Mechanics of fluids

ONE WEEK LOAN

Mechanics of Fluids

Seventh edition

Bernard Massey

Reader Emeritus in Mechanical Engineering
University College London

Revised by

John Ward-Smith

Senior Research Fellow in Biomechanics
Brunel University

Stanley Thornes (Publishers) Ltd

Sixth edition published by Chapman & Hall in 1989
(ISBN 0 412 34280 4)

Seventh edition published in 1998 by
Stanley Thornes (Publishers) Ltd
Ellenborought House
Wellington Street
CHELTENHAM
GL50 1YW
United Kingdom

98 99 00 01 02 / 10 9 8 7 6 5 4 3 2 1

A catalogue record for this book is available from the British Library

ISBN 0-7487-4043-0

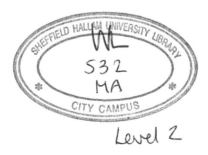
Typeset by Best-set Typesetter Ltd, Hong Kong
Printed and bound in Great Britain at the Alden Press, Oxford

Contents

Preface to the sixth edition

Goethe thought that prefaces were useless, and I am not disposed to disagree. Yet there may be value in stating, somewhere between the covers of this book, that its purpose is to present the basic principles of mechanics of fluids and to illustrate them by application to a variety of problems in different branches of engineering. Emphasis, however, is on principles rather than engineering practice. Attention is also given to the assumptions on which the principles rest, for only thus can the limits of their validity be appreciated.

Although the book is intended primarily for students taking honours degrees in engineering, it should also serve the needs of those studying for the CEI examinations in Fluid and Particle Mechanics.

It is a book for engineers rather than mathematicians. Stress is laid on physical concepts rather than mathematics; specialized mathematical techniques are avoided and algebraic manipulations are kept to a minimum. For the majority of students, the occasional small sacrifice of rigour is, I believe, more than justified by the greater ease of visualizing and understanding the physical circumstances.

The book is reasonably self-contained, but references are given to more detailed discussions of certain topics. These references are not intended as an exhaustive bibliography but rather as useful clues to start the search for further information.

The choice of algebraic symbols largely follows the recommendations of the International Organization for Standardization, but, as our requirements strain even the combined resources of the Roman and Greek alphabets, occasional departures from orthodoxy have been made where confusion might otherwise result. Printers can widen the range of available symbols by using different typefaces, but as this book will be used chiefly by students, whose own use of symbols is in handwriting that may leave something to be desired, these subtler distinctions have been avoided.

The end-of-chapter problems are to be regarded as an essential part of the book. Each problem is designed to further the student's grasp of the subject, and is not to be solved simply by inserting figures in a 'standard' formula, or by a line-by-line following of a worked example. Some of these problems are taken from, or adapted from, past examination papers of the University of London, and I gratefully acknowledge the University's permission for their use here. The majority, however, have been specially devised for this book. I have provided what I hope are correct answers to all numerical problems; for satisfactory checking with students' answers, mine are normally given to three or four significant figures, whether or not this is warranted by the accuracy of the data or by the validity of assumptions involved in the solution.

It is a pleasure to express my gratitude to many who have helped in the preparation of this book. I owe an inestimable debt to several colleagues who by personal kindness have cheerfully shared with me their wisdom. Responsibility for any errors is mine alone, but I should particularly like to acknowledge helpful discussions I have had with Professor H. Billett, Dr J. W. Fox, Professor K. J. Ives and Mr L. A. Wigglesworth. They may not recognize their contributions in these pages, but I assure them that for what I have received I am truly thankful. I am, of course, much indebted to previous authors, even when I have dissented from them, and to students who have pointed, sometimes unwittingly, to the real difficulties of the subject. My sincere appreciation goes also to Miss C. R. Clewer for her expert typing of the manuscript, and to the staff of the publishers for much ready help and encouragement.

B.S.M.

Preface to the seventh edition

In this seventh edition the aim has been to retain unchanged the broad ethos of the book, established in earlier editions. However, some important detailed changes have been introduced. Most obvious are the changes in typeface and page layout, which are intended to make the book yet more reader-friendly. Various sections of text have been reorganized, changes having been made to the content both within the main chapters and in the appendices. As a consequence of this reorganization, the introductory chapter on fundamental concepts has been expanded. Some new material has also been added. There are a substantial number of new worked examples, bringing the total to over 70, distributed throughout the book. A short introduction to computational fluid dynamics has also been provided. The index has been greatly expanded. Data on conversion factors, physical properties of fluids, gas flow functions and algebraic symbols are now all collected together in the appendices for ease of use.

In conclusion, Dr Massey and I wish to record our thanks to those readers of previous editions whose correspondence has helped shape the changes incorporated here.

J.W.S.

Fundamental concepts $\boxed{1}$

1.1 THE CHARACTERISTICS OF FLUIDS

A fluid is a substance which may flow; that is, its constituent particles   *Fluid*
may continuously change their positions relative to one another.
Moreover, it offers no lasting resistance to the displacement, however
great, of one layer over another. This means that, if the fluid is at rest,
no shear force (that is, a force tangential to the surface on which it acts)
can exist in it. A solid, on the other hand, can resist a shear force while
at rest; the shear force may cause some displacement of one layer over
another, but the material does not continue to move indefinitely. In a
fluid, however, shear forces are possible only while relative movement
between layers is actually taking place. A fluid is further distinguished
from a solid in that a given amount of it owes its shape at any particular
time to that of a vessel containing it, or to forces that in some way
restrain its movement.

The distinction between solids and fluids is usually clear, but there are
some substances not easily classified. Some fluids, for example, do not
flow easily: thick tar or pitch may at times appear to behave like a solid.
A block of such a substance may be placed on the ground, and, although
its flow would take place very slowly, over a period of time – perhaps
several days – it would spread over the ground by the action of gravity,
that is, its constituent particles would change their relative positions. On
the other hand, certain solids may be made to 'flow' when a sufficiently
large force is applied; these are known as *plastic solids*.

Even so, the essential difference between solids and fluids remains.
Any fluid, no matter how 'thick' or viscous it is, begins to flow, even if
imperceptibly, under the action of the slightest net shear force.
Moreover, a fluid continues to flow as long as such a force is applied. A
solid, however, no matter how plastic it is, does not flow unless the net
shear force on it exceeds a certain value. For forces less than this value
the layers of the solid move over one another only by a certain amount.
The more the layers are displaced from their original relative positions,
however, the greater are the forces resisting the displacement. Thus, if
a steady force is applied, a state will be reached in which the forces
resisting the movement of one layer over another balance the force
applied and so no further movement of this kind can occur. If the

applied force is then removed, the resisting forces will tend to restore the solid body to its original shape.

In a fluid, however, the forces opposing the movement of one layer over another exist only while the movement is taking place, and so static equilibrium between applied force and resistance to shear never occurs. Deformation of the fluid takes place continuously so long as a shear force is applied. But if this applied force is removed the shearing movement subsides and, as there are then no forces tending to return the particles of fluid to their original relative positions, the fluid keeps its *new* shape.

Liquid

Fluids may be sub-divided into liquids and gases. A fixed amount of a liquid has a definite volume which varies only slightly with temperature and pressure. If the capacity of the containing vessel is greater than this definite volume, the liquid occupies only part of the container, and it forms an interface separating it from its own vapour, the atmosphere or any other gas present.

Gas

On the other hand, a fixed amount of a gas, by itself in a closed container, will always expand until its volume equals that of the container. Only then can it be in equilibrium. In the analysis of the behaviour of fluids the most important difference between liquids and gases is that, whereas under ordinary conditions liquids are so difficult to compress that they may for most purposes be regarded as incompressible, gases may be compressed much more readily. Where conditions are such that an amount of gas undergoes a negligible change of volume, its behaviour is similar to that of a liquid and it may then be regarded as incompressible. If, however, the change in volume is not negligible, the compressibility of the gas must be taken into account in examining its behaviour.

1.1.1 Molecular structure

The different characteristics of solids, liquids and gases result from differences in their molecular structure. All substances consist of vast numbers of molecules separated by empty space. The molecules have an attraction for one another, but when the distance between them becomes very small (of the order of the diameter of a molecule) there is a force of repulsion between them which prevents them all gathering together as a solid lump.

The molecules are in continual movement, and when two molecules come very close to one another the force of repulsion pushes them vigorously apart, just as though they had collided like two billiard balls. In solids and liquids the molecules are much closer together than in a gas. A given volume of a solid or a liquid therefore contains a much larger number of molecules than an equal volume of a gas, so solids and liquids have a greater density (i.e. mass per unit volume).

In a solid the movement of individual molecules is slight – just a vibration of small amplitude – and they do not readily move relative to one another. In a liquid the movement of the molecules is greater, but they continually attract and repel one another so that they move in curved, wavy paths rather than in straight lines. The force of attraction between the molecules is sufficient to keep the liquid together in a definite volume although, because the molecules can move past one another, the substance is not rigid. In a gas the molecular movement is very much greater; the number of molecules in a given space is much less, and so any molecule travels a much greater distance before meeting another. The forces of attraction between molecules – being inversely proportional to the square of the distance between them – are, in general, negligible and so molecules are free to travel away from one another until they are stopped by a solid or liquid boundary.

The activity of the molecules increases as the temperature of the substance is raised. Indeed, the temperature of a substance may be regarded as a measure of the average kinetic energy of the molecules.

When an external force is applied to a substance the molecules tend to move relative to one another. A solid may be deformed to some extent as the molecules change position, but the strong forces between molecules remain, and they bring the solid back to its original shape when the external force is removed. Only when the external force is very large is one molecule wrenched away from its neighbours; removal of the external force does not then result in a return to the original shape, and the substance is said to have been deformed beyond its elastic limit.

In a liquid, although the forces of attraction between molecules cause it to 'hold together', the molecules can move past one another and find new neighbours. Thus a force applied to an unconfined liquid causes the molecules to slip past one another until the force is removed.

If a liquid is in a confined space and is compressed it exhibits elastic properties like a solid in compression. Because of the close spacing of the molecules, however, the resistance to compression is great. A gas, on the other hand, with its molecules much farther apart, offers much less resistance to compression.

1.1.2   The continuum

An absolutely complete analysis of the behaviour of a fluid would have to account for the action of each individual molecule. In most engineering applications, however, interest centres on the *average* conditions of velocity, pressure, temperature, density and so on. Therefore, instead of the actual conglomeration of separate molecules, we suppose that the fluid is a *continuum*, that is a continuous distribution of matter with no empty space. This assumption is normally justifiable because the number of molecules involved in the situation is so vast and the distances between them are so small. The assumption fails, of course, when these conditions are not satisfied as, for example, in a gas at extremely low pressure. The average distance between molecules may then be appreciable in comparison with the smallest significant length in the fluid boundaries. However, as this situation is well outside the range of normal engineering work, we shall in this book regard a fluid as a continuum. Although it is often necessary to postulate a small element (or 'particle') of fluid, this is supposed large enough to contain very many molecules.

The properties of a fluid, although molecular in origin, may be adequately accounted for in their overall effect by ascribing to the continuum such attributes as temperature, thermal conductivity, pressure, viscosity and so on. Quantities such as velocity, acceleration and the properties of the fluid are assumed to vary continuously (or remain constant) from one point to another in the fluid.

1.1.3   The development of 'mechanics of fluids'

The *mechanics of fluids* is the study in which the fundamental principles of general mechanics are applied to liquids and gases. These principles are those of the conservation of matter, the conservation of energy and Newton's laws of motion. To the study of compressible fluids we also bring some of the laws of thermodynamics. By the use of these principles we are enabled not only to explain and bring into relation observed phenomena but also to predict, at least approximately, the behaviour of fluids under a set of specified conditions.

Ever since the first attempts to move water from one place to another without filling a container and carrying it, people have been interested in mechanics of fluids. For centuries, however, their knowledge of the subject was gained solely from simple observations and the tedious processes of trial and error. Thus, from the time of the ancient Greeks and Romans, the largely empirical subject of *hydraulics* grew up. Against such a background, the recognition of fundamental principles governing the observed phenomena was not easy.

Since about 1750 mathematicians and mathematical physicists have attempted to obtain solutions of many of the problems of fluid motion, solutions not relying on the results of experimental measurement. Most of these attempts have been made possible, however, only by the introduction of major simplifying assumptions, so the results have sometimes borne little relation to practical problems. The solutions, which refer to an 'ideal' fluid not to be found in the physical world, constitute a body of knowledge now known as *classical hydrodynamics*. Although this ideal fluid lacks some of the attributes of real fluids, in particular viscosity, the descriptions of its behaviour given by classical hydrodynamics are not without value in those circumstances where the neglected properties have only a small effect on the behaviour of a real fluid. The mathematicians' outlook, moreover, was the praiseworthy one of attempting to find solutions of as wide an application as possible rather than just the formulation of rules obtained from a limited range of observed conditions and therefore strictly applicable only to those conditions. The historical development of the subject is admirably described by Hunter Rouse and Simon Ince in *History of Hydraulics*, Iowa Institute of Hydraulic Research, State University of Iowa (1957). There is a shorter account by W. R. Durand, 'The development of our knowledge of the laws of fluid mechanics', *Science* **78**, 343–51 (1933).

It is not always easy to decide whether a particular simplifying assumption is justifiable, and ultimately experimental verification of theoretical results is always required. Conversely, experimental results themselves sometimes suggest assumptions that may profitably be made in further theoretical work. Modern mechanics of fluids, which is one of the sciences on which many important branches of engineering are founded, combines both theoretical and experimental approaches to problems. In this book we shall deal chiefly with those fundamental principles underlying all logical analysis of the behaviour of fluids. A sure grasp of these principles is essential to the understanding of more detailed and advanced work.

1.2 UNITS AND RELATED MATTERS

In order to introduce some basic ideas about units, let us consider the following. A speed is reported as 30 cm/s. In this statement, 30 is described as the *numeric*, and cm/s are the *units* of measurement of speed. There are 10^5 cm in 1 km, and 3600 seconds in 1 hour. Hence the speed of 30 cm/s is equivalent to 1.08 km/h. In the latter case, the numeric is 1.08 and the units are km/h. This simple example illustrates the importance of specifying both the *units* and the *numeric* when quoting the magnitude of physical quantities.

1.2.1 Units of the Système International d'Unités (SI units)

This system of units is an internationally agreed version of the metric system; it is rapidly being adopted throughout most of the world and will no doubt eventually come into universal use. There are seven base units, and not only their names but also their symbols have been internationally agreed (see Table 1.1).

From these base units, all others are derived (e.g. metre per second as a unit of velocity). Two forms of notation are in common use for derived units, e.g. $kg/(s\,m^2)$ or $kg\,s^{-1}\,m^{-2}$; both forms are used in this book. Certain combinations of the base units are given internationally agreed special names. Table 1.2 lists only those used in this book. Some other special names have been proposed and may be adopted in the future.

1.2.2 Prefixes

To avoid inconveniently large or small numbers, prefixes may be put in front of the unit names (see Table 1.3). Especially recommended are prefixes that refer to factors of 10^{3n}, where n is a positive or negative integer.

Table 1.1 Base SI units

Quantity	Unit	Symbol
length	metre	m
mass	kilogram	kg
time interval	second	s
electric current	ampere	A
temperature	kelvin (formerly 'degree Kelvin')	K (formerly °K)
luminous intensity	candela	cd
amount of substance	mole	mol

Table 1.2 Names of some derived units

Quantity	Unit	Symbol	Equivalent combination of other units
force	newton	N	$kg\,m/s^2$
pressure and stress	pascal	Pa	N/m^2 ($\equiv kg\,m^{-1}\,s^{-2}$)
work, energy, quantity of heat (but *not* torque)	joule (pronounced 'jool')	J	$N\,m$ ($\equiv kg\,m^2\,s^{-2}$)
power	watt	W	J/s ($\equiv kg\,m^2\,s^{-3}$)
frequency	hertz	Hz	s^{-1}
plane angle	radian (actually unity)	rad	m/m (i.e. length of circular arc ÷ radius)

Table 1.3 Prefixes for multiples and submultiples of SI units

Prefix	Symbol	Factor by which unit is multiplied
exa	E	10^{18}
peta	P	10^{15}
tera	T	10^{12}
giga	G	10^{9}
mega	M	10^{6}
kilo	k	10^{3}
hecto	h	10^{2}
deca	da	10
deci	d	10^{-1}
centi	c	10^{-2}
milli	m	10^{-3}
micro	μ	10^{-6}
nano	n	10^{-9}
pico (pronounced 'peeko')	p	10^{-12}
femto	f	10^{-15}
atto	a	10^{-18}

Care is needed in using these prefixes. The symbol for a prefix should always be written close to the symbol of the unit it qualifies, e.g. kilometre (km), megawatt (MW), microsecond (μs). On the other hand, the symbols for the units themselves should be spaced apart. Only one prefix at a time may be applied to a unit; thus 10^{-6} kilogram is one milligram (mg), *not* one microkilogram.

The symbol 'm' stands both for the basic unit 'metre' and for the prefix 'milli', so especial care is needed in using it. For example, mN means millinewton whereas m N denotes metre newton. In this case, the difference of spacing is hardly sufficient safeguard against confusion, and so if the 'm' denotes 'metre' it is better to reverse the order of the unit symbols: N m.

When a unit with a prefix is raised to a power, the exponent applies to the *whole multiple* and not just to the original unit. Thus $1\,mm^2$ means $1(mm)^2 = (10^{-3}m)^2 = 10^{-6}m^2$, and *not* $1\,m(m^2) = 10^{-3}m^2$.

If a derived unit is in the form of a quotient, any prefix should normally be applied only to the numerator and not to the denominator. For example, kN/m, not N/mm, should be used for $10^3\,N/m$.

The symbols for units refer not only to the singular but also to the plural. For instance, the symbol for kilometres is km, not kms. Dots, to denote either multiplication or abbreviation, are not used at all.

Capital or lower case (small) letters are used strictly in accordance with the definitions, no matter in what combination the letters may appear. In print, ordinary upright type is used for unit symbols and prefixes, whereas italic type (*thus*) is used for algebraic symbols.

1.2.3 Units for particular quantities

'Conventional' temperatures are expressed in 'degrees Celsius', °C (formerly known as 'degrees Centigrade'). The zero of the Celsius scale is so defined that the triple-point of water substance is exactly 0.01 °C. 'Absolute' temperatures are expressed in kelvins, K.   *Temperature*

$$t\,°C = \left(273.15 + t\right)K$$

For many purposes the 273.15 can be rounded off to 273 without significant loss of accuracy. Differences of temperature, whether 'conventional' or 'absolute', are also expressed in kelvins.

Note that 1 newton is the net force required to give a body of mass 1 kg an acceleration of $1\,m/s^2$. The force required to give a body of mass 1 kg an acceleration of (approximately) $9.81\,m/s^2$ is sometimes called one kilogram-force (1 kgf). This therefore equals 9.81 N and must be carefully distinguished from 1 N, and also from 1 kilogram mass (1 kg).   *Force*

Note that 1 pascal is the pressure induced by a force of 1 N acting on an area of $1\,m^2$. The pascal, Pa, is small for most purposes, and thus   *Pressure and stress*

multiples are often used. The unit 'bar', equal to 10^5 Pa, has been in use for many years, but as it breaks the 10^{3n} convention it is not an SI unit.

Volume

In the measurement of fluids the name 'litre' is commonly given to 10^{-3} m^3. The agreed abbreviation for litre has been 'l', but as this is easily mistaken for 1 (one) 'L' is now recommended.

1.2.4 Conversion factors

It is often desirable to change data involving units of one size so as to involve units of the same kind but of a different size. This may be done by using 'conversion factors' which relate the sizes of different units of the same kind.

As an example, consider the identity

$$1 \text{ inch} \equiv 25.4 \text{ mm}$$

(The use of three lines ($\equiv$), instead of the two lines of the usual 'equals' sign, indicates not simply that one inch equals or is equivalent to 25.4 mm but that one inch *is* 25.4 mm. At all times and in all places one inch and 25.4 mm are precisely the same magnitude.) The identity may be rewritten as

$$1 \equiv \frac{25.4 \text{ mm}}{1 \text{ inch}}$$

and this ratio *equal to unity* is a conversion factor. Because unity may be introduced as a factor into any expression without altering the value of the expression, changing the size of units cannot essentially affect any equation in 'physical' algebra. To change the size of units (though not their nature) is equivalent to multiplying by unity. Moreover, as the reciprocal of unity is also unity, any conversion factor may be used in reciprocal form when the desired result requires it.

This simple example may be extended indefinitely:

$$1 \equiv \frac{25.4 \text{ mm}}{1 \text{ inch}} \equiv \frac{1 \text{ inch}}{25.4 \text{ mm}} \equiv \frac{304.8 \text{ mm}}{1 \text{ ft}} \equiv \frac{0.4536 \text{ kg}}{1 \text{ pound mass (lbm)}}$$

$$\equiv \frac{10^{-5} \text{ N}}{1 \text{ dyne}} \equiv \frac{4.448 \text{ N}}{1 \text{ pound-force (lbf)}}$$

$$\equiv \frac{32.174 \text{ poundal (pdl)}}{1 \text{ lbf}} \equiv \ldots$$

Application of such conversion factors enables us to express, for example, 936 ft lbf/min in terms of watts. Set out fully, the successive operations are

$$936 \frac{\text{ft lbf}}{\text{min}} \equiv 936 \frac{\text{ft lbf}}{\text{min}} \times \frac{0.3048\,\text{m}}{1\,\text{ft}} \times \frac{4.448\,\text{N}}{1\,\text{lbf}} \times \frac{1\,\text{min}}{60\,\text{s}}$$

$$\equiv \frac{936 \times 0.3048 \times 4.448}{60} \frac{\text{N m}}{\text{s}}$$

$$\equiv 21.15 \frac{\text{N m}}{\text{s}} \equiv 21.15\,\text{W}$$

Such detail of setting out is not normally necessary, of course, but this pattern of operations, this successive multiplication by unity, is the logical basis of the use of conversion factors and is therefore implicit in all correct conversions of data.

A conversion factor equal to unity may be applied to a magnitude that has been defined in one particular way. But a unity factor does not apply when different definitions are used, e.g. plane angle expressed in radian measure instead of 'protractor' measure involving revolutions, degrees, etc. Also, if magnitudes are expressed on scales with different zeros (e.g. the Fahrenheit and Celsius scales of temperature) then unity conversion factors may be used only for *differences* of the quantity, not for individual points on a scale. For instance,

$$\text{a temperature } \textit{difference} \text{ of } 36\,°\text{F} = 36\,°\text{F} \times \frac{1\,°\text{C}}{1.8\,°\text{F}} = 20\,°\text{C},$$

but a temperature of 36 °F corresponds to 2.22 °C, not 20 °C.

1.2.5 Dimensional formulae

It is possible to define the magnitudes of some quantities (those of mass and time interval, for example) without reference to any other kind of quantity. For most magnitudes, however, this is not practicable and so they are defined in terms of others. We cannot of course define *all* magnitudes in terms of others, because the definitions would then just be going round in circles. We therefore begin by considering a small number of magnitudes as specified independently of any others; these are termed *reference* (or *primary* or *fundamental*) magnitudes. From them all the remaining magnitudes are then defined; consequently these are referred to as *derived* (or *secondary*) magnitudes.

For example, although velocity may be determined in various ways, the method normally used to define its magnitude is to divide the magnitude of a length by the magnitude of a time interval. That is, the magnitude of velocity is defined in terms of those of length and time interval, and, if either the unit of length or the unit of time interval were altered in size, the expression for the magnitude of a velocity would be changed. Velocity, therefore, is considered to have a derived magnitude.

The classification of magnitudes as reference or derived is not a rigid one imposed by restrictions of nature, but is largely arbitrary. For example, the magnitude of a force is normally defined in terms of units of mass and acceleration. But it is quite possible to compare the magnitudes of two forces directly, that is to define one entirely in terms of the other, and so force could be regarded as having a reference magnitude rather than a derived one. We could then regard mass as having a derived instead of a reference magnitude (mass = force ÷ acceleration). In fact, the number of reference magnitudes is arbitrary: there is no reason why, with sufficient ingenuity, any magnitude should not be definable independently of any other magnitude.

However, let us regard the magnitudes of length and time interval as reference ones, and the magnitude of velocity as derived directly from them. The magnitude u of velocity is then given by the equation $u = l/t$ where l represents the magnitude of the length traversed by a moving object and t the magnitude of the corresponding time interval (assumed small enough for the velocity to be considered constant during that interval). Suppose that l is measured as $n_1[L]$ and that t is measured as $n_2[T]$, where n_1, n_2 are numerics, and [L] and [T] represent suitable units of length and of time interval respectively. Therefore the magnitude u is given by $n_1[L] \div n_2[T] = n_3[L]/[T]$, where the ratio n_1/n_2 is simply another numeric n_3. Since velocity, like other physical quantities, has its magnitude expressed by the product of a numeric and a unit it follows that [L]/[T] is a possible unit of velocity. The expression [L]/[T] is usually known as the *dimensional formula* of velocity – although it should be noted that the dimensional formula is characteristic not of velocity itself but of the units with which its magnitude is expressed.

A further step gives the dimensional formula of acceleration. Taking the magnitude of acceleration as defined by (increase of velocity) ÷ (corresponding short time interval), we obtain, for a particular value of acceleration, n_4[unit of velocity] ÷ $n_5[T]$

$$= n_6 \frac{[L]}{[T][T]} = n_6 [LT^{-2}]$$

where the ns again represent numerics and, for brevity, one pair of square brackets is used to enclose the dimensional formula. Note that the symbol [X] refers simply to a *possible* unit of the quantity X; it is not restricted to any particular unit.

Since the magnitude of force is defined by the equation $F = ma$, the dimensional formula of force is $[M] \times [LT^{-2}] = [MLT^{-2}]$. By similar reasoning the dimensional formulae of other magnitudes may be built up from those of the reference magnitudes. The power to which a reference unit is raised in the dimensional formula of any magnitude is said to be the dimension in respect to that reference magnitude in the formula. For example, the magnitude of force is said to have dimensions of 1 in respect to mass, 1 in respect to length and −2 in respect to time interval. (Although this usage is generally understood it is not always followed. What is here termed the dimensional formula is often loosely called 'the dimensions'.)

A dimensional formula for the magnitude of a particular quantity is not unique: it depends on which magnitudes are regarded as reference ones and on how the derived magnitudes are defined. It tells us nothing about the intrinsic nature of the quantity concerned (that information must come from a definition of the *quantity*, not a dimensional formula); it simply defines a possible unit of the quantity. Any magnitude that may be expressed by only a numeric has no dimensions and is therefore said to be 'dimensionless' or 'non-dimensional'.

Although the magnitudes of length, mass and time interval are usually regarded as reference ones, they are in no way superior to other magnitudes. Nevertheless, for most applications it is convenient to consider the magnitudes of length, mass and time interval as reference ones, and other magnitudes in mechanics as derived from them. Many magnitudes occurring in thermodynamics cannot be expressed in terms of those of length, mass and time interval only. An additional reference magnitude is required, and that of temperature is usually chosen. Other reference magnitudes are often required when quantities connected with electricity, magnetism or light are under discussion. However, as we shall not need them in this book we shall not consider them here.

1.2.6 Dimensional homogeneity

For a given choice of reference magnitudes, quantities of the same kind have magnitudes with the same dimensional formulae. (The converse, however, is not necessarily true: identical dimensional formulae are no guarantee that the corresponding quantities are of the same kind.) Since adding, subtracting or equating magnitudes makes sense only if the magnitudes refer to quantities of the same kind, it follows that all terms added, subtracted or equated must have identical dimensional formulae; that is, an equation must be *dimensionally homogeneous*.

In addition to the variables of major interest, equations in physical algebra may contain constants. These may be numerics, like the $\frac{1}{2}$ in Kinetic energy $= \frac{1}{2}mu^2$, and therefore dimensionless, but in general they are not dimensionless; their dimensional formulae are determined from those of the other magnitudes in the equation, so that dimensional homogeneity is achieved. For instance, in Newton's Law of Universal Gravitation, $F = Gm_1m_2/r^2$, the constant G must have the same dimensional formula as Fr^2/m_1m_2, that is, $[MLT^{-2}][L^2]/[M][M] \equiv [L^3M^{-1}T^{-2}]$, otherwise the equation would not be dimensionally homogeneous. The fact that G is a 'universal constant' is beside the point: dimensions are associated with it, and in analysing the equation they must be accounted for.

An important consequence of the principle of dimensional homogeneity is that the truth of a correct relation in physical algebra in which all relevant factors are *explicitly* included does not depend on the size of units used in expressing the various magnitudes. If a relation is dimensionally homogeneous, the same basic units appear to the same power in the dimensional formulae of each side of the relation. Thus if the size of any of the basic units is changed by some factor, each side of the relation is similarly affected and the relation is no less true than before.

Dimensional homogeneity, as we have seen, is a condition for a relation in 'physical' algebra to have physical significance. Although a relation may be dimensionally homogeneous yet wrong in some other respect, a relation that is *not* dimensionally homogeneous cannot have physical meaning, and an error has probably been made in its derivation. If the dimensional formulae of terms are checked at frequent intervals in algebraic working, many slips may be detected and corrected.

1.3 PROPERTIES OF FLUIDS

1.3.1 Density

The basic definition of the density of a substance is the ratio of the mass of a given amount of the substance to the volume it occupies. For liquids, this definition is generally satisfactory. However, since gases are compressible, further clarification is required.

Mean density

The ratio of the mass of a given amount of a substance to the volume that this amount occupies. If the mean density in all parts of a substance is the same then the density is said to be *uniform*.

Density at a point

The limit to which the mean density tends as the volume considered is indefinitely reduced, that is $\lim_{V \to 0}(m/V)$. As a mathematical definition this is satisfactory; since, however, all matter actually consists of separate molecules, we should think of the volume reduced not absolutely to zero, but to an exceedingly small amount that is nevertheless large enough to contain a considerable number of molecules. The concept of a continuum is thus implicit in the definition of density at a point.

Relative density

The ratio of the density of a substance to some standard density. The standard density chosen for comparison with the density of a solid or a liquid is invariably that of water at $4\,°C$. For a gas, the standard density may be that of air or that of hydrogen, although for gases the term is little used. (The term 'specific gravity' has also been used for the relative density of a solid or a liquid, but 'relative density' is much to be preferred.) As relative density is the ratio of two magnitudes of the same kind it is merely a numeric without units.

1.3.2 Pressure

Pressure

A fluid always has pressure. As a result of innumerable molecular 'collisions', any part of the fluid must experience forces exerted on it by adjoining fluid or by adjoining solid boundaries. If, therefore, part of the fluid is arbitrarily divided from the rest by an imaginary plane, there will be forces that may be considered as acting at that plane.

The term 'pressure intensity' is usually abbreviated to 'pressure', but it is not uncommon for careless people to use 'pressure' as a synonym for 'thrust', that is force, not force/area. This dangerous misuse of the term should be carefully avoided.

Pressure intensity cannot be measured directly; all instruments said to measure it in fact indicate a difference of pressure. This difference is frequently that between the pressure of the fluid under consideration and the pressure of the surrounding atmosphere. The pressure of the atmosphere is therefore commonly used as the reference or datum pressure, that is the starting point of the scale of measurement. The difference in pressure recorded by the measuring instrument is then termed the *gauge pressure*.

Gauge pressure

The *absolute pressure*, that is the pressure considered relative to that of a perfect vacuum, is then given by $p_{abs} = p_{gauge} + p_{atm}$. (See also Section 2.3.)

Absolute pressure

The pressure of the atmosphere is not constant. For many engineering purposes the variation of atmospheric pressure (and therefore the variation of absolute pressure for a given gauge pressure, or vice versa) is of no consequence. In other cases, however – especially for the flow of gases – it is necessary to consider absolute pressures rather than gauge pressures, and a knowledge of the pressure of the atmosphere is then required.

Pressure (intensity), we have seen, is defined as force/area (and so the corresponding dimensional formula is $[F]/[L^2] = [MLT^{-2}]/[L^2] = [ML^{-1}T^{-2}]$). Now although the force has direction, the pressure has not. The direction of the force also specifies the direction of the imaginary plane surface, since the latter is defined by the direction of a line perpendicular to the surface. Here, then, the force and the surface have the same direction and so in the equation

$$\overrightarrow{\text{Force}} = \text{Pressure} \times \overrightarrow{\text{Area of plane}} \text{ surface}$$

pressure must be a scalar quantity. Pressure is a property of the fluid at the point in question. Similarly, temperature and density are properties of the fluid and it is just as illogical to speak of 'downward pressure', for example, as of 'downward temperature' or 'downward density'. To say that pressure acts in any direction, or even in all directions, is meaningless; pressure is a scalar quantity.

Suitable units for pressure are N/m^2, lbf/ft^2, lbf/in^2. By international agreement the unit N/m^2 is now termed *pascal*, with the abbreviation 'Pa'. (In America the abbreviations 'psf' and 'psi' are frequently used for 'lbf/ft²' and 'lbf/in²' respectively.) Pressures of large magnitude have often been expressed in 'atmospheres' (abbreviated 'atm'). For precise definition, one atmosphere is taken as 1.01325×10^5 Pa (approximately

14.7 lbf/in^2). A pressure of 10^5 Pa has been called 1 *bar*. The thousandth part of this unit, called a *millibar* (abbreviated 'mbar'), has commonly been used by meteorologists.

For pressures less than that of the atmosphere the units normally used are millimetres (or inches) of mercury vacuum. These units refer to the difference between the height of a vertical column of mercury supported by the pressure considered, and the height of one supported by the atmosphere.

In the absence of shear forces, the direction of the plane over which the force due to the pressure acts has no effect on the magnitude of the pressure at a point. The fluid may even be accelerating in a particular direction provided that shear forces are absent – a condition that requires no *relative* motion between different particles of fluid.

Imagine a small prism, with plane faces and triangular section, surrounding the point in question. Figure 1.1 shows one end ABC of the prism; a parallel end face $A'B'C'$ is at a perpendicular distance l from ABC. The rectangular face $ABB'A'$ is assumed vertical and the rectangular face $BCC'B'$ horizontal, but the face $ACC'A'$ is at any angle. The mean density of the fluid inside the prism is ϱ and the average pressures at each face are p_1, p_2 and p_3, respectively.

If there is no relative motion between particles of the fluid, the forces on the end faces ABC and $A'B'C'$ act only perpendicular to those faces. The net force towards the right is given by resolving horizontally (and parallel to the plane ABC):

$$p_1\, ABl - p_3\, ACl\cos A = (p_1 - p_3)ABl$$

since $AC\cos A = AB$. By Newton's Second Law, this net force equals the product of the mass of the fluid and its mean acceleration (say a_x) in that direction:

$$(p_1 - p_3)ABl = \frac{1}{2} BC\, ABl\,\varrho a_x$$

i.e.

$$p_1 - p_3 = \frac{1}{2} BC\, \varrho a_x \qquad (1.1)$$

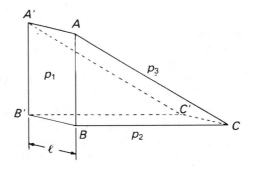

Fig. 1.1

If the prism is made exceedingly small, the right-hand side of eqn 1.1 tends to zero and so, at the point considered,

$$p_1 = p_3 \qquad (1.2)$$

The weight of the fluid in the prism is g times its mass, so the net force vertically downwards is

$$p_3 \, AC \, l \cos C + \frac{1}{2} BC \, AB l \varrho g - p_2 \, BC \, l$$

$$= BC \, l \left(p_3 + \frac{1}{2} AB \varrho g - p_2 \right)$$

since $AC \cos C = BC$.

Again by Newton's Second Law, this net force equals the product of the mass of the fluid and its mean acceleration vertically downwards (say a_y):

$$BC \, l \left(p_3 + \frac{1}{2} AB \varrho g - p_2 \right) = \frac{1}{2} BC \, AB l \varrho a_y$$

or, after division by $BC \, l$ and rearrangement:

$$p_3 - p_2 = \frac{1}{2} AB \varrho (a_y - g)$$

If the size of the prism is reduced $AB \to 0$ and, at the point considered,

$$p_3 = p_2 \qquad (1.3)$$

So, combining eqns 1.2 and 1.3, we have

$$p_1 = p_2 = p_3 \qquad (1.4)$$

We remember that the direction of the face $ACC'A'$ was not specified at all, and so the result is valid for any value of the angle ACB. Moreover, the plane $ABB'A'$ may face any point of the compass and therefore the pressure intensity is quite independent of the direction of the surface used to define it. This result is frequently known as *Pascal's Law* after the French philosopher Blaise Pascal (1623–62), although the principle had previously been deduced by G. B. Benedetti (1530–90) and Simon Stevin (1548–1620) in about 1586. The only restrictions are that the fluid is a continuum, i.e. that the prism, even when made very small, contains a large number of molecules, and that, if it is moving, there is no relative motion between adjacent particles.

If, however, there is relative motion between adjacent layers in the fluid, then shear forces are set up and eqn 1.4 is not, in general, true. Nevertheless, the quantity referred to as the pressure at a point in such circumstances is taken as the mean of the perpendicular force per unit area on three mutually perpendicular planes. The perpendicular forces per unit area are, in any case, usually large compared with the shear forces per unit area, so Pascal's Law is not far from the truth even then.

1.3.3 Vapour pressure

All liquids tend to evaporate (or vaporize). This is because there is at the free surface a continual movement of molecules out of the liquid. Some of these molecules return to the liquid, so there is, in fact, an interchange of molecules between the liquid and the space above it. If the space above the surface is enclosed, the number of liquid molecules in the space will – if the quantity of liquid is sufficient – increase until the rate at which molecules escape from the liquid is balanced by the rate at which they return to it.

Saturation pressure

Just above the liquid surface the molecules returning to the liquid create a pressure known as the partial pressure of the vapour. This partial pressure and the partial pressures of other gases above the liquid together make up the total pressure there. Molecules leaving the liquid give rise to the *vapour pressure*, the magnitude of which corresponds to the rate at which molecules escape from the surface. When the vapour pressure equals the partial pressure of the vapour above the surface, the rates at which molecules leave and enter the liquid are the same, and the gas above the surface is then said to be *saturated* with the vapour. The value of the vapour pressure for which this is so is the *saturation pressure*.

Since the velocity of the molecules, and hence their ability to escape through the liquid surface, increases with temperature, so does the vapour pressure. If the total pressure of the gas above the liquid becomes less than the saturation pressure, molecules escape from the liquid very rapidly in the phenomenon known as boiling. Bubbles of vapour are formed in the liquid itself and then rise to the surface. For pure water the saturation pressure at 100 °C is approximately 10^5 Pa, which is the total pressure of the atmosphere at sea level, so water subject to this atmospheric pressure boils at this temperature. If, however, the external pressure to which the liquid is subjected is lower, then boiling commences at a lower value of the saturation pressure, that is at a lower temperature. Water therefore boils even at room temperature if the pressure is reduced to the value of the saturation vapour pressure at that temperature (for numerical data see Appendix 2).

Cavitation

Effects very similar to boiling occur if a liquid contains dissolved gases. When the pressure of the liquid is sufficiently reduced the dissolved gases are liberated in the form of bubbles; a smaller reduction of pressure is, however, required for the release of dissolved gases than for the boiling of the liquid. A subsequent increase of pressure may cause bubbles, whether of vapour or of other gases, to collapse; very high impact forces may then result. The latter phenomenon is known as cavitation, and has serious consequences in fluid machinery (see Section 13.3.6).

There is a wide variation in vapour pressure among liquids, as shown in Appendix 2. These figures clearly indicate that it is not only its high density that makes mercury valuable in a barometer; the vapour pressure is so low that there is an almost perfect vacuum above the liquid column. It will also be seen why a liquid such as petrol evaporates much more readily than water at the same temperature.

1.4   THE PERFECT GAS: EQUATION OF STATE

No actual gas is 'perfect' although the assumed properties of a perfect gas are closely matched by those of actual gases in many circumstances. The molecules of a perfect gas would behave like tiny, perfectly elastic spheres in random motion, and would influence one another only when they collided. Their total volume would be negligible in comparison with the space in which they moved. From these hypotheses the kinetic theory of gases indicates that, for equilibrium conditions, the absolute pressure p, the volume V occupied by mass m, and the absolute temperature T would be related by the expression

$$pV = mRT$$

i.e.

$$p = \varrho RT \qquad\qquad (1.5)$$

where ϱ represents the density and R the 'gas constant', the value of which depends on the gas concerned.

Equation of state

Any equation that relates p, ϱ and T is known as an *equation of state* and eqn 1.5 is therefore termed the *equation of state of a perfect gas*. Most gases, if at temperatures and pressures well away both from the liquid phase and from dissociation, obey this relation closely and so their pressure, density and (absolute) temperature may, to a good approximation, be related by eqn 1.5. For example, air at normal temperatures and pressures behaves closely in accordance with the equation. But gases near to liquefaction – which are then usually termed vapours – depart markedly from the behaviour of a perfect gas. Equation 1.5 therefore does not apply to substances such as non-superheated steam and the vapours used in refrigerating plants. For such substances, corresponding values of pressure, temperature and density must be obtained from tables or charts.

Thermally perfect gas

A gas for which

$$p/\varrho T = R = \text{constant}$$

is said to be *thermally perfect*.

It is usually assumed that the equation of state is valid not only when the fluid is in mechanical equilibrium and neither giving nor receiving heat, but also when it is not in mechanical or thermal equilibrium. This assumption seems justified because deductions based on it have been found to agree with experimental results.

It should be noted that R is defined by eqn 1.5 as $p/\varrho T$: its dimensional formula is therefore

$$\left[\frac{F}{L^2}\right]\bigg/\left[\frac{M}{L^3}\theta\right] = \frac{[FL]}{[M\theta]}$$

where $[F]$ is the dimensional symbol for force and $[\theta]$ that for temperature. For air the value of R is 287 J/kg K. (Some – particularly American – writers define the gas constant as p/wT, where w represents the weight per unit volume; this form has the dimensional formula

$$\left[F/L^2\right] \div \left[F\theta/L^3\right] = \left[L/\theta\right]$$

The dependence of a gas 'constant' on the weight of a given volume of gas rather than on its mass is illogical, and eqn 1.5 – used throughout this book – is the preferred form of the equation of state for a perfect gas.)

■

Example 1.1   A mass of air, at a pressure of 200 kPa and a temperature of 300 K, occupies a volume of $3\,m^3$. Determine:

(a) the density of the air
(b) its mass

Solution

□

$(a)\quad \varrho = \dfrac{p}{RT} = \dfrac{2 \times 10^5\,N/m^2}{287\,J/(kg\,K)\times 300\,K} = 2.32\,kg/m^3$

$(b)\quad m = \varrho V = 2.32\,kg/m^3 \times 3\,m^3 = 6.96\,kg$

Universal gas constant

For a given temperature and pressure eqn 1.5 indicates that $\varrho R =$ constant. By Avogadro's hypothesis, all pure gases at the same temperature and pressure have the same number of molecules per unit volume. The density is proportional to the mass of an individual molecule and so the product of R and the relative molecular mass M is constant for all perfect gases. This product MR is known as the *universal gas constant*, R_0; for real gases it is not strictly constant but for monatomic and diatomic gases its variation is slight. If M is the ratio of the mass of the molecule to the mass of a normal hydrogen atom, $MR = 8314\,J/kg\,K$.

Calorically perfect gas

A gas for which the specific heat capacity at constant volume, c_v, is a constant is said to be *calorically perfect*. The term *perfect gas*, used without qualification, generally refers to a gas that is both thermally and calorically perfect. (Some writers use semi-perfect as a synonym for thermally perfect.)

Example 1.2 Find the gas constant for the following gases: CO, CO_2, NO, N_2O. The relative atomic masses are: C = 12, N = 14, O = 16.

Solution
For CO, the relative molecular mass $M = 12 + 16 = 28$.
Hence, for CO

$$R = \frac{R_0}{M} = \frac{MR}{M} = \frac{8314\,J/(kg\,K)}{28} = 297\,J/(kg\,K)$$

Similarly

For CO_2, $M = 12 + (2 \times 16) = 44$ and
$$R = 8314/44 = 189\,J/(kg\,K)$$

For NO, $M = 14 + 16 = 30$ and
$$R = 8314/30 = 277\,J/(kg\,K)$$

For N_2O, $M = (2 \times 14) + 16 = 44$ and
$$R = 8314/44 = 189\,J/(kg\,K)$$

1.4.1 Changes of state

A change of density may evidently be achieved both by a change of pressure and by a change of temperature. If the process is one in which the temperature is held constant, it is known as *isothermal*.

Isothermal process

On the other hand, the pressure may be held constant while the temperature is changed. In either of these two cases there must be a transfer of heat to or from the gas so as to maintain the prescribed conditions. If the density change occurs with no heat transfer to or from the gas, the process is said to be *adiabatic*.

Adiabatic process

If, in addition, no heat is generated within the gas (for example by friction) then the process is described as isentropic, and the absolute pressure and density of a perfect gas are related by the additional expression (developed in Section 11.2):

Isentropic process

$$p/\varrho^\gamma = \text{constant} \qquad (1.6)$$

where γ (Greek 'gamma') $= c_p/c_v$ and c_p and c_v represent the specific heat capacities at constant pressure and constant volume respectively. For air and other diatomic gases in the usual ranges of temperature and pressure $\gamma = 1.4$.

1.5 COMPRESSIBILITY

All matter is to some extent compressible. That is to say, a change in the compressive stress applied to a certain amount of a substance always produces some change in its volume. Although the compressibility of different substances varies widely, the *proportionate* change in volume of a particular material that does not change its phase (e.g. from liquid to solid) during the compression is directly related to the change in the compressive stress.

Bulk modulus of elasticity

The degree of compressibility of a substance is characterized by the *bulk modulus of elasticity*, K, which is defined by the equation

$$K = -\frac{\delta p}{\delta V/V} \qquad (1.7)$$

Here δp represents a small increase in pressure applied to the material and δV the corresponding small increase in the original volume V. Since a rise in pressure always causes a *decrease* in volume, δV is always negative, and the minus sign is included in the equation to give a positive value of K. As $\delta V/V$ is simply a ratio of two volumes it is dimensionless and thus K has the same dimensional formula as pressure. In the limit, as $\delta p \to 0$, eqn 1.7 becomes $K = -V(\partial p/\partial V)$. As the density ϱ is given by mass/volume $= m/V$

$$d\varrho = d(m/V) = -\frac{m}{V^2}dV = -\varrho\frac{dV}{V}$$

so K may also be expressed as

$$K = \varrho(\partial p/\partial \varrho) \qquad (1.8)$$

Compressibility

The reciprocal of bulk modulus is sometimes termed the *compressibility*.

The value of the bulk modulus, K, depends on the relation between pressure and density for the conditions under which the compression takes place. Two sets of conditions are especially important. If the compression occurs while the temperature is kept constant, the value of K is the *isothermal* bulk modulus. On the other hand, if no heat is added to or taken from the fluid during the compression, and there is no friction, the corresponding value of K is the *isentropic* bulk modulus. The ratio of the isentropic to the isothermal bulk modulus is γ, the ratio of the specific heat capacity at constant pressure to that at constant volume. For liquids the value of γ is practically unity, so the isentropic and isothermal bulk moduli are almost identical. Except in work of high accuracy it is not usual to distinguish between the bulk moduli of a liquid.

For liquids the bulk modulus is very high, so the change of density with increase of pressure is very small even for the largest pressure changes encountered. Accordingly, the density of a liquid can normally be regarded as constant, and the analysis of problems involving liquids is thereby simplified. In circumstances where changes of pressure are either very large or very sudden, however – as in 'water hammer' (see Section 12.3) – the compressibility of liquids must be taken into account.

As a liquid is compressed its molecules become closer together, so its resistance to further compression increases, i.e. K increases. The bulk modulus of water, for example, roughly doubles as the pressure is raised from 1 atm to 3500 atm. There is also a decrease of K with increase of temperature.

Unlike liquids, gases are easily compressible. However, the term 'bulk modulus of elasticity' is seldom used of a gas because, as will be seen in Section 11.2, the quantity $\varrho(\partial p/\partial \varrho)$ is usually directly proportional to the pressure, so K may not be regarded as even approximately constant.

Where gases undergo only very small changes of density (as in ventilation systems, for example) the effects of compressibility may be disregarded. These effects, however, come into prominence when the relative velocity between the gas and a solid body approaches the velocity at which sound is propagated through the gas. Serious errors would arise if relations for the flow of an incompressible fluid were applied to such situations.

1.6   VISCOSITY

Although all real fluids resist any force tending to cause one layer to move over another, the resistance is offered only while the movement is taking place. Thus when the external force is removed, the flow subsides because of the resisting forces, but when the flow stops the particles of fluid stay in the positions they have reached and have no tendency to revert to their original positions. This resistance to the movement of one layer of fluid over an adjoining one is ascribed to the *viscosity* of the fluid. Since relative motion between layers requires shearing forces, that is, forces parallel to the surfaces over which they act, the resisting forces must be in the exactly opposite directions and so they too are parallel to the surfaces.

It is a matter of common experience that, under particular conditions, one fluid offers greater resistance to flow than another. Such liquids as tar, treacle and glycerine cannot be rapidly poured or easily stirred, and are commonly spoken of as 'thick'; on the other hand, so-called 'thin' liquids such as water, petrol and paraffin flow much more readily. (Lubricating oils with small viscosity are sometimes referred to as 'light', and those with large viscosity as 'heavy'; but viscosity is not related to density.)

Gases as well as liquids have viscosity, although the viscosity of gases is less evident in everyday life.

1.6.1 Quantitative definition of viscosity

Consider the motion of fluid illustrated in Fig. 1.2. All particles are moving in the same direction, but different layers of the fluid move with different velocities (as indicated here by the lengths of the arrows). Thus one layer moves relative to another. We assume for the moment that the parallel movements of the layers are in straight lines. A particular small portion of the fluid will be deformed from its original rectangular shape *PQRS* to *P'Q'R'S'* as it moves along. However, it is not the displacement of *P'Q'* relative to *S'R'* that is important, so much as the angle *α*. The right-hand diagram of Fig. 1.3 represents a smaller degree of deformation than does the left-hand diagram, although the relative movement between top and bottom of the portion considered is the same in each case. The *linear* displacement is a matter of the difference of velocity between the two planes *PQ* and *SR* but the *angular* displacement depends also on the distance between the planes. Thus the important factor is the *velocity gradient*, that is the rate at which the velocity changes with the distance across the flow.

Suppose that the velocity *u* varies with distance *y* (measured from some fixed reference plane) in such a manner as in Fig. 1.4 (such a curve

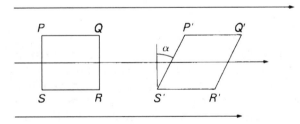

Fig. 1.2

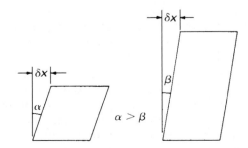

Fig. 1.3

is termed *velocity profile*). The velocity gradient is given by $\delta u/\delta y$ or, in the limit as $\delta y \rightarrow 0$, by $\partial u/\partial y$. The partial derivative $\partial u/\partial y$ is used because in general the velocity varies also in other directions. Only the velocity gradient in the y direction concerns us here.

Figure 1.5 represents two adjoining layers of the fluid, although they are shown slightly separated for the sake of clarity. The upper layer, supposed the faster of the two, tends to draw the lower one along with it by means of a force F on the lower layer. At the same time, the lower layer (by Newton's Third Law) tends to retard the faster, upper, one by an equal and opposite force acting on that. If the force F acts over an area of contact A the stress τ (Greek letter 'tau') is given by F/A.

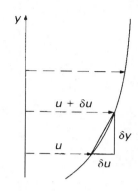

Fig. 1.4

Newton (1642–1727) postulated that, for the straight and parallel motion of a given fluid, the tangential stress between two adjoining layers is proportional to the velocity gradient in a direction perpendicular to the layers. That is

$$\tau = F/A \propto \partial u/\partial y$$

or

$$\tau = \mu \frac{\partial u}{\partial y} \qquad (1.9)$$

where μ (Greek 'mu') is a constant for a particular fluid at a particular temperature. This coefficient of proportionality μ is now known as the *coefficient of viscosity* or, more simply, the *viscosity* of the fluid. It is also termed absolute viscosity or dynamic viscosity to distinguish it from kinematic viscosity (Section 1.6.4). The symbols μ and η (Greek 'eta') are both widely used for viscosity; in this book μ will be used. The restriction of eqn 1.9 to straight and parallel flow is necessary because only in these circumstances does the increment of velocity δu necessarily represent the rate at which one layer of fluid slides over another.

It is important to note that eqn 1.9 strictly concerns the velocity gradient and the stress *at a point*: the change of velocity considered is that occurring over an infinitesimal thickness and the stress is given by the force acting over an infinitestimal area. Thus the relation $\tau = \mu \Delta u/\Delta y$, where Δu represents the change of velocity occurring over a larger, finite distance Δy, is not necessarily true.

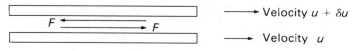

Fig. 1.5

To remove the restriction to straight and parallel flow, we may substitute 'the rate of relative movement between adjoining layers of the fluid' for δu, and 'rate of shear' for 'velocity gradient'. As will be shown in Section 6.6.4, if angular velocity is involved then the rate of shear is not necessarily identical with the velocity gradient and, in general, the rate of shear represents only part of the velocity gradient. With this modification, eqn 1.9 may be used to define viscosity as the shear stress, at any point in a flow, divided by the rate of shear at that point in the direction perpendicular to the surface over which the stress acts.

The absolute viscosity μ is a property of the fluid and a scalar quantity. The other terms in eqn 1.9, however, refer to vector quantities, and it is important to relate their directions. We have already seen that the surface over which the stress τ acts is (for straight and parallel flow) perpendicular to the direction of the velocity gradient. (With the notation of eqn 1.9 the surface is perpendicular to the y coordinate or, in other words, parallel to the x–z plane.) We have seen too that the line of action of the force F is parallel to the velocity component u. Yet what of the *sense* of this force? In Fig. 1.5, to which of the two forces each labelled F does eqn 1.9 strictly apply?

If the velocity u increases with y, then $\partial u/\partial y$ is positive and eqn 1.9 gives a positive value of τ. For simplicity the positive sense of the force or stress is defined as being the same as the positive sense of velocity. Thus, referring again to Fig. 1.5, the value of τ given by the equation refers to the stress acting on the lower layer. In other words, both velocity and stress are considered positive in the direction of increase of the coordinate parallel to them; and the stress given by eqn 1.9 acts over the surface facing the direction in which the perpendicular coordinate (e.g. y) increases.

For many fluids the magnitude of the viscosity is independent of the rate of shear, and although it may vary considerably with temperature it may be regarded as a constant for a particular fluid and temperature. Such fluids are known as Newtonian fluids. Those fluids that behave differently are discussed in Section 1.6.6.

Equation 1.9 shows that, irrespective of the magnitude of μ, the stress is zero when there is no relative motion between adjoining layers. Moreover, it is clear from the equation that $\partial u/\partial y$ must nowhere be infinite, since such a value would cause an infinite stress and this is physically impossible. Consequently, if the velocity varies across the flow, it must do so continuously and not change by abrupt steps between adjoining elements of the fluid. This condition of continuous variation must be met also at a solid boundary; the fluid immediately in contact with the boundary does not move relative to it because such motion would constitute an abrupt change. (This fluid cannot slide past the surface *en bloc* because, no matter how smooth the surface, its inevitable irregularities are large compared with the molecules of the fluid and so some fluid is held in the depressions on the surface. Molecules of the fluid may also be held on the solid surface by adsorption.) In a viscous fluid, then, a condition that must always be satisfied is that there should be no 'slipping' at solid boundaries. (Gases at extremely low pressures – for which the mean free path of the molecules becomes comparable with the size of the irregularities of the surface – may no longer be regarded as continuous dispersions of matter and they consequently lose some of the characteristics of fluids. In these conditions appreciable slip at a boundary may occur. Such conditions, however, are outside normal engineering practice.)

It will be seen that there is a certain similarity between the coefficient of viscosity in a fluid and the shear modulus of elasticity in a solid. Whereas, however, a solid continues to deform only until equilibrium is reached between the internal resistance to shear and the external force producing it, a fluid continues to deform indefinitely, provided that the external force remains in action. In a fluid it is the *rate* of deformation, not the deformation itself, that provides the criterion for equilibrium of force. For this reason J. C. Maxwell called viscosity 'fugitive elasticity'.

To maintain relative motion between adjoining layers of a fluid, work must be done continuously against the viscous forces of resistance. In other words, energy must be continuously supplied. Whenever a fluid flows there is a loss of mechanical energy, often ascribed to 'fluid friction', which is that used to overcome the viscous forces. The energy is dissipated as heat, and for practical purposes may usually be regarded as lost for ever. On some occasions the energy needed to overcome 'fluid friction' may be neglected without serious error; at other times this energy may be far from negligible, but the way in which viscosity enters a problem is frequently so intricate that a solution is possible only if the effects of viscosity are first neglected, and then an overall correction is applied to the results.

1.6.2   The causes of viscosity

For one possible cause of viscosity we may consider the forces of attraction between molecules. Yet there is evidently also some other explanation, because gases have by no means negligible viscosity although their molecules are in general so far apart that no appreciable inter-molecular force exists. We know, however, that the individual molecules of a fluid are continuously in motion and this motion makes possible a process of exchange of momentum between different layers of the fluid. Suppose that, in straight and parallel flow, a layer *aa* (Fig. 1.6) in the fluid is moving more rapidly than an adjacent layer *bb*. Some molecules from the layer *aa*, in the course of their continuous thermal agitation, migrate into the layer *bb*, taking with them the momentum they have as a result of the overall velocity of layer *aa*. By 'collisions' with other molecules already in layer *bb* this momentum is shared among the occupants of *bb*, and thus layer *bb* as a whole is speeded up. Similarly, molecules from the slower layer *bb* cross to *aa* and tend to retard the layer *aa*. Every such migration of molecules, then, causes forces of acceleration

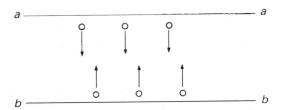

Fig. 1.6

or deceleration in such directions as to tend to eliminate the differences of velocity between the layers.

In gases this interchange of molecules forms the principal cause of viscosity, and the kinetic theory of gases (which deals with the random motions of the molecules) allows the predictions, borne out by experimental observations, that (a) the viscosity of a gas is independent of its pressure (except at very high or very low pressure), and (b) because the molecular motion increases with a rise of temperature, the viscosity also increases with a rise of temperature (unless the gas is so highly compressed that the kinetic theory is invalid).

Although the process of momentum exchange occurs also in liquids, the molecules of a liquid are sufficiently close together for there to be appreciable forces between them. Relative movement of layers in a liquid modifies these inter-molecular forces, thereby causing a net shear force which resists the relative movement. The exact way in which the inter-molecular forces are modified is, however, a matter for debate[1]. At all events, the viscosity of a liquid is the resultant of two mechanisms, each of which depends on temperature, and so the variation of viscosity with temperature is much more complex than for a gas. The viscosity of nearly all liquids decreases with rise of temperature, but the rate of decrease also falls. Except at very high pressures, however, the viscosity of a liquid is independent of pressure.

The variation of the viscosity of a few common fluids with temperature is given in Appendix 2.

1.6.3 The dimensional formula and units of absolute viscosity

Absolute viscosity is defined as the ratio of a shear stress to a velocity gradient. Since stress is defined as the ratio of a force to the area over which it acts, its dimensional formula is $[F/L^2]$. Velocity gradient is defined as the ratio of increase of velocity to the distance across which the increase occurs, thus giving the dimensional formula $[L/T] \div [L] \equiv [1/T]$. Consequently the dimensional formula of viscosity is $[F/L^2] \div [1/T] \equiv [FT/L^2]$. Since $[F] \equiv [ML/T^2]$ the expression is equivalent to $[M/LT]$. Although this last form is usually more suitable in dimensional analysis, the version involving $[F]$ is also useful in defining units of viscosity.

The basic SI unit of viscosity is Pa s, or $kg\,m^{-1}\,s^{-1}$, but no special name for it has yet found international agreement. (The name *poiseuille*, abbreviated Pl, has been used in France but must be carefully distinguished from poise – see below. 1 Pl = 10 poise.) Water at 20°C has a viscosity of almost exactly 10^{-3} Pa s.

Data for absolute viscosity are still commonly found in terms of units from the c.g.s. system where the dyne was the unit of force, the centimetre the unit of length and the second the unit of time interval. Thus the resulting unit of absolute viscosity was second $\times$ dyne/cm^2. This was termed the *poise* (abbreviated P) in honour of J. L. M. Poiseuille (1799– 1869) whom we shall meet again in Chapter 6.

Smaller units, the centipoise, cP, i.e. 10^{-2} poise, the millipoise, mP (10^{-3} poise) and the micropoise, μP (10^{-6} poise) were also used.

Many units based on those of the Imperial system have been used for absolute viscosity, but none has had a well-known and unambiguous name. The units lbf s/ft^2 and pdl s/ft^2 [respectively identical to slug/(ft s) and lbm/(ft s)] have both been widely used. Unfortunately these units have not always been expressed in such unequivocal form. Terms such as 'British units', 'engineers' units', 'FPS units' and so on not only fail to distinguish between units based on the pound force and those based on the poundal, but have sometimes been used with different meanings by different people.

1.6.4 Kinematic viscosity and its units

In perhaps the majority of problems involving viscosity, we are concerned with the magnitude of the viscous forces compared with the magnitude of the inertia forces, that is, those forces causing acceleration of particles of the fluid. Since the viscous forces are proportional to the absolute viscosity μ and the inertia forces to the density ϱ, the ratio μ/ϱ is frequently involved, and in fact has come to be regarded almost as a property of the fluid in its own right. The ratio of absolute viscosity to density is known as the *kinematic viscosity* and is denoted by the symbol ν (Greek 'nu') so that

$$\nu = \frac{\mu}{\varrho} \tag{1.10}$$

(Care should be taken in writing the symbol ν: it is easily confused with v.)

The dimensional formula of ν is given by

$$\left[M/LT\right] \div \left[M/L^3\right] \equiv \left[L^2/T\right]$$

It will be noticed that [M] cancels and so ν is independent of the units of mass. Only magnitudes of length and time interval are involved. Kinematics is defined as the study of motion without regard to the causes of the motion, and so is concerned with lengths and time intervals only, not with masses. That is why the name kinematic viscosity, now in universal use, has been given to the ratio μ/ϱ.

The basic unit m^2/s is too large for most purposes, so the mm^2/s $(=10^{-6}\,m^2/s)$ is generally employed. The c.g.s. unit, cm^2/s, was termed the *stokes* (abbreviated S or St), thus honouring the Cambridge physicist, Sir George Stokes (1819–1903), who contributed much to the theory of viscous fluids. This unit too was rather large, and the centistokes (cS), i.e. 10^{-2} stokes = mm^2/s, was in wide use. Water has a kinematic viscosity of exactly $1\,mm^2/s$ at $20.2\,°C$.

Since data expressed in Imperial units are still commonly encountered, it is well to point out that errors frequently arise when units are used that are based on different primary units. If the unit used for (absolute) viscosity is $lbf\,s/ft^2$ [identical with $slug/(ft\,s)$] then the unit used for density must be $slug/ft^3$ if the implied unit of kinematic viscosity is to be ft^2/s. Or if density is expressed in terms of lbm/ft^3 the unit of viscosity must also involve lbm in the form $lbm/ft\,s$ (i.e. $pdl\,s/ft^2$) if hybrid units are to be avoided. With either combination of consistent units, ft^2/s may be obtained as the unit of kinematic viscosity whereas the *inconsistent* pair $lbf\,s/ft^2$ and lbm/ft^3 yield $32.1740\,ft^2/s$ as the implied unit of kinematic viscosity. The inclusion of the units as well as the numerics in every stage of calculations is strongly recommended.

Although, as Appendix 2 shows, the absolute viscosity of air at ordinary temperatures is only about one-sixtieth that of water, yet because of its much smaller density its *kinematic* viscosity is 13 times that of water.

Measurement of absolute and kinematic viscosities is discussed in Chapter 6.

1.6.5 Fluidity

Instead of viscosity the inverse property of *fluidity* is sometimes considered. It is defined as the reciprocal of viscosity; its usual symbol is ϕ (i.e. $\phi = 1/\mu$) and the usual unit has been the 'reciprocal poise', i.e. $cm\,s/g \equiv cm^2/(s\,dyne)$, occasionally termed 'rhe'. Fluidity, however, is seldom used in engineering practice.

1.6.6 Non-Newtonian liquids

For most fluids the absolute viscosity is independent of the velocity gradient in straight and parallel flow, so Newton's hypothesis is fulfilled. Equation 1.9 indicates that a graph of stress against rate of shear is a straight line through the origin with slope equal to μ (Fig. 1.7). There is, however, a fairly large category of liquids for which the viscosity is not independent of the rate of shear, and these liquids are referred to as non-Newtonian. Solutions (particularly of colloids) often have a reduced viscosity when the rate of shear is large, and such liquids are said to be *pseudo-plastic*. Gelatine, clay, milk, blood and liquid cement come in this category.

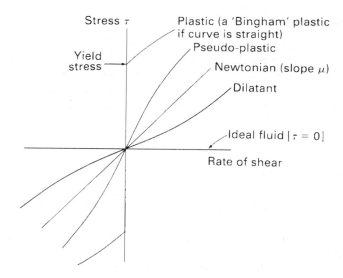

Fig. 1.7

A few liquids exhibit the converse property of *dilatancy*; that is, their effective viscosity *increases* with increasing rate of shear. Concentrated solutions of sugar in water and aqueous suspensions of rice starch (in certain concentrations) are examples.

Additional types of non-Newtonian behaviour may arise if the apparent viscosity changes with the time for which the shearing forces are applied. Liquids for which the apparent viscosity increases with the duration of the stress are termed *rheopectic*; those for which the apparent viscosity decreases with the duration are termed *thixotropic*.

A number of materials have the property of *plasticity*. Metals when strained beyond their elastic limit or when close to their melting points can deform continuously under the action of a constant force, and thus in some degree behave like liquids of high viscosity. Their behaviour, however, is non-Newtonian, and most of the methods of mechanics of fluids are therefore inapplicable to them.

Viscoelastic materials possess both viscous and elastic properties; bitumen, nylon and flour dough are examples. In steady flow, that is, flow not changing with time, the rate of shear is constant and may well be given by τ/μ where μ represents a constant coefficient of viscosity as in a Newtonian fluid. Elasticity becomes evident when the shear stress is changed. A rapid increase of stress from τ to $\tau + \delta\tau$ causes the material to be sheared through an additional angle $\delta\tau/G$ where G represents an elastic modulus; the corresponding rate of shear is $(1/G)\partial\tau/\partial t$ so the total rate of shear in the material is $(\tau/\mu) + (1/G)\partial\tau/\partial t$.

The fluids with which engineers most often have to deal are Newtonian, that is, their viscosity is not dependent on either the rate of shear or its duration, and the term 'mechanics of fluids' generally refers only to Newtonian fluids. (The study of non-Newtonian liquids is termed 'rheology'.)

1.6.7 Ideal fluid

For mathematical simplicity in describing flow in which the influence of viscosity is small a hypothetical fluid having zero viscosity may be postulated. This is known as an *ideal fluid*. Although 'ideal' is sometimes a synonym for 'perfect', the 'idealizing' referred to here concerns only the neglect of viscosity: a perfect gas (Section 1.4) need not be considered an ideal fluid.

1.7 SURFACE TENSION

Although any detailed consideration of surface tension is more in the province of the physicist than of the engineer, this property of liquids deserves brief mention because it does enter certain engineering problems.

It arises from the forces between the molecules of a liquid and the forces (generally of different magnitude) between the liquid molecules and those of any adjacent substance. There is, in consequence, a resultant force which opposes any increase in the area of a liquid surface. Such an increase therefore requires the expenditure of mechanical energy, and the existence of a free surface implies the presence of *free surface energy*, which equals the work that was done when the surface was formed. (Only stored mechanical energy is referred to. The *total* surface energy also includes thermal energy.)

Any system tends to move towards a condition of stable equilibrium in which its potential energy is a minimum. Thus a quantity of a liquid will (in the absence of any other constraint) adjust its shape until its surface area, and consequently its free surface energy, is a minimum. For example, a drop of liquid, free from all other forces, takes on a spherical form since, for a given volume, the sphere is the geometrical shape having the least surface area.

Free surface energy necessarily implies the existence of a tensile force in the surface. A surface requires mechanical energy for its formation, and if it contracts it loses mechanical energy, that is, it does mechanical work. Hence it exerts on its surroundings a force in the direction in which it moves when contracting. In other words, the surface is in a state of tension.

If a line is imagined drawn in the surface, then the liquid on one side of the line pulls that on the other side. The magnitude of surface tension is defined as that of the tensile force acting across and perpendicular to a short, straight element of the line, divided by the length of that line element. The dimensional formula of surface tension, therefore, is given by [Force/Length], i.e. $[MLT^{-2}]/[L] = [MT^{-2}]$. As symbols for surface tension, γ (Greek 'gamma') and σ (Greek 'sigma') are both widely used; γ will be used in this book.

Surface tension thus defined, however, is equivalent to the free surface energy per unit area as may be readily shown. Suppose that the

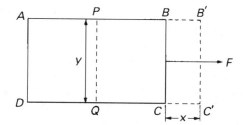

Fig. 1.8

small rectangle of surface $ABCD$ (Fig. 1.8), associated with a given volume of liquid, is increased to the area $AB'C'D$ by a force applied to the side BC. From the definition of surface tension γ, the tensile force acting across any line PQ, of length y parallel to AD, equals γy. This may be regarded as the force with which the section $APQD$ holds the section $PBCQ$. Provided that the movement of BC to $B'C'$ (a distance x) is carried out very slowly and with no acceleration, the necessary force F equals in magnitude the resisting force γy, and the work done in the process is γyx. Thus the mechanical work done per unit increase of area, that is, the free surface energy per unit area is $\gamma yx \div xy = \gamma$.

It is perhaps prudent to mention that some writers have adduced fallacious arguments to deny the existence of surface tension while admitting the idea of surface energy. One concept, however, implies the other: surface tension does exist.

Water in contact with air has a surface tension of about 0.073 N/m at usual temperature; most organic liquids have values between 0.020 and 0.030 N/m and mercury about 0.48 N/m, the liquid in each case being in contact with air. For all liquids the surface tension decreases as the temperature rises. The surface tension of water may be considerably reduced by the addition of small quantities of organic solutes such as soap and detergents. Salts such as sodium chloride in solution raise the surface tension of water. That tension which exists in the surface separating two immiscible liquids is usually known as *interfacial tension*.

1.7.1 Cohesion and adhesion

As we have seen, the molecules of a liquid are bound to one another by forces of molecular attraction, and it is these forces that give rise to *cohesion*, that is, the tendency of the liquid to remain as one assemblage of particles rather than to behave as a gas and fill the entire space within which it is confined. Forces between the molecules of a fluid and the molecules of a solid boundary surface give rise to *adhesion* between the fluid and the boundary.

If the forces of adhesion between the molecules of a particular liquid and a particular solid are greater than the forces of cohesion among the liquid molecules themselves, the liquid molecules tend to crowd towards the solid surface, and the area of contact between liquid and solid tends to increase. Given the opportunity, the liquid then spreads over the solid surface and 'wets' it. Water will wet clean glass, but mercury will not. Water, however, will not wet wax or a greasy surface.

The interplay of these various forces explains the capillary rise or depression that occurs when a free liquid surface meets a solid boundary. Unless the attraction between molecules of the liquid exactly equals that between molecules of the liquid and molecules of the solid, the surface near the boundary becomes curved. Now if the surface of a liquid is curved the surface tension forces have a resultant towards the concave side. For equilibrium this resultant must be balanced by a greater pressure at the concave side of the surface. It may readily be shown that if the surface has radii of curvature R_1 and R_2 in two perpendicular planes the pressure at the concave side of the surface is greater than that at the convex side by

$$\gamma\left(\frac{1}{R_1} + \frac{1}{R_2}\right) \qquad (1.11)$$

Inside a spherical drop, for instance, the pressure exceeds that outside by $2\gamma/R$ (since here $R_1 = R_2 = R$). However, the excess pressure inside a soap bubble is $4\gamma/R$; this is because the very thin soap film has two surfaces, an inner and an outer, each in contact with air. Applying expression 1.11 and the principles of statics to the rise of a liquid in a vertical capillary tube yields

$$h = \frac{4\gamma\cos\theta}{\varrho g d} \qquad (1.12)$$

where h represents the 'capillary rise' of the liquid surface (see Fig. 1.9), θ represents the 'angle of contact' between the wall of the tube and the interface between the liquid and air, ϱ the density of the liquid, g the weight per unit mass, and d the diameter of the tube. (For two liquids in contact γ represents the interfacial tension, and ϱ the *difference* of their densities.) However, the assumption of a spherical interface between the liquid and air (and other approximations) restricts the application of

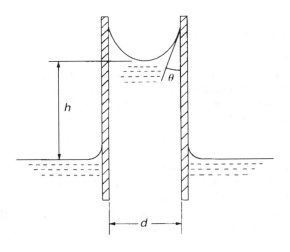

Fig. 1.9

the formula to tubes of small bore (say less than 3 mm). Moreover, although for pure water in a completely clean glass tube $\theta = 0$, the value may well be different in engineering practice, where cleanliness of a high order is seldom found, or with tubes of other materials. Equation 1.12 therefore overestimates the actual capillary rise. Mercury, which has an angle of contact with clean glass of about 130° in air, and therefore a negative value of cos θ, experiences a capillary depression.

Surface tension becomes important when solid boundaries of a liquid surface are close together or when the surface separating two immiscible fluids has a very small radius of curvature. The forces due to surface tension then become comparable with other forces and so may appreciably affect the behaviour of the liquid. Such conditions may occur, for example, in small-scale models of rivers or harbours. The surface tension forces may be relatively much more significant in the model than in the full-size structure; consequently a simple scaling-up of measurements made on the model may not yield results accurately corresponding to the full-size situation.

In apparatus of small size the forces due to surface tension can completely stop the motion of a liquid if they exceed the other forces acting on it. It is well known, for example, that a cup or tumbler may be carefully filled until the liquid surface is perhaps 3 mm above the rim before overflowing begins. Among other instances in which surface tension plays an important role are the formation of bubbles, the break-up of liquid jets and the formation of drops, the rise of water in soil above the level of the water table, and the flow of thin sheets of liquid over a solid surface.

In most engineering problems, the distances between boundaries are sufficient for surface tension forces to be negligible compared with the other forces involved. The consequent customary neglect of the surface tension forces should not, however, lead us to forget their importance in small-scale work.

1.8 BASIC CHARACTERISTICS OF FLUID FLOW

1.8.1 Variation of flow parameters in time and space

In general the parameters such as velocity, pressure and density, which describe the behaviour and state of a fluid, are not constant in a particular set of circumstances. They may vary from one point to another, or from one instant of time to another, or they may vary with both position and time.

Steady flow is defined as that in which the various parameters at any point do not change with time. Flow in which changes with time do occur is termed *unsteady* or *non-steady*. Not surprisingly, steady flow is simpler to analyse than unsteady flow and we shall confine our attention mainly to steady conditions except in Chapter 12. In practice absolutely

Steady flow

steady flow is the exception rather than the rule, but many problems may be studied effectively by assuming that the flow is steady, for, even though minor fluctuations of velocity and other quantities with time do in fact occur, the *average* value of any quantity over a reasonable interval of time remains unchanged.

A particular flow may appear steady to one observer but unsteady to another. This is because all movement is relative; any motion of one body can be described only by reference to another 'body', often a set of coordinate axes. For example, the movement of water past the sides of a motor-boat travelling at constant velocity would (apart from small fluctuations) appear steady to an observer in the boat. Such an observer would compare the water flow with an imaginary set of reference axes fixed to the boat. To someone on a bridge, however, the same flow would appear to change with time as the boat passed underneath the bridge. This observer would be comparing the flow with reference axes fixed relative to the bridge.

Since steady flow is usually much easier to examine than unsteady flow, reference axes are chosen, where possible, so that the flow with respect to them is steady. It should be remembered, however, that Newton's Laws of Motion are valid only if any movement of the coordinate axes takes place uniformly in a straight line.

So much for variations with time, or *temporal* variations as they are sometimes called. If, at a particular instant, the various quantities do not change from point to point over a specified region, then the flow is said to be *uniform* over that region. If however, changes do occur from one point to another, the flow is said to be *non-uniform*. These changes with position may be found in the direction of the flow or in directions perpendicular to it. This latter kind of non-uniformity is always encountered near solid boundaries past which the fluid is flowing. This is because all fluids possess viscosity which reduces the relative velocity to zero at a solid boundary. In a river, for example, the velocity of the water close to the sides and the base is less than in the centre of the cross-section. Nevertheless it may be possible to treat the flow as uniform over the cross-section, provided that only a region well removed from the boundaries is considered.

Steadiness of flow and uniformity of flow do not necessarily go together. Any of four combinations is possible, as shown in Table 1.4.

1.8.2 Describing the pattern of flow

Streamline

For a particular instant of time we may consider in the fluid an imaginary curve across which, at that instant, no fluid is flowing. Such a line is called a *streamline* (sometimes also known as a *flowline* or *line of flow*). At that instant, therefore, the direction of the velocity of every particle on the line is along the line. If a number of streamlines is considered at a particular instant, the pattern they form gives a good indication of the flow then occurring. For steady flow the pattern is unchanging, but for unsteady flow it changes with time. Consequently, streamlines must be

Table 1.4 Types of flow

Type	Example
1. Steady uniform flow	Flow at a high constant rate through a long straight pipe of constant cross-section. (The region close to the walls of the pipe is, however, disregarded.)
2. Steady non-uniform flow	Flow at constant rate through a tapering pipe.
3. Non-steady uniform flow	High-velocity flow accelerating or decelerating through a long straight pipe of constant cross-section. (Again the region close to the walls of the pipe is ignored.)
4. Non-steady, non-uniform flow	Accelerating or decelerating flow through a tapering pipe.

thought of as instantaneous, and the pattern they form may be regarded as corresponding to an instantaneous photograph of the flow.

The boundaries of the flow are always composed of streamlines because there is no flow across them. Provided that the flow is continuous, every streamline must be a continuous line, either extending to infinity both upstream and downstream or forming a closed curve.

Stream-tube

A bundle of neighbouring streamlines may be imagined which form a passage through which the fluid flows (Fig. 1.10), and this passage (not necessarily circular) is known as a *stream-tube*. A stream-tube with a cross-section small enough for the variation of velocity over it to be negligible is sometimes termed a *stream filament*. Since the stream-tube is bounded on all sides by streamlines and since, by definition, there can be no velocity across a streamline, no fluid may enter or leave a stream-tube except through its ends. The entire flow may be imagined to be composed of stream-tubes arranged in some arbitrary pattern.

Path-line

An individual particle of fluid does not necessarily follow a streamline, but traces out a *path-line*. In distinction to a streamline, a path-line may be likened, not to an instantaneous photograph of a procession of particles, but to a time exposure showing the direction taken by the *same* particle at successive instants of time.

Streak-line

In experimental work a dye or some other indicator is often injected into the flow, and the resulting stream of colour is known as a *streak-line* or *filament line*. It gives an instantaneous picture of the positions of all particles which have passed through the point of injection.

In general, the patterns of streamlines, path-lines and streak-lines for a given flow differ; apart from a few special cases it is only for steady flow that all three patterns coincide.

Fig. 1.10 A stream-tube.

1.8.3 One-, two- and three-dimensional flow

Three-dimensional flow

In general, fluid flow is three-dimensional in the sense that the flow parameters – velocity, pressure, and so on – vary in all three coordinate directions. Considerable simplification in analysis may often be achieved, however, by selecting the coordinate directions so that appreciable variation of the parameters occurs in only two directions, or even in only one.

One-dimensional flow

So-called *one-dimensional flow* is that in which all the flow parameters may be expressed as functions of time and one space coordinate only. This single space coordinate is usually the distance measured along the centre-line (not necessarily straight) of some conduit in which the fluid is flowing. For instance, the flow in a pipe is frequently considered one-dimensional: variations of pressure, velocity and so on may occur along the length of the pipe, but any variation over the cross-section is assumed negligible. In reality flow is never truly one-dimensional because viscosity causes the velocity to decrease to zero at the boundaries. Figure 1.11 compares the hypothetical 'one-dimensional' flow with actual flow in, say, a pipe. If, however, the non-uniformity of the actual flow is not too great, valuable results may often be obtained from a 'one-dimensional analysis'. In this the average values of the parameters at any given section (perpendicular to the flow) are assumed to apply to the entire flow at that section.

Two-dimensional flow

In *two-dimensional flow* the flow parameters are functions of time and two rectangular space coordinates (say x and y) only. There is no variation in the z direction and therefore the same streamline pattern could at any instant be found in all planes in the fluid perpendicular to the z direction. The flow past an aerofoil of uniform cross-section and infinite span (width), for instance, is the same in all planes perpendicular to its span. Water flow over a weir of uniform cross-section and infinite width is likewise two-dimensional. In practice, where the width is not infinite it may be satisfactory to assume two-dimensional flow over most of the aerofoil or weir, that is, to consider flow in a single plane; the results may then be modified by 'end corrections' accounting for the three-dimensional flow at the ends.

The flow illustrated in Fig. 1.12a is one-*directional* since flow occurs in the x direction only, but it is two-*dimensional* because the pressure

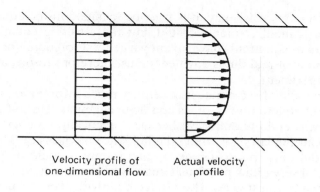

Velocity profile of   Actual velocity
one-dimensional flow   profile

Fig. 1.11

varies with x and the velocity varies with y. In Fig. 1.12b the velocity varies with both x and y; the pressure varies only with x when the streamlines are parallel, but with both x and y where the streamlines are curved.

Axi-symmetric flow, although not two-dimensional in the sense just defined, may be analysed more simply with the use of two cylindrical coordinates (x and r).

1.9   TWO KINDS OF FLOW

The influence of viscosity on flow is frequently very complex and not amenable to rigorous mathematical analysis. The shear stresses

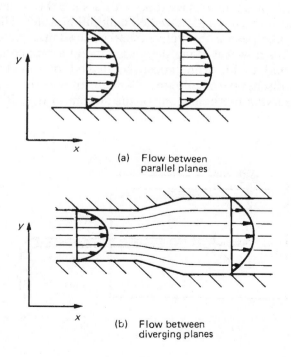

(a)   Flow between
parallel planes

(b)   Flow between
diverging planes

Fig. 1.12

resulting from viscosity may, it is true, be included in the equations of motion for a small element of fluid, but the resulting equations (the Navier–Stokes equations) have no known general solution. For studying the behaviour of real fluids engineers must therefore resort, at least in part, to experiment.

By the middle of the nineteenth century much information had been obtained relating to flow in pipes and open channels. This information, however, was entirely empirical because practically nothing was then known about the laws governing the influence of viscosity. The behaviour of a fluid in a particular set of circumstances could be predicted only from observations previously made of the same fluid in a closely similar set of circumstances. The results of individual experiments could not be brought together to form a universal law.

For some years, however, it had been realized that the flow of a fluid could be of two quite different kinds. Although the work of Hagen in about 1840 served to emphasize certain consequences of this difference, it was not until the experiments of Reynolds in the early 1880s that the essential nature of the two types of flow was clearly demonstrated. In fact, the distinction between them is most easily understood by reference to the work of Osborne Reynolds (1842–1912), Professor of Engineering at Manchester University, so it is to his experiments that we now turn on attention.

1.9.1 Reynolds's demonstration of the different kinds of flow

The apparatus used by Reynolds was basically as shown in Fig. 1.13. A straight length of circular glass tube with a smoothly rounded, flared inlet was placed in a large glass-walled tank full of water. The other end of the glass tube passed through the end of the tank. Water from the tank could thus flow out along the glass tube at a rate controlled by a valve at the outlet end. A fine nozzle connected to a small reservoir of a liquid dye discharged a coloured filament into the inlet of the glass tube. By observing the behaviour of the stream of dye, Reynolds was

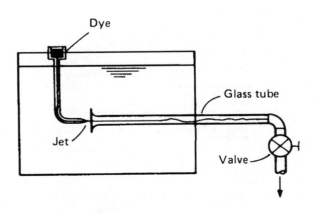

Fig. 1.13

able to study the way in which the water was flowing along the glass tube.

If the velocity of the water remained low and especially if the water in the tank had previously been allowed to settle for some time so as to eliminate all disturbances as far as possible, the filament of dye would pass down the tube without mixing with the water, and often so steadily as almost to seem stationary (Fig. 1.14a). As the valve was opened further and the velocity of the water thereby increased, this type of flow would persist until the velocity reached a value at which the stream of dye began to waver (Fig. 1.14b). Further increase in the velocity of the water made the fluctuations in the stream of dye more intense, particularly towards the outlet end of the tube, until a state was reached in which the dye no longer travelled the whole length of the tube as a distinct, unbroken thread, but, quite suddenly, mixed more or less completely with the water in the tube. Thus, except for a region near the inlet, the water in the tube became evenly coloured by the dye. Still further increases of velocity caused no more alteration in the type of flow, but the dye mixed even more readily with the water and complete mixing was achieved nearer the inlet of the tube. The original type of flow, in which the dye remained as a distinct streak, could be restored by reducing the velocity.

It is of particular interest that the disturbed flow always began far from the inlet (in Reynolds's tests, usually at a length from the inlet equal to about 30 times the diameter of the tube); also that the complete mixing occurred suddenly.

Although Reynolds used water in his original tests, subsequent experiments have shown beyond any doubt that the phenomenon is exhibited by all fluids, gases as well as liquids. Moreover, the two types of flow are to be found whatever the shape of the solid boundaries: there is no restriction to circular tubes.

In the first kind of flow, that occurring at the lower velocities, the particles of fluid are evidently moving entirely in straight lines even though the velocity with which particles move along one line is not

Laminar flow

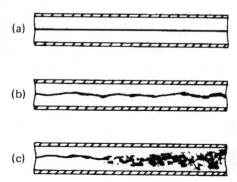

Fig. 1.14

necessarily the same as that along another line. Since the fluid may therefore be considered as moving in layers, or laminae (in this example, parallel to the axis of the glass tube), this kind of flow is now usually called *laminar flow*. (In a more general case of laminar flow the concept of 'laminae' is less appropriate because the velocity may vary in all directions and not just in the direction perpendicular to the 'laminae'.)

Turbulent flow

The second type of flow is known as *turbulent flow*. As indicated in Fig. 1.15a, the paths of individual particles of fluid are no longer everywhere straight but are sinuous, intertwining and crossing one another in a disorderly manner so that a thorough mixing of the fluid takes place. When turbulent flow takes place in a cylindrical tube, for example, only the *average* motion of the fluid is parallel to the axis of the tube. Turbulent flow, in short, is characterized by the fact that superimposed on the principal motion of the fluid are countless, irregular, haphazard secondary components. A single particle would thus follow an erratic path involving movements in three dimensions (Fig. 1.15b).

Turbulent motion is essentially irregular motion on a small scale. The movements of individual particles have no definite frequency, nor is there any ordered pattern as may be observed in large-scale eddies. Movements, even irregular movements, of sizeable portions of fluid are not a necessary feature of turbulence. Turbulence may – in fact, frequently does – exist in a fluid that appears to be flowing very smoothly and in which there is no obvious source of disturbance at all. Indeed, the fluctuations in the magnitude of the velocity of individual particles are often so small compared with the average magnitude, and the changes in direction so short-lived, that they are difficult to detect. These small fluctuations nevertheless have far-reaching effects.

Although the particles taking part in these random movements are small they are larger than individual molecules. Even in laminar flow there is a continual exchange of *molecules* between adjacent layers; this exchange, it will be remembered (Section 1.6.2), contributes to the phenomenon of viscosity. In turbulent flow not only do these tiny motions of molecules occur, but in addition there are random movements of larger particles.

Thus turbulent flow is very complex. Whereas laminar flow may be studied mathematically without the need for additional information obtainable only from experiments, a rigid mathematical analysis of turbulent flow is impossible because the erratic motion of one particle

Fig. 1.15

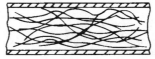

(a)

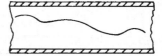

(b)

in the flow is not exactly repeated by any other particle. Statistical techniques have been called on, although they have as yet brought little benefit to practical calculations. We are forced therefore to use experimentally determined figures in calculations relating to turbulent flow.

Only laminar flow can strictly be described as steady. In turbulent flow there are continual variations of velocity (and pressure) at every point. If, however, the *average* velocity and pressure over a reasonable time interval remain constant the flow is nevertheless termed 'steady'. The term 'mean steady flow' would perhaps be a more accurate description.

'Mean steady flow', as just defined, does not include periodic motion such as tidal movements, ocean waves, or other oscillations of fluids. In such instances not only are the deviations of quantities such as velocity from their mean values much greater than the variations found in turbulent flow, but each oscillation takes place over a much greater time than do the random variations in turbulence.

A similar relaxing of the strict definition for *uniform* flow becomes necessary when turbulence is present. In turbulent flow the uniformity is understood to refer to the values at particular points averaged over a short interval of time.

The strict meaning of the word 'streamline' may also be relaxed. A streamline in turbulent flow is taken as a line across which there is no *net* flow during a short interval of time.

In engineering practice, fluid flow is nearly always turbulent. There are, however, some important instances of wholly laminar flow, for example in lubrication, and there are also many instances in which part of the flow is laminar.

1.9.2 The criterion of flow

As laminar and turbulent flow are widely different in their nature and effect, it is of prime importance to know the conditions under which each may be expected to exist, and the laws which govern them. We have already seen that the transition from laminar to turbulent flow depends on the velocity with which the fluid is moving. It was, it seems, the German engineer G. H. L. Hagen (1797–1884) who first noticed that the law governing the flow of a fluid through a pipe changed when a certain velocity was exceeded. He later discovered that the change from one type of flow to the other depended also on the temperature of the fluid, and so deduced that, since viscosity is the property that principally varies with temperature, the transition depended on the viscosity of the fluid. But Hagen was not apparently able to generalize his results into a law for all fluids and all the circumstances in which fluids might move. It was left to Reynolds to find the fundamental criterion that determines whether the flow is to be laminar or turbulent.

Reynolds's own experiments showed further that the velocity at which the transition occurred depended inversely on the diameter of the tube. He reasoned, however, that since all measurements are essentially comparisons, any apparent dependence of a phenomenon on absolute size or absolute velocity must actually be a dependence on the size or velocity compared with some other size or velocity. Any general law governing the flow of fluids must therefore be concerned with the criteria that make the behaviour of the fluid *similar* to the behaviour of the same or a different fluid when similar conditions are imposed on it. In the case of flow through a circular pipe, for example, the general law should enable us to say that if the fluid behaves in a particular manner when flowing in one circular pipe under a particular set of conditions then a fluid would behave in an *exactly similar* manner when flowing in another circular pipe under another set of conditions appropriately similar to the first.

The subject of similarity is treated more fully in Chapter 5. However, at this stage, we can confirm that, for the conditions under consideration similar behaviour of the fluid requires equality of $\varrho l u/\mu$. Here l and u represent a characteristic length and velocity, respectively, and ϱ and μ represent density and absolute viscosity. The quantity $\varrho l u/\mu$ is dimensionless. This may readily be verified by considering the dimensional formulae of the constituent terms:

$$\left[\frac{\varrho l u}{\mu}\right] \equiv \frac{[M/L^3][L][L/T]}{[M/LT]} = \frac{[M][L^2][L][T]}{[L^3][T][M]} = [1]$$

As will be shown in Chapter 5, the ratio $\varrho l u/\mu$ may be interpreted as the ratio of inertia to viscous forces in a flow. In calculating values of $\varrho l u/\mu$ we may use any consistent set of units, that is, such units that when the various magnitudes are inserted in the expression all the 'units' parts of the magnitudes cancel, leaving only the numerics. It is again strongly recommended that all the units be written beside the corresponding numerics: any inconsistencies will then be revealed.

Reynolds number

The conclusion that flow in which viscous and inertia forces are important is governed by the ratio of their magnitudes as represented by $\varrho l u/\mu$ was reached by Reynolds, a result of the greatest importance. The ratio $\varrho l u/\mu$ is a fundamental characteristic of the flow, and is now universally known as Reynolds number. The symbol now most commonly used for it is *Re*, and it is often an advantage to put the letters in parentheses – (*Re*) – to show that it is a single symbol and does not represent the product of *R* and *e*.

1.9.3 The significance of Reynolds number

The length l and velocity u appearing in Reynolds number are, as we have seen, quantities chosen as representative ones. Which length and which velocity are selected naturally affect the numerical value of *Re* but they do not affect its fundamental significance. In many applications convention has standardized the length and velocity to be considered; for flow in a circular pipe, for example, the representative length measurement is the diameter and the representative velocity is the mean velocity (that is, volume flow rate divided by cross-sectional area); for flow past a single plane boundary the length is usually measured from the forward edge in the direction of flow; in more unusual applications the length and velocity to be used must be specified.

The Reynolds number is essentially a means of comparing one flow with another, and provided that *corresponding* lengths and *corresponding* velocities are compared in the two flows the particular choices of length and velocity do not matter. For turbulent flow, the velocity considered is inevitably an *average* velocity. In such flow the instantaneous velocity at a point is in continual fluctuation, but if the flow is 'steady', an average of the velocity at a particular point, taken over a sufficient time interval, is constant in magnitude and direction. These average velocities are characteristic of the given pattern of flow and they are readily measurable.

The essential condition of geometric similarity must not be forgotten. It would be quite wrong, for example, to take a value of *Re* that refers to the onset of turbulence in a circular pipe and to apply this value to predict the onset of turbulence in flow past a flat plate.

We must also remember that the Reynolds number concerns only the forces due to viscosity and inertia (and, by implication, those due to differences of pressure). Inertia forces are present even when the flow as a whole is steady and not changing direction. Individual particles of fluid may not move entirely in straight paths, and if the path-line followed by a particle has even the slightest degree of curvature the particle must undergo an acceleration. When forces arising from other causes (those due, for example, to gravity or the compressibility of the fluid) play an important part in the flow, its general character may be determined by other criteria. The influence of these other forces will be the concern of later chapters. In situations where viscous and inertia forces are the most significant, however, provided that the necessary condition of geometric similarity is met, the Reynolds number is the parameter which may be used to compare experimental observations and to assemble even apparently unrelated data into comprehensive laws.

It is evident from the form of the expression $\varrho l u / \mu$ that a high value of ϱ, l or u, or a small value of μ, gives a high value of *Re*. Conversely, a low value of *Re* is brought about by high viscosity, or low density, low velocity or small size. A high value of *Re* indicates that inertia forces dominate the flow while viscous forces play only a small part; when *Re* is small in value, the viscous forces have the upper hand and inertia forces take second place.

Under normal engineering conditions, flow through pipes at a Reynolds number, $\varrho u d/\mu$ below 2000 may be regarded as laminar, and flows for $Re > 4000$ may be taken as turbulent.

■

Example 1.3 Water, at 20 °C, flows through a pipe of diameter 4 mm at 3 m/s. Determine whether the flow is laminar or turbulent.

□

Solution From Appendix 2, at 20 °C, water has a density of $10^3 \, \mathrm{kg \, m^{-3}}$ and an absolute viscosity $\mu = 1 \times 10^{-3} \, \mathrm{kg \, m^{-1} s^{-1}}$. Hence

$$Re = \frac{\varrho u d}{\mu} = \frac{10^3 \, \mathrm{kg/m^3} \times 3 \, \mathrm{m/s} \times 0{\cdot}004 \, \mathrm{m}}{10^{-3} \, \mathrm{kg/(m \, s)}} = 12\,000$$

The Reynolds number is well in excess of 4000, so the flow is turbulent.

REFERENCE

1. Brush, S. G. 'Theories of liquid viscosity', *Chem. Rev.* **62**, 513–48 (1962).

FURTHER READING

Blair, G. W. Scott *Elementary Rheology*, Academic Press, New York (1969).

Brown, R. C. 'The Fundamental Concepts concerning Surface Tension and Capillarity', *Proc. phys. Soc.* **59**, 429–48 (1947).

Harris, J. *Rheology and non-Newtonian flow*, Longman, London (1977).

von Kármán, 'Turbulence', *Jl. R. aeronaut. Soc.* **41**, 1109–41 (1937).

Reid, R. C., Prausnitz, J. M. and Poling, B. E. *The Properties of Gases and Liquids* (4th edn), McGraw-Hill, New York (1987).

Reynolds, O. 'An experimental investigation of the circumstances which determine whether the motion of water shall be direct or sinuous, and the law of resistance in parallel channels', *Phil. Trans. R. Soc.* **174**, 935–82 (1883).

Tanner, R. I. *Engineering Rheology*, Clarendon Press, Oxford (1985).

PROBLEMS

1.1   A hydrogen-filled balloon is to expand to a sphere 20 m diameter at a height of 30 km where the absolute pressure is 1100 Pa and the temperature −40 °C. If there is to be no stress in the fabric of the balloon what volume of hydrogen must be added at ground level where the absolute pressure is 101.3 kPa and the temperature 15 °C?

1.2   Calculate the density of air when the absolute pressure and the temperature are respectively 140 kPa and 50 °C and $R = 287$ J/kg K.

1.3   Eight kilometres below the surface of the ocean the pressure is 81.7 MPa. Determine the density of sea-water at this depth if the density at the surface is 1025 kg/m^3 and the average bulk modulus of elasticity is 2.34 GPa.

1.4   At an absolute pressure of 101.3 kPa and temperature of 20 °C the absolute viscosity of a certain diatomic gas is 2×10^{-5} Pa s and its kinematic viscosity is 15 mm^2/s. Taking the universal gas constant as 8310 J/kg K and assuming the gas to be 'perfect', calculate its approximate relative molecular mass.

1.5   A hydraulic ram 200 mm in diameter and 1.2 m long moves wholly within a concentric cylinder 200.2 mm in diameter, and the annular clearance is filled with oil of relative density 0.85 and kinematic viscosity 400 mm^2/s. What is the viscous force resisting the motion when the ram moves at 120 mm/s?

1.6   The space between two large flat and parallel walls 25 mm apart is filled with a liquid of absolute viscosity 0.7 Pa s. Within this space a thin flat plate 250 mm × 250 mm is towed at a velocity of 150 mm/s at a distance of 6 mm from one wall, the plate and its movement being parallel to the walls. Assuming linear variations of velocity between the plate and the walls, determine the force exerted by the liquid on the plate.

1.7   A uniform film of oil 0.13 mm thick separates two discs, each of 200 mm diameter, mounted co-axially. Ignoring edge effects, calculate the torque necessary to rotate one disc relative to the other at a speed of 7 rev/s if the oil has a viscosity of 0.14 Pa s.

1.8   By how much does the pressure in a cylindrical jet of water 4 mm in diameter exceed the pressure of the surrounding atmosphere if the surface tension of water is 0.073 N/m?

1.9   What is the approximate capillary rise of water in contact with air (surface tension 0.073 N/m) in a clean glass tube 5 mm in diameter?

1.10 What is the approximate capillary rise of mercury (relative density 13.56, interfacial tension 0.377 N/m, angle of contact approximately 140°) in contact with water in a clean glass tube 6 mm in diameter? (*Note:* As the mercury moves it displaces water, the density of which is not negligible.)

1.11 Calculate the Reynolds number for a fluid of density 900 kg/m^3 and viscosity 0.038 Pa s flowing in a 50 mm diameter pipe at the rate of 2.5 litres/s. Estimate the mean velocity above which laminar flow would be unlikely.

1.12 A liquid of kinematic viscosity 370 mm^2/s flows through an 80 mm diameter pipe at 0.01 m^3/s. What type of flow is to be expected?

Fluid statics | 2

2.1 INTRODUCTION

Fluid statics is that branch of mechanics of fluids that deals primarily with fluids at rest. Problems in fluid statics are much simpler than those associated with the motion of fluids, and exact analytical solutions are possible. Since individual elements of fluid do not move relative to one another, shear forces are not involved and all forces due to the pressure of the fluid are normal to the surfaces on which they act. Fluid 'statics' may thus be extended to cover instances in which elements of the fluid do not move relative to one another even though the fluid as a whole may be moving. With no relative movement between the elements, the viscosity of the fluid is of no concern.

In this chapter we shall first examine the variation of pressure intensity throughout an expanse of fluid. We shall then study the forces caused by pressure on solid surfaces in contact with the fluid, and also the effects (such as buoyancy) of these forces in certain circumstances.

2.2 VARIATION OF PRESSURE WITH POSITION IN A FLUID

Consider a small cylinder of fluid PQ as illustrated in Fig. 2.1. If the fluid is at rest, the cylinder must be in equilibrium and the only forces acting on it are those on its various faces (due to pressure), and gravity. The cross-sectional area δA is very small and the variation of pressure over it therefore negligible. Let the pressure at the end P be p and that at the end Q be $p + \delta p$ where δp may be positive or negative. The force on the end P is therefore $p\,\delta A$ and the force on the end Q is $(p + \delta p)\delta A$. If the length of the cylinder is δl its volume is $\delta A\,\delta l$ and its weight $\varrho g\,\delta A\,\delta l$ where ϱ represents the density and g the weight per unit mass. Since no shear forces are involved in a fluid at rest, the only forces acting on the sides of the cylinder are perpendicular to them and therefore have no component along the axis.

For equilibrium, the algebraic sum of the forces in any direction must be zero. Resolving in the direction QP:

$$(p + \delta p)\delta A - p\,\delta A + \varrho g\,\delta A\,\delta l\cos\theta = 0 \qquad (2.1)$$

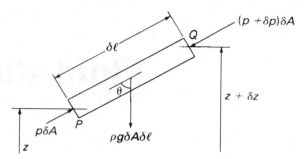

Fig. 2.1

Now if P is at a height z above some suitable horizontal datum plane and Q is at height $z + \delta z$, then the vertical difference in level between the ends of the cylinder is δz and $\delta l \cos \theta = \delta z$. Equation 2.1 therefore simplifies to

$$\delta p + \varrho g \, \delta z = 0$$

and in the limit as $\delta z \to 0$

$$\frac{\partial p}{\partial z} = -\varrho g \qquad (2.2)$$

The minus sign indicates that the pressure decreases in the direction in which z increases, i.e. upwards.

If P and Q are in the same horizontal plane, then $\delta z = 0$, and consequently δp is also zero whatever the value of ϱ. The argument may readily be extended to cover any two points in the same horizontal plane by considering a series of cylinders of fluid, PQ, QR, RS, etc. Then, in any fluid in equilibrium, the pressure is the same at any two points in the same horizontal plane, provided that they can be connected by a line in that plane and wholly in the fluid. In other words, a surface of equal pressure (an *isobar*) is a horizontal plane. More precisely, the surface is everywhere perpendicular to the direction of gravity and so is approximately a spherical surface concentric with the earth. Our concern is usually only with very small portions of that surface, however, and they may therefore be considered plane.

A further deduction is possible from eqn 2.2. If everywhere in a certain horizontal plane the pressure is p, then in another horizontal plane, also in the fluid and at a distance δz above, the pressure will be $p + (\partial p / \partial z)\delta z$. Since this pressure also must be constant throughout a horizontal plane, $\partial p / \partial z$ does not vary horizontally, and so, by eqn 2.2, ϱg does not vary horizontally. For a homogeneous incompressible fluid this is an obvious truth because the density is the same everywhere and g does not vary horizontally. But the result tells us that a condition for equilibrium of any fluid is that the density as well as the pressure must be constant over any horizontal plane. This is why immiscible fluids of different densities have a horizontal interface when they are in equilibrium (except very close to solid boundaries where surface tension usually causes curvature of the interface).

There are, then, three conditions for equilibrium of any fluid:

1. the pressure must be the same over any horizontal plane;
2. the density must be the same over any horizontal plane;
3. $\mathrm{d}p/\mathrm{d}z = -\varrho g$. (Since the pressure varies only in the vertical (z) direction, the partial derivative in eqn 2.2 may give way to the full derivative.)

To determine the pressure at any point in a fluid in equilibrium, eqn 2.2 must be integrated:

$$p = \int -\varrho g \, \mathrm{d}z$$

Evaluation of the integral is not possible, however, unless the variation of ϱ with z is known.

2.2.1   The equilibrium of a fluid of constant density

Since for all practical differences in height the variation of g is negligible, integration of eqn 2.2 for a homogeneous fluid of constant density gives

$$p + \varrho g z = \text{constant} \qquad (2.3)$$

This result is valid throughout a continuous expanse of the same fluid since, in deriving eqn 2.2, no restriction at all was placed on the value of θ. The value of the constant in eqn 2.3 is determined by the value of p at a point where z is specified. If the fluid is a liquid with a horizontal free surface at which the pressure is atmospheric (p_a) this free surface may be taken as the datum level $z = 0$. For equilibrium of the surface the pressure immediately below it must equal that immediately above it, and so the pressure in the liquid at the surface is p_a. Then, for a point at a depth h below the surface, $h = -z$ (since h is measured downwards whereas z is measured upwards) and, from eqn 2.3,

$$p = p_a + \varrho g h \qquad (2.4)$$

The pressure therefore increases linearly with the depth, whatever the shape of any solid boundaries may be.

Equation 2.4 shows that the pressure at a point in a liquid in equilibrium is due partly to atmospheric pressure at the free surface and partly to the weight of the liquid. Thus atmospheric pressure is usually effective, even if indirectly, on all surfaces, and over the differences of height normally encountered it is sensibly constant. Consequently it is often simpler to regard atmospheric pressure as the zero of the pressure scale. A pressure expressed above atmospheric pressure as zero is known as a *gauge pressure*. Equation 2.4 then reduces to $p = \varrho g h$. That is, the gauge pressure at any point in a liquid in equilibrium is equal to the product of

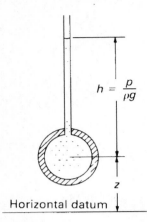

$$h = \frac{p}{\rho g}$$

Horizontal datum

Fig. 2.2

Piezometric pressure

ϱg and the depth vertically below the plane free surface in contact with the atmosphere. As we shall see in Section 2.3, this relation forms the basis of a number of methods of measuring pressure.

The direct proportionality between gauge pressure and h for a fluid of constant density enables the pressure to be simply visualized in terms of the vertical distance $h = p/\varrho g$. The quotient $p/\varrho g$ is termed the *pressure head* corresponding to p. So useful is the concept of pressure head that it is employed whether or not an actual free surface exists above the point in question. For a liquid without a free surface, as for example in a closed pipe, $p/\varrho g$ corresponds to the height above the pipe to which a free surface would rise if a small vertical tube of sufficient length and open to atmosphere – known as a *piezometer tube* – were connected to the pipe (Fig. 2.2). Provided that ϱ is constant, all pressures may be expressed as heads. Thus pressures are sometimes quoted in terms of *millimetres of mercury* or *metres of water*.

Equation 2.3 may be divided by ϱg to give $(p/\varrho g) + z = $ constant. That is, the sum of the pressure head and the elevation above the chosen horizontal datum plane is constant. This constant is known as the *piezometric head* and corresponds to the height of the free surface above the datum plane. The quantity $p + \varrho g z$ is termed the *piezometric pressure*.

The fact that an increase of pressure in any part of a confined fluid is transmitted uniformly throughout the fluid is utilized in such devices as the hydraulic press and the hydraulic jack. A large force F may be produced on a piston of area A by subjecting it to a pressure $p = F/A$. This pressure p, however, may be produced by applying a smaller force f to a smaller piston of area a (see Fig. 2.3). If the pistons move very slowly viscous and inertia forces may be neglected, and, if the pistons are at the same level, the pressure at one equals that at the other. Then $F/A = f/a$, i.e. $F = fA/a$. By a suitable choice of piston areas a considerable multiplication of the force may be achieved. The work done by each force, however, is the same (apart from the effects of friction); since the compressibility of the liquid used is extremely small its volume remains practically unchanged, so the smaller force moves through a correspondingly greater distance.

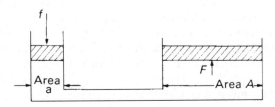

Fig. 2.3

2.2.2 The equilibrium of the atmosphere

Equation 2.2 expresses the conditions for equilibrium of any fluid. For a compressible fluid, however, the density varies with the pressure, so, unless the manner of this variation is known, the equation cannot be integrated to give the value of the pressure at a particular position. In connection with the atmosphere the problem is encountered in aeronautics and meteorology; it is also of concern in oceanography, since at great depths there is a small increase in the density of sea-water.

Let us consider a perfect gas. The density may be obtained from the equation of state $p = \varrho RT$ (in which p represents the *absolute* pressure). Then, from eqn 2.2,

$$\frac{dp}{dz} = -\varrho g = -\frac{pg}{RT}$$

i.e.

$$\frac{dp}{p} = -\frac{g}{R}\frac{dz}{T} \tag{2.5}$$

Since the value of g decreases by only about 0.03 % for a 1000 m increase in altitude it may be assumed constant. If conditions are isothermal, T = constant, and eqn 2.5 may be integrated to give

$$\ln\left(\frac{p}{p_0}\right) = -\frac{g}{RT}(z - 0)$$

where p_0 represents the (absolute) pressure when $z = 0$. That is,

$$p/p_0 = \exp\left(-gz/RT\right) \tag{2.6}$$

However, in the atmosphere the temperature varies with altitude. For the first 11 km above the ground there is a uniform decrease, that is, $\partial T/\partial z$ = constant = $-\lambda$ where λ (Greek 'lambda') is known as the *temperature lapse rate*. The observed value of λ in this region is about 0.0065 K/m. From 11 km to 20 km the temperature is constant at $-56.5\,°C$ and then beyond 20 km the temperature rises again.

If the temperature lapse rate is constant, $T = T_0 - \lambda z$ where T_0 represents the temperature at $z = 0$. Substituting this relation into eqn 2.5 and integrating we obtain

$$\ln\left(\frac{p}{p_0}\right) = \frac{g}{R\lambda}\ln\left(\frac{T_0 - \lambda z}{T_0}\right)$$

i.e.

$$\frac{p}{p_0} = \left(1 - \frac{\lambda z}{T_0}\right)^{g/R\lambda} \tag{2.7}$$

If the right-hand sides of eqns 2.6 and 2.7 are expanded in series form and then, for small values of z, all but the first two terms are neglected, the result in each case is

$$\frac{p}{p_0} = 1 - \frac{gz}{RT_0}$$

i.e.

$$p = p_0 - \frac{p_0}{RT_0} gz = p_0 - \varrho_0 gz$$

This corresponds to the relation $p + \varrho gz$ = constant (eqn 2.3) for a fluid of constant density. Thus, for small differences of height (less than, say, 300 m in air), sufficient accuracy is usually obtained by considering the fluid to be of constant density. If changes of z are appreciable, however, the full relation (eqn 2.6 or 2.7) is required.

Certain values of p_0, T_0 and λ are used to define a 'standard atmosphere' which provides a set of data reasonably representative of the actual atmosphere. Aircraft instruments and performance are related to these standard conditions, and figures are subject to error if the actual conditions differ appreciably from the standard. A knowledge of λ is important in practice, not only for predicting conditions in the atmosphere at various altitudes, but also for the calibration of altimeters which depend on the atmospheric pressure for their operation. Information on the International Standard Atmosphere is contained in Appendix 2.

In fact, however, the atmosphere is not in perfect equilibrium. The lower part is continually being mixed by convection and winds and there are varying amounts of water vapour in it. A little more about the equilibrium of the atmosphere will be said in Section 2.7.3.

■
Example 2.1 A spherical balloon, of diameter 1.5 m and total mass 1.2 kg, is released in the atmosphere. Assuming that the balloon does not expand and that the temperature lapse rate in the atmosphere is 0.0065 K/m, determine the height above sea-level to which the balloon will rise. Atmospheric temperature and pressure at sea-level are 15 °C and 101 kPa respectively; for air, R = 287 J/kg K.

Solution

$$\text{Density of balloon} = 1.2\,\text{kg} \bigg/ \frac{\pi}{6}(1.5\,\text{m})^3 = 0.679\,\text{kg/m}^3$$

∴ Balloon rises until atmospheric density = 0.679 kg/m³

By eqn 2.7

$$\ln\left(\frac{p}{p_0}\right) = \frac{g}{R\lambda} \ln \frac{T_0 - \lambda z}{T_0}$$

$$\therefore \quad \text{since } \varrho = p/RT$$

$$\ln\frac{\varrho}{\varrho_0} = \ln\frac{p}{p_0} - \ln\frac{T}{T_0} = \left(\frac{g}{R\lambda} - 1\right)\ln\left(\frac{T_0 - \lambda z}{T_0}\right)$$

$$\varrho_0 = \frac{101 \times 10^3\,\text{kg}}{287 \times 288.15\,\text{m}^3} = 1.221\,\text{kg/m}^3$$

$$\therefore \quad \ln\frac{0.679}{1.221} = \left(\frac{9.81}{287 \times 0.0065} - 1\right)\ln\left(\frac{288.15 - 0.0065\,z}{288.15}\right)$$

Whence $z = 5708\,\text{m}$ □

2.3 THE MEASUREMENT OF PRESSURE

In practice, pressure is always measured by the determination of a pressure difference. If the difference is that between the pressure of the fluid in question and that of a vacuum then the result is known as the *absolute pressure* of the fluid. More usually, however, the difference determined is that between the pressure of the fluid concerned and the pressure of the surrounding atmosphere. This is the difference normally recorded by pressure gauges and so is known as *gauge pressure*. If the pressure of the fluid is below that of the atmosphere it is termed *vacuum* or *suction* (see Fig. 2.4). (The term 'high vacuum' refers to a low value of the absolute pressure.) The absolute pressure is always positive but gauge pressures are positive if they are greater than atmospheric and negative if less than atmospheric.

Most of the properties of a gas are functions of its absolute pressure and consequently values of the absolute pressure are usually required in problems concerning gases. Frequently it is the gauge pressure that is measured and the atmospheric pressure must be added to this to give the value of the absolute pressure. The properties of liquids, on the other hand, are little affected by pressure and the pressure of a liquid is

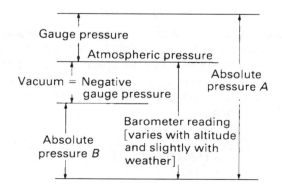

Fig. 2.4

therefore usually expressed as a gauge value. There is little point in using absolute pressures for liquids because in an equation, for example, the contribution of atmospheric pressure would cancel from both sides. The absolute pressure of a liquid may, however, be of concern when the liquid is on the point of vaporizing. In this book, values of pressure will be understood to be gauge pressures unless there is a specific statement that they are absolute values.

We now consider some of the means of measuring pressure.

2.3.1 The barometer

We have already seen that there is a simple relation (eqn 2.3) between the height of a column of liquid and the pressure at its base. Indeed, if the pressure of a liquid is only slightly greater than that of the atmosphere a simple way of measuring it is to determine the height of the free surface in a piezometer tube as illustrated in Fig. 2.2. (The diameter of the tube must be large enough for the effect of surface tension to be negligible.) If such a piezometer tube – of sufficient length – were closed at the top and the space above the liquid surface were a perfect vacuum the height of the column would then correspond to the absolute pressure of the liquid at the base. This principle is used in the well-known mercury barometer.

Mercury is employed because its density is sufficiently high for a fairly short column to be obtained, and also because it has, at normal temperatures, a very small vapour pressure. A perfect vacuum at the top of the tube is not in practice possible; even when no air is present the space is occupied by vapour given off from the surface of the liquid. The mercury barometer was invented in 1643 by the Italian Evangelista Torricelli* (1608–47) and the near-vacuum above the mercury is often known as the Torricellian vacuum. All air and other foreign matter is removed from the mercury, and a glass tube full of it is then inverted with its open end submerged in pure mercury. The pressure at A (Fig. 2.5) equals that at B (atmospheric because the surface curvature is here negligible) since these points are at the same level and connected by a path wholly in the mercury. Therefore, by eqn 2.3

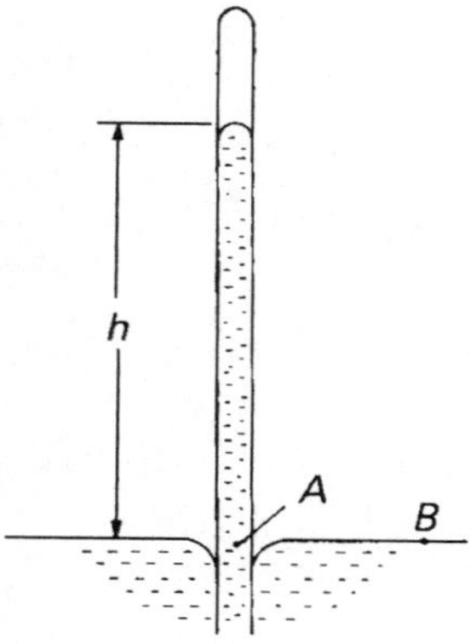

$$p_a = p_v + \varrho g h$$

where p_a represents the absolute pressure of the atmosphere, ϱ the density of the mercury and h the height of the column above A. The pressure of the mercury vapour in the space at the top of the tube is represented by p_v. However, at 20 °C, p_v is only 0.16 Pa and so may normally be neglected in comparison with p_a, which is usually about 10^5 Pa at sea level. Then

$$h = p_a/\varrho g = \frac{10^5 \, \text{N/m}^2}{\left(13\,560\,\text{kg/m}^3\right)\left(9.81\,\text{N/kg}\right)} = 0.752\,\text{m}$$

Fig. 2.5

* pronounced *Tor'ry-chell'y*.

For accurate work small corrections are necessary to allow for the variation of ϱ with temperature, the thermal expansion of the (usually brass) scale, and surface-tension effects.

If water were used instead of mercury the corresponding height of the column would be about 10.4 m provided that a perfect vacuum could be achieved above the water. However, the vapour pressure of water at ordinary temperatures is appreciable and so the actual height at, say, 15 °C would be about 180 mm less than this value. With a tube smaller in diameter than about 15 mm, surface-tension effects become significant.

In the aneroid barometer, a metal bellows containing a near-perfect vacuum is expanded or contracted according to variations in the pressure of the atmosphere outside it. This movement is transmitted by a suitable mechanical linkage to a pointer moving over a calibrated scale.

2.3.2 Manometers

Manometers are devices in which columns of a suitable liquid are used to measure the difference in pressure between a certain point and the atmosphere, or between two points neither of which is necessarily at atmospheric pressure. For measuring small gauge pressures of liquids simple piezometer tubes (Fig. 2.2) may be adequate, but for larger pressures some modification is necessary. A common type of manometer is that employing a transparent 'U-tube' set in a vertical plane as shown in Fig. 2.6. This is connected to the pipe or other container in which is the fluid (A) whose pressure is to be measured. The lower part of the U-tube contains a liquid (B) immiscible with A and of greater density. Within a *continuous expanse of the same fluid* the pressure is the same at any two points in a horizontal plane when equilibrium is achieved. Therefore, since points P and Q are in the same horizontal plane and are joined by a continuous expanse of liquid B, the pressures at P and Q are equal when the system is in equilibrium. Let the pressure

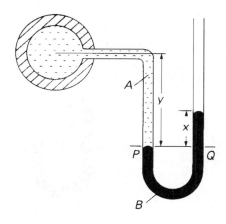

Fig. 2.6

in the pipe at its centreline be p. Then, provided that the fluid A is of constant density, the pressure at P is $p + \varrho_A g y$ (from eqn 2.3), where ϱ_A represents the density of fluid A. If the other side of the U-tube is open to atmosphere the (gauge) pressure at Q is $\varrho_B g x$ where ϱ_B represents the density of liquid B.

Therefore

$$p + \varrho_A g y = \varrho_B g x$$

and if ϱ_A, ϱ_B are known and y and x are measured, the value of p may be calculated. If A is a gas, ϱ_A is negligible compared with ϱ_B, and then $p = \varrho_B g x$. The arrangement is suitable for measuring pressures below atmospheric as illustrated in Fig. 2.7. Application of the same principles then gives $p + \varrho_A g y + \varrho_B g x = 0$.

U-tube manometers are also frequently used for measuring the difference between two unknown pressures, say p_1 and p_2. Figure 2.8 shows such an arrangement for measuring the pressure difference across a restriction in a horizontal pipe. (When fluid is flowing past the connections to a manometer it is very important for the axis of each connecting tube to be perpendicular to the direction of flow and also for the edges of the connections to be smooth, flush with the main surface and free from burrs. To reduce the risk of errors resulting from imperfect connections several openings round the periphery of the pipe are often used, all connected to the manometer via a common annulus; an average pressure is thus obtained in which the individual, probably random, errors tend to cancel.) Again applying the principle that pressures at P and Q must be equal for equilibrium, we have

$$p_1 + (y + x)\varrho_A g = p_2 + \varrho_A g y + \varrho_B g x$$

$$\therefore \qquad p_1 - p_2 = (\varrho_B - \varrho_A)g x \qquad (2.8)$$

If it is desired to express this pressure difference as a head difference for fluid A, then

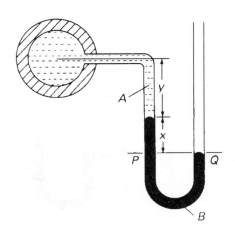

Fig. 2.7

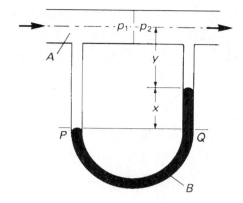

Fig. 2.8

$$h_1 - h_2 = \frac{p_1 - p_2}{\varrho_A g} = \left(\frac{\varrho_B}{\varrho_A} - 1\right)x \qquad (2.9)$$

If, for example, fluid A is water and B is mercury, then a difference x in manometer levels corresponds to a difference in head of water $=$ $(13.56 - 1)x$. A common error is to use simply ϱ_B/ϱ_A instead of $\{(\varrho_B/\varrho_A) - 1\}$ in eqn 2.9. It should not be forgotten that the pressure at P includes a contribution made by the column of fluid A above it. More generally, it may be shown that a differential manometer such as this records the difference of *piezometric pressure* $p + \varrho g z$: only when points (1) and (2) are at the same level does the manometer reading correspond to the difference of actual pressure p.

Many modifications of the U-tube manometer have been developed for particular purposes. A common modification is to make one limb of the 'U' very much greater in cross-section than the other. When a pressure difference is applied across the manometer the movement of the liquid surface in the wider limb is practically negligible compared with that occurring in the narrow limb. If the level of the surface in the wide limb is assumed constant the height of the meniscus in only the narrow limb need be measured, and only this limb need therefore be transparent.

For accurate measurements reasonable values of x are desirable. For small pressure differences such values may be obtained by selecting liquid B so that the ratio ϱ_B/ϱ_A is close to unity. If fluid A is a gas, however, this is not possible, and a sloping manometer may then be used. For example, if the transparent tube of a manometer is set not vertically, but at an angle θ to the horizontal, then a pressure difference corresponding to a vertical difference of levels x gives a movement $s = x/\sin\theta$ along the slope (see Fig. 2.9). If θ is small, a considerable magnification of the movement of the meniscus may be achieved. Angles less than $5°$, however, are not usually satisfactory, because the exact position of the meniscus is difficult to determine, and also small changes in the surface tension forces, arising from imperfect cleanliness of the walls of the tube, may considerably affect the accuracy.

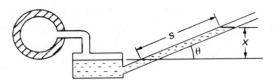

Fig. 2.9

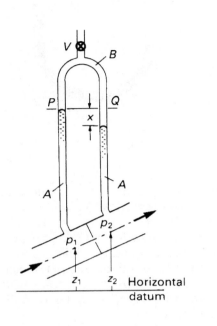

Fig. 2.10

When large pressure differences are to be measured a number of U-tube manometers may be connected in series. The required pressure difference can be calculated by the application of the basic principles: (1) the pressure within a continuous expanse of the same fluid in equilibrium is the same at any two points in a horizontal plane; (2) the hydrostatic equation $p + \varrho gz =$ constant for a homogeneous fluid of constant density.

For the measurement of small pressure differences in liquids an inverted U-tube manometer, as illustrated in Fig. 2.10, is often suitable. Here $\varrho_B < \varrho_A$ and it is the *upper* fluid which is in equilibrium. The horizontal line PQ is therefore taken at the level of the higher meniscus. By equating the pressures at P and Q it may readily be shown that, for a manometer in a vertical plane, $p_1^* - p_2^* = (\varrho_A - \varrho_B)gx$ where p^* represents the *piezometric pressure $p + \varrho_A gz$*. If $\varrho_A - \varrho_B$ is sufficiently small a large value of x may be obtained for a small value of $p_1^* - p_2^*$. The interface between liquids of closely similar densities, however, is very sensitive to changes in surface tension and therefore to traces of grease and other impurities. Air may be used as fluid B: it may be pumped through the valve V at the top of the manometer until the liquid menisci are at a suitable level. Then, of course, ϱ_B is negligible compared with ϱ_A.

Certain practical considerations arise in the use of manometers. (1) Since the densities of liquids depend on temperature the temperature of the liquids should be known for accurate results. (2) Some liquids, otherwise suitable for use in manometers, give ill-defined menisci. (3) Fluctuations of menisci reduce accuracy; such movements may be reduced by restrictions in the manometer connections (e.g. lengths of small-diameter pipe) which, under equilibrium conditions, do not affect the pressure. (4) The density of the fluid in the connecting tubes must be uniform; for example, water must not contain air bubbles, nor air contain 'blobs' of water. The layout of the connecting tubes should be such as to minimize the possibility of trapping air bubbles, and means should be provided for flushing the connecting tubes through before the manometer is used. A valve by which the pressure difference may be reduced to zero and the zero reading thus checked is also desirable. (5) In tubes of less than about 15 mm diameter surface tension effects may be appreciable and the meniscus is either raised above or lowered below its 'correct' position. For example, for pure water in a clean, vertical, glass tube, 6 mm diameter, the 'capillary rise' is about 5 mm. The corresponding depression for mercury is about 1.25 mm. Because of the uncertain degree of cleanliness of tubes used in practice, however, it is difficult to allow for surface tension effects. Fortunately these effects can be nullified, for example in a U-tube manometer where the limbs are of equal diameter and cleanliness, or where measurements are made of the movement of a single meniscus in a uniform tube. Alcohol, being a solvent of grease, is less sensitive to the cleanliness of the tube and so is frequently preferred to water in manometers.

Micro-manometers

For measuring very small pressure differences, a wide variety of special manometers has been developed. Several devices may be used to increase the accuracy of a reading. For example, a meniscus may be observed through a small telescope containing a horizontal cross-wire, and the assembly may be raised or lowered by a slow-motion screw with a micrometer scale. Or a scale floating on the surface of a liquid may be optically magnified.

When an additional gauge liquid is used in a U-tube a large difference of meniscus level may be produced by a small pressure difference. One arrangement is illustrated in Fig. 2.11. The appropriate equilibrium equation is

$$p_1 + \varrho_A g(h + \Delta z) + \varrho_B g\left(z - \Delta z + \frac{y}{2}\right)$$

$$= p_2 + \varrho_A g(h - \Delta z) + \varrho_B g\left(z + \Delta z - \frac{y}{2}\right) + \varrho_C gy \quad (2.10)$$

The amount of liquid B on each side remains constant. Therefore

$$a_1 \Delta z = a_2 \frac{y}{2} \quad (2.11)$$

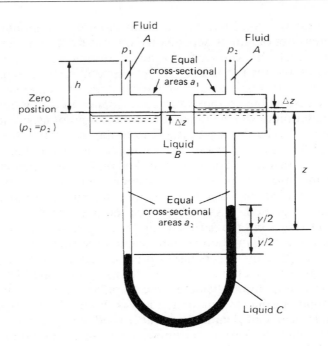

Fig. 2.11

Substituting for Δz in eqn 2.10 we obtain

$$p_1 - p_2 = gy\left\{\varrho_C - \varrho_B\left(1 - \frac{a_2}{a_1}\right) - \varrho_A\frac{a_2}{a_1}\right\}$$

Since a_2 is usually very small compared with a_1, $p_1 - p_2 \simeq (\varrho_C - \varrho_B)gy$, so when ϱ_C and ϱ_B are closely similar a reasonable value of y may be achieved for a small pressure difference.

In several micro-manometers the pressure difference to be measured is balanced by the slight raising or lowering (on a micrometer screw) of one arm of the manometer whereby a meniscus is brought back to its original position. Well-known micro-manometers of this type are those invented by Chattock, Small and Krell, and they are sensitive to pressure differences down to less than 0.002 mm of water. They suffer from the disadvantage that an appreciable time is required to make a reading and they are therefore suitable only for completely steady pressures.

2.3.3 The Bourdon gauge

Where high precision is not required a pressure difference may be indicated by the deformation of an elastic solid. For example, in an engine indicator the pressure to be measured acts at one side of a small piston, the other side being subject to atmospheric pressure. The difference between these pressures is then indicated by the movement of the piston against the resistance of a calibrated spring. The principle of the aneroid barometer (Section 2.3.1) may also be adapted for the

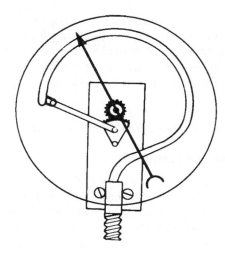

Fig. 2.12

measurement of pressures other than atmospheric. The most common type of pressure gauge – compact, reasonably robust and simple to use – is that invented by Eugène Bourdon (1808–84). A curved tube of elliptical cross-section is closed at one end; this end is free to move, but the other end – through which the fluid enters – is rigidly fixed to the frame as shown in Fig. 2.12. When the pressure inside the tube exceeds that outside (which is usually atmospheric) the cross-section tends to become circular, thus causing the tube to uncurl slightly. The movement of the free end of the tube is transmitted by a suitable mechanical linkage to a pointer moving over a scale. Zero reading is of course obtained when the pressure inside the tube equals the local atmospheric pressure. By using tubes of appropriate stiffness, gauges for a wide range of pressures may be made. If, however, a pressure higher than the intended maximum is applied to the tube, even only momentarily, the tube may be strained beyond its elastic limit and the calibration invalidated.

All gauges depending on the elastic properties of a solid require calibration. For small pressures this may be done by using a column of mercury; for higher pressures the standard, calibrating, pressure is produced by 'weights' of known magnitude exerting a downward force on a piston of known area.

2.3.4   Other types of pressure gauge

For very high pressures, piezo-electric gauges may be used in which a crystal of quartz or other material, when subjected to the pressure of the fluid, produces across itself a small but measurable difference of electrical potential. Other gauges utilize the increase of electrical resistance exhibited by metals under very high pressures. As there are no moving parts these electrical gauges respond practically instantaneously to changes of pressure.

In a 'pressure transducer' the pressure of the fluid acts at one side of a thin diaphragm; movements of the diaphragm caused by changes of pressure are indicated by an electrical strain gauge on the diaphragm. Alternatively, the change of electrical capacitance between the moving diaphragm and a fixed plate may be measured.

2.4 FIRST AND SECOND MOMENTS OF AREA

2.4.1 First moments and centroids

Figure 2.13 shows a plane area A of which an infinitesimal element is δA. The first moment of the elemental area δA about an axis in the plane is defined as the product of δA and its perpendicular distance from that axis. Consequently the first moment of the elemental area about the y axis is given by $x\delta A$ and the first moment of the entire area about the y axis is therefore $\int_A x\,dA$. (The symbol $\int_A$ indicates that the integration is performed over the entire area.) Since individual values of x may be positive or negative according as the element is to the right or left of the y axis, so the integral may be positive or negative or zero. The first moment of area about an axis at $x = k$ is $\int_A (x - k)dA = \int_A x\,dA - kA$. This moment is zero when

$$k = \frac{1}{A}\int_A x\,dA = \bar{x} \qquad (2.12)$$

and the axis is then known as a *centroidal axis*.

Similarly, moments may be taken about the x axis, and another centroidal axis is then found at

$$y = \bar{y} = \frac{1}{A}\int_A y\,dA \qquad (2.13)$$

The intersection C of these two centroidal axes is known as the *centroid* of the area, and it may be shown that the first moment of the area is zero about any axis through the centroid. An axis of symmetry is evidently a

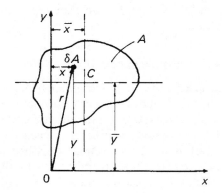

Fig. 2.13

centroidal axis since for every element on one side of the axis and contributing $x\delta A$ to the total moment there is an element on the other side contributing $-x\delta A$. Equations 2.12 and 2.13 show that the first moment of an area about any axis may be written as $A\bar{z}$ where $\bar{z}$ is the perpendicular distance of the centroid from that axis. Both the area and the axis must of course be specified.

The position of the centroid of a volume may be determined similarly. For example, the x coordinate of the centroid is found by summing moments in which x is the perpendicular distance of the element δV from the yz plane. Then

$$\bar{x} = \frac{1}{V}\int_V x\,\mathrm{d}V$$

Or the first moment of mass of a body about the yz plane may be calculated as $\int_M x\,\mathrm{d}M$, and the x coordinate of the *centre of mass* is then given by

$$\frac{1}{M}\int_M x\,\mathrm{d}M$$

By taking elements of weight $\delta(Mg)$, rather than of mass, we may determine the position of the centre of gravity. For bodies small compared with the earth, however, the variations of g are negligible and thus the centre of gravity and the centre of mass coincide.

All the moments just considered are termed first moments because each element of area, volume and so on is multiplied by the first power of the appropriate distance.

2.4.2   Second moment of area

The second moment of the plane area illustrated in Fig. 2.13 about the y axis is $\int_A x^2\,\mathrm{d}A$. Similarly, the second moment of the area about the x axis is $\int_A y^2\,\mathrm{d}A$. A second moment of area may alternatively be written as Ak^2 (a suitable suffix being used to indicate the axis concerned), so in $(Ak^2)_{Oy}$, for example, k^2 represents the mean value of x^2. The dimensional formula of a second moment of area is evidently $[\mathrm{L}^4]$ and a suitable unit is m^4.

The value of a second moment of area about a particular axis may always be found by performing the appropriate integration, but a 'short cut' is often possible when the second moment about another axis is known. Consider an axis through the centroid of the area and parallel to the given axis (say Oy in Fig. 2.13). The second moment of the area about this new axis is

$$\left(Ak^2\right)_C = \int_A \left(x - \bar{x}\right)^2 \mathrm{d}A = \int_A x^2\mathrm{d}A - 2\bar{x}\int_A x\,\mathrm{d}A + \bar{x}^2\int_A \mathrm{d}A$$
$$= \left(Ak^2\right)_{Oy} - 2\bar{x}\left(A\bar{x}\right) + A\bar{x}^2 = \left(Ak^2\right)_{Oy} - A\bar{x}^2$$

where the suffixes C and Oy indicate the axes used. Therefore

$$\left(Ak^2\right)_{Oy} = \left(Ak^2\right)_C + A\bar{x}^2$$

The direction of the y axis was arbitrary; hence it may be said that the second moment of a plane area about any axis equals the sum of the second moment about a parallel axis *through the centroid* and the product of the area and the square of the perpendicular distance between the axes. This result is frequently known as the *parallel axes theorem*. Moreover, by definition,

$$\left(Ak^2\right)_{Oy} + \left(Ak^2\right)_{Ox} = \int_A x^2 \,\mathrm{d}A + \int_A y^2 \,\mathrm{d}A$$
$$= \int_A \left(x^2 + y^2\right)\mathrm{d}A = \int_A r^2 \,\mathrm{d}A$$

The last term corresponds to the second moment of the area about an axis perpendicular to the plane of the area at the origin. Since the origin was arbitrarily chosen we have the *perpendicular axes theorem*: the second moment of a plane area about an axis meeting the plane perpendicularly at any point P equals the sum of the second moments of that area about two axes in the plane that intersect perpendicularly at P.

The second moment of mass about a particular axis is $\int_M z^2 \,\mathrm{d}M$, where z represents the perpendicular distance of an element from the axis in question. If the mean value of z^2 is represented by k^2, the second moment of mass may alternatively be written Mk^2. The second moment of mass is also known as the moment of inertia, and k is termed the radius of gyration. The dimensional formula of moment of inertia is $[\mathrm{ML}^2]$ and a suitable unit is $\mathrm{kg\,m^2}$.

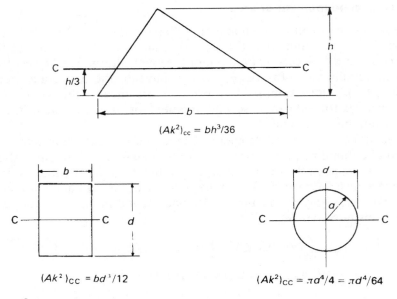

$(Ak^2)_{cc} = bh^3/36$

$(Ak^2)_{cc} = bd^3/12$

$(Ak^2)_{cc} = \pi a^4/4 = \pi d^4/64$

Fig. 2.14 Second moments about centroidal axes.

Unfortunately the second moment of area is sometimes referred to wrongly as moment of inertia. Inertia is a property of matter and has nothing to do with area. Second moment of area Ak^2, a purely geometric quantity, and moment of inertia Mk^2 are fundamentally different. In this book we use Ak^2 as the symbol for second moment of area, a suitable suffix indicating the axis about which the moment is taken.

Examples of second moments about centroidal axes are given in Fig. 2.14.

2.5 HYDROSTATIC THRUSTS ON SUBMERGED SURFACES

The pressure of a fluid causes a thrust to be exerted on every part of any surface with which the fluid is in contact. The individual forces distributed over the area have – in general – a resultant, and determination of the magnitude, direction and position of this resultant force is frequently important. For a plane *horizontal* surface at which the fluid is in equilibrium the matter is simple: the pressure does not vary over the plane and the total force is given by the product of the pressure and the area. Its direction is perpendicular to the plane – downwards on the upper face, upwards on the lower face – and its position is at the centroid of the plane. But if the surface is not horizontal the pressure varies (in general) from one point of the surface to another and the calculation of the total thrust is a little less simple.

2.5.1 Thrust on a plane surface

Figure 2.15 shows a plane surface of arbitrary shape, wholly submerged in a liquid in equilibrium. The plane of the surface makes an angle θ with the horizontal, and the intersection of this plane with the plane of the free surface (where the pressure is atmospheric) is taken as the x-axis. The y-axis is taken down the sloping plane. Every element of the area is subjected to a force due to the pressure of the liquid. At any element of area δA, at a depth h below the free surface, the (gauge) pressure is $p = \varrho g h$ and the corresponding force is

$$\delta F = p\,\delta A = \varrho g h\,\delta A = \varrho g\, y \sin\theta\,\delta A \qquad (2.14)$$

As the fluid is not moving relative to the plane there are no shear stresses. Thus the force is perpendicular to the element, and since the surface is plane all the elemental forces are parallel. The total force on one side of the plane is therefore

$$F = \int_A \varrho g y \sin\theta\,\mathrm{d}A = \varrho g \sin\theta \int_A y\,\mathrm{d}A$$

But $\int_A y\,\mathrm{d}A$ is the first moment of the area about the x-axis and may be represented by $A\bar{y}$ where A represents the total area and $(\bar{x}, \bar{y})$ is the position of its centroid C. Therefore

$$F = \varrho g \sin\theta\, A\bar{y} = \varrho g A\bar{h} \qquad (2.15)$$

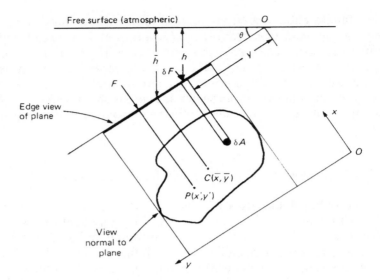

Fig. 2.15

Now $\varrho g \overline{h}$ is the pressure at the centroid, so, whatever the slope of the plane, the total force exerted on it by the static fluid is given by the product of the area and the pressure at the centroid. Whether the fluid actually has a free surface in contact with the atmosphere is of no consequence: for a fluid of uniform density in equilibrium the result is true however the pressure is produced.

Centre of pressure for a plane surface

In addition to the magnitude of the total force we need to know its line of action. Since all the elemental forces are perpendicular to the plane, their total is also perpendicular to the plane. It remains to determine the point at which its line of action meets the plane. This point is known as the *centre of pressure* (although 'centre of thrust' might be a better term).

For the resultant force to be equivalent to all the individual forces, its moment about any axis must be the same as the sum of the moments of the individual forces about the axis. The x- and y-axes are most suitable to our purpose. From eqn 2.14 the force on an element of the area is $\varrho g y \sin\theta\,\delta A$ and the moment of this force about Ox is therefore $\varrho g y^2 \sin\theta\,\delta A$. Let the centre of pressure P be at (x', y'). Then the total moment about Ox is

$$Fy' = \int_A \varrho g y^2 \sin\theta\,\mathrm{d}A$$

Substituting for the total force F from eqn 2.15 we obtain

$$y' = \frac{\varrho g \sin\theta \int_A y^2 \mathrm{d}A}{\varrho g \sin\theta\, A\overline{y}} = \frac{\left(Ak^2\right)_{Ox}}{A\overline{y}} \tag{2.16}$$

where $(Ak^2)_{Ox}$ is the second moment of the area about Ox. In other words, the *slant* depth (i.e. measured down the plane) of the centre of pressure equals:

Second moment of the area about the intersection of
its plane with that of the free (i.e. atmospheric) surface
_____
First moment of the area about the intersection of
its plane with that of the free surface

The centre of pressure is always lower than the centroid (except when
the surface is horizontal). From the parallel axes theorem:

$$\left(Ak^2\right)_{Ox} = \left(Ak^2\right)_C + A\bar{y}^2$$

so eqn 2.16 becomes

$$y' = \frac{\left(Ak^2\right)_C + A\bar{y}^2}{A\bar{y}} = \bar{y} + \left(Ak^2\right)_C \big/ A\bar{y} \qquad (2.17)$$

Since a second moment of area is always positive $y' > \bar{y}$.

We see also that the more deeply the surface is submerged, i.e. the
greater the value of $\bar{y}$, the smaller is the contribution made by the last
term in eqn 2.17 and the closer is the centre of pressure to the centroid.
This is because, as the pressure becomes greater with increasing depth,
its variation over a given area becomes smaller *in proportion*, so making
the distribution of pressure more uniform. Thus where the variation of
pressure is negligible the centre of pressure may be taken as approximately at the centroid. This is justifiable in liquids only if the depth is
very large and the area small, and in gases because in them the pressure
changes very little with depth.

The expressions 2.16 and 2.17, it is re-emphasized, give the distance to
the centre of pressure measured *down the plane from the level of the free
surface* and not vertically.

The x-coordinate of the centre of pressure may be determined by
taking moments about Oy. Then the moment of δF is $\varrho gy \sin\theta\, \delta A\, x$ and
the total moment is

$$Fx' = \int_A \varrho gxy \sin\theta\, dA = \varrho g \sin\theta \int_A xy\, dA$$

so

$$x' = \frac{\int_A xy\, dA}{A\bar{y}} \qquad (2.18)$$

When the area has an axis of symmetry in the y direction, this axis
may be taken as Oy and then $\int_A xy\, dA$ is zero, i.e. the centre of pressure
lies on the axis of symmetry. It will be noted from eqns 2.16 and 2.18 that
the position of the centre of pressure is independent of the angle θ and
of the density of the fluid. However, a constant value of ϱ was used; the
relations are therefore valid *only for a single homogeneous fluid*.

For the plane lamina of negligible thickness illustrated in Fig. 2.15, the
force on one face would exactly balance the force on the other if both
faces were in contact with the fluid. In most cases of practical interest,
however, there is no continuous path in the fluid from one face of the

plane to the other and therefore the pressures at corresponding points on the two faces are not necessarily the same. For example, the surface may be that of a plate covering a submerged opening in the wall of a reservoir, or a canal lock-gate which has different depths of water on the two sides. The surface may extend up through the liquid and into the atmosphere; only the part below the free surface then has a net hydrostatic thrust exerted on it.

The pressures we have considered have been expressed as gauge pressures. It is unnecessary to use absolute pressures because the effect of atmospheric pressure at the free surface is to provide a uniform addition to the gauge pressure throughout the liquid, and therefore to the force on any surface in contact with the liquid. Normally atmospheric pressure also provides a uniform force on the other face of the plane, and so it has no effect on either the magnitude or position of the resultant net force.

It should be particularly noted that, although the total force acts at the *centre of pressure*, its magnitude is given by the product of the area and the pressure at the *centroid*.

Example 2.2 A cylindrical tank 2 m diameter and 4 m long, with its axis horizontal, is half filled with water and half filled with oil of density 880 kg/m³. Determine the magnitude and position of the net hydrostatic force on one end of the tank.

Solution
We assume that the tank is only just filled, i.e. the pressure in the fluids is due only to their weight, and thus the (gauge) pressure at the top is zero. Since two immiscible fluids are involved we must consider each separately. In equilibrium conditions the oil covers the upper semicircular half of the end wall.

Since the centroid of a semicircle of radius a is on the central radius and $4a/3\pi$ from the bounding diameter, the centroid C_o of the upper semicircle is $4(1\,\text{m})/3\pi = 0.4244\,\text{m}$ above the centre of

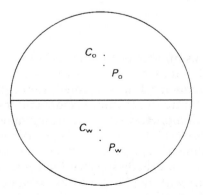

the tank, i.e. $(1 - 0.4244)\,\text{m} = 0.5756\,\text{m}$ from the top. The pressure of the oil at this point is

$$\varrho g \overline{h} = \left(880\,\text{kg/m}^3\right)\left(9.81\,\text{N/kg}\right)\left(0.5756\,\text{m}\right) = 4969\,\text{Pa}$$

and thus

the force exerted by the oil on the upper half of the wall

$$= 4969\,\text{Pa} \times \left(\frac{1}{2}\pi 1^2\right)\text{m}^2 = 7805\,\text{N}.$$

By eqn 2.17 the centre of pressure is $(Ak^2)_c/A\overline{y}$ below the centroid. Now Ak^2 about the bounding diameter $= \pi a^4/8$ (see Fig. 2.14). So, by the parallel axes theorem, for a horizontal axis through C_o,

$$\left(Ak^2\right)_c = \frac{\pi a^4}{8} - \left(\frac{1}{2}\pi a^2\right)\left(\frac{4a}{3\pi}\right)^2 = a^4\left(\frac{\pi}{8} - \frac{8}{9\pi}\right) = 0.1098\,\text{m}^4$$

Therefore, the centre of pressure P_o is $0.1098\,\text{m}^4/\frac{1}{2}\pi(1\,\text{m})^2$ $(0.5756\,\text{m}) = 0.1214\,\text{m}$ below the centroid, i.e. $(0.5756 + 0.1214)\,\text{m}$ $= 0.6970\,\text{m}$ below the top.

For the lower semicircle, in contact with water, the centroid C_w is $0.4244\,\text{m}$ below the central diameter. The pressure here is that due to $1\,\text{m}$ of oil together with $0.4244\,\text{m}$ of water, i.e.

$$\left(800\,\text{kg/m}^3\right)\left(9.81\,\text{N/kg}\right)\left(1\,\text{m}\right)$$
$$+ \left(1000\,\text{kg/m}^3\right)\left(9.81\,\text{N/kg}\right)\left(0.4244\,\text{m}\right) = 12\,796\,\text{Pa}.$$

Thus the force on the lower semicircle is $(12\,796\,\text{Pa})(\frac{1}{2}\pi 1^2\,\text{m}^2) = 20\,100\,\text{N}.$

$(Ak^2)_c$ is again $0.1098\,\text{m}^4$ but we must be very careful in calculating $\overline{y}$ since there is not a *single* fluid between this centroid and the zero-pressure position. However, conditions in the water are the same as if the pressure at the oil–water interface $[(880\,\text{kg/m}^3)(9.81\,\text{N/kg})(1\,\text{m})]$ were produced instead by $0.88\,\text{m}$ of water $[(1000\,\text{kg/m}^3)(9.81\,\text{N/kg})(0.88\,\text{m})]$. In that case the vertical distance from the centroid C_w to the zero-pressure position would be $0.4244\,\text{m} + 0.88\,\text{m} = 1.3044\,\text{m}.$

∴ Centre of pressure P_w for lower semicircle is

$$0.1098\,\text{m}^4 \Big/ \left(\frac{1}{2}\pi\,1^2\,\text{m}^2 \times 1.3044\,\text{m}\right)$$

$= 0.0536\,\text{m}$ below the centroid C_w,

i.e. $\left(1 + 0.4244 + 0.0536\right)\text{m} = 1.478\,\text{m}$ below top of cylinder.

The total force on the circular end is $(7805 + 20\,100)\,\text{N} = 27\,905\,\text{N}$, acting horizontally. Its position may be determined by taking moments about, for example, a horizontal axis at the top of the cylinder:

$$(7805\,\text{N})(0.697\,\text{m}) + (20\,100\,\text{N})(1.478\,\text{m}) = (27\,905\,\text{N})x$$

where x = distance of line of action of total force
from top of cylinder
= 1.260 m.

By symmetry, the centre of pressure is on the vertical diameter of the circle.

An alternative, though algebraically more tiresome, technique would be to consider horizontal strips of the surface of vertical thickness, say, δy; and then to integrate, over each semicircle, expressions for forces on the strips and their moments about a horizontal axis at, say, the top of the cylinder. However, *there is no escape from dealing <u>separately</u> with the surfaces in contact with each fluid.*

2.5.2 Hydrostatic thrusts on curved surfaces

On a curved surface the forces $p\,\delta A$ on individual elements differ in direction, so a simple summation of them may not be made. Instead, the resultant thrusts in certain directions may be determined, and these forces may then be combined vectorially. It is simplest to calculate horizontal and vertical components of the total thrust.

Horizontal component

Any curved surface may be projected on to a vertical plane. Take, for example, the curved surface illustrated in Fig. 2.16. Its projection on to the vertical plane shown is represented by the trace MN and the horizontal projection lines may be supposed in the x-direction. Let F_x represent the component in this direction of the total thrust exerted by the fluid on the curved surface. By Newton's Third Law the surface

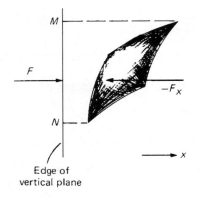

Fig. 2.16

Edge of
vertical plane

exerts a force $-F_x$ on the fluid. Consider the fluid enclosed by the curved surface, the projection lines and the vertical plane. For this fluid to be in equilibrium the force $-F_x$ must be equal in magnitude to the force F on the fluid at the vertical plane. Also the two forces must be in line, that is, $-F_x$ must act through the centre of pressure of the vertical projection.

In any given direction, therefore, the horizontal force on any surface equals the force on the projection of that surface on a vertical plane perpendicular to the given direction. The line of action of the horizontal force on the curved surface is the same as that of the force on the vertical projection.

The vertical component of the force on a curved surface may be determined by considering the fluid enclosed by the curved surface and vertical projection lines extending to the free surface (see Fig. 2.17). (We assume for the moment that a free surface does exist above the curved surface in question.) As the sides of this volume are vertical the forces acting on them must everywhere be horizontal if the fluid is in equilibrium. If the pressure at the free surface is taken as zero then there are only two vertical forces acting on the fluid in the space considered: (1) its own weight W; (2) the reaction $-F_y$ to the vertical component F_y of the total force exerted on the curved surface. Hence $W = F_y$. Moreover, W acts at G, the centre of gravity of the fluid in that space, and for equilibrium the line of action of F_y must also pass through G. Thus the vertical force acting on any surface equals the weight of the fluid extending above that surface to the free (zero-pressure) surface, and it acts through the centre of gravity of that fluid.

Vertical component

In some instances it is the underside of a curved surface that is subjected to the hydrostatic pressure, whereas the upper side is not. The vertical component of the thrust on the surface then acts upwards and equals the weight of an imaginary amount of fluid extending from the surface up to the level of the free (zero-pressure) surface. This is

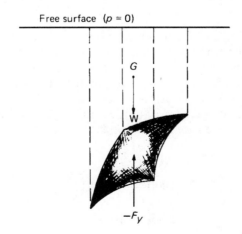

Free surface $(p = 0)$

$-F_y$

Fig. 2.17

because, if the imaginary fluid were in fact present, pressures at the two sides of the surface would be identical and the net force reduced to zero.

If a free surface does not actually exist, an imaginary free surface may be considered at a height $p/\varrho g$ above any point at which the pressure p is known. The density of the imaginary fluid must, of course, be supposed the same as that of the actual fluid so that the variation of pressure over the surface is correctly represented. The vertical component of the total force is then equal to the weight of the imaginary fluid vertically above the curved surface.

Resultant thrust

In general the components of the total force must be considered in three mutually perpendicular directions, two horizontal and one vertical. These three components need not meet at a single point, so there is, in general, no single resultant force. In many instances, however, two forces lie in the same plane and may then be combined into a single resultant by the parallelogram of forces. If there is a vertical plane on which the surface has no projection (for example, the plane perpendicular to the horizontal axis of a cylindrical surface) there is no component of hydrostatic force perpendicular to that plane. The only horizontal component then needing consideration is the one parallel to that plane.

When the two sides of a surface are wholly in contact with a single fluid of uniform density but the level of the free (atmospheric) surface on one side is different from that on the other, the net effective pressure at any point depends only on the difference in free surface levels. The effective pressure is therefore uniform over the area and so the components of the resultant force then pass through the centroids of the vertical and horizontal projections respectively.

Example 2.3 A sector gate, of radius 4 m and length 5 m, controls the flow of water in a horizontal channel. For the (equilibrium) conditions shown in Fig. 2.18, determine the total thrust on the gate.

Solution
Since the curved surface of the gate is part of a cylinder, the water exerts no thrust along its length, so we consider the horizontal and vertical components in the plane of the diagram.

The horizontal component is the thrust that would be exerted by the water on a vertical projection of the curved surface. The depth d of this projection is $4 \sin 30° \text{m} = 2 \text{m}$ and its centroid is $1 \text{m} + d/2 = 2 \text{m}$ below the free surface. Therefore

$$\text{Horizontal force} = \varrho g \bar{h} A$$

$$= 1000 \frac{\text{kg}}{\text{m}^3} \times 9.81 \frac{\text{N}}{\text{kg}} \times 2 \text{m} (5 \times 2) \text{m}^2$$

$$= 1.962 \times 10^5 \text{N}$$

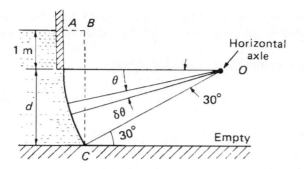

Fig. 2.18

Its line of action passes through the centre of pressure of the vertical projection, i.e. at a distance $(Ak^2)_0/A\bar{h}$ below the free surface, given by:

$$\frac{\left(Ak^2\right)_0}{A\bar{h}} = \frac{\left(Ak^2\right)_C + A\bar{h}^2}{A\bar{h}} = \frac{bd^3/12}{bd\bar{h}} + \bar{h}$$

$$= \left\{ \frac{(2\,\text{m})^2}{12} \middle/ 2\,\text{m} \right\} + 2\,\text{m}$$

$$= 2.167\,\text{m}$$

The vertical component of the total thrust = weight of imaginary water ABC.

$$AB = \left(4 - 4\cos 30° \right)\text{m} = 0.536\,\text{m}$$

Vertical force $= \varrho g V$

$$= 1000\frac{\text{kg}}{\text{m}^3} \times 9.81\frac{\text{N}}{\text{kg}} \times 5\,\text{m}\left\{ (0.536 \times 1) \right.$$

$$\left. + \pi \times 4^2 \times \frac{30}{360} - \frac{1}{2} \times 2 \times 4\cos 30° \right\}\text{m}^2$$

$$= 6.18 \times 10^4\,\text{N}$$

The centre of gravity of the imaginary fluid ABC may be found by taking moments about BC. It is 0.237 m to the left of BC.

The horizontal and vertical components are co-planar and therefore combine to give a single resultant force of magnitude

$$\left\{ \left(1.962 \times 10^5 \right)^2 + \left(6.18 \times 10^4 \right)^2 \right\}^{1/2}\,\text{N} = 2.057 \times 10^5\,\text{N}$$

at an angle arctan $(61\,800/196\,200) \simeq 17.5°$ to the horizontal.

It is instructive to obtain the result in an alternative way. Consider an element of the area of the gate subtending a small angle $\delta\theta$ at O. Then the thrust on this element $= \varrho g h\,\delta A$, and the horizontal component of this thrust $= \varrho g h \, \delta A \cos\theta$ where θ is the angle between the horizontal and the radius to the element. Now $h = (1 + 4\sin\theta)\,\mathrm{m}$ and $\delta A = (4\,\delta\theta \times 5)\,\mathrm{m}^2 = 20\,\delta\theta\,\mathrm{m}^2$, so the total horizontal component $= \varrho g \int h \cos\theta \, \mathrm{d}A =$

$$1000\frac{\mathrm{kg}}{\mathrm{m}^3} \times 9.81\frac{\mathrm{N}}{\mathrm{kg}} \times 20\,\mathrm{m}^2 \int_0^{\pi/6} (1 + 4\sin\theta)\cos\theta\,\mathrm{d}\theta\,\mathrm{m}$$

$$= 1.962 \times 10^5\,\mathrm{N}$$

as before.

The vertical component of the thrust on an element is $\varrho g h\,\delta A\sin\theta$ and the total vertical component $= \varrho g \int h \sin\theta\,\mathrm{d}A =$

$$1000\frac{\mathrm{kg}}{\mathrm{m}^3} \times 9.81\frac{\mathrm{N}}{\mathrm{kg}} \times 20\,\mathrm{m}^2 \int_0^{\pi/6} (1 + 4\sin\theta)\sin\theta\,\mathrm{d}\theta\,\mathrm{m}$$

$$= 6.18 \times 10^4\,\mathrm{N}$$

as before.

Since all the elemental thrusts are perpendicular to the surface their lines of action all pass through O and that of the resultant force therefore also passes through O.

When variations of pressure with depth may be neglected – for example, when the fluid is a gas – the magnitude of the force exerted on a curved surface in any direction is given by the product of the (uniform) pressure and the projected area of the surface perpendicular to that direction.

2.6   BUOYANCY

Because the pressure in a fluid in equilibrium increases with depth, the fluid exerts a resultant upward force on any body wholly or partly immersed in it. This force is known as the *buoyancy* and it may be determined by the methods of Section 2.5.

The buoyancy has no horizontal component, because the horizontal thrust in any direction is the same as on a vertical projection of the surface of the body perpendicular to that direction, and the thrusts on the two faces of such a vertical projection exactly balance. Consider the body *PQRS* of Fig. 2.19. The upward thrust on the lower surface *PSR* corresponds to the weight of the fluid, real or imaginary, vertically above that surface, that is, that corresponding to the volume *PSRNM*. The downward thrust on the upper surface *PQR* equals the weight of the fluid *PQRNM*. The resultant upward force exerted by the fluid on the body is therefore

weight of fluid   weight of fluid   weight of fluid
corresponding to $-$ corresponding to $=$ corresponding to
PSRNM   *PQRNM*   *PQRS*.

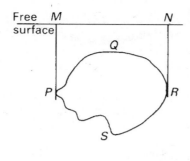

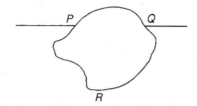

Fig. 2.19

Fig. 2.20

It may be noted that in this case there is no restriction to a fluid of uniform density.

Since the fluid is in equilibrium we may imagine the body removed and its place occupied by an equal volume of the fluid. This extra fluid would be in equilibrium under the action of its own weight and the thrusts exerted by the surrounding fluid. The resultant of these thrusts (the buoyancy) must therefore be equal and opposite to the weight of the fluid taking the place of the body, and must also pass through the centre of gravity of that fluid. This point, which corresponds to the centroid of the volume if the fluid is of uniform weight per unit volume, is known as the *centre of buoyancy*. Its position depends on the shape of the volume considered and it should be carefully distinguished from the centre of gravity *of the body* which depends on the way in which the weight of the body is distributed.

For a body only partly immersed in the fluid (as in Fig. 2.20) similar considerations show that the buoyancy corresponds to the weight of fluid equal in volume to *PRQ*. In general, then, the buoyancy is the resultant upward force exerted by the fluid on the body, and is equal in magnitude to $\varrho g V$ where ϱ represents the mean density of the fluid and V the immersed volume of the body. (In calculating V for a partly immersed body, any meniscus due to surface tension must be disregarded. Except for very small bodies the vertical force directly due to surface tension is negligible.) This result is often known as the Principle of Archimedes* (*c.* 287–212 B.C.).

The buoyancy is not related to – and indeed may even exceed – the weight of fluid actually present. For example, the mirrors of astronomical telescopes are sometimes floated in mercury; the buoyancy

* pronounced *Ark-im-mee'-deez.*

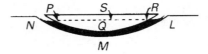

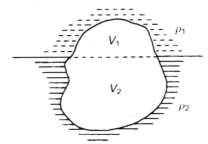

Fig. 2.21

Fig. 2.22

corresponds to the weight of mercury having a volume equal to *PQRS* (Fig. 2.21) and this may be many times greater than the volume of mercury present, *PQRLMN*.

A body may be partly immersed in each of two immiscible fluids, as shown in Fig. 2.22. The total buoyancy is then $\varrho_1 g V_1 + \varrho_2 g V_2$. In general, however, the centres of buoyancy of the volumes V_1 and V_2 are not on the same vertical line and the total buoyancy force then does not pass through the centroid of the entire volume. Where the lower fluid is a liquid and the upper a gas, the buoyancy provided by the gas may, except in very accurate work, be neglected and the total buoyancy assumed to be $\varrho_2 g V_2$ only, acting at the centroid of the volume V_2. Buoyancy due to the atmosphere is also usually neglected when a body is weighed on a balance in air, although in very accurate work a correction is applied to account for it.

If a body is otherwise unsupported it is in equilibrium in the fluid only when its buoyancy balances its weight. If the buoyancy exceeds the weight – as for a balloon in air or an air bubble in water – the body rises until its average density equals that of the surrounding fluid. If the body is more compressible than the surrounding fluid its own average density decreases faster than that of the fluid and, unless the height of the fluid has a definite limit, the body rises indefinitely.

For a floating body to be in vertical equilibrium, the volume immersed in the liquid must be such as to provide a buoyancy exactly equal to the weight of the body.

2.7 THE STABILITY OF BODIES IN FLUIDS

2.7.1 The stability of submerged bodies

For a body not otherwise restrained it is important to know not only whether it will rise or fall in the fluid, but also whether an originally

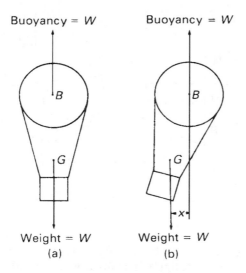

Fig. 2.23

vertical axis in the body will remain vertical. We are not here concerned with effects of a fluid in motion but with states of equilibrium. We must, however, distinguish three types of equilibrium. A body in *stable equilibrium* will, if given a small displacement and then released, return to its original position. If, on the other hand, the equilibrium is *unstable* the body will not return to its original position but will move further from it. In *neutral equilibrium*, the body, having been given a small displacement and then released, will neither return to its original position nor increase its displacement; it will simply adopt its new position.

For a body wholly immersed in a single fluid – as, for example the balloon and gondola illustrated in Fig. 2.23 – the conditions for stability of equilibrium are simple. An angular displacement from the normal position (a) brings into action the couple Wx which tends to restore the system to position (a). This, then, is a stable arrangement. If, however, the centre of gravity G were above the centre of buoyancy B the couple arising from a small angular displacement would be such as to cause the assembly to topple over. So for a completely immersed body the condition for stability is simply that G be below B. If B and G coincide, neutral equilibrium is obtained.

2.7.2 The stability of floating bodies

The condition for angular stability of a body floating in a liquid is a little more complicated. This is because, when the body undergoes an angular displacement about a horizontal axis, the shape of the immersed volume in general changes, so the centre of buoyancy moves relative to the body. As a result stable equilibrium can be achieved even when G is above B.

Figure 2.24a illustrates a floating body – a boat, for example – in its equilibrium position. The net force is zero, so the buoyancy is equal in

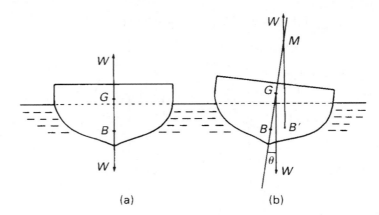

Fig. 2.24

(a)　　　　(b)

magnitude to the weight W of the body. There must be no moment on the body, so the weight acting vertically downwards through the centre of gravity G must be in line with the buoyancy acting vertically upwards through the centre of buoyancy B. Figure 2.24b shows the situation after the body has undergone a small angular displacement (or 'angle of heel') θ. It is assumed that the position of the centre of gravity G remains unchanged relative to the body. (This is not always a justifiable assumption for a ship since some of the cargo may shift during an angular displacement.) The centre of buoyancy B, however, does not remain fixed relative to the body. During the movement, the volume immersed on the right-hand side increases while that on the left-hand side decreases, so the centre of buoyancy (i.e. the centroid of the immersed volume) moves to a new position B'. Suppose that the line of action of the buoyancy (which is always vertical) intersects the axis BG at M. For small values of θ, the point M is practically constant in position and is known as the *metacentre*. For the body shown in the figure, M is above G, and the couple acting on the body in its displaced position is a *restoring couple*, that is, it tends to restore the body to its original position. If M were below G the couple would be an *overturning couple* and the original equilibrium would have been unstable.

The distance of the metacentre above G is known as the *metacentric height*, and for stability of the body it must be positive (i.e. M above G). Neutral equilibrium is of course obtained when the metacentric height is zero and G and M coincide. For a floating body, then, stability is not determined simply by the relative positions of B and G.

The magnitude of the restoring couple is $W(GM)\sin\theta$ and the magnitude of GM therefore serves as a measure of the stability of a floating body. A simple experiment may be conducted to determine GM. Suppose that for the boat illustrated in Fig. 2.25 the metacentric height corresponding to 'roll' about the longitudinal axis is required. If a body of weight P is moved transversely across the deck (which is initially horizontal) the boat moves through a small angle θ – which may be measured by the movement of a plumb line over a scale – and comes to

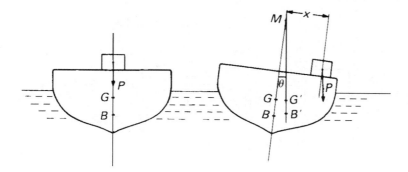

Fig. 2.25

rest in a new position of equilibrium. The centres of gravity and buoy-ancy are therefore again vertically in line. Now the movement of the weight P through a distance x causes a parallel shift of the 'total' centre of gravity (that is, the centre of gravity of the whole boat *including P*) from G to G' such that $Px = W(GG')$, W being the total weight includ-ing P. But $(GG') = (GM)\tan\theta$, so

$$(GM) = \frac{Px}{W}\cot\theta \qquad (2.19)$$

Since the point M corresponds to the metacentre for small angles of heel only, the true metacentric height is the limiting value of GM as $\theta \to 0$. This may be determined from a graph of nominal values of GM calculated from eqn 2.19 for various values of θ (positive and negative).

It is desirable, however, to be able to determine the position of the metacentre and the metacentric height before a boat is constructed. Fortunately this may be done simply by considering the shape of the hull. Figure 2.26 shows that cross-section, perpendicular to the axis of rotation, in which the centre of buoyancy B lies. At (a) is shown the equilibrium position: after displacement through a small angle θ (here exaggerated for the sake of clarity) the body has the position shown at (b). The section on the left, indicated by cross-hatching, has emerged from the liquid whereas the cross-hatched section on the right has moved down into the liquid. We assume that there is no overall vertical movement; thus the *vertical* equilibrium is undisturbed. As the total weight of the body remains unaltered so does the volume immersed, and therefore the volumes corresponding to the cross-hatched sections are equal. This is so if the planes of flotation for the equilibrium and dis-placed positions intersect along the centroidal axes of the planes.

We choose coordinate axes through O as origin: Ox is perpendicular to the plane of diagrams (a) and (b), Oy lies in the original plane of flotation and Oz is vertically downwards in the equilibrium position. As the body is displaced the axes move with it. (The axis Ox may move sideways during the rotation: thus Ox is not necessarily the axis of rotation.) The entire immersed volume V may be supposed to be made

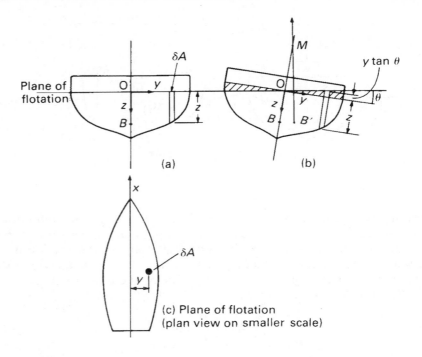

Plane of flotation

(a)

(b)

(c) Plane of flotation
(plan view on smaller scale)

Fig. 2.26

up of elements like that shown – each underneath an area δA in the plane of flotation. Now the centre of buoyancy B by definition corresponds to the centroid of the immersed volume (the liquid being assumed homogeneous). Its y-coordinate $(\bar{y}_0)$ may therefore be determined by taking moments of volume about the xz plane:

$$V\bar{y}_0 = \int (z\,\mathrm{d}A)y \qquad (2.20)$$

(For a symmetrical body $\bar{y}_0 = 0$.) After displacement the centre of buoyancy is at B' whose y-coordinate is $\bar{y}$. (For ships rotating about a longitudinal axis the centre of buoyancy may not remain in the plane shown in Fig. 2.26a because the underwater contour is not, in general, symmetrical about a transverse section. One should therefore regard B' as a projection of the new centre of buoyancy on to the transverse plane represented in the diagram.) The depth of each element of volume is now $z + y \tan\theta$, so

$$V\bar{y} = \int y(z + y \tan\theta)\mathrm{d}A \qquad (2.21)$$

Subtraction of eqn 2.20 from eqn 2.21 gives

$$V(\bar{y} - \bar{y}_0) = \int y^2 \tan\theta \, \mathrm{d}A = \tan\theta \left(Ak^2 \right)_{Ox}$$

where $(Ak^2)_{Ox}$ represents the second moment of area of the plane of flotation about the axis Ox (see Fig. 2.26c).

But, for *small* angular displacements, $\bar{y} - \bar{y}_0 = (BM)\tan\theta$ and therefore

$$V(BM) = (Ak^2)_{Ox}$$

or

$$(BM) = \frac{(Ak^2)_{Ox}}{V}$$

$$= \frac{\text{Second moment of area of Plane of Flotation about}}{\text{Immersed Volume}} \quad (2.22)$$

The length BM is sometimes known as the metacentric radius; it must not be confused with the metacentric height GM.

For rolling (i.e. side to side) movements of a ship the centroidal axis about which the second moment is taken is the longitudinal one. Stability in this direction is normally by far the most important. For pitching movements (i.e. stern up, bow down, or vice versa) the appropriate axis is the transverse one. The metacentres corresponding to different axes of rotation in general have different positions. The position of B can be calculated since the contours of the hull at various levels are normally known, and hence the position of the metacentre may be determined from eqn 2.22.

The equation strictly applies only to very small angular displacements, and this limitation is more important if the body does not have vertical sides (although for ships the sides are usually approximately vertical at the water-line). The result may legitimately be used to indicate the *initial* stability of the body. It is nevertheless sufficiently accurate for most calculations involving angles up to about 8°.

The value of BM for a ship is of course affected by a change of loading whereby the immersed volume alters. If the sides are not vertical at the water-line the value of Ak^2 may also change as the vessel rises or falls in the water. Naval architects must design vessels so that they are stable under all conditions of loading and movement. Wide ships are stable in rolling movements because $(Ak^2)_{Ox}$ is then large and the metacentre high.

Example 2.4 A uniform, closed cylindrical buoy, 1.5 m high, 1.0 m diameter, and of mass 80 kg is to float with its axis vertical in sea water of density 1026 kg/m³. A body of mass 10 kg is attached to the centre of the top surface of the buoy. Show that, if the buoy floats freely, initial instability will occur.

Solution
Moments of mass about horizontal axis through O:

$$(10\,\text{kg})(1.5\,\text{m}) + (80\,\text{kg})\left(\frac{1.5}{2}\,\text{m}\right) = \{(80 + 10)\text{kg}\}\,(OG)$$

$$\therefore \quad OG = 0.8333\,\text{m}$$

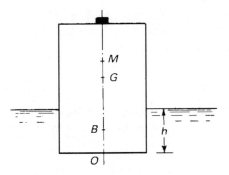

For vertical equilibrium, buoyancy = weight.

$$\therefore \ \frac{\pi}{4}(1\,\mathrm{m})^2 h \times 1026\frac{\mathrm{kg}}{\mathrm{m}^3}g = (80 + 10)\mathrm{kg}\,g$$

whence $h = 0.1117\,\mathrm{m}$.

From Fig. 2.14, Ak^2 of a circle about a centroidal axis $= \pi d^4/64$.

$$\therefore \ BM = Ak^2/V = \frac{\pi}{64}d^4 \Big/ \frac{\pi}{4}d^2 h = \frac{d^2}{16h}$$

$$= \frac{1^2}{16 \times 0.1117}\mathrm{m} = 0.560\,\mathrm{m}$$

and

$$GM = OB + BM - OG = \left(\frac{0.1117}{2} + 0.560 - 0.8333\right)\mathrm{m}$$

$$= -0.2175\,\mathrm{m}$$

Since this is negative (i.e. M is below G) *buoy is unstable*.

Floating bodies containing a liquid

If a floating body carries liquid with a free surface, this contained liquid will move in an attempt to keep its free surface horizontal when the body undergoes angular displacement. Thus not only does the centre of buoyancy move, but also the centre of gravity of the floating body and its contents. The movement of G is in the same direction as the movement of B and consequently the stability of the body is reduced. For this reason a liquid (e.g. oil) which has to be carried in a ship is put into a number of separate compartments so as to minimize its movement within the ship. It may be shown that, for small angular movements, the effective metacentric height is reduced by an amount $\varrho_i(Ak^2)'/\varrho V$ for each compartment, where $(Ak^2)'$ represents the second moment of area of the free surface of the liquid in the compartment about its centroidal axis parallel to the axis of rotation, ϱ_i represents the density of the liquid in the compartment and ϱV the total mass of the vessel and its cargo.

As we have seen, the restoring couple caused by the hydrostatic forces acting on a floating body displaced from its equilibrium position is $W(GM)\sin\theta$ (see Fig. 2.24). Since torque equals moment of inertia (i.e. second moment of mass) multiplied by angular acceleration we may write

Period of oscillation

$$W(GM)\sin\theta = -\left(Mk^2\right)_R \frac{\mathrm{d}^2\theta}{\mathrm{d}t^2}$$

if we assume that the torque caused by the hydrostatic forces is the only one acting on the body. (In practice a certain amount of liquid moves with the body, but the effect of this is slight.) Here $(Mk^2)_R$ represents the moment of inertia of the body about its axis of rotation. The minus sign arises because the torque acts so as to decrease θ, that is, the angular acceleration $\mathrm{d}^2\theta/\mathrm{d}t^2$ is negative. Thus for small angular movements $\sin\theta$ is proportional to $-\mathrm{d}^2\theta/\mathrm{d}t^2$ as for a simple pendulum. If there is no relative movement (e.g. of a liquid) within the body $(Mk^2)_R$ is constant and if θ is small so that $\sin\theta \simeq \theta$ (in radian measure) the equation may be integrated to give $2\pi k_R/\{g(GM)\}^{1/2}$ as the time of a complete oscillation from one side to the other and back again.

If the only forces acting are the weight of the body and the buoyancy – both of which are vertical – then G does not move horizontally. The instantaneous axis of rotation therefore lies in a horizontal plane through G. Moreover, for a body symmetrical about the axis Ox (of Fig. 2.26) the instantaneous axis of rotation must lie in a vertical plane through Ox so that Ox does not move vertically out of the free surface. For small angular displacements, however, the line of intersection of these horizontal and vertical planes is very close to G, so the axis of rotation is considered to pass through G. The moment of inertia is consequently calculated for an axis through G.

The oscillation of the body results in some flow of the liquid round it and this flow has been disregarded here. Reasonable agreement between theoretical and experimental values of the period of oscillation has been found for the rolling motion of ships but the agreement is less good for pitching movements. In practice, of course, viscosity in the water introduces a damping action which quickly suppresses the oscillation unless further disturbances such as waves cause new angular displacements.

The metacentric height of ocean-going vessels is, for rotation about a longitudinal axis, usually of the order of 300 mm to 1.2 m. Increasing the metacentric height gives greater stability but reduces the period of roll, so the vessel is less comfortable for passengers and is subjected to strains which may damage its structure. In cargo vessels the metacentric height varies with the loading although some control of its value is possible by adjusting the position of the cargo. Some control of the period of roll is also possible: if the cargo is placed further from the centre-line the moment of inertia of the vessel, and consequently the period, may be increased with little sacrifice of stability. On the other hand, in warships and racing yachts, for example, stability is more important than comfort, and such vessels have larger metacentric heights.

2.7.3 Stability of a body subject to an additional force

When an unconstrained body is in equilibrium in a fluid the only forces relevant to its stability are the weight of the body and its buoyancy. If, however, an additional force is provided – by, for example, an anchor chain – stability is determined by the lines of action of the buoyancy and the *resultant* downward force.

Example 2.5   For the buoy considered in example 2.4, calculate the least vertical downward force applied at the centre of the base that would just keep the buoy upright. What would then be the depth of immersion?

Solution
A vertically downward force F applied at O increases the total downward force from W (the total weight of the buoy) to $W + F$. To maintain vertical equilibrium the buoyancy too is increased to $W + F$, and so the new depth of immersion h' is given by

$$\varrho g \frac{\pi}{4} d^2 h' = W + F$$

Taking moments of forces about a horizontal axis through O gives the requirement for the restoring couple to be just zero:

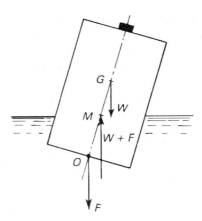

$$W(OG) = (W + F)(OB + BM) = \varrho g \frac{\pi}{4} d^2 h' \left(\frac{h'}{2} + \frac{d^2}{16h'} \right)$$

i.e.

$$(90\,\text{kg})g(0.8333\,\text{m}) = \left(1026\,\frac{\text{kg}}{\text{m}^3} \right) g \frac{\pi}{4} (1\,\text{m})^2 \left\{ \frac{(h')^2}{2} + \frac{(1\,\text{m})^2}{16} \right\}$$

whence

$$h' = 0.2473\,\text{m}$$

and

$$F = \varrho g \frac{\pi}{4} d^2 h' - W = \left(1026 \times 9.81 \frac{\pi}{4} 1^2 \times 0.2473 - 90 \times 9.81 \right) \text{N}$$

$$= 1072\,\text{N}$$

2.7.4 Stability of a fluid itself

In the preceding sections we have considered the stability of separate, identifiable, bodies wholly or partly immersed in a fluid. We now turn attention to the stability of parts of the fluid itself which, perhaps because of uneven heating or cooling, have a density slightly different from that of neighbouring fluid. These differences of density are the cause of 'convection currents' frequently encountered in both liquids and gases.

If, for example, only the lower layers of a certain bulk of fluid are heated, an unstable condition results. This is because if some of the warmer fluid is displaced upwards it finds itself surrounded by cooler, and therefore denser, fluid. The buoyancy force exerted on the warmer fluid by its surroundings is equal in magnitude to the weight of an equal volume of the surrounding denser fluid. As this buoyancy is greater than the weight of the newly arrived fluid there is a net upward force on the warmer fluid which therefore continues to rise. Heavier fluid then flows downwards to take the place of the less dense fluid which has moved up and thus 'free convection' is started.

If, however, the lower layers of fluid are cooled the conditions are stable. Fluid displaced downwards would be surrounded by cooler, denser, fluid; it would therefore experience a buoyancy force greater than its own weight and would return upwards to its original position.

Such movements occur on a large scale in the atmosphere. The lower part of the atmosphere is continually being mixed by convection which is largely due to the unequal heating of the earth's surface. When air is heated more in one locality than in another, it rises and then, as its pressure falls with increase of altitude, it cools. Because air is a poor conductor of heat the cooling takes place approximately adiabatically according to eqn 1.6. The adiabatic temperature lapse rate in a dry atmosphere is approximately 0.01 K/m whereas the temperature change normally found in nature is of the order of 0.0065 K/m. Rising air, cooling adiabatically, therefore becomes cooler and denser than its surroundings and tends to fall back to its original level. Normally, then, the atmosphere is stable. If, however, the natural temperature lapse rate exceeds the adiabatic, the equilibrium is unstable – a condition frequently responsible for thunderstorms.

2.8 EQUILIBRIUM OF MOVING FLUIDS

In certain instances the methods of hydrostatics may be used to study the behaviour of fluids in motion. For example, if all the fluid concerned moves uniformly in a straight line, there is no acceleration and there are no shear forces. Thus no force acts on the fluid as a result of the motion and, in these circumstances, the hydrostatic equations apply without change.

If all the fluid concerned is undergoing uniform acceleration in a straight line, no layer moves relative to another, so there are still no shear forces. There is, however, an additional force acting to cause the acceleration. Nevertheless, provided that due allowance is made for this additional force the system may be studied by the methods of hydrostatics. Fluids in such motion are said to be in *relative equilibrium*.

Consider the small rectangular element of fluid of size $\delta x \times \delta y \times \delta z$ shown in Fig. 2.27 (δy being measured perpendicularly to the paper). The pressure at the centre is p, so the mean pressure over the left-hand face is $p - (\partial p/\partial x)\frac{1}{2}\delta x$ and the mean pressure over the right-hand face $p + (\partial p/\partial x)\frac{1}{2}\delta x$. If the fluid is in relative equilibrium there are no shear forces. The net force in the (horizontal) x direction is therefore

$$\left\{ \left(p - \frac{\partial p}{\partial x}\frac{1}{2}\delta x \right) - \left(p + \frac{\partial p}{\partial x}\frac{1}{2}\delta x \right) \right\}\delta y\,\delta z = -\frac{\partial p}{\partial x}\delta x\,\delta y\,\delta z$$

$$= \text{(by Newton's Second Law)} \; \varrho\,\delta x\,\delta y\,\delta z\,a_x$$

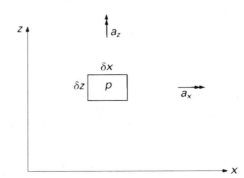

Fig. 2.27

where ϱ represents the mean density of the fluid in the element and a_x the component of acceleration in the x direction. Therefore the *pressure gradient* in the x direction, $\partial p/\partial x$, is given by

$$\frac{\partial p}{\partial x} = -\varrho a_x \qquad (2.23)$$

Similarly, the net force in the (vertical) z direction due to pressure is given by

$$-\frac{\partial p}{\partial z}\,\delta x\,\delta y\,\delta z$$

The weight of the element acting vertically downwards is $\varrho g\,\delta x\,\delta y\,\delta z$, so

$$-\frac{\partial p}{\partial z}\,\delta x\,\delta y\,\delta z - \varrho g\,\delta x\,\delta y\,\delta z = \varrho\,\delta x\,\delta y\,\delta z\,a_z$$

$$\therefore \quad \frac{\partial p}{\partial z} = -\varrho\left(g + a_z\right) \qquad (2.24)$$

In general there would also be a component of acceleration in the y direction and a corresponding pressure gradient in the y direction $\partial p/\partial y = -\varrho a_y$. For simplicity, however, we shall consider the total acceleration to be in the x–z plane.

From eqns 2.23 and 2.24 the pressure variation throughout the fluid may be determined. A surface of constant pressure in the fluid is one along which

$$dp = \frac{\partial p}{\partial x}\,dx + \frac{\partial p}{\partial z}\,dz = 0$$

i.e. along which

$$\frac{dz}{dx} = -\frac{\partial p/\partial x}{\partial p/\partial z} = \frac{-a_x}{g + a_z} \qquad (2.25)$$

For constant acceleration, therefore, dz/dx is constant and a surface of constant pressure has a constant slope relative to the x direction of $-a_x/(g + a_z)$. One such surface is a free surface; other constant pressure planes are parallel to it.

For example, consider the tank illustrated in Fig. 2.28. It contains a liquid and is given a uniform horizontal acceleration a_x (the vertical acceleration a_z is zero). Once the liquid has adjusted itself to uniform conditions the free surface settles at a slope as shown. (During the period when the liquid is moving into its new position shear forces are involved, so the methods of 'hydrostatics' do not then apply.) Here $\tan\theta = dz/dx = -a_x/g$.

If, however, the acceleration is only in the vertical direction eqn 2.25 shows that $dz/dx = 0$, so planes of constant pressure are horizontal. The variation of pressure through the fluid is then given simply by eqn 2.24. A tank of fluid allowed to fall freely would have an acceleration in the

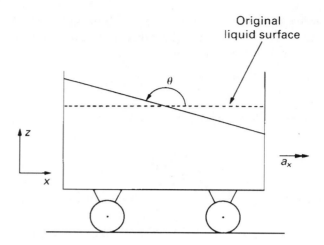

Fig. 2.28

z direction (upwards) of $-g$ and thus uniform pressure would be obtained throughout the fluid. But if a tank of liquid were accelerated upwards the hydrostatic pressure variation would be intensified.

Fluid completely filling a closed tank would have no free surface, but planes of constant pressure would still be inclined at an angle of arctan $\{-a_x/(g + a_z)\}$ to the x direction. Pressures at particular points in the fluid may be determined by integrating eqns 2.23 and 2.24:

$$p = \int dp = \int \frac{\partial p}{\partial x} dx + \int \frac{\partial p}{\partial z} dz$$
$$= -\varrho a_x x - \varrho(g + a_z)z + \text{constant} \qquad (2.26)$$

for a constant-density fluid. The integration constant is determined by the conditions of the problem – for example, that $p = p_{\text{atm}}$ at a free surface.

All the foregoing results refer *only* to a horizontal x (and y) axis and a vertical z axis; it should be remembered too that z is measured *upwards* from a suitable horizontal datum.

Once the direction of the constant-pressure planes is known, alternative expressions may be obtained by considering, say, ξ and η axes, respectively parallel to and perpendicular to the constant-pressure planes (see Fig. 2.29). Then $\partial p/\partial \xi = 0$ by definition of ξ but

$$\frac{\partial p}{\partial \eta} = \frac{\partial p}{\partial x} \bigg/ \frac{\partial \eta}{\partial x} = \frac{-\varrho a_x}{\sin \theta} = -\varrho\left\{a_x^2 + (g + a_z)^2\right\}^{1/2} = \frac{dp}{d\eta}$$

since p depends only on η. Comparison with $dp/dz = -\varrho g$, the equilibrium equation for zero acceleration, shows that pressures for relative equilibrium may be calculated by hydrostatic principles provided that $\{a_x^2 + (g + a_z)^2\}^{1/2}$ takes the place of g; and η the place of z. However, it is usually simpler and certainly safer to work with horizontal and vertical axes only.

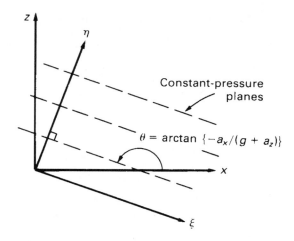

Constant-pressure planes

$\theta = \arctan \{-a_x/(g + a_z)\}$

Fig. 2.29

Example 2.6 A thin-walled, open-topped tank in the form of a cube of 500 mm side is initially full of oil of relative density 0.88. It is accelerated uniformly at 5 m/s² up a long straight slope at arctan (1/4) to the horizontal, the base of the tank remaining parallel to the slope, and the two side faces remaining parallel to the direction of motion. Calculate (a) the volume of oil left in the tank when no more spilling occurs, and (b) the pressure at the lowest corners of the tank.

Solution
The forward acceleration causes the free surface to slope at angle $(180° - \theta)$ to the forward horizontal, and the oil therefore spills over the corner B until conditions are as shown in the diagram.

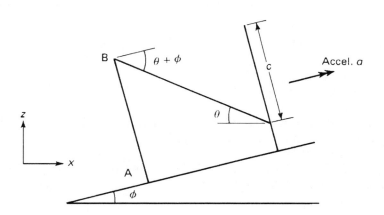

Horizontal component of acceleration

$$= a_x = a\cos\phi = \left(5\,\text{m/s}^2\right) \times 4/\sqrt{17}$$

Vertical component of acceleration

$$= a_z = a\sin\phi = \left(5\,\text{m/s}^2\right) \times 1/\sqrt{17}$$

From eqn 2.25

$$\tan\left(180° - \theta\right) = -\frac{a_x}{a_z + g}$$

i.e.

$$\tan\theta = \frac{a_x}{a_z + g} = \frac{20/\sqrt{17}}{\left(5/\sqrt{17}\right) + 9.81} = 0.440$$

(a) $\therefore$ $\tan\left(\phi + \theta\right) = \dfrac{\tan\phi + \tan\theta}{1 - \tan\phi\tan\theta} = \dfrac{0.25 + 0.440}{1 - 0.25 \times 0.440} = 0.775$

$\therefore$ $c = 0.5\,\text{m} \times 0.775 = 0.3875\,\text{m}$

Then volume of oil left $= 0.5\left(0.5^2 - \dfrac{1}{2}0.5 \times 0.3875\right)\text{m}^3$

$$= 0.0765\,\text{m}^3 = 76.5\,\text{litres}$$

(b) From eqn 2.26 $\quad p = -\varrho a_x x - \varrho\left(a_z + g\right)z + \text{constant}$

Pressure at B is atmospheric and if B is at $(0, 0)$ then constant $= 0$

Point A will be at $\left[\left(0.5\,\text{m}\right)\sin\phi, -\left(0.5\,\text{m}\right)\cos\phi\right]$

$\therefore$ $p_A = -\varrho a\cos\phi\left(0.5\,\text{m}\right)\sin\phi + \varrho\left(a\sin\phi + g\right)\left(0.5\,\text{m}\right)\cos\phi$

$$= \varrho g\left(0.5\,\text{m}\right)\cos\phi = 880\frac{\text{kg}}{\text{m}^3}9.81\frac{\text{N}}{\text{kg}}0.5\,\text{m}\frac{4}{\sqrt{17}} = 4190\,\text{Pa}$$

FURTHER READING

(Manometers, etc.)

Doyle, F. E. *Instrumentation: Pressure and Liquid Level*, Blackie, Glasgow and London (1969).

Ower, E. and Pankhurst, R. C. *The Measurement of Air Flow* (5th edn), Chapter X, Pergamon, Oxford (1977).

(Metacentric height, etc.)

Clayton, B. R. and Bishop, R. E. D. *Mechanics of Marine Vehicles*, Spon, London (1982).

PROBLEMS

2.1 To what head of carbon tetrachloride (relative density 1.59) is a pressure of 200 kPa equivalent?

2.2 A tank 3.5 m long and 2.5 m wide contains alcohol of relative density 0.82 to a depth of 3 m. A 50 mm diameter pipe leads from the bottom of the tank. What will be the reading on a gauge calibrated in Pa connected at a point (a) 150 mm above the bottom of the tank; (b) in the 50 mm diameter pipe, 2 m below the bottom of the tank; (c) at the upper end of a 25 mm diameter pipe, connected to the 50 mm pipe 2 m below the bottom of the tank, sloping upwards at 30° to the horizontal for 1.2 m and then rising vertically for 600 mm? What is the load on the bottom of the tank?

2.3 To what head of air ($R = 287$ J/kg K) at an absolute pressure of 101.3 kPa and temperature of 15 °C is a pressure of 75 mm of water equivalent?

2.4 A spherical air bubble rises in water. At a depth of 9 m its diameter is 4 mm. What is its diameter just as it reaches the free surface where the absolute pressure is 101.3 kPa? (Surface tension effects are negligible.)

2.5 Two small vessels are connected to a U-tube manometer containing mercury (relative density 13.56) and the connecting tubes are filled with alcohol (relative density 0.82). The vessel at the higher pressure is 2 m lower in elevation than the other. What is the pressure difference between the vessels when the steady difference in level of the mercury menisci is 225 mm? What is the difference of piezometric head? If an inverted U-tube manometer containing a liquid of relative density 0.74 were used instead, what would be the manometer reading for the same pressure difference?

2.6 A manometer consists of two tubes A and B, with vertical axes and uniform cross-sectional areas 500 mm^2 and 800 mm^2 respectively, connected by a U-tube C of cross-sectional area 70 mm^2 throughout. Tube A contains a liquid of relative density 0.8; tube B contains one of relative density 0.9. The surface of separation between the two liquids is in the vertical side of C connected to tube A. What additional pressure, applied to the tube B, would cause the surface of separation to rise 60 mm in the tube C?

2.7 Assuming that atmospheric temperature decreases with increasing altitude at a uniform rate of 0.0065 K/m, determine the atmospheric pressure at an altitude of 7.5 km if the temperature and pressure at sea level are 15 °C and 101.5 kPa respectively. ($R = 287$ J/kg K)

2.8 At the top of a mountain the temperature is $-5\,^\circ$C and a mercury barometer reads 566 mm, whereas the reading at the foot of the mountain is 749 mm. Assuming a temperature lapse rate of 0.0065 K/m and $R = 287\,$J/kg K, calculate the height of the mountain. (Neglect thermal expansion of mercury.)

2.9 A rectangular plane, 1.2 m by 1.8 m is submerged in water and makes an angle of 30° with the horizontal, the 1.2 m sides being horizontal. Calculate the magnitude of the net force on one face and the position of the centre of pressure when the top edge of the plane is (a) at the free surface, (b) 500 mm below the free surface, (c) 30 m below the free surface.

2.10 What is the position of the centre of pressure for a vertical semicircular plane submerged in a homogeneous liquid and with its diameter d at the free surface?

2.11 An open channel has a cross-section in the form of an equilateral triangle with 2.5 m sides and a vertical axis of symmetry. Its end is closed by a triangular vertical gate, also with 2.5 m sides, supported at each corner. Calculate the horizontal thrust on each support when the channel is brim-full of water.

2.12 A circular opening 1.2 m in diameter in the vertical side of a reservoir is closed by a disc which just fits the opening and is pivoted on a shaft along its horizontal diameter. Show that, if the water level in the reservoir is above the top of the disc, the turning moment on the shaft required to hold the disc vertical is independent of the head of water. Calculate the amount of this moment.

2.13 A square aperture in the vertical side of a tank has one diagonal vertical and is completely covered by a plane plate hinged along one of the upper sides of the aperture. The diagonals of the aperture are 2 m long and the tank contains a liquid of relative density 1.15. The centre of the aperture is 1.5 m below the free surface. Calculate the net hydrostatic thrust on the plate, the moment of this thrust about the hinge and the position of the centre of pressure.

2.14 A canal lock is 6 m wide and has two vertical gates which make an angle of 120° with each other. The depths of water on the two sides of the gates are 9 m and 2.7 m respectively. Each gate is supported on two hinges, the lower one being 600 mm above the bottom of the lock. Neglecting the weight of the gates themselves, calculate the thrust between the gates and the height of the upper hinges if the forces on them are to be half those on the lower hinges.

2.15 The profile of the inner face of a dam takes the form of a parabola with the equation $18y = x^2$, where y m is the height above the base and x m is the horizontal distance of the face from the vertical reference plane. The water level is 27 m above the base. Determine the thrust on the dam (per metre width) due to the water pressure, its inclination to the vertical and the point where the line of action of this force intersects the free water surface.

2.16 A tank with vertical sides contains water to a depth of 1.2 m and a layer of oil 800 mm deep which rests on top of the water. The relative density of the oil is 0.85 and above the oil is air at atmospheric pressure. In one side of the tank, extending through its full height, is a protrusion in the form of a segment of a vertical circular cylinder. This is of radius 700 mm and is riveted across an opening 500 mm wide in the plane wall of the tank. Calculate the total horizontal thrust tending to force the protrusion away from the rest of the tank and the height of the line of action of this thrust above the base of the tank.

2.17 A vertical partition in a tank has a square aperture of side a, the upper and lower edges of which are horizontal. The aperture is completely closed by a thin diaphragm. On one side of the diaphragm there is water with a free surface at a distance b ($> a/2$) above the centre-line of the diaphragm. On the other side there is water in contact with the lower half of the diaphragm, and this is surmounted by a layer of oil of thickness c and relative density σ. The free surfaces on each side of the partition are in contact with the atmosphere. If there is no net force on the diaphragm, determine the relation between b and c, and the position of the axis of the couple on the diaphragm.

2.18 In the vertical end of an oil tank is a plane rectangular inspection door 600 mm wide and 400 mm deep which closely fits an aperture of the same size. The door can open about one vertical edge by means of two hinges, respectively 125 mm above and below the horizontal centre-line, and at the centre of the opposite vertical edge is a locking lever. Determine the forces exerted on each hinge and on the locking lever when the tank contains an oil of relative density 0.9 to a depth of 1 m above the centre of the door and the air above the oil surface is at a gauge pressure of 15 kPa.

2.19 A vessel of water of total mass 5 kg stands on a parcel balance. An iron block of mass 2.7 kg and relative density 7.5 is suspended by a fine wire from a spring balance and is lowered into the water until it is completely immersed. What are the readings on the two balances?

2.20 A cylindrical tank of diameter $3d$ contains water in which a solid circular cylinder of length l and diameter d floats with its axis vertical. Oil is poured into the tank so that the length of the float finally protruding above the oil surface is $l/20$. What vertical movement of the float has taken place? (Relative density of oil 0.8, of cylinder 0.9.)

2.21 A hollow cylinder with closed ends is 300 mm diameter and 450 mm high, has a mass of 27 kg and has a small hole in the base. It is lowered into water so that its axis remains vertical. Calculate the depth to which it will sink, the height to which the water will rise inside it and the air pressure inside it. Disregard the effect of the thickness of the walls but assume that it is uniform and that the compression of the air is isothermal. (Atmospheric pressure = 101.3 kPa.)

2.22 A spherical, helium-filled balloon of diameter 800 mm is to be used to carry meteorological instruments to a height of 6000 m above sea level. The instruments have a mass of 60 g and negligible volume, and the balloon itself has a mass of 100 g. Assuming that the balloon does not expand and that atmospheric temperature decreases with increasing altitude at a uniform rate of 0.0065 K/m, determine the mass of helium required. Atmospheric pressure and temperature at sea level are 15 °C and 101 kPa respectively; for air, $R = 287$ J/kg K.

2.23 A uniform wooden cylinder has a relative density of 0.6. Determine the ratio of diameter to length so that it will just float upright in water.

2.24 A rectangular pontoon 6 m by 3 m in plan, floating in water, has a uniform depth of immersion of 900 mm and is subjected to a torque of 7600 N m about the longitudinal axis. If the centre of gravity is 700 mm up from the bottom, estimate the angle of heel.

2.25 A solid uniform cylinder of length 150 mm and diameter 75 mm is to float upright in water. Between what limits must its mass be?

2.26 A sea-going vessel, containing a quantity of ballast, has its centre of gravity 300 mm, and its centre of buoyancy 1.6 m, below the water line. The total displacement is 80 MN and the reaction of the screw causes a heel of 0.53° when the shaft speed is 1.4 rev/s and the shaft power is 3.34 MW. After the removal of 400 Mg of ballast, resulting in the centre of buoyancy being lowered by 75 mm relative to the boat, the same screw reaction produces a heel of 0.75°. By how much has the centre of gravity been raised relative to the boat? (The sides of the vessel may be assumed vertical at the water line.)

2.27 A buoy, floating in sea water of density 1025 kg/m³, is conical in shape with a diameter across the top of 1.2 m and a vertex angle of 60°. Its mass is 300 kg and its centre of gravity is 750 mm from the vertex. A flashing beacon is to be fitted to the top of the buoy. If this unit is of mass 55 kg what is the maximum height of its centre of gravity above the top of the buoy if the whole assembly is not to be unstable? (The centroid of a cone of height h is at $3h/4$ from the vertex.)

2.28 A solid cylinder, 1 m diameter and 800 mm high, is of uniform relative density 0.85 and floats with its axis vertical in still water. Calculate the periodic time of small angular oscillations about a horizontal axis.

2.29 An open-topped tank, in the form of a cube of 900 mm side, has a mass of 340 kg. It contains 0.405 m³ of oil of relative density 0.85 and is accelerated uniformly up a long slope at arctan $(\frac{1}{3})$ to the horizontal. The base of the tank remains parallel to the slope, and the side faces are parallel to the direction of motion. Neglecting the thickness of the walls of the tank, estimate the net force (parallel to the slope) accelerating the tank if the oil is just on the point of spilling.

2.30 A test vehicle contains a U-tube manometer for measuring differences of air pressure. The manometer is so mounted that, when the vehicle is on level ground, the plane of the U is vertical and in the fore-and-aft direction. The arms of the U are 60 mm apart, and contain alcohol of relative density 0.79. When the vehicle is accelerated forwards down an incline at 20° to the horizontal at 2 m/s² the difference in alcohol levels (measured parallel to the arms of the U) is 73 mm, that nearer the front of the vehicle being the higher. What is the difference of air pressure to which this reading corresponds?

3 The principles of fluid motion

3.1 INTRODUCTION

The flow of a real fluid is usually very complex and, as a result, complete solutions of problems can seldom be obtained without recourse to experiment. Mathematical analysis of problems of fluid flow is generally possible only if certain simplifying assumptions are made. One of the chief of these is that the fluid is an ideal one without viscosity. In situations where the effect of viscosity is small this assumption often yields results of acceptable accuracy, although where viscosity plays a major part the assumption is clearly untenable. Simplification is often obtained also by assuming that the flow does not change with time.

In this chapter we lay the foundations of the analysis of flow by considering first the description of motion in terms of displacement, velocity and acceleration but without regard to the forces causing it. The Principle of Conservation of Mass is introduced; then the inter-relation between different forms of energy associated with the fluid flow is examined; and finally some simple applications of these results are considered.

3.2 ACCELERATION OF A FLUID PARTICLE

In general, the velocity of a fluid particle is a function both of position and of time. As the particle moves from, say, point A to point B, its velocity changes for two reasons. One is that particles at B have a different velocity from particles at A, even at the same instant of time; the other reason is that during the time the given particle moves from A to B the velocity at B changes. If B is at only a small distance δs from A the particle's total increase of velocity δu is the sum of the increase due to its change of position and the increase due to the passing of a time interval δt:

$$\delta u = \frac{\partial u}{\partial s}\delta s + \frac{\partial u}{\partial t}\delta t$$

and so, in the limit, as $\delta t \to 0$, the acceleration a_s in the direction of flow is given by:

$$a_s = \frac{du}{dt} = \frac{\partial u}{\partial s}\frac{ds}{dt} + \frac{\partial u}{\partial t}$$

or, since $ds/dt = u$,

$$a_s = \frac{du}{dt} = u\frac{\partial u}{\partial s} + \frac{\partial u}{\partial t} \qquad (3.1)$$

The full rate of increase du/dt for a given particle is often termed the *substantial* acceleration. The term $\partial u/\partial t$ represents only the *local* or *temporal* acceleration, i.e. the rate of increase of velocity with respect to time at a particular point in the flow. The term $u(\partial u/\partial s)$ is known as the *convective* acceleration, i.e. the rate of increase of velocity due to the particle's change of position. Although in steady flow $\partial u/\partial t$ is zero, the convective acceleration is not necessarily zero, so the substantial acceleration is not necessarily zero.

A particle may also have an acceleration in a direction perpendicular to the direction of flow. When a particle moves in a curved path, it changes direction and so has an acceleration towards the centre of curvature of the path, whether or not the magnitude of the velocity is changing. If the radius of the path-line is r_p the particle's acceleration towards the centre of curvature is u^2/r_p. Alternatively, if the streamline has a radius of curvature r_s, the particle's acceleration a_n towards the centre of curvature of the streamline has in general a convective part u^2/r_s and a temporal part $\partial u_n/\partial t$, where u_n represents the component of velocity of the particle towards the centre of curvature. Although, at that moment, u_n is zero it is, unless the flow is steady, increasing at the rate $\partial u_n/\partial t$. Thus

$$a_n = \frac{u^2}{r_s} + \frac{\partial u_n}{\partial t} \qquad (3.2)$$

3.3   CONTINUITY

The Equation of Continuity is really a mathematical statement of the Principle of Conservation of Mass. To consider first its most general expression we turn to Fig. 3.1. Here we bring under observation any

Fig. 3.1

fixed region within the fluid and since, in the absence of nuclear reaction, matter is neither created nor destroyed within that region:

Rate at which mass = Rate at which mass leaves the region
enters the region   + Rate of accumulation of mass in the region

If the flow is steady (i.e. unchanging with time) the rate at which mass is accumulated within the region is zero. The expression then reduces to:

Rate at which mass enters the region = Rate at which mass leaves the region

This relation may now be applied to a stream-tube whose cross-section is small enough for there to be no significant variation of velocity over it. A length δs of the stream-tube is considered between the cross-sectional planes B and C (Fig. 3.2), δs being so small that any variation in the cross-sectional area ΔA along that length is negligible. Then, if we assume that the fluid is a continuum, the volume of fluid contained in that small piece of the stream-tube is $(\Delta A)\delta s$. (We recall that cross-section by definition is perpendicular to the length.) If the fluid initially between planes B and C passes through the plane C in a short time interval δt, then the rate at which fluid volume passes through C is $(\Delta A)\delta s/\delta t$, or in the limit $(\Delta A)\mathrm{d}s/\mathrm{d}t$. But $\mathrm{d}s/\mathrm{d}t$ is the linear velocity there, say u, so the rate of volume flow is $(\Delta A)u$. As in calculating a volume a length must be multiplied by the area of a surface *perpendicular to* that length, so in calculating the rate of volume flow (frequently termed the *discharge* and represented by the symbol Q) the velocity must be multiplied by the area of a surface perpendicular to it. The rate of *mass* flow is given by the product of the discharge and the density.

The rate at which mass of fluid enters a selected portion of a stream-tube where the cross-sectional area is ΔA_1, the velocity of the fluid u_1, and its density ϱ_1, is therefore $\varrho_1(\Delta A_1)u_1$. For steady flow there is no accumulation of mass within the stream-tube, so the same mass must pass through all cross-sections of the tube in unit time. Thus

$$\varrho_1\left(\Delta A_1\right)u_1 = \varrho_2\left(\Delta A_2\right)u_2 = \ldots = \text{constant} \qquad (3.3)$$

For the entire collection of stream-tubes occupying the cross-section of a passage through which the fluid flows, eqn 3.3 may be integrated to give

$$\int \varrho u \,\mathrm{d}A = \text{constant} \qquad (3.4)$$

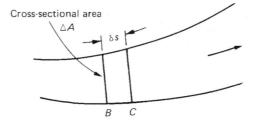

Fig. 3.2

where u is everywhere perpendicular to the elemental area δA. If ϱ and u are constant over the entire cross-section the equation becomes

$$\varrho A u = \text{constant} \qquad (3.5)$$

For a fluid of constant density the continuity relation reduces to

$$\int u \, dA = \text{constant}$$

which may be written

$$A\bar{u} = \text{constant}$$

where $\bar{u}$ represents the mean velocity.

Equation 3.3 indicates that, for flow of an *incompressible* fluid along a stream-tube, $u \Delta A = \text{constant}$, so as the cross-sectional area ΔA decreases, the velocity increases. This fact at once allows a partial interpretation of the pattern formed by streamlines in steady flow: in regions where the streamlines are close together the velocity is high, but where the same streamlines are more widely spaced the velocity is lower. This conclusion, however, does not necessarily apply to the flow of compressible fluids.

Other topics in the kinematics of fluids – for example, stream function, velocity potential, vorticity and circulation – will be discussed in Chapter 9.

3.4 BERNOULLI'S EQUATION

The velocity of a fluid in general varies from one point to another even in the direction of flow. Since, by Newton's First Law, a change of velocity must be associated with a force, it is to be expected that the pressure of the fluid also changes from point to point.

The relation between these changes may be studied by applying Newton's Second Law to a small element of the fluid over which the changes of velocity and pressure are very small. Although small, however, the element consists of a large number of molecules, so that the characteristic property of a fluid continuum is retained. It is so chosen that it occupies part of a stream-tube of small cross-section (see Fig. 3.3). The ends of the element are plane and perpendicular to the central streamline, but may be of any geometrical shape.

The forces under investigation are those due to the pressure of the fluid all round the element, and to gravity. Other forces, such as those due to viscosity, surface tension, electricity, magnetism, chemical or nuclear reactions are assumed negligibly small. Even the assumption of negligible viscosity is less restrictive than it may at first seem. The fluids more frequently encountered have small values of viscosity, and except when eddies are present viscous forces are significant only very close to solid boundaries. The behaviour of an actual fluid is thus often remarkably similar to that of an ideal, inviscid one. In the absence of shearing

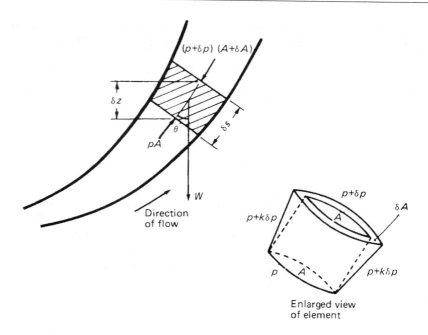

Fig. 3.3

Enlarged view
of element

forces, any force acting on a surface is perpendicular to it, whether the surface is that of a solid boundary or that of an element of fluid.

It is also assumed that the flow is steady.

The element is of length δs where s represents the distance measured along the stream-tube in the direction of flow. The length δs is so small that curvature of the streamlines over this distance may be neglected.

The pressure, velocity and so on will (in general) vary with s, but, as the flow is steady, quantities at a particular point do not change with time and so, for the stream-tube considered, each variable may be regarded as a function of s only.

At the upstream end of the element the pressure is p, and at the downstream end $p + \delta p$ (where δp may of course be negative). At the sides of the element the pressure varies along the length, but a mean value of $p + k\delta p$ may be assumed where k is a fraction less than unity. The pressure at the upstream end (where the cross-sectional area is A) results in a force pA on the element in the direction of flow; the pressure at the downstream end (where the cross-sectional area is $A + \delta A$) causes a force $(p + \delta p)(A + \delta A)$ on the element in the opposite direction.

Unless the element is cylindrical, the forces due to the pressure at its sides also have a component in the flow direction. Since the force in any direction is given by the product of the pressure and the *projected* area perpendicular to that direction, the net axial force downstream due to the pressure at the sides of the element is $(p + k\delta p)\delta A$ since δA is the *net* area perpendicular to the flow direction.

The weight of the element, W, equals $\varrho g A\,\delta s$ (the second order of small quantities being neglected) and its component in the direction of motion is $-\varrho g A \,\delta s \cos\theta$, where ϱ represents the density of the fluid and θ the angle shown between the vertical and the direction of motion. Thus in the absence of other forces, such as those due to viscosity, the total force acting on the element in the direction of flow is

$$pA - (p + \delta p)(A + \delta A) + (p + k\,\delta p)\delta A - \varrho g A \,\delta s \cos\theta$$

When the second order of small quantities is neglected, this reduces to

$$-A\delta p - \varrho g A \,\delta s \cos\theta \qquad (3.6)$$

Since the mass of the element is constant, this net force must, by Newton's Second Law, equal the mass multiplied by the acceleration in the direction of the force, that is, $\varrho A \,\delta s \times (\mathrm{d}u/\mathrm{d}t)$.

We may write $\delta s \cos\theta$ as δz where z represents height above some convenient horizontal datum plane and δz the increase in level along the length of the element. Then dividing by $\varrho A \,\delta s$ and taking the limit $\delta s \to 0$ we obtain .

$$\frac{1}{\varrho}\frac{\mathrm{d}p}{\mathrm{d}s} + \frac{\mathrm{d}u}{\mathrm{d}t} + g\frac{\mathrm{d}z}{\mathrm{d}s} = 0 \qquad (3.7)$$

From eqn 3.1

$$\frac{\mathrm{d}u}{\mathrm{d}t} = u\frac{\partial u}{\partial s} + \frac{\partial u}{\partial t}$$

however for steady flow the local acceleration $\partial u/\partial t = 0$ and so $\mathrm{d}u/\mathrm{d}t = u(\mathrm{d}u/\mathrm{d}s)$ (the full derivative now taking the place of the partial because for this stream-tube u is a function of s only). We then have

$$\frac{1}{\varrho}\frac{\mathrm{d}p}{\mathrm{d}s} + u\frac{\mathrm{d}u}{\mathrm{d}s} + g\frac{\mathrm{d}z}{\mathrm{d}s} = 0 \qquad (3.8)$$

as the required equation in differential form. This is often referred to as Euler's equation, after the Swiss mathematician Leonhard Euler* (1707–83). It cannot be completely integrated with respect to s unless ϱ is either constant or a known function of p. For a fluid of *constant density*, however, the result of integration is

$$\frac{p}{\varrho} + \frac{u^2}{2} + gz = \text{constant} \qquad (3.9)$$

or, if we divide by g,

$$\frac{p}{\varrho g} + \frac{u^2}{2g} + z = \text{constant} \qquad (3.10)$$

This result (in either form) is usually known as *Bernoulli's equation* in honour of another Swiss mathematician, Daniel Bernoulli (1700–82)

* pronounced *Oy'-ler.*

who in 1738 published one of the first books on fluid flow. (Equations 3.9 and 3.10, however, were not developed until some years later.)

The quantity z represents the elevation above some horizontal plane arbitrarily chosen as a base of measurement. The level of this plane is of no consequence: if it were moved, say, one metre higher all the values of z for the stream-tube considered would be reduced by one metre, so the sum of the three quantities in eqn 3.10 would still be constant.

The assumption that the flow is steady must not be forgotten; the result does not apply to unsteady motion. Moreover, in the limit the cross-sectional area of the stream-tube considered tends to zero and the tube becomes a single streamline. Thus the sum of the three terms is constant along a single streamline but, in general, the constant on the right-hand side of either eqn 3.9 or eqn 3.10 has different values for different streamlines. For those special cases in which all the streamlines start from, or pass through, the same conditions of pressure, velocity and elevation, the constants for the several streamlines are of course equal, but not every example of fluid motion meets these conditions.

To sum up, these are the conditions to which Bernoulli's equation applies. The fluid must be *frictionless* (inviscid) and of *constant density*; the flow must be *steady*, and the relation holds in general only for a *single streamline*.

For describing the behaviour of liquids, especially when there is a free surface somewhere in the system considered, eqn 3.10 is usually the most suitable form of the expression. The equation may be applied to gases in those circumstances where changes of density are not appreciable. Then eqn 3.9 has certain advantages. A simplification of the equation is frequently possible for describing the behaviour of gases in such conditions; because the density of gases is small, changes in the values of z from one point to another in the flow may well have negligible effect compared with the term p/ϱ in eqn 3.9, and so the gz term may be omitted without appreciable error. The equation then becomes

$$\frac{p}{\varrho} + \frac{u^2}{2} = \text{constant}$$

or, in the form more usually employed,

$$p + \frac{1}{2}\varrho u^2 = \text{constant} \tag{3.11}$$

3.4.1 The significance of the terms in Bernoulli's equation

Equation 3.9 states that the sum of three quantities is constant. Consequently the separate quantities must be interchangeable and thus of the same kind. The second term, $u^2/2$, represents the kinetic energy of unit mass of the fluid, or more precisely the ratio of the kinetic energy of a small element of the fluid to the mass of the element.

The third term, gz, also represents energy/mass and corresponds to the work that would be done on unit mass of the fluid in raising it from datum level to the height z.

Similarly the term p/ϱ must also represent an amount of work associated with unit mass of the fluid. We see from the expression 3.6 that the contribution of the 'pressure forces' to the net force acting on the element is $-A\,\delta p$ in the direction of motion. Therefore the work done by this force on the element as it moves a distance δs (that is, from a point where the pressure is p to a point where it is $p + \delta p$) is given by $(-A\,\delta p)\delta s$. But the mass of the element is $\varrho A\,\delta s$, so the work done by the force per unit mass of fluid is

$$-A\,\delta p\,\delta s/\varrho A\delta s = -\delta p/\varrho$$

If the element moves from a point where the pressure is p_1 to one where the pressure is p_2, then the work done by the 'pressure forces' per unit mass of fluid is $\int_{p_1}^{p_2} - \mathrm{d}p/\varrho$, that is, if ϱ is constant, $(p_1 - p_2)/\varrho$. The term p/ϱ in Bernoulli's equation therefore corresponds to the work that would be done on unit mass of the fluid by 'pressure forces' if the fluid moved from a point where the pressure was p to one where the pressure was zero. The work is done simply because the fluid moves. Consequently it is often known as the 'flow work' or 'displacement work'.

Thus each of the terms p/ϱ, $u^2/2$ and gz represents energy/mass. In the alternative form, eqn 3.10, each term represents energy/weight and so has the dimensional formula $[ML^2/T^2] \div [ML/T^2] = [L]$. The quantities in eqn 3.10 are therefore usually referred to respectively as 'pressure head' (or 'static head'), 'velocity head', and 'gravity head' or 'elevation', and their sum as the 'total head'. In eqn 3.11 each term corresponds to energy/volume.

The quantity p/ϱ is sometimes misleadingly termed 'pressure energy'. It has, however, nothing to do with the elastic energy given to a fluid when it is compressed – even when it is easily compressible. The fluid in fact does not even *possess* the 'pressure energy' (as it possesses kinetic energy, for example). A transmission belt transmits energy between two pulleys simply because it is under stress; the transmission of energy is in fact in the *opposite* direction to the movement of the belt (see Fig. 3.4) and so it is clearly absurd to regard the energy as being carried along in the belt. Likewise a fluid under stress (pressure) can transmit energy without necessarily possessing it. The terms in Bernoulli's equation, then, do not represent energy *stored* in unit mass of fluid but rather the total mechanical energy transmitted by this amount of fluid. The equation may be likened to the cash-book of an honest treasurer keeping account of the mechanical energy transactions of a small society, Unit Mass of Fluid, during its steady, frictionless travel along a streamline without change of density.

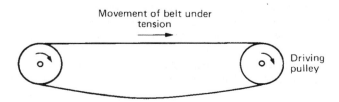

Movement of belt under tension

Driving pulley

Fig. 3.4

3.5 GENERAL ENERGY EQUATION FOR STEADY FLOW OF ANY FLUID

The application of Newton's Second Law of Motion to an element of fluid yields eqn 3.8 which may be integrated to relate the pressure, velocity and elevation of the fluid. This result, it will be remembered, is subject to a number of restrictions, among which is that there are no viscous forces in the fluid. In many instances, however, viscous forces are appreciable. Moreover, transfers of energy to or from the fluid may occur. These situations are investigated by a general energy equation which may be developed from the First Law of Thermodynamics.

3.5.1 The First Law of Thermodynamics

One of the fundamental generalizations of science is that, in the absence of nuclear reactions, energy can be neither created nor destroyed. The First Law of Thermodynamics expresses this principle thus:

> For any mass system (that is, any identified and unchanging collection of matter) the net heat supplied to the system equals the increase in energy of the system plus all the energy that leaves the system as work is done.

Or, in algebraic terms,

$$\Delta Q = \Delta E + \Delta W \tag{3.12}$$

where E represents the energy of the system, ΔQ the heat transferred to the system and ΔW the work done by the system while the change ΔE occurs.

The energy content of the system consists of:

1. Energy which may be ascribed to the substance considered as a continuum: that is, kinetic energy associated with its motion and potential energy associated with its position in fields of external forces. The latter is usually gravitational energy, but may also be electrical or magnetic. Although arising from intermolecular forces, free surface energy and elastic energy may also conveniently be included in this category.
2. Internal energy. This consists of the kinetic and potential energies of individual molecules and atoms and is thus, in general, a function of temperature and density. For a perfect gas, however, the potential energy arising from the attractive forces between molecules is assumed zero and the internal energy is then a function of temperature only. In any case, the internal energy depends only on the internal state of the matter constituting the system and not on the position or velocity of the system as a whole relative to a set of coordinate axes.

The First Law of Thermodynamics, as applied to the flow of fluids, keeps account of the various interchanges of energy that occur.

3.5.2 Derivation of the Steady-Flow Energy Equation

Let us consider the steady flow of a fluid through the device illustrated in Fig. 3.5. Energy in the form of heat is supplied steadily to this device and mechanical work is done by it, for example by means of a rotating shaft. (This external mechanical work is usually termed 'shaft work' whether or not the mechanism actually involves a rotating shaft; 'machine work' might be a better term.) The arrangement might be, say, a steam engine or a turbine. (In practice, most machines involve reciprocating parts of rotating blades, close to which the fluid flow cannot be strictly steady. However, we assume that such unsteadiness, if present, is only local and is simply a series of small fluctuations superimposed on steady mean flow.) We assume that the heat is supplied at a constant net rate, the shaft work is performed at a constant net rate, and the mass of the fluid entering in unit time is constant and equal to that leaving in unit time. Fluid at pressure p_1 and with velocity u_1 enters at a section where the (average) elevation is z_1 and leaves with pressure p_2 and velocity u_2 where the (average) elevation is z_2.

Although, as the fluid moves from inlet to outlet, its properties, in general, change from one point to another they do not (we assume) change with time.

We fix attention on the 'batch' of fluid bounded originally by the device itself and the planes A (at entry) and C (at exit). After a short time interval δt this fluid has moved forward, and is then bounded by the device itself, entry plane B and exit plane D. During this time interval an elemental mass δm (originally between planes A and B) enters the

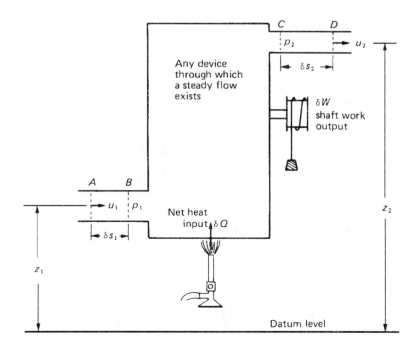

Fig. 3.5

device and, by the principle of continuity, a mass δm also leaves to occupy the space between planes C and D. The elements are assumed small enough for their properties to be uniform. The element at entry has internal energy $\delta m\, e_1$ (where e represents internal energy per unit mass), kinetic energy $\frac{1}{2}\delta m\, u_1^2$ and gravitational energy $\delta m\, gz_1$. (Changes of electrical, chemical, nuclear or free surface energy are disregarded here.) If the energy of the fluid in the device itself (i.e., between B and C) totals E then the energy of the fluid between A and C is $E + \delta m$ $(e_1 + \frac{1}{2}u_1^2 + gz_1)$.

After this fluid has moved to the position between B and D its energy is that of the fluid between B and C plus that of the element between C and D. In other words the total is $E + \delta m(e_2 + \frac{1}{2}u_2^2 + gz_2)$. Consequently the increase in energy which this particular 'batch' of fluid receives is

$$\left\{ E + \delta m\left(e_2 + \frac{1}{2}u_2^2 + gz_2 \right) \right\} - \left\{ E + \delta m\left(e_1 + \frac{1}{2}u_1^2 + gz_1 \right) \right\}$$

$$= \delta m \left\{ \left(e_2 - e_1 \right) + \frac{1}{2}\left(u_2^2 - u_1^2 \right) + g\left(z_2 - z_1 \right) \right\}$$

During this same interval of time a net amount of heat δQ is supplied to the system and a net amount of work δW is done by the fluid on, for example, a rotating shaft. (If the fluid were at a higher temperature than its surroundings heat would be transferred *from* the fluid *to* the surroundings and so this heat would be regarded as negative. Also, δW would be negative if work were done *on* the fluid, for example, by a pump.) The work δW, however, is not the only work done by the fluid. In moving from its position between A and C to that between B and D the fluid does work against the forces due to pressure. At the outlet, where the cross-sectional area is A_2, the fluid we are considering exerts a force $p_2 A_2$ on the material in front of it. During the short time interval δt, this end of the fluid system moves from C to D, a distance δs_2. The force, in moving in its own direction through the distance δs_2, therefore does an amount of work $p_2 A_2 \delta s_2$. Similarly, the force $p_1 A_1$ at the inlet does work $-p_1 A_1 \delta s_1$. (The minus sign arises because the force $p_1 A_1$ exerted by this 'batch' of fluid is in the opposite direction to the displacement δs_1.) The total work done by the fluid considered is therefore

$$\delta W + p_2 A_2 \delta s_2 - p_1 A_1 \delta s_1$$

Substitution into eqn 3.12 now yields

$$\delta Q = \delta m \left\{ \left(e_2 - e_1 \right) + \frac{1}{2}\left(u_2^2 - u_1^2 \right) + g\left(z_2 - z_1 \right) \right\}$$

$$+ \delta W + p_2 A_2 \delta s_2 - p_1 A_1 \delta s_1$$

Now $\varrho_1 A_1 \delta s_1 = \delta m = \varrho_2 A_2 \delta s_2$ and so division by δm gives

$$\frac{\delta Q}{\delta m} = \left(e_2 - e_1 \right) + \frac{1}{2}\left(u_2^2 - u_1^2 \right) + g\left(z_2 - z_1 \right) + \frac{\delta W}{\delta m} + \frac{p_2}{\varrho_2} - \frac{p_1}{\varrho_1}$$

or

$$q = \left(\frac{p_2}{\varrho_2} + \frac{1}{2}u_2^2 + gz_2 \right) - \left(\frac{p_1}{\varrho_1} + \frac{1}{2}u_1^2 + gz_1 \right) + e_2 - e_1 + w$$

(3.13)

where q represents the net heat transferred *to* the fluid per unit mass, and w represents the net shaft work done *by* the fluid per unit mass.

The relation 3.13 is known as the *Steady-Flow Energy Equation*. It may be expressed in words as follows:

> In steady flow through any region the net heat transferred to the fluid per unit mass equals the net shaft work performed by the fluid per unit mass plus the increase per unit mass in flow work, kinetic energy, gravitational energy and internal energy.

We recall that, apart from the First Law of Thermodynamics, the result is based on the following assumptions:

1. The flow is steady and continuous, that is, the rate at which mass enters the region considered equals that at which mass leaves the region and neither varies with time.
2. Conditions at any point between the inlet and outlet sections 1 and 2 do not vary with time.
3. Heat and shaft work are transferred to or from the fluid at a constant net rate.
4. Quantities are uniform over the inlet and outlet cross-sections 1 and 2.
5. Energy due to electricity, magnetism, surface tension or nuclear reaction is absent. If energy due to any of these phenomena is, in fact, involved appropriate additional terms will appear in the equations.

No assumptions are made about details of the flow pattern between inlet and outlet and no assumption is made about the presence or absence of 'friction' between inlet and outlet. The restrictions of assumptions 1 and 2 may in practice be slightly relaxed. Fluctuations in conditions are permissible if they occur through a definite cycle so that identical conditions are again reached periodically. This happens in fluid machinery operating at constant speed and torque. Flow in the neighbourhood of the moving blades or pistons of the machine is cyclic rather than absolutely steady. In other words, the conditions at any particular point in the fluid vary with time in a manner which is regularly repeated at a certain frequency. In such a case the equation may be used to relate values of the quantities averaged over a time considerably longer than the period of one cycle.

In practice, assumption 4 is never completely justified since viscous forces cause the velocity to fall rapidly to zero at a solid boundary. Thermodynamic properties may also vary somewhat over the cross-section. Appropriate correction factors may be introduced – for example, the kinetic energy correction factor α we shall mention in Section

3.5.3. However, the use of mean values of the velocity and other quantities normally yields results of sufficient accuracy. Developments of a more general equation not involving our assumption 4 may be found in, for example, the books by Shames[1] and Brenkert[2].

3.5.3 The kinetic energy correction factor

In investigating many problems of fluid dynamics it is frequently assumed that the flow is 'one-dimensional'; in other words, all the fluid is regarded as being within a single large stream-tube in which the velocity is uniform over the cross-section. The value of the kinetic energy per unit mass is then calculated as $\bar{u}^2/2$ where $\bar{u}$ represents the mean velocity, that is, the total discharge divided by the cross-sectional area of the flow. The only situation in which use of this mean velocity would be completely justified is that represented by the relation:

$$\left(\Sigma m\right)\bar{u}^2 = \Sigma\left(mu^2\right)$$

where m represents the mass and u the velocity of fluid in a short length of a small individual stream-tube while $\bar{u}$ represents the mean velocity over the entire cross-section of the flow $(= \Sigma(mu)/\Sigma m)$.

This equation, it may be shown, is satisfied only when all the us are equal, a condition never reached in practice because of the action of viscosity. The error involved in using the mean velocity to calculate the kinetic energy per unit mass may be estimated as follows.

Instead of the entire cross-section, consider first a small element of it whose area δA is small enough for there to be no appreciable variation of velocity u over it. The discharge through this small element is therefore $u\,\delta A$ and the mass flow rate $\varrho u\,\delta A$. The kinetic energy passing through the element in unit time is $\frac{1}{2}(\varrho u\,\delta A)u^2$ and consequently the total kinetic energy passing through the whole cross-section is $\int \frac{1}{2}\varrho u^3\,\mathrm{d}A$ in unit time.

We may also integrate the mass flow rate through an element of cross-section so as to obtain the total mass flow rate $= \int \varrho u\,\mathrm{d}A$. The kinetic energy per unit mass is then:

$$\frac{\text{Total kinetic energy in unit time}}{\text{Total mass in unit time}} = \frac{\int \frac{1}{2}\varrho u^3\,\mathrm{d}A}{\int \varrho u\,\mathrm{d}A}$$

For a fluid of constant density this reduces to

$$\frac{\int u^3\,\mathrm{d}A}{2\int u\,\mathrm{d}A} \tag{3.14}$$

Now unless u is constant over the entire cross-section this expression does not correspond to $\bar{u}^2/2$.

The factor by which the term $\bar{u}^2/2$ should be multiplied to give the true mean kinetic energy per unit mass is often known as the 'Kinetic

Energy Correction Factor', α. It can never be less than unity because the mean of different cubes is always greater than the cube of the mean.

Example 3.1 Consider fully-developed turbulent flow in a circular pipe (to be discussed more fully in Chapter 7). The velocity over the cross-section of the pipe varies approximately in accordance with Prandtl's 'one-seventh power law'

$$\frac{u}{u_{\text{max}}} = \left(\frac{y}{R}\right)^{1/7} \tag{3.15}$$

where R represents the radius of the pipe and u the velocity of the fluid at a distance y from the wall of the pipe. The maximum velocity u_{max} occurs at the centre of the pipe where $y = R$. Calculate the kinetic energy per unit mass.

Solution

Assuming the variation of density over the cross-section is negligible, the integrals in the expression 3.14 may be evaluated by the use of eqn 3.15. Because of axial symmetry the element of area δA may be taken as an annulus of radius r and area $2\pi r\,\delta r$ (see Fig. 3.6). Then

$$\int u^3\,\mathrm{d}A = \int u_{\text{max}}^3 \left(\frac{y}{R}\right)^{3/7} \mathrm{d}A = \frac{u_{\text{max}}^3}{R^{3/7}} \int_0^R y^{3/7}\,2\pi r\,\mathrm{d}r$$

An increase δr of r corresponds to a *decrease* $(-\delta y)$ of y. With $r = R - y$ the integral becomes

$$\frac{u_{\text{max}}^3}{R^{3/7}} \int_R^0 y^{3/7}\,2\pi(R - y)(-\mathrm{d}y) = \frac{2\pi u_{\text{max}}^3}{R^{3/7}} \int_0^R \left(Ry^{3/7} - y^{10/7}\right)\mathrm{d}y$$

$$= \frac{2\pi u_{\text{max}}^3}{R^{3/7}} \left[\frac{7}{10}Ry^{10/7} - \frac{7}{17}y^{17/7}\right]_0^R = \frac{98}{170}\pi R^2 u_{\text{max}}^3$$

Similarly

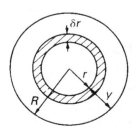

Fig. 3.6

$$\int u\,\mathrm{d}A = \int_0^R u_{max}\left(\frac{y}{R}\right)^{1/7} 2\pi r\,\mathrm{d}r = \frac{2\pi u_{max}}{R^{1/7}}\int_R^0 y^{1/7}(R-y)(-\mathrm{d}y)$$

$$= \frac{2\pi u_{max}}{R^{1/7}}\int_0^R \left(Ry^{1/7} - y^{8/7}\right)\mathrm{d}y = \frac{2\pi u_{max}}{R^{1/7}}\left[\frac{7}{8}Ry^{8/7} - \frac{7}{15}y^{15/7}\right]_0^R$$

$$= \frac{49}{60}\pi R^2 u_{max}$$

The mean velocity

$$\bar{u} = \frac{\text{Total discharge}}{\text{Area}} = \frac{\int u\,\mathrm{d}A}{\pi R^2} = \frac{49}{60}u_{max}$$

whence

$$u_{max} = \frac{60}{49}\bar{u}$$

Hence kinetic energy per unit mass

$$= \frac{\int u^3\,\mathrm{d}A}{2\int u\,\mathrm{d}A} = \frac{\dfrac{98}{170}\pi R^2 u_{max}^3}{2\times\dfrac{49}{60}\pi R^2 u_{max}}$$

$$= \frac{12}{17}\frac{u_{max}^2}{2} = \frac{12}{17}\left(\frac{60}{49}\right)^2\frac{\bar{u}^2}{2} = 1.058\frac{\bar{u}^2}{2}$$

Further discussion
With this particular distribution of velocity over the cross-section the term $\bar{u}^2/2$ is therefore about 6% too low to represent the mean kinetic energy per unit mass. It is usual, however, to disregard such discrepancy except where great accuracy is aimed at. In any case, the exact value of the correction to be applied is much influenced by conditions upstream and is scarcely ever known. The correction should nevertheless be remembered if the Steady-Flow Energy Equation is applied to fully-developed laminar flow in a circular pipe (see Chapter 6), for then the mean kinetic energy per unit mass $= 2\bar{u}^2/2$. Even so, as laminar flow is generally associated only with very low velocities the kinetic energy term would in these circumstances probably be negligible.

3.5.4 The Steady-Flow Energy Equation in practice

The Steady-Flow Energy Equation applies to liquids, gases and vapours, and to real fluids having viscosity, no less than to ideal fluids. In many applications it is considerably simplified because some of the terms are zero or cancel with others. If no heat energy is supplied to the fluid from outside the boundaries, and if the temperature of the fluid and that of its surroundings are practically identical (or if the boundaries are well insulated) q may be taken as zero. If there is no machine between sections (1) and (2) the shaft work per unit mass w is zero. And for fluids of constant density $\varrho_1 = \varrho_2$.

If an incompressible fluid with zero viscosity flows in a stream-tube across which there is no transfer of heat or work, the temperature of the fluid remains constant. Therefore the internal energy is also constant and the equation reduces to

$$0 = \left(\frac{p_2}{\varrho_2} + \frac{1}{2}u_2^2 + gz_2 \right) - \left(\frac{p_1}{\varrho_1} + \frac{1}{2}u_1^2 + gz_1 \right)$$

This is seen to be identical with Bernoulli's equation (3.9).

Real fluids, however, have viscosity, and the work done in overcoming the viscous forces corresponds to the so-called fluid 'friction'. The energy required to overcome the friction is transformed into thermal energy. The temperature of the fluid rises above the value for frictionless flow; the internal energy increases and, in general, the heat transferred *from* the fluid to its surroundings, $-q$ per unit mass, is increased. The increase of temperature, and consequently of internal energy, is generally of no worth (the temperature rise is normally only a very small fraction of a degree) and thus corresponds to a loss of *useful* energy. Moreover, $-q$ represents a loss of useful energy from the system and so the total loss (per unit mass of the fluid) is $e_2 - e_1 - q$. For a fluid of constant density it is usual to express this loss of useful energy, resulting from 'friction', as energy per unit weight, that is, 'head lost to friction', h_f. Therefore

$$h_f = \left(e_2 - e_1 - q \right)/g$$

Then for a *constant-density* fluid with no other heat transfer and no shaft work performed the Steady-Flow Energy Equation reduces to

$$\frac{p_1}{\varrho g} + \frac{u_1^2}{2g} + z_1 - h_f = \frac{p_2}{\varrho g} + \frac{u_2^2}{2g} + z_2 \tag{3.16}$$

Here u_1 and u_2 represent mean velocities over the cross-sections (1) and (2) respectively and the Kinetic Energy Correction Factor α is assumed negligibly different from unity. If we assume further that the flow occurs in a horizontal pipe of uniform cross-section then $u_1 = u_2$ and $z_1 = z_2$ and so $(p_1 - p_2)/\varrho g = h_f$. That is, the 'displacement work' done on the fluid in the pipe is entirely used in overcoming friction.

For an incompressible fluid the values of $(e_2 - e_1)$ and the heat transfer resulting from friction are in themselves rarely of interest, and so combining the magnitudes of these quantities into the single term h_f is a useful simplification. We can see that the head loss h_f represents, not the entire disappearance of an amount of energy, but the conversion of mechanical energy into thermal energy. This thermal energy, however, cannot normally be recovered, and so h_f refers to a loss of *useful* energy.

For a compressible fluid, on the other hand, that statement would not, in general, be true since the internal energy is then included in the total of useful energy.

We shall consider the flow of compressible fluids in more detail in Chapter 11. For the moment we look more particularly at the behaviour of incompressible fluids.

Example 3.2 A pump delivers water through a pipe 150 mm in diameter. At the pump inlet A, which is 225 mm diameter, the mean velocity is 1.35 m/s and the pressure 150 mmHg *vacuum*. The pump outlet B is 600 mm above A and is 150 mm diameter. At a section C of the pipe, 5 m above B, the gauge pressure is 35 kPa. If friction in the pipe BC dissipates energy at the rate of 2.5 kW and the power required to drive the pump is 12.7 kW, calculate the overall efficiency of the pump. (Relative density of mercury = 13.56)

Solution

Mean velocity at $A = u_A = 1.35$ m/s

$$\therefore \text{ by continuity, } u_B = u_C = u_A \times \frac{(\text{Area})_A}{(\text{Area})_{B,C}}$$

$$= 1.35 \left(\frac{225}{150} \right)^2 \text{ m/s} = 3.038 \text{ m/s}$$

Steady-Flow Energy Equation:

$$\frac{p_A}{\varrho} + \frac{1}{2}u_A^2 + gz_A + \frac{\text{Energy added by pump/time}}{\text{Mass/time}}$$

$$- \frac{\text{Energy loss to friction/time}}{\text{Mass/time}} = \frac{p_C}{\varrho} + \frac{1}{2}u_C^2 + gz_C$$

$$\therefore \quad \frac{\text{Energy added by pump}}{\text{Time}}$$

$$= \frac{\text{Mass}}{\text{Time}} \left\{ \frac{p_C - p_A}{\varrho} + \frac{u_C^2 - u_A^2}{2} + g(z_C - z_A) \right\}$$

$$+ \frac{\text{Energy loss to friction}}{\text{Time}}$$

$$= \frac{\text{Volume}}{\text{Time}} \left\{ p_C - p_A + \frac{1}{2}\varrho(u_C^2 - u_A^2) + \varrho g(z_C - z_A) \right\}$$

$$+ \frac{\text{Energy loss to friction}}{\text{Time}}$$

$$= \frac{\pi}{4}(0.225\,\text{m})^2 1.35\frac{\text{m}}{\text{s}}\left\{35000 - \left[13\,560 \times 9.81(-0.150)\right]\right.$$

$$\left. + \frac{1}{2} \times 1000\!\left(3.038^2 - 1.35^2\right) + 1000 \times 9.81 \times 5.6\right\}\frac{\text{N}}{\text{m}^2} + 2.5\,\text{kW}$$

$$= \frac{\pi}{4}(0.225)^2 1.35\{35000 + 19950 + 3702 + 54900\}\frac{\text{N}\,\text{m}}{\text{s}}$$

$$+ 2.5\,\text{kW} = 8.6\,\text{kW}$$

$\therefore$ Overall efficiency of pump $= 8.6/12.7 = \underline{67.7\%}$

Notice that p_A, a vacuum pressure, is negative. $\qquad\qquad$ □

3.5.5 Energy transformations in a constant-density fluid

The concept of 'head', that is, energy per unit weight of a constant-density fluid, is of great value in allowing a geometrical representation of energy changes. We recall from Section 2.2.1 that steady pressures not greatly in excess of atmospheric pressure may be measured by the rise of liquid in a glass tube. We may therefore imagine, for example, the system depicted in Fig. 3.7, in which such piezometer tubes are connected at certain points to a pipe conveying liquid from a large reservoir. At a point where the (gauge) pressure in the pipe is p the liquid will rise in the piezometer tube to a height $p/\varrho g$.

At points in the reservoir far from the outlet the velocity of the liquid is so small as to be negligible. At such a point 1 at a depth h_1 below the free surface, the pressure is therefore given by the hydrostatic relation $p_1 = \varrho g h_1$, so the sum of the three terms in Bernoulli's expression is

$$\varrho g h_1/\varrho g + 0^2/2g + z_1 = h_1 + z_1 = H$$

Thus H is the total head for the streamline on which the point 1 lies. If no energy is dissipated by 'friction' the total head H is constant along that streamline and may therefore be represented by a line parallel to the datum plane.

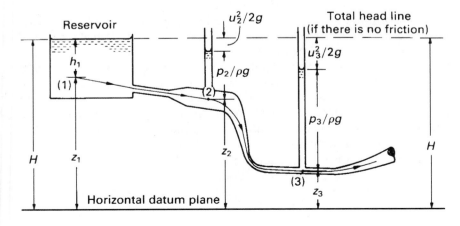

Fig. 3.7

At a point 2 in the pipe the pressure is indicated by the rise $p_2/\varrho g$ of the liquid in the piezometer tube. (For reasons that will become apparent in Section 3.6, there should be no appreciable curvature of the streamlines at positions 2 and 3.) The amount by which the sum of $p_2/\varrho g$ and z_2 falls short of the total head corresponds to the velocity head $u_2^2/2g$ for the streamline considered. There is a similar state of affairs at point 3, although here the cross-section of the pipe is smaller and so the mean velocity is greater than at 2 by virtue of the continuity equation $A\bar{u}$ = constant.

In practice, friction leads to a loss of mechanical energy, so the *total head line* (sometimes known as the *total energy line*) does not remain horizontal, but drops. The height of any point on this line above the datum plane always represents the total head $(p/\varrho g) + (u^2/2g) + z$ of the fluid at the point in question. Another line that may be drawn is that representing the sum of the pressure head and elevation only: $(p/\varrho g) + z$. This line, which would pass through the surface levels in the piezometer tubes of Fig. 3.7, is known as the *pressure line* or *hydraulic grade line* and is always a distance $u^2/2g$ vertically below the total head line. The geometrical representation that these lines afford is frequently useful, and it is therefore important to distinguish clearly between them.

Strictly speaking, each streamline has its own total head and pressure lines. When 'one-dimensional' flow is assumed, however, it is usual to consider only the streamline in the centre of the pipe, so that the z measurements are taken to the centre line and the static head $p/\varrho g$ is measured upwards from there. The mean total head line is then a distance $\alpha\bar{u}^2/2g$ vertically above the pressure line. Other conventional assumptions about these lines are mentioned in Section 7.7.

The continuity relation shows that, for a fluid of constant density, a reduction in the cross-sectional area causes an increase in the mean velocity, and the energy equation (3.16) indicates that unless additional energy is given to the fluid an increase of velocity is accompanied by a decrease of pressure (provided that the change of elevation z is small). Conversely, an increase of cross-sectional area of the flow gives rise to a decrease of velocity and an increase of pressure.

The energy equation (3.16) further indicates, however, that for a given elevation, the velocity cannot be increased indefinitely by reducing the cross-sectional area. Apart from exceptional circumstances, not encountered in normal engineering practice, the absolute pressure can never become less than zero; thus a maximum velocity is reached when the pressure has been reduced to zero. Any further reduction of the cross-sectional area would not bring about an increase of velocity, and therefore the discharge (i.e. area × mean velocity) would be reduced. There would then be a consequent decrease in the velocity at other sections. This phenomenon is known as *choking*.

With liquids, however, difficulties arise before the pressure becomes zero. At low pressures liquids vaporize and pockets of vapour may thus be formed where the pressure is sufficiently low. These pockets may suddenly collapse – either because they are carried along by the liquid until they arrive at a region of higher pressure, or because the pressure increases again at the point in question. The forces then exerted by the liquid rushing into the cavities cause very high localized pressures, which can lead to serious erosion of the boundary surfaces. This action is known as *cavitation*. Furthermore, the flow may be considerably disturbed when cavitation occurs.

In ordinary circumstances, liquids contain some dissolved air. This air is released as the pressure is reduced, and it too may form pockets in the liquid which are often known as 'air locks'. To avoid these undesirable effects, the absolute pressure head in water, for example, should not be allowed to fall below about 2 m (equivalent to about 20 kPa).

Choking is important in the study of the flow of compressible fluids and will be considered further in Chapter 11.

3.6 PRESSURE VARIATION PERPENDICULAR TO STREAMLINES

Euler's equation (3.8) (or, for a constant-density fluid, Bernoulli's equation) expresses the way in which pressure varies along a streamline in steady flow with no friction. Now although, from the definition of a streamline, fluid particles have no velocity component perpendicular to it, an acceleration perpendicular to a streamline is possible if the streamline is curved. Since any acceleration requires a net force in the same direction it follows that a variation of pressure (other than the hydrostatic one) occurs across curved streamlines.

Consider two streamlines sufficiently close together to be regarded as having the same centre of curvature (Fig. 3.8). Between them is a small cylindrical element of fluid of length δr normal to the streamlines, cross-sectional area δA and therefore weight $W = \varrho g \delta A\, \delta r$. At radius r the pressure is p; at radius $r + \delta r$ the pressure is $p + \delta p$. As in the development of Euler's equation, forces other than those due to pressure and gravity are neglected. In any case, for steady flow viscous forces have no

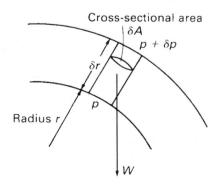

Fig. 3.8

components perpendicular to the streamlines. The net force on the element acting inwards along the streamline radius is therefore

$$(p + \delta p)\delta A - p\,\delta A + W\cos\theta = \delta p\,\delta A + \varrho g\,\delta A\,\delta r\cos\theta$$

where θ is the angle between the radius and the vertical.

By Newton's Second Law this force equals the product of the mass of the element and its centripetal acceleration, a_n. Noting that $\delta r\cos\theta = \delta z$, z being measured vertically upwards from a suitable datum level, and using eqn 3.2, we then have

$$\delta p\,\delta A + \varrho g\,\delta A\,\delta z = \varrho\,\delta A\,\delta r\,a_n = \varrho\,\delta A\,\delta r\left(\frac{u^2}{r} + \frac{\partial u_n}{\partial t}\right)$$

Dividing by $\delta A\,\delta r$ and taking the limit as $\delta r \to 0$ now gives

$$\frac{\partial p}{\partial r} + \varrho g\frac{\partial z}{\partial r} = \varrho\left(\frac{u^2}{r} + \frac{\partial u_n}{\partial t}\right) \tag{3.17}$$

If the streamlines are straight and not changing direction with time, the right-hand side of eqn 3.17 is zero since $r = \infty$ while $u_n = 0$ and is not changing. For a constant-density fluid, integration in the direction r then gives $p + \varrho gz = $ constant, that is, the piezometric pressure is constant normal to the streamlines. Where r is not infinite the exact manner in which p varies across the streamlines, even for steady flow, depends on the way in which u varies with r. Two special cases, the free and forced vortex, are discussed in Sections 11.7.4 and 11.7.5.

An important consequence of the pressure variation perpendicular to curved streamlines is the tendency of a jet of fluid to attach itself to a convex solid body. This is known as the Coanda effect, after the Romanian engineer Henri Coanda (1885–1972) who made use of it in various aeronautical applications. It may be simply demonstrated by using a solid cylinder (e.g. a finger) to deflect the flow from a water tap (Fig. 3.9). The curvature of the streamlines between sections AA' and BB' requires a net force towards the centre of curvature, and, as the outer edge of the stream is at atmospheric pressure, the pressure at the surface of the cylinder must be below atmospheric. Consequently the flow does not continue vertically downwards from BB', but bends towards the cylinder. The sub-atmospheric pressure in the neighbourhood of B results in a net force F on the cylinder (as may be demonstrated by suspending the cylinder on strings so that it can move freely).

The deflection of a jet because of a pressure difference across the streamlines can also arise in another way. When a jet goes into fluid of the same kind (e.g. when an air jet escapes into the atmosphere), nearby particles of fluid are dragged along with the jet. This is a process called *entrainment*. If, however, one side of the jet is close to a large solid surface, the supply of particles for entrainment there is restricted. Thus, particularly if the flow pattern is essentially two-dimensional, a partial vacuum is created between the jet and the surface, so the jet tends to attach itself to the surface.

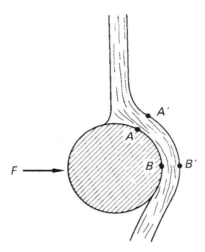

Fig. 3.9

Use is made of the Coanda effect on aircraft control flaps and in various 'fluid logic' devices.

3.7 SIMPLE APPLICATIONS OF BERNOULLI'S EQUATION

3.7.1 Flow through a sharp-edged orifice

An *orifice* is an aperture through which fluid passes and its thickness (in the direction of flow) is very small in comparison with its other measurements. An orifice used for measuring purposes has a sharp edge (the bevelled side facing downstream as in Fig. 3.10) so that there is the minimum contact with the fluid and consequently the minimum 'frictional' effects. If a sharp edge is not provided, the flow depends on the thickness of the orifice and the roughness of its boundary surface.

The diagram illustrates an orifice in one side of an open reservoir containing a liquid. The reservoir and the free surface are so large in comparison with the orifice that the velocity of the fluid at the free surface is negligibly small. The liquid issues from the orifice as a *free jet*, that is, a jet unimpeded by other liquid, and therefore under the influence of gravity.

Fluid approaching the orifice converges towards it. Because an instantaneous change of direction is impossible, the streamlines continue to converge beyond the orifice until they become parallel at the section *cc*. Strict parallelism of the streamlines would be approached asymptotically in an ideal fluid; in practice, however, frictional effects produce parallel flow at only a short distance (about half the diameter if the orifice is circular) from the orifice. The jet may diverge again beyond section *cc*, so this is then the section of minimum area. It is termed the *vena contracta* (Latin: contracted vein).

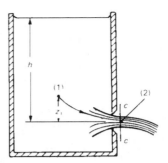

Fig. 3.10

(At low velocities some curvature of the streamlines results from the downward deflection of the jet by gravity and the vena contracta may be ill-defined. This curvature, however, is generally negligible close to the orifice. When a jet of liquid is discharged vertically downwards, gravity causes further acceleration of the liquid and so, by the principle of continuity $\varrho A u = $ constant, a further reduction of the cross-sectional area; the vena contracta is then defined as the section at which marked contraction from the orifice ceases.)

Since the streamlines are parallel and, we assume, sensibly straight at the vena contracta, the pressure in the jet there is uniform. (A non-uniform pressure over the section would cause accelerations perpendicular to the axis, and thus curved or non-parallel streamlines.) Apart from a small difference due to surface tension the pressure in the jet at the vena contracta therefore equals that of the fluid – for example, the atmosphere – surrounding the jet. The vena contracta, however, is the only section of the jet at which the pressure is completely known.

If the flow is steady and frictional effects are negligible Bernoulli's equation may be applied between two points on a particular streamline. Taking a horizontal plane through the centre of the orifice as datum level and considering the points (1) and (2) in Fig. 3.10, we have

$$\frac{p_1}{\varrho g} + \frac{u_1^2}{2g} + z_1 = \frac{p_{\text{atm}}}{\varrho g} + \frac{u_2^2}{2g} + 0$$

We suppose that the reservoir is sufficiently large and the point (1) sufficiently far from the orifice for the velocity u_1 to be negligible. Subject to this proviso the actual position of the point (1) is immaterial. Then, since hydrostatic conditions prevail there, p_1 corresponds to the depth of the point (1) below the free surface. With p_{atm} taken as zero (its variation from the free surface of the reservoir to the orifice being negligible) $p_1 = \varrho g(h - z_1)$. Consequently $(p_1/\varrho g) + z_1 = h$

$$\therefore h = \frac{u_2^2}{2g} \quad \text{i.e. } u_2 = \sqrt{(2gh)} \tag{3.18}$$

If the orifice is small in comparison with h, the velocity of the jet is uniform across the vena contracta. Evangelista Torricelli (1608–47), a pupil of Galileo, demonstrated experimentally in 1643 that the velocity with which a jet of liquid escapes from a small orifice is proportional to the square root of the head above the orifice, so eqn 3.18 is often known as Torricelli's formula. The equation refers to the velocity at the vena contracta: in the plane of the orifice itself neither the pressure nor the velocity is uniform and the average velocity is less than that at the vena contracta.

In the foregoing analysis friction and surface tension have been neglected, so eqn 3.18 represents the ideal velocity. The velocity actually attained at the vena contracta is slightly less, and a *coefficient of velocity* C_v is defined as the ratio of the actual (mean) velocity to the ideal. In other words, actual mean velocity $= C_v\sqrt{(2gh)}$.

Coefficient of velocity

The *coefficient of contraction* C_c is defined as the ratio of the area of the vena contracta to the area of the orifice itself.

Coefficient of contraction

Because of the two effects of friction and contraction the discharge from the orifice is less than the ideal value and the *coefficient of discharge* C_d is defined as the ratio of the actual discharge to the ideal value.

Coefficient of discharge

$$
\begin{aligned}
C_d &= \frac{\text{Actual discharge}}{\text{Ideal discharge}} \\
&= \frac{\text{Area of vena contracta} \times \text{Actual velocity there}}{\text{Ideal cross-sectional area} \times \text{Ideal velocity}} \\
&= \frac{\text{Area of vena contracta} \times \text{Actual velocity}}{\text{Area of orifice} \times \text{Ideal velocity}} = C_c \times C_v \quad (3.19)
\end{aligned}
$$

For a large vertical orifice the velocity in the plane of the vena contracta varies with the depth below the level of the free surface in thereservoir. The total discharge is therefore not calculated simply as $A_c u$ (where A_c represents the area of the vena contracta) but has to be determined by integrating the discharges through small elements of the area.

Consider, for example, a large, vertical, rectangular orifice of breadth b and depth d discharging into the atmosphere. A vena contracta of breadth b_c and depth d_c is formed as shown in Fig. 3.11. The streamlines here are parallel and practically straight; thus the pressure at any point in the plane of the vena contracta is atmospheric and for steady conditions the velocity at a depth h below the free surface in the reservoir is $C_v\sqrt{(2gh)}$. Through a small element of the vena contracta at this depth, the discharge equals velocity $\times$ area, which equals $C_v\sqrt{(2gh)}b_c\delta h$. Therefore if H_c represents the depth at the *centre* of the vena contracta, the total discharge is:

$$
\begin{aligned}
Q &= C_v b_c \sqrt{(2g)} \int_{H_c-d_c/2}^{H_c+d_c/2} h^{1/2}\, dh \\
&= \frac{2}{3} C_v b_c \sqrt{(2g)} \left\{ \left(H_c + \frac{d_c}{2} \right)^{3/2} - \left(H_c - \frac{d_c}{2} \right)^{3/2} \right\}
\end{aligned}
$$

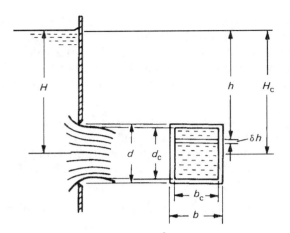

Fig. 3.11

(The value of C_v may not be exactly the same for all streamlines; the value in the equation must therefore be regarded as an average one.)

The difficulty now arises of determining the values of b_c, d_c and H_c. They vary with the corresponding values of b, d and H for the orifice itself, but the relation is not a simple one. To circumvent the difficulty we may write b, d and H respectively in place of b_c, d_c and H_c and introduce a coefficient of contraction C_c. Then

$$Q = \frac{2}{3}C_cC_vb\sqrt{(2g)}\left\{\left(H + \frac{d}{2}\right)^{3/2} - \left(H - \frac{d}{2}\right)^{3/2}\right\}$$

$$= \frac{2}{3}C_db\sqrt{(2g)}\left\{\left(H + \frac{d}{2}\right)^{3/2} - \left(H - \frac{d}{2}\right)^{3/2}\right\}$$

It is important to note that the integration is performed, not across the plane of the orifice, but across the plane of the vena contracta. The latter is the only plane across which the pressure is sensibly uniform and the velocity at every point known. When $H > d$, however, the orifice may usually be regarded as 'small' so that the ideal velocity at the vena contracta has the uniform value $\sqrt{(2gH)}$. The simpler formula $Q = C_dbd\sqrt{(2gH)}$ then gives an error of less than 1%.

Of the three coefficients the one most easily determined is the coefficient of discharge. For a liquid the amount emerging from the orifice in a known time interval may be determined by weighing or by direct measurement of volume, and the discharge then compared with the ideal value. The coefficient of contraction may be determined by direct measurement of the jet with calipers, although the accuracy is not usually high. One way of determining the actual velocity of a small jet is to allow it to describe a trajectory in the atmosphere, under the influence of gravity. If air resistance is negligible the horizontal component u_x of the jet velocity remains unchanged, and after a time t a particle leaving the vena contracta has travelled a horizontal distance $x = u_x t$. Since there is a uniform downward acceleration g, and the vertical component of velocity is initially zero for horizontal discharge, the vertical distance y travelled in the same time is $\frac{1}{2}gt^2$. Elimination of t from these expressions gives $u_x = u = x(g/2y)^{1/2}$ and so u may be determined from the coordinates of a point in the trajectory (which is a parabola). Although this result does not account for air resistance or for the possible influence of one particle on the trajectory of another, it is sufficiently near the truth for most purposes.

Example 3.3   Water flows through a sharp-edged orifice in the side of a tank and discharges as a jet into the atmosphere.

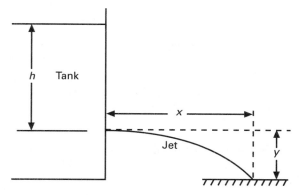

(a) The free surface in the tank is a height h above the centre of the orifice. The jet of water, on leaving the tank, strikes a surface distance y vertically below and at a horizontal distance x from the orifice. Derive an expression for the coefficient of velocity in terms of x, y and h.

(b) Water discharges through a sharp-edged orifice of diameter 11 mm into a pond, the surface of which is 0.6 m below the centre of the orifice. The jet strikes the surface at a horizontal distance of 2 m from the orifice. At the vena contracta plane the diameter of the jet is 8.6 mm. If the free surface in the tank is 1.75 m above the orifice, determine the rate of discharge.

Solution
(a) We consider first the horizontal and vertical components of the motion of the fluid in the jet. Assuming no effects of air resistance the trajectory is parabolic. Hence

$$x = u_2 t \quad \text{and} \quad y = \frac{1}{2}gt^2$$

which can be combined on eliminating t to yield

$$u_2 = x \sqrt{\frac{g}{2y}}$$

On substitution of this relation in the definition of coefficient of velocity there results

$$C_v = \frac{u_2}{\sqrt{2gh}} = x \sqrt{\frac{g}{2y}} \frac{1}{\sqrt{2gh}} = \frac{x}{2\sqrt{yh}}$$

(b) From the definition of the coefficient of contraction

$$C_c = \left(\frac{8.6}{11}\right)^2 = 0.611$$

The coefficient of velocity is evaluated from the equation derived in part (a). Thus

$$C_v = \frac{2\,\text{m}}{2\sqrt{0.6\,\text{m} \times 1.75\,\text{m}}} = 0.976$$

Hence the coefficient of discharge is

$$C_d = C_v \times C_c = 0.976 \times 0.611 = 0.596$$

The flow rate is calculated as

$$Q = C_d A_2 \sqrt{2gh} = 0.596 \times \frac{\pi}{4}(0.011\,\text{m})^2 \sqrt{2 \times 9.81\,\text{m/s}^2 \times 1.75\,\text{m}}$$

$$= 3.32 \times 10^{-4}\,\text{m}^3/\text{s}$$

Equation 3.18 strictly applies only to a fluid of constant density. It may, however, be used for a gas, provided that the drop in pressure across the orifice is small compared with the absolute pressure, so that the change of density is small. For h we may substitute $p/\varrho g$ and then

$$u = \sqrt{\left(\frac{2p}{\varrho}\right)} \tag{3.20}$$

Values of orifice coefficients

The values of the coefficients must be determined experimentally. For well-made, sharp-edged, circular orifices producing free jets, the coefficient of velocity is usually in the range 0.97 to 0.99, although slightly smaller values may be obtained with small orifices and low heads. For orifices not sharp-edged nor of negligible thickness the coefficient may be markedly lower.

The coefficient of contraction for a circular sharp-edged orifice is about 0.61 to 0.66. For low heads and for very small orifices, however, the effects of surface tension may raise the value to as much as 0.72. If the orifice is near the corner of a tank, for example, or if there are obstructions that prevent the full convergence of the streamlines approaching the orifice, then the coefficient is increased. Roughness of the walls near the orifice may reduce the velocity of the fluid approaching it and so curtail the contraction. The contraction after the orifice can be eliminated almost entirely by a 'bell-mouthed' approach to the orifice (Fig. 3.12) or the provision of a short length with uniform diameter immediately before the final exit. But although the coefficient of contraction is then unity, friction materially reduces the coefficient of velocity. It also reduces the velocity at the edge of the jet, and so the velocity over the cross-section is non-uniform.

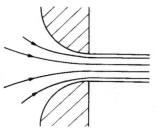

Fig. 3.12

The coefficient of discharge for a small sharp-edged orifice is usually in the range 0.6 to 0.65.

For a particular orifice and a particular fluid the coefficients are practically constant above a certain value of the head (the coefficients are in fact functions of Reynolds number). A good deal of information about the coefficients has been obtained and may be found in various hydraulics handbooks.

A submerged orifice is one that discharges a jet into fluid of the same kind. The orifice illustrated in Fig. 3.13, for example, discharges liquid into more of the same liquid. A vena contracta again forms, and the pressure there corresponds to the head h_2. Application of Bernoulli's equation between points 1 and 2 gives   *Submerged orifice*

$$\frac{p_1}{\varrho g} + z_1 + \frac{u_1^2}{2g} = h_1 + \frac{0}{2g} = \frac{p_2}{\varrho g} + z_2 + \frac{u_2^2}{2g} = h_2 + 0 + \frac{u_2^2}{2g}$$

so the ideal velocity $u_2 = \sqrt{\{2g(h_1 - h_2)\}}$. In other words, Torricelli's formula is still applicable provided that h refers to the *difference* of head across the orifice. Except for very small orifices, the coefficients for a

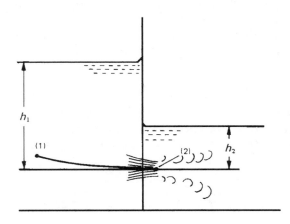

Fig. 3.13

submerged orifice are little different from those for an orifice producing a free jet.

The kinetic energy of a submerged jet is usually dissipated in turbulence in the receiving fluid.

Quasi-steady flow through an orifice

The use of Bernoulli's equation is strictly permissible only for steady flow. Unless the reservoir of Fig. 3.10, for example, is continuously replenished, the level of the free surface falls as fluid escapes through the orifice. Provided that the free surface is large compared with the orifice, however, the rate at which it falls is small, and the error involved in applying Bernoulli's equation is negligible. Such conditions may be termed *quasi-steady*. But an assumption of quasi-steady flow should always be checked to see whether the rate of change of h is negligible in comparison with the velocity of the jet.

Quasi-steady flow will be further considered in Section 7.10. Flow through an orifice for which the upstream velocity is *not* negligible will be discussed in Section 3.7.4.

3.7.2 The Pitot tube

Stagnation point

A point in a fluid stream where the velocity is reduced to zero is known as a *stagnation point*. Any non-rotating obstacle placed in the stream produces a stagnation point next to its upstream surface. Consider the symmetrical object illustrated in Fig. 3.14 as an example. On each side of the central streamline OX the flow is deflected round the object. The divergence of the streamlines indicates that the velocity along the central streamline decreases as the point X is approached. The contour of the body itself, however, consists of streamlines (since no fluid crosses it) and the fluid originally moving along the streamline OX cannot turn both left and right on reaching X. The velocity at X is therefore zero: X is a stagnation point.

Stagnation pressure

By Bernoulli's equation (3.9) the quantity $p + \frac{1}{2}\varrho u^2 + \varrho gz$ is constant along a streamline for the steady frictionless flow of a fluid of constant density. Consequently, if the velocity u at a particular point is brought to zero the pressure there is increased from p to $p + \frac{1}{2}\varrho u^2$. For a constant-density fluid the quantity $p + \frac{1}{2}\varrho u^2$ is therefore known as the *stagnation pressure* of that streamline.

Dynamic pressure

That part of the stagnation pressure due to the motion – $\frac{1}{2}\varrho u^2$ – is termed the *dynamic pressure*. (If heads rather than pressures are used the term *total head* is often preferred to 'stagnation head'.) A manometer connected to the point X would record the stagnation pressure, and

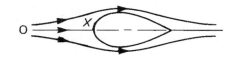

Fig. 3.14

if the *static* pressure p were also known $\frac{1}{2}\varrho u^2$ could be obtained by subtraction, and hence u calculated.

Henri Pitot (1695–1771) adopted this principle in 1732 for measuring velocities in the River Seine, and Fig. 3.15 shows the sort of device he used. A right-angled glass tube, large enough for capillary effects to be negligible, has one end (A) facing the flow. When equilibrium is attained the fluid at A is stationary and the pressure in the tube exceeds that of the surrounding stream by $\frac{1}{2}\varrho u^2$. The liquid is forced up the vertical part of the tube to a height

$$h = \Delta p / \varrho g = \frac{1}{2}\varrho u^2 / \varrho g = u^2 / 2g$$

above the surrounding free surface. Measurement of h therefore enables u to be calculated.

Such a tube is termed a Pitot tube* and provides one of the most accurate means of measuring the velocity of a fluid. For an open stream of liquid only this single tube is necessary, since the difference between stagnation and static pressures (or heads) is measured directly. (In practice, however, it is difficult to measure the height h above the surface of a moving liquid.) But for an enclosed stream of liquid, or for a gas, the Pitot tube indicates simply the stagnation pressure and so the static pressure must be measured separately.

Measurement of the static pressure may be made at the boundary of the flow, as illustrated in Fig. 3.16a, provided that the axis of the piezometer is perpendicular to the boundary and the connection is free from burrs, that the boundary is smooth and that the streamlines adjacent to it are not curved. A tube projecting into the flow (as at Fig. 3.16c) does not give a satisfactory reading because the fluid is accelerating round the end of the tube. The Pitot tube and that recording the static pressure may be connected to a suitable differential manometer: piezometer tubes are shown in Fig. 3.16 only for the sake of illustration.

The tubes recording static pressure and stagnation pressure are frequently combined into one instrument known as a Pitot-static tube (Fig.

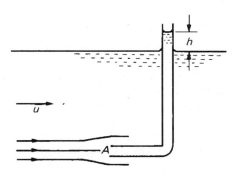

Fig. 3.15 Simple Pitot tube.

*pronounced *Pee'-toe*.

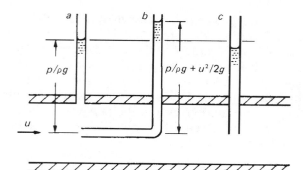

Fig. 3.16

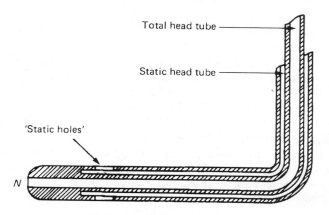

Fig. 3.17 Pitot-static tube.

3.17). The 'static' tube surrounds the 'total head' tube and two or more small holes are drilled radially through the outer wall into the annular space. The position of these 'static holes' is important. Downstream of the nose N the flow is accelerated somewhat with consequent reduction of static pressure; in front of the supporting stem there is a reduction of velocity and increase of pressure; the 'static holes' should therefore be at the position where these two opposing effects are counterbalanced and the reading corresponds to the undisturbed static pressure. Standard proportions of Pitot-static tubes have been determined that give very accurate results. If other proportions are used, a correction factor C, determined by calibration, has to be introduced: $u = C\sqrt{(2\Delta p/\varrho)}$. Misalignment of the tube with the flow leads to error: fortunately, however, since 'total head' and 'static' readings are both reduced, their difference is less seriously affected, and refinements of design reduce the sensitivity to changes of direction. A good Pitot-static tube gives errors less than 1% in velocity for misalignments up to about 15°.

When the flow is highly turbulent, individual particles of the fluid have velocities that fluctuate both in magnitude and direction. In such circumstances a Pitot tube records a value of Δp rather higher than that corresponding to the time-average component of velocity in the direction of the tube axis. This is partly because the mean pressure difference corresponds to the mean value of u^2 rather than to the square of the mean velocity. Although these errors are not large they should not be overlooked in accurate work.

An adaptation of the Pitot-static tube is the so-called 'Pitometer' (Fig. 3.18). The 'static' tube faces backwards into the wake behind the instrument, where the pressure is usually somewhat lower than the undisturbed static pressure. Such an instrument therefore requires calibration to determine the correction factor C (which may not be constant over more than a limited range of velocities), but it has the advantages of cheapness and compactness.

The use of a Pitot tube in the flow of a compressible fluid is discussed in Section 11.7.

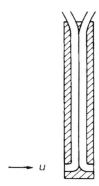

Fig. 3.18

3.7.3 The venturi-meter

The principle of the venturi-meter was demonstrated in 1797 by the Italian Giovanni Battista Venturi (1746–1822) but it was not until 1887 that the principle was applied, by the American Clemens Herschel (1842–1930), to a practical instrument for measuring the rate of flow of a fluid. The device consists essentially of a convergence in a pipe-line, followed by a short parallel-sided 'throat' and then a divergence (see Fig. 3.19). Continuity requires a greater velocity at the throat than at the inlet; there is consequently a difference of pressure between inlet and throat, and measurement of this pressure difference enables the rate of flow through the meter to be calculated.

We suppose for the moment that the fluid is ideal (so that no energy is dissipated by friction) and that the velocities u_1 at the inlet and u_2 at the throat are uniform and parallel over the cross-sections (which have areas A_1 and A_2 respectively). For the steady flow of a constant-density fluid we may apply Bernoulli's equation to a streamline along the axis between sections 1 and 2. Into this equation we may then substitute values of u_1 and u_2 from the continuity relation $Q = A_1 u_1 = A_2 u_2$ to give

$$\frac{p_1}{\varrho g} + \frac{Q^2}{2gA_1^2} + z_1 = \frac{p_2}{\varrho g} + \frac{Q^2}{2gA_2^2} + z_2 \qquad (3.21)$$

Hence the ideal discharge

$$Q_{\text{ideal}} = \left[\frac{2g\{(p_1/\varrho g + z_1) - (p_2/\varrho g + z_2)\}}{(1/A_2^2) - (1/A_1^2)} \right]^{1/2}$$

$$= A_1 \left[\frac{2g\,\Delta h}{(A_1/A_2)^2 - 1} \right]^{1/2} \qquad (3.22)$$

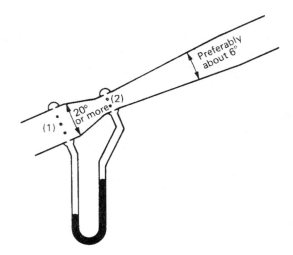

Fig. 3.19

The change in piezometric head $\{(p_1/\varrho g + z_1) - (p_2/\varrho g + z_2)\} = \Delta h$ may be measured directly by a differential manometer. It will be recalled from Section 2.3.2 that the difference of levels in the manometer is directly proportional to the difference of piezometric head, regardless of the difference of level between the manometer connections.

In practice, friction between sections 1 and 2, although small, causes p_2 to be slightly less than for an ideal fluid and so Δh is slightly greater. As the use of this value of Δh in eqn 3.22 would give too high a value of Q a *coefficient of discharge* C_d is introduced. Actual discharge

$$Q = C_d A_1 \left\{ \frac{2g\Delta h}{\left(A_1/A_2\right)^2 - 1} \right\}^{1/2} = \frac{C_d A_1 A_2}{\left(A_1^2 - A_2^2\right)^{1/2}} \left(2g\Delta h\right)^{1/2} \quad (3.23)$$

(Remember that Δh is the difference of piezometric head of the fluid in the meter, *not* the difference of levels of the manometer liquid.)

The coefficient of discharge also accounts for effects of non-uniformity of velocity over sections 1 and 2. Although C_d varies somewhat with the rate of flow, the viscosity of the fluid and the surface roughness, a value of about 0.98 is usual with fluids of low viscosity (British Standard 1042).

To ensure that the pressure measured at each section is the average, connections to the manometer are made via a number of holes into an annular ring. The holes are situated where the walls are parallel so that there is no variation of piezometric pressure across the flow. To discourage the formation of eddies, sharp corners at the joins between the conical sections and the parallel-sided ones are avoided.

Rapidly converging flow, as between the inlet and throat of a venturimeter, almost always causes the velocity to become more uniform over the cross-section. Over the short length involved the loss of head to friction, h_f, is negligible in comparison with $(p_1 - p_2)/\varrho g$. For a single streamline Bernoulli's equation then gives

$$u_2^2 = u_1^2 + 2g\left(\frac{p_1^* - p_2^*}{\varrho g}\right)$$

where p^* represents the piezometric pressure $p + \varrho gz$. On a nearby streamline, slightly different values $u_1 + \delta u_1$ and $u_2 + \delta u_2$ may be found instead of u_1 and u_2. If the streamlines are straight and parallel at these sections, however, no difference of piezometric pressure across the flow can be sustained and so all streamlines have the same values of p_1^* and of p_2^*. Subtracting the Bernoulli equations for the two streamlines therefore gives

$$\left(u_2 + \delta u_2\right)^2 - u_2^2 = \left(u_1 + \delta u_1\right)^2 - u_1^2$$

i.e. $2u_2\delta u_2 = 2u_1\delta u_1$, higher orders of the small quantities being neglected.

$$\therefore \quad \frac{\delta u_2}{u_2} = \left(\frac{u_1}{u_2}\right)^2 \frac{\delta u_1}{u_1}$$

Since $u_1 < u_2$,

$$\frac{|\delta u_2|}{u_2} < \frac{|\delta u_1|}{u_1}$$

in other words, the proportionate variation of velocity is less after the contraction than before it. This is why a rapid contraction is placed before the working section of a wind tunnel or water tunnel, where a uniform velocity is especially important.

In a few extreme cases, where friction between inlet and throat of a venturi-meter is very small, the uniformity of velocity may be so much improved at the throat as to outweigh the effect of friction and give a value of C_d slightly greater than unity. (A C_d greater than unity may also result from faulty manometer connections.) For accurate results, α_1, the kinetic energy correction factor at inlet, should not be large (and therefore unpredictable) and it is desirable that the venturi-meter be approached by a sufficient length of straight pipe for the fluid to be free from large eddies and similar disturbances.

The function of the diverging part of the meter is to reduce the velocity again gradually and restore the pressure as nearly as possible to its original value. Unfortunately, in a diverging tube the flow tends to separate from the walls, eddies are formed and energy is dissipated as heat. The greater the angle of divergence the greater this dissipation of energy. On the other hand, a small angle results in a large overall length and therefore a large loss by 'ordinary' friction at the walls. The best

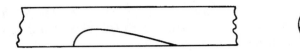

Fig. 3.20

compromise has been found for a total angle of divergence about 6° (see Section 7.6.3). Larger angles have, however, been used, with a consequent reduction in overall length (and cost). Of the drop in pressure between inlet and throat only 80–90% is recovered in the diverging part.

A common ratio of diameters d_2/d_1 is $\frac{1}{2}$. (Thus $A_2/A_1 = \frac{1}{4}$ and $u_2/u_1 = 4$.) Although a smaller throat area gives a greater and more accurately measured difference of pressure, the subsequent dissipation of energy in the diverging part is greater. Moreover, the pressure at the throat may become low enough for dissolved gases to be liberated from the liquid, or even for vaporization to occur.

To save expense, large venturi-meters are sometimes made by welding a plate of sheet metal to the inside of a pipe, thus producing a D-shaped throat (Fig. 3.20).

Example 3.4 A horizontal venturi tube, 280 mm diameter at the entrance and 140 mm diameter at the throat, has a discharge coefficient of 0.97. A differential U-tube manometer, using mercury as the manometric fluid, is connected between pressure tappings at the entrance and at the throat. The venturi tube is used to measure the flow of water, which fills the leads to the U-tube and is in contact with the mercury. Calculate the flow rate when the difference in the mercury levels is 50 mm. Take the densities of water and mercury as 10^3 kg/m³ and 13.6×10^3 kg/m³ respectively.

Solution Denote conditions at the entrance and in the throat by suffixes 1 and 2, respectively, and the difference in levels in the manometer by x. Also represent the densities of the water and mercury by ϱ_w and ϱ_m respectively.

Considering the pressure balance in the two limbs of the manometer (see Section 2.3.2) it follows that

$$p_1 + \varrho_w x g = p_2 + \varrho_m x g$$

Hence

$$p_1 - p_2 = 9.81\,\text{m/s}^2 \times \frac{50\,\text{mm}}{1000\,\text{mm/m}} \times 10^3\,\text{kg/m}^3 \times \left(13.6 - 1\right)$$

$$= 6180\,\text{N/m}$$

Since

$$\Delta h = \frac{p_1 - p_2}{\varrho_w g}$$

the flow rate is obtained by substituting in the relation

$$Q = C_d A_1 \left\{ \frac{2g\Delta h}{\left(A_1/A_2\right)^2 - 1} \right\}^{1/2}$$

$$= 0.97 \times \frac{\pi}{4}\left(\frac{280}{1000}\,\text{m}\right)^2 \left\{ \frac{2 \times 6180\,\text{N/m}^2}{10^3\,\text{kg/m}^3 \times \left[\left(28/14\right)^4 - 1\right]} \right\}^{1/2}$$

$$= 0.0542\,\text{m}^3/\text{s}$$

□

3.7.4 The flow nozzle and orifice meter

The nozzle meter or 'flow nozzle' illustrated in Fig. 3.21 is essentially a venturi-meter with the divergent part omitted, and the basic equations are the same as those for the venturi-meter. The dissipation of energy downstream of the throat is greater than for a venturi-meter but this disadvantage is often offset by the lower cost of the nozzle. The pressure in the manometer connection in the wall of the pipe at section 2 may not be that at the throat of the nozzle because of the eddying motion surrounding the jet from the nozzle. Nor may the upstream connection be made at a point sufficiently far from the nozzle for the flow to be uniform. These deviations, however, are allowed for in values of C_d (see, for example, British Standard 1042). The coefficient depends on the shape of the nozzle, the diameter ratio d_2/d_1 and the Reynolds number of the flow, but as it does not depend on what happens beyond the throat it is little different from that for a venturi-meter.

A still simpler and cheaper arrangement is a sharp-edged orifice fitted concentrically in the pipe (Fig. 3.22). Application of Bernoulli's equation between a point 1 upstream of the orifice and the vena contracta (2) gives, for an ideal fluid and uniform velocity distribution:

$$\frac{u_2^2}{2g} = \frac{p_1^*}{\varrho g} - \frac{p_2^*}{\varrho g} + \frac{u_1^2}{2g} = \Delta h + \frac{u_1^2}{2g}, \text{ say.}$$

For a real fluid we introduce a coefficient of velocity:

$$u_2 = C_v\left\{2g\left(\Delta h + u_1^2/2g\right)\right\}^{1/2} \qquad (3.24)$$

We now put $u_1 = Q/A_1$ and $u_2 = Q/A_c = Q/C_c A_o$ where Q represents the discharge, A_c the cross-sectional of the vena contracta, A_o the area of the orifice itself and C_c the coefficient of contraction. Then

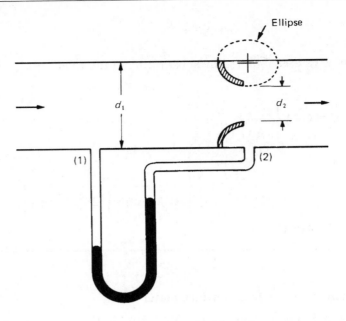

Fig. 3.21

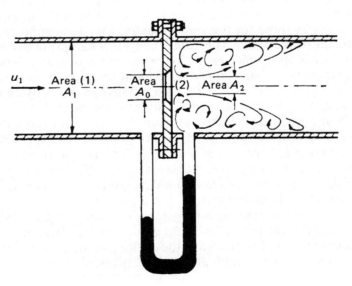

Fig. 3.22

$$Q = \frac{C_v C_c A_o A_1}{\left(A_1^2 - C_v^2 C_c^2 A_o^2\right)^{1/2}} \left(2 g \Delta h\right)^{1/2} = \frac{C_d A_o \left(2 g \Delta h\right)^{1/2}}{\left\{1 - C_d^2 \left(A_o / A_1\right)^2\right\}^{1/2}}$$

since $C_v C_c = C_d$ (eqn 3.19).

Unless the orifice diameter is less than about one-fifth of the pipe diameter the contraction of the jet is affected by the proximity of the pipe walls and C_d is higher than for an orifice in a large reservoir (Section 3.7.1). In any case it is usual to simplify the expression by using a coefficient

$$C = C_d \left\{ \frac{1 - (A_o/A_1)^2}{1 - C_d^2(A_o/A_1)^2} \right\}^{1/2}$$

so that

$$Q = \frac{C A_o (2g\Delta h)^{1/2}}{\left\{1 - (A_o/A_1)^2\right\}^{1/2}}$$

The downstream manometer connection should strictly be made to the section where the vena contracta occurs, but this is not feasible as the vena contracta is somewhat variable in position. In practice, various positions are used for the manometer connections and C is thereby affected (again, see British Standard 1042).

Much of the kinetic energy of the jet from either a flow nozzle or an orifice is dissipated in eddies downstream and so the overall loss of useful energy is considerably larger than for a venturi-meter.

The accuracy of any measuring device may be affected by swirling motion or non-uniformity of the flow approaching it. Therefore if pipe-bends, valves and so on which cause such disturbances are not at least 50 times the pipe diameter upstream, straightening grids are often fitted in front of the metering device.

Example 3.5 A fluid of relative density 0.86 flows through a pipe of diameter 120 mm. The flow rate is measured using a 6 cm diameter orifice plate with corner tappings, which are connected to the two limbs of a differential U-tube manometer using mercury as the manometric fluid. The discharge coefficient is 0.62. Calculate the mass flow rate when the difference in the mercury levels in the U-tube is 100 mm.

Solution
Denote conditions at the measurement points upsteam and downstream of the orifice plate by suffixes 1 and 2, respectively, and the difference in levels in the manometer by x. Also represent the densities of the flowing fluid and mercury by ϱ_f and ϱ_m respectively.

Considering the pressure balance in the two limbs of the manometer (see Section 2.3.2) it follows that

$$p_1 + \varrho_f x g = p_2 + \varrho_m x g$$

Hence

$$p_1 - p_2 = 9.81\,\text{m/s}^2 \times \frac{100\,\text{mm}}{1000\,\text{mm/m}} \times 10^3\,\text{kg/m}^3 \times \left(13.6 - 0.86\right)$$

$$= 12\,500\,\text{N/m}^2$$

Since

$$\Delta h = \frac{p_1 - p_2}{\varrho_f g}$$

the volumetric flow rate is obtained by substituting in the relation

$$Q = CA_0 \left\{ \frac{2g\Delta h}{1 - \left(A_o/A_1\right)^2} \right\}^{1/2}$$

$$= 0.62 \times \frac{\pi}{4}\left(\frac{6}{100}\,\text{m}\right)^2 \left\{ \frac{2 \times 12\,500\,\text{N/m}^2}{0.86 \times 10^3\,\text{kg/m}^3 \times \left[1 - \left(6/12\right)^4\right]} \right\}^{1/2}$$

$$= 0.00976\,\text{m}^3/\text{s}$$

Hence

$$m = \varrho_f Q = 0.86 \times 10^3\,\text{kg/m}^3 \times 0.00976\,\text{m}^3/\text{s} = 8.39\,\text{kg/s}$$

3.7.5 Notches and sharp-crested weirs

A notch may be defined as a sharp-edged obstruction over which flow of a liquid occurs. As the depth of flow above the base of the notch is related to the discharge, the notch forms a useful measuring device. It is formed in a smooth, plane, vertical plate and its edges are bevelled on the downstream side so as to give minimum contact with the fluid. The area of flow is most commonly either rectangular or V-shaped. A large rectangular notch is more often termed a sharp-crested weir.

The pattern of flow over a notch is a very complex one and no rigorous analytical method of studying it is possible. Owing to the curvature of the streamlines (Fig. 3.23) there is no cross-section of the flow over which the pressure is uniform (as in the vena contracta associated with an orifice). The variation of velocity within the stream therefore cannot be accurately determined. Turbulence and frictional effects also enter the problem in an unpredictable way. At low rates of flow surface tension can affect the discharge appreciably. Secondary motions superimposed on the main flow just upstream of the notch plate sometimes influence the flow.

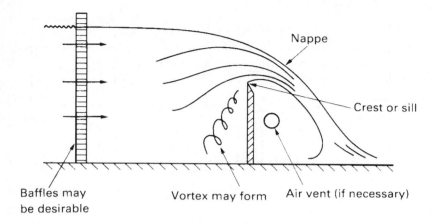

Nappe

Crest or sill

Baffles may
be desirable

Vortex may form   Air vent (if necessary)

Fig. 3.23

Any attempt, therefore, to discover analytically the relation between the rate of flow and the depth at the notch can be based only on drastic simplifying assumptions. But although the formula so obtained will have to be modified by an experimentally determined coefficient it is nevertheless useful in showing the essential form of the relation between depth and discharge.

The sheet of liquid escaping over the notch or weir is known as the *nappe*. If the pressure underneath it is atmospheric, the nappe (except at very low rates of flow) springs clear of the notch plate. For a notch extending across the entire width of a channel, atmospheric air may not be able to get under the nappe, and the liquid then clings to the downstream side of the notch plate and the discharge is unpredictable. So the space underneath the nappe must be 'ventilated', if necessary by providing an air vent as shown in Fig. 3.23.

Consider a sharp-edged, rectangular notch as shown in Fig. 3.24. The crest is horizontal and normal to the general direction of flow. The classical analysis, usually ascribed to the German engineer Julius Weisbach (1806–71), requires these admittedly radical assumptions:

1. Upstream of the notch, the velocities of particles in the stream are uniform and parallel; thus the pressure there varies according to the hydrostatic equation $p = \varrho gh$. (In practice it is often necessary to install baffles to achieve reasonably steady and uniform conditions.)
2. The free surface remains horizontal as far as the plane of the notch, and all particles passing through the notch move horizontally, and perpendicular to its plane.
3. The pressure throughout the nappe is atmospheric.
4. The effects of viscosity and surface tension are negligible.

These assumptions give the idealized pattern of flow shown in Fig. 3.24. At section 1, $(p_1/\varrho g) + z_1 =$ height H of the free surface. So for a typical streamline Bernoulli's equation gives

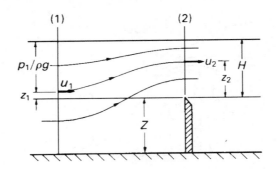

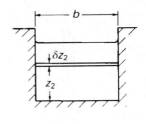

Fig. 3.24

$$H + u_1^2/2g = 0 + u_2^2/2g + z_2$$

$$\therefore u_2 = \left\{2g\left(H - z_2 + u_1^2/2g\right)\right\}^{1/2}$$

This shows that u_2 varies with z_2. In the plane of the notch the discharge through a horizontal element of depth δz_2 is $u_2 b\, \delta z_2$ and so the idealized total discharge

$$Q_{\text{ideal}} = b\int_0^H u_2\, dz_2$$

(the datum of z_2 being taken at the crest of the notch)

$$= b\sqrt{(2g)}\int_0^H \left(H - z_2 + u_1^2/2g\right)^{1/2} dz_2$$

$$= -\frac{2}{3}b\sqrt{(2g)}\left[\left(H - z_2 + u_1^2\ 2g\right)^{3/2}\right]_0^H$$

$$= \frac{2}{3}b\sqrt{(2g)}\left\{\left(H + u_1^2\ 2g\right)^{3/2} - \left(u_1^2/2g\right)^{3/2}\right\} \qquad (3.25)$$

Since u_1 depends on Q, solution of eqn 3.25 is troublesome except by trial and error. However, $u_1^2/2g$ is frequently negligible in comparison with H and then the equation becomes

$$Q_{\text{ideal}} = \frac{2}{3}b\sqrt{(2g)}H^{3/2}$$

An experimentally determined coefficient of discharge now has to be inserted to account for the inadequacy of the assumptions made. The contraction of the nappe as it passes through the notch is a significant factor and the coefficient is considerably less than unity. Its value depends primarily on H and H/Z where Z is the height of the crest above the bed of the approach channel[3]. The effects of viscosity and surface tension are marked only when H is small.

A *suppressed weir* is one for which the breadth b is the same as the width of the approach channel. The nappe then contracts in the vertical direction only and not horizontally (see Fig. 3.25).

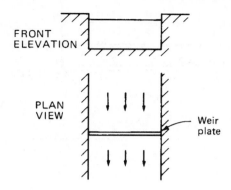

Fig. 3.25 Suppressed weir.

When the notch is symmetrically placed in the width of a channel and has the proportions shown in Fig. 3.26, full contraction of the nappe takes place in the horizontal direction. The contraction at each side of the nappe is about $H/10$; thus the width of the nappe and consequently the coefficient of discharge vary with H.

The V notch scores over the rectangular notch in producing a nappe with the same shape whatever the value of H, and thus it has a less variable coefficient.

With the same assumptions used in deriving the expression for the discharge through a rectangular notch, we have

$$u_2 = \left\{2g\left(H - z_2 + u_1^2/2g\right)\right\}^{1/2}$$

For a V notch, however, the cross-sectional area of the approach channel is usually so much greater than that of the notch that $u_1^2/2g$ may be neglected. The idealized discharge through an element of the notch (as in Fig. 3.27) is therefore $b\delta z_2\{2g(H - z_2)\}^{1/2}$ and, if each side makes an angle $\theta/2$ with the vertical, $b = 2z_2\tan(\theta/2)$. The total idealized discharge is therefore

$$Q_{ideal} = 2\tan\frac{\theta}{2}\sqrt{(2g)}\int_0^H z_2\left(H - z_2\right)^{1/2} dz_2$$

$$= 2\tan\frac{\theta}{2}\sqrt{(2g)}\int_0^H \left(H - h\right)h^{1/2} dh$$

(where $h = H - z_2$)

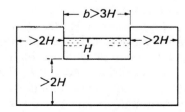

Fig. 3.26

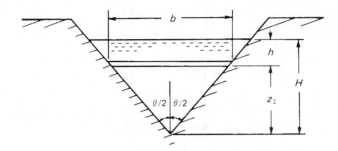

Fig. 3.27

$$= 2\tan\frac{\theta}{2}\sqrt{(2g)}\left[\frac{2}{3}Hh^{3/2} - \frac{2}{5}h^{5/2}\right]_0^H$$

$$= \frac{8}{15}\tan\frac{\theta}{2}\sqrt{(2g)}H^{5/2} \tag{3.26}$$

The actual discharge is

$$\frac{8}{15}C_d \tan\frac{\theta}{2}\sqrt{(2g)}H^{5/2}$$

The angle θ is seldom outside the range 30°–90°. If the head H is sufficient for the nappe to spring clear of the notch plate and the width of the approach channel is at least four times the maximum width of the nappe, C_d is about 0.59 for water flow. As a result of viscosity and surface tension effects the value increases somewhat as the head falls[4].

The classical notch formulae 3.25 and 3.26 are obtained by using an approximate relation for velocity (which ignores effects of curvature and non-uniform pressure) and integrating the discharge between approximate limits in the plane of the notch. In practice the 'head' H is *not* the depth of flow over the crest but the *upstream* level of the free surface measured relative to the crest. As the liquid approaches the notch the free surface level falls appreciably. (This does not, however, affect the result of the integration markedly since $H - z_2 + u_1^2/2g$ is small when z_2 approaches H.) The head should be measured at a point upstream before the fall of the surface has begun and ideally where the velocity of the stream is negligible. When this ideal is not possible, allowance must be made for the velocity at the point of measurement. As a first approximation $u_1^2/2g$, the head corresponding to the 'velocity of approach', is assumed negligible and a value of Q obtained. From this approximate value of Q, u_1 is calculated as Q divided by the cross-sectional area of flow at the point where H is measured. Hence $u_1^2/2g$ is calculated and, for a rectangular notch for example, may be substituted in eqn 3.25 to determine a more accurate value of Q. Since $u_1^2/2g$ is small in comparison with H, $(H + u_1^2/2g)^{3/2}$ is often used rather than $(H + u_1^2/2g)^{3/2} - (u_1^2/2g)^{3/2}$. The formula is in any case based on the assumption of a uniform value of u_1 although in fact the discharge over the notch can be appreciably affected by non-uniform conditions upstream. In comparison with such indeterminate errors the omission of $(u_1^2/2g)^{3/2}$ is of no account.

Example 3.6   Water flows over a sharp-crested weir 600 mm wide. The measured head (relative to the crest) is 155 mm at a point where the cross-sectional area of the stream is 0.26 m² (see Fig. 3.28). Calculate the discharge, assuming that $C_d = 0.61$.

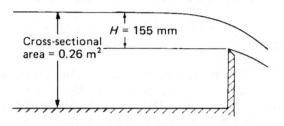

Fig. 3.28

Solution
As a first approximation,

$$Q = \frac{2}{3} C_d \sqrt{(2g)} b H^{3/2}$$

$$= \frac{2}{3} 0.61 \sqrt{(19.62 \,\text{m/s}^2)} 0.6 \,\text{m} (0.155 \,\text{m})^{3/2}$$

$$= 0.0660 \,\text{m}^3/\text{s}$$

$$\therefore \text{ Velocity of approach} = \frac{0.0660 \,\text{m}^3/\text{s}}{0.26 \,\text{m}^2} = 0.254 \,\text{m/s}$$

$$\therefore \frac{u_1^2}{2g} = \frac{(0.254 \,\text{m/s})^2}{19.62 \,\text{m/s}^2} = 3.28 \times 10^{-3} \,\text{m}$$

$$\therefore H + u_1^2/2g = (0.155 + 0.00328) \,\text{m} = 0.1583 \,\text{m}$$

Second approximation: $Q = \frac{2}{3} 0.61 \sqrt{(19.62)} 0.6 (0.1583)^{3/2} \,\text{m}^3/\text{s}$

$$= 0.0681 \,\text{m}^3/\text{s}$$

Further refinement of the value could be obtained by a new calculation of u_1 (0.0681 m³/s ÷ 0.26 m²), a new calculation of $H + u_1^2/2g$ and so on. One correction is usually sufficient, however, to give a value of Q acceptable to three significant figures.   □

Since an empirical coefficient of discharge has to be used there may seem little point in deriving formulae from Bernoulli's equation. The subject has been treated here in some detail, however, to demonstrate the care necessary in applying Bernoulli's equation and the importance of a full realization of all the assumptions involved in an analysis.

REFERENCES

1. Shames, I. H. *Mechanics of Fluids* (3rd edn), McGraw-Hill, New York (1992).
2. Brenkert, K. *Elementary Theoretical Fluid Mechanics*, Wiley, New York (1960).
3. Kindsvater, C. E. and Carter, R. W. 'Discharge characteristics of rectangular thin-plate weirs', *Trans. Am. Soc. civ. Engrs* **124**, 722–822 (1959).
4. Lenz, A. T. 'Viscosity and surface tension effects on V notch weir coefficients', *Trans. Am. Soc. civ. Engrs* **108**, 759–802 (1943).
5. BS 1042: Part 1: Section 1.6: 1993. Method of measurement of pulsating fluid flow in a pipe by means of orifice plates, nozzles or venturi tubes.

FURTHER READING

Ackers, P., White, W. R., Perkins, J. A. and Harrison, A. J. M. *Weirs and Flumes for Measurement*, Wiley, Chichester (1978).
Folsom, R. G. 'Review of the Pitot tube', *Trans. Am. Soc. mech. Engrs* **78**, 1447–60 (1956).

FILMS

Kline, S. J. *Flow Visualization*, Encyclopaedia Britannica Films or from Central Film Library.
Shapiro, A. H. *Pressure Fields and Fluid Acceleration*, Encyclopaedia Britannica Films or from Central Film Library.

PROBLEMS

3.1 A pipe carrying water tapers from a cross-section of $0.3\,\text{m}^2$ at A to $0.15\,\text{m}^2$ at B which is 6 m above the level of A. At A the velocity, assumed uniform, is 1.8 m/s and the pressure 117 kPa gauge. If frictional effects are negligible, determine the pressure at B.

3.2 A long bridge with piers 1.5 m wide, spaced 8 m between centres, crosses a river. The depth of water before the bridge is 1.6 m and that between the piers is 1.45 m. Calculate the volume rate of flow under one arch assuming that the bed of the river is horizontal, that its banks are parallel and that frictional effects are negligible.

3.3 A pipe takes water from a reservoir where the temperature is 12 °C to a hydro-electric plant 600 m below. At the 1.2 m diameter inlet to the power house the gauge pressure of the water is 5.5 MPa and its mean velocity 2 m/s. If its temperature there is 13.8 °C at what rate has heat passed into the pipe as a result of hot sunshine? (*Note:* The change of atmospheric pressure over 600 m must be accounted for: mean atmospheric density = 1.225 kg/m^3. Specific heat capacity of water = 4.187 kJ/kg K.)

3.4 In an open rectangular channel the velocity, although uniform across the width, varies linearly with depth, the value at the free surface being twice that at the base. Show that the value of the kinetic energy correction factor is 10/9.

3.5 From a point 20 m away from a vertical wall a fireman directs a jet of water through a window in the wall at a height of 15 m above the level of the nozzle. Neglecting air resistance determine the angle at which the nozzle must be held when the supply pressure at the nozzle is only just sufficient, and calculate this minimum pressure if C_v for the nozzle is 0.95. (The velocity head in the hose may be neglected.)

3.6 The air supply to an oil-engine is measured by being taken directly from the atmosphere into a large reservoir through a sharp-edged orifice 50 mm diameter. The pressure difference across the orifice is measured by an alcohol manometer set at a slope of arcsin 0.1 to the horizontal. Calculate the volume flow rate of air if the manometer reading is 271 mm, the relative density of alcohol is 0.80, the coefficient of discharge for the orifice is 0.602 and atmospheric pressure and temperature are respectively 755 mm Hg and 15.8 °C. (You may assume $R = 287$ J/kg K.)

3.7 Oil of relative density 0.85 issues from a 50 mm diameter orifice under a pressure of 100 kPa (gauge). The diameter of the vena contracta is 39.5 mm and the discharge is 18 litres/s. What is the coefficient of velocity?

3.8 A submarine, submerged in sea-water, travels at 16 km/h. Calculate the pressure at the front stagnation point situated 15 m below the surface. (Density of sea-water = 1026 kg/m^3.)

3.9 A vertical venturi-meter carries a liquid of relative density 0.8 and has inlet and throat diameters of 150 mm and 75 mm respectively. The pressure connection at the throat is 150 mm above that at the inlet. If the actual rate of flow is 40 litres/s and the coefficient of discharge is 0.96, calculate (*a*) the pressure difference between inlet and throat, and (*b*) the difference of levels in a vertical U-tube mercury manometer connected between these points, the tubes above the mercury being full of the liquid. (Relative density of mercury = 13.56.)

3.10 A servo-mechanism is to make use of a venturi contraction in a horizontal 350 mm diameter pipe carrying a liquid of relative density 0.95. The upper end of a vertical cylinder 100 mm diameter is connected by a pipe to the throat of the venturi and the lower end of the cylinder is connected to the inlet. A piston in the cylinder is to be lifted when the flow rate through the venturi exceeds $0.15 \, \text{m}^3/\text{s}$. The piston rod is 20 mm diameter and passes through both ends of the cylinder. Neglecting friction, calculate the required diameter of the venturi throat if the gross effective load on the piston rod is 180 N.

3.11 A sharp-edged notch is in the form of a symmetrical trapezium. The horizontal base is 100 mm wide, the top is 500 mm wide and the depth is 300 mm. Develop a formula relating the discharge to the upstream water level and estimate the discharge when the upstream water surface is 228 mm above the level of the base of the notch. Assume that $C_d = 0.6$ and that the velocity of approach is negligible.

3.12 The head upstream of a rectangular weir 1.8 m wide is 80 mm. If $C_d = 0.6$ and the cross-sectional area of the upstream flow is $0.3 \, \text{m}^2$, estimate, to a first approximation, the discharge allowing for the velocity of approach.

The momentum equation $\boxed{4}$

4.1 INTRODUCTION

It is often important to determine the force produced on a solid body by fluid flowing steadily over it. For example, there is the force exerted on a solid surface by a jet of fluid impinging on it; there are also the aerodynamic forces (lift and drag) on an aircraft wing, the force on a pipe-bend caused by the fluid flowing within it, the thrust on a propeller and so on. All these forces are *hydrodynamic* forces and they are associated with a change in the momentum of the fluid.

The magnitude of such a force is determined essentially by Newton's Second Law. However, the law usually needs to be expressed in a form particularly suited to the steady flow of a fluid: this form is commonly known as the steady-flow momentum equation and may be applied to the whole bulk of fluid within a prescribed space. Only forces acting at the boundaries of this fluid concern us; any force within this fluid is involved only as one half of an 'action-and-reaction' pair and so does not affect the overall behaviour. Use of the momentum equation therefore happily does not require a knowledge of the flow pattern in detail. Moreover, the fluid may be compressible or incompressible, and the flow with or without friction.

4.2 THE MOMENTUM EQUATION FOR STEADY FLOW

In its most general form, Newton's Second Law states that the net force acting on a body in any fixed direction is equal to the rate of increase of momentum of the body in that direction. Since force and momentum are both vector quantities it is essential to specify the direction. Where we are concerned with a collection of bodies (which we shall here term a *system*) the law may be applied (for a given direction) to each body individually. If the resulting equations are added, the total force in the given fixed direction corresponds to the net force acting in that direction at the boundaries of the system. Only these external, boundary forces are involved because any internal forces between the separate bodies occur in pairs of action and reaction and therefore cancel in the total. For a fluid, which is a continuum of particles, the same result applies: the net force in any fixed direction on a certain 'batch' of fluid equals the total rate of increase of momentum of that fluid in that direction.

Our aim now is to derive a relation by which force may be related to the fluid within a given space rather than to the 'batch' consisting of a given collection of particles. We begin by applying Newton's Second Law to a small element in a stream-tube (shown in Fig. 4.1). The flow is steady and so the stream-tube remains stationary with respect to the fixed coordinate axes. The cross-section of this stream-tube is sufficiently small for the velocity to be considered uniform over the plane AB and over the plane CD. After a short interval of time δt the fluid that formerly occupied the space $ABCD$ will have moved forward to occupy the space $A'B'C'D'$. In general, its momentum changes during this short time interval.

If u_x represents the component of velocity in the x direction then the element (of mass δm) has a component of momentum in the x direction equal to $u_x \delta m$. The total 'x-momentum' of the fluid in the space $ABCD$ at the beginning of the time interval δt is therefore

$$\sum_{ABCD} u_x \delta m$$

The same fluid at a time δt later will have a total 'x-momentum'

$$\sum_{A'B'C'D'} u_x \delta m$$

This last expression may be expanded as

$$\sum_{ABCD} u_x \delta m - \sum_{ABB'A'} u_x \delta m + \sum_{DCC'D'} u_x \delta m$$

The net increase of 'x-momentum' during the time interval δt is therefore

$$\left(\sum_{A'B'C'D'} u_x \delta m \right)_{\text{after } \delta t} - \left(\sum_{ABCD} u_x \delta m \right)_{\text{before } \delta t}$$

$$= \left(\sum_{ABCD} u_x \delta m - \sum_{ABB'A'} u_x \delta m + \sum_{DCC'D'} u_x \delta m \right)_{\text{after } \delta t} - \left(\sum_{ABCD} u_x \delta m \right)_{\text{before } \delta t}$$

$$= \left(\sum_{DCC'D'} u_x \delta m - \sum_{ABB'A'} u_x \delta m \right)_{\text{after } \delta t}$$

since, as the flow is assumed steady, $(\sum u_x \delta m)_{ABCD}$ is the same after δt as before δt. Thus, during the time interval δt, the increase of x-momentum of the 'batch' of fluid considered is equal to the x-momentum leaving the stream-tube in that time minus the x-momentum entering in that time:

$$\left(\sum_{DCC'D'} u_x \delta m \right) - \left(\sum_{ABB'A'} u_x \delta m \right)$$

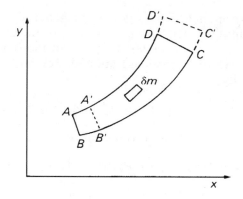

Fig. 4.1

For a very small value of δt the distances AA', BB' are very small, so the values of u_x for all the particles in the space $ABB'A'$ are substantially the same. Similarly, all particles in the space $DCC'D'$ have substantially the same value of u_x, although this may differ considerably from the value for particles in $ABB'A'$. The u_x terms may consequently be taken outside the summations.

Therefore the increase of x-momentum during the interval δt is

$$\left(u_x \sum \delta m\right)_{DCC'D'} - \left(u_x \sum \delta m\right)_{ABB'A'} \tag{4.1}$$

Now $(\Sigma \delta m)_{DCC'D'}$ is the mass of fluid which has crossed the plane CD during the interval δt and so is expressed by $\dot{m}\,\delta t$, where $\dot{m}$ denotes the rate of mass flow. Since the flow is steady, $(\Sigma \delta m)_{ABB'A'}$ also equals $\dot{m}\,\delta t$. Thus expression 4.1 may be written $\dot{m}(u_{x2} - u_{x1})\delta t$, where suffix 1 refers to the inlet section of the stream-tube, suffix 2 to the outlet section. The *rate* of increase of x-momentum is obtained by dividing by δt and the result, by Newton's Second Law, equals the net force F_x on the fluid in the x direction:

$$F_x = \dot{m}\left(u_{x_2} - u_{x_1}\right) \tag{4.2}$$

The corresponding force in the x direction exerted *by* the fluid *on* its surroundings is, by Newton's Third Law, $-F_x$.

A similar analysis for the relation between force and rate of increase of momentum in the y direction gives

$$F_y = \dot{m}\left(u_{y_2} - u_{y_1}\right) \tag{4.3}$$

In steady flow $\dot{m}$ is constant and so $\dot{m} = \varrho_1 A_1 u_1 = \varrho_2 A_2 u_2$ where ϱ represents the density of the fluid and A the cross-sectional area of the stream-tube (A being perpendicular to u).

We have so far considered only a single stream-tube with a cross-sectional area so small that the velocity over each end face (AB, CD) may be considered uniform. Let us now consider a bundle of adjacent stream-tubes, each of cross-sectional area δA, which together carry all the flow being examined. The velocity, in general, varies from one stream-tube to another. The space enclosing all these stream-tubes is often known as the *control volume* and it is to the boundaries of this volume that the external forces are applied. For one stream-tube the 'x-force' is given by

$$\delta F_x = \dot{m}\left(u_{x_2} - u_{x_1}\right) = \varrho_2 \delta A_2 u_2 u_{x_2} - \varrho_1 \delta A_1 u_1 u_{x_1}$$

The total force in the x direction is therefore

$$F_x = \int dF_x = \int \varrho_2 u_2 u_{x_2}\, dA_2 - \int \varrho_1 u_1 u_{x_1}\, dA_1 \qquad (4.4a)$$

(The elements of area δA must everywhere be perpendicular to the velocities u.) Similarly

$$F_y = \int \varrho_2 u_2 u_{y_2}\, dA_2 - \int \varrho_1 u_1 u_{y_1}\, dA_1 \qquad (4.4b)$$

and

$$F_z = \int \varrho_2 u_2 u_{z_2}\, dA_2 - \int \varrho_1 u_1 u_{z_1}\, dA_1 \qquad (4.4c)$$

These equations are required whenever the force exerted on a flowing fluid has to be calculated. They express the fact that for steady flow the net force on the fluid in the control volume equals the net rate at which momentum flows out of the control volume, the force and the momentum having the same direction. It will be noticed that conditions only at inlet 1 and outlet 2 are involved. The details of the flow between positions 1 and 2 therefore do not concern us for this purpose. Such matters as friction between inlet and outlet, however, may affect the magnitudes of quantities at outlet.

It will also be noticed that eqns 4.4 take account of variation of ϱ and so are just as applicable to the flow of compressible fluids as to the flow of incompressible ones.

The integration of the terms on the right-hand side of the eqns 4.4 is not possible without further information. By judicious choice of the control volume, however, it is often possible to use sections 1 and 2 over which ϱ, u, u_x and so on do not vary significantly, and then the equations reduce to a form such as:

$$F_x = \varrho_2 u_2 A_2 u_{x_2} - \varrho_1 u_1 A_1 u_{x_1} = \dot{m}\left(u_{x_2} - u_{x_1}\right)$$

It should never be forgotten, however, that such simplified forms involve the assumption of uniform values of the quantities over the inlet and outlet cross-sections of the control volume: the validity of these assumptions should therefore always be checked. (See Section 4.2.1.)

A further assumption is frequently involved in the calculation of F. A contribution to the total force acting at the boundaries of the control volume comes from the force due to the pressure of the fluid at a cross-section of the flow. If the streamlines at this cross-section are sensibly straight and parallel, the pressure over the section varies uniformly with depth as for a static fluid; in other words, $p^* = p + \varrho gz$ is constant. If, however, the streamlines are not straight and parallel, there are accelerations perpendicular to them and consequent variations of p^*. Ideally, then, the control volume should be so selected that at the sections where fluid enters or leaves it the streamlines are sensibly straight and parallel and, for simplicity, the density and the velocity (in both magnitude and direction) should be uniform over the cross-section.

Newton's Laws of Motion, we remember, are limited to describing motions with respect to coordinate axes that are not themselves accelerating. Consequently the momentum relations for fluids, being derived from these Laws, are subject to the same limitation. That is to say, the coordinate axes used must either be at rest or moving with uniform velocity in a straight line.

Here we have developed relations only for steady flow in a stream-tube. More general expressions are beyond the scope of this book.

4.2.1 Momentum correction factor for one-dimensional analysis

By methods analogous to those of Section 3.5.3 it may be shown that where the velocity of a constant-density fluid is not uniform (although essentially parallel) over a cross-section, the true rate of momentum flow perpendicular to the cross-section is not $\varrho \bar{u}^2 A$ but $\varrho \int u_\perp^2 dA = \beta \varrho \bar{u}^2 A$. Here $\bar{u} = (1/A) \int u_\perp dA$, the mean velocity over the cross-section, and β is the 'momentum correction factor'. The suffix $\perp$ is used as a reminder that the velocity must always be perpendicular to the element of area dA. With constant ϱ, the value of β for the velocity distribution postulated in Section 3.5.3 is $100/98 = 1.02$ which for most purposes differs negligibly from unity. Disturbances upstream, however, may give a markedly higher value. For fully-developed laminar flow in a circular pipe (see Section 6.2) $\beta = 4/3$. In a given situation β is always less than α, the kinetic energy correction factor.

4.3 APPLICATIONS OF THE MOMENTUM EQUATION

4.3.1 The force caused by a jet striking a surface

When a steady jet strikes a solid surface it does not rebound from the surface as a rubber ball would rebound. Instead, a stream of fluid is formed which moves over the surface until the boundaries are reached, and the fluid then leaves the surface tangentially. (It is assumed that the surface is large compared with the cross-sectional area of the jet.)

Consider a jet striking a large plane surface as shown in Fig. 4.2. A suitable control volume is that indicated by dotted lines on the diagram. If the x direction is taken perpendicular to the plane, the fluid, after passing over the surface, will have no component of velocity and therefore no momentum in the x direction. (It is true that the thickness of the

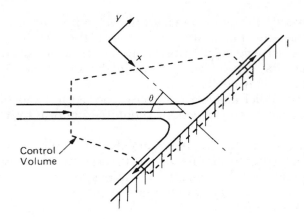

Fig. 4.2

stream probably changes as the fluid moves over the surface, but this change of thickness corresponds to a negligible movement in the x direction.) The rate at which x-momentum enters the control volume is $\int \varrho_1 u_1 u_{x1} \mathrm{d}A_1 = \cos\theta \int \varrho_1 u_1^2 \mathrm{d}A_1$ and so the rate of *increase* of x-momentum is $-\cos\theta \int \varrho_1 u_1^2 \mathrm{d}A_1$ and this equals the net force on the fluid in the x direction. If the fluid on the solid surface were stationary and at atmospheric pressure there would of course be a force between the fluid and the surface due simply to the static (atmospheric) pressure of the fluid. However, the change of fluid momentum is produced by a *hydrodynamic* force additional to this static force. By regarding atmospheric pressure as zero we can determine the hydrodynamic force directly.

Since the pressure is atmospheric both where the fluid enters the control volume and where it leaves, the hydrodynamic force on the fluid can be provided only by the solid surface (effects of gravity being neglected). The hydrodynamic force exerted by the fluid *on* the surface is equal and opposite to this and is thus $\cos\theta \int \varrho_1 u_1^2 \mathrm{d}A_1$ in the x direction. If the jet has uniform density and velocity over its cross-section the integral becomes

$$\varrho_1 u_1^2 \cos\theta \int \mathrm{d}A_1 = \varrho_1 Q_1 u_1 \cos\theta$$

(where Q_1 is the volume flow rate at inlet)

In the y direction the jet carries a component of momentum equal to $\sin\theta \int \varrho_1 u_1^2 \mathrm{d}A_1$ per unit time. For this component to undergo a change, a net force in the y direction would have to be applied to the fluid. Such a force, being parallel to the surface, would be a shear force exerted by the surface on the fluid. For an ideal fluid moving over a smooth surface no shear force is possible, so the component $\sin\theta \int \varrho_1 u_1^2 \mathrm{d}A_1$ would be unchanged and equal to the rate at which y-momentum leaves the control volume. Except when $\theta = 0$, the spreading of the jet over the surface is not symmetrical, and for a real fluid the rate at which y-momentum leaves the control volume differs from the rate at which it enters. In general, the force in the y direction may be calculated if the final velocity of the fluid is known. This, however, requires further experimental data.

When the fluid flows over a curved surface similar techniques of calculation may be used as the following example will show.

Example 4.1   A jet of water flows smoothly on to a stationary curved vane which turns it through 60°. The initial jet is 50 mm in diameter, and the velocity, which is uniform, is 36 m/s. As a result of friction, the velocity of the water leaving the surface is 30 m/s. Neglecting gravity effects, calculate the hydrodynamic force on the vane.

Solution
Taking the x direction as parallel to the initial velocity (Fig. 4.3) and assuming that the final velocity is uniform, we have

Force *on fluid* in x direction = Rate of increase of x-momentum

$$= \varrho Q u_2 \cos 60° - \varrho Q u_1$$

$$= \left(1000\, \frac{\text{kg}}{\text{m}^3} \right) \left\{ \frac{\pi}{4}(0.05)^2\, \text{m}^2 \times 36\, \frac{\text{m}}{\text{s}} \right\} \left(30\, \cos 60° \frac{\text{m}}{\text{s}} - 36\, \frac{\text{m}}{\text{s}} \right)$$

$$= -1484\, \text{N}$$

Similarly, force *on fluid* in y direction

$$= \varrho Q u_2 \sin 60° - 0$$

$$= \left\{ 1000\, \frac{\pi}{4}(0.05)^2\, 36\, \frac{\text{kg}}{\text{s}} \right\} \left(30\, \sin 60° \frac{\text{m}}{\text{s}} \right)$$

$$= 1836\, \text{N}$$

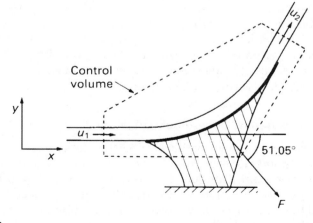

Fig. 4.3

The resultant force *on the fluid* is therefore $\sqrt{(1484^2 + 1836^2)}\,\text{N} = 2361\,\text{N}$ acting in a direction arctan $\{1836/(-1484)\} = 180° - 51.05°$ to the x direction. Since the pressure is atmospheric both where the fluid enters the control volume and where it leaves, the hydrodynamic force on the fluid can be provided only by the vane. The force exerted by the fluid on the vane is opposite to the force exerted by the vane on the fluid.

Therefore the hydrodynamic force F on the vane acts in the direction shown on the diagram.

□

If the vane is moving with a uniform velocity in a straight line the problem is not essentially different. To meet the condition of steady flow (and only to this does the equation apply) coordinate axes moving with the vane must be selected. Therefore the velocities concerned in the calculation are velocities *relative to* these axes, that is, relative to the vane. The volume flow rate Q must also be measured *relative* to the vane. As a simple example we may suppose the vane to be moving at velocity c in the same direction as the jet. If c is greater than u_1, that is, if the vane is receding from the orifice faster than the fluid is, no fluid can act on the vane at all. If, however, c is less than u_1, the mass of fluid reaching the vane in unit time is given by $\varrho A(u_1 - c)$ where A represents the cross-sectional area of the jet, and uniform jet velocity and density are assumed. (Use of the *relative* incoming velocity $u_1 - c$ may also be justified thus. In a time interval δt the vane moves a distance $c\,\delta t$, so the jet lengthens by the same amount; as the mass of fluid in the jet increases by $\varrho A c\,\delta t$ the mass actually reaching the vane is only $\varrho A u_1\,\delta t - \varrho A c\,\delta t$, i.e. $\varrho A(u_1 - c)$ per unit time.) The direction of the exit edge of the vane corresponds to the direction of the velocity of the fluid there *relative* to the vane.

The action of a stream of fluid on a single body moving in a straight line has little practical application. To make effective use of the principle a number of similar vanes may be mounted round the circumference of a wheel so that they are successively acted on by the fluid. In this case, the system of vanes *as a whole* is considered. No longer does the question arise of the jet lengthening so that not all the fluid from the orifice meets a vane; the entire mass flow rate $\varrho A u_1$ from the orifice is intercepted *by the system* of vanes. Such a device is known as a turbine, and we shall consider it further in Chapter 13.

4.3.2 Force caused by flow round a pipe-bend

When the flow is confined within a pipe the static pressure may vary from point to point and forces due to differences of static pressure must be taken into account. Consider the pipe-bend illustrated in Fig. 4.4 in which not only the direction of flow but also the cross-sectional area is changed. The 'control volume' selected is that bounded by the inner surface of the pipe and sections 1 and 2. For simplicity we here assume that the axis of the bend is in the horizontal plane: changes of elevation

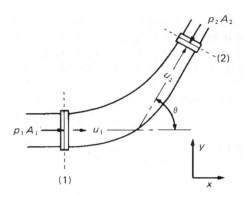

Fig. 4.4

are thus negligible; moreover, the weights of the pipe and fluid act in a direction perpendicular to this plane and so do not affect the changes of momentum. We assume too that conditions at sections 1 and 2 are uniform and that the streamlines there are straight and parallel.

If the mean pressure and cross-sectional area at section 1 are p_1 and A_1 respectively, the fluid adjacent to this cross-section exerts a force $p_1 A_1$ on the fluid in the control volume. Similarly, there is a force $p_2 A_2$ acting at section 2 on the fluid in the control volume. Let the pipe-bend exert a force F *on the fluid*, with components F_x and F_y in the x and y directions indicated. The force F is the resultant of all forces acting over the inner surface of the bend. Then the total force in the x direction *on the fluid* in the control volume is

$$p_1 A_1 - p_2 A_2 \cos\theta + F_x$$

This total 'x-force' must equal the rate of increase of x-momentum

$$\varrho Q (u_2 \cos\theta - u_1)$$

Equating these two expressions enables F_x to be calculated.

Similarly, the total y-force acting *on the fluid* in the control volume is

$$- p_2 A_2 \sin\theta + F_y = \varrho Q (u_2 \sin\theta - 0)$$

and F_y may thus be determined. From the components F_x and F_y the magnitude and direction of the total force exerted by the bend on the fluid can readily be calculated. The force exerted by the fluid on the bend is equal and opposite to this.

If the bend were empty (except for stationary atmosphere) there would be a force exerted by the atmosphere on the inside surfaces of the bend. In practice we are concerned with the amount by which the force exerted by the moving fluid exceeds the force that would be exerted by a stationary atmosphere. Thus we use gauge values for the pressures p_1 and p_2 in the above equations. The force due to the atmospheric part of the pressure is counterbalanced by the atmosphere surrounding the bend: if absolute values were used for p_1 and p_2 separate account would have to be taken of the force, due to atmospheric pressure, on the outer surface.

Where only one of the pressures p_1 and p_2 is included in the data of the problem, the other may be deduced from the energy equation.

Particular care is needed in determining the signs of the various terms in the momentum equation. It is again emphasized that the principle used is that the resultant force *on the fluid* in a particular direction is equal to the rate of *increase* of momentum *in that direction*.

The force on a bend tends to move it and a restraint must be applied if movement is to be prevented. In many cases the joints are sufficiently strong for that purpose, but for large pipes (e.g. those used in hydro-electric installations) large concrete anchorages are usually employed to keep the pipe-bends in place.

The force F includes any contribution made by friction forces on the boundaries. Although it is not necessary to consider friction forces separately they do influence the final result, because they affect the relation between p_1 and p_2.

Example 4.2 A 45° reducing pipe-bend (in a horizontal plane) tapers from 600 mm diameter at inlet to 300 mm diameter at outlet (see Fig. 4.5). The gauge pressure at inlet is 140 kPa and the rate of flow of water through the bend is 0.425 m³/s. Neglecting friction, calculate the net resultant horizontal force exerted by the water on the bend.

Solution
Assuming uniform conditions with straight and parallel stream-lines at inlet and outlet, we have:

$$u_1 = \frac{0.425\,\text{m}^3/\text{s}}{\frac{\pi}{4}(0.6\,\text{m})^2} = 1.503\,\text{m/s}$$

$$u_2 = \frac{0.425\,\text{m}^3/\text{s}}{\frac{\pi}{4}(0.3\,\text{m})^2} = 6.01\,\text{m/s}$$

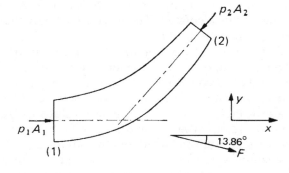

Fig. 4.5

By the energy equation

$$p_2 = p_1 + \frac{1}{2}\varrho\left(u_1^2 - u_2^2\right)$$

$$= 1.4 \times 10^5\,\text{Pa} + 500\frac{\text{kg}}{\text{m}^3}\left(1.503^2 - 6.01^2\right)\frac{\text{m}^2}{\text{s}^2}$$

$$= 1.231 \times 10^5\,\text{Pa}$$

In the x direction, force on water in control volume =

$$p_1A_1 - p_2A_2\cos 45° + F_x = \varrho Q\left(u_2\cos 45° - u_1\right)$$

= Rate of increase of x-momentum, where F_x represents x-component of force exerted *by* bend *on* water. Therefore

$$1.4 \times 10^5\,\text{Pa}\,\frac{\pi}{4}0.6^2\,\text{m}^2 - 1.231 \times 10^5\,\text{Pa}\,\frac{\pi}{4}0.3^2\,\text{m}^2\,\cos 45° + F_x$$

$$= 1000\frac{\text{kg}}{\text{m}^3}0.425\frac{\text{m}^3}{\text{s}}\left(6.01\cos 45° - 1.503\right)\frac{\text{m}}{\text{s}}$$

i.e. $\left(39\,580 - 6150\right)\text{N} + F_x = 1168\,\text{N}$ whence $F_x = -32\,260\,\text{N}$.

In the y direction, force on water in control volume

$$= -p_2A_2\sin 45° + F_y = \varrho Q\left(u_2\sin 45° - 0\right)$$

= Rate of increase of y-momentum, whence

$$F_y = 1000 \times 0.425\left(6.01\sin 45°\right)\text{N} + 1.231 \times 10^5\,\frac{\pi}{4}0.3^2\,\sin 45°\text{N}$$

$$= 7960\,\text{N}$$

Therefore total net force exerted on water $= \sqrt{(32\,260^2 + 7960^2)}\text{N}$ $= 33\,230\,\text{N}$ acting in direction arctan $\{7960/(-32\,260)\} = 180° - 13.86°$ to the x direction.

Force F exerted on bend is equal and opposite to this, i.e. in the direction shown on Fig. 4.5.

□

For a pipe-bend with a centre-line not entirely in the horizontal plane the weight of the fluid in the control volume contributes to the force causing the momentum change. It will be noted, however, that detailed information is not required about the shape of the bend or the conditions between the inlet and outlet sections.

4.3.3 Force at a nozzle and reaction of a jet

As a special case of the foregoing we may consider the horizontal nozzle illustrated in Fig. 4.6. Assuming uniform conditions with streamlines straight and parallel at the sections 1 and 2 we have

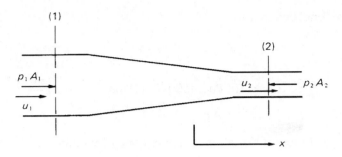

(1)

$p_1 A_1$

u_1

(2)

u_2 ← $p_2 A_2$

x

Fig. 4.6

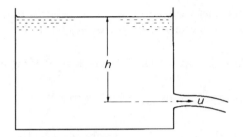

h

u

Fig. 4.7

Force exerted in the x direction on the fluid between planes 1 and 2

$$= p_1 A_1 - p_2 A_2 + F_x = \varrho Q (u_2 - u_1)$$

If a small jet issues from a reservoir large enough for the velocity within it to be negligible (except close to the orifice) then the velocity of the fluid is increased from zero in the reservoir to u at the vena contracta (see Fig. 4.7). Consequently the force exerted *on* the fluid to cause this change is $\varrho Q(u - 0) = \varrho Q C_v \sqrt{(2gh)}$. An equal and opposite reaction force is therefore exerted by the jet on the reservoir.

The existence of the reaction may be explained in this way. At the vena contracta the pressure of the fluid is reduced to that of the surrounding atmosphere and there is also a smaller reduction of pressure in the neighbourhood of the orifice, where the velocity of the fluid becomes appreciable. On the opposite side of the reservoir, however, and at the same depth, the pressure is expressed by ϱgh and the difference of pressure between the two sides of the reservoir gives rise to the reaction force.

Such a reaction force may be used to propel a craft – aircraft, rocket, ship or submarine – to which the nozzle is attached. The jet may be formed by the combustion of gases within the craft or by the pumping of fluid through it. For the *steady* motion of such a craft in a straight line the propelling force may be calculated from the momentum equation. For steady flow the reference axes must move with the craft, so all velocities are measured relative to the craft. If fluid (air, for example) is taken in at the front of the craft with a uniform velocity c and spent fluid (for example, air plus fuel) is ejected at the rear with a velocity u_r, then, for a control volume closely surrounding the craft,

The net rate of increase of fluid momentum backwards (*relative to the craft*) is

$$= \int \varrho u_r^2 dA_2 - \int \varrho c^2 dA_1 \qquad (4.5)$$

where A_1, A_2 represent the cross-sectional areas of the entry and exit orifices respectively. (In some jet-propelled boats the intake faces downwards in the bottom of the craft, rather than being at the front. This, however, does not affect the application of the momentum equation since, wherever the water is taken in, its initial momentum relative to the boat is ϱQc per unit time. Nevertheless, a slightly better efficiency can be expected with a forward-facing inlet because the pressure there is increased – as in a Pitot tube – so the pump has to do less work to produce a given outlet jet velocity.)

Equation 4.5 is restricted to a craft moving steadily in a straight line because Newton's Second Law is valid only for a non-accelerating set of reference axes.

In practice the evaluation of the integrals in eqn 4.5 is not readily accomplished because the assumption of a uniform velocity – particularly over the area A_2 – is seldom justified. Moreover, the tail pipe is not infrequently of diverging form and thus the velocity of the fluid is not everywhere perpendicular to the cross-section.

In a jet-propelled aircraft the spent gases are ejected to the surroundings at high velocity – usually greater than the velocity of sound in the fluid. Consequently (as we shall see in Chapter 11) the pressure of the gases at discharge does not necessarily fall immediately to the ambient pressure. If the mean pressure p_2 at discharge is greater than the ambient pressure p_a then a force $(p_2 - p_a)A_2$ contributes to the propulsion of the aircraft.

The relation 4.5 represents the propulsive force exerted by the engine on the fluid in the backward direction. There is a corresponding forward force exerted by the fluid on the engine, and the total thrust available for propelling the aircraft at uniform velocity is therefore

$$\left(p_2 - p_a\right)A_2 + \int \varrho u_r^2 dA_2 - \int \varrho c^2 dA_1 \qquad (4.6)$$

It might appear from this expression that, to obtain a high value of the total thrust, a high value of p_2 is desirable. When the gases are not fully expanded, however, that is, when $p_2 > p_a$, the exit velocity u_r relative to the aircraft is reduced and the total thrust is in fact decreased. This is a matter about which the momentum equation itself gives no information and further principles must be drawn upon to decide the optimum design of a jet-propulsion unit.

Although not applying the Steady-Flow Momentum Equation, a few paragraphs on rocket propulsion are interpolated here. They are

Rocket propulsion

included partly because of current interest in the topic but principally to correct some common misconceptions.

A rocket is driven forward by the reaction of its jet. The gases constituting the jet are produced by the combustion of a fuel and appropriate oxidant; no air is required, so a rocket can operate satisfactorily in a vacuum. The penalty of this independence of the atmosphere, however, is that a large quantity of oxidant has to be carried along with the rocket. At the start of a journey the fuel and oxidant together form a large proportion of the total load carried by the rocket. Work done in raising the fuel and oxidant to a great height before they are burnt is wasted. Therefore the most efficient use of the materials is achieved by accelerating the rocket to a high velocity in a short distance. It is this period during which the rocket is accelerating that is of principal interest. We note that the simple relation $F = ma$ is not directly applicable here because, as fuel and oxidant are being consumed, the mass of the rocket is not constant.

In examining the behaviour of an accelerating rocket particular care is needed in selecting the coordinate axes to which measurements of velocities are referred. We here consider our reference axes fixed to the earth and *all* velocities must be expressed with respect to these axes. We may not consider our reference axes attached to the rocket, because the rocket is accelerating and Newton's Laws of Motion are applicable only when the reference axes are *not* accelerating.

If Newton's Second Law is applied to all the particles in a system and the resulting equations are added, the result is:

Rate of increase of total momentum of the system of particles
= Vector sum of the external forces.

(When the vector sum of the external forces is zero, the total momentum of the system of particles is consequently constant: this is the Principle of Conservation of Linear Momentum.) The system must be so defined that, whatever changes occur, it always consists of the same collection of particles. Here the system comprises the rocket and its fuel and oxidant at a particular time t.

If, at time t, the total mass of rocket, fuel and oxidant $= M$ and the velocity of the rocket (relative to the earth) $= v$, then the total momentum of the system $= Mv$. Let the spent gases be discharged from the rocket at a rate $\dot{m}$ (mass/time). Then, at a time δt later (δt being very small) a mass $\dot{m}\,\delta t$ of spent gases has left the rocket. These gases then have an average velocity u *relative to the earth*, and therefore momentum $\dot{m}\,\delta t u$. For consistency u is considered positive in the same direction as v, i.e. forwards.

At time $t + \delta t$ the velocity of the rocket has become $v + \delta v$, so the momentum of the whole system (relative to the earth) is now

$$\left(M - \dot{m}\delta t\right)\left(v + \delta v\right) + \dot{m}\delta t u$$

Since all velocities are in the same (unchanging) direction the momentum has increased by an amount

$$\left(M - \dot{m}\delta t\right)\left(v + \delta v\right) + \dot{m}\delta tu - Mv = M\delta v + \dot{m}\delta t\left(u - v - \delta v\right)$$

This increase has occurred in a time interval δt, so the rate of increase = $M(\delta v/\delta t) + \dot{m}(u - v - \delta v)$ which, in the limit as δt and δv both tend to zero, becomes

$$M\frac{\mathrm{d}v}{\mathrm{d}t} + \dot{m}\left(u - v\right) \qquad (4.7)$$

But, when $\delta t \to 0$, u represents the (mean) absolute velocity of the gases at the moment they leave the rocket. So $u - v$ is the difference between the absolute velocity of the jet and the absolute velocity of the rocket, both being considered positive in the forward direction. Therefore $u - v$ = the velocity of the jet *relative* to the rocket (forwards) = $-u_r$, where u_r is the (mean) rearward velocity of the jet relative to the rocket. We may usefully express the relation 4.7 in terms of u_r: though not necessarily constant, it is, for a given fuel and oxidant and shape of nozzle, a quantity normally known to rocket designers. Therefore

$$\text{Rate of increase of momentum} = M\frac{\mathrm{d}v}{\mathrm{d}t} - \dot{m}u_r$$

$$= \text{vector sum of the external forces on}$$

$$\text{the system }\left(\text{in the forward direction}\right) = \sum F \qquad (4.8)$$

Now the vector sum of the external forces on the system is not the same as the net force on the rocket. The system, we recall, consists of the rocket plus the gases that escape from it. The jet reaction on the rocket and the force exerted on the escaping gases constitute an action-and-reaction pair and their sum therefore cancels in the total. The propulsive force may readily be deduced, however. The force (forwards) applied to the spent gases to change their momentum from $\dot{m}\delta tv$ to $\dot{m}\delta tu$ is $\dot{m}\delta t(u - v)/\delta t = \dot{m}(u - v) = -\dot{m}u_r$. Therefore the corresponding reaction (forwards) on the rocket = $\dot{m}u_r$. In addition, however, a contribution to the propulsive force is made by $(p_2 - p_a)A_2$, where p_2 represents the mean pressure of the gases at discharge, p_a the ambient pressure and A_2 the cross-sectional area of the discharge orifice. The total propulsive force is therefore $\dot{m}u_r + (p_2 - p_a)A_2$. In the absence of gravity, air or other resistance $(p_2 - p_a)A_2$ is the sole external force. In these circumstances eqn 4.8 shows that the propulsive force is $\dot{m}u_r + \Sigma F = M(\mathrm{d}v/\mathrm{d}t)$.

Since M represents the mass of the rocket and the remaining fuel and oxidant at the time t it is a function of t. Consequently the rocket acceleration is not constant even if $\dot{m}$, u_r and the propulsive force are constant.

This method of analysis is not the only one possible, but of the *correct* methods it is probably the most simple, direct and general. It is *incorrect* to argue that force equals rate of increase of momentum = $(\mathrm{d}/\mathrm{d}t)Mv = M(\mathrm{d}v/\mathrm{d}t) + v(\mathrm{d}M/\mathrm{d}t)$, where M and v refer to the mass and velocity of the rocket. This is wrong because the 'system' to which the argument is applied (the rocket) does *not* always consist of the same collection of particles.

4.3.4   Force on a solid body in a flowing fluid

The momentum equation may be used to determine the hydrodynamic force exerted on a solid body by fluid flowing past it. The equal and opposite force exerted by the body on the fluid corresponds to a change in the momentum of the fluid.

Figure 4.8 shows a stationary body immersed in a fluid stream that is otherwise undisturbed. The steady uniform velocity well upstream of the body is U_∞. Downstream of the body the velocity in general differs from U_∞ and is no longer uniform. We consider, for simplicity, a body round which the flow is 'two-dimensional', that is, the flow pattern is the same in all planes parallel to that of the diagram. A suitable 'control volume' is one with an inner boundary on the solid body and an outer rectangular boundary $ABCD$, where AD and BC are parallel to the direction of U_∞, and where each of the planes AB, BC, AD is far enough from the body for the flow at these planes to be unaffected by the presence of the body. If u_x, the component of velocity in the x direction (that is, parallel to U_∞), is measured at the plane CD then the mass flow rate through an element of that plane, of thickness δy and of unit breadth, is $\varrho u_x \delta y$. The rate of increase of x-momentum of this fluid is therefore $\varrho u_x \delta y(u_x - U_\infty)$ and the total rate of increase of x-momentum is

$$\int_C^D \varrho u_x\left(u_x - U_\infty\right) \mathrm{d}y$$

By the momentum equation the rate of increase of x-momentum of the fluid equals the sum of all the forces acting on it in the x direction

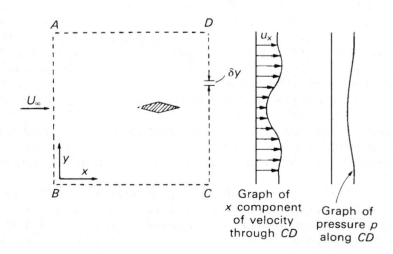

Fig. 4.8

between the upstream plane AB and the plane CD. The forces concerned are F_x, the x-component of the force exerted by the body directly on the fluid, and also any force resulting from the pressure p in the plane CD being different from the upstream pressure p_∞. (Another force might be gravity, but this need not be considered separately if we use piezometric pressure $p^* = p + \varrho gz$ in place of p.) The pressure downstream of a body is frequently less than the upstream pressure because of the turbulence in the wake. At the plane CD, the total 'pressure force' on the fluid in the control volume is $\int_C^D p\,\mathrm{d}y$ (per unit breadth) acting upstream whereas at plane AB the total force is $\int_B^A p_\infty\,\mathrm{d}y$ acting downstream. The net pressure force in the x direction (i.e. downstream) is therefore

$$\int_B^A p_\infty\,\mathrm{d}y - \int_C^D p\,\mathrm{d}y = \int_C^D \left(p_\infty - p\right)\mathrm{d}y$$

since $AB = DC$. Thus the momentum equation is

$$F_x + \int_C^D \left(p_\infty - p\right)\mathrm{d}y = \int_C^D \varrho u_x\left(u_x - U_\infty\right)\mathrm{d}y$$

Beyond C and D, however, $u_x = U_\infty$ and $p = p_\infty$ and so we may put $-\infty$ and $+\infty$ as limits of integration instead of C and D since the regions beyond C and D would make no contribution to the integrals. Hence

$$F_x = \int_{-\infty}^{\infty} \left\{\varrho u_x\left(u_x - U_\infty\right) - \left(p_\infty - p\right)\right\}\mathrm{d}y \qquad (4.9)$$

If the body is not symmetrical about an axis parallel to U_∞ there may be a change of momentum in the y-direction and consequently a component of force in that direction. Originally the fluid had no component of momentum in the y direction. If the stream is deflected by the body so that a component of velocity u_y is produced, the corresponding rate at which y-momentum passes through an element in the plane CD is $\varrho u_x \delta y\, u_y$ and the total rate of increase of y-momentum experienced by the fluid is $\int_{-\infty}^{\infty} \varrho u_x u_y\,\mathrm{d}y$. This equals the component of force on the fluid in the y direction.

Planes AD and BC are far enough from the body for the (piezometric) pressures at them to be equal and so there is no net 'pressure force' on the control volume in the y direction. The rate of increase of y-momentum therefore directly equals the component of force on the body in the $-y$ direction.

The hydrodynamic force on a body in a stream of fluid may thus be deduced from measurements of velocity and pressure in the wake downstream of the body. In many cases, of course, the force on a body is more suitably determined by direct measurement. In other instances, however, the direct method may be impracticable because of the size of the body. It may even be ruled out because the fluid would exert a force not only on the body itself, but on the members supporting it from the measuring balance, and this latter force might falsify the result. Nevertheless, measurements of velocity and pressure downstream of the body are not without difficulties. The chief of these is that close to the body the flow is frequently highly turbulent, so accurate values of the magnitude and direction of the velocity are not easy to obtain.

4.3.5 Momentum theory of a propeller

A propeller uses the torque of a shaft to produce axial thrust. This it does by increasing the momentum of the fluid in which it is submerged: the reaction to the necessary force on the fluid provides a forward force on the propeller itself, and this force is used for propulsion. A good deal more than the momentum and energy equations is needed for the complete design of a propeller. Nevertheless, the application of these equations produces some illuminating results, and we shall here make a simple analysis of the problem.

Figure 4.9 shows a propeller and its 'slipstream' (that is, the fluid on which it directly acts) and we assume that it is 'unconfined' (that is, not in a duct, for example). So that we can consider the flow steady, we shall suppose the propeller in a fixed position while fluid flows past it. Far upstream the flow is undisturbed as at section 1 where the pressure is p_1 and the velocity u_1. Just in front of the propeller, at section 2, the pressure is p_2 and the mean axial velocity u_2. Across the propeller the pressure increases to p_3. Downstream of the propeller the axial velocity of the fluid increases further, and, for a constant-density fluid, continuity therefore requires that the cross-section of the slipstream be reduced. At section 4 the streamlines are again straight and parallel; there is thus no variation in piezometric pressure across them and the pressure is again that of the surrounding undisturbed fluid.

This is a simplified picture of what happens. For one thing, it suggests that the boundary of the slipstream is a surface across which there is a

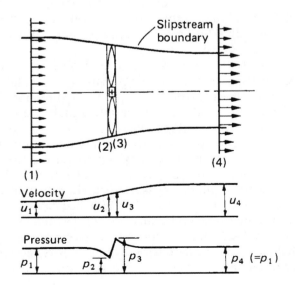

Fig. 4.9

discontinuity of pressure and velocity. In reality the pressure and velocity at the edge of the slipstream tail off into the values outside it. In practice too, there is some interaction between the propeller and the craft to which it is attached, but this is not amenable to simple analysis and allowance is usually made for it by empirical corrections.

The fluid leaving the propeller has rotary motion about the propeller axis, in addition to its axial motion. The rotary motion, however, has no contribution to make to the propulsion of the craft and represents a waste of energy. It may be eliminated by the use of guide vanes placed downstream of the propeller or by the use of a pair of contra-rotating propellers.

Certain assumptions are made for the purpose of analysis. In place of the real propeller we imagine an ideal one termed an *actuator disc*. This is assumed to have the same diameter as the actual propeller; it gives the fluid its rearward increase of momentum but does so without imparting any rotational motion. Conditions over each side of the disc are assumed uniform. This means, for example, that all elements of fluid passing through the disc undergo an equal increase of pressure. (This assumption could be realized in practice only if the propeller had an infinite number of blades.) It is also assumed that changes of pressure do not significantly alter the density and that the disc has negligible thickness in the axial direction. Consequently, the cross-sectional areas of the slipstream on each side of the disc are equal and so $u_2 = u_3$ by continuity. (At the disc the fluid has a small component of velocity radially inwards but this is small enough to be neglected and all fluid velocities are assumed axial.) The fluid is assumed frictionless.

Consider the space enclosed by the slipstream boundary and planes 1 and 4 as a control volume. The pressure all round this volume is the same, and, for the ideal fluid assumed, shear forces are absent. Consequently the only net force F on the fluid in the axial direction is that produced by the actuator disc. Therefore, for steady flow,

$$F = \varrho Q(u_4 - u_1) \tag{4.10}$$

This is equal in magnitude to the net force on the disc. Since there is no change of velocity across the disc this force is given by $(p_3 - p_2)A$. Equating this to eqn 4.10 and putting $Q = Au_2$, where A represents the cross-sectional area of the disc, we obtain

$$p_3 - p_2 = \varrho u_2(u_4 - u_1) \tag{4.11}$$

Bernoulli's equation may not be applied between sections 2 and 3 because the flow through the propeller is unsteady. However, applying the equation between sections 1 and 2 gives

$$p_1 + \frac{1}{2}\varrho u_1^2 = p_2 + \frac{1}{2}\varrho u_2^2 \tag{4.12}$$

the axis being assumed horizontal for simplicity. Similarly, between sections 3 and 4:

$$p_3 + \frac{1}{2}\varrho u_3^2 = p_4 + \frac{1}{2}\varrho u_4^2 \qquad (4.13)$$

Now $u_2 = u_3$ and also $p_1 = p_4$ = pressure of undisturbed fluid. Therefore, adding eqns 4.12 and 4.13 and rearranging gives

$$p_3 - p_2 = \frac{1}{2}\varrho\left(u_4^2 - u_1^2\right) \qquad (4.14)$$

Eliminating $p_3 - p_2$ from eqns 4.11 and 4.14 we obtain

$$u_2 = \frac{u_1 + u_4}{2} \qquad (4.15)$$

The velocity through the disc then is the arithmetic mean of the velocities upstream and downstream from it; in other words, half the change of velocity occurs before the disc and half after it (as shown in Fig. 4.9). This result, known as Froude's theorem after William Froude* (1810–79), is one of the principal assumptions in propeller design.

If the undisturbed fluid be considered stationary, the propeller advances through it at velocity u_1. The rate at which useful work is done by the propeller is given by the product of the thrust and the velocity:

$$\text{Power output} = Fu_1 = \varrho Q\left(u_4 - u_1\right)u_1 \qquad (4.16)$$

In addition to the useful work, kinetic energy is given to the slipstream which is wasted. Consequently the power input is

$$\varrho Q\left(u_4 - u_1\right)u_1 + \frac{1}{2}\varrho Q\left(u_4 - u_1\right)^2 \qquad (4.17)$$

since $u_4 - u_1$ is the velocity of the downstream fluid relative to the earth. The ratio of the expressions 4.16 and 4.17 is sometimes known as the Froude efficiency:

$$\eta_{\text{Fr}} = \frac{\text{Power output}}{\text{Power input}} = \frac{u_1}{u_1 + \frac{1}{2}\left(u_4 - u_1\right)} \qquad (4.18)$$

This efficiency, it should be noted, does not account for friction or for the effects of the rotational motion imparted to the fluid. A propulsive force requires a non-zero value of $u_4 - u_1$ (see eqn 4.10) and so even for an ideal fluid a Froude efficiency of 100% could not be achieved. Equation 4.18 in fact represents an upper limit to the efficiency. It does, however, show that a higher efficiency may be expected as the velocity increase $u_4 - u_1$ becomes smaller. The actual efficiency of an aircraft propeller is, under optimum conditions, about 0.85 to 0.9 times the value given by eqn 4.18. At speeds above about 650 km/h, however, effects of compressibility of the air (at the tips of the blades where the relative velocity is highest) cause the efficiency to decline. Ships' propellers are

* pronounced *Frood.*

usually less efficient, mainly because of restrictions in diameter, and interference from the hull of the ship.

The thrust of a propeller is often expressed in terms of a dimensionless thrust coefficient $C_T = F / \frac{1}{2} \varrho u_1^2 A$. It may readily be shown that

$$\eta_{Fr} = \frac{2}{1 + \sqrt{(1 + C_T)}}$$

Since the derivation of eqn 4.18 depends only on eqns 4.10, 4.16 and 4.17 no assumption about the form of the actuator is involved. Equation 4.18 may therefore be applied to any form of propulsion unit that works by giving momentum to the fluid surrounding it. The general conclusion may be drawn that the best efficiency is obtained by imparting a relatively small increase of velocity to a large quantity of fluid. A large velocity $(u_4 - u_1)$ given to the fluid by the actuator evidently produces a poor efficiency if u_1, the forward velocity of the craft relative to the undisturbed fluid, is small. This is why jet propulsion for aircraft is inefficient at low speeds.

For a stationary propeller, as on an aircraft before take-off, the approach velocity u_1 is zero and the efficiency is therefore zero. This is also true for helicopter rotors when the machine is hovering. No effective work is being done on the machine and its load, yet there must be a continuous input of energy to maintain the machine at constant height. With $u_1 = 0$, eqn 4.15 gives $u_2 = \frac{1}{2} u_4$ and eqn 4.10 becomes $F = \varrho Q u_4 = \varrho A u_2 \times 2 u_2$. From eqn 4.17, power input $= \frac{1}{2} \varrho Q u_4^2 = \frac{1}{2} F u_4 = F u_2 = \sqrt{(F^3 / 2 \varrho A)}$. This result shows that, for a helicopter rotor to support a given load, the larger the area of the rotor, the smaller is the power required. The weight of the rotor itself, however, increases rapidly with its area, and so the rotor diameter is, in practice, limited.

The foregoing analysis of the behaviour of propellers is due to W. J. M. Rankine (1820–72) and William Froude. Although it provides a valuable picture of the way in which velocity changes occur in the slipstream, and indicates an upper limit to the propulsive efficiency, the basic assumptions – particularly those of lack of rotary motion of the fluid and the uniformity of conditions over the cross-section – lead to inaccuracy when applied to actual propellers. To investigate the performance of an actual propeller having a limited number of blades, a more detailed analysis is needed. This, however, is outside the scope of the present book.

4.3.6 Momentum theory of a wind turbine

A wind turbine is similar to a propeller but it takes energy *from* the fluid instead of giving energy *to* it. Whereas the thrust on a propeller is made as large as possible, that on a wind turbine is, for structural reasons, ideally small. The flow pattern for the wind turbine is the opposite of that for the propeller: the slipstream widens as it passes the disc. Again, however, $u_2 = \frac{1}{2}(u_1 + u_4)$. The rate of loss of kinetic energy by the air $= \frac{1}{2}\varrho Q(u_1^2 - u_4^2)$ and, for a frictionless machine, this would equal the output of the wind turbine. The efficiency is customarily defined as the ratio of this output to the power in the wind that would pass through the area A if the disc were not present.

$$\therefore \quad \eta_{th} = \frac{\frac{1}{2}\varrho Q(u_1^2 - u_4^2)}{\frac{1}{2}\varrho(Au_1)u_1^2} = \frac{Au_2(u_1^2 - u_4^2)}{Au_1^3} = \frac{(u_1 + u_4)(u_1^2 - u_4^2)}{2u_1^3}$$

This expression has a maximum value when $u_4/u_1 = 1/3$ and the theoretical efficiency η_{th} then $= 16/27 = 59.3\%$. Efficiencies achieved in practice are much less than this, that of the traditional windmill with a small number of sail-like blades being not much over 5%. Since wind is normally available in practically unlimited quantity, however, this 'efficiency' is of little practical concern and the cost of construction is much more important.

Example 4.3
(a) Using actuator disc theory, show that half the change of velocity of the wind passing through a wind turbine occurs upstream and half in the wake of the turbine.
(b) A wind turbine, 12 m in diameter, operates at sea level in a wind of 20 m/s. The wake velocity is measured as 8 m/s. Estimate the thrust on the turbine.
(c) Calculate the power being generated by the turbine.

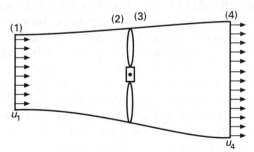

Assume a sea level air density of $1.2\,\text{kg/m}^3$.

Solution
(a) Applying Bernoulli's equation upstream of the turbine, between sections 1 and 2

$$p_1 + \frac{1}{2}\varrho u_1^2 = p_2 + \frac{1}{2}\varrho u_2^2$$

Applying Bernoulli's equation downstream of the turbine, between sections 3 and 4

$$p_3 + \frac{1}{2}\varrho u_3^2 = p_4 + \frac{1}{2}\varrho u_4^2$$

Since $u_2 = u_3$ and $p_1 = p_4$ = atmospheric pressure, these two equations can be combined to yield

$$p_2 - p_3 = \frac{1}{2}\varrho(u_1^2 - u_4^2)$$

Loss of momentum between sections 1 and 4 is equal to the thrust on the wind turbine. Thus

$$F = \varrho Q(u_1 - u_4) = \varrho A u_2(u_1 - u_4)$$

But the thrust can also be expressed as

$$F = (p_2 - p_3)A$$

So, from the two expressions for F

$$F/A = p_2 - p_3 = \varrho u_2(u_1 - u_4)$$

Equating the two expressions for $(p_2 - p_3)$

$$\frac{1}{2}\varrho(u_1^2 - u_4^2) = \varrho u_2(u_1 - u_4)$$

from which it follows that

$$u_2 = \frac{(u_1 + u_4)}{2} \qquad \text{QED}$$

(b) $F = \varrho A u_2(u_1 - u_4) = \varrho A \dfrac{(u_1 + u_4)}{2}(u_1 - u_4)$

$$= 1.2\,\text{kg/m}^3 \times \frac{\pi}{4} \times (12\,\text{m})^2 \times \frac{(20 + 8)\text{m/s}}{2} \times 12\,\text{m/s}$$

$$= 22.8 \times 10^3\,\text{N}$$

(c) $P = \varrho Q\left(\dfrac{u_1^2}{2} - \dfrac{u_4^2}{2}\right) = \varrho A u_2\left(\dfrac{u_1^2}{2} - \dfrac{u_4^2}{2}\right)$

$$= \varrho A \frac{(u_1 + u_4)}{2}\left(\frac{u_1^2}{2} - \frac{u_4^2}{2}\right)$$

$$= 1.2\,\text{kg/m}^3 \times \frac{\pi}{4}(12\,\text{m})^2 \times \frac{(20 + 8)\text{m/s}}{2}$$

$$\times \left(\frac{(20\,\text{m/s})^2}{2} - \frac{(8\,\text{m/s})^2}{2}\right)$$

$$= 319\,000\,\text{W} = 319\,\text{kW} \qquad \square$$

FURTHER READING

Seifert, H. S., Mills, M. W. and Summerfield, M. 'The physics of rockets', *Amer. J. Phys.* **15**, 1–21, 121–40, 255–72 (1947).

PROBLEMS

4.1 A stationary curved vane deflects a 50 mm diameter jet of water through 150°. Because of friction over the surface, the water leaving the vane has only 80% of its original velocity. Calculate the volume flow rate necessary to produce a hydro-dynamic force of 2000 N on the vane.

4.2 The diameter of a pipe-bend is 300 mm at inlet and 150 mm at outlet and the flow is turned through 120° in a vertical plane. The axis at inlet is horizontal and the centre of the outlet section is 1.4 m below the centre of the inlet section. The total volume of fluid contained in the bend is 0.085 m^3. Neglecting friction, calculate the magnitude and direction of the net force exerted on the bend by water flowing through it at 0.23 m^3/s when the inlet gauge pressure is 140 kPa.

4.3 Air at constant density 1.22 kg/m^3 flows in a duct of internal diameter 600 mm and is discharged to atmosphere. At the outlet end of the duct, and coaxial with it, is a cone with base diameter 750 mm and vertex angle 90°. Flow in the duct is controlled by moving the vertex of the cone into the duct, the air then escaping along the sloping sides of the cone. The mean velocity in the duct upstream of the cone is 15 m/s and the air leaves the cone (at the 750 mm diameter) with a mean velocity of 60 m/s parallel to the sides. Assuming temperature changes and frictional effects to be negligible, calculate the net axial force exerted by the air on the cone.

4.4 Two adjacent parallel and horizontal rectangular conduits A and B, of cross-sectional areas 0.2 m^2 and 0.4 m^2 respectively, discharge water axially into another conduit C of cross-sectional area 0.6 m^2 and of sufficient length for the individual streams to become thoroughly mixed. The rates of flow through A and B are 0.6 m^3/s and 0.8 m^3/s respectively and the pressures there are 31 kPa and 30 kPa respectively. Neglecting friction at the boundaries, determine the energy lost by each entry stream (per unit mass) and the total power dissipated.

4.5 A boat is driven at constant velocity c (relative to the undisturbed water) by a jet propulsion unit which takes in water at the bow and pumps it astern, beneath the water surface, at velocity u relative to the boat. Show that the efficiency of the propulsion, if friction and other losses are neglected, is $2c/(c + u)$.

Such a boat moves steadily up a wide river at 8 m/s (relative to the land). The river flows at 1.3 m/s. The resistance to motion of the boat is 1500 N. If the velocity of the jet relative to the boat is 17.5 m/s, and the overall efficiency of the pump is 65%, determine the total area of the outlet nozzles, and the engine power required.

4.6   A toy balloon of mass 86 g is filled with air of density 1.29 kg/m^3. The small filling tube of 6 mm bore is pointed vertically downwards and the balloon is released. Neglecting frictional effects calculate the rate at which the air escapes if the initial acceleration of the balloon is 15 m/s^2.

4.7   A rocket sled of 2.5 Mg (tare) burns 90 kg of fuel a second and the uniform exit velocity of the exhaust gases relative to the rocket is 2.6 km/s. The total resistance to motion at the track on which the sled rides and in the air equals KV, where $K = 1450$ N per m/s and V represents the velocity of the sled. Assuming that the exhaust gases leave the rocket at atmospheric pressure, calculate the quantity of fuel required if the sled is to reach a maximum velocity of 150 m/s.

4.8   A boat travelling at 12 m/s in fresh water has a 600 mm diameter propeller which takes 4.25 m^3 of water per second between its blades. Assuming that the effects of the propeller hub and the boat hull on flow conditions are negligible, calculate the thrust on the boat, the theoretical efficiency of the propulsion, and the power input to the propeller.

4.9   To propel a light aircraft at an absolute velocity of 240 km/h against a head wind of 48 km/h a thrust of 10.3 kN is required. Assuming a theoretical efficiency of 90% and a constant air density of 1.2 kg/m^3 determine the diameter of ideal propeller required and the power needed to drive it.

4.10 An ideal wind turbine, 12 m diameter, operates at a theoretical efficiency of 50% in a 14 m/s wind. If the air density is 1.235 kg/m^3 determine the thrust on the wind turbine, the air velocity through the disc, the mean pressures immediately in front of and behind the disc, and the shaft power developed.

5 | Physical similarity and dimensional analysis

5.1 INTRODUCTION

Physical similarity is of very wide relevance – in fact, it is applicable whenever we wish to compare the magnitudes of physical quantities in one situation with those in another. Here, however, we shall confine attention to its application in problems of fluid flow.

Theoretical analysis alone seldom solves such problems, and it is thus frequently necessary to turn to experimental results to complete the study. Even if a complete quantitative theory has been worked out, experiments are still necessary to verify it, because theories are invariably based on certain assumptions and these may not be precisely satisfied by actual fluids.

Much of this experimental work is, of course, done either on the apparatus for which the results are required, or on an exact duplicate of it. But a large part of the progress made in the study of mechanics of fluids and in its engineering applications has come from experiments conducted on scale models. No aircraft is now built before exhaustive tests have been carried out on small models in a wind-tunnel; the behaviour and power requirements of a ship are calculated in advance from results of tests in which a small model of the ship is towed through water. Flood control of rivers, spillways of dams, harbour works and similar large-scale projects are studied in detail with small models, and the performance of turbines, pumps, propellers and other machines is investigated with smaller, model, machines. There are clearly great economic advantages in testing and probably subsequently modifying small-scale equipment; not only is expense saved, but also time. Sometimes, tests are conducted with one fluid – water, perhaps – and the results applied to situations in which another fluid – air, steam, oil, for example – is used.

In all these examples, results taken from tests performed under one set of conditions are applied to another set of conditions. This procedure is made possible and justifiable by the laws of similarity. By these laws, the behaviour of a fluid in one set of circumstances may be related to the behaviour of the same, or another, fluid in other sets of circumstances. Comparisons are usually made between the *prototype*, that is, the full-size aircraft, ship, river, turbine, or other device, and the *model* apparatus. As already indicated, the use of the same fluid with both prototype and model is not necessary. Nor is the model necessarily smaller than the prototype. The flow of fluid through an injection nozzle or a carburettor, for example, would be more easily studied by using a model much larger than the prototype. So would the flow of gas between small turbine blades. Indeed, model and prototype may even be of identical size, although the two may then differ in regard to other factors, such as velocity and viscosity of the fluid.

For any comparison between prototype and model to be valid, the sets of conditions associated with each must be *physically similar*. 'Physical similarity' is a general term covering several different kinds of similarity. We shall first define physical similarity as a general proposition, and then consider separately the various forms it may take.

Two systems are said to be physically similar in respect to certain specified physical quantities when the ratio of corresponding magnitudes of these quantities between the two systems is everywhere the same.

If the specified physical quantities are *lengths*, the similarity is called geometric similarity. This is probably the type of similarity most commonly encountered and, from the days of Euclid, most readily understood.

5.2 TYPES OF PHYSICAL SIMILARITY

5.2.1 Geometric similarity

Geometric similarity is similarity of shape. The characteristic property of geometrically similar systems, whether plane figures, solid bodies or patterns of fluid flow, is that the ratio of any length in one system to the corresponding length in the other system is everywhere the same. This ratio is usually known as the *scale factor*.

Scale factor

Geometric similarity is perhaps the most obvious requirement in a model system designed to correspond to a given prototype system. Yet perfect geometric similarity is not always easy to attain. Not only must the overall shape of the model be geometrically similar to that of the prototype, but the inevitable roughness of the surfaces should also be geometrically similar. For a small model the surface roughness might not be reduced according to the scale factor – unless the model surfaces can be made very much smoother than those of the prototype. And in the study of the movement of sediment in rivers, for example, a small model might require – according to the scale factor – the use of a powder of impossible fineness to represent sand.

If for any reason the scale factor is not the same throughout, a distorted model results. For example, a prototype and its model may have overall shapes that are geometrically similar, but surface finishes that are not. In the case of very large prototypes, such as rivers, the size of the model is probably limited by the available floor space: but if the scale factor used in reducing the horizontal lengths is also used for the vertical lengths, the result may be a stream so shallow that surface tension has a considerable effect and, moreover, the flow may be laminar instead of turbulent. Here, then, a distorted model may be unavoidable.

The extent to which perfect geometric similarity should be sought naturally depends on the problem being investigated, and the accuracy required from the solution. We shall consider specific instances later.

5.2.2 Kinematic similarity

Kinematic similarity is similarity of motion. This implies similarity of lengths (i.e. geometric similarity) and, in addition, similarity of time intervals. Then, since corresponding lengths in the two systems are in a fixed ratio and corresponding time intervals are also in a fixed ratio, the velocities of corresponding particles must be in a fixed ratio of magnitude at corresponding times. Moreover, accelerations of corresponding particles must be similar. If the ratio of corresponding lengths is r_l and the ratio of corresponding time intervals is r_t, then the magnitudes of corresponding velocities are in the ratio r_l/r_t and the magnitudes of corresponding accelerations in the ratio r_l/r_t^2.

A well-known example of kinematic similarity is found in a planetarium. Here the heavens are reproduced in accordance with a certain length scale factor, and in copying the motions of the planets a fixed ratio of time intervals (and hence velocities and accelerations) is used.

When fluid motions are kinematically similar the patterns formed by streamlines are geometrically similar (at corresponding times). Since the boundaries consist of streamlines, kinematically similar flows are possible only past geometrically similar boundaries. This condition, however, is not sufficient to ensure geometric similarity of the streamline patterns at a distance from the boundaries. Geometrically similar boundaries therefore do not necessarily imply kinematically similar flows.

5.2.3 Dynamic similarity

Dynamic similarity is similarity of forces. If two systems are dynamically similar then the magnitudes of forces at similarly located points in each system are in a fixed ratio. Consequently the magnitude ratio of any two forces in one system must be the same as the magnitude ratio of the corresponding forces in the other system. In a system involving fluids, forces may be due to many causes: viscosity, gravitational attraction, differences of pressure, surface tension, elasticity and so on. For perfect dynamic similarity, therefore, there are many requirements to be met, and it is usually impossible to satisfy all of them simultaneously. Fortunately, in many instances, some of the forces do not apply at all, or have negligible effect, so it becomes possible to concentrate on the similarity of the most important forces.

Now the justification for comparing results from one flow system with those for another system is that the behaviour of the fluid is similar in the two systems, that is, the flows are kinematically similar. What, then, are the conditions in which kinematically similar flows are obtained? As we have seen, one necessary condition is that the boundaries be geometrically similar. In addition, however, similarity of forces is necessary because the direction taken by any particle is determined by the resultant force acting on it. Consequently, complete similarity of two flows can be achieved only when corresponding particles in the two flows are acted on by resultant forces that have the same direction and are in a fixed ratio of magnitude. Moreover, the same conditions apply to the components of these resultant forces. The directions of component forces are determined either by external circumstances (as for gravity forces, for example) or by the flow pattern itself (as for viscous forces). In dynamically similar flows, therefore, the force polygons for corresponding individual particles are geometrically similar, and so the component forces too have the same ratio of magnitude between the two flows. Dynamic similarity, then, produces geometric similarity of the flow patterns. It should be noted, however, that the existence of geometric similarity does not, in general, imply dynamic similarity.

Before examining dynamic similarity in more detail we may note in passing that we have not exhausted the list of types of similarity, even for fluids. For example, some investigations may call for thermal similarity in which differences of temperature are in a fixed ratio between model and prototype. In chemical similarity there is a fixed ratio of concentrations of reactants at corresponding points.

One important feature, common to all kinds of physical similarity, is that for the two systems considered certain ratios of *like* magnitudes are fixed. Geometric similarity requires a fixed ratio of lengths, kinematic similarity a fixed ratio of velocities, dynamic similarity a fixed ratio of forces, and so on. Whatever the quantities involved, however, the ratio of their magnitudes is dimensionless.

Now let us suppose that the behaviour of two systems, such as a prototype and model, is governed by an equation which may be put into the form

$$\phi(\Pi_1, \Pi_2, \Pi_3 \ldots) = 0$$

where $\Pi_1, \Pi_2, \Pi_3 \ldots$ represent dimensionless groups of variables. Any correct equation in which the magnitudes of all relevant quantities are explicitly included may be rearranged in this way. If the two systems are to be physically similar, then corresponding Πs must be the same in the two cases. The fact that the precise form of the function $\phi(\Pi_1, \Pi_2, \Pi_3 \ldots)$ may be unknown is of no importance: if each Π is the same for one system as for the other then the function must have the same value in each case and the equation is satisfied in each case. For physical similarity between two systems, then, the individual magnitudes for one system must bear such ratios to those for the other system that any Π containing the magnitudes has the same value in one system as in the other.

In the study of mechanics of fluids the type of similarity with which we are almost always concerned is dynamic similarity. We now turn attention to some of the force ratios that enter that study.

5.3 RATIOS OF FORCES ARISING IN DYNAMIC SIMILARITY

Forces controlling the behaviour of fluids arise in several ways. Not every kind of force enters every problem, but a list of the possible types is usefully made at the outset of the discussion.

1. Forces due to differences of piezometric pressure between different points in the fluid. The phrase 'in the fluid' is worth emphasis. Dynamic similarity of flow does not necessarily require similarity of thrusts on corresponding parts of the boundary surfaces and so the magnitude of piezometric pressure relative to the surroundings is not important. For the sake of brevity, the forces due to differences of piezometric pressure will be termed 'pressure forces'.
2. Forces resulting from the action of viscosity.
3. Forces acting from outside the fluid – gravity, for example.
4. Forces due to surface tension.
5. Elastic forces, i.e. those due to the compressibility of the fluid.

Now any of these forces, acting in combination on a particle of fluid, in general have a resultant, which, in accordance with Newton's Second Law, $F = ma$, causes an acceleration of the particle in the same direction as the force. And the accelerations of individual particles together determine the pattern of the flow. Let us therefore examine a little further these accelerations and the forces causing them.

If, in addition to the resultant force F, an extra force $(-ma)$ were applied to the particle in the same direction as F, the net force on the particle would then be $F - ma$, i.e. zero. With zero net force on it, the particle would, of course, have zero acceleration. This hypothetical force $(-ma)$, required to bring the acceleration of the particle to zero, is termed the *inertia force*: it represents the reluctance of the particle to be accelerated. The inertia force is, however, in no way independent of the other forces on the particle; it is, as we have seen, equal and opposite to their resultant, F. Nevertheless, since our concern is primarily with the pattern of the flow, it is useful to add inertia forces to our list as a separate item:

(6. Inertia forces.)

If the forces on any particle are represented by the sides of a force polygon, then the inertia force corresponds to the side required to close the polygon. Now a polygon can be completely specified by the magnitude and direction of all the sides except one. The remaining, unspecified, side is fixed automatically by the condition that it must just close the polygon. Consequently, if for any particular particle this hypothetical inertia force is specified, one of the other forces may remain unspecified; it is fixed by the condition that the force polygon must be completely closed, in other words, that the addition of the inertia force would give zero resultant force.

For dynamic similarity between two systems, the forces on any particle in one system must bear the same ratios of magnitude to one another as the forces on the corresponding particle in the other system. In most cases several ratios of pairs of forces could be selected for consideration. But, because the accelerations of particles (and hence the inertia forces) play an important part in practically every problem of fluid motion, it has become conventional to select for consideration the ratios between the magnitude of the inertia force and that of each of the other forces in turn. For example, in a problem such as that studied in Section 1.9.2, the only relevant forces are inertia forces, viscous forces and the forces due to differences of pressure. The ratio chosen for consideration in this instance is that of |Inertia force| to |Net viscous force| and this ratio must be the same for corresponding particles in the two systems if dynamic similarity between the systems is to be realized. (There is no need to consider separately the ratio of |Inertia force| to |Pressure force| since, once the inertia force and net viscous force are fixed, the 'pressure force' is determined automatically by the condition that the resultant of all three must be zero.) In a case where the forces involved are weight, pressure force and inertia force, the ratio chosen is |Inertia force| to |Weight|.

We shall now consider the various force ratios in turn.

5.3.1 Dynamic similarity of flow with viscous forces acting

There are many instances of flow that is affected only by viscous, pressure and inertia forces. If the fluid is in a full, completely closed conduit, gravity cannot affect the flow pattern; surface tension has no effect since there is no free surface, and if the velocity is well below the speed of sound in the fluid the compressibility is of no consequence. These conditions are met also in the flow of air past a low-speed aircraft and the flow of water past a submarine deeply enough submerged to produce no waves on the surface.

Now for dynamic similarity between two systems, the magnitude ratio of any two forces must be the same at corresponding points of the two systems (and, if the flow is unsteady, at corresponding times also). There are three possible pairs of forces of different kinds but, by convention, the ratio of |Inertia force| to |Net viscous force| is chosen to be the same in each case.

The inertia force acting on a particle of fluid is equal in magnitude to the mass of the particle multiplied by its acceleration. The mass is equal to the density ϱ times the volume (and the latter may be taken as proportional to the cube of some length l which is characteristic of the geometry of the system). The mass, then, is proportional to ϱl^3. The acceleration of the particle is the rate at which its velocity in that direction changes with time and so is proportional in magnitude to some particular velocity divided by some particular interval of time, that is, to u/t, say. The time interval, however, may be taken as proportional to the chosen characteristic length l divided by the characteristic velocity, so that finally the acceleration may be set proportional to $u \div (l/u) = u^2/l$. The magnitude of the inertia force is thus proportional to $\varrho l^3 u^2/l = \varrho l^2 u^2$.

The shear stress resulting from viscosity is given by the product of viscosity μ and the rate of shear; this product is proportional to $\mu u/l$. The magnitude of the area over which the stress acts is proportional to l^2 and thus the magnitude of viscous force is proportional to $(\mu u/l) \times l^2 = \mu u l$.

Consequently, the ratio

$$\frac{|\text{Inertia force}|}{|\text{Net viscous force}|} \text{ is proportional to } \frac{\varrho l^2 u^2}{\mu u l} = \frac{\varrho l u}{\mu}$$

As we recall from Section 1.9.2, the ratio $\varrho l u/\mu$ is known as the *Reynolds number*. For dynamic similarity of two flows past geometrically similar boundaries and affected only by viscous, pressure and inertia forces, the magnitude ratio of inertia and viscous forces at corresponding points must be the same. Since this ratio is proportional to Reynolds number, the condition for dynamic similarity is satisfied when the Reynolds numbers based on corresponding characteristic lengths and velocities are identical for the two flows.

The length l in the expression for Reynolds number may be any length that is significant in determining the pattern of flow. For a circular pipe completely full of the fluid the diameter is now invariably used – at least in Great Britain and North America. (Except near the inlet and outlet of the pipe the length along its axis is *not* relevant in determining the pattern of flow. Provided that the cross-sectional area of the pipe is constant and that the effects of compressibility are negligible, the flow pattern does not change along the direction of flow – except near the ends as will be discussed in Section 7.9.) Also by convention the mean velocity over the pipe cross-section is chosen as the characteristic velocity u.

For flow past a flat plate, the length taken as characteristic of the flow pattern is that measured along the plate from its leading edge, and the characteristic velocity is that well upstream of the plate. The essential point is that, in all comparisons between two systems, lengths similarly defined and velocities similarly defined must be used.

Example 5.1 The flow rate through differential-pressure flow-metering devices, such as venturi and orifice-plate meters, can be calculated when the value of the discharge coefficient is known. Information on the discharge coefficient is conventionally presented as a function of the Reynolds number.

(a) Show that this method of presenting the data is inconvenient for many purposes.

(b) By using different dimensionless parameters, show that these difficulties can be avoided.

Solution
Define the mean velocities in the pipe and at the throat by u_1 and u_2, and the upstream pipe diameter and throat diameter by D and d, respectively.

(a) Reference to Chapter 3 shows that the discharge coefficient, C, is defined as

$$C = \frac{Q}{E(\pi/4)d^2} \sqrt{\frac{1}{2g\Delta h}} = \frac{Q}{E(\pi/4)d^2} \sqrt{\frac{\varrho}{2\Delta p}}$$

$$\text{where} \quad E = \left\{1 - (d/D)^4\right\}^{-1/2}$$

The Reynolds number can be defined in terms of conditions in the pipe or at the throat, so that

$$Re_D = \frac{\varrho u_1 D}{\mu} \quad \text{and} \quad Re_d = \frac{\varrho u_2 d}{\mu}$$

Data are usually presented in the form

$$C = \text{function }(\beta,\, Re) \quad \text{or} \quad \alpha = \text{function }(\beta,\, Re),$$

where $\alpha = CE$ and $\beta = d/D$.

Whether Re_d or Re_D is used is unimportant in the context of the present arguments.

In order to evaluate the flow rate Q corresponding to a measured differential pressure or head, the value of C must first be established. But C is, in general, a function of Re, and Re itself depends upon the value of Q, since, from the continuity condition,

$$u_1 = \frac{4Q}{\pi D^2} \quad \text{and} \quad u_2 = \frac{4Q}{\pi d^2}$$

Hence it is demonstrated that it is not possible to determine Q directly from this method of presentation of the data, whether in chart or equation form. In this situation, the use of a process of successive approximations to determine Q is often recommended, but it is rather cumbersome and is, in any event, unnecessary.

(b) A presentation which allows a direct approach to the calculation of the flow rate is to define a new non-dimensional parameter, based on the known value Δh or Δp, rather than the unknown u_1 or u_2.

Define

$$Np_d = \frac{(\varrho \Delta p)^{1/2} d}{\mu} = \frac{(\varrho^2 g \Delta h)^{1/2} d}{\mu}$$

and

$$Np_D = \frac{(\varrho\Delta p)^{1/2} D}{\mu} = \frac{(\varrho^2 g\Delta h)^{1/2} D}{\mu}$$

These new variables are related to Re by

$$Np_d = \frac{Re_d}{\sqrt{2\alpha}} \quad \text{and} \quad Np_D = \frac{Re_D}{\sqrt{2\alpha\beta^2}}$$

A plot or correlating equation, with C as the dependent variable, and β and Np_d (or Np_D) as independent variables, allows the flow rate to be determined directly from measurements of differential pressure or head. The method is as follows: First, β and Np_d (or Np_D) are evaluated. Second, C is now determined. Finally, Q is calculated from the equation

$$Q = \frac{C(\pi/4)d^2(2g\Delta h)^{1/2}}{(1 - \beta^4)^{1/2}}$$

5.3.2 Dynamic similarity of flow with gravity forces acting

We now consider flow in which the significant forces are gravity forces, pressure forces and inertia forces. Motion of this type is found when a free surface is present or when there is an interface between two immiscible fluids. One example is the flow of a liquid in an open channel; another is the wave motion caused by the passage of a ship through water. Other instances are the flow over weirs and spillways and the flow of jets from orifices into the atmosphere.

The condition for dynamic similarity of flows of this type is that the magnitude ratio of inertia to gravity forces should be the same at corresponding points in the systems being compared. The pressure forces, as in the previous case where viscous forces were involved, are taken care of by the requirement that the force polygon must be closed. The magnitude of the inertia force on a fluid particle is, as shown in Section 5.3.1, proportional to $\varrho l^2 u^2$ where ϱ represents the density of the fluid, l a characteristic length and u a characteristic velocity. The gravity force on the particle is its weight, that is, ϱ (volume) g which is proportional to $\varrho l^3 g$ where g represents the weight per unit mass. Consequently the ratio

$$\frac{|\text{Inertia force}|}{|\text{Gravity force}|} \quad \text{is proportional to} \quad \frac{\varrho l^2 u^2}{\varrho l^3 g} = \frac{u^2}{lg}$$

In practice it is often more convenient to use the square root of this ratio so as to have the first power of the velocity. This is quite permissible: equality of $u/(lg)^{1/2}$ implies equality of u^2/lg.

The ratio $u/(lg)^{1/2}$ is known as the *Froude number* after William Froude* (1810–79), a pioneer in the study of naval architecture, who first introduced it. Some writers have termed the square of this the Froude number, but the definition Froude number $= u/(lg)^{1/2}$ is now more usual.   *Froude number*

Dynamic similarity between flows of this type is therefore obtained by having values of Froude number (based on corresponding velocities and corresponding lengths) the same in each case. The boundaries for the flows must, of course, be geometrically similar, and the geometric scale factor should be applied also to the depths of corresponding points below the free surface.

Gravity forces are important in any flow with a free surface. Since the pressure at the surface is constant (usually atmospheric) only gravity forces can under steady conditions cause flow. Moreover, any disturbance of the free surface, such as wave motion, involves gravity forces because work must be done in raising the liquid against its weight. The Froude number is thus a significant parameter in determining that part of a ship's resistance which is due to the formation of surface waves.

5.3.3 Dynamic similarity of flow with surface tension forces acting

In most examples of flow occurring in engineering work, surface tension forces are negligible compared with other forces present, and the engineer is not often concerned with dynamic similarity in respect to surface tension. However, surface tension forces are important in certain problems such as those in which capillary waves appear, in the behaviour of small jets formed under low heads, and in flow of a thin sheet of liquid over a solid surface.

Here the significant force ratio is that of |Inertia force| to |Surface tension force|. (Again, pressure forces, although present, need not be separately considered.) The force due to surface tension is tangential to the surface and has the same magnitude perpendicular to any line element of unit length along the surface. If the line element is of length Δl then the surface tension force is $\gamma(\Delta l)$ where γ represents the surface tension. Since inertia force is proportional to $\varrho l^2 u^2$ (Section 5.3.1) and Δl is proportional to the characteristic length l, the ratio

$$\frac{|\text{Inertia force}|}{|\text{Surface tension force}|} \text{ is proportional to } \frac{\varrho l^2 u^2}{\gamma l} = \frac{\varrho l u^2}{\gamma}$$

The square root of this ratio, $u(\varrho l/\gamma)^{1/2}$, is now usually known as the *Weber number* after the German naval architect Moritz Weber† (1871–1951) who first suggested the use of the ratio as a relevant parameter. Sometimes, however, the ratio $\varrho l u^2/\gamma$ and even its reciprocal are also given this name.   *Weber number*

---

*pronounced *Frood*.
†pronounced *Vay'-ber*.

5.3.4 Dynamic similarity of flow with elastic forces acting

Where the compressibility of the fluid is important the elastic forces must be considered along with the inertia and pressure forces, and the magnitude ratio of inertia force to elastic force is the one considered for dynamic similarity. Equation 1.7 shows that for a given degree of compression the increase of pressure is proportional to the bulk modulus of elasticity, K. Therefore, if l again represents a characteristic length of the system, the pressure increase acts over an area of magnitude proportional to l^2, and the magnitude of the force is proportional to Kl^2. Hence the ratio

$$\frac{|\text{Inertia force}|}{|\text{Elastic force}|} \text{ is proportional to } \frac{\varrho l^2 u^2}{Kl^2} = \frac{\varrho u^2}{K}$$

Cauchy number

The parameter $\varrho u^2/K$ is known as the *Cauchy number*, after the French mathematician A. L. Cauchy (1789–1857). However, as we shall see in Chapter 11, the velocity with which a sound wave is propagated through the fluid (whether liquid or gas) is $a = \sqrt{(K_s/\varrho)}$ where K_s represents the *isentropic* bulk modulus. If we assume for the moment that the flow under consideration is isentropic, the expression $\varrho u^2/K$ becomes u^2/a^2.

Mach number

In other words, dynamic similarity of two isentropic flows is achieved if, along with the prerequisite of geometric similarity of the boundaries, u^2/a^2 is the same for corresponding points in the two flows. This condition is equivalent to the simpler one that u/a must be the same at corresponding points. This latter ratio is known as the *Mach number* in honour of Ernst Mach* (1838–1916), the Austrian physicist and philosopher. It is very important in the study of the flow of compressible fluids. It should be remembered that a represents the *local* velocity of sound, which, for a given fluid, is determined by the values of absolute pressure and density at the point where u is measured.

If the change of density is not small compared with the mean density, then thermodynamic considerations arise. In particular, the ratio of principal specific heat capacities γ must be the same in the two cases considered. (This point is usually taken care of because, in cases where compressibility of the fluid is important, experiments on models are almost invariably performed with the same fluid as for the prototype.) Where appreciable changes of temperature occur, the ways in which viscosity and thermal conductivity vary with temperature may also be important. These matters are outside the scope of this book, but it is well to remember that equality of the Mach numbers is not in every case a sufficient criterion for dynamic similarity of the flow of compressible fluids.

Effects of compressibility usually become important in practice when the Mach number exceeds about 0.4. Apart from its well-known

---

*pronounced *Mahk*.

significance in connection with high-speed aircraft and missiles, the Mach number also enters the study of propellers and rotary compressors.

It is appropriate at this point to summarize the principal ratios arising in dynamic similarity. (See Table 5.1.)

In every case l represents a length that is characteristic of the flow pattern. Except in the few cases where it is determined by convention, it must always be specified. Also u represents a velocity that is characteristic of the flow pattern and must usually be specified.

5.3.5 Other types of dynamic similarity

The foregoing ratios are not the only significant ones that could be devised. But since the inertia forces are usually the most important, each ratio is by convention constructed from inertia force and one other kind of force.

Pressure forces are always present and are therefore represented in any complete equation describing the flow. When the equation is in dimensionless form, the ratio of these forces to other types of forces appears. For example, the ratio

$$\frac{|\text{Pressure force}|}{|\text{Inertia force}|} \text{ is proportional to } \frac{\Delta p * l^2}{\varrho l^2 u^2} = \frac{\Delta p *}{\varrho u^2}$$

where $\Delta p*$ represents the difference in piezometric pressure between two points in the flow. This ratio (sometimes known as the Newton number) is often supplanted by $\Delta p*/\frac{1}{2}\varrho u^2$, the $\frac{1}{2}$ being inserted so that the denominator represents kinetic energy per unit volume or, for an

Table 5.1 Principal dynamic similarity ratios

Definition of ratio	Name of ratio	Represents magnitude ratio of these forces	Recommended symbol
$ul\varrho/\mu$	Reynolds number	$\dfrac{|\text{Inertia force}|}{|\text{Viscous force}|}$	Re
$u/(lg)^{1/2}$	Froude number	$\dfrac{|\text{Inertia force}|}{|\text{Gravity force}|}$	Fr
$u(l\varrho/\gamma)^{1/2}$	Weber number	$\dfrac{|\text{Inertia force}|}{|\text{Surface tension force}|}$	We
$\dfrac{u}{\sqrt{(K/\varrho)}} = \dfrac{u}{a}$ $= \dfrac{u}{local \text{ velocity of sound}}$	Mach number (isentropic conditions)	$\dfrac{|\text{Inertia force}|}{|\text{Elastic force}|}$	Ma (but M much more usual)

incompressible fluid, the dynamic pressure of the stream (see Section 3.7.2). This modified form is sometimes known as the *pressure coefficient*.

Cavitation number

In some instances of liquid flow the pressure at certain points may become so low that vapour cavities form – this is the phenomenon of cavitation (see Section 13.3.6). Pressures are then usefully expressed relative to p_v, the vapour pressure of the liquid at the temperature in question. A significant dimensionless parameter is the *cavitation number*, $(p - p_v)/\frac{1}{2}\varrho u^2$ (which may be regarded as a special case of the pressure coefficient). For fluid machines a special definition due to D. Thoma is more often used (see Section 13.3.6).

Ratios involving electrical and magnetic forces may arise if the fluid is permeable to electrical and magnetic fields. These topics, however, are outside the scope of this book.

5.4 THE APPLICATION OF DYNAMIC SIMILARITY

We have already seen that, for the solution of many problems in the mechanics of fluids, experimental work is required, and that such work is frequently done with models of the prototype. To take a well-known example, the development of an aircraft is based on experiments made with small models of the aircraft held in a wind-tunnel. Not only would tests and subsequent modifications on an actual aircraft prove too costly: there could be considerable danger to human life. The tests on the model, however, will have little relevance to the prototype unless they are carried out under conditions in which the flow of air round the model is dynamically similar to the flow round the prototype.

Any quantities measured in the course of testing the model will depend on the values of the independent variables involved. The application of dimensional analysis to the problem will yield a result in the form:

$$\Pi_1 = \phi(\Pi_2, \Pi_3, \Pi_4, \ldots)$$

in which the Πs are dimensionless groups. To discover the value of Π_1 which corresponds to particular values of Π_2, Π_3, Π_4 etc., a test may be made on the model, and if each of the independent groups Π_2, Π_3, Π_4 etc., has the same value for the prototype as for the model then the result is equally applicable to the prototype.

If those conditions are achieved the model and prototype are said to be *completely similar*. Complete similarity includes geometrical similarity since some of the Πs will represent ratios of significant lengths.

It is rarely feasible to obtain complete similarity between model and prototype. This is because the conditions required to make, say, Π_2 the same in each case conflict with the requirements for Π_3 to be the same. Fortunately, in many problems not all the force magnitude ratios are relevant, and some may be known to have a negligible effect.

Consider the testing of an aircraft model in a wind-tunnel in order to find the drag force F on the prototype. The model and prototype are, of course, geometrically similar. Viscous and inertia forces are involved; gravity forces, however, may be disregarded. This is because the fluid concerned – air – is of small density, and the vertical movements of the air particles as they flow past the aircraft are small; thus the work done by gravity is negligible. Surface tension does not concern us because there is no interface between a liquid and another fluid, and we shall assume that the aircraft is a low-speed type so that effects of compressibility are negligible. Apart from the inevitable pressure forces, then, only viscous and inertia forces are involved, and thus the relevant dimensionless parameter appearing as an independent variable is the Reynolds number. The usual processes of dimensional analysis, in fact, yield the result

$$\frac{F}{\varrho l^2 u^2} = \phi\left(\frac{ul\varrho}{\mu}\right) = \phi(Re) \tag{5.1}$$

where ϱ and μ respectively represent the density and viscosity of the fluid used, u represents the relative velocity between the aircraft (prototype or model) and the fluid at some distance from it, and l some characteristic length – the wing span, for instance.

Now eqn 5.1 is true for both model and prototype. For true dynamic similarity between model and prototype, however, the ratio of force magnitudes at corresponding points must be the same for both, and so, for this case, the Reynolds number must be the same. Consequently the function $\phi(Re)$ has the same value for both prototype and model and so $F/\varrho l^2 u^2$ is the same in each case. Using suffix m for the model and suffix p for the prototype, we may write

$$\frac{F_p}{\varrho_p l_p^2 u_p^2} = \frac{F_m}{\varrho_m l_m^2 u_m^2} \tag{5.2}$$

This result is valid only if the test on the model is carried out under such conditions that the Reynolds number is the same as for the prototype. Then

$$\frac{u_m l_m \varrho_m}{\mu_m} = \frac{u_p l_p \varrho_p}{\mu_p} \quad \text{and so} \quad u_m = u_p\left(\frac{l_p}{l_m}\right)\left(\frac{\varrho_p}{\varrho_m}\right)\left(\frac{\mu_m}{\mu_p}\right) \tag{5.3}$$

The velocity u_m for which this is true is known as the *corresponding velocity*. Only when the model is tested at the corresponding velocity does the formation of eddies and so on take place at points corresponding to those on the prototype, and only then are the overall flow patterns exactly similar.

Although the corresponding velocity u_m must be determined by equating the Reynolds number (or other relevant parameter) for model and prototype, there may be other considerations that limit the range of u_m. When a model aircraft is tested, the size of the model is naturally less than that of the prototype. In other words, $l_p/l_m > 1.0$. If the model is tested in the same fluid (atmospheric air) as is used for the prototype we have $\varrho_m = \varrho_p$ and $\mu_m = \mu_p$. Thus, from eqn 5.3, $u_m = u_p(l_p/l_m)$ and since $l_p/l_m > 1$, u_m is larger than u_p. Even for a prototype intended to fly at only 300 km/h, for example, a model constructed to one-fifth scale would have to be tested with an air speed of $300 \times 5 = 1500$ km/h. At velocities as high as this, the effects of the compressibility of air become very important and the pattern of flow round the model will be quite different from that round the prototype, even though the Reynolds number is kept the same. With present-day high-speed aircraft, such difficulties in the testing of models are of course accentuated.

One way of obtaining a sufficiently high Reynolds number without using inconveniently high velocities is to test the model in air of higher density. In the example just considered, if the air were compressed to a density five times that of the atmosphere, then $u_m = u_p(5)(1/5)(1/1) = u_p$, that is, model and prototype could be tested at the same relative air velocity. (Since the temperature of the compressed air would not depart greatly from that of atmospheric air, the absolute viscosity μ would not be significantly different for model and prototype. The sonic velocity $a = \sqrt{(p\gamma/\varrho)} = \sqrt{(\gamma RT)}$ would also be unchanged. Undesirable compressibility effects could thus be avoided.) Such a procedure of course involves much more complicated apparatus than a wind-tunnel using atmospheric air. If such a 'variable density' tunnel is not available, complete similarity has to be sacrificed and a compromise solution sought. It is in fact possible in this instance to extrapolate the results of a test at moderate velocity to the higher Reynolds number encountered with the prototype. The basis of the extrapolation is that the relationship between the Reynolds number and the forces involved has been obtained from experiments in which other models have been compared with their prototypes.

For dynamic similarity in cases where the effects of compressibility are important for the prototype (as for a higher-speed aircraft, for example) the Mach numbers also must be identical (Section 5.3.4). For complete similarity, therefore, we require

$$\frac{u_m l_m \varrho_m}{\mu_m} = \frac{u_p l_p \varrho_p}{\mu_p} \quad \text{(equality of Reynolds number)}$$

and

$$\frac{u_m}{a_m} = \frac{u_p}{a_p} \quad \text{(equality of Mach number)}$$

For both conditions to be satisfied simultaneously

$$\left(\frac{l_p}{l_m}\right)\left(\frac{\varrho_p}{\varrho_m}\right)\left(\frac{\mu_m}{\mu_p}\right) = \frac{u_m}{u_p} = \frac{a_m}{a_p} = \left(\frac{K_m/\varrho_m}{K_p/\varrho_p}\right)^{1/2}$$

Unfortunately, the range of values of ϱ, μ and K for available fluids is limited, and no worthwhile size ratio (l_p/l_m) can be achieved. The compressibility phenomena, however, may readily be made similar by using the same Mach number for both model and prototype. This condition imposes no restriction on the scale of the model, since no characteristic length is involved in the Mach number. If the same fluid at the same temperature and pressure is used for both model and prototype, then $K_m = K_p$ and $\varrho_m = \varrho_p$ (at corresponding points) and so $a_m = a_p$ and $u_m = u_p$. Thus the model has to be tested at the same speed as the prototype no matter what its size. In a test carried out under these conditions the Reynolds number would not be equal to that for the prototype, and the viscous forces would consequently be out of scale. Fortunately, in the circumstances considered here, the viscous forces may be regarded as having only a secondary effect, and may be allowed to deviate from the values that complete similarity requires. It has been found in practice that, for most purposes, eqn 5.2 is an adequate approximation, even though u_m is not the corresponding velocity giving equality of Reynolds number.

The important point is this. If forces of only one kind are significant, apart from inertia and pressure forces, then complete dynamic similarity is achieved simply by making the values of the appropriate dimensionless parameter (Reynolds, Froude, Weber or Mach number) the same for model and prototype. Where more than one such parameter is relevant it may still be possible to achieve complete similarity, but it is usually necessary to depart from it. It is essential that these departures be justified. For example, those forces for which dynamic similarity is not achieved must be known to have only a small influence, or an influence that does not change markedly with an alteration in the value of the appropriate dimensionless parameter. If possible, corrections to compensate for these departures from complete similarity should be made. An example of such a procedure will be given in Section 5.5.

One further precaution in connection with the testing of models should not be overlooked. It may happen that forces having a negligible effect on the prototype do materially affect the behaviour of the model. For example, surface tension has a negligible effect on the flow in rivers, but if a model of a river is of small scale the surface tension forces may have a marked effect. In other words, the Weber number may have significance for the model although it may safely be disregarded for the prototype. The result of this kind of departure from complete similarity – an effect negligible for the prototype being significant in the model (or vice versa) – is known as a *scale effect*. The roughness of the surface of solid boundaries frequently gives rise to another scale effect. Even more serious is a discrepancy of Reynolds number whereby laminar flow exists in the model system although the flow in the prototype is turbulent. Scale effects are minimized by using models that do not differ in size from the prototype more than necessary.

5.5 SHIP RESISTANCE

We now turn attention to a particularly important and interesting example of model testing in which complete similarity is impossible to achieve, and we shall examine the methods by which the difficulty has been surmounted. The testing of models is the only way at present known of obtaining a reliable indication of the power required to drive a ship at a particular speed, and also other important information.

Any solid body moving through a fluid experiences a drag, that is, a resistance to its motion. Part of the drag results directly from the viscosity of the fluid: at the surface of the body the fluid moves at the same velocity as the surface (the condition of 'no slip' at a boundary) and thus viscous forces are set up between layers of fluid there and those farther away. This part of the total drag is usually termed *skin friction*. In addition, part of the flow towards the rear of the body breaks away from the surface to form a 'wake' of eddying motion. The eddies give rise to a distribution of pressure round the body different from that to be expected in an ideal fluid, and so provide another contribution to the total drag force.

These two types of resistance – the skin friction and eddy-making resistance – are experienced by any solid body moving through any fluid. A ship, however, is only partly immersed in a liquid, and its motion through liquid gives rise to waves on the surface. The formation of these waves requires energy, and, since this energy must be derived from the motion, the ship experiences an increased resistance to its passage through the liquid.

Waves on the surface of a liquid may be of two kinds. Those of the first type are due to surface tension forces, and are known as capillary waves or ripples; they are of little importance except for bodies that are small in size compared with the waves. In the case of ship resistance it is waves of the second type that are important. These result from the action of gravity on the water that tends to accumulate around the sides of the hull. Usually two main sets of waves are produced, one originating at the bow and the other at the stern of the ship. These both diverge from each side of the hull, and there are also smaller waves whose crests are perpendicular to the direction of motion.

In the formation of waves, some water is raised above the mean level, while some falls below it. When particles are raised work must be done against their weight, so gravity, as well as viscosity, plays a part in the resistance to motion of a ship. In a dimensional analysis of the situation the quantities considered must therefore include the weight per unit mass, g. The complete list of relevant quantities is therefore the total resistance force F, the velocity of the ship u, the viscosity of the liquid μ, its density ϱ, some characteristic length l to specify the size of the ship (the overall length, for example) and g. As already indicated, surface tension forces are negligible (except for models so small as to be most suitable as children's bath-time toys). We assume that the shape is specified, so that quoting a single length is sufficient to indicate all other lengths for a particular design. The geometric similarity of a model should extend to the relative roughness of the surfaces if complete similarity is to be achieved. (Usually the depth of water is sufficiently large compared with the size of the ship not to affect the resistance.

Where this is not so, however, the depth of the water must be included among the lengths that have to satisfy geometric similarity. Likewise, the distance from a boundary, such as a canal bank, might have to be considered.) The most suitable form of the result of dimensional analysis is

$$F = \varrho u^2 l^2 \phi \left\{ \frac{ul\varrho}{\mu}, \frac{u^2}{lg} \right\}$$

that is

$$F = \varrho u^2 l^2 \phi \left\{ Re, \left(Fr \right)^2 \right\} \tag{5.4}$$

The total resistance therefore depends in some way both on the Reynolds number and on the Froude number. We remember that dimensional analysis tells us nothing about either the form of the function $\phi\{ \ \}$, or how Re and Fr are related within it. For complete similarity between a prototype and its model the Reynolds number must be the same for each, that is

$$\frac{u_p l_p \varrho_p}{\mu_p} = \frac{u_m l_m \varrho_m}{\mu_m} \tag{5.5}$$

and also the Froude number must be the same, that is

$$\frac{u_p}{\left(l_p g_p \right)^{1/2}} = \frac{u_m}{\left(l_m g_m \right)^{1/2}} \tag{5.6}$$

Equation 5.5 gives $u_m/u_p = (l_p/l_m)(\nu_m/\nu_p)$ where $\nu = \mu/\varrho$, and eqn 5.6 gives $u_m/u_p = (l_m/l_p)^{1/2}$ since, in practice, g_m cannot be significantly different from g_p. For testing small models these conditions are incompatible; together they require $(l_m/l_p)^{3/2} = \nu_m/\nu_p$ and, since both model and prototype must operate in water, l_m cannot be less than l_p. (There is no practicable liquid that would enable ν_m to be much less than ν_p.) Thus similarity of viscous forces (represented by Reynolds number) and similarity of gravity forces (represented by Froude number) cannot be achieved simultaneously between model and prototype.

The way out of the difficulty is basically that suggested by Froude. The *assumption* is made that the total resistance is the sum of three distinct parts: (a) the wave-making resistance; (b) skin friction; (c) the eddy-making resistance. It is then further *assumed* that part (a) is uninfluenced by viscosity and is therefore independent of the Reynolds number; also that (b) depends only on Reynolds number. Item (c) cannot be readily estimated, but in most cases it is only a small proportion of the total resistance, and varies little with Reynolds number. Therefore it is usual to lump (c) together with (a).

These assumptions amount to expressing the function of Re and Fr in eqn 5.4 as the sum of two separate functions, $\phi_1(Re) + \phi_2(Fr)$. Now the skin friction, (b), may be estimated by assuming that it has the same value as that for a thin flat plate, with the same length and wetted surface area, moving end-on through the water at the same velocity. Froude and others have obtained experimental data for the drag on flat plates, and this information has been systematized by boundary-layer theory. The difference between the result for skin friction so obtained and the total resistance must then be the resistance due to wave-making and eddies. Since the part of the resistance that depends on the Reynolds number is separately determined, the test on the model is conducted at the corresponding velocity giving equality of the Froude number between model and prototype; thus dynamic similarity for the wave-making resistance is obtained.

■

Example 5.2 A production torpedo has a maximum speed of 11 m/s as originally designed. By introducing a series of design changes, the following improvements were achieved:

1. the cross-sectional area was reduced by 12%;
2. the overall drag coefficient was reduced by 15%;
3. the propulsion power was increased by 20%.

What was the maximum speed of the redesigned torpedo?

Solution

Use suffix 1 to denote the original design, and suffix 2 to denote the revised design. Then

$$A_2 = A_1 - 0.12A_1 = 0.88\,A_1$$
$$C_{D2} = C_{D1} - 0.15C_{D1} = 0.85C_{D1}$$
$$P_2 = P_1 + 0.20P_1 = 1.20\,P_1$$

Since

$$P = D\dot{V} \quad \text{and} \quad D = \frac{1}{2}\varrho V^2 A C_D$$

it follows that

$$\frac{P_2}{P_1} = \frac{D_2 V_2}{D_1 V_1} = \frac{\left(\frac{1}{2}\varrho V^2 A C_D\right)_2 V_2}{\left(\frac{1}{2}\varrho V^2 A C_D\right)_1 V_1} = \left(\frac{V_2}{V_1}\right)^3\left(\frac{A_2}{A_1}\right)\left(\frac{C_{D2}}{C_{D1}}\right)$$

Hence

$$\frac{V_2}{V_1} = \left(\frac{P_2 A_1 C_{D1}}{P_1 A_2 C_{D2}}\right)^{1/3} \quad \text{and} \quad V_2 = 11\,\text{m/s} \times \left(\frac{(1.2)}{(0.88)(0.85)}\right)^{1/3}$$

□

$$V_2 = 12.88\,\text{m/s}$$

Example 5.3 A ship 125 m long (at the water-line) and having a wetted surface of 3500 m^2 is to be driven at 10 m/s in sea-water. A model ship of 1/25th scale is to be tested to determine its resistance.

Solution
The velocity at which the model must be tested is that which gives dynamic similarity of the wave-making resistance. That is, the Froude number of prototype and model must be the same.

$$\frac{u_m}{(gl_m)^{1/2}} = \frac{u_p}{(gl_p)^{1/2}} \quad \therefore u_m = u_p \left(\frac{l_m}{l_p}\right)^{1/2} = \frac{10\,\text{m/s}}{\sqrt{25}} = 2\,\text{m/s}$$

To be tested, the model must therefore be towed through water at 2 m/s. There will then be a pattern of waves geometrically similar to that with the prototype.

Suppose that, at this velocity, the total resistance of the model (in fresh water) is 54.2 N and that its skin friction is given by $\frac{1}{2}\varrho u_m^2 A_m \times C_F$, where A_m represents the wetted surface area of the model and C_F its skin-friction coefficient which is given by

$$C_F = \frac{0.075}{\left(\log_{10} Re - 2\right)^2}$$

This is a widely used empirical formula for ships with smooth surfaces. (Model ships are usually constructed of, e.g., polyurethane so that modifications to the shape may easily be made.) The Reynolds number Re in the formula is based on the water-line length, which for the model is 125 m/25 = 5 m. The kinematic viscosity of fresh water at, say, 12 °C is 1.235 mm^2/s and so Re for the model is

$$\frac{2\,\text{m/s} \times 5\,\text{m}}{1.235 \times 10^{-6}\,\text{m}^2/\text{s}} = 8.10 \times 10^6$$

Hence

$$C_F = \frac{0.075}{\left(4.9085\right)^2} = 3.113 \times 10^{-3}$$

and the skin-friction resistance is

$$\frac{1}{2} \times 1000 \frac{\text{kg}}{\text{m}^3} (2\,\text{m/s})^2 \left(\frac{3500\,\text{m}^2}{25^2}\right) 3.113 \times 10^{-3} = 34.87\,\text{N}$$

However, the total resistance was 54.2 N, so (54.2 − 34.87) N = 19.33 N must be the wave resistance + eddy-making resistance of the model. This quantity is usually known as the 'residual resistance'.

Now if the residual resistance is a function of Froude number but not of Reynolds number, eqn 5.4 becomes, *for the residual resistance only*, $F_{resid} = \varrho u^2 l^2 \phi(Fr)$. Therefore

$$\frac{\left(F_{resid}\right)_p}{\left(F_{resid}\right)_m} = \frac{\varrho_p u_p^2 l_p^2}{\varrho_m u_m^2 l_m^2}$$

(Since *Fr* is the same for both model and prototype the function $\phi(Fr)$ cancels.)

Hence

$$\left(F_{resid}\right)_p = \left(F_{resid}\right)_m \left(\frac{\varrho_p}{\varrho_m}\right)\left(\frac{u_p}{u_m}\right)^2\left(\frac{l_p}{l_m}\right)^2 = \left(F_{resid}\right)_m \left(\frac{\varrho_p}{\varrho_m}\right)\left(\frac{l_p}{l_m}\right)\left(\frac{l_p}{l_m}\right)^2$$

$$= 19.33\left(\frac{1025}{1000}\right)25^3 \, \text{N} = 3.096 \times 10^5 \, \text{N}$$

For the prototype ship *Re* is $10 \, \text{m/s} \times 125 \, \text{m}/(1.188 \times 10^{-6} \, \text{m}^2/\text{s})$ $= 1.025 \times 10^9$ (using a value of kinematic viscosity for sea-water of standard temperature and salinity). For the prototype, which, even in new condition, has a rougher surface than the model, C_F is given by

$$\frac{0.075}{\left(\log_{10} Re - 2\right)^2} + 0.0004 = \frac{0.075}{\left(7.0220\right)^2} + 0.0004 = 1.921 \times 10^{-3}$$

However, the magnitude of the skin friction is not easy to determine accurately for the prototype ship because the condition of the surface is seldom known exactly. The surface of a new hull is quite smooth, but after some time in service it becomes encrusted with barnacles and coated with slime, so the roughness is rather indeterminate. The naval architect has to make allowance for these things as far as possible. For, say, six months in average conditions C_F would be increased by about 45%. Taking C_F in the present example as $1.45 \times 1.921 \times 10^{-3} = 2.785 \times 10^{-3}$ we then find that the ship's frictional resistance is

$$\frac{1}{2}\varrho_p u_p^2 A_p C_F = \frac{1}{2} \times 1025\frac{\text{kg}}{\text{m}^3}(10 \, \text{m/s})^2 3500 \, \text{m}^2 \times 2.785 \times 10^{-3}$$

$$= 5.00 \times 10^5 \, \text{N}$$

The total resistance of the prototype is therefore $(3.096 + 5.00)10^5 \, \text{N} = 810 \, \text{kN}$. □

We may note that, for a submarine travelling at such a depth below the surface that no surface waves are formed, the total resistance is that due only to skin friction and eddies.

Details of methods used for calculating the skin friction vary somewhat, but all methods are in essence based on the assumption that the

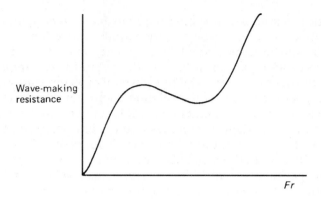

Wave-making resistance

Fr

Fig. 5.1

drag is equal to that for a flat plate of the same length and wetted area as the hull, and moving, parallel to its own length, at the same velocity. This assumption cannot be precisely true: the fact that eddy resistance is present indicates that the flow breaks away from the surface towards the stern of the vessel, and thus there must be some discrepancy between the skin friction of the hull and that of a flat plate. Indeed, the entire method is based on a number of assumptions, all of which involve a certain simplification of the true state of affairs.

In particular, the complete independence of skin friction and wave-making resistance is, we recall, an assumption. The wave-making resistance is in fact affected by the viscous flow round the hull, which in turn is dependent on the Reynolds number. Moreover, the formation of eddies influences the waves generated near the stern of the vessel. We may note in passing that the wave-making resistance of a ship bears no simple relation to the Froude number (and thus to the speed). At some speeds the waves generated from the bow of the ship reinforce those produced near the stern. At other speeds, however, the effect of one series of waves may almost cancel that of the other because the crests of one series are superimposed on the troughs of the other. As a result the graph of wave-making resistance against speed often has a sinuous form similar to that of Fig. 5.1. Nevertheless, the overall error introduced by the assumptions is sufficiently small to make the testing of models extremely valuable in the design of ships.

5.6 DIMENSIONAL ANALYSIS

Complete solutions to engineering problems can seldom be obtained by analytical methods alone and experiments are usually necessary to determine fully the way in which one variable depends on others. A technique that has proved very useful in reducing to a minimum the number of experiments required is the branch of applied mathematics known as *dimensional analysis*. Although not producing analytical solutions to problems, it does yield information about the form of mathematical relations connecting the relevant variables, and suggests the most effective way of grouping the variables together.

The principle of dimensional homogeneity enables us to test the consistency of an equation in 'physical' algebra: it also imposes conditions on the quantities involved in a physical problem, and so provides valuable clues to the form of the relation connecting their magnitudes. This is because a relation can, in general, be dimensionally homogeneous only if the individual magnitudes enter it in certain combinations. Knowledge of the form of the relation is valuable not only in interpreting the results of experiments, but also in suggesting the pattern that a series of experiments could most usefully take. The search for the correct form of the relation is known as dimensional analysis.

The methods of dimensional analysis are of very wide application and are especially valuable in the study of phenomena that are too complex for complete theoretical treatment. Such complexity may result from a large number of independent variables affecting the phenomenon, or from a mathematically complicated form of the expressions relating these variables. Dimensional analysis alone can never give the complete solution of a problem. It does, however, usually permit considerable simplifications in investigating complex phenomena and may show the effect of particular variables, especially when the effects of some of the other variables are known.

5.6.1 The process of analysis

Our first task in dimensional analysis is to decide what the quantities whose magnitudes enter the problem are. It is at this point particularly that there is risk of error. No quantity that might affect the result must be overlooked, even though, in the conditions of the problem considered, that quantity might not vary in magnitude. For instance, if the mass of a particular body is a relevant factor and the gravitational force tending to pull that body towards the earth also enters the problem, then the weight per unit mass, g, must be included in the list of significant quantities – even though its magnitude is constant at, say, 9.81 N/kg, and so it is not, in the ordinary sense, a 'variable'. In some problems dimensional constants appear – such as the gravitational constant G, or the velocity of light in a vacuum. If they are relevant to the problem being investigated they must be included in the analysis because their magnitudes are not merely numbers. The fact that their magnitudes do not vary is beside the point.

On the other hand, there is no virtue in bringing in quantities that have no bearing on the situation. For example, in a problem concerned with the equilibrium of a fluid, the viscosity of the fluid – which is in evidence only when there is relative motion between different particles of fluid – need not be considered. Moreover, some quantities have only an indirect effect on a problem. Temperature, for instance, may affect the magnitude of some of the quantities that directly enter a physical relation – such as density, electrical resistance, viscosity and other properties of substances. But the effect of a variation in the magnitude of the property could just as well be achieved by changing the substance instead of the temperature. In such cases temperature may be left out of account in a dimensional analysis because its effect is confined to determining values of variables that directly influence the phenomenon being investigated; consequently, the way in which the magnitude of temperature might be expressed is here irrelevant. In problems of heat transmission, however, the temperature of a body can have a direct effect on the phenomenon and so must be included in the list of relevant quantities.

In deciding which quantities are relevant it is often helpful to ask: If I were planning an experiment to investigate this phenomenon, what quantities should I need to specify? Would it matter if the experiment were performed in an orbiting satellite rather than in a laboratory where the weight per unit mass is 9.81 N/kg? Would it matter whether the experiment were performed in a hot or a cold environment, provided that values of all temperature-dependent properties were known? Would it matter if a wetting agent were added to a liquid to reduce its surface tension? Or if treacle or mercury were used instead of water?

It is not wise to proceed until one is sure that all relevant quantities have been listed. However, once we have completed the list of relevant quantities we may begin the analysis. As illustration, let us consider the problem of determining the force exerted on a solid body by the flow of a single, homogeneous, fluid of constant density past it. We suppose that all the conditions are steady, that is, not changing with time, and that the fluid is infinite in extent in all directions. We shall disregard the effects of gravity: these are not necessarily negligible, but they would be present even if the fluid were stationary and so would simply provide a force additional to that caused by the flow. The quantities involved in the problem are listed in Table 5.2, together with the exponents of the reference magnitudes in their dimensional formulae.

In the neighbourhood of the body the velocity of the fluid relative to the body will not be uniform, but the velocity referred to here is the

Table 5.2 Dimensional analysis

Quantity	Symbol	Dimensional formula		
		[M]	[L]	[T]
Force on body	F	1	1	−2
Size of body (e.g. diameter)	d	0	1	0
Velocity of fluid	u	0	1	−1
Density of fluid	ϱ	1	−3	0
Absolute viscosity of fluid (see Section 1.6.3)	μ	1	−1	−1

(Dimensional formulae are listed in Appendix 4.)

value far upstream where the fluid is not yet affected by the presence of the body. The body is not necessarily circular in section, but, for a given shape, we may select one length measurement as representative of its size. If we assume that any roughness of the body's surface is insufficient to affect the result, the only relevant quantities are those listed in Table 5.2. (If the fluid were a liquid having a free surface near to the body, then the flow round the body might distort the free surface, that is, raise or lower it by the formation of waves. Such vertical motion would involve work being done against gravity and so the weight per unit mass, g, would have to be added to the list of relevant quantities. If the compressibility of the fluid were significant its bulk modulus of elasticity, K, would have to be included.)

Now the object of dimensional analysis is to assemble the original variables (which in general are not dimensionless) into a smaller number of dimensionless groups. This is most simply achieved by putting the variables together in such a way as to remove the dependence on the individual reference magnitudes one after another.

In the present example let us first eliminate the dependence on [M]. This can be done by dividing each of the variables having dimensions in respect to [M] by, say, ϱ. Thus we obtain

	[M]	[L]	[T]
F/ϱ	$1 - 1 = 0$	$1 - (-3) = 4$	$-2 - 0 = -2$
d	0	1	0
u	0	1	-1
$(\varrho/\varrho$	0	0	$0)$
μ/ϱ	$1 - 1 = 0$	$-1 - (-3) = 2$	$-1 - 0 = -1$

Dividing ϱ by itself gives unity, which is not a variable and so need not be included in the list. Hence both the number of reference magnitudes and the number of separate variables have been reduced by one.

In a similar way we can now remove the dependence on [T]. Each variable with dimensions in respect to [T] in our new table is multiplied or divided by an appropriate power of, say, u so as to make the [T] exponent zero:

	[M]	[L]	[T]
$F/\varrho u^2$	0	$4 - 2 = 2$	$-2 - (-2) = 0$
d	0	1	0
$(u/u$	0	0	$0)$
$\mu/\varrho u$	0	$2 - 1 = 1$	$-1 - (-1) = 0$

Finally, we can use appropriate powers of, say, d to remove the dependence on [L]:

	[M]	[L]	[T]
$F/\varrho u^2 d^2$	0	$2 - 2 = 0$	0
$(d/d$	0	0	0)
$\mu/\varrho ud$	0	$1 - 1 = 0$	0

This leaves the two dimensionless groups $F/\varrho u^2 d^2$ and $\mu/\varrho ud$. The second of these is the reciprocal of the Reynolds number (Re) (see Section 1.9.2). So the sought-for relation connecting the five original variables is in fact one connecting those two dimensionless groups: that is, one is a function of the other:

$$F\big/\varrho u^2 d^2 = \phi(\mu/\varrho ud) = \phi_1(Re) \qquad (5.7)$$

(ϕ_1 is of course not the same function as ϕ.)

This is as far as dimensional analysis will take us. To discover the form of the function $\phi_1(Re)$ experimental results will be required. But the dimensional analysis has achieved a considerable simplification. Whereas we began with five separate variables (F, d, u, ϱ, μ) we now have only two ($F/\varrho u^2 d^2$ and Re), and the relation between them can be represented by a single graph. (The word 'variable' here simply means the magnitude of a relevant physical quantity: it is not implied that the magnitude varies; it could even be a 'universal constant', such as the velocity of light in a vacuum.)

This reduction from five parameters to two illustrates the general principle (the 'Pi' theorem – where the title 'Pi' refers to the Greek capital letter Π used to denote a product; it has nothing to do with $3.14159\ldots$) that if r is the number of distinct reference magnitudes required to express the dimensional formulae of all the n magnitudes directly affecting a physical phenomenon, then these n magnitudes may be grouped into $(n - r)$ independent dimensionless parameters (often referred to as Πs).

If, in the elimination of say [M] in this example, we had used μ as a divisor instead of ϱ the groups obtained would have been $F/\mu ud$ and $\varrho ud/\mu$ ($= Re$), giving

$$F/\mu ud = \phi_2(Re) \qquad (5.8)$$

This result is just as correct as the other. It merely presents the variables in a different way. Again experimental results could be depicted on a single graph: this time $F/\mu ud$ would be plotted against Reynolds number. The resulting curve would look different from the previous one but would be just as valid. The choice of dimensionless parameters is partly a matter of convention; it depends too on the use to be made of the results. For example, if the effect of viscosity is to be studied, eqn 5.8 would be inconvenient because μ appears both in $F/\mu ud$ and in Re ($= \varrho ud/\mu$): for this purpose eqn 5.7, in which only Re incorporates μ, would be better.

If we do not wish a particular variable to appear in more than one Π – for example, because we wish to consider it the dependent variable

– then clearly we should not use that variable (or any power of it) as a multiplier or divisor when eliminating a reference magnitude.

The order in which reference magnitudes are eliminated does not matter, although some slight economy of mental arithmetic is usually achieved if columns already containing a number of zeros are dealt with first.

Time can often be saved if there are obvious dimensionless combinations of variables. For instance, if two of the original variables have the same dimensional formula, their ratio will be dimensionless and so this ratio can be one of the Πs in the final result. Then, in obtaining the remaining Πs, only one of these two variables need be included in our table. Or if the original variables include μ, ϱ, u and an appropriate length measure (as in the example just considered) we can immediately write down Reynolds number ($= \varrho u \times \text{length}/\mu$) as one of the Πs, and omit from the variables tabulated one (but not more than one!) of these four.

The smaller the number $(n - r)$ of dimensionless groups appearing in the final result the more precise and informative it will be. Therefore, the number of separate variables n should be kept as small as possible without omitting any which are indeed relevant. As an instance of the rather rare possibility of reducing n let us suppose that, in the example discussed here, the velocity of the fluid relative to the body is very low. Thus the forces required to accelerate individual particles of fluid around the body are negligible compared with the viscous forces. We may then say that (most unusually) the relation Force = mass × acceleration does not enter the problem. Consequently, although we are concerned with forces we are not concerned with mass, so the density ϱ may be left out of account. Then $n = 4$ instead of 5 and the number of dimensionless groups $n - r$ is one instead of two. The final result for these conditions of very low velocity is $F/du\mu = \text{constant}$, that is, $\phi_1(Re)$ in eqn 5.7 is a constant × $(Re)^{-1}$, and $\phi_2(Re)$ in eqn 5.8 is a constant.

In brief, dimensional analysis is a study of the restrictions placed on the form of an algebraic function by the requirement of dimensional homogeneity. After such an analysis has been made, additional restrictions may sometimes be imposed by applying any available special information about the situation. For example, it may be known that certain effects become negligible when a particular quantity is very large or very small. Such information may provide further clues to the form which an unknown function takes.

Results obtained from dimensional analysis do of course depend on which quantities are at the outset considered to affect the phenomenon being studied. If the original list of quantities fails to include one that is, in fact, relevant then the result obtained from the analysis will be incorrect. However, a quantity not relevant to the phenomenon might be included in the list: this redundant quantity might be rejected by the analysis itself, but more probably the analysis would yield an extra Π. Whether the original list of relevant variables is too short or too long, only experiment can usually indicate that something is wrong.

By itself, dimensional analysis will not provide a complete solution to a problem, but the partial solution it yields will indicate that the individual magnitudes can be grouped only in certain ways. It thus enables the experimenter to obtain the maximum of information from the minimum number of experiments. The grouping of individual variables into a smaller number of dimensionless parameters greatly simplifies the presentation of experimental results. And since dimensionless parameters are not restricted to particular units, it is an advantage to use a dimensionless group even if only one of its constituent variables actually changes.

There is a good deal more to dimensional analysis and the interpretation of its results than we have space for here. For further details more specialized works should be consulted (e.g. Massey, *Measures in Science and Engineering*).

5.7 ASPECTS OF LAMINAR AND TURBULENT FLOW IN PIPES

The flow of an incompressible fluid in a closed conduit such as a pipe is, we have seen, subject to inertia forces and viscous forces but, apart from forces due to differences of piezometric pressure between different parts of the fluid, it is not normally subject to any other forces. It is to be expected therefore that the nature of the flow will be determined by the magnitude of the Reynolds number, $\varrho l u / \mu$. Reynolds himself deduced this and conducted a series of experiments to verify his deduction.

For a circular pipe it is usual to take the diameter d as the linear measurement representative of the flow pattern. (Notice particularly that, for a pipe of uniform diameter, the *length* of the pipe is not relevant in determining the flow pattern – provided that the length is sufficient for the appropriate pattern to have been reached. If, for example, the total length were doubled by adding an identical length of pipe to an existing one, the flow pattern for a given velocity in the existing pipe would be unchanged.) Thus Reynolds number for a circular pipe may be written as $\varrho d u / \mu$ or $u d / \nu$ where ν denotes the *kinematic* viscosity μ/ϱ. The representative velocity is invariably the mean velocity in the axial direction.

Several things may at once be deduced from Reynolds's tests with the injected stream of dye (see Section 1.9.1). One is that laminar flow, in which all the particles travel parallel to the axis of the tube, occurs at low velocities and therefore low values of Reynolds number, whereas turbulent flow takes place at high values of *Re*. Thus in laminar flow the viscous forces (which exert a stabilizing influence) are predominant, but in turbulent flow it is the inertia forces that take the lead.

Further, the fact that when the velocity is increased eddies begin suddenly rather than gradually indicates that laminar flow is then unstable and so only a slight disturbance is sufficient to bring on fully turbulent flow. Indeed, Reynolds found – as have many other investigators since – that the critical velocity at which turbulence begins is very sensitive to any initial disturbances. Not only does careless handling of the apparatus reduce this critical velocity, but so does insufficient stilling of the fluid before the flow begins, too fast a flow of dye, vibration of the apparatus, convection currents in the tank, an insufficiently smooth bellmouth at the pipe inlet, or sudden adjustment – especially partial closing – of the valve.

By experimenting with tubes of different diameters, and with water at different temperatures (and therefore different viscosities) Reynolds verified that the ratio ud/ν was the factor chiefly determining the onset of turbulence. Because the transition from laminar to turbulent flow is caused by disturbances such as vibration, it may occur in a range of Reynolds numbers. Indeed, by taking extreme precautions to avoid disturbances of any kind, later investigators have achieved laminar flow in pipes at values of Re much higher than Reynolds's own figures. Under normal engineering conditions, however, where disturbances are always present, transition occurs at values between 2000 and 4000.

There is, then, apparently no upper limit to the value of Re at which the change from laminar flow to turbulent flow occurs. There is, however, a definite lower limit. When Re is less than this lower limit any disturbances in the flow are damped out by the viscous forces, which are then predominant.

The technique using a stream of dye is not suitable for demonstrating the change from turbulent to laminar flow. Reynolds therefore employed another method to study the changes that occur as the velocity is reduced. It had been known for many years that the law governing the resistance to the flow of liquid along a pipe is different for the two kinds of flow. At the lower velocities, at which the flow is laminar, steady flow of a given liquid through a straight horizontal pipe requires, over a given length of the pipe, a difference of pressure directly proportional to the velocity. In turbulent flow, however, the pressure difference increases at a greater rate than the velocity.

Reynolds accordingly set up an apparatus of which the essential part is indicated in Fig. 5.2. The velocity of water through a uniform horizontal pipe was controlled by a valve at its downstream end, and for various values of the mean velocity, the drop in pressure over a given length of the pipe was measured by a differential manometer. The first manometer connection was situated at a great distance from the inlet to the pipe (at least 250 times its diameter) so that any disturbances from the inlet could be damped out and the flow pattern could settle down to its final form before the measuring section was reached. Figure 5.3 illustrates the result obtainable from such an experiment.

Here values of the drop in (piezometric) pressure Δp divided by the length l over which it occurs are plotted against values of the mean velocity on log scales. For the lower velocities, which correspond to laminar flow, a straight line is obtained with a slope of one. This result shows that in laminar flow the pressure drop per unit length is directly proportional to the velocity, as predicted by the theory of laminar flow.

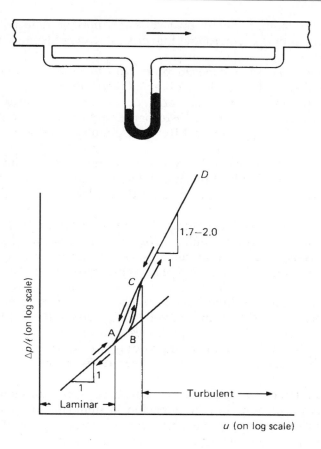

Fig. 5.2

Fig. 5.3

As the velocity is slowly increased, however, an abrupt increase in the pressure drop is found (point *B* on the diagram). This point in fact corresponds to the change from laminar to turbulent flow when a thin stream of dye would mix with the fluid. There is a region in which it is difficult to discern any simple law connecting the two variables, and then another line (*CD* on the diagram), which has a slope greater than the first. The slope of such lines as *CD* varies from test to test: for pipes with very smooth walls it may be as low as 1.7; as the roughness of the walls increases the slope approaches a maximum value of 2.0. The lines *CD* may be only approximately straight; that is, the slope may alter somewhat as *u* increases.

If, however, the velocity is carefully reduced, the previous route is not exactly retraced. The line *DC* is followed but, instead of the path *CB* being taken, the line *CA* is traced out. Then, from the point *A*, the line corresponding to laminar flow is followed.

These results clearly confirm that there are two critical velocities: one, corresponding to the change from laminar to turbulent flow, is known as the *upper* or *higher critical velocity* (point *B*); the other (point *A*) is known as the *lower critical velocity*, where the flow changes from turbulent to laminar. The critical velocities may be determined from the graph, and the corresponding critical Reynolds numbers then calculated.

Reynolds, and other experimenters who have taken great care to eliminate all disturbances, have obtained high values of the upper critical Reynolds number. In normal practice, however, extreme figures are never attainable because small disturbances are always present, and the usual value of the upper critical Reynolds number is between 2500 and 4000.

It is the lower critical value which is of much greater interest and importance because it is at this point that laminar flow becomes unstable. It is usual to refer to this lower critical value simply as the *critical Reynolds number*.

The experiments of Reynolds and more particularly the later, more detailed, ones of Ludwig Schiller (1882–1961) have shown that for very smooth, straight, uniform circular pipes this lower critical value of Reynolds number is about 2300. The value is slightly lower for pipes with the usual degree of roughness of the walls, and for ordinary purposes it is usual to take it as 2000. Curvature of the pipe axis also reduces the figure. If the pipe converges, the critical Reynolds number is higher, whereas divergence produces a lower value. (These values of critical Reynolds number apply only to Newtonian fluids. No satisfactory general criterion has yet been devised for non-Newtonian fluids.)

A calculation or two will help to illustrate the conditions under which the two types of flow may be expected in uniform, straight, circular pipes. First we consider the sort of flow perhaps most frequently encountered by the engineer, that of water at about 15 °C. At this temperature the kinematic viscosity of water is about $1.15 \, mm^2/s$. Since $Re = ud/\nu$, the velocity is given by $\nu \, Re/d$ and the critical velocity by $\nu(Re_{crit})/d$. Thus for a pipe 25 mm in diameter

$$u_{crit} = \frac{1.15 \times 2000}{25} \frac{mm^2}{s \, mm} = 99 \, mm/s$$

Such a low velocity is seldom of interest in practice. A critical velocity 10 times as great would require a diameter 10 times smaller, i.e. 2.5 mm. This, too, is a figure outside normal engineering practice. The velocity of water is generally far greater than the critical value and so the flow is normally assumed fully turbulent. This is not to say that laminar flow of water may not be found in laboratory experiments or other small-scale work. Indeed, in the testing of small-scale models of hydraulic structures, the presence of laminar flow that does not correspond to the flow occurring in the full-size prototype may constitute a considerable difficulty.

Water, however, has a low viscosity. Oil having a kinematic viscosity of, say, 200 times that of water would, in a pipe of 25 mm diameter, have a critical velocity of $200 \times 0.092 \, m/s = 18.4 \, m/s$. This velocity is far in excess of any that one would expect in practice, so it would be necessary to treat the flow of such oil as laminar. Even in a pipe of 250 mm diameter, a mean velocity of 1.84 m/s could be reached without the flow becoming turbulent.

By similar considerations we may see that the wholly laminar flow of air or steam in a pipe is not likely to be found often.

There is a great temptation to suppose that if the Reynolds number of flow in any pipe is 1999 the flow will certainly be entirely laminar, whereas if $Re = 2001$ the flow will certainly be turbulent. All that may definitely be said is that, irrespective of the length of the pipe, if the Reynolds number is less than the critical, an originally laminar flow will remain laminar, and if the flow at the entrance to the pipe is turbulent the flow will become purely laminar if the pipe is straight, reasonably smooth and sufficiently long and the Reynolds number remains less than the critical. On the other hand, if Re exceeds the critical value, an originally turbulent flow will remain turbulent, but if the flow is originally laminar it will tend to become turbulent if any disturbances are introduced. The higher the Reynolds number and the greater the disturbances, the shorter will be the distance the fluid has to travel before turbulence sets in.

5.7.1 Flows neither wholly laminar nor wholly turbulent

A Reynolds number greater than the critical indicates that laminar flow is unstable, and consequently that the tendency is to turbulence. The fluid may have to travel a length equal to 60 times the diameter of the pipe before the stable pattern of flow corresponding to the particular Reynolds number is established. The actual conditions of flow before that are influenced considerably by the state of the flow preceding the entry to the pipe, disturbances in approach and other characteristics of the entry, and to some extent by the roughness of the walls of the pipe.

In a short tube, therefore, an intermediate state may well exist in which part of the flow (that near the periphery) is laminar and the other part (near the centre) is turbulent. With decreasing Reynolds number the thickness of the laminar envelope increases, while the turbulent core decreases in size until it finally disappears. Thus in a short tube the change from turbulent to laminar flow occurs gradually rather than abruptly. The semi-turbulent state can exist over a wide range of Reynolds numbers, the effect of changes in Reynolds number being only to alter the relative thickness of the turbulent core and laminar envelope.

This explains the necessity, in the experiment described in Section 5.7, for providing a long length of straight pipe through which the fluid passes before entering the section under test. The flow characteristics and the laws governing laminar and turbulent flow hold only for 'fully developed' patterns of flow. In the region close to the inlet of the pipe, and in other places where the flow is disturbed (as, for example, at bends in the pipe or changes in its cross-sectional area), the pattern of flow may be notably different from that of 'fully developed' flow, whether laminar or turbulent, and, in consequence, the laws governing the pressure drop may also differ. These matters will be more fully discussed in later chapters.

Figure 5.3 shows that for values of Reynolds number greater than the critical value the pressure drop per unit length for a particular velocity is greater for turbulent flow than for laminar. This is hardly surprising. The drop in pressure as the fluid flows along represents a loss of mechanical energy. In laminar flow neighbouring layers of fluid remain distinct from one another and mechanical energy is required only to overcome the viscous stresses between layers moving at different velocities. In turbulent flow, however, there is a continuous interchange of fluid between different parts of the flow, and innumerable small variations of velocity both in magnitude and direction. Differences of velocity between adjacent particles of fluid abound, and the opportunities for viscosity to exercise its restraining hand are therefore greatly increased. The dissipation of energy in turbulent flow consequently occurs at a greater rate than in laminar flow, even at the same mean velocity.

This observation explains the phenomenon of *intermittent turbulence*. As the rate of dissipation of energy in turbulent flow is greater than in laminar flow at the same mean velocity, the pressure drop over the whole length of the pipe increases when part of the flow within it becomes turbulent. That is, the resistance to the flow increases and thus – since the overall pressure difference causing the flow is fixed by external conditions – the mean velocity is reduced. This reduction of the mean velocity may bring the Reynolds number down below the critical value again and so restore laminar flow. When, however, the turbulent fluid has escaped from the end of the pipe the resistance to flow drops to its former value and the mean velocity consequently rises again to a value slightly above the critical, and so the sequence is repeated.

This phenomenon of intermittent turbulence occurs more particularly in long, narrow pipes. Its effects are frequently noticed when apparatus similar to that depicted in Fig. 5.2 is used to study the transition from one type of flow to the other; when the Reynolds number of the flow is near the critical value, the levels in the manometer tubes tend to oscillate up and down and thus taking readings is rendered difficult. (Usually the manometer connections provide considerable 'damping', that is, resistance to motion of the fluid through them, and so the fluctuations of manometer levels do not follow faithfully the more rapid changes occurring in the pipe itself; the oscillations in the manometer are thus much slower.)

5.7.2 The more general occurrence of turbulence

We have considered in some detail the change from laminar to turbulent flow and vice versa for circular pipes. Similar changes between laminar and turbulent flow are found for other shapes of boundary, although it must be emphasized that the values of the critical Reynolds numbers are not the same. For flow between very wide parallel plane walls, for example, the lower critical Reynolds number is about 1000 (where the representative linear measurement is the distance between the planes and the representative velocity is the mean velocity). Such values have to be determined by experiment; in the present state of knowledge they cannot be calculated theoretically.

Fine-scale turbulence is made up of many small eddies in which energy is rapidly dissipated. *Large-scale turbulence* consists of large eddies – which usually break down into fine scale turbulence. Since all turbulence consists of fluctuations of velocity superimposed on the main velocity, the root mean square value of the fluctuations gives a quantitative measure of *intensity* of turbulence – which, in general, increases as Reynolds number increases. The frequency with which the fluctuations at a particular point change sign is an indication of the size of the eddies, and therefore of the scale of the turbulence.

FURTHER READING

Kline, S. J. *Similitude and Approximation Theory*, McGraw-Hill, New York (1965). Reprinted 1986.

Sedov, L. I. *Similarity and Dimensional Methods in Mechanics*, Academic Press (Infosearch), London (1959). Reprinted 1996.

On particular applications:

Clayton, B. R. and Bishop, R. E. D. *Mechanics of Marine Vehicles*, Spon, London (1982).

Keith, H. D. 'Simplified theory of ship waves', *Am. J. Phys.* **25**, 466–74 (1957).

Massey, B. S. *Measures in Science and Engineering*, Ellis Horwood, Chichester (1986).

Novak, P. and Čábelka, J. *Models in Hydraulic Engineering*, Pitman, London (1981).

Sharp, J. J. *Hydraulic Modelling*, Butterworths, London (1981).

Quantities, Units and Symbols (2nd edn), The Royal Society, London (1975).

PROBLEMS

5.1 A pipe of 40 mm bore conveys air at a mean velocity of 21.5 m/s. The density of the air is 1.225 kg/m^3 and its viscosity is 1.8×10^{-5} Pa s. Calculate that volume flow rate of water through the pipe which would correspond to the same value of friction factor f if the viscosity of water is 1.12×10^{-3} Pa s. Compare the drop in piezometric pressure per unit length of pipe in the two cases.

5.2 Derive an expression for the volume flow rate of a liquid (of viscosity μ, density ϱ and surface tension γ) over a V notch of given angle θ. Experiments show that for water flowing over a 60° V notch a useful practical formula is $Q = 0.762h^{2.47}$ for metre-second units. What limitation would you expect in the validity of this formula? Determine the head over a similar notch when a liquid with a kinematic viscosity 8 times that of water flows over it at the rate of 20 litres/s.

5.3 A disc of diameter D immersed in a fluid of density ϱ and viscosity μ has a constant rotational speed N. The power required to drive the disc is P. Show that $P = \varrho N^3 D^5 \phi$ $(\varrho ND^2/\mu)$. A disc 225 mm diameter rotating at 23 rev/s in water requires a driving torque of 1.1 N m. Calculate the corresponding speed and the torque required to drive a similar disc 675 mm diameter rotating in air. (Viscosities: air 1.86×10^{-5} Pa s; water 1.01×10^{-3} Pa s. Densities: air 1.20 kg/m³; water 1000 kg/m³.)

5.4 The flow through a closed, circular-sectioned pipe may be metered by measuring the speed of rotation of a propeller having its axis along the pipe centre-line. Derive a relation between the volume flow rate and the rotational speed of the propeller, in terms of the diameters of the pipe and the propeller and of the density and viscosity of the fluid. A propeller of 75 mm diameter, installed in a 150 mm pipe carrying water at 42.5 litres/s, was found to rotate at 20.7 rev/s. If a geometrically similar propeller of 375 mm diameter rotates at 10.9 rev/s in air flow through a pipe of 750 mm diameter, estimate the volume flow rate of the air. The density of the air is 1.28 kg/m³ and its viscosity 1.93×10^{-5} Pa s. The viscosity of water is 1.145×10^{-3} Pa s.

5.5 A torpedo-shaped object 900 mm diameter is to move in air at 60 m/s and its drag is to be estimated from tests in water on a half-scale model. Determine the necessary speed of the model and the drag of the full-scale object if that of the model is 1140 N. (Fluid properties as in Problem 5.3.)

5.6 What types of force (acting on particles of fluid) would you expect to influence the torque needed to operate the rudder of a deeply submerged mini-submarine? To investigate the operation of such a rudder, tests are conducted on a quarter-scale model in a fresh-water tunnel. If, for the relevant temperatures, the density of sea water is 2.5% greater than that of fresh water, and the absolute viscosity 7% greater, what velocity should be used in the water tunnel to correspond to a velocity of 3.5 m/s for the prototype submarine? If the measured torque on the model rudder is 20.6 N m, what would be the corresponding torque on the full-size rudder?

5.7 Show that, for flow governed only by gravity, inertia and pressure forces, the ratio of volume flow rates in two dynamically similar systems equals the 5/2 power of the length ratio.

5.8 The drag on a stationary hemispherical shell with its open, concave side towards an oncoming airstream is to be investigated by experiments on a half-scale model in water. For a steady air velocity of 30 m/s determine the corresponding velocity of the water relative to the model, and the drag on the prototype shell if that on the model is 152 N. (Fluid properties as in Problem 5.3.)

5.9 The flow rate over a spillway is 120 m³/s. What is the maximum length scale factor for a dynamically similar model if a flow rate of 0.75 m³/s is available in the laboratory? On a part of such a model a force of 2.8 N is measured. What is the corresponding force on the prototype spillway? (Viscosity and surface tension effects are here negligible.)

5.10 An aircraft is to fly at a height of 9 km (where the temperature and pressure are −45 °C and 30.2 kPa respectively) at 400 m/s. A 1/20th-scale model is tested in a pressurized wind-tunnel in which the air is at 15 °C. For complete dynamic similarity what pressure and velocity should be used in the wind-tunnel? (For air at T K, $\mu \propto T^{3/2}/(T + 117)$.)

5.11 In a 1/100th-scale model of a harbour what length of time should correspond to the prototype tidal period of 12.4 hours?

5.12 Cavitation is expected in an overflow siphon where the head is −7 m, the water temperature 10 °C and the rate of flow 7 m³/s. The conditions are to be reproduced on a 1/12th-scale model operating in a vacuum chamber with water at 20 °C. If viscous and surface tension effects may be neglected, calculate the pressure required in the vacuum chamber and the rate of flow in the model. (Hint: In the absence of friction the velocity in the siphon is determined only by the overall head difference, and the pressure at a point by the fluid velocity and the elevation of the point. Saturated vapour pressure of water = 1230 Pa at 10 °C; 2340 Pa at 20 °C.)

5.13 A ship has a length of 100 m and a wetted area of 1200 m². A model of this ship of length 4 m is towed through fresh water at 1.5 m/s and has a total resistance of 15.5 N. For the model the skin resistance/wetted area is 14.5 N/m² at 3 m/s, and the skin resistance is proportional to (velocity)^{1.9}. The prototype ship in sea water has a skin resistance of 43 N/m² at 3 m/s and a velocity exponent of 1.85. Calculate the speed and total resistance of the prototype ship in conditions corresponding to the model speed of 1.5 m/s. The relative density of sea water is 1.026.

6 Laminar flow between solid boundaries

6.1 INTRODUCTION

Laminar flow may occur in many situations. Its distinguishing features, however, are always the same: individual particles of fluid follow paths that do not cross those of neighbouring particles. There is nevertheless a velocity gradient across the flow, and so laminar flow is not normally found except in the neighbourhood of a solid boundary, the retarding effect of which causes the transverse velocity gradient. Laminar flow occurs at velocities low enough for forces due to viscosity to predominate over inertia forces, and thus, if any individual particle attempts to stray from its prescribed path, viscosity firmly restrains it, and the orderly procession of particles continues.

We recall from Section 1.6.1 that viscous stresses are set up whenever there is relative movement between adjacent particles of fluid and that these stresses tend to eliminate the relative movement. The basic law of viscous resistance was described by Newton in 1687:

$$\tau = \mu \frac{\partial u}{\partial y} \tag{6.1}$$

Here $\partial u / \partial y$ is the rate at which the velocity u (in straight and parallel flow) increases with coordinate y perpendicular to the velocity, μ represents the coefficient of (absolute) viscosity and τ the resulting shear stress on a surface perpendicular to, and facing the direction of increase of y. The partial derivative $\partial u / \partial y$ is used because u may vary not only with y but also in other directions.

We now consider a number of cases of laminar flow which are of particular interest.

6.2 STEADY LAMINAR FLOW IN CIRCULAR PIPES: THE HAGEN–POISEUILLE LAW

The law governing laminar flow in circular pipes was one of the first examples to be studied. About 1840, experimental investigations of flow in straight pipes of circular cross-section were carried out independently by two men. The first results, published in 1839, were the work of the

German engineer G. H. L. Hagen* (1797–1884). He had experimented
with the flow of water through small brass tubes, and his figures showed
that the loss of head experienced by the water as it flowed through
a given length of the tube was directly proportional to the rate of flow,
and inversely proportional to the fourth power of the diameter of the
tube.

At the same time a French physician, J. L. M. Poiseuille[†] (1799–1869)
was engaged in a long series of careful and accurate experiments with
the object of studying the flow of blood through veins. In his experiments he used water in fine glass capillary tubes and arrived at the same
conclusions as Hagen (although the reason for an apparent discrepancy
found with very short tubes escaped him). His first results were made
public in 1840. Subsequently the law governing laminar flow in circular
pipes was derived theoretically. It is of interest to note that, invaluable
though Poiseuille's results have been in pointing the way to the theory
of laminar flow in circular tubes, they are not really applicable to the
flow of blood in veins; for one thing the walls of veins are not rigid, and
also blood is not a Newtonian fluid, that is, it does not have a constant
viscosity, even at a fixed temperature.

Figure 6.1 shows, on the left, a side view of a straight pipe of constant
internal radius R. On the right is shown the circular cross-section.
We assume that the part of the pipe considered is far enough from
the inlet for conditions to have become settled. In laminar flow, the
paths of individual particles of fluid do not cross, and so the pattern
of flow may be imagined as a number of thin, concentric cylinders
which slide over one another like the tubes of a pocket telescope. The
diagram shows a cylinder, of radius r, moving from left to right
with velocity u inside a slightly larger cylinder, of radius $r + \delta r$, moving
in the same direction with velocity $u + \delta u$. (δu may, of course, be
negative.)

The difference of velocity between the two cylinders brings viscosity
into play, and thus there is a stress along the interface between the two
layers of fluid so as to oppose the relative motion. The force balance on
the cylinder of radius r and length δx is given by $p^* \pi r^2 - (p^* + \delta p^*)\pi r^2
+ \tau 2\pi r \delta x = 0$. Hence in the limit $\delta x \to 0$, for *steady flow*, the shear stress
τ at radius r is given by

$$\tau = \frac{r}{2}\frac{\mathrm{d}p^*}{\mathrm{d}x}$$

where $p^* = p + \rho g z$, the piezometric pressure. In laminar flow the stress
τ is due entirely to viscous action and so is given by eqn 6.1, in which
r takes the place of y:

$$\tau = \mu\frac{\partial u}{\partial r}$$

*pronounced *Har'-gen.*
[†] pronounced *Pwah-zoy'-yuh.*

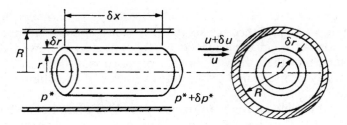

Fig. 6.1

As the flow is steady and 'fully developed' the velocity varies only with the radius, and so in place of the partial derivative $\partial u/\partial r$ we may write the full derivative $\mathrm{d}u/\mathrm{d}r$. Hence

$$\mu \frac{\mathrm{d}u}{\mathrm{d}r} = \frac{r}{2} \frac{\mathrm{d}p^*}{\mathrm{d}x} \quad \text{i.e.} \quad \frac{\mathrm{d}u}{\mathrm{d}r} = \frac{r}{2\mu} \frac{\mathrm{d}p^*}{\mathrm{d}x} \tag{6.2}$$

If μ is constant, integration with respect to r gives

$$u = \frac{r^2}{4\mu} \frac{\mathrm{d}p^*}{\mathrm{d}x} + A \tag{6.3}$$

Now the constant of integration A is determined from the boundary conditions. As there is no slip at the wall of the pipe, $u = 0$ where $r = R$. Consequently $A = -(R^2/4\mu)(\mathrm{d}p^*/\mathrm{d}x)$, so the velocity at any point is given by

$$u = -\frac{1}{4\mu}\left(\frac{\mathrm{d}p^*}{\mathrm{d}x}\right)\left(R^2 - r^2\right) \tag{6.4}$$

(Since p^* falls in the direction of flow, $\mathrm{d}p^*/\mathrm{d}x$ is negative.)

From eqn 6.4 it is clear that the maximum velocity occurs at the centre of the pipe, where $r = 0$. The distribution of velocity over the cross-section may be represented graphically by plotting u against r as in Fig. 6.2. The shape of the graph is parabolic; in other words, the velocity 'profile' has the shape of a paraboloid of revolution.

Equation 6.4 by itself is of limited application. Of far more interest than the velocity at a particular point in the pipe is the total discharge through it. Now the discharge δQ through the annular space between radii r and $r + \delta r$ is (velocity × perpendicular area) $= u\,2\pi r\,\delta r$. Using eqn 6.4 this may be written:

$$\delta Q = -\frac{1}{4\mu}\left(\frac{\mathrm{d}p^*}{\mathrm{d}x}\right)\left(R^2 - r^2\right)2\pi r\,\delta r = -\frac{\pi}{2\mu}\left(\frac{\mathrm{d}p^*}{\mathrm{d}x}\right)\left(R^2 r - r^3\right)\delta r \tag{6.5}$$

The discharge through the entire cross-section is therefore

$$Q = -\frac{\pi}{2\mu}\left(\frac{\mathrm{d}p^*}{\mathrm{d}x}\right)\int_0^R\left(R^2 r - r^3\right)\mathrm{d}r$$

$$= -\frac{\pi}{2\mu}\left(\frac{\mathrm{d}p^*}{\mathrm{d}x}\right)\left(R^2\frac{R^2}{2} - \frac{R^4}{4}\right) = -\frac{\pi R^4}{8\mu}\left(\frac{\mathrm{d}p^*}{\mathrm{d}x}\right) \tag{6.6}$$

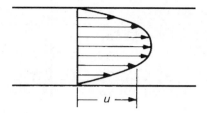

Fig. 6.2

This equation is known usually as Poiseuille's formula and sometimes as the Hagen–Poiseuille formula, although neither Hagen nor Poiseuille derived an equation in this form. (The first complete theoretical derivation appears to have been made by G. H. Wiedermann (1826–99) in 1856.) For a length l of the pipe over which the piezometric pressure drops from p_1^* to p_2^* the equation may be written

$$Q = \frac{\pi R^4}{8\mu l}\left(p_1^* - p_2^*\right) \tag{6.7}$$

An expression in terms of diameter d rather than radius is often more suitable, and eqns 6.6 and 6.7 may be written

$$Q = -\frac{\pi d^4}{128\mu}\frac{dp^*}{dx} = \frac{\pi d^4}{128\mu l}\left(p_1^* - p_2^*\right) \tag{6.8}$$

Equation 6.6 applies to both incompressible and compressible fluids since it concerns only an infinitesimal length δx of the pipe and any change of density of the fluid in this distance would be negligible. When a compressible fluid flows through an appreciable length of pipe, however, the density changes as the pressure changes and so, although the total *mass* flow rate $m = \rho Q$ is constant, the volume flow rate Q is not constant. (The laminar flow of compressible fluids will be discussed in Section 11.10.3.) Thus eqn 6.7 is strictly applicable only to the laminar flow of constant-density fluids; it may, however, be used for compressible fluids if the change of density is negligible.

A further restriction on these equations is that they apply only to conditions in which laminar flow is 'fully developed'. From the entrance to the pipe the fluid has to traverse a certain distance before the parabolic pattern of velocity distribution depicted in Fig. 6.2 is established.

If the fluid enters the pipe from a much larger section, for example, the velocity is at first practically uniform (as at A in Fig. 6.3). The retarding effect of the walls, where the velocity must always be zero, at once operates, so that at B more of the layers nearer the walls are slowed down. As the cross-sectional area of the pipe is constant the mean velocity over the whole section must remain unchanged so as to satisfy the equation of continuity; thus, to compensate for the reduction in velocity experienced by the fluid near the walls, that near the centre must be accelerated. Not until the position C is reached, however, is the

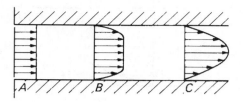

Fig. 6.3

full parabolic distribution established. It is this final, 'fully developed' form to which the equations we have derived apply.

The parabolic velocity profile is theoretically reached only after an infinite distance; but it is usual to regard the so-called 'entry length' as the distance along the pipe to the point at which the maximum velocity is only 1% different from the final value. The way in which the flow thus 'settles down' in the entry length is discussed in greater detail in Section 7.9. It may, however, be said here that except in short pipes (that is, with a length up to about 200 times the diameter) the different conditions at the inlet to the pipe do not significantly affect the change of pressure over the whole length. A deviation from the parabolic profile is also found at the outlet end of the pipe but, except for extremely short pipes, the effect of this modification on the value of $(p_1^* - p_2^*)$ is negligible.

From eqn 6.6 the mean velocity of the fluid may be calculated.

$$\text{Mean velocity} = \frac{Q}{\text{Area}} = -\frac{\pi R^4}{8\mu}\frac{dp^*}{dx}\bigg/\pi R^2 = -\frac{R^2}{8\mu}\frac{dp^*}{dx} \quad (6.9)$$

From eqn 6.4 we see that the maximum velocity occurs in the centre of the pipe, where $r = 0$.

$$\therefore \ u_{\max} = -\frac{dp^*}{dx}\left(\frac{R^2}{4\mu}\right) \quad (6.10)$$

Thus the mean velocity is here $\frac{1}{2}u_{\max}$.

The Hagen–Poiseuille formula has been amply verified by experiment and it is interesting to note that this agreement between theory and experimental results is perhaps the principal justification for the assumption that a fluid continuum does not 'slip' past a solid boundary. The integration constant in eqn 6.3 had to be determined from the conditions at the wall: if a velocity other than zero had been assumed, say u_0, the value of Q given by eqn 6.6 would be increased by an amount $\pi R^2 u_0$. It is only when u_0 is zero that agreement between theory and experiment is obtained. Newton's hypothesis (eqn 6.1) is also of course vindicated by the agreement with experimental results.

Another result which the Hagen–Poiseuille formula verifies is that in laminar flow the drop in piezometric pressure is proportional directly to the mean velocity.

Moreover, the formula is completely determined by our analysis and does not involve any additional coefficients that have to be obtained experimentally – or estimated – for a particular pipe. Thus we should expect moderate roughness of the walls of the pipe not to affect laminar flow, and this conclusion too is confirmed by experiment.

The Hagen–Poiseuille formula was developed on the assumption that the centre-line of the pipe was straight. Slight curvature of the centre-line, in other words, a radius of curvature large compared with the radius of the pipe, does not appreciably affect the flow through the pipe. For smaller radii of curvature, however, the flow is not accurately described by the Hagen–Poiseuille formula.

Example 6.1 Oil, of relative density 0.83 and absolute viscosity $0.08\,\mathrm{kg\,m^{-1}s^{-1}}$, passes through a circular pipe of 12 mm diameter with a mean velocity of $2.3\,\mathrm{m\,s^{-1}}$. Determine:
(a) the Reynolds number
(b) the maximum velocity
(c) the volumetric flow rate
(d) the change of pressure per unit length of pipe.

Solution
Denote the mean velocity by $\bar{u}$.

(a) $Re = \dfrac{\rho \bar{u} d}{\mu} = \dfrac{0.83 \times 1000\,\mathrm{kg/m^3} \times 2.3\,\mathrm{m/s} \times 12\,\mathrm{mm}}{0.08\,\mathrm{kg/(m\,s)} \times 1000\,\mathrm{mm/m}}$

$\qquad = 286$

This value of Reynolds number is well within the laminar range, so the relations for laminar flow may be used throughout the remainder of the question.

(b) $u_{\max} = 2\bar{u} = 2 \times 2.3\,\mathrm{m/s} = 4.6\,\mathrm{m\,s^{-1}}$

(c) $Q = \dfrac{\pi}{4}d^2 \bar{u} = \dfrac{\pi}{4}\left(\dfrac{12}{10^3}\,\mathrm{m}\right)^2 \times 2.3\,\mathrm{m/s} = 260 \times 10^{-6}\,\mathrm{m^3 s^{-1}}$

(d) $\dfrac{dp*}{dx} = -\dfrac{128 Q \mu}{\pi d^4}$

$\qquad = \dfrac{-128 \times \left(260 \times 10^{-6}\right)\mathrm{m^3/s} \times 0.08\,\mathrm{kg/(m\,s)} \times \left(10^3\,\mathrm{mm/m}\right)^4}{\pi \times 12^4\,\mathrm{mm^4}}$

$\qquad = -40.9 \times 10^3\,\mathrm{kg\,m^{-2}\,s^{-2}} = -40.9 \times 10^3\,\mathrm{Pa/m}$

The negative pressure gradient indicates that the pressure decreases with distance along the pipe axis. □

6.2.1 Laminar flow of a non-Newtonian liquid in a circular pipe

Results corresponding to those in Section 6.2 may be obtained for a non-Newtonian liquid. We again consider steady flow and thus suppose that the rate of shear is a function of τ only, say $f(\tau)$. Then, for fully-developed flow in a circular pipe,

$$\frac{\mathrm{d}u}{\mathrm{d}r} = f(\tau) \tag{6.11}$$

The expression for the total discharge may be integrated by parts:

$$Q = \int_0^R u\, 2\pi r\, \mathrm{d}r = \pi \int_{r=0}^{r=R} u\, \mathrm{d}(r^2) = \pi\left[ur^2 - \int r^2\, \mathrm{d}u\right]_{r=0}^{r=R}$$

$$= -\pi \int_{u_{\max}}^{0} r^2\, \mathrm{d}u$$

since, with no slip at the walls, $u = 0$ when $r = R$. Define $\tau = \tau_0$, when $r = R$. It will be shown, eqn 7.11, that $r = R\tau/\tau_0$. Substituting $\mathrm{d}u = f(\tau)\mathrm{d}r$ from eqn 6.11 then gives

$$Q = -\pi \int_0^R r^2 f(\tau)\mathrm{d}r = -\frac{\pi R^3}{\tau_0^3} \int_0^{\tau_0} \tau^2 f(\tau)\mathrm{d}\tau \tag{6.12}$$

Hence $Q/\pi R^3$ is a function of τ_0 and since $\tau_0 = (R/2)(\mathrm{d}p^*/\mathrm{d}x)$, eqn 7.9, the relation between $Q/\pi R^3$ and τ_0 may be determined experimentally.

For many pseudo-plastic and dilatant liquids the relation between the magnitudes of shear stress and rate of shear (see Fig. 1.7) may be represented to an acceptable degree of approximation by a power law

$$|\tau| = k\left(A|\partial u/\partial y|\right)^n$$

where k, A and n are constants (though possibly dependent on pressure and temperature). For a pseudo-plastic liquid $n < 1$; for a dilatant liquid $n > 1$. Noting that τ_0 is negative because $\mathrm{d}p^*/\mathrm{d}x$ is negative and substituting

$$\left|\frac{\mathrm{d}u}{\mathrm{d}y}\right| = f(\tau) = \frac{1}{A}\left(\frac{|\tau|}{k}\right)^{1/n}$$

into eqn 6.12, we obtain

$$Q = \frac{\pi R^3}{|\tau_0|^3} \int_0^{\tau_0} \frac{|\tau|^{(2n+1)/n}}{A k^{1/n}}\, \mathrm{d}\tau$$

$$= \frac{\pi R^3}{|\tau_0|^3 A k^{1/n}}\left(\frac{n}{3n+1}\right)\left[|\tau|^{(3n+1)/n}\right]_0^{\tau_0}$$

$$= \frac{n\pi R^3}{A(3n+1)}\left(\frac{|\tau_0|}{k}\right)^{1/n} = \frac{n\pi R^3}{A(3n+1)}\left(-\frac{R}{2k}\frac{\mathrm{d}p^*}{\mathrm{d}x}\right)^{1/n} \tag{6.13}$$

The constants in the power-law relation may thus be determined from measurements of Q and dp^*/dx for a circular pipe.

For a Newtonian fluid $n = 1$ and eqn 6.13 then reduces to eqn 6.6.

For a Bingham plastic (see Fig. 1.7) the relation between shear stress and rate of shear is

$$\tau = \tau_y + \mu_p \frac{\partial u}{\partial y} \quad \text{when } |\tau| > |\tau_y|$$

and

$$\frac{\partial u}{\partial y} = 0 \quad \text{when } |\tau| < |\tau_y|$$

Here τ_y represents the yield stress of the material and μ_p the slope of the graph of stress against rate of shear when $|\tau| > |\tau_y|$. Therefore

$$\partial u / \partial y = f(\tau) = (\tau - \tau_y)/\mu_p \quad \text{for } |\tau| > |\tau_y|$$

and

$$f(\tau) = 0 \quad \text{for } |\tau| < |\tau_y|$$

In the central region of the pipe, for all radii where the shear stress is smaller in magnitude than the yield value, the substance moves as a solid plug. Splitting the integration range of eqn 6.12 into two parts and then making the appropriate substitutions for $f(\tau)$ we obtain

$$Q = -\frac{\pi R^3}{\tau_0^3} \int_0^{\tau_y} \tau^2 f(\tau) d\tau - \frac{\pi R^3}{\tau_0^3} \int_{\tau_y}^{\tau_0} \tau^2 f(\tau) d\tau$$

$$= 0 - \frac{\pi R^3}{\mu_p \tau_0^3} \int_{\tau_y}^{\tau_0} \tau^2 (\tau - \tau_y) d\tau = -\frac{\pi R^3}{\mu_p \tau_0^3} \left[\frac{\tau^4}{4} - \frac{\tau^3 \tau_y}{3} \right]_{\tau_y}^{\tau_0}$$

$$= -\frac{\pi R^3}{\mu_p \tau_0^3} \left[\frac{\tau_0^4}{4} - \frac{\tau_0^3 \tau_y}{3} + \frac{\tau_y^4}{12} \right]$$

$$= -\frac{\pi R^3 \tau_0}{\mu_p} \left[\frac{1}{4} - \frac{1}{3} \left(\frac{\tau_y}{\tau_0} \right) + \frac{1}{12} \left(\frac{\tau_y}{\tau_0} \right)^4 \right] \quad (6.14)$$

The substitution $\tau_0 = (R/2)(dp^*/dx)$ may again be made and the equation is seen to reduce to that for a Newtonian fluid when $\tau_y = 0$.

A solution for Q is obtainable directly from eqn 6.14, but when Q is known and τ_0 (and hence dp^*/dx) is to be determined we are faced with a fourth-order equation. However, the equation may be put in the form

$$m = B - \frac{1}{3m^3} \quad (6.15)$$

where

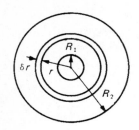

Fig. 6.4

$$m = \tau_0/\tau_y \quad \text{and} \quad B = \frac{4Q\mu_p}{\pi R^3 |\tau_y|} + \frac{4}{3}$$

With $m = B$ as a first approximation in the right-hand side of eqn 6.15, a value of m may be calculated which is then used for a second approximation, and so on. Sufficient accuracy is usually attained after three or four such steps.

6.3 STEADY LAMINAR FLOW THROUGH AN ANNULUS

The analysis developed for steady laminar flow in circular pipes can be readily extended to flow through an annulus of inner radius R_1 and outer radius R_2.

Between the concentric boundary surfaces of radii R_1 and R_2 (see Fig. 6.4) an annular shell of thickness δr is considered, and the axial force on this, due to the difference of piezometric pressure, is equated to the difference between the resisting viscous forces on the inner and outer surfaces of the shell. The subsequent development follows the pattern already used for flow in a circular pipe: the integration constants are determined from the boundary conditions, and then the discharge through an elemental ring can be integrated to give the total discharge. Thus the result

$$Q = \frac{\pi}{8\mu l}\left(p_1^* - p_2^*\right)\left(R_2^2 - R_1^2\right)\left\{R_2^2 + R_1^2 - \frac{R_2^2 - R_1^2}{\ln(R_2/R_1)}\right\}$$

is obtained. It will be seen, however, that as the value of R_1 approaches that of R_2 the last term in the bracket, $(R_2^2 - R_1^2)/\ln(R_2/R_1)$, becomes difficult to determine accurately (and the simpler formula of eqn 6.22 is then far better).

6.4 STEADY LAMINAR FLOW BETWEEN PARALLEL PLANES

Another example of laminar flow that may be studied quite simply is that in which fluid is forced to flow, at a velocity less than the critical velocity, between parallel plane solid boundaries.

Figure 6.5 represents a cross-section viewed in a direction perpendicular to the flow. The boundary planes are assumed to extend to a great distance, both in the direction left to right across the page and in that perpendicular to the page. This assumption is necessary in order that 'end effects' may be neglected; in other words, the edges of the planes are so far distant from the portion of the fluid being considered that they have no effect on its behaviour.

The flow is caused by a difference of piezometric pressure between the two ends of the system. As the flow is laminar, there is no movement

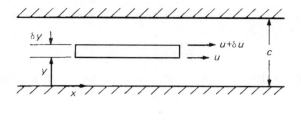

Fig. 6.5

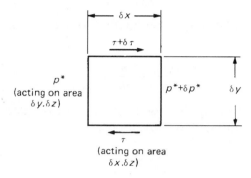

Fig. 6.6

of fluid in any direction perpendicular to the flow, and thus p^* varies only in the direction of flow.

As the origin of coordinates we may select a point on the lower plane in Fig. 6.5 and take the x axis along the plane parallel to the flow, and the y axis perpendicular to the plane. Let the distance separating the planes be c.

The requirement of no slip at each boundary produces a variation of velocity in the y direction. Viscous stresses are set up, and these may be related to the forces due to the difference of piezometric pressure by considering a small rectangular element of the fluid, with sides parallel to the coordinate axes, as shown in Fig. 6.5. Let the lower face of the element be a distance y from the lower plane and here let the velocity be u. At the upper face of the element, a distance $y + \delta y$ from the lower plane, the velocity is $u + \delta u$.

If δu is positive, the faster-moving fluid just above the element exerts a forward force on the upper face. Similarly, the slower-moving fluid adjacent to the lower face tends to retard the element. Thus there are stresses of magnitude τ on the lower face and $\tau + \delta \tau$, say, on the upper face in the directions shown in Fig. 6.6. (The indicated directions of the stresses follow the convention mentioned in Section 1.6.1 by which a stress in the x direction acts on the surface facing the direction of increase of y.) Let the piezometric pressure be p^* at the left-hand end face, and $p^* + \delta p^*$ at the right-hand end face.

Then, if the width of the element in the direction perpendicular to the page is δz, the total force acting on the element towards the right is

$$\{p^* - (p^* + \delta p^*)\}\delta y\, \delta z + \{(\tau + \delta \tau) - \tau\}\delta x\, \delta z$$

But for steady, fully developed flow there is no acceleration and so this total force must be zero.

$$\therefore\ -\delta p^* \, \delta y + \delta\tau \, \delta x = 0$$

Dividing by $\delta x \, \delta y$ and proceeding to the limit $\delta y \rightarrow 0$, we get

$$\frac{\delta p^*}{\delta x} = \frac{\partial \tau}{\partial y} \tag{6.16}$$

For laminar flow of a Newtonian fluid the stress $\tau = \mu\, \partial u/\partial y$. Hence eqn 6.16 becomes

$$\frac{\delta p^*}{\delta x} = \frac{\partial}{\partial y}\left(\mu \frac{\partial u}{\partial y}\right) \tag{6.17}$$

As p^* nowhere varies in the y direction, $\delta p^*/\delta x$ is independent of y and eqn 6.17 may be integrated with respect to y:

$$\frac{\delta p^*}{\delta x} y = \mu \frac{\partial u}{\partial y} + A$$

If μ is constant, a further integration with respect to y gives

$$\left(\frac{\delta p^*}{\delta x}\right)\frac{y^2}{2} = \mu u + Ay + B \tag{6.18}$$

Since the portion of fluid studied is very far from the edges of the planes, A and B are constants, independent of both x and z.

To determine these constants the boundary conditions must be considered. If both the planes are stationary the velocity of the fluid in contact with each must be zero so as to meet the requirement of no slip. Thus $u = 0$ at the lower plane where $y = 0$, and substituting these values in eqn 6.18 gives $B = 0$. Further, $u = 0$ at the upper plane where $y = c$ and the substitution of these values gives

$$A = \left(\frac{\delta p^*}{\delta x}\right)\frac{c}{2}$$

Inserting the values of A and B in eqn 6.18 and rearranging gives the value of u at any distance y from the lower plane:

$$u = \frac{1}{2\mu}\left(\frac{\delta p^*}{\delta x}\right)(y^2 - cy) \tag{6.19}$$

As shown in Fig. 6.7, the velocity profile is in the form of a parabola with its vertex (corresponding to the maximum velocity) mid-way between the planes as is to be expected from symmetry. Putting $y = c/2$ in eqn 6.19 gives the maximum velocity as $-(c^2/8\mu)(\delta p^*/\delta x)$, the minus sign indicating, not that the fluid flows backwards, but that $(\delta p^*/\delta x)$ is negative, that is, p^* decreases in the direction of flow.

The total volume flow rate between the two planes may readily be calculated. Consider an elemental strip, of thickness δy and fixed

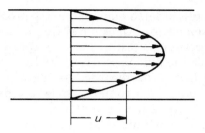

Fig. 6.7

breadth b, perpendicular to the page. The breadth b is assumed very large so that 'end effects' associated with it are negligible. In other words, there is a region, at each end of the strip, in which the velocity u may be somewhat different from the velocity in the centre; if b is sufficiently large this region is small compared with the total breadth, and the velocity u may then fairly be taken as the average velocity across the breadth. With this proviso, the discharge through the strip is $ub\ \delta y$ and the total discharge is:

$$Q = \int_0^c ub\ \mathrm{d}y = b \int_0^c \frac{1}{2\mu}\left(\frac{\delta p^*}{\delta x}\right)(y^2 - cy)\mathrm{d}y$$

$$= \frac{b}{2\mu}\left(\frac{\delta p^*}{\delta x}\right)\left[\frac{y^3}{3} - \frac{cy^2}{2}\right]_0^c$$

$$= -\frac{bc^3}{12\mu}\left(\frac{\delta p^*}{\delta x}\right) \qquad (6.20)$$

Dividing this result by the area of the cross-section bc gives the mean velocity as $-c^2\,(\delta p^*/\delta x)/12\mu$ which is two-thirds of the maximum value.

Equation 6.20 shows that, provided the width b and the density and viscosity of the fluid are constant, the term $\delta p^*/\delta x$ is independent of x. Thus where the piezometric pressure changes from p_1^* to p_2^* over a finite length l in the direction of flow $-\delta p^*/\delta x$ may give place to $-(p_2^* - p_1^*)/l = (p_1^* - p_2^*)/l$ and eqn 6.20 then becomes

$$Q = \frac{\left(p_1^* - p_2^*\right)bc^3}{12\mu l} \qquad (6.21)$$

(Corresponding results for non-Newtonian liquids may be derived by methods similar to those used in Section 6.2.1.)

The formula applies strictly only to conditions in which the flow is 'fully developed'. In practice, just as for laminar flow in circular tubes, there is an 'entry length', and also a small region near the outlet end of the passage, in which the pattern of flow differs from that described by the formula. If the length of the flow passage is small the effect of this modified flow on the overall change of piezometric pressure $(p_1^* - p_2^*)$ may be significant.

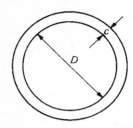

Fig. 6.8

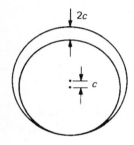

Fig. 6.9

Equation 6.21 is often called on in discussing problems of the leakage of a fluid through small gaps. An example frequently encountered is the leakage occurring between a piston and a concentric cylinder. Here the space through which the fluid passes is in the form of a narrow annulus (see Fig. 6.8) and it may be regarded as the space between parallel planes which have been bent round a circle, the breadth b previously considered having become the circumference πD. This supposition of planes bent round is legitimate if the clearance c is considerably less than the diameter D so that there is a negligible difference between the inner and outer circumferences.

The requirement that the breadth b should be so large that 'end effects' are negligible is here happily met. As b now represents the circumference of a circle there are no ends and thus no end effects. (Nevertheless there may still be 'end effects' in the direction of l.) Substituting πD for b in eqn 6.21 we get

$$Q = \left(p_1^* - p_2^*\right)\pi D c^3 \big/ 12\mu l \qquad (6.22)$$

This result, it should be emphasized, applies to situations where c is small compared with D. If this is not so, the analysis for flow through an annulus can be used.

The importance of the concentricity of the piston and the cylinder should not be overlooked, for these formulae apply to those instances in which the clearance c is constant. Even a slight degree of eccentricity between the two boundaries can affect the rate of flow considerably. Indeed it may be shown that in the extreme case depicted in Fig. 6.9, where there is a point of contact between the two surfaces, the flow rate is about 2.5 times the value when the same two surfaces are concentric with a constant clearance c.

This result explains the difficulty of making a reliable control device in which the flow rate is determined by the position of an adjustable tapered needle: a small amount of sideways movement of the needle may cause a considerable alteration in the flow rate.

Example 6.2 Two stationary, parallel, concentric discs of external radius R_2 are a distance c apart. Oil is supplied at a gauge pressure p_1^* from a central source of radius R_1 in the lower disc. From there it spreads radially outwards between the two discs and escapes to atmosphere. Assume the flow is laminar.

(a) Starting from the equation for flow between fixed surfaces

$$u = \frac{1}{2\mu}\left(\frac{\mathrm{d}p^*}{\mathrm{d}x}\right)\left(y^2 - cy\right)$$

show that the pressure distribution gives rise to a vertical force F on the upper disc given by

$$F = \pi p_1^* \left[R_1^2 + 2\int_{R_1}^{R_2} x \left(1 - \frac{\log_e(x/R_1)}{\log_e(R_2/R_1)} \right) dx \right]$$

(b) Calculate the rate at which oil of viscosity $0.6\,\mathrm{kg\,m^{-1}s^{-1}}$ must be supplied to maintain a pressure p_1^* of $15\,\mathrm{kPa}$ when $R_2/R_1 = 6$ and the clearance c is $1\,\mathrm{mm}$.

Solution

(a)

$$Q = \int_0^c 2\pi x(u)\,dy = \int_0^c 2\pi x \frac{1}{2\mu}\left(\frac{dp^*}{dx}\right)(y^2 - cy)\,dy = -\frac{\pi x c^3}{6\mu}\left(\frac{dp^*}{dx}\right)$$

Hence

$$dp^* = -\frac{6\mu Q}{\pi c^3}\frac{dx}{x}$$

Integrating with respect to x:

$$p^* = -\frac{6\mu Q}{\pi c^3}\log_e x + A$$

The boundary conditions are

$$p^* = p_1^*, \quad x = R_1; \quad p^* = 0, \quad x = R_2$$

So

$$A = p_1^* + \frac{6\mu Q}{\pi c^3}\log_e R_1 \quad \text{and} \quad A = \frac{6\mu Q}{\pi c^3}\log_e R_2$$

Combining these expressions we obtain

$$p^* - p_1^* = -\frac{6\mu Q}{\pi c^3}\log_e \frac{x}{R_1} \quad \text{and} \quad p_1^* = \frac{6\mu Q}{\pi c^3}\log_e \frac{R_2}{R_1}$$

Eliminating $(6\mu Q)/(\pi c^3)$ between these two relations produces

$$p^* = p_1^* - p_1^* \frac{\log_e(x/R_1)}{\log_e(R_2/R_1)}$$

$$F = \pi R_1^2 p_1^* + \int_{R_1}^{R_2} 2\pi x p^*\,dx$$

$$= \pi R_1^2 p_1^* + 2\pi \int_{R_1}^{R_2} \left(p_1^* - p_1^* \frac{\log_e(x/R_1)}{\log_e(R_2/R_1)} \right) x\,dx$$

$$= \pi p_1^* \left[R_1^2 + 2\int_{R_1}^{R_2} x\left(1 - \frac{\log_e(x/R_1)}{\log_e(R_2/R_1)} \right) dx \right]$$

(b) From the above analysis

$$p_1^* = \frac{6\mu Q}{\pi c^3}\log_e \frac{R_2}{R_1}$$

So

$$Q = \frac{\pi c^3 p_1^*}{6\mu \log_e(R_2/R_1)} = \frac{\pi \times 1 \, \text{mm}^3 \times 15 \times 10^3 \, \text{Pa}}{6 \times 0.6 \, \text{kg}/(\text{m s}) \times 1.792 \times \left(10^3 \, \text{mm/m}\right)^3}$$

$$= 7.3 \times 10^{-6} \, \text{m}^3 \, \text{s}^{-1}$$

6.5 STEADY LAMINAR FLOW BETWEEN PARALLEL PLANES, ONE OF WHICH IS MOVING

We now consider steady laminar flow between plane boundaries that move relative to one another while still remaining parallel and the same distance apart. We may conveniently assume that one of the boundaries is stationary and that the other moves with a velocity V in the direction of flow. Even if both are moving, this assumption still serves: all velocities are considered relative to one of the boundaries which is then supposed at rest.

Here we may assume the lower plane in Fig. 6.5 to be the stationary one. Again we may consider the flow from left to right and, for consistency, we consider the velocity of the upper plane as positive from left to right. The analysis proceeds as before, and again we arrive at eqn 6.18. The boundary conditions, however, are now different. The velocity at the lower plane where $y = 0$ is again zero and hence $B = 0$. But the fluid at $y = c$ must have the velocity V of the upper plane. Hence, substituting the values in eqn 6.18:

$$\left(\frac{\delta p^*}{\delta x}\right)\frac{c^2}{2} = \mu V + Ac \quad \text{and so} \quad A = \left(\frac{\delta p^*}{\delta x}\right)\frac{c}{2} - \frac{\mu V}{c}$$

Thus for any value of y the velocity of the fluid may now be expressed by

$$u = \left(\frac{\delta p^*}{\delta x}\right)\frac{y^2}{2\mu} - \frac{y}{\mu}\left\{\left(\frac{\delta p^*}{\delta x}\right)\frac{c}{2} - \frac{\mu V}{c}\right\}$$

$$= \left(\frac{\delta p^*}{\delta x}\right)\frac{1}{2\mu}(y^2 - cy) + \frac{Vy}{c}$$

For a fixed breadth b perpendicular to the page, and with provisos as before concerning 'end effects', the total volume flow rate Q may be calculated by the same method as before.

$$Q = \int_0^c ub \, \mathrm{d}y = b\left[\left(\frac{\delta p^*}{\delta x}\right)\frac{1}{2\mu}\left(\frac{y^3}{3} - \frac{cy^2}{2}\right) + \frac{Vy^2}{2c}\right]_0^c$$

$$= b\left[-\left(\frac{\delta p^*}{\delta x}\right)\frac{c^3}{12\mu} + \frac{Vc}{2}\right] \tag{6.23}$$

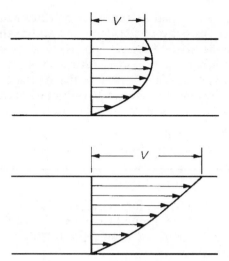

Fig. 6.10

Fig. 6.11

It is important to remind ourselves here that V must be measured in the same direction as the flow. If the movement of the boundary is in fact opposite to the direction of flow, then, in eqn 6.23, V is negative. The velocity profile (Fig. 6.10) is modified by the movement of one of the boundary surfaces and the maximum velocity no longer occurs in the centre of the section. Indeed, if V is sufficiently large the greatest velocity may occur at the moving surface as in Fig. 6.11.

Equation 6.23 shows that flow may occur even without a difference of piezometric pressure provided that one boundary is moving. In these circumstances $\delta p^*/\delta x$ is zero and $Q = \frac{1}{2}bVc$. Such flow, caused only by the movement of a boundary, is known as Couette flow (after M. F. A. Couette (1858–1943)). Couette flow, however, is not necessarily laminar.

Couette flow

Just as eqn 6.21 may be adapted to apply to laminar flow in a narrow annular space when both boundaries are stationary, so may eqn 6.23 be adapted for instances where one boundary is moving. As an example we may consider the application of the result to describe the operation of a simple cylindrical dashpot.

Example 6.3 ■

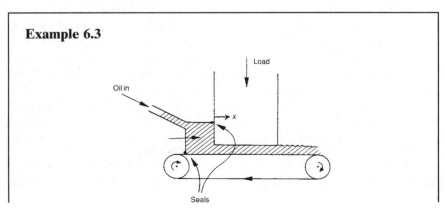

The diagram shows a pad and moving belt lubricated by oil supplied at a gauge pressure of 12 kPa at one end of the pad. The oil flows through the space between the two surfaces, emerging at atmospheric pressure. The pad is 120 mm long and the gap between the two surfaces is 0.18 mm. If the belt speed is 5 m/s and assuming the flow may be taken as two-dimensional, estimate (per unit span of pad):

(a) the load the pad will support
(b) the rate at which oil of viscosity $0.5\,\mathrm{kg\,m^{-1}\,s^{-1}}$ must be supplied.

Solution

(a) $\mathrm{d}p^*/\mathrm{d}x$ is constant and independent of x. Hence the average pressure of 6 kPa is applied over an area of $0.12\,\mathrm{m^2}$ per unit span, giving $f = F/b = 6 \times 10^3\,\mathrm{Pa} \times 0.12\,\mathrm{m} = 720\,\mathrm{N/m}$.

(b) $$q = \frac{Q}{b} = \left[-\left(\frac{\mathrm{d}p^*}{\mathrm{d}x}\right)\frac{c^3}{12\mu}\right] + \frac{Vc}{2}$$

$$= \left[\left(\frac{12 \times 10^3\,\mathrm{N/m^2}}{0.12\ \mathrm{m}}\right) \times \frac{(0.18)^3\ \mathrm{mm^3} \times \left(10^{-3}\ \mathrm{m/mm}\right)^3}{12 \times 0.5\ \mathrm{kg/(m\ s)}}\right]$$

$$+ \frac{5\ \mathrm{m/s} \times \left(0.18 \times 10^{-3}\right)\mathrm{m}}{2}$$

$$= 4.5 \times 10^{-4}\ \mathrm{m^2\ s^{-1}}$$

In this example, almost the entire flow is generated by the movement of the belt. The flow due to the pressure gradient (the bracketed term) is insignificant.

6.5.1 The simple cylindrical dashpot

A dashpot is essentially a device for damping vibrations of machines, or rapid reciprocating motions. This aim may readily be achieved by making use of a fluid of fairly high viscosity. A dashpot of the simplest kind is illustrated in Fig. 6.12. A piston P, connected to the mechanism whose movement is to be restrained, may move in a concentric cylinder C, the diameter of which is only slightly greater than that of the piston. The cylinder contains a viscous oil, and the quantity of oil should be sufficient to cover the top of the piston. If the piston is caused to move downwards, oil is displaced from underneath it. This displaced oil must move to the space above the piston and its only route is through the small annular clearance between the piston and the wall of the cylinder. If the viscosity of the oil is great enough, its flow upwards through the

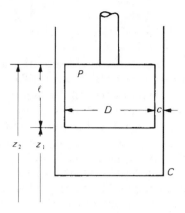

Fig. 6.12

clearance space is laminar and occurs simply as a result of the pressure developed underneath the piston. The more viscous the oil and the smaller the clearance between the piston and the cylinder, the greater is the pressure required to produce a particular movement of oil; thus the greater is the resistance to the motion of the piston.

For an upward movement of the piston, oil must flow *downwards* through the clearance space, otherwise a void will form underneath the piston. That is to say, an upward movement of the piston reduces the pressure underneath it, and the resulting difference of piezometric pressure between the top and bottom of the piston causes the flow of oil.

In applying eqn 6.23 we may consider the downward movement of the piston. The oil flow is therefore upwards through the clearance space, so upward velocities are suitably taken as positive. The velocity of the piston, which here forms one of the boundaries of the flow, is then negative and may be written $-V_p$. The planes between which the flow represented by eqn 6.23 takes place may be regarded as bent round a circle: one becomes the surface of the piston, the other the inner wall of the cylinder; the breadth b becomes the circumferential length πD.

The expression for the steady rate of flow of oil becomes

$$Q = \pi D \left\{ -\left(\frac{\delta p^*}{\delta x}\right) \frac{c^3}{12\mu} - \frac{V_p c}{2} \right\}$$

and this must exactly equal the rate at which oil is being displaced by the piston, $(\pi D^2/4)V_p$. If end effects are neglected $(-\delta p^*/\delta x)$ may be considered constant for a passage of uniform cross-section and may be written $\Delta p^*/l$, where Δp^* represents the difference of piezometric pressure from bottom to top of the piston.

$$\therefore \; \pi D \left[\frac{\Delta p^*}{l} \frac{c^3}{12\mu} - \frac{V_p c}{2} \right] = \frac{\pi}{4} D^2 V_p$$

whence

$$V_p\left(\frac{D}{2} + c\right) = \Delta p^* c^3/6\mu l \qquad (6.24)$$

The pressure underneath the piston exceeds that above it by Δp; thus there is an upward force of $\Delta p(\pi D^2/4)$ on it. There is also an upward force on the piston as a result of the oil flowing past it in the clearance space. The shear stress in a viscous fluid is given by $\mu(\partial u/\partial y)$. Here y is the coordinate perpendicular to the flow and so may be supplanted by r. By taking the value of $\mu \partial u/\partial r$ at the moving boundary, the shear stress on the piston surface may be calculated, and thus the upward shear force exerted by the oil on the piston. It may be shown, however, that this shear force is usually negligible compared with the other forces.

For steady, i.e. non-accelerating, movement, the sum of the forces on the piston must be zero.

$$\therefore \ \Delta p \frac{\pi}{4} D^2 - F - W = 0 \qquad (6.25)$$

where W denotes the weight of the piston and F the downward force exerted on it by the mechanism to which it is connected.

Now $p^* = p + \rho g z$ and so

$$p_1^* - p_2^* = p_1 - p_2 + \rho g(z_1 - z_2)$$
$$= \Delta p + \rho g(-l)$$

The minus sign appears in front of the l because z must be measured upwards (see Fig. 6.12). Substituting for Δp from eqn 6.25 and then putting the result in eqn 6.24 we obtain

$$V_p\left(\frac{D}{2} + c\right) = \frac{c^3}{6\mu l}\left(\frac{F+W}{\pi D^2/4} - \rho g l\right) \qquad (6.26)$$

The clearance c is normally small compared with the radius of the piston $D/2$, so the left-hand side of eqn 6.26 may be simplified to $V_p D/2$. Rearrangement then gives

$$F + W - \rho g l\, \pi D^2/4 = \frac{3}{4}\pi\mu l\left(\frac{D}{c}\right)^3 V_p \qquad (6.27)$$

The term $\rho g l\pi D^2/4$ represents the buoyancy of the piston, and in some instances this is negligible compared with F. This, however, is not always so and the buoyancy term should not be omitted without investigation.

For *upward* movement of the piston the signs of F and V_p are of course changed.

Equation 6.27 is a formula commonly used, but even for steady conditions it is only approximate. The clearance c has been neglected in comparison with the radius of the piston; the shear force on the piston has been neglected; the circumferences of piston and cylinder are not exactly equal; end effects have been neglected. It may be shown, however, that the accuracy is much improved if the mean diameter $2(D/2 + c/2) = (D + c)$ is used in eqn 6.27 in place of the piston diameter D.

In a dashpot of this kind the piston has to be maintained concentric with the cylinder by external means. If it is free to move laterally the piston tends to move to one side of the cylinder and, having once touched the side of the cylinder, is reluctant to leave it. Under such conditions the relation between the load on the piston and the rate at which oil flows through the clearance space is drastically altered and the effectiveness of the dashpot in restraining the movement of the piston is greatly reduced. Three buttons, equally spaced round the circumference of the piston, are sometimes used to maintain its concentricity with the cylinder.

Many other forms of dashpot are in use, but an account of these is beyond the scope of this book.

Example 6.4 A simple dashpot consists of a piston of diameter 50 mm and length 130 mm positioned concentrically in a cylinder of 50.4 mm diameter. If the dashpot contains oil of specific gravity 0.87 and kinematic viscosity $10^{-4}\,\text{m}^2\text{s}^{-1}$, determine the velocity of the dashpot if the difference in pressures Δp^* is 1.4 MPa.

Solution

Clearance:

$$c = \left(50.4 - 50\right)/2 = 0.2\ \text{mm}$$

Absolute viscosity:

$$\mu = \rho v = \left(0.87 \times 10^3\right)\text{kg/m}^3 \times 10^{-4}\ \text{m}^2/\text{s} = 0.087\ \text{kg m}^{-1}\text{s}^{-1}$$

From equation 6.24:

$$V_\text{p} = \frac{\Delta p^* c^3}{6\mu l\left(D/2 + c\right)}$$

$$= \frac{1.4 \times 10^6\ \text{Pa} \times \left(0.2 \times 10^{-3}\right)^3 \text{m}^3}{6 \times 0.087\ \text{kg/(m s)} \times 0.13\ \text{m} \times \left(25 + 0.2\right) \times 10^{-3}\ \text{m}}$$

$$\times 10^{-2}\ \text{m/cm}$$

$$= 6.55 \times 10^{-3}\ \text{m s}^{-1} = 6.55\ \text{mm/s}$$

6.6 THE MEASUREMENT OF VISCOSITY

The viscosity of a fluid cannot be measured directly, but its value can be calculated from some equation relating it to quantities that are directly measurable. A piece of apparatus specially suitable for the necessary measurements is known as a *viscosimeter* or, more usually, a *viscometer*

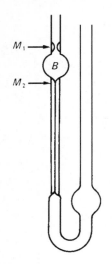

Fig. 6.13

and the study of methods of determining viscosity is known as *viscometry*.

In an ideal viscometer, the flow of the fluid under test would be completely determined by its viscosity. For practical viscometers, however, it is always necessary to introduce into the equations corrections to account for other effects or to calibrate the instrument with a fluid whose viscosity is already accurately known.

A few methods used in determining viscosity will be mentioned here but for details of these and other techniques more specialist works, e.g. those by Van Wazer[1] and Dinsdale and Moore[2], should be consulted.

6.6.1 Transpiration methods

Many types of viscometer involve laminar flow through a circular tube. Poiseuille's Law (eqn 6.8), rearranged as

$$\mu = \frac{\pi\left(p_1^* - p_2^*\right)d^4}{128Ql},$$

is therefore called upon. The difference of piezometric pressure between the ends of a capillary tube of known length and diameter, connecting two constant-level reservoirs, may be measured by a manometer. When the fluid under test is a liquid the volume rate of flow, Q, may be determined simply by collecting and measuring the quantity passing through the tube in a certain time. For gases, however, special arrangements must be made for measuring the flow.

An adaptation of this method is used in the Ostwald viscometer (invented by Wilhelm Ostwald (1853–1932)). The instrument (Fig. 6.13) is mounted vertically, a fixed volume of liquid is placed in it and drawn up into the upper bulb B and beyond the mark M_1. It is then allowed to flow back and the passage of the liquid level between two marks M_1 and M_2 is timed. It is difficult to determine accurately the dimensions of the capillary section and there are, moreover, surface tension effects and 'end effects' that are not negligible: this viscometer must therefore be calibrated with a liquid of known viscosity.

6.6.2 Industrial viscometers

Several instruments are in industrial and technical use for measuring viscosity – particularly the viscosity of oils – and most require the measurement of the time taken by a certain quantity of the liquid to flow through a short capillary tube. In many viscometers this capillary tube is so short that it is more like an orifice, and in any case fully developed laminar flow is scarcely achieved before the liquid reaches the end of the capillary. Thus Poiseuille's formula does not strictly apply. The rate of flow does not bear a simple relation to the viscosity and so such a viscometer requires calibration with a liquid of known viscosity.

In Great Britain the Redwood viscometer is widely used; America favours the Saybolt viscometer; Germany and other Continental coun-

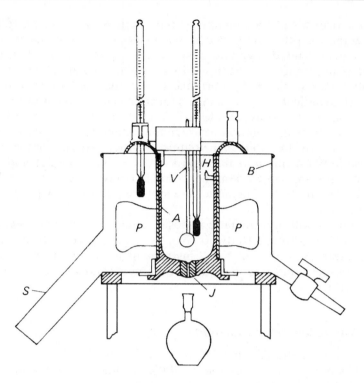

Fig. 6.14 Redwood No. 1 viscometer.

tries the Engler. These instruments differ in detail but not in principle and it is sufficient to refer to the Redwood instrument – invented by Sir Boverton Redwood (1846–1919) – as typical of its kind.

The liquid under test is placed in the container A (Fig. 6.14) and may escape through a capillary tube in the block of agate J, except when prevented by the valve V which fits the concave upper surface of J. The surrounding water bath B serves to control the temperature; heat may be applied to the side tube S (or by an electrical immersion heater) and the water may be stirred by the paddles P. A hook H indicates the correct starting level of the test liquid and the time is recorded for 50 ml of liquid to escape into a flask.

If Poiseuille's formula is assumed valid here, at least approximately, the rate of flow for a particular value of the head h (i.e., the level of the free surface in A above the outlet of J) is given by substituting $\rho g h$ for the difference of piezometric pressure in eqn 6.8:

$$Q = \frac{\pi \rho g h d^4}{128 \mu l}$$

The rate of flow is thus inversely proportional to the kinematic viscosity $\mu/\rho = v$. As the liquid runs out, h, and therefore Q, decreases; the proportionality between Q and $1/v$, however, remains, and so the time required for a fixed volume of liquid to escape is directly proportional to v.

In practice this proportionality is only approximate. The difference of piezometric pressure between the ends of the tube is used not only to overcome the viscous resistance represented by Poiseuille's formula, but also to give the liquid its kinetic energy as it flows through the tube. Moreover, at the entrance to the tube $\partial p^*/\partial x$ is somewhat greater than indicated by Poiseuille's formula because the flow there has not yet taken the 'fully developed' laminar form. Because of these additional effects the relation between the kinematic viscosity v and the time t taken by the standard volume of liquid to run out is better expressed by a formula of the type $v = At - B/t$, where A and B are constants for the instrument concerned. Conversion tables relating 'Redwood seconds' to kinematic viscosity are published by the Institute of Petroleum.

For liquids with a kinematic viscosity greater than about $500\,\text{mm}^2/\text{s}$ the time of efflux from a Redwood No. 1 viscometer would be more than half an hour. A Redwood No. 2 instrument is then more suitable because it has a capillary tube of greater diameter and so a much reduced time of efflux.

6.6.3 The falling sphere method

Transpiration methods (that is, those requiring the flow of the fluid through a tube) are not suitable for fluids of high viscosity because of the very low rate of flow. For liquids of high viscosity (such as treacle) a more satisfactory method makes use of Stokes's Law, which describes the steady motion of a sphere through a large expanse of a fluid at conditions of very low Reynolds number. An expression for the force exerted by the fluid on the sphere as a result of such motion was first obtained by Sir George G. Stokes (1819–1903). Although the mathematical details are beyond the scope of this book, it is important to give attention to the fundamental assumptions on which the solution is based.

First, it is assumed that the motion is such that the inertia forces on the particles of fluid (that is, the forces required to accelerate or decelerate them) may be neglected in comparison with the forces due to viscosity. As we have seen, the ratio of inertia forces to viscous forces is represented by Reynolds number, and so this condition is met if the Reynolds number of the flow is very small.

Stokes's Law

Other assumptions are that no other boundary surface is sufficiently near to affect the flow round the sphere, that the sphere is rigid, that the motion is steady and that there is no slip between the fluid and the sphere. Using these hypotheses, Stokes found that the force opposing the motion equals $3\pi\mu u d$, where μ represents the coefficient of absolute viscosity, u the velocity of the sphere relative to the undisturbed fluid and d the diameter of the sphere. This result is now known as Stokes's Law. It has been found that, to obtain good agreement with experimental results, the Reynolds number (expressed as $ud\rho/\mu$) must be less than about 0.1. The result of this restriction is that, for ordinary fluids such as water or air, the sphere must be almost microscopic in size. If Stokes's Law is to be valid for larger spheres, then either the viscosity must be very large or the velocity exceedingly small.

If a small solid particle is falling through a fluid under its own weight, the particle accelerates until the net downward force on it is zero. No further acceleration is then possible and the particle is said to have reached its 'terminal velocity'. This may be calculated for a small sphere on the assumption that the Reynolds number is small enough for Stokes's Law to be valid. If the density of the fluid is ρ and the mean density of the sphere is ρ_s, then, when the terminal velocity has been reached,

Downward force = Weight of sphere − Buoyancy − Resisting force
caused by flow round sphere

$$= \frac{\pi}{6}d^3\rho_s g - \frac{\pi}{6}d^3\rho g - 3\pi\mu u d = 0$$

whence

$$u = \frac{d^2(\rho_s - \rho)g}{18\mu} \tag{6.28}$$

For the determination of viscosity a small solid sphere of known weight is allowed to fall vertically down the centre of a cylinder containing the liquid under test. The velocity with which the sphere falls is measured; it does not, however, quite coincide with the terminal velocity as given by eqn 6.28. In practice the liquid cannot be of infinite extent as assumed in the derivation of Stokes's Law, so corrections are necessary to allow for the effect of the walls of the cylinder. (For this effect to be negligible, the diameter of the cylinder must be more than about 100 times the diameter of the sphere.) Moreover, the measurement of velocity must not be begun until the sphere has reached its terminal velocity, and should not be continued when the sphere nears the base of the cylinder since this influences the rate of fall of objects in its vicinity.

Example 6.5 The viscosity of an oil is to be measured using the falling sphere method. The oil, of relative density 0.88, is contained in a vertical tube of diameter $D = 19\,\text{mm}$. A sphere of relative density 1.151 and diameter $d = 4.75\,\text{mm}$ is dropped into the oil along the axis of the tube and reaches a terminal velocity of $6\,\text{mm/s}$. For $d/D = 0.25$, the viscous drag force acting on the sphere may be taken as 1.80 times the value in a fluid of infinite extent. Determine:

(a) the absolute and kinematic viscosities of the oil
(b) the Reynolds number of the sphere.

Solution

(a) Denote the viscous drag force acting on the sphere by F. Then the force balance on the sphere yields

$$F = \frac{\pi}{6} d^3 g (\rho_s - \rho)$$

Hence

$$F = \frac{\pi}{6} \times (4.75 \text{ mm})^3 \times (10^{-3} \text{ m/mm})^3 \times 9.81 \text{ m/s}^2$$
$$\times (1.151 - 0.88) \times 10^3 \text{ kg/m}^3$$
$$= 149 \times 10^{-6} \text{ N}$$

Also for $d/D = 0.25$, $F/F_0 = 1.80$, where $F_0 = 3\pi\mu u d$. So:

$$\mu = \frac{F}{1.8 \times 3\pi u d}$$
$$= \frac{149 \times 10^{-6} \text{ N}}{1.8 \times 3\pi \times (6 \times 10^{-3} \text{ m/s}) \times (4.75 \times 10^{-3} \text{ m})}$$
$$= 0.308 \text{ kg m}^{-1} \text{s}^{-1}$$

and

$$v = \frac{\mu}{\rho} = \frac{0.308 \text{ kg/(m s)}}{0.88 \times 10^3 \text{ kg/m}^3} = 3.50 \times 10^{-4} \text{ m}^2 \text{s}^{-1}$$

(b) Reynolds number:

$$Re = \frac{\rho u d}{\mu} = \frac{880 \text{ kg/m}^3 \times (6 \times 10^{-3} \text{ m/s}) \times (4.75 \times 10^{-3} \text{ m})}{0.308 \text{ kg/(m s)}}$$
$$= 0.08$$

The Reynolds number is below the upper limit for the application of this method of measuring viscosity, and experience shows that reliable results are obtained.

6.6.4 Rotary viscometers

A simple method of applying a known rate of shear to a fluid and of measuring the viscous stress thus produced is illustrated in Fig. 6.15. The annular gap between two concentric cylinders is filled with the fluid under test. If one cylinder is rotated at a constant, known, angular velocity the fluid tends to rotate at the same angular velocity and thus exerts a torque on the other cylinder. The balancing couple necessary to keep the second cylinder at rest may be measured, and the viscosity of the fluid then calculated. If the inner cylinder is the stationary one, the torque may be effectively determined by measuring the torsion in a wire suspending the cylinder.

If there is no slip at either boundary surface, the fluid in contact with the rotating cylinder has the same velocity as the periphery of the cylinder, and the fluid in contact with the stationary cylinder is at rest. The resulting velocity gradient across the layer of fluid brings the viscous forces into play.

It is perhaps worth developing in detail a formula applicable to such a viscometer, if only because formulae are not infrequently produced that are in error. More important, analysis of the problem illustrates the application of Newton's formula for viscosity when *angular* velocity is involved.

If the speed of rotation is not so high that turbulence is generated, the fluid in the annular space rotates in layers concentric with the cylinders. We may consider a small element of fluid, between two such layers distance δr apart and subtending an angle $\delta\theta$ at the centre of rotation (see Fig. 6.16: the rotation is considered in the plane of the paper). As a result of the relative movement of fluid particles at different radii, a stress τ is exerted at the interface between adjacent layers at radius r. Similarly, there is a stress $\tau + \delta\tau$ at radius $r + \delta r$. Since (in the absence of turbulence) the fluid moves in a tangential direction only, and not radially, there are no forces due to viscosity on the end faces of the element.

At radius r the area over which the stress acts is $r\,\delta\theta$ per unit thickness of fluid perpendicular to the plane of rotation. Therefore the force on this area is $\tau r\,\delta\theta$, and the corresponding torque $\tau r^2\delta\theta$. Similarly the viscous forces on the other side of the element produce a torque of $(\tau + \delta\tau)(r + \delta r)^2\,\delta\theta$. The forces on the two sides of the element are in opposite directions: if, for example, the angular velocity increases with radius then the force on the outer face of the element tends to accelerate it, while the force on the inner face tends to retard it. Thus the *net* torque on the element is $(\tau + \delta\tau)(r + \delta r)^2\delta\theta - \tau r^2\delta\theta$.

Under steady conditions the element does not undergo angular acceleration, so the net torque on it is zero. (Forces due to pressure do not contribute to the torque because they are identical at the two ends of the element and thus balance each other in the tangential direction.) Therefore

$$(\tau + \delta\tau)(r + \delta r)^2 - \tau r^2 = 0$$

i.e.

$$2\tau r\,\delta r + r^2\,\delta\tau = 0$$

higher orders of small quantities being neglected. Dividing by τr^2 we obtain

$$\frac{\delta\tau}{\tau} = -2\frac{\delta r}{r}$$

which may be integrated to give

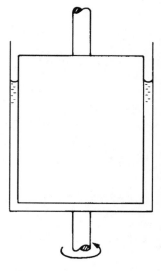

Fig. 6.15

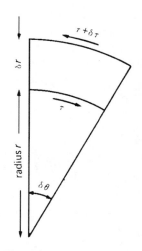

Fig. 6.16

$$\tau = A/r^2 \qquad (6.29)$$

where A is a constant.

We now have to evaluate τ in terms of the viscosity μ. Since we are here concerned with the viscous stress over an area perpendicular to the radius, the velocity gradient must be calculated along the radius. The tangential velocity at radius r is given by ωr where ω represents the angular velocity of the fluid, and the full velocity gradient is therefore

$$\frac{\partial u}{\partial r} = \frac{\partial}{\partial r}(\omega r) = \omega + r\frac{\partial \omega}{\partial r}$$

In this expression, however, only the second term contributes to relative motion between particles. Suppose that the angular velocity of the fluid does not vary with the radius. Then $\partial \omega/\partial r$ is zero and $\partial u/\partial r$ reduces to ω. In this case there is no relative motion between the particles of fluid, even though $\partial u/\partial r$ is not zero; the entire quantity of fluid rotates as if it were a solid block. (One may imagine a cylinder of liquid placed on a gramophone turntable: when conditions are steady the liquid will have the same angular velocity as the turntable and there will be no relative motion between particles at different radii, even though the peripheral velocity increases with radius.) Therefore the *rate of shear*, which represents relative motion between particles, is simply $r\partial\omega/\partial r$ and so the stress τ is given by $\mu r\partial\omega/\partial r$.

$$\therefore \ \mu\frac{\partial \omega}{\partial r} = \frac{\tau}{r} = \frac{A}{r^3} \qquad (6.30)$$

(from eqn 6.29). Since ω is here a function of r alone, eqn 6.30 may be integrated to give

$$\mu\omega = -\frac{A}{2r^2} + B \quad \text{where } B \text{ is a constant} \qquad (6.31)$$

Now if the rotating cylinder has radius a and angular velocity Ω and the stationary cylinder has radius b, the condition of no slip at a boundary requires ω to be zero when $r = b$. Substituting these simultaneous values in eqn 6.31 gives $B = A/2b^2$. The same no slip requirement makes $\omega = \Omega$ at $r = a$, and these values substituted in eqn 6.31 give

$$\mu\Omega = -\frac{A}{2a^2} + \frac{A}{2b^2} = \frac{A(a^2 - b^2)}{2a^2b^2}$$

whence

$$A = \frac{2a^2b^2\mu\Omega}{a^2 - b^2}$$

The torque T on the cylinder of radius b

$$= \text{stress} \times \text{area} \times \text{radius}$$

$$= \left(\mu r \, \partial\omega/\partial r\right)_{r=b} \times 2\pi b h \times b$$

where h represents the height of the cylinder in contact with the fluid, and end effects are assumed negligible.

$$\therefore\ T = 2\pi\mu b^3 h \left(\frac{\partial \omega}{\partial r}\right)_{r=b} = 2\pi\mu b^3 h \left(\frac{A}{\mu b^3}\right) \text{(from eqn 6.30)}$$

$$= 2\pi h A = \frac{4\pi h a^2 b^2 \mu \Omega}{a^2 - b^2} = k\mu\Omega \tag{6.32}$$

where k is a constant for any given apparatus.

It may readily be shown that the torque on the rotating cylinder is the same. Equation 6.32 applies whether the inner cylinder is stationary while the outer one rotates, or the outer cylinder is stationary while the inner one rotates.

In the derivation of eqn 6.32 it was assumed that h was large enough to render negligible any special effects at the ends. In practice, however, the cylinders are of moderate length and some account must therefore be taken of effects produced by the ends. The end effects are very similar to that of an additional length of the cylinder in contact with the fluid, in other words $(h + l)$ rather than h would arise in eqn 6.32. By using two, or preferably more, values of the liquid depth, and therefore of h, simultaneous equations are obtained from which l may be eliminated.

If the radii of the two cylinders are closely similar eqn 6.32 may be slightly simplified. Putting the annular clearance $c = a - b$ and $(a + b) = 2$ (mean radius) = mean diameter = D, we obtain

$$T = 4\pi h\mu\Omega \frac{a^2 b^2}{(a + b)(a - b)} = 4\pi h\mu\Omega \left(\frac{D + c}{2}\right)^2 \left(\frac{D - c}{2}\right)^2 \Big/ Dc$$

$$= \frac{\pi h\mu\Omega}{4Dc}\left(D^2 - c^2\right)^2$$

Then, neglecting c^2 in comparison with D^2, we have

$$T = \frac{\pi h\mu\Omega D^3}{4c} \tag{6.33}$$

A simple laboratory viscometer of this type is that devised by G. F. C. Searle (1864–1954).

It is important to remember that neither eqn 6.32 nor eqn 6.33 is applicable if the motion of the fluid is turbulent. Moreover, the assumption has been made throughout that the two cylinders are concentric. The formulae are therefore not applicable (except in very rare instances) to journal bearings, for, as we shall see in Section 6.7.3, a journal bearing supports a load only if the journal and bearing are *not* concentric.

The rotary type of viscometer may be modified so that the moving cylinder is allowed to oscillate about a mean position instead of rotating steadily. Measurements are then taken of the damping imposed by the fluid. However, the number of methods of determining viscosity is legion and for details of them reference must be made to specialist works[1,2].

In any method of measuring viscosity it is important to maintain the fluid at a constant and known temperature. The viscosity of both liquids and gases varies markedly with temperature and, indeed, the relevant temperature should always be quoted alongside any value for the viscosity of a fluid.

Example 6.6 In a rotary viscometer the gap between two concentric cylinders is filled with a fluid under test, as shown in the diagram. The outer cylinder is rotated with constant angular velocity Ω and the torque T required to hold the inner cylinder at rest is measured. The fluid motion is everywhere laminar.

The torque T consists of two components T_a and T_b. The torque T_a is due to the flow in the annular clearance c, where $c << D$. The torque T_b is due to the fluid motion generated in the clearance t at the base of the viscometer.

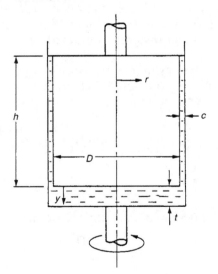

(a) Show that the torque T is related to D by an equation of the form

$$T = T_a + T_b = \left(K_1 D^3 + K_2 D^4\right)\mu\Omega$$

where K_1 and K_2 are constants of the viscometer. Derive expressions for K_1 and K_2.

(b) The dimensions of the viscometer are $D = 120\,\text{mm}$, $h = 80\,\text{mm}$, $c = 1\,\text{mm}$, $t = 18.75\,\text{mm}$. Evaluate the viscosity of the liquid if a torque of $4 \times 10^{-3}\,\text{N}\,\text{m}$ is required to hold the inner cylinder stationary when the outer cylinder is rotated at 65 revolutions per minute.

Solution

(a) In the annular clearance c the torque T_a is given by the analysis leading to equation (6.33). Thus

$$T_a = \frac{\pi h \mu \Omega D^3}{4c}$$

The torque T_b due to the fluid motion in the gap t at the base of the viscometer is determined as follows.

At radius r the velocity of the outer cylinder is Ωr. The shear stress on the inner cylinder at radius r is given by

$$\tau_b = \mu \frac{du}{dy} = \mu \frac{\Omega r}{t}$$

At radius r the area of an element is $2\pi r dr$, the shear force on the element is

$$2\pi r \tau_b dr = \frac{2\pi \mu \Omega}{t} r^2 dr$$

and the torque due to the elemental shear force is

$$\frac{2\pi \mu \Omega}{t} r^3 dr$$

Hence torque

$$T_b = \int_0^{D/2} \frac{2\pi \mu \Omega}{t} r^3 dr = \frac{\pi \mu \Omega D^4}{32t}$$

and total torque

$$T = T_a + T_b = \left(\frac{\pi h}{4c} D^3 + \frac{\pi}{32t} D^4 \right) \mu \Omega$$

Hence

$$T = T_a + T_b = \left(K_1 D^3 + K_2 D^4 \right) \mu \Omega$$

where

$$K_1 = \frac{\pi h}{4c}, \quad K_2 = \frac{\pi}{32t} \quad \text{QED.}$$

(b) Substituting above $T = 4 \times 10^{-3}\,\mathrm{N\,m}$

Hence μ is evaluated as μ $\left(\frac{\pi}{4} \frac{8.36\,\mathrm{mm}}{1\,\mathrm{mm}} 10^{-3}\,\mathrm{N\,s\,m^{-2}} \times (120\,\mathrm{mm})^3 \right.$

$\left. + \frac{\pi}{32 \times 18.75\,\mathrm{mm}} \times (120\,\mathrm{mm})^4 \right) \times \frac{1}{\left(10^3\,\mathrm{mm/m}\right)^3}$

$$\mu \Omega = \frac{65\,\mathrm{rev/min} \times 2\pi\,\mathrm{rad/rev}}{60\,\mathrm{s/min}} \mu$$

The calculations show that, for the particular design of viscometer considered here, if the contribution from T_b is ignored, a 1% error is incurred in the magnitude of μ.

☐

6.7 FUNDAMENTALS OF THE THEORY OF HYDRODYNAMIC LUBRICATION

Another important application of laminar flow arises in the lubrication of various types of bearings. In these, laminar motion may almost always be assumed because, although high velocities may be involved, the thickness of the film of lubricant is so small that the Reynolds number is usually well below its critical value for that system.

The primary function of the lubricant is to separate the bearing surfaces, and so long as the lubrication is effective there is no direct contact between properly finished surfaces. If the film of lubricant is to keep the bearing surfaces apart it must be capable of sustaining a load. One way of achieving this is to supply the fluid lubricant to the space between the surfaces at a sufficiently high pressure from some external source. This provides *hydrostatic* lubrication. But in many instances a high pressure may be more readily produced in the lubricant as a result of the shape and relative motion of the bearing surfaces themselves. This action gives *hydrodynamic* lubrication. The theory of it can give rise to considerable mathematical complexity and no attempt will be made here to consider more than very simple examples. Nevertheless, we shall see how the clearances that must be allowed between the surfaces may be determined; also the degree of smoothness to which they must be finished, the viscosity and the rate of flow of the lubricant necessary to prevent the bearing surfaces coming into direct contact.

The simplest form of bearing is the slipper or slide-block moving over a horizontal plane surface as illustrated in Fig. 6.17. For the purpose of our analysis we shall assume that the bearing plate is infinite in extent and that the slipper is infinitely wide in the horizontal direction perpendicular to its motion (that is, perpendicular to the diagram). Thus the flow of the lubricant may be considered two-dimensional; in other words, there is no component of velocity in the direction perpendicular to the diagram.

If the further assumption is made that the surfaces of the slipper and bearing plate are parallel, the behaviour of the lubricant corresponds to

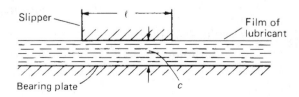

Fig. 6.17

laminar flow between parallel boundaries, one of which is moving. Assuming steady conditions and a constant viscosity, we may therefore use eqn 6.23:

$$\frac{Q}{b} = -\frac{\delta p^*}{\delta x}\frac{c^3}{12\mu} + \frac{Vc}{2}$$

In view of the negligible differences of elevation within the lubricant film, the asterisk may be dropped from p^*. Then, with the volume flow rate per unit width Q/b written as q, the equation may be rearranged to give (in the limit as $\delta x \to 0$)

$$\frac{dp}{dx} = \frac{12\mu}{c^3}\left(\frac{Vc}{2} - q\right) \tag{6.34}$$

We have assumed c constant, and from the principle of continuity and the assumption of two-dimensional constant-density flow, q is constant. So, from eqn 6.34, dp/dx = constant. But this constant must be zero because the pressure has the same value at each end of the slipper, say p_0 (usually atmospheric). In other words, there is no variation of pressure throughout the space between slipper and plate. The same pressure, p_0, also acts uniformly over the outer surface of the slipper and on the exposed part of the lubricant film, so the lubricant exerts no resultant force normal to the boundaries. We are therefore led to conclude that the bearing surfaces are incapable of supporting any load – except when they actually touch each other, but such contact of course defeats the purpose of lubrication.

This conclusion, however, depends on the assumptions of constant density and viscosity. If a film of lubricant could in fact be maintained between the slipper and the plate, the energy needed to overcome the viscous resistance would be dissipated as heat, and the temperature of the lubricant would increase in the direction of flow. Consequently there would be not only a variation of viscosity but also a decrease of density in the flow direction and thus an increase of q. (We may disregard any variation of temperature, and therefore viscosity, across the very small clearance c, and, for a liquid lubricant, the effect of pressure on density may also be neglected.) In such circumstances a parallel slider bearing could in fact support a load. This would require the pressure in the lubricant to increase from the ambient value p_0 at one end of the slipper, pass through a maximum, and return to p_0 at the other end of the slipper, as shown in Fig. 6.18. At the peak value $dp/dx = 0$ and there, by eqn 6.34, $q = Vc/2$; but downstream of this position q would be greater, thereby making dp/dx negative, whereas upstream dp/dx would be positive. (Although the derivation of eqn 6.34 involves the assumption of constant density, a small, non-zero, value of $\delta p/\delta x$ has a negligible effect on that equation, and does not undermine the present argument.) In addition, the small clearance c might not remain uniform even for surfaces initially parallel. Significant variations of c may be caused by quite tiny distortions of the surfaces, particularly as a result of temperature changes.

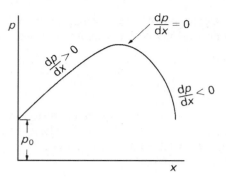

Fig. 6.18

Thus any ability of a parallel bearing to support a load depends on the wasteful dissipation of energy as heat! A much more economical and reliable way of obtaining a pressure graph like Fig. 6.18 is to make the clearance c vary in some suitable way with x.

6.7.1 The inclined slipper bearing

The simplest form of bearing in which the clearance varies with x is one in which the two surfaces are plane but inclined to one another in the direction of motion. A bearing of this type is illustrated in Fig. 6.19. The slipper is at an angle, arctan δ, to the bearing plate. (In practice, this angle is very small and it is greatly exaggerated in the diagram.) It is again assumed that the bearing plate is infinite in extent, and that the slipper has an infinite width so that the flow may be considered two-dimensional. To give steady conditions, the slipper is considered fixed and the bearing plate moving. The origin of coordinates is chosen so that the left-hand end of the slipper is at $x = 0$, and the bearing plate moves relative to the slipper with constant velocity V in the x direction. The clearance between slipper and plate at a distance x from the origin is h, and the plane of the slipper surface intersects that of the bearing plate at a distance a from the origin.

Over a small length δx the clearance h may be assumed unchanged, and from eqn 6.34 we have

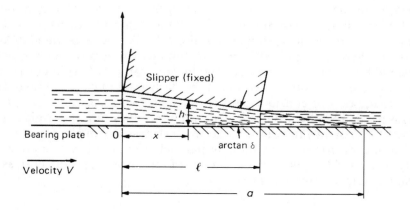

Fig. 6.19

$$\frac{\mathrm{d}p}{\mathrm{d}x} = 12\mu\left(\frac{V}{2h^2} - \frac{q}{h^3}\right) \tag{6.35}$$

Equation 6.35 involves two noteworthy assumptions: first, that the variation of velocity with y is the same as if the boundaries were parallel; second, that the acceleration of the fluid as h decreases requires inertia forces that are negligible compared with the viscous forces. For the usual very small inclination of the slipper, however, these assumptions are justified. Furthermore, components of velocity in the y direction are negligible.

If μ is assumed constant throughout, and $(a - x)\delta$ is substituted for h, eqn 6.35 may be integrated:

$$p = 12\mu\left\{\frac{V}{2(a-x)\delta^2} - \frac{q}{2(a-x)^2\delta^3}\right\} + A$$

The integration constant A may be determined from the condition that $p = p_0$, the ambient pressure, when $x = 0$.

The discharge per unit width, q, still remains to be determined. However, p again equals p_0 when $x = l$, and inserting this condition into the equation enables q to be calculated. After simplification the final result is

$$p = p_0 + \frac{6\mu Vx(l - x)}{\delta^2(a - x)^2(2a - l)} \tag{6.36}$$

This equation expresses the relation between p, the pressure in the lubricant, and x. Since, under the slipper, x is always less than l and l is less than $2a$, the last term of eqn 6.36 is positive; p is thus greater than p_0 and the bearing can sustain a load. If the distance a approaches infinity, that is, if the surfaces become parallel, then $p = p_0$ for all values of x – as we found in Section 6.7. The inclination of the slipper must be such that the fluid is forced into a passage of decreasing cross-section: if V, and therefore the fluid velocity, were reversed the last term of eqn 6.36 would be made negative, p would become less than p_0, and the bearing would collapse.

Thus the fluid is, as it were, used as a wedge. This 'wedge' principle is, in fact, one of the fundamental features of any hydrodynamically lubricated bearing, whether plane or not.

The load that such a bearing can support is determined by the total net thrust which the fluid exerts on unit width of either bearing surface. This is obtained from the integral

$$\int_0^l (p - p_0)\mathrm{d}x \tag{6.37}$$

Substituting from eqn 6.36 we have

$$\text{Net thrust per unit width} = T = \frac{6\mu V}{\delta^2(2a - l)}\int_0^l \frac{x(l - x)}{(a - x)^2}\,\mathrm{d}x$$

and the result after simplifying is

$$T = \frac{6\mu V}{\delta^2}\left(\ln\frac{a}{a-l} - \frac{2l}{2a-l}\right) \tag{6.38}$$

The position of the centre of pressure (that is, the point of application of this resultant force) may be determined by calculating the total moment of the distributed force over the slipper about some convenient axis and then dividing the result by the total force. The axis $x = 0$, $y = 0$ is a suitable one about which to take moments. Thus if the centre of pressure is at $x = x_p$,

$$Tx_p = \int_0^l (p - p_0)x\,dx = \frac{6\mu V}{\delta^2(2a-l)}\int_0^l \frac{x^2(l-x)}{(a-x)^2}dx \tag{6.39}$$

After carrying out the integration and substituting the value of T from eqn 6.38 we finally obtain

$$x_p = \frac{a(3a - 2l)\ln\left\{a/(a-l)\right\} - l\left\{3a - (l/2)\right\}}{(2a - l)\ln\left\{a/(a-l)\right\} - 2l} \tag{6.40}$$

This, it may be shown, is always greater than $l/2$; that is, the centre of pressure is always behind the geometrical centre of the slipper. When the plates are parallel, a is infinite, and the centre of pressure and the geometrical centre coincide. As the inclination of the slipper is increased the centre of pressure moves back. At the maximum possible inclination, the 'heel' of the slipper touches the bearing plate; thus $a = l$ and the centre of pressure is at the heel. Equation 6.36 shows that the pressure at this point is then infinite. However, under these conditions flow between the slipper and plate is impossible, and so this limiting case has no physical significance.

It will be noted from eqn 6.40 that the position of the centre of pressure in no way depends on the value of T, μ or V. It depends in fact only on the geometry of the slipper. Since a given small angle is difficult to maintain accurately in practice, it is usual to pivot the slipper on its support, the axis of the pivot being towards the rear of the slipper. The slipper then so adjusts its inclination that the line of action of the resultant force passes through the pivot axis.

Since the centre of pressure moves towards the rear of the slipper as the inclination is increased, the arrangement is stable; if the inclination increases, the point of application of the force moves back beyond the axis of the pivot, and so provides a restoring moment to reduce the inclination again. Conversely, a reduction of the inclination brings the centre of pressure forward and so produces the necessary restoring moment.

The pivoted plane slipper bearing is sometimes known as a Michell bearing after the Australian engineer A. G. M. Michell (1870–1959). (The principle was discovered independently by the U.S. engineer

Albert Kingsbury (1862–1943), so in North America such bearings are more often termed Kingsbury bearings.)

Although the theory suggests that the pivot axis should be nearer the rear of the bearing, centrally pivoted slippers have been found to work satisfactorily. This is in part explained by the fact that the energy required to overcome the viscous resistance is dissipated as heat, and the consequent variations in the viscosity and density of the lubricant cause the centre of pressure to move forward. A centrally pivoted slipper is of course suitable for movement in either direction.

We have seen that the 'heel' of the slipper should not touch the bearing plate because lubrication would then break down entirely. In practice there is a lower limit, greater than zero, to the clearance at the heel. This limit is largely governed by the inevitable lack of perfect smoothness of the surfaces and the probable size of solid particles (e.g. grit) in the lubricant. If these particles are unable to escape under the heel of the bearing, they collect at this point and score the surfaces.

The tangential force (that is, the resistance to the relative movement between the slipper and the bearing plate) may also be calculated. The viscous shearing stress at any point in the film of lubricant is given by $\tau = \mu(\partial u/\partial y)$. The velocity u is given by eqn 6.18

$$\left(\frac{\delta p^*}{\delta x}\right)\frac{y^2}{2} = \mu u + Ay + B$$

It is again permissible to drop the asterisk from the symbol p^* since we consider a horizontal slipper bearing, and again the inclination of the slipper is assumed so small that any component of velocity in the vertical direction is negligible. Since the rate of change of pressure with x is not constant, the differential dp/dx must be used instead of the ratio $\delta p/\delta x$. The integration constants A and B are determined by the conditions at the boundaries. Referring to Fig. 6.19 we see that the velocity of the lubricant must equal the velocity of the bearing plate when $y = 0$. Substituting $u = V$ and $y = 0$ gives $B = -\mu V$. If the clearance is h at a particular value of x, the other boundary condition is that $u = 0$ when $y = h$ (since the slipper is considered stationary).

$$\therefore \quad \frac{dp}{dx}\frac{h^2}{2} = 0 + Ah - \mu V \quad \text{whence } A = \frac{dp}{dx}\frac{h}{2} + \frac{\mu}{h}V$$

Thus the expression for velocity at any point becomes

$$\frac{dp}{dx}\frac{y^2}{2} = \mu u + \frac{dp}{dx}\frac{h}{2}y + \frac{\mu}{h}Vy - \mu V$$

and differentiation with respect to y then gives an expression for the viscous stress in the x direction at any value of y

$$\mu\frac{\partial u}{\partial y} = -\frac{dp}{dx}\left(\frac{h}{2} - y\right) - \frac{\mu V}{h} \tag{6.41}$$

The stress at the bearing plate is given by the value of this expression when $y = 0$, that is, $-(dp/dx)(h/2) - \mu V/h$. (The sign convention for stress shows that the stress on the bearing plate is given by $\mu(\partial u/\partial y)$ since the plate faces the direction of increase of y.) Thus the total tangential force on the bearing plate (per unit width) is

$$-\int_0^l \left(\frac{dp}{dx}\frac{h}{2} + \frac{\mu V}{h}\right)dx$$

Substituting $(a - x)\delta$ for h, we have

Total tangential force per unit width $= F_p$

$$= -\frac{\delta}{2}\int_0^l (a - x)\frac{dp}{dx}\,dx - \frac{\mu V}{\delta}\int_0^l \frac{dx}{a - x}$$

To evaluate the first of these integrals we may obtain an expression for dp/dx by differentiating eqn 6.36. It is simpler, however, to integrate by parts

$$F_p = -\frac{\delta}{2}\left\{\left[(a - x)p\right]_0^l - \int_0^l p\frac{d}{dx}(a - x)dx\right\} + \frac{\mu V}{\delta}\ln\left(\frac{a - l}{a}\right)$$

$$= -\frac{\delta}{2}\left\{(a - l)p_0 - ap_0 + \int_0^l p\,dx\right\} + \frac{\mu V}{\delta}\ln\left(\frac{a - l}{a}\right)$$

But since $T = \int_0^l (p - p_0)dx$ (eqn 6.37) the term

$$\int_0^l p\,dx = T + \int_0^l p_0\,dx = T + p_0 l$$

Thus the total tangential force per unit width of bearing plate

$$F_p = -\frac{\delta}{2}\left(-p_0 l + T + p_0 l\right) + \frac{\mu V}{\delta}\ln\left(\frac{a - l}{a}\right)$$

$$= -\frac{T\delta}{2} + \frac{\mu V}{\delta}\ln\left(\frac{h_2}{h_1}\right)$$

$$= -\left\{\frac{T\delta}{2} + \frac{\mu V}{\delta}\ln\left(\frac{h_1}{h_2}\right)\right\} \tag{6.42}$$

where h_1, h_2 are the clearances under the 'toe' and 'heel' of the slipper respectively $(h_1 > h_2)$. The minus sign indicates that the force acts in the direction opposite to x, i.e. towards the left in Fig. 6.19.

To calculate the horizontal force on the slipper we may proceed similarly. The viscous stress in the lubricant at the surface of the slipper is given by putting $y = h$ in eqn 6.41. The stress on the boundary is given by $-\mu\,\partial u/\partial y$ for $y = h$. The minus sign here arises from the sign convention for stress: the surface in question faces the direction *opposite* to that in which y increases. The stress on the surface of the slipper is therefore

$$-\frac{dp}{dx}\frac{h}{2} + \frac{\mu V}{h}$$

The total horizontal force per unit width arising from the viscous stress is, since the inclination of the slipper is again assumed to be exceedingly small,

$$\int_0^l \left(-\frac{dp}{dx}\frac{h}{2} + \frac{\mu V}{h} \right) dx$$

This expression may be integrated like the corresponding one for the bearing plate, and the result is

$$-\frac{T\delta}{2} + \frac{\mu V}{\delta} \ln\left(\frac{h_1}{h_2} \right)$$

In addition, however, the normal force on the slipper has a component $T\delta$ which helps to resist the motion. If the total normal force on the slipper is F, then F has a component, towards the right of Fig. 6.20, equal to $F \sin\theta$. But if the separating force is T (perpendicular to the bearing plate) then, resolving vertically, we see that $T = F\cos\theta$. Thus the component towards the right becomes $(T/\cos\theta) \sin\theta = T\tan\theta = T\delta$. Hence the total force resisting the movement of the slipper is

$$F_s = -\frac{T\delta}{2} + \frac{\mu V}{\delta} \ln\left(\frac{h_1}{h_2} \right) + T\delta = \frac{T\delta}{2} + \frac{\mu V}{\delta} \ln\left(\frac{h_1}{h_2} \right) \qquad (6.43)$$

Comparison of the expressions 6.42 and 6.43 shows that $F_p = -F_s$. This, of course, is to be expected from Newton's Third Law.

A further point of interest is this. In the case of friction between dry solid surfaces the 'coefficient of friction', that is, the ratio of the friction force to the normal force, is largely independent of the load and speed of sliding. This is not so for lubricated surfaces, as eqns 6.42 and 6.43 show.

The load that such a bearing can support is determined by the inclination of the slipper. Using the geometrical relations

$$\delta = (h_1 - h_2)/l \quad \text{and} \quad a/l = h_1/(h_1 - h_2)$$

eqn 6.38 may be written in the alternative form

$$T = \frac{6\mu V l^2}{(h_1 - h_2)^2} \left\{ \ln\left(\frac{h_1}{h_2} \right) - 2\left(\frac{h_1 - h_2}{h_1 + h_2} \right) \right\} \qquad (6.44)$$

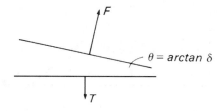

$$\theta = \arctan \delta$$

Fig. 6.20

For given values of l and h_2, the condition for T to be a maximum is $h_1/h_2 = 2.189$. Then $T = 0.1602\mu Vl^2/h_2^2$ and $F_\mathrm{s} = -F_\mathrm{p} = 0.754\mu Vl/h_2$. For these conditions the ratio T/F_s, which may be regarded as a measure of the efficiency of the bearing, is $0.212l/h_2$ and since h_2 may be very small (0.02 mm perhaps) this ratio may be very large.

The theory of the simple slider bearing, as here developed, is broadly that first put forward by Osborne Reynolds in 1886. Considerable extensions of the theory have since been made, but these further developments are beyond the scope of the present volume. It should, however, be realized that in the foregoing analysis the problem has been idealized in a number of respects. In practice the slipper would not be infinite in width: there would thus be some loss of lubricant from the sides of the bearing and, as a result, the pressure in the film of lubricant would be reduced. The flow, moreover, would not be truly two-dimensional.

A further complication is that the energy required to overcome the viscous resistance in the bearing is dissipated as heat. Some of this heat is taken up by the lubricant, with the result that its temperature rises and its viscosity falls. Indeed, in a number of applications the lubricant is used as much for cooling as for supporting a load. The variation of viscosity is accounted for – at least approximately – in some of the more advanced theories.

6.7.2 The Rayleigh step

An analysis by Lord Rayleigh (1842–1919) suggested that a greater load-carrying capacity would be obtained with a fixed pad providing two constant clearances h_1 and h_2 (Fig. 6.21). If we again assume steady conditions, constant density and viscosity, and an infinite width (perpendicular to the diagram), eqn 6.34 shows that in each part of the bearing the volume rate of flow of lubricant per unit width is

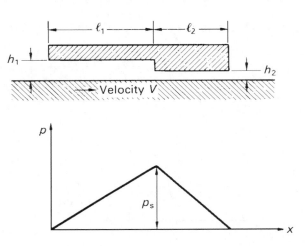

Fig. 6.21

$$q = \frac{Vc}{2} - \frac{c^3}{12\mu}\frac{dp}{dx} \tag{6.45}$$

Since q is independent of x, so too is dp/dx if the clearance is constant; that is, the pressure varies linearly with x in each part of the bearing, and hence

$$\left(\frac{dp}{dx}\right)_1 = \frac{p_s}{l_1} \quad \text{and} \quad \left(\frac{dp}{dx}\right)_2 = -\frac{p_s}{l_2}$$

where p_s represents the pressure at the step and the ambient pressure is for simplicity taken as zero. Using eqn 6.45 to equate the values of q for each part of the bearing then gives

$$\frac{Vh_1}{2} - \frac{h_1^3}{12\mu}\frac{p_s}{l_1} = \frac{Vh_2}{2} - \frac{h_2^3}{12\mu}\left(-\frac{p_s}{l_2}\right)$$

whence

$$\frac{p_s}{6\mu V} = \frac{h_1 - h_2}{\left(h_1^3/l_1\right) + \left(h_2^3/l_2\right)} \tag{6.46}$$

Since p_s must be positive, $V(h_1 - h_2)$ must be positive; that is, the lubricant must flow from the larger clearance to the smaller. The load carried by unit width of the bearing corresponds to the area of the pressure graph (Fig. 6.21)

$$\int_0^l p\,dx = \frac{1}{2}p_s l$$

where $l = l_1 + l_2$. The maximum load is determined by the maximum value of the expression 6.46 and this occurs when $h_1/h_2 = 1 + \sqrt{3}/2 = 1.866$ and $l_1/l_2 = (5 + 3\sqrt{3})/4 = 2.549$. The maximum load per unit width is then

$$\frac{4}{9}\left(2\sqrt{3} - 3\right)\mu Vl^2 \ h_2^2 = 0.2063\mu Vl^2/h_2^2$$

an appreciable increase above the maximum value, $0.1602\mu Vl^2/h_2^2$, obtained with the inclined-plane type of bearing. Unfortunately, for bearings that are not infinite in width, sideways leakage is greater with a Rayleigh step, and thus in practice its superiority in load-carrying is lessened. Refinements have been devised, however, to improve the performance.

Example 6.7   Consider a step bearing of breadth b, with the step centrally positioned.

(a) Show that the volumetric flow rate Q through the bearing is given by

$$Q = \frac{V\left(1 + H^2\right)}{2\left(1 + H^3\right)}bh_1$$

where $H = h_1/h_2$.

(b) A bearing has the following dimensions: $h_1 = 0.5 \, \text{mm}$, $h_2 = 0.25 \, \text{mm}$, $l = 100 \, \text{mm}$, $b = 100 \, \text{mm}$. The bearing is used in conjunction with an oil of relative density 0.87 and kinematic viscosity $2 \times 10^{-4} \, \text{m}^2\text{s}^{-1}$. The relative velocity between the bearing surfaces is $10 \, \text{ms}^{-1}$. Determine the volumetric flow rate of oil.

(c) Determine the load supported by the bearing.

Solution

(a) Since $Q = qb$ and $(dp/dx)_1 = -(dp/dx)_2$, we may write

$$\left(\frac{V}{2} - \frac{Q}{bh_1}\right)\frac{12\mu}{h_1^2} = -\left(\frac{V}{2} - \frac{Q}{bh_2}\right)\frac{12\mu}{h_2^2}$$

or

$$\left(\frac{V}{2} - \frac{Q}{bHh_2}\right)\frac{12\mu}{H^2 h_2^2} = -\left(\frac{V}{2} - \frac{Q}{bh_2}\right)\frac{12\mu}{h_2^2}$$

which may be rewritten as

$$Q = \frac{V}{2}\frac{(1+H^2)}{(1+H^3)}bHh_2 = \frac{V}{2}\frac{(1+H^2)}{(1+H^3)}bh_1 \quad \text{QED.}$$

(b) $Q = \dfrac{V}{2}\dfrac{(1+H^2)}{(1+H^3)}bh_1$

$$= \frac{10 \, \text{m/s}}{2} \times \frac{(1+4)}{(1+8)} \times 0.1 \, \text{m} \times \left(0.5 \times 10^{-3} \, \text{m}\right)$$

$$= 139 \times 10^{-6} \, \text{m}^3\,\text{s}^{-1}$$

(c) The load F supported by the bearing is given by

$$F = \left(\frac{dp}{dx}\right)_1 \frac{L}{2}\frac{L}{2}b = \frac{V}{2}\left(1 - \frac{(1+H^2)}{(1+H^3)}\right)\frac{12\mu}{h_1^2}\frac{L^2}{4}b$$

$$= \frac{10 \, \text{m/s}}{2}\left(1 - \frac{(1+4)}{(1+8)}\right)\frac{12 \times 870 \, \text{kg/m}^3 \times \left(2 \times 10^{-4}\right)\text{m}^2/\text{s}}{\left(0.5 \, \text{mm}\right)^2 \times \left(10^{-3} \, \text{m/mm}\right)^2}$$

$$\times \frac{\left(0.1 \, \text{m}\right)^2}{4} \times 0.1 \, \text{m}$$

$$= 4640 \, \text{N}$$

6.7.3 Journal bearings

The basic 'wedge' principle again applies in a journal bearing on a rotating shaft. This can carry a transverse load only if the journal (i.e. the part of the shaft within the bearing) and the bush (i.e. the lining of the bearing) are not exactly coaxial. Under load, the journal (centre C in Fig. 6.22) takes up an eccentric position in the bush (centre O). The distance CO between the centres is known as the *eccentricity, e,* and the ratio of this to the average clearance c ($= R_b - R$) is termed the *eccentricity ratio*, ε. The thickness h of the lubricant film (greatly exaggerated in the diagram: in practice c is of order $0.001R$) varies with the angle θ which is measured, in the direction of shaft rotation, from the line CO. (The magnitude and direction of CO will be determined by our analysis.) From Fig. 6.22

$$R + h = e \cos\theta + R_b \cos\xi$$

but, because e is exceedingly small (compared with R_b), so is the angle ξ, and we can take $\cos\xi = 1$. Then

$$h = R_b - R + e \cos\theta = c + e \cos\theta = c(1 + \varepsilon \cos\theta) \quad (6.47)$$

We consider first a bearing of infinite length, so that the lubricant flow may be supposed two-dimensional (in the plane of the diagram). We assume also that conditions are steady, that the viscosity and density are constant throughout, and that the axes of journal and bush are exactly straight and parallel. As the film thickness h is so small compared with the radii, the effect of curvature is negligible, and, for a small length δx in the direction of flow, the flow is governed essentially by eqn 6.35:

Bearing of infinite length

$$\frac{\mathrm{d}p}{\mathrm{d}x} = \frac{12\mu}{h^3}\left(\frac{\Omega R h}{2} - q\right) = \frac{6\mu\Omega R}{h^3}(h - h_0) \quad (6.48)$$

where $h_0 = 2q/\Omega R$ which, for steady conditions, is constant and may be regarded as representing the value of h at which $\mathrm{d}p/\mathrm{d}x = 0$. By putting $x = R\theta$ and $h = c(1 + \varepsilon \cos\theta)$ we transform eqn 6.48 to

$$\frac{1}{R}\frac{\mathrm{d}p}{\mathrm{d}\theta} = \frac{6\mu\Omega R}{c^2}\left\{\frac{1}{(1 + \varepsilon \cos\theta)^2} - \frac{h_0}{c(1 + \varepsilon \cos\theta)^3}\right\} \quad (6.49)$$

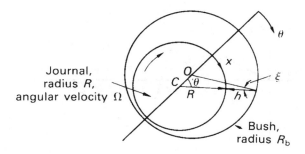

Journal, radius R, angular velocity Ω

Bush, radius R_b

Fig. 6.22

The pressure p may now be determined by integrating eqn 6.49 with respect to θ. This may be achieved by using a new variable α such that

$$1 + \varepsilon \cos \theta = \frac{1 - \varepsilon^2}{1 - \varepsilon \cos \alpha} \qquad (6.50)$$

and thus $d\theta = (1 - \varepsilon^2)^{1/2} d\alpha / (1 - \varepsilon \cos \alpha)$. Then $pc^2/6\mu\Omega R^2 = C_p$, say

$$= \frac{\alpha - \varepsilon \sin \alpha}{\left(1 - \varepsilon^2\right)^{3/2}} - \frac{h_0}{c\left(1 - \varepsilon^2\right)^{5/2}}$$

$$\times \left\{ \alpha\left(1 + \frac{\varepsilon^2}{2}\right) - 2\varepsilon \sin \alpha + \frac{\varepsilon^2}{4} \sin 2\alpha \right\} + C_0 \qquad (6.51)$$

where C_0 is an integration constant.

Now if the lubricant occupies the entire space between journal and bush, eqn 6.51 must give the same value for p at the maximum clearance whether we set $\theta = 0$ or 2π, i.e. $\alpha = 0$ or 2π. Hence

$$0 + C_0 = \frac{2\pi}{\left(1 - \varepsilon^2\right)^{3/2}} - \frac{h_0}{c\left(1 - \varepsilon^2\right)^{5/2}} \left\{ 2\pi\left(1 + \frac{\varepsilon^2}{2}\right) \right\} + C_0$$

and so

$$h_0 = \frac{c\left(1 - \varepsilon^2\right)}{1 + \varepsilon^2/2} \qquad (6.52)$$

Substituting this in eqn 6.51 we obtain

$$C_p = \frac{\varepsilon\left(2 - \varepsilon^2 - \varepsilon \cos \alpha\right) \sin \alpha}{\left(1 + \varepsilon^2\right)^{3/2}\left(2 + \varepsilon^2\right)} + C_0$$

$$= \frac{\varepsilon\left(2 + \varepsilon \cos \theta\right) \sin \theta}{\left(2 + \varepsilon^2\right)\left(1 + \varepsilon \cos \theta\right)^2} + C_0 \qquad (6.53)$$

This expression is plotted in Fig. 6.23, which shows that when θ is between π and 2π, the pressure drops considerably below the value at $\theta = \pi$. The result was first obtained by the German physicist Arnold

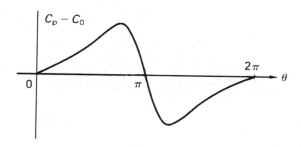

Fig. 6.23

Sommerfeld (1868–1951), and so the existence of these symmetrical high- and low-pressure sections of the graph is termed the 'Sommerfeld boundary condition'.

The condition frequently cannot be achieved in practice. This is because a film of liquid lubricant will break when the pressure falls below the vapour pressure of the liquid, and the bearing will then behave largely as though no lubricant were present in the low pressure region. In any case, appreciable quantities of dissolved air are likely to be released as soon as the pressure falls much below atmospheric. A continuous film of lubricant can of course be preserved if the value of the constant C_0 in eqn 6.53 is sufficiently large. Normally the lubricant is introduced under pressure through a small hole at some point in the bush, and eventually leaves via the ends of the bearing. For any position of the inlet hole, C_0 could be increased by raising the supply pressure, but, as this would often require unacceptably high values, a better means (in theory) of maintaining continuity of the lubricant film would be to locate the inlet hole at the position of minimum pressure. Unfortunately, since this position depends on ε and thus on the load, it is seldom accurately known in practice, and, to ensure that the inlet hole is not in the high pressure region ($\theta < \pi$), the hole is for safety's sake placed beyond the expected position of minimum pressure.

However, if we assume that a continuous film of lubricant can in fact be achieved, the friction torque and the load carried by the bearing may be calculated by using the full Sommerfeld boundary condition. From eqn 6.41 the shear stress at the surface of the journal is

$$\tau_0 = \mu \left(\frac{\partial u}{\partial y} \right)_{y=0} = -\frac{dp}{dx} \left(\frac{h}{2} \right) - \frac{\mu \Omega R}{h}$$

$$= -\frac{3\mu \Omega R}{h^2} (h - h_0) - \frac{\mu \Omega R}{h}$$

$$= \frac{\mu \Omega R}{h^2} (3h_0 - 4h) \tag{6.54}$$

in which substitution for dp/dx was made from eqn 6.48. The torque on unit length of the journal is given by

$$\int_0^{2\pi} R\tau_0 \left(R\,d\theta \right) = \mu \Omega R^3 \int_0^{2\pi} \left(\frac{3h_0 - 4h}{h^2} \right) d\theta$$

$$= \frac{\mu \Omega R^3}{c} \int_0^{2\pi} \left\{ \frac{3h_0}{c(1 + \varepsilon \cos \theta)^2} - \frac{4}{1 + \varepsilon \cos \theta} \right\} d\theta$$

Again using the substitution 6.50, we may evaluate the integral and obtain

$$\text{Torque per unit length} = -\frac{4\pi\mu\Omega R^{3}\left(1 + 2\varepsilon^{2}\right)}{c\left(1 - \varepsilon^{2}\right)^{1/2}\left(2 + \varepsilon^{2}\right)} \qquad (6.55)$$

By the sign convention for viscous stress (see Section 1.6.1), the direction of this torque is that of increasing θ, that is, the direction of the journal's rotation. The minus sign in eqn 6.55 thus indicates that the torque opposes the rotation. It will be noticed that for very small eccentricity ratios eqn 6.55 reduces to eqn 6.33, sometimes known as Petroff's Law. (The name – that of the Russian engineer Nikolaĭ Pavlovich Petroff (1836–1920) – may also be transliterated Petrov.)

The load that the journal will bear may be determined by calculating its components perpendicular and parallel to OC. On a small element of the journal surface subtending an angle $\delta\theta$ at the centre C, the force in the direction perpendicular to OC is

$$pR\,\delta\theta\,\sin\theta + \tau_{0}R\,\delta\theta\,\cos\theta$$

per unit length of the bearing, and hence the total force per unit length is

$$F_{\perp oc} = R\int_{0}^{2\pi} p \sin\theta\,d\theta + R\int_{0}^{2\pi}\tau_{0}\cos\theta\,d\theta$$

$$= R\left[-p\cos\theta\right]_{0}^{2\pi} - R\int_{0}^{2\pi}(-\cos\theta)\frac{dp}{d\theta}d\theta + R\int_{0}^{2\pi}\tau_{0}\cos\theta\,d\theta$$

Since the first term has identical values at both limits we have

$$F_{\perp oc} = R\int_{0}^{2\pi}\left(\frac{dp}{d\theta} + \tau_{0}\right)\cos\theta\,d\theta \qquad (6.56)$$

However, it may be shown that the contribution of τ_{0} is less than c/R times that of $dp/d\theta$ and so for simplicity we may justifiably neglect τ_{0} in the expression 6.56. Obtaining $dp/d\theta$ from eqn 6.49 and again using the substitution 6.50, we finally arrive at

$$F_{\perp oc} = \frac{12\pi\mu\Omega R^{3}\varepsilon}{c^{2}\left(1 - \varepsilon^{2}\right)^{1/2}\left(2 + \varepsilon^{2}\right)} \qquad (6.57)$$

Similarly, the total force, per unit length, parallel to OC is

$$R\int_{0}^{2\pi} p\cos\theta\,d\theta + R\int_{0}^{2\pi}\tau_{0}\sin\theta\,d\theta$$

$$= R\left[p\sin\theta\right]_{0}^{2\pi} + R\int_{0}^{2\pi}\left(\tau_{0} - \frac{dp}{d\theta}\right)\sin\theta\,d\theta$$

but this may be shown to be exactly zero. Consequently the *attitude angle* (that is, the angle between OC and the direction of the load) is 90°, and the total load is given by eqn 6.57. (The attitude angle is often known simply as *attitude*. In America, however, 'attitude' sometimes refers to the eccentricity ratio ε.) It is more usual to put eqn 6.57 in dimensionless form:

$$\frac{Fc^2}{\mu\Omega R^3} = \frac{12\pi\varepsilon}{\left(1 - \varepsilon^2\right)^{1/2}\left(2 + \varepsilon^2\right)} \qquad (6.58)$$

where F represents the total load per unit length of the bearing. The left-hand side of eqn 6.58 is often termed the Sommerfeld number but unfortunately various modified forms are sometimes given the same name. It is therefore essential to check the definition a writer uses.

Equation 6.58 is plotted in Fig. 6.24, from which it will be seen that the greater the load the greater the eccentricity. For a heavily loaded bearing, ε approaches 1.0 and the minimum clearance $= c(1 - \varepsilon)$ becomes extremely small: unless the surfaces are very well finished and the lubricant is completely free from solid particles, it may not then be possible to maintain the film of lubricant. (Designers commonly aim at $\varepsilon \simeq 0.6$.)

There are two important limitations to this analysis. One is that the value of the constant C_0 in eqn 6.51 may not be sufficient to ensure a continuous film of lubricant throughout the bearing and the limits 0 and 2π applied to the various integrals are then not appropriate. Many assumptions about the extent of the film have been suggested, none entirely satisfactory, but for discussion of these reference must be made to more specialized works[3,4,5].

The other principal limitation is that the bearing is assumed to be infinitely long so that there is no flow in the axial direction. For lengths more than about four times the diameter the assumption is not

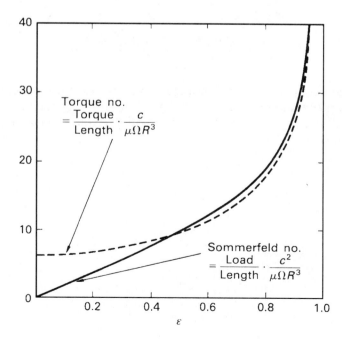

Fig. 6.24

unreasonable but in practice it is rare for journal bearings to have lengths that much exceed the diameter. To account for axial flow in addition to that already considered involves great mathematical complexity, but we can readily investigate the behaviour, under steady conditions, of a very short bearing.

Very short bearing

Here there is, in general, a large pressure gradient in the axial (z) direction. Again, because the clearance is very small compared with the radius, the curvature of the lubricant film may be neglected, and at a particular position within it we consider a small box-like element over an area $\delta x \times \delta z$ of the journal surface where the clearance is h (Fig. 6.25). In the circumferential (x) direction we suppose the volume rate of flow to be Q_x within a space δz wide, and so the net rate at which fluid flows out of the box in this direction is

$$\left(Q_x + \frac{\partial Q_x}{\partial x} \delta x \right) - Q_x = \frac{\partial Q_x}{\partial x} \delta x$$

Similarly there is a net outflow $(\partial Q_z/\partial z)\,\delta z$ from the box in the z direction. If, at the particular position considered, h does not vary with time (that is, if both the speed and the load are steady), the total net outflow of lubricant from the box must be zero.

$$\therefore \quad \frac{\partial Q_x}{\partial x} \delta x + \frac{\partial Q_z}{\partial z} \delta z = 0 \tag{6.59}$$

To obtain $\partial Q_x/\partial x$ we make appropriate substitutions in eqn 6.23:

$$\frac{\partial Q_x}{\partial x} = \frac{\partial}{\partial x} \left[\delta z \left\{ -\frac{\partial p}{\partial x} \frac{h^3}{12\mu} + \frac{\Omega R h}{2} \right\} \right]$$

$$= \left\{ \frac{\Omega R}{2} \frac{\partial h}{\partial x} - \frac{1}{12} \frac{\partial}{\partial x} \left(\frac{h^3}{\mu} \frac{\partial p}{\partial x} \right) \right\} \delta z \tag{6.60}$$

where the negligible changes of elevation allow us to drop the asterisk from p^* and δz takes the place of b. The corresponding equation for the z (axial) direction, in which there is no relative velocity between journal and bush, is

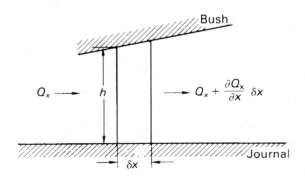

Fig. 6.25

$$\frac{\partial Q_z}{\partial z} = -\frac{1}{12}\frac{\partial}{\partial z}\left(\frac{h^3}{\mu}\frac{\partial p}{\partial z}\right)\delta x \tag{6.61}$$

Substituting eqns 6.60 and 6.61 into eqn 6.59, we obtain

$$\frac{\Omega R}{2}\frac{\partial h}{\partial x}\delta x\,\delta z - \frac{1}{12}\frac{\partial}{\partial x}\left(\frac{h^3}{\mu}\frac{\partial p}{\partial x}\right)\delta x\,\delta z - \frac{1}{12}\frac{\partial}{\partial z}\left(\frac{h^3}{\mu}\frac{\partial p}{\partial z}\right)\delta x\,\delta z = 0$$

If the axes of journal and bush are parallel, h will be independent of z, and dependent only on x. With μ again assumed constant throughout, the equation thus becomes

$$6\mu\Omega R\frac{dh}{dx} - \frac{\partial}{\partial x}\left(h^3\frac{\partial p}{\partial x}\right) - h^3\frac{\partial^2 p}{\partial z^2} = 0 \tag{6.62}$$

The position $z = 0$ may be taken mid-way between the ends of the bearing; for a given value of x, the pressure varies from, say, p_c at $z = 0$ to atmospheric at each end of the bearing (where $z = \pm L/2$). That is, p/p_c varies symmetrically along the length of the bearing from zero to 1 and back to zero. It is reasonable to suppose that the form of this variation is the same whatever the value of x; that is, that $f = p/p_c$ is a function of z only. (We shall show later that for a short bearing this assumption is justified.) Substituting $p_c f$ for p in eqn 6.62 gives

$$6\mu\Omega R\frac{dh}{dx} - \frac{d}{dx}\left(h^3\frac{dp_c}{dx}\right)f - h^3 p_c\frac{d^2 f}{dz^2} = 0 \tag{6.63}$$

and division by $h^3 p_c$ then yields

$$a - b^2 f - \frac{d^2 f}{dz^2} = 0 \tag{6.64}$$

where

$$a = \frac{6\mu\Omega R}{h^3 p_c}\frac{dh}{dx} \quad \text{and} \quad b^2 = \frac{1}{h^3 p_c}\frac{d}{dx}\left(h^3\frac{dp_c}{dx}\right)$$

Both a and b are independent of z and so the solution of eqn 6.64 is

$$f = \frac{a}{b^2} + A\sin(bz + B)$$

Symmetry requires $df/dz = Ab\cos(bz + B)$ to be zero when $z = 0$; hence $\cos B = 0$. Consequently $\sin(bz + B) = \pm\cos bz$ and

$$f = \frac{a}{b^2} + (\pm A)\cos bz$$

Putting $f = 0$ at $z = \pm L/2$ then gives $\pm A = -(a/b^2)\sec(bL/2)$, and so

$$f = \frac{a}{b^2}\left\{1 - \sec\left(\frac{bL}{2}\right)\cos bz\right\} \tag{6.65}$$

Moreover, $f = 1$ when $z = 0$; thus $a/b^2 = \{1 - \sec(bL/2)\}^{-1}$ and so

$$f = \frac{1 - \sec(bL/2)\cos bz}{1 - \sec(bL/2)} = \frac{\cos(bL/2) - \cos bz}{\cos(bL/2) - 1}$$

For small values of L (and thus of z), expansion of the cosines yields $f = 1 - 4z^2/L^2$ when higher orders of the small quantities are neglected. This confirms our hypothesis that, for a short bearing, f is independent of x. (In parts of the bearing b^2 may be negative, but the solution is no less valid when b is imaginary.)

Substituting for a in eqn 6.65 we now obtain

$$p = p_c f = p_c \frac{6\mu\Omega R}{h^3 p_c} \frac{dh}{dx} \frac{1}{b^2} \left\{ 1 - \sec\left(\frac{bL}{2}\right)\cos bz \right\}$$

which, for small values of L, becomes

$$p = \frac{3\mu\Omega R}{h^3} \frac{dh}{dx} \left(z^2 - \frac{L^2}{4} \right) \tag{6.66}$$

We now put $x = R\theta$ and $h = c(1 + \varepsilon\cos\theta)$ (eqn 6.47) whence

$$p = \frac{3\mu\Omega}{c^2} \left(\frac{L^2}{4} - z^2 \right) \frac{\varepsilon\sin\theta}{(1 + \varepsilon\cos\theta)^3}$$

This result indicates that the pressure is negative between $\theta = \pi$ and $\theta = 2\pi$. Unless the bearing is so lightly loaded that ε is very small, the reduction of pressure below atmospheric is sufficient to cause the film to break in this region. Then only the lubricant between $\theta = 0$ and $\theta = \pi$ is effective in sustaining a load. (This is known as the *half Sommerfeld condition*.)

Again the contribution of the shear stress τ_0 to supporting a load is negligible, and the force resulting from pressure at a small element $R\,\delta\theta\,\delta z$ of the journal surface is $pR\,\delta\theta\,\delta z$ with components $pR\cos\theta\,\delta\theta\,\delta z$ and $pR\sin\theta\,\delta\theta\,\delta z$ respectively parallel and perpendicular to OC. These expressions may be integrated with respect to z between the limits $z = -L/2$ and $z = L/2$, and (with the help of the substitution 6.50) with respect to θ between the limits $\theta = 0$ and $\theta = \pi$. Then

$$\text{Force parallel to } OC = -\frac{\mu\Omega R L^3 \varepsilon^2}{c^2(1 - \varepsilon^2)^2}$$

$$\text{Force perpendicular to } OC = \frac{\mu\Omega R L^3 \pi\varepsilon}{4c^2(1 - \varepsilon^2)^{3/2}}$$

The resultant of these two components is

$$\frac{\pi\mu\Omega R L^3 \varepsilon}{4c^2(1 - \varepsilon^2)^2} \left\{ \frac{16\varepsilon^2}{\pi^2} + (1 - \varepsilon^2) \right\}^{1/2} = \frac{\pi\mu\Omega R L^3 \varepsilon}{4c^2(1 - \varepsilon^2)^2} \left\{ \left(\frac{16}{\pi^2} - 1 \right)\varepsilon^2 + 1 \right\}^{1/2}$$

This resultant force exerted by the lubricant on the journal is equal and opposite to the load F which the journal supports. The attitude angle ψ (Fig. 6.26) is given by the ratio of the components

$$\tan\psi = \frac{\pi\left(1 - \varepsilon^2\right)^{1/2}}{4\varepsilon}$$

Under steady conditions the rate at which lubricant must be supplied to the bearing equals the rate at which it escapes from the sides. According to the theory of the very short bearing, pressure is positive only between $\theta = 0$ and $\theta = \pi$, so lubricant escapes only from this half of the bearing. (In practice, some which has escaped may then be drawn back into the other half and thus recirculated, but this amount is likely to be small.) By eqn 6.20 the volume flow rate in the z (axial) direction for a short element dx of the circumference is

$$-\frac{\partial p}{\partial z}\frac{h^3}{12\mu}dx$$

which, on substitution from eqn 6.66, becomes

$$-\frac{6\mu\Omega R}{h^3}\frac{dh}{dx}\,z\,\frac{h^3}{12\mu}dx = -\frac{\Omega Rz}{2}dh$$

Integrating this between $\theta = 0$ and $\theta = \pi$, that is, from $h = c(1 + \varepsilon)$ to $h = c(1 - \varepsilon)$, gives a total volume flow rate $\Omega R\varepsilon cz$. Consequently, at one end of the bearing, where $z = L/2$, the lubricant escapes at the rate $\Omega R\varepsilon cL/2$; a similar amount escapes at the other end, and thus the total volume to be supplied in unit time is $\Omega R\varepsilon cL$.

In conclusion we may remark that the hydrodynamics of lubrication is not, in any event, the whole story. Two oils may have the same viscosity yet not be equally effective as lubricants in a particular application. There are other characteristics – mainly in the realm of physical chemistry – that are of great importance. Also, conditions of low speed or high load may reduce the thickness of the lubricant film to only a few molecules and it may then take on properties different from those of the normal fluid. And, the dissipation of energy as heat may not only result

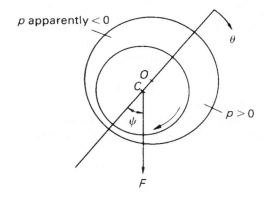

Fig. 6.26

in non-uniform viscosity of the fluid, but may also cause distortion of the journal and the bush. For these further aspects of the subject, more specialist works[3,4,5] should be consulted.

6.8   LAMINAR FLOW THROUGH POROUS MEDIA

There are many important instances of the flow of fluids through porous materials. For example, the movement of water, oil and natural gas through the ground, seepage underneath dams and large buildings, flow through packed towers in some chemical processes, filtration – all these depend on this type of flow. The velocity is usually so small and the flow passages so narrow that laminar flow may be assumed without hesitation. Rigorous analysis of the flow is not possible because the shape of the individual flow passages is so varied and so complex. Several approximate theories have, however, been formulated, and we shall briefly examine the principal ones.

In 1856 the French engineer Henri Darcy (1803–58) published experimental results from which he deduced what is now known as Darcy's Law:

$$\bar{u} = -\frac{C\partial p^*}{\partial x} \tag{6.67}$$

In this expression x refers to the average direction of flow, $\bar{u}$ and $\partial p^*/\partial x$ represent respectively the steady mean velocity and the rate of increase of piezometric pressure in this direction, and C is a constant at a given temperature for a particular fluid (free from suspended solid particles) and for the piece of porous medium concerned. To calculate the mean velocity we consider a cross-sectional area ΔA (perpendicular to the x direction) through which a volume of fluid ΔQ passes in unit time. Then, if ΔA is large enough to contain several flow passages, $\bar{u} = \Delta Q/\Delta A$. The minus sign appears in eqn 6.67 because p^* decreases in the mean flow direction, i.e. $\partial p^*/\partial x$ is negative. The direct proportionality between the mean velocity and $\partial p^*/\partial x$ is characteristic of steady laminar flow. Further experiments have shown that the mean velocity is inversely proportional to the viscosity μ of a Newtonian fluid, and this result too is characteristic of steady laminar flow.

The value of C in eqn 6.67 depends not only on the viscosity of the fluid but also on the size and geometrical arrangement of the solid particles in the porous material. It is thus difficult to predict the value of C appropriate to a particular set of conditions. One of many attempts to obtain a more precise expression is that made by the Austrian Josef Kozeny* (1889–1967) in 1927.

He considered the porous material to be made up of separate small solid particles, and the flow to be, broadly speaking, in one direction only: thus, although the fluid follows a somewhat tortuous path through the material, there is no net flow across the block of material in any

---

* pronounced *Kozz-ay'-nee.*

direction other than (say) the x direction. Whatever the actual shape of the individual flow passages the overall result is the same as if, in place of the porous material, there were a large number of parallel, identical, capillary tubes with their axes in the x direction. Kozeny then reasoned that, since the resistance to flow results from the requirement of no slip at any solid boundary, the capillary tubes and the porous medium would be truly equivalent only if, for a given volume occupied by the fluid, the total surface area of the solid boundaries were the same in each case.

In the porous medium the ratio of the volume of voids (i.e. the spaces between the solid particles) to the total volume is known as the *voidage* or *porosity*, ε. That is

$$\varepsilon = \frac{\text{Volume of voids}}{\text{Volume of voids} + \text{Volume of solids}} = \frac{V_v}{V_v + V_s}$$

whence

$$V_v = \frac{V_s \varepsilon}{1 - \varepsilon} \tag{6.68}$$

Therefore

$$\frac{\text{Volume of voids}}{\text{Total surface area}} = \frac{V_v}{S} = \frac{V_s \varepsilon}{(1 - \varepsilon)S} \tag{6.69}$$

The corresponding value for a capillary tube of internal diameter d and length l is $(\pi/4)d^2l/\pi dl = d/4$. Thus if the set of capillary tubes is to be equivalent to the porous medium then each must have an internal diameter

$$d = \frac{4V_s \varepsilon}{(1 - \varepsilon)S} \tag{6.70}$$

Now if all the flow passages in the porous material were entirely in the x direction, the mean velocity of the fluid in them would be $\bar{u}/\varepsilon$ (because only a fraction ε of the total cross-section is available for flow). The actual paths, however, are sinuous and have an average length l_e which is greater than l, the thickness of the porous material. Philip C. Carman later pointed out that the mean velocity in the passages is therefore greater than if the passages were straight, and is given by $(\bar{u}/\varepsilon)(l_e/l)$. Flow at this mean velocity in capillary tubes of length l_e would require a drop of piezometric pressure Δp^* given by Poiseuille's equation (6.8)

Kozeny–Carman equation

$$\frac{\Delta p^*}{l_e} = \frac{32\mu}{d^2}\frac{Q}{\pi d^2/4} = \frac{32\mu}{d^2}\frac{\bar{u}}{\varepsilon}\frac{l_e}{l} \tag{6.71}$$

Rearrangement of eqn 6.71 and substitution from eqn 6.70 yields

$$\bar{u} = \frac{\Delta p^*}{l_e}\frac{\varepsilon d^2}{32\mu}\frac{l}{l_e} = \frac{\Delta p^*}{l}\frac{\varepsilon^3}{(1 - \varepsilon)^2}\frac{V_s^2}{2\mu S^2}\frac{l^2}{l_e^2}$$

$$= \frac{\Delta p^*}{l}\frac{\varepsilon^3}{(1 - \varepsilon)^2}\frac{1}{\mu k(S/V_s)^2} \tag{6.72}$$

where $k = 2(l_e/l)^2$. Equation 6.72 is known as the Kozeny–Carman equation, and k as the Kozeny constant or Kozeny function.

The porosity ε is seen to be an important factor in determining the mean velocity of flow and small changes in ε markedly affect the value of $\varepsilon^3/(1 - \varepsilon)^2$. However, for given particles randomly packed, ε is approximately constant. Most granular materials have values between 0.3 and 0.5, but values as high as 0.9 may be obtained with specially designed packings such as rings. For packings of spheres of similar size ε is usually between 0.4 and 0.55. An increase in the size range of the particles generally reduces the porosity, and in the neighbourhood of solid boundaries ε is usually increased because the boundaries prevent the close interlocking of particles.

Kozeny's function k depends to some extent on ε and it depends also on the arrangement of particles, their shape and their size range. Nevertheless, experiment shows that for many materials k ranges only between about 4.0 and 6.0, and a mean value of 5.0 is commonly adopted.

Because the solid particles touch one another, part of the total surface area S is ineffective in causing resistance to flow – especially if there are flat-sided particles. Moreover, any re-entrant parts of the surface make little, if any, contribution to the resistance. Consequently S would be better defined as the *effective* surface area. In practice, however, the difference between the total and the effective surface area is absorbed into the experimentally determined value of k.

The concept of flow paths of average length l_e is not quite so simple as may at first appear because individual paths are interconnected in a complex way and there may be numerous abrupt changes in cross-section at which additional pressure drops can occur. Also, as the flow paths are not straight there are accelerations and thus inertia forces which do not arise in flow through straight capillary tubes.

Yet in spite of such uncertainties in its derivation, the Kozeny–Carman equation is quite well supported by experiment. For particles of given specific surface S/V_s, randomly packed, similar values of k are obtained in different ranges of ε. The equation becomes unreliable, however, if the packing of particles is not random, if the porosity is very high ($\varepsilon \to 1$), or if local values of the ratio given by eqn 6.69 vary widely from the mean.

6.8.1 Fluidization

The force (other than buoyancy) exerted by a fluid on the solid particles in a porous medium increases as the velocity of the fluid increases. Thus, if flow occurs upwards through a stationary bed of solid particles, a condition may be reached in which the upward force on the particles equals their weight. The particles are then supported by the fluid and the bed is said to be fluidized.

If ρ_s and ρ represent the density of the solid particles and the fluid respectively and ε the porosity, then, for a thickness l of the bed, the effective weight of the particles (i.e. actual weight minus buoyancy) is $(\rho_s - \rho)gl(1 - \varepsilon)$ per unit cross-sectional area. This must be exactly balanced, when fluidization begins, by the drop in piezometric pressure through the bed as given by eqn 6.72. Hence the minimum fluidizing velocity is given by

$$\bar{u}_f = \frac{(\rho_s - \rho)g\varepsilon^3}{(1 - \varepsilon)\mu k(S/V_s)^2}$$

If the velocity is increased beyond this value, there is little further change in the pressure drop (although some fluctuation may occur as fluidization begins because movement of the particles produces non-uniform porosity). When ρ_s and ρ are not widely different the bed expands uniformly; this is known as *particulate* fluidization. For a large density difference, as with a gas, 'bubbles' of fluid, mostly devoid of solid particles, may rise through the bed; the upper surface then looks rather like that of a boiling liquid. This type of fluidization is called *aggregative* and the bed is described as a boiling bed.

When the fluid velocity exceeds the settling velocity for a particle the particle is carried away. Where the particles are not of uniform size, this process is of course gradual, the smaller particles being removed at lower velocities.

Fluidization is a valuable method for handling solids such as grain and pulverized coal: also, because it allows the maximum area of contact between solid and fluid, it finds wide application in catalytic and heat transfer processes in the chemical industry.

REFERENCES

1. Van Wazer, J. R. et al. *Viscosity and Flow Measurement*, Interscience, New York (1963).
2. Dinsdale, A. and Moore, F. *Viscosity and its Measurement*, The Institute of Physics and The Physical Society, London (1962).
3. Cameron, A. et al. *Principles of Lubrication* (2nd edn), Longman, London (1983).
4. Tipei, N. *Theory of Lubrication*, Stanford University Press, California (1962).
5. Barwell, F. T. *Bearing Systems*, Oxford University Press (1979).

FURTHER READING

Papers by Hagen, Poiseuille and Hagenbach were reprinted in *Ostwald's Klassiker der exakten Wissenschaften*, No. 237 (1933).
Bear, J. *Dynamics of Fluids in Porous Media*, Elsevier, Amsterdam (1984).
Reynolds, O. 'On the theory of lubrication' (etc.) *Phil. Trans. R. Soc.* **177**, 157–234 (1886).

PROBLEMS

6.1 Glycerin (relative density 1.26, viscosity 0.9 Pa s) is pumped at 20 litres/s through a straight, 100 mm diameter pipe, 45 m long, inclined at 15° to the horizontal. The gauge pressure at the lower, inlet, end of the pipe is 590 kPa. Verify that the flow is laminar and, neglecting 'end effects', calculate the pressure at the outlet end of the pipe and the average shear stress at the wall.

6.2 Show that when laminar flow occurs with mean velocity u_m between extensive stationary flat plates the mean kinetic energy per unit mass of the fluid is $1.543\, u_m^2/2$.

6.3 A Bingham plastic with a yield stress of 120 Pa and apparent viscosity 1.6 Pa s is to be pumped through a horizontal pipe, 100 mm diameter and 15 m long at 10 litres/s. Neglecting 'end effects' and assuming that the flow is laminar, determine the pressure difference required.

6.4 A cylindrical drum of length l and radius r can rotate inside a fixed concentric cylindrical casing, the clearance space c between the drum and the casing being very small and filled with a liquid of viscosity μ. To rotate the drum with angular velocity ω requires the same power as to pump the liquid axially through the clearance space while the drum is stationary, and the pressure difference between the ends of the drum is p. The motion in both cases is laminar. Neglecting 'end effects', show that

$$p = \frac{2\mu l r \omega \sqrt{3}}{c^2}.$$

6.5 Two circular plane discs of radius R_2 have their axes vertical and in line. They are separated by an oil film of uniform thickness. The oil is supplied continuously at a pressure p_1 to a well of radius R_1 placed centrally in the lower disc, from where it flows radially outwards and escapes to atmosphere. The depth of the well is large compared with the clearance between the discs. Show that the total force tending to lift the upper disc is given by

$$\frac{\pi p_1 \left(R_2^2 - R_1^2\right)}{2 \ln\left(R_2/R_1\right)}$$

and determine the clearance when the viscosity of the oil is 0.08 Pa s, the flow 0.85 litre/s, $p_1 = 550$ kPa, $R_1 = 12.5$ mm and $R_2 = 50$ mm. Assume laminar flow and neglect 'end effects'.

6.6 Two coaxial flat discs of radius a are separated by a small distance h. The space between them is filled with fluid, of viscosity μ, which is in direct contact with a large volume of the same fluid at atmospheric pressure at the periphery of the discs. Working from first principles for the theory of laminar flow between stationary flat plates of indefinite extent, show that the force necessary to separate the discs in the axial direction is given by $(3\pi\mu a^4/2h^3)(dh/dt)$ where dh/dt represents the rate of separation of the discs.

6.7 A piston 113 mm diameter and 150 mm long has a mass of 9 kg. It is placed in a vertical cylinder 115 mm diameter containing oil of viscosity 0.12 Pa s and relative density 0.9 and it falls under its own weight. Assuming that piston and cylinder are concentric, calculate the time taken for the piston to fall steadily through 75 mm.

6.8 What is the maximum diameter of a spherical particle of dust of density 2500 kg/m³ which will settle in the atmosphere (density 1.225 kg/m³, kinematic viscosity 14.9 mm²/s) in good agreement with Stokes's Law? What is its settling velocity?

6.9 A steel sphere, 1.5 mm diameter and of mass 13.7 mg, falls steadily in oil through a vertical distance of 500 mm in 56 s. The oil has a density of 950 kg/m³ and is contained in a drum so large that any 'wall effects' are negligible. What is the viscosity of the oil? Verify any assumptions made.

6.10 In a rotary viscometer the radii of the cylinders are respectively 50 mm and 50.5 mm, and the outer cylinder is rotated steadily at 30 rad/s. For a certain liquid the torque is 0.45 N m when the depth of the liquid is 50 mm and the torque is 0.81 N m when the depth is 100 mm. Estimate the viscosity of the liquid.

6.11 A plane bearing plate is traversed by a very wide, 150 mm long, plane inclined slipper moving at 1.5 m/s. The clearance between slipper and bearing plate is 0.075 mm at the 'toe' and 0.025 mm at the 'heel'. If the load to be sustained by the bearing is 500 kN per metre width, determine the viscosity of the lubricant required, the power consumed per metre width of the bearing, the maximum pressure in the lubricant and the position of the centre of pressure.

6.12 In a wide slipper bearing of length l the clearance is given by

$$h = h_1 \exp(-x/l)$$

in which h_1 denotes the clearance at the inlet (where $x = 0$). Assuming constant density and viscosity of the lubricant, determine the position of the maximum pressure in the bearing.

6.13 A journal bearing of length 60 mm is to support a steady load of 20 kN. The journal, of diameter 50 mm, runs at 10 rev/s. Assuming $c/R = 0.001$ and $\varepsilon = 0.6$, determine a suitable viscosity of the oil, using the theory of (a) the very long bearing with the full Sommerfeld condition, (b) the very short bearing with the half Sommerfeld condition. For (a) determine the power required to overcome friction, and the best position for the oil supply hole relative to the load line. Would this position also be suitable according to theory (b)?

Turbulent flow in pipes | 7

7.1 INTRODUCTION

Flow in pipes is usually turbulent and therefore highly complex. Random fluctuating components are superimposed on the main flow, and as these haphazard movements are unpredictable no complete theory has yet been developed for the analysis of turbulent flow. Even the most advanced theories depend at some point on experimentally derived information. Much of this information about turbulent flow has been gained from the study of flow in pipes. Not only is such flow of considerable practical importance but the study of it has brought better understanding of turbulent flow in general.

The main concern of this chapter will be to introduce those experimental results needed to calculate the dissipation of energy in pipelines. A more detailed analysis of turbulent flow, however, requires boundary-layer theory, and some consideration will be given to this in Chapter 8. We shall here confine our attention to homogeneous fluids of constant viscosity and constant density. (The results are applicable to gases provided that density changes are small.) Pipes or conduits that are not completely full of the fluid concerned (e.g. sewers and culverts) are in effect open channels and these are treated separately in Chapter 10. The flow of compressible fluids is discussed in Chapter 11.

7.2 HEAD LOST TO FRICTION IN A PIPE

One of the most important items of information the engineer needs is the difference of piezometric head required to force fluid at a certain steady rate through a pipe. About the middle of the nineteenth century, therefore, many experimenters devoted attention to this topic. Among them was the French engineer Henri Darcy (1803–58) who investigated the flow of water, under turbulent conditions, in long, unobstructed, straight pipes of uniform diameter. The dissipation of energy by fluid friction results in a fall of piezometric head in the direction of flow, and if the pipe is of uniform cross-section and roughness, and the flow is 'fully developed', that is, if it is sufficiently far from the inlet of the pipe for conditions to have become settled, the piezometric head falls uniformly. Darcy's results suggest the formula (now commonly named after him):

$$h_f = \frac{\Delta p^*}{\varrho g} = \frac{4fl}{d} \frac{\bar{u}^2}{2g} \tag{7.1}$$

which, with certain provisos (see Section 7.5), may be generalized to

$$h_f = \frac{fl}{m} \frac{\bar{u}^2}{2g} \tag{7.2}$$

Hydraulic mean depth

In these equations h_f represents the 'head lost to friction' corresponding (in steady flow) to the drop Δp^* of piezometric pressure over length l of the pipe, ϱ represents the density of the fluid, $\bar{u}$ the mean velocity (that is, discharge divided by cross-sectional area), f is a coefficient, g the weight per unit mass, d the pipe diameter and m the *hydraulic mean depth*, that is, the ratio of the cross-sectional area of the flow to the perimeter in contact with the fluid. For a circular section *completely full of the fluid* $m = A/P = (\pi/4)d^2/\pi d = d/4$ and so eqns 7.1 and 7.2 are seen to be equivalent. (The hydraulic mean depth has unfortunately sometimes been termed the *hydraulic radius*, even though it is not equal to the actual radius.) The head loss, it will be noted, is expressed in terms of the velocity head $\bar{u}^2/2g$ corresponding to the mean velocity and this is why the 4 in the numerator of eqn 7.1 is left uncancelled by the 2 in the denominator.

Friction factor

The coefficient f is usually known as the *friction factor*. Comparison of the dimensional formulae of the two sides of eqn 7.1 shows that f is simply a numeric without units. Observations show that its value depends primarily on the relative roughness of the pipe surface, but also on the Reynolds number of the flow. In America the friction factor commonly used is equal to 4 times that defined by eqns 7.1 and 7.2. Although the use of the symbol λ for $4f$ is now encouraged (giving $h_f = \lambda l \bar{u}^2/2gd$) the symbol f is unfortunately most often employed and care should therefore be taken to check whether f is that defined by eqn 7.1 or 4 times that value. In this book f always denotes the value defined by eqn 7.1 (or 7.2).

The value of f is simply related to the mean stress $\bar{\tau}_0$ at the wall of the pipe. Over a short length δx of the pipe the head lost to friction $\delta h_f = f \delta x \, \bar{u}^2/2gm$ (from eqn 7.2) and so $f = (2gm/\bar{u}^2)dh_f/dx$. We shall show in Section 7.4 that $\bar{\tau}_0$ and dp^*/dx are related by the expression

$$\bar{\tau}_0 = \frac{A}{P} \frac{dp^*}{dx} = m \frac{dp^*}{dx}$$

The drop in piezometric pressure $-\delta p^* = \varrho g \, \delta h_f$ and so

$$\frac{dh_f}{dx} = -\frac{1}{\varrho g} \frac{dp^*}{dx} = -\frac{\bar{\tau}_0}{\varrho g m} \tag{7.3}$$

and

$$f = -\frac{2gm}{\bar{u}^2}\frac{\bar{\tau}_0}{\varrho gm} = \frac{|\bar{\tau}_0|}{\frac{1}{2}\varrho\bar{u}^2} \tag{7.4}$$

(The minus sign in eqn 7.4 arises because we define it as a forward stress on the fluid, whereas it is actually in the opposite direction. It is usual to consider only the magnitude of $\bar{\tau}_0$.)

As we have said, the value of f depends on the roughness of the walls of the pipe. Many attempts have therefore been made to express the roughness quantitatively. All surfaces, no matter how well polished, are to some extent rough. The irregularities of the surface usually vary greatly in shape, size and spacing. To specify the roughness quantitatively is therefore extremely difficult. Nevertheless, one feature that may be expected to influence the flow appreciably is the average height of the 'bumps' on the surface. Following Reynolds's reasoning (Section 5.3), however, we realize that it is not the *absolute* size of the bumps that would be significant, but their size in relation to some other characteristic length of the system. It is to be expected, then, that it is the *relative* roughness that affects the flow. For a circular pipe, the relative roughness may be suitably expressed as the ratio of the average height k of the surface 'bumps' to the pipe diameter d.

Dimensional analysis may be used to show that f is a function both of Reynolds number and of the relative roughness k/d. True, other features of the roughness, such as the spacing of the bumps, may influence the flow. Yet the use of k/d as a simple and measurable indication of the relative roughness has made possible notable progress towards the solution and understanding of a very complex problem.

7.3 VARIATION OF FRICTION FACTOR

Since f is a function of Reynolds number and of the relative roughness k/d, experimental data on friction in pipes may be represented diagrammatically in the form shown in Fig. 7.1. One of the first to present results in this way was Sir Thomas E. Stanton (1865–1931) who, with J. R. Pannell, conducted experiments on a number of pipes of various diameters and materials, and with various fluids. Figure 7.1 is based on results obtained by the German engineer Johann Nikuradse using circular pipes that had been artificially roughened.

At the left-hand side of the diagram is a single line corresponding to Reynolds numbers $\bar{u}d/\nu$ less than 2000. In this range the flow is laminar and is governed by Poiseuille's equation (6.7). A rearrangement of that equation gives

$$h_f = \frac{\Delta p^*}{\varrho g} = \frac{8Ql\mu}{\pi R^4 \varrho g} = \frac{8\bar{u}l\mu}{R^2\varrho g}$$

since the mean velocity $\bar{u} = Q/\pi R^2$. But a rearrangement of eqn 7.1 gives $f = (2gd/4l\bar{u}^2)h_f$ and so, for laminar flow,

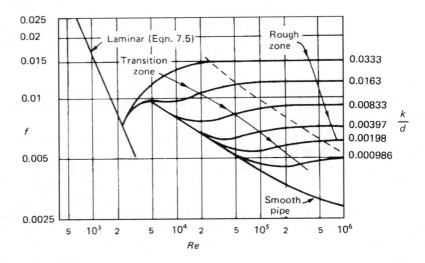

Fig. 7.1

$$f = \frac{2gd}{4l\bar{u}^2}\frac{8\bar{u}l\mu}{(d/2)^2\varrho g} = \frac{16\mu}{\bar{u}d\varrho} = \frac{16}{Re} \tag{7.5}$$

Calculations for laminar flow in circular pipes are of course most readily made with Poiseuille's formula, and eqn 7.5 is used only for the sake of including the laminar flow region in the graph of f. With the logarithmic scales used in Fig. 7.1 eqn 7.5 is represented by a straight line. Experimental results confirm the equation and the fact that laminar flow is independent of the roughness of the pipe walls (unless the roughness is so great that the irregularities constitute an appreciable change of diameter).

Beyond a Reynolds number of about 2000, that is, for turbulent flow, the flow does depend on the roughness of the pipe, and different curves are obtained for different values of the relative roughness. In Nikuradse's experiments, grains of sand of uniform size were glued to the walls of pipes of various diameters, which were initially very smooth. Thus a value of the relative roughness was readily deduced since k could be said to correspond to the diameter of the sand grains. It must be admitted at once that such uniform artificial roughness is quite different from the roughness of ordinary commercial pipes, which is random is both size and spacing. Sand grains, however, do provide a definite, measurable value of k which serves as a reliable basis for gauging the effect of this uniform surface roughness on the flow. Values of the relative roughness in Nikuradse's experiments are marked against each curve. It will be noticed that, close to the critical Reynolds number, all the curves are coincident, but for successively higher Reynolds numbers the curves separate in sequence from the curve for 'smooth' pipes, and the greater the relative roughness the sooner the corresponding curve branches off. Eventually, each curve flattens out to a straight line parallel to the Reynolds number axis; this indicates that f has become independent of Re.

Nikuradse's results confirm the significance of *relative* rather than absolute roughness: the same value of k/d was obtained with different values of k and d individually, and yet points for the same value of k/d lie on a single curve.

For moderate degrees of roughness a pipe acts as a smooth pipe up to that value of Re at which its curve separates from the 'smooth-pipe' line. The region in which the curve is coincident with the 'smooth-pipe' line is known as the *smooth zone of flow*. Where the curve becomes horizontal, showing that f is independent of Re, the flow is said to be in the *rough zone* and the region between the two is known as the *intermediate* or *transition zone*. The position and extent of these zones depends on the relative roughness of the pipe.

This behaviour may be explained by reference to the *viscous sublayer*. The random movements perpendicular to the pipe axis, which occur in turbulent flow, must die out as the wall of the pipe is approached, so even for the most highly turbulent flow there is inevitably a very thin layer, immediately adjacent to the wall, in which these random motions are negligible. The higher the Reynolds number, the more intense are the secondary motions that constitute the turbulence, and the closer they approach to the boundary. So the very small thickness of the viscous sub-layer becomes smaller still as the Reynolds number increases.

In the smooth zone of flow the viscous sub-layer is thick enough to cover completely the irregularities of the surface. Consequently the size of the irregularities has no effect on the main flow (just as when the entire flow is laminar) and all the curves for the smooth zone coincide. With increasing Reynolds number, however, the thickness of the sublayer decreases and so the surface bumps can protrude through it. The rougher the pipe, the lower the value of Re at which this occurs. In the rough zone of flow the thickness of the sub-layer is negligible compared with the height of the surface irregularities. The turbulent flow round each bump then generates a wake of eddies giving rise to a resistance force known as *form drag* (see Chapter 8). Energy is dissipated by the continual production of these eddies; their kinetic energy is proportional to the square of their velocities, and these, in turn, are proportional to the general velocity. Form drag is thus proportional to the square of the mean velocity of flow. In the complete turbulence of the rough zone of flow simple viscous effects are negligible and so $h_f \propto \bar{u}^2$ and (from eqn 7.1) f is constant. In the transition zone the surface bumps partly protrude through the viscous sub-layer. Thus form drag and viscous effects are both present to some extent.

The foregoing explanation is somewhat simplified. For one thing, there is no sharp demarcation between the viscous sub-layer and the rest of the flow. Moreover, a bump on the surface can affect the flow to some degree before the peak emerges from the sub-layer. Nevertheless, this idealized picture of the way in which the surface irregularities influence the flow provides a useful qualitative explanation of the phenomena. It also permits the definition of a hydrodynamically smooth surface as one on which the protuberances are so far submerged in the viscous sublayer as to have no effect on the flow. (Thus a surface that is 'smooth' at low values of Re may be 'rough' at higher values of Re.)

Now Nikuradse's results were obtained for uniform roughness – not for that of pipes encountered in practice. Even though the *average* height of the irregularities on pipe surfaces may be determined, Nikuradse's diagram (Fig. 7.1) is not suitable for actual pipes. Where the surface irregularities are of various heights they begin to protrude through the viscous sub-layer at differing values of Re, and the transition zones of Nikuradse's curves do not correspond at all well to those for actual pipes. However, at a high enough Reynolds number the friction factor of many ordinary pipes becomes independent of Re, and under these conditions a comparison of the value of f with Nikuradse's results enables an *equivalent* uniform size of sand grain k to be specified for the pipe. For instance, uncoated cast iron has an equivalent grain size of about 0.25 mm, galvanized steel 0.15 mm and drawn brass 0.0015 mm. Using these equivalent grain sizes – which, in the present state of knowledge, cannot be deduced from measurements of the actual roughness – the American engineer Lewis F. Moody (1880–1953) prepared a modified diagram (Fig. 7.2) for use with ordinary commercial pipes. (The diagram is based largely on the results of C. F. Colebrook, discussed in Section 8.12.3.)

Moody's diagram is now widely employed, and is the best means at present available for predicting values of f. Nevertheless, the concept of an equivalent grain size is open to serious objection. For instance, it implies that only the height of surface irregularities significantly affects the flow. There is evidence, however, that in the rough zone of flow the spacing of the irregularities is of greater importance. If the irregularities are far apart the wake of eddies formed by one bump may die away before the fluid encounters the next bump. When the bumps are closer, however, the wake from one may interfere with the flow round the next. And if they are exceptionally close together the flow may largely skim over the peaks while eddies are trapped in the valleys. In Nikuradse's experiments the sand grains were closely packed, and so the spacing may be supposed approximately equal to the grain diameter. His results could therefore just as validly be taken to demonstrate that f depends on s/d where s represents the average spacing of the grains. The equivalent grain size, moreover, takes no account of the shape of the irregularities. Another factor that may appreciably affect the value of f in large pipes is 'waviness' of the surface, that is, the presence of transverse ridges on a larger scale than the normal roughness.

These and other arguments indicate that Moody's diagram is at best an approximation. In addition, the roughness of many materials is very variable and it frequently increases with age, particularly if the surface becomes dirty or corroded. (In pipes of small diameter the effective bore may be altered by dirt and corrosion.) Accurate prediction of friction losses is thus difficult to achieve.

A number of empirical formulae have been put forward to describe

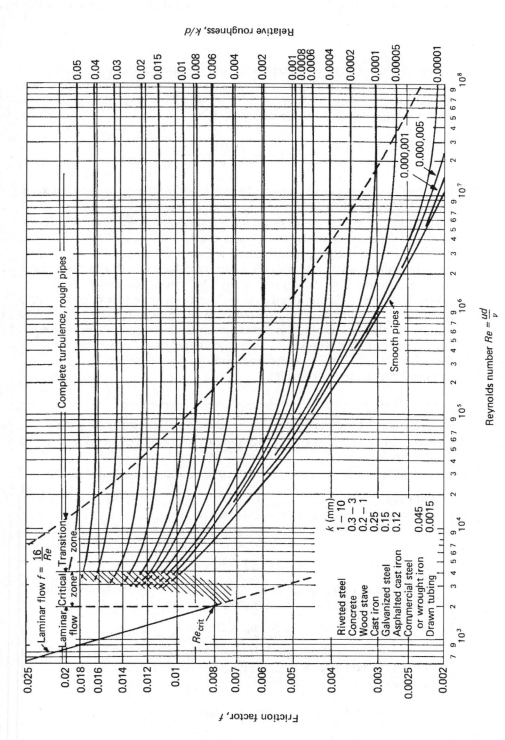

Fig. 7.2

certain parts of Fig. 7.2. There is, for example, Blasius's formula for the smooth-pipe curve:

$$f = 0.079(Re)^{-1/4} \tag{7.6}$$

which agrees closely with experimental results for Reynolds numbers between 3000 and 10^5. Many formulae have been proposed for giving f directly for the entire range of k/d and Re. The best yet produced is probably that by S. E. Haaland[1]:

$$\frac{1}{\sqrt{f}} = -3.6 \log_{10} \left\{ \frac{6.9}{Re} + \left(\frac{k}{3.71d} \right)^{1.11} \right\}$$

It combines reasonable simplicity with acceptable accuracy (within about 1.5%) and is useful if an algebraic expression is required.

A fuller discussion of friction factor relationships for flow in smooth or rough pipes is to be found in Section 8.12.

A common practical problem is to determine the head lost to friction for a given mean velocity in a pipe of given diameter. From these data the appropriate Reynolds number may be calculated and a value of f taken from Fig. 7.2. The head loss is then determined from eqn 7.1. If, however, the velocity or the diameter is unknown then the Reynolds number is also unknown. Nevertheless, since the value of f changes but slowly with Reynolds number, assumed values of Re and f may be used for a first trial. Better approximations to these values can then be obtained from the trial results.

Example 7.1 Determine the head lost to friction when water flows through 300 m of 150 mm diameter galvanized steel pipe at 50 litres/s.

Solution
For water at, say, 15 °C, $v = 1.14 \, \text{mm}^2/\text{s}$.

$$u = \frac{50 \times 10^{-3} \, \text{m}^3/\text{s}}{(\pi/4)(0.15)^2 \, \text{m}^2} = 2.83 \, \text{m/s}$$

(From here onwards u represents the mean velocity over the cross-section, and for simplicity the bar over the u is omitted.)

$$Re = \frac{ud}{v} = \frac{2.83 \, \text{m/s} \times 0.15 \, \text{m}}{1.14 \times 10^{-6} \, \text{m}^2/\text{s}} = 3.72 \times 10^5$$

For galvanized steel $k = 0.15$ mm, say. $\therefore$ $k/d = 0.001$
From Fig. 7.2 $f = 0.00515$ so

$$h_f = \frac{4 \times 0.00515 \times 300 \, \text{m}}{0.15 \, \text{m}} \frac{(2.83 \, \text{m/s})^2}{19.62 \, \text{m/s}^2} = 16.81 \, \text{m, say 17 m}$$

Example 7.2 Calculate the steady rate at which oil ($v = 10^{-5}\,\text{m}^2/\text{s}$) will flow through a cast-iron pipe 100 mm diameter and 120 m long under a head difference of 5 m.

Solution
As yet Re is unknown since the velocity is unknown. For cast iron (in new condition) $k = 0.25\,\text{mm}$, say. $\therefore k/d = 0.0025$ and Fig. 7.2 suggests $f = 0.0065$ as a first trial. Then

$$5\,\text{m} = \frac{4 \times 0.0065 \times 120\,\text{m}}{0.10\,\text{m}} \frac{u^2}{19.62\,\text{m/s}^2}$$

whence $u = 1.773\,\text{m/s}$. Therefore

$$Re = \frac{1.773\,\text{m/s} \times 0.10\,\text{m}}{10^{-5}\,\text{m}^2/\text{s}} = 1.773 \times 10^4$$

These values of Re and k/d give $f = 0.0079$ from Fig. 7.2. A recalculation of u gives 1.608 m/s, hence $Re = 1.608 \times 10^4$. The corresponding change of f is insignificant. The value $u = 1.608\,\text{m/s}$ is accepted and

$$Q = 1.608\,\text{m/s} \times (\pi/4)(0.10)^2\,\text{m}^2 = 0.01263\,\text{m}^3/\text{s}.$$

[Alternatively, use may be made of the expression

$$Q = -\frac{\pi}{2}d^{5/2}\left(\frac{2gh_f}{l}\right)^{1/2} \log\left\{\frac{k}{3.71d} + \frac{2.51v}{d^{3/2}\left(2gh_f/l\right)^{1/2}}\right\}$$

which is derived from the formula (eqn 8.56) on which Fig. 7.2 is based. In this example

$$Q = -\frac{\pi}{2}(0.10\,\text{m})^{5/2}\left(\frac{19.62 \times 5}{120}\frac{\text{m}}{\text{s}^2}\right)^{1/2} \log\left\{\frac{0.0025}{3.71}\right.$$

$$\left. + \frac{2.51 \times 10^{-5}\,\text{m}^2/\text{s}}{(0.10\,\text{m})^{3/2}\left(\frac{19.62 \times 5\,\text{m}}{120\,\text{s}^2}\right)^{1/2}}\right\} = 0.01262\,\text{m}^3/\text{s}.$$

□

■ **Example 7.3** Determine the size of galvanized steel pipe needed to carry water a distance of 180 m at 85 litres/s with a head loss of 9 m.

Solution

$$h_f = 9\,\text{m} = \frac{4f(180\,\text{m})}{d}\left(\frac{0.085\,\text{m}^3/\text{s}}{\pi d^2/4}\right)^2 \frac{1}{19.62\,\text{m/s}^2}$$

whence $d^5 = 0.0478f\,\text{m}^5$.

Again taking $v = 1.14\,\text{mm}^2/\text{s}$ for water at 15 °C,

$$Re = \frac{ud}{v} = \frac{0.085\,\text{m}^3/\text{s}}{\pi d^2/4}\frac{d}{1.14 \times 10^{-6}\,\text{m}^2/\text{s}}$$

$$= \frac{9.49 \times 10^4\,\text{m}}{d}$$

$$k = 0.15\,\text{mm}$$

Try $f = 0.006$. Then the foregoing expressions successively give $d = 0.1956$ m, $Re = 4.85 \times 10^5$ and $k/d = 0.00077$. These figures give $f = 0.00475$ from Fig. 7.2. From this value $d = 0.1867$ m, $Re = 5.08 \times 10^5$ and $k/d = 0.00080$. Then $f = 0.0048$, which differs negligibly from the previous value. The calculated diameter $= 0.1867$ m, but the nearest pipe size available is probably 200 mm (nominal). In view of this discrepancy of diameters it can be seen that great accuracy in the calculations is not warranted.

Specially prepared charts[2] enable solutions of acceptable accuracy to be obtained without trial-and-error methods.

For restricted ranges of Reynolds number and pipe diameter, f is sufficiently constant for tables of values to have been compiled for use in engineering calculations. Although the use of such values often allows problems to be solved more simply, limitation of the values to a particular range of conditions should never be forgotten.

Darcy's formula (eqn 7.1), together with Moody's chart, provides the best data at present available on pipe friction in turbulent flow, but for some problems the variation of f necessitates trial-and-error solutions. Many empirical formulae, admittedly limited to particular conditions and fluids but more convenient for certain problems, have therefore been developed. Such formulae are often numerical ones, applying only to particular units and to restricted ranges of the variables.

7.4 DISTRIBUTION OF SHEAR STRESS IN A CIRCULAR PIPE

If fluid passes with steady velocity along a pipe, the loss of mechanical energy brought about by viscosity results in a decrease of the piezometric pressure $p + \varrho g z$. This decrease in piezometric pressure is related directly to the shear stresses at the boundaries of the flow. Consider flow within a straight, completely closed conduit, such as a pipe, which the fluid fills entirely. Figure 7.3 depicts a short length δx of this conduit of uniform cross-sectional area A. The mean pressure at section 1 is p; that at section 2 is $p + \delta p$. (Where the flow is turbulent there are small fluctuations of pressure at any point in the fluid just as there are small fluctuations of velocity, but time-average values are here considered.) Noting that the weight of the fluid between sections 1 and 2 is $\varrho g A \delta x$, where ϱ represents the density of the fluid, we obtain for the net force on the fluid in the direction of flow

$$pA - (p + \delta p)A - \varrho g A \delta x \cos\theta + \bar{\tau}_0 P \delta x \qquad (7.7)$$

where P represents the perimeter of the section in contact with the fluid and $\bar{\tau}_0$ the mean stress on the fluid at the boundary. If the flow is steady and 'fully developed', that is, the velocity distribution over the cross-section does not change with distance x, there is no increase of momentum in this direction and therefore the net force is zero. Setting the expression 7.7 equal to zero and putting $\delta x \cos\theta = \delta z$, where z is measured vertically upwards from some horizontal datum, we obtain

$$\bar{\tau}_0 P \delta x = A(\delta p + \varrho g \delta z)$$

For a fluid of constant density the right-hand side may be written $A\delta p^*$ where $p^* = p + \varrho g z$. Then, in the limit as $\delta x \to 0$,

$$\bar{\tau}_0 = \frac{A}{P} \frac{\mathrm{d}p^*}{\mathrm{d}x} \qquad (7.8)$$

The piezometric pressure p^* is constant over the cross-section. If differences of p^* between different parts of the cross-section did exist they

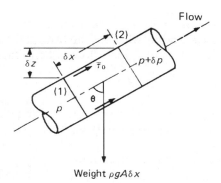

Weight $\rho g A \delta x$

Fig. 7.3

would give rise to movements of the fluid perpendicular to the main flow. Even in turbulent motion, however, there is no *net* flow perpendicular to the main flow.

Equation 7.8 is substantially true even for a gas, because when $\bar{\tau}_0$ is large enough to be of interest $\varrho g \, \delta z$ is negligible compared with δp and the asterisk may be dropped from p^* anyway.

In practice the boundary stress resists flow, that is, it acts on the fluid in the direction opposite to that shown in Fig. 7.3. Consequently both $\bar{\tau}_0$ and dp^*/dx are negative, so p^* decreases in the direction of flow.

Equation 7.8 applies to any shape of cross-section provided that its area does not change along the length, and to any sort of steady flow. We note that $\bar{\tau}_0$ is an *average* stress; for cross-sections other than circular, the stress varies in magnitude round the perimeter. For a circular section, however, the symmetry of a circle requires the stress to have the same value at all points on the circumference (if the roughness of the surface is uniform) and we may then drop the bar from the symbol $\bar{\tau}_0$. For a pipe of radius R, eqn 7.8 becomes

$$\tau_0 = \frac{\pi R^2}{2\pi R}\frac{dp^*}{dx} = \frac{R}{2}\frac{dp^*}{dx} \tag{7.9}$$

Similar reasoning can also be applied to a smaller, concentric, cylinder of fluid having radius r. Again for 'fully developed' steady flow, the net force on the fluid is zero and the result

$$\tau = \frac{r}{2}\frac{dp^*}{dx} \tag{7.10}$$

is obtained.

Thus τ, the internal shear stress in the axial direction, varies with r, the distance from the centre-line of the pipe. Dividing eqn 7.10 by eqn 7.9 gives

$$\frac{\tau}{\tau_0} = \frac{r}{R}, \quad \text{that is, } \tau = \tau_0\frac{r}{R} \tag{7.11}$$

Consequently τ varies linearly with r from a value of zero at the centre-line of the pipe, where $r = 0$, to a maximum at the wall, where $r = R$. This distribution of stress is represented graphically at the right-hand side of Fig. 7.4.

The law of linear distribution of shear stress over a circular section, represented by eqn 7.11, holds whether the flow in the pipe is laminar or turbulent. In steady, fully-developed flow, all the factors affecting the

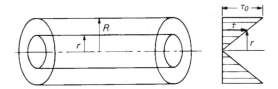

Fig. 7.4

magnitude of the stress must in fact so adjust themselves that this law is obeyed.

7.5 FRICTION IN NON-CIRCULAR CONDUITS

The majority of closed conduits used in engineering practice are of circular cross-section, but the friction loss in non-circular passages – for example, rectangular air ducts – quite often has to be estimated. Experiment shows that for many shapes the relations developed for turbulent flow in circular sections may be applied to non-circular ones if, in place of the diameter of the circle, an 'equivalent diameter' is used, such that the hydraulic mean depth m (see Section 7.2) is the same. For a circular section flowing full

$$m = \frac{A}{P} = \frac{(\pi/4)d^2}{\pi d} = \frac{d}{4}$$

so the equivalent diameter must equal $4m$.

Equation 7.3 shows that this 'equivalent diameter' involves the assumption that the mean shear stress at the boundary is the same as for a circular section. For the circular section the stress is uniform (unless the roughness varies round the circumference) but for non-circular sections it is not. Contours of equal velocity are entirely parallel to the perimeter only in the circular section (see Fig. 7.5); in a rectangular section, for example, the velocity gradient is highest at the mid-point of a side, and least in the corners, and the shear stress varies accordingly. It is therefore to be expected that the less the shape deviates from a circle the more reliable will be the use of the equivalent diameter. The assumptions may be quite invalid for odd-shaped sections, but reasonable results are usually obtained with ovals, triangles and rectangles (if the longer side is not greater than about 8 times the shorter). For annuli between concentric cylinders the larger diameter must be at least 3 times the smaller. (Note that for an annulus the relevant perimeter used in determining the hydraulic mean depth is the total for inner and outer surfaces together.)

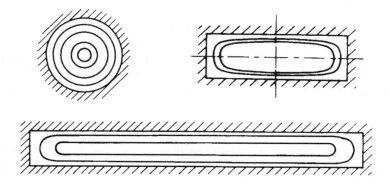

Fig. 7.5 Contours of equal velocity.

O. C. Jones[3,4] has suggested that better correlation between eqn 7.2 and measured values of h_f is obtainable if the hydraulic mean depth m is modified by factors derived from laminar flow results for the cross-section concerned. Experimental results for rectangular sections and concentric annuli seem to support this procedure, but firm theoretical justification is lacking.

The concept of equivalent diameter is not applicable at all to laminar flow. Moreover, if part of a cross-section is very narrow, laminar flow may exist there even if the main flow is turbulent; consequently the equivalent diameter as defined here is not valid.

7.6   OTHER HEAD LOSSES IN PIPES

Not only is there a loss of head caused by friction in a uniform straight pipe, but additional losses may be incurred at changes in the cross-section, bends, valves and fittings of all kinds. In long pipes these extra losses may, without serious error, be neglected in comparison with the 'ordinary' friction loss. Although they are often termed 'minor' (or 'secondary') losses they may, however, actually outweigh the ordinary friction loss in short pipes. The losses invariably arise from sudden changes of velocity (either in magnitude or direction). These changes generate large-scale turbulence in which energy is dissipated as heat.

The source of the loss is usually confined to a very short length of the pipe, but the turbulence produced may persist for a considerable distance downstream. The flow after the sudden change of velocity is exceedingly complicated, and the processes of 'ordinary' pipe friction are inevitably affected by the additional turbulence. For the purposes of analysis, however, it is assumed that the effects of ordinary friction and of the additional large-scale turbulence can be separated, and that the additional loss is concentrated at the device causing it. The total head lost in a pipe may then be calculated as the sum of the normal friction for the length of pipe considered and the additional losses.

A theoretical determination of the additional losses is seldom possible, and so experimentally-determined figures must be called on. Since the losses have been found to vary as the square of the mean velocity, they are frequently expressed in the form

$$\text{Head loss} = k \frac{u^2}{2g}$$

and for high Reynolds numbers the value of k is practically constant.

For further information on head losses in pipe flow, Reference 8 should be consulted.

7.6.1   Loss at abrupt enlargement

One 'minor loss' that can be subjected to analysis is that at an abrupt enlargement of the cross-section, such as that illustrated in Fig. 7.6. The

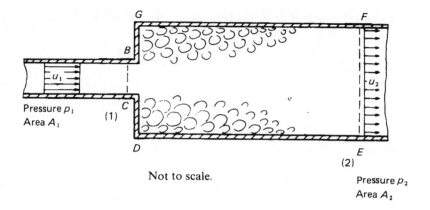

G   F

B

u_1

Pressure p_1
Area A_1
(1)

C

D   E
(2)

Not to scale.

u_2

Pressure p_2
Area A_2

Fig. 7.6

pipes run full and the flow is assumed steady. Fluid emerging from the smaller pipe is unable to follow the abrupt deviation of the boundary; consequently turbulent eddies form in the corners and result in the dissipation of energy as heat. By means of a few simplifying assumptions, an estimate of the head lost may be made.

For the high values of Reynolds number usually found in practice, the velocity in the smaller pipe may be assumed sensibly uniform over the cross-section. At section 1 the streamlines are straight and parallel and so the piezometric pressure here is uniform. Downstream of the enlargement the vigorous mixing caused by the turbulence helps to even out the velocity and it is assumed that at a section 2 sufficiently far from the enlargement (about 8 times the larger diameter) the velocity (like the piezometric pressure) is again uniform over the cross-section. For simplicity the axes of the pipes are assumed horizontal. Continuity requires the velocity u_2 to be less than u_1 and the corresponding momentum change requires a net force to act on the fluid between sections 1 and 2. On the fluid in the 'control volume' $BCDEFG$ the net force acting towards the right is

$$p_1 A_1 + p'(A_2 - A_1) - p_2 A_2$$

where p' represents the mean pressure of the eddying fluid over the annular face GD. (Shear forces on the boundaries over the short length between sections 1 and 2 are neglected.) Since radial accelerations over the annular face GD are very small, we assume (with the support of experimental evidence) that p' is sensibly equal to p_1. The net force on the fluid is thus $(p_1 - p_2)A_2$. From the steady-flow momentum equation (see Section 4.2) this force equals the rate of increase of momentum in the same direction:

$$(p_1 - p_2)A_2 = \varrho Q(u_2 - u_1)$$

where ϱ represents the density and Q the volume flow rate.

$$\therefore p_1 - p_2 = \varrho \frac{Q}{A_2}(u_2 - u_1) = \varrho u_2(u_2 - u_1) \qquad (7.12)$$

From the energy equation for a constant density fluid we have

$$\frac{p_1}{\varrho g} + \frac{u_1^2}{2g} + z - h_l = \frac{p_2}{\varrho g} + \frac{u_2^2}{2g} + z$$

where h_l represents the loss of total head between sections 1 and 2.

$$\therefore h_l = \frac{p_1 - p_2}{\varrho g} + \frac{u_1^2 - u_2^2}{2g}$$

and substitution from eqn 7.12 gives

$$h_l = \frac{u_2(u_2 - u_1)}{g} + \frac{u_1^2 - u_2^2}{2g} = \frac{(u_1 - u_2)^2}{2g} \qquad (7.13)$$

Since by continuity $A_1 u_1 = A_2 u_2$ eqn 7.13 may be alternatively written

$$h_l = \frac{u_1^2}{2g}\left(1 - \frac{A_1}{A_2}\right)^2 = \frac{u_2^2}{2g}\left(\frac{A_2}{A_1} - 1\right)^2 \qquad (7.14)$$

This result was first obtained by J.-C. Borda (1733–99) and L. M. N. Carnot (1753–1823) and is sometimes known as the Borda–Carnot head loss. In view of the assumptions made, eqns 7.13 and 7.14 are subject to some inaccuracy, but experiments show that for coaxial pipes they are within only a few per cent off the truth.

Exit loss

If $A_2 \to \infty$, then eqn 7.14 shows that the head loss at an abrupt enlargement tends to $u_1^2/2g$. This happens at the submerged outlet of a pipe discharging into a large reservoir, for example (Fig. 7.7). The velocity head in the pipe, corresponding to the kinetic energy of the fluid per unit weight, is thus lost in turbulence in the reservoir. In such circumstances the loss is usually termed the exit loss for the pipe.

7.6.2 Loss at abrupt contraction

Although an abrupt contraction (Fig. 7.8) is geometrically the reverse of an abrupt enlargement it is not possible to apply the momentum equation to a control volume between sections 1 and 2. This is because, just

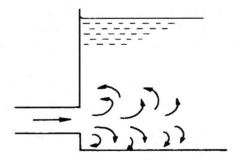

Fig. 7.7

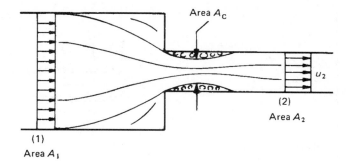

Area A_c

u_2

(2)

Area A_2

(1)

Area A_1

Fig. 7.8

upstream of the junction, the curvature of the streamlines and the acceleration of the fluid cause the pressure at the annular face to vary in an unknown way. However, immediately downstream of the junction a vena contracta is formed, after which the stream widens again to fill the pipe. Eddies are formed between the vena contracta and the wall of the pipe, and it is these which cause practically all the dissipation of energy. Between the vena contracta and the downstream section 2 – where the velocity has again become sensibly uniform – the flow pattern is similar to that after an abrupt enlargement, and so the loss of head is assumed to be given by eqn 7.14:

$$ h_l = \frac{u_2^2}{2g}\left(\frac{A_2}{A_c} - 1\right)^2 = \frac{u_2^2}{2g}\left(\frac{1}{C_c} - 1\right)^2 \qquad (7.15) $$

where A_c represents the cross-sectional area of the vena contracta, and the coefficient of contraction $C_c = A_c/A_2$.

Although the area A_1 is not explicitly involved in eqn 7.15, the value of C_c depends on the ratio A_2/A_1. For coaxial circular pipes and fairly high Reynolds numbers Table 7.1 gives representative values of the coefficient k in

$$ h_l = \frac{ku_2^2}{2g} \qquad (7.16) $$

As $A_1 \rightarrow \infty$ the value of k in eqn 7.16 tends to 0.5, and this limiting case corresponds to the flow from a large reservoir into a sharp-edged pipe, provided that the end of the pipe does not protrude into the reservoir. (A protruding pipe, as at Fig. 7.9b, causes a greater loss of head.) For a

Entry loss

Table 7.1 Loss coefficient k for abrupt contraction

d_2/d_1	0	0.2	0.4	0.6	0.8	1.0
k		0.5	0.45	0.38	0.28	0.14	0

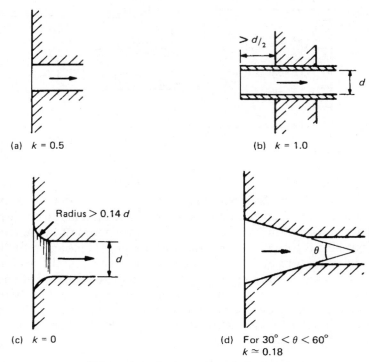

(a) $k = 0.5$

(b) $k = 1.0$

(c) $k = 0$

(d) For $30° < \theta < 60°$ $k \simeq 0.18$

Fig. 7.9

Values given for k are the usually accepted values.

non-protruding, sharp-edged pipe the loss $0.5u_2^2/2g$ is known as the entry loss. If the inlet to the pipe is well rounded, as in Fig. 7.9c, the fluid can follow the boundary without separating from it, and the entry loss is much reduced. A tapered entry, as in Fig. 7.9d, also gives a much lower loss than the abrupt entry.

7.6.3 Diffusers

The head lost at an abrupt enlargement (or at the exit from a pipe) can be considerably reduced by the substitution of a gradual, tapered, enlargement usually known as a *diffuser* or *recuperator*. Its function is to reduce the velocity of the fluid gradually, and thus eliminate, as far as possible, the eddies responsible for the dissipation of energy.

The loss of head that does occur in a diffuser depends on the angle of divergence θ and on the ratio of the upstream and downstream areas. One contribution to the loss is made by ordinary pipe friction. This decreases as the angle θ increases since, for a given ratio of areas A_2/A_1, a larger angle gives a smaller length. For all but the smallest angles, however, energy is also dissipated by eddies caused by the separation of the flow from the walls. This loss increases with θ. There is an optimum angle for which the sum of the two types of loss is a minimum and, for a conical diffuser with a smooth surface, the optimum value of the total angle is about 6°. With greater surface roughness the ordinary friction increases without affecting the eddies significantly and so the optimum angle is then somewhat larger.

The loss of head in a diffuser may be expressed as

$$h_l = k\frac{\left(u_1 - u_2\right)^2}{2g} = k\left(1 - \frac{A_1}{A_2}\right)^2 \frac{u_1^2}{2g} \qquad (7.17)$$

where A_1, A_2 represent the cross-sectional areas at inlet and outlet respectively and u_1, u_2 the corresponding mean velocities. Values of the factor k for conical diffusers are indicated in Fig. 7.10 where it will be noted that for angles greater than about 40° the total loss exceeds that for an abrupt enlargement (for which $\theta = 180°$) and the maximum loss occurs at about $\theta = 60°$. For $\theta = 180°$, $k \simeq 1.0$, and eqn 7.17 then corresponds to eqn 7.13 for an abrupt enlargement.

Diffusers are commonly used to obtain an increase of pressure in the direction of flow. In a perfect diffuser there would be a rise of piezometric pressure given by Bernoulli's equation (3.9):

$$p_2^* - p_1^* = \frac{1}{2}\varrho\left(u_1^2 - u_2^2\right) = \frac{1}{2}\varrho u_1^2\left\{1 - \left(\frac{A_1}{A_2}\right)^2\right\} \qquad (7.18)$$

if steady flow and uniform conditions over inlet and outlet cross-sections are assumed. Because of the energy losses, however, the actual rise in pressure is less than this, and the 'pressure efficiency' of a diffuser is commonly defined as the ratio of the actual rise in piezometric pressure to that predicted by eqn 7.18.

As a result of the flow separation from the walls of a diffuser, the velocity at the section where the area A_2 is reached is far from uniform.

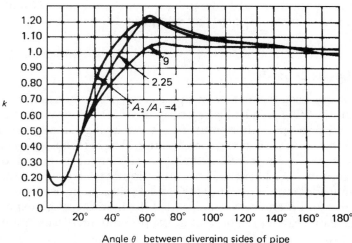

Fig. 7.10 Loss of head in conical diffuser.

If a straight pipe of uniform section follows the diffuser, the velocity distribution gradually reverts to the more nearly uniform pattern of normal pipe flow and, in consequence, a further pressure rise may be observed for some distance downstream from the diffuser itself. (For a circular pipe this distance is about 4 to 6 times the diameter.) The phenomenon may be deduced by including kinetic energy correction factors α (see Section 3.5.3) in the energy equation between section 2 (diffuser outlet) and section 3 further downstream:

$$\frac{p_2^*}{\varrho g} + \alpha_2 \frac{u_2^2}{2g} - h_f = \frac{p_3^*}{\varrho g} + \alpha_3 \frac{u_3^2}{2g}$$

that is

$$\frac{p_3^* - p_2^*}{\varrho g} = \alpha_2 \frac{u_2^2}{2g} - \alpha_3 \frac{u_3^2}{2g} - h_f = (\alpha_2 - \alpha_3)\frac{u^2}{2g} - h_f$$

since, for a constant cross-section, the mean velocity $u = u_2 = u_3$. If the uniformity of velocity is poor at section 2, so that α_2 is large, and yet is much improved at section 3, then $(\alpha_2 - \alpha_3)u^2/2g$ is positive and may outweigh h_f in a smooth pipe. Then $p_3^* > p_2^*$.

Losses in a diffuser are also affected by the conditions at inlet: the less uniform the velocity the more readily will eddies form in the diffuser.

Diverging flow, such as occurs in a diffuser, is always subject to separation from the boundaries, and the consequent formation of eddies. In general, therefore, the dissipation of energy in diverging flow is greater than in converging flow. Indeed, a gradual contraction without sharp corners causes a loss so small that it may usually be neglected. In other words, the conversion of velocity head to piezometric head is inherently a less efficient process than the conversion of piezometric head to velocity head.

Much research has been devoted to the design of efficient diffusers. Examples of diffusers are the outlets from wind tunnels and water tunnels, the volutes of centrifugal pumps and the draft tubes of water turbines.

7.6.4 Losses in bends

Energy losses occur when flow in a pipe is caused to change its direction. Consider the pipe bend illustrated in Fig. 7.11. Now whenever fluid flows in a curved path there must be a force acting radially inwards on the fluid to provide the inward acceleration. There is thus an increase of pressure near the outer wall of the bend, starting at point A and rising to a maximum at B. There is also a reduction of pressure near the inner wall giving a minimum pressure at C and a subsequent rise from C to D. Between A and B and between C and D the fluid therefore experiences an adverse pressure gradient, that is, the pressure increases in the direction of flow.

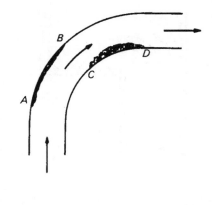

Fig. 7.11

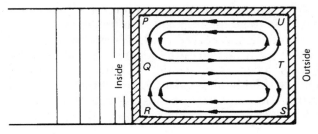

Fig. 7.12

These conditions are similar to those in a diffuser and, unless the radius of curvature of the bend is very large, they lead, as in the diffuser, to separation of the flow from the boundary and thus energy losses in turbulence. The magnitude of these losses depends mainly on the radius of curvature of the bend.

Energy losses also arise from *secondary flow*. This phenomenon may most easily be explained by reference to a pipe of rectangular cross-section as shown in Fig. 7.12. Adjacent to the upper and lower walls the velocity is reduced by the viscous action in the boundary layers there and, as a result, the increase in pressure from the inner to the outer radius is less in the boundary layers (*PU* and *RS*) than along the centre line *QT*. Since the pressure at *T* is greater than at *U* and *S* and the pressure at *Q* less than at *P* and *R* a secondary flow in the radial plane takes place as indicated in the figure. Similar twin eddies are produced in circular pipes. In association with the main flow, the secondary flow produces a double spiral motion which may persist for a distance downstream as much as 50 to 75 times the pipe diameter. The spiralling of the fluid increases the local velocity, so the loss to friction is greater than for the same rate of flow without the secondary motion.

A pipe bend thus causes a loss of head additional to that which would occur if the pipe were of the same total length but straight. This extra loss is conveniently expressed as $ku^2/2g$. The value of k depends on the total angle of the bend and on the relative radius of curvature R/d, where R represents the radius of curvature of the pipe centre line and d the diameter of the pipe. The factor k varies only slightly with Reynolds

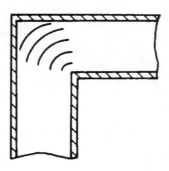

Fig. 7.13

number in the usual range and increases with surface roughness. Where space does not permit a bend of large radius a so-called mitre bend with $R/d = 0$ may have to be used. Then k is approximately 1.1. If, however, a series of well-designed curved guide vanes (known as a *cascade*) is installed (see Fig. 7.13) much of the separation and secondary flow that would otherwise occur is prevented, and the loss is greatly reduced, even though the total boundary surface is thereby increased.

Much information on flow in bends has been provided by Ito[5,6] and Ward-Smith[8].

7.6.5 Losses in pipe fittings

All pipe fittings – valves, couplings and so on – cause additional losses of head and, in general, the more intricate the passage through which the fluid has to pass the greater the head loss. For turbulent flow the head lost may be represented by $ku^2/2g$ where u represents the mean velocity in the pipe. Values of the factor k depend critically on the exact shape of the flow passages. A few typical values are given in Table 7.2 but should be regarded as very approximate.

Because the eddies generated by fittings persist for some distance downstream, the total loss of head caused by two fittings close together is not necessarily the same as the sum of the losses that each alone would cause. Usually the total loss is less than the sum of the individual losses but if, in design calculations, individual losses are simply added, any error thereby introduced will at least be on the safe side.

The losses are sometimes expressed in terms of an equivalent length of unobstructed straight pipe in which an equal loss would occur. That is

$$\frac{ku^2}{2g} = \frac{4fl_e}{d}\frac{u^2}{2g}$$

where l_e represents the equivalent length for the fitting. Where f is known, l_e can be expressed as 'n diameters', i.e. $n = l_e/d$. The value of l_e thus depends on the value of f, and therefore on the Reynolds number and the roughness of the pipe, but the error made by considering n and k constant for a particular fitting is usually small in comparison with

Table 7.2 Approximate loss coefficient k for commercial pipe fittings

Globe valve, wide open	10
Gate valve,	
wide open	0.2
three-quarters open	1.15
half-open	5.6
quarter open	24
Pump foot valve	1.5
90° elbow (threaded)	0.9
45° elbow (threaded)	0.4
Side outlet of T junction	1.8

other uncertainties. Adding the equivalent length to the actual length gives the effective length of the pipe concerned, and the effective length can be used in eqn 7.1 to obtain the expression relating h_f, u and d.

7.7 TOTAL HEAD AND PRESSURE LINES

In the study of flow in pipe systems the concepts of the total head line and the pressure line are often useful. The quantity $(p/\varrho g) + (u^2/2g)$ at any point in the pipe may be represented by a vertical ordinate above the centre-line of the pipe. (The kinetic energy correction factor is assumed negligibly different from unity.) The line linking the tops of such ordinates is termed the *total head line* or *energy line*, and its height above a horizontal datum represents the total head of the fluid relative to that datum. (In a diagram it is usual to plot the total head as ordinate against length along the pipe as abscissa.) At an open reservoir of still liquid the total head line coincides with the free surface; where there is a uniform dissipation of energy (such as that due to friction in a long uniform pipe) the total head line slopes uniformly downwards in the direction of flow; where there is a concentrated dissipation of energy (as at an abrupt change of section) there is an abrupt step downwards in the total head line. If mechanical energy is added to the fluid – by means of a pump, for example – the total head line has a step up. A turbine, on the other hand, which takes energy from the fluid, causes a step down in the total head line.

In drawing a total head line certain conventional approximations are made. We have already noted (Section 7.6) that the losses of head arising from changes of section, bends or other fittings in a pipe are in reality not concentrated at one point since they are largely caused by the continuing turbulence downstream. Nevertheless it is customary to represent the loss by an abrupt step in the total head line. Also the velocity distribution across the pipe, which is appropriate to the Reynolds number of the flow and to the surface roughness, is not achieved immediately the fluid enters the pipe. The 'entry length' in which the normal pattern of velocity is developed is often about 50 times the pipe diameter (see Section 7.9) and over this length the value of f appropriate in Darcy's equation (7.1) varies somewhat. For many pipes, however, the entry length is a small proportion of the total length; little error is thus involved by the usual assumption that the slope of the total head line is uniform for the entire length of a uniform pipe.

The *pressure line* (sometimes known as the *piezometric line* and more usually in America as the *hydraulic grade line*) is obtained by plotting values of $p/\varrho g$ vertically above the pipe centre-line. It is therefore a distance $u^2/2g$ below the total head line. If the pressure line and the axis of the pipe itself coincide at any point then the gauge pressure in the pipe is zero, i.e. atmospheric. A pipe that rises above its pressure line has a negative pressure within it and is known as a *siphon*. The gauge pressure in any event cannot fall below $-100\,\mathrm{kPa}$ (if the atmospheric pressure is $100\,\mathrm{kPa}$) because that limit would correspond to a perfect vacuum and the flow would stop. However, the flow of a liquid in a pipe would almost certainly stop before the pressure fell to this value because dissolved air or other gases would come out of solution and collect in the highest part of the siphon in sufficient quantity to form an *air-lock*. Even if the liquid contained no dissolved gases, it would itself vaporize when the pressure fell to the vapour pressure of the liquid at that temperature. For these reasons the pressure of water in a pipe-line should not fall below about $-75\,\mathrm{kPa}$ gauge. This suggests that the pressure line should not be more than about

$$75\,000\,\frac{\mathrm{N}}{\mathrm{m^2}}\bigg/1000\,\frac{\mathrm{kg}}{\mathrm{m^3}}9.81\,\frac{\mathrm{N}}{\mathrm{kg}}=7.64\,\mathrm{m}$$

below the pipe, but siphons with somewhat lower pressure lines are possible because the emerging gas bubbles reduce the mean density of the liquid.

The air in a siphon must of course be extracted by some means in order to start the flow of liquid, and air subsequently collecting at the highest point must be removed if flow is to continue. Automatic float valves may be used to do this. But sub-atmospheric pressure in pipe-lines should where possible be avoided by improving the design of the system.

Figure 7.14 illustrates the total head and pressure lines for the steady flow of a liquid in a pipe-line connecting two reservoirs. All the junctions are supposed abrupt. Thus, if u_1 represents the mean velocity in pipe 1, there is a head loss of approximately $0.5u_1^2/2g$ at the inlet (see Section 7.6.2) and this is represented by the step down in the total head line at that point. Along pipe 1, supposed uniform, head is lost to friction at a uniform rate and the total head line therefore has a constant downward slope. At the abrupt enlargement (see Section 7.6.1) the loss $(u_1 - u_2)^2/2g$ is represented (conventionally) by an abrupt step down in the total head line. Uniform head loss to friction in pipe 2 is represented by the uniform downward slope of the corresponding part of the total head line and, finally, the exit loss $u_2^2/2g$ (see Section 7.6.1) is represented by an abrupt step down at the end of the line.

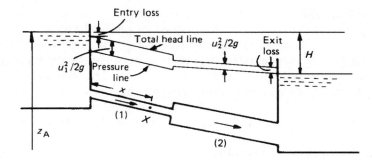

Fig. 7.14

The pressure line is everywhere a distance $u^2/2g$ below the total head line. A step *up* in the pressure line occurs at the abrupt enlargement because, by the energy equation,

$$\frac{p_1 - p_2}{\varrho g} = \frac{u_2^2 - u_1^2}{2g} + h_l = \frac{1}{2g}\left\{u_2^2 - u_1^2 + \left(u_1 - u_2\right)^2\right\}$$

$$= \frac{u_2}{g}\left(u_2 - u_1\right)$$

and, since this expression is negative, p_2 exceeds p_1.

For steady flow – and only to this is the foregoing applicable – the diagram shows that the difference in reservoir levels equals the sum of all the head losses along the pipe-line. The rate of flow so adjusts itself that this balance is achieved. In the example just considered the total head losses are

$$\underset{\substack{\text{Entry}\\\text{loss}}}{0.5\frac{u_1^2}{2g}} + \underset{\substack{\text{Friction in}\\\text{pipe 1}}}{\left(\frac{4fl}{d}\frac{u^2}{2g}\right)_1} + \underset{\substack{\text{Abrupt}\\\text{enlargement}}}{\frac{\left(u_1 - u_2\right)^2}{2g}} + \underset{\substack{\text{Friction in}\\\text{pipe 2}}}{\left(\frac{4fl}{d}\frac{u^2}{2g}\right)_2} + \underset{\substack{\text{Exit}\\\text{loss}}}{\frac{u_2^2}{2g}}$$

This total equals H and, since u_1 and u_2 are related by the continuity condition $A_1u_1 = A_2u_2$, either u_1 or u_2 (and hence the discharge Q) may be determined if H and the pipe lengths, diameters and friction factors are known. It may be noted that, provided that the outlet is submerged, the result is independent of the position of either end of the pipe; the head determining the flow is simply the difference of the two reservoir levels.

Once the rate of flow is known the pressure at any point along the pipe may be determined by applying the energy equation between the upper reservoir and the point in question. At point X, for example, the pressure p_X is given by

$$z_A - \underbrace{\left(0.5\frac{u_1^2}{2g} + \frac{4f_1x}{d_1}\frac{u_1^2}{2g}\right)}_{\text{losses up to } X} = \frac{p_X}{\varrho g} + \frac{u_1^2}{2g} + z_X$$

If the fluid discharges freely into the atmosphere from the outlet end of the pipe, the pressure at that end must be atmospheric and in steady flow the sum of the head losses (including the exit velocity head $u_e^2/2g$) equals the vertical difference in level between the pipe outlet and the free surface in the inlet reservoir (Fig. 7.15). This is readily verified by applying the energy equation between the reservoir and the pipe outlet.

Further examples of total head and pressure lines are given in Fig. 7.16.

In cases where the Reynolds number of the flow cannot be directly

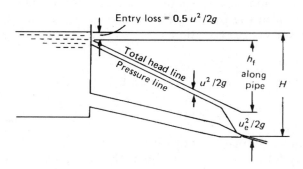

Fig. 7.15

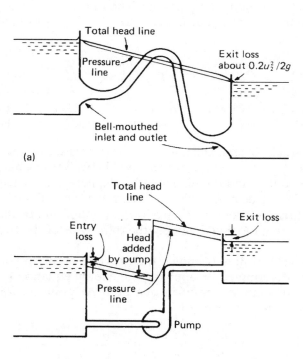

Fig. 7.16

determined from the data of a problem (because u or d is unknown) a solution is generally obtainable only by trial because of the variation of f with Reynolds number. Many problems in engineering, however, involve flow at high Reynolds numbers in rough pipes and so f is sensibly constant (see Fig. 7.2). In any case, since the inexact determination of relative roughness leads to uncertainty in the value of f, solutions with great arithmetical accuracy are not warranted. Trial-and-error methods are thus seldom called for.

For pipes of reasonable length (say, more than 1000 times the diameter) the 'minor losses' may often be neglected and the calculations thereby simplified. It should, however, be checked that such simplification is justifiable. For example, a valve which may constitute a negligible 'minor loss' when fully open may, when partly closed, provide the largest head loss in the system. And since the function of the nozzle illustrated in Fig. 7.15 is to produce a high velocity jet the 'exit loss' $u_e^2/2g$ here may well be more than the friction loss along the entire pipe.

Example 7.4

(a) Derive an expression for the power P required to pump a volumetric flow rate Q through a horizontal pipeline of constant diameter D, length l and friction factor f. Assume that the friction in the pipeline is the only source of dissipation, other than that within the pump itself.

(b) For the purpose of project calculations the total cost of moving a fluid over a distance by pipeline at a steady flow rate, Q, can be broken down into two items. First, the manufacture, laying and maintenance of the pipeline, including interest charges, are represented by the cost C_1, which is proportional to D^3. The second item, C_2, depends solely upon the energy required to pump the fluid.

A preliminary design study for a particular project showed that the total cost was a minimum for $D = 600\,\text{mm}$. If fuel prices are increased by 150%, and assuming only C_2 is affected, make a revised estimate of the optimum pipe diameter.

Solution

(a) Power required, $P = \dfrac{Q\Delta p^*}{\eta}$

where η = pump efficiency and from equation (7.1)

$$\Delta p^* = \frac{4fl}{D}\frac{1}{2}\varrho\bar{u}^2$$

From the continuity equation

$$\bar{u} = \frac{Q}{(\pi/4)D^2}$$

Eliminating $\bar{u}$ and substituting for Δp^* the relation for P becomes

$$P = \frac{32fl\varrho Q^3}{\pi^2 \eta D^5}$$

(b) From part (a), $C_2 \propto D^{-5}$. Hence

$$C = C_1 + C_2 = aD^3 + bD^{-5}$$

where a and b are constant coefficients.

C is a minimum at $dC/dD = 0$, and for initial fuel costs write $b = b_1$, $D = D_1$.

$$\frac{dC}{dD} = 3aD^2 - 5bD^{-6} = 0$$

Hence

$$D^8 = \frac{5b}{3a} \quad \text{or} \quad D = \sqrt[8]{\frac{5b}{3a}}$$

The diameter of the original pipe is given by

$$D_1 = \sqrt[8]{\frac{5b_1}{3a}}$$

and the diameter of the new pipe, D_2, corresponding to $b = b_2$, is given by

$$D_2 = \sqrt[8]{\frac{5b_2}{3a}} = \sqrt[8]{\frac{(5)(2.5)b_1}{3a}}$$

Hence $\quad D_2 = \sqrt[8]{2.5}D_1 = \sqrt[8]{2.5} \times 600\,\text{mm} = 673\,\text{mm}$

7.8 COMBINATION OF PIPES

7.8.1 Pipes in series

If two or more pipes are connected in *series*, that is, end to end (as in Fig. 7.14), the same flow passes through each in turn and, as the total head line shows, the total head loss is the sum of the losses in all the individual pipes and fittings.

7.8.2 Pipes in parallel

When two or more pipes are connected as shown in Fig. 7.17, so that the flow divides and subsequently comes together again, the pipes are said to be in *parallel*. Referring to the figure, the continuity equation is

$$Q = Q_A + Q_B$$

Now at any point there can be only one value of the total head. So all the fluid passing point 1 has the same total head $(p_1/\varrho g) + (u_1^2/2g) + z_1$. Similarly, at point 2, after passing through the parallel pipes, all the fluid must have the same total head $(p_2/\varrho g) + (u_2^2/2g) + z_2$ no matter which path it took between points 1 and 2. Therefore, whatever changes occur in the total head lines for pipes A and B, these lines must join common ones at sections 1 and 2 because there cannot be more than one total head line for the single pipe upstream or downstream of the junctions.

The steady-flow energy equation may be written

$$\frac{p_1}{\varrho g} + \frac{u_1^2}{2g} + z_1 - h_l = \frac{p_2}{\varrho g} + \frac{u_2^2}{2g} + z_2$$

Consequently all the fluid must suffer the same loss of head h_l whether it goes via pipe A or pipe B. Flow takes place in pipes A and B as a result of the difference of head between sections 1 and 2 and, once steady conditions are established, the velocities must be such as to give

$$\left(h_l\right)_A = \left(h_l\right)_B$$

If, as is often the case, minor losses are negligible compared with ordinary pipe friction, then

$$h_l = \frac{4 f_A l_A}{d_A} \frac{u_A^2}{2g} = \frac{4 f_B l_B}{d_B} \frac{u_B^2}{2g}$$

In a parallel-pipe system, therefore, the total flow rate is the sum of the flow rates through the individual pipes, but the overall head loss is the same as that through any *one* of the individual pipes. (A similarity

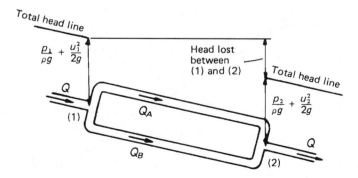

Fig. 7.17

with d.c. electrical circuits will be noted. The total current through resistors in parallel is the sum of the currents through the individual resistors whereas the drop in potential is the same as that across any one resistor.)

Simultaneous solution of the continuity and head loss equations enables the distribution of the total flow rate Q between the individual pipes to be determined.

Problems involving parallel pipes require for their solution an estimate of the value of f for each pipe. From the results of the trial solution, the Reynolds number of the flow in each pipe may be calculated and the assumed values of f checked; if necessary, new values of f may be used for an improved solution. In view of the approximate nature of the data usually available, however, the variation of f with Reynolds number is commonly neglected and the velocity head $u^2/2g$ is also neglected in comparison with the friction loss $4flu^2/2gd$. (Thus the total head line and the pressure line are assumed to coincide.)

7.8.3 Branched pipes

When a pipe system consists of a number of pipes meeting at a junction (as in the simple example illustrated in Fig. 7.18) the basic principles which must be satisfied are:

1. Continuity. At any junction the total mass flow rate towards the junction must equal the total mass flow rate away from it.
2. There can be only one value of head at any point.
3. The friction equation (e.g. Darcy's) must be satisfied for each pipe.

'Minor losses' and the velocity head are usually negligible in comparison with the pipe friction but a difficulty which remains is that formulae for friction loss take no account of the direction of flow. The total head always falls in the direction of flow and yet Darcy's formula, for example, gives a positive value of h_f whether u is positive or negative. It

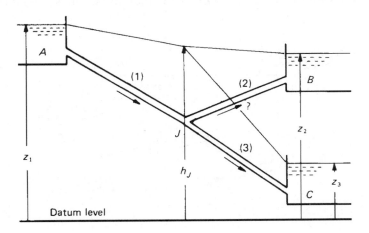

Fig. 7.18

follows that the direction of flow must be specified separately. The direction of flow in a pipe is not often in doubt. When, however, it is unknown it must be assumed and if that assumption yields no physically possible solution the assumption has to be revised.

This is illustrated by the system shown in Fig. 7.18. Three points A, B, C are connected to a common junction J by pipes 1, 2, 3 in which the head losses are respectively h_{f1}, h_{f2}, h_{f3}. The heads at A, B, C are known and may be suitably represented by the surface levels in open reservoirs. Details of the pipes are known and the flow rate in each has to be determined. The head at J is unknown, although its value is evidently between z_1 and z_3.

Since the head at A is the highest and that at C the lowest the direction of flow in pipes 1 and 3 is as indicated by the arrows. The direction of flow in pipe 2, however, is not immediately evident. If h_J, the head at J, is intermediate between the heads at A and B then flow occurs from J to B and for steady conditions the following equations apply:

$$\left. \begin{aligned} z_1 - h_J &= h_{f1} \\ h_J - z_2 &= h_{f2} \\ h_J - z_3 &= h_{f3} \\ Q_1 &= Q_2 + Q_3 \end{aligned} \right\} \quad (7.19)$$

Since h_f is a function of Q, these four equations involve the four unknowns h_J, Q_1, Q_2, Q_3. Even when f is assumed constant and minor losses are neglected so that

$$h_f = \frac{4fl}{d} \frac{Q^2}{\left(\pi d^2/4\right)^2 2g} = \frac{32flQ^2}{\pi^2 gd^5}$$

algebraic solution is tedious (and for more than four pipes impossible). Trial values of h_J substituted in the first three equations, however, yield values of Q_1, Q_2, Q_3 to be checked in the fourth equation. If the calculated value of Q_1 exceeds $Q_2 + Q_3$, for example, the flow rate towards J is too great and a larger trial value of h_J is required. Values of $Q_1 - (Q_1 + Q_3)$ may be plotted against h_J as in Fig. 7.19 and the value of h_J for which $Q_1 - (Q_2 + Q_3) = 0$ readily found. If, however, the direction of flow in pipe 2 was incorrectly assumed no solution is obtainable.

For the opposite direction of flow in pipe 2 the equations are:

$$\left. \begin{aligned} z_1 - h_J &= h_{f1} \\ z_2 - h_J &= h_{f2} \\ h_J - z_3 &= h_{f3} \\ Q_1 + Q_2 &= Q_3 \end{aligned} \right\} \quad (7.20)$$

It will be noticed that the two sets of equations, 7.19 and 7.20, become identical when $z_2 = h_J$ and $Q_2 = 0$. A preliminary trial with $h_J = z_2$ may

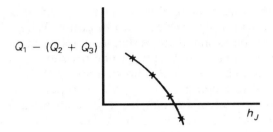

Fig. 7.19

therefore be used to determine the direction of flow in pipe 2. If the trial value of Q_1 is greater than that of Q_3, that is, if the flow rate towards J exceeds that leaving J, then a greater value of h_J is required to restore the balance. On the other hand, if $Q_1 < Q_3$ when h_J is set equal to z_2, then h_J is actually less than z_2.

Similar trial-and-error methods are used for more complex branched pipe problems. An assumption is made for the value of one of the variables and the other quantities are then calculated in turn from that assumption. Adjustments of the initial trial value are made as necessary.

7.8.4 Pipe networks

Complicated problems are posed by pipe networks such as are used for municipal water distribution systems. Here, in addition to a large number of pipes connected in a variety of ways, there may be pumps, check valves to prevent reverse flow, pressure-reducing valves to obviate excessive pressures in the lower parts of the network, and other devices. Nevertheless, the fundamental principles of continuity and uniqueness of the head at any point are again the basis of solution. (As a result of the second principle, the *net* head loss round any closed loop in the network must be zero.)

Ordinary trial-and-error methods are prohibitively time-consuming, particularly as the direction of flow in many of the pipes may be unknown. Until recently, various iterative methods of calculation were used, dealing with one loop in the network at a time[7,8]. Usually, flow rates in the individual pipes would be guessed and then adjusted until the net head loss round that loop was reduced to zero. However, even with good initial guesses, iterations sometimes converged slowly, if at all, particularly when flow rates in some pipes were small and changed sign from one iteration to the next. Now that computers with very large storage capacities are available, preference is given to techniques in which the heads at junctions are guessed (and then adjusted until the resulting calculated flow rates satisfy the continuity relation at each junction), and even to methods in which the many simultaneous equations are solved directly without initial guesses. Analogue computers have also been developed for pipe networks. These highly specialized techniques of calculation are, however, outside the scope of this book.

Some simplification for the purposes of analysis may often be obtained by substituting 'equivalent single pipes' for sets of pipes effectively in series or parallel. For example, for two pipes in series, with negligible minor losses and constant and equal friction factors,

$$h_f = \frac{4f}{2g}\left(\frac{l_1 u_1^2}{d_1} + \frac{l_2 u_2^2}{d_2}\right) = \frac{32 f Q^2}{\pi^2 g}\left(\frac{l_1}{d_1^5} + \frac{l_2}{d_2^5}\right)$$

An equivalent pipe of length $l_1 + l_2$ would therefore have a diameter d such that

$$\frac{l_1 + l_2}{d^5} = \frac{l_1}{d_1^5} + \frac{l_2}{d_2^5}$$

For two pipes in parallel, and similar assumptions,

$$h_f = \frac{32 f Q_1^2 l_1}{\pi^2 g d_1^5} = \frac{32 f Q_2^2 l_2}{\pi^2 g d_2^5}$$

so

$$Q_1 + Q_2 = \left(\frac{\pi^2 g h_f}{32f}\right)^{1/2}\left\{\left(\frac{d_1^5}{l_1}\right)^{1/2} + \left(\frac{d_2^5}{l_2}\right)^{1/2}\right\}$$

$$= \left(\frac{\pi^2 g h_f}{32f}\right)^{1/2}\left(\frac{d^5}{l}\right)^{1/2}$$

for the equivalent pipe.

For a first approximation, pipes in which the flow rate is obviously small may be supposed closed.

7.8.5 Pipe with side tappings

Fluid may be withdrawn from a pipe by side tappings (or 'laterals') along its length. Thus, for a constant diameter, the velocity, and hence the slope of the total head line, varies along the length. If the side tappings are very close together the loss of head over a given length of the main pipe may be obtained by integration of the equation

$$dh_f = 4f\frac{dl}{d}\frac{u^2}{2g} \qquad (7.21)$$

between appropriate limits. In the general case, integration might require, for example, a graphical method in which values of $4fu^2/2gd$ are plotted as a function of l. However, if the tappings are uniformly and closely spaced and are assumed to remove fluid at a uniform rate q per unit length of the main pipe, the volume flow rate at a distance l from the inlet is $Q_0 - ql$, where Q_0 denotes the initial value. For a uniform cross-sectional area A:

$$\frac{Q_0}{u_0} = A = \frac{Q_0 - ql}{u}$$

so

$$u = \left(1 - \frac{ql}{Q_0}\right)u_0$$

Substitution for u in eqn 7.21 and integration from $l = 0$ to $l = l$ gives

$$h_f = \frac{4f}{d}\frac{u_0^2}{2g}\frac{Q_0}{3q}\left\{1 - \left(1 - \frac{ql}{Q_0}\right)^3\right\} \tag{7.22}$$

where f is assumed constant.

If the end of the pipe is closed so that $Q_0 - ql$ is there zero, eqn 7.22 becomes

$$h_f = \frac{4f}{d}\frac{u_0^2}{2g}\frac{l}{3} \tag{7.23}$$

That is, the total loss of head is one-third what it would be if the inlet flow rate Q_0 were the same but no fluid were drawn off along the length.

■

> **Example 7.5** A horizontal water main is 10 cm diameter and 4.5 km long. Supplies are taken uniformly from the main at the rate of $10^{-6}\,\mathrm{m^3\,s^{-1}}$ per metre of length.
>
> Calculate the difference in pressure between the point where the feed enters the main and the remotest point of supply, when the feed is:
>
> (a) at the end of the main
> (b) at the centre of the main.
>
> Assume the friction factor f is constant throughout at 0.006.

Solution
(a) Total water supply $Q_0 = 10^{-6}\,\mathrm{m^2/s} \times (4.5 \times 10^3)\mathrm{m} = 0.0045\,\mathrm{m^3\,s^{-1}}$. Hence

$$u_0 = \frac{Q_0}{(\pi/4)d^2} = \frac{0.0045\,\mathrm{m^3/s}}{(\pi/4)(0.1\,\mathrm{m})^2} = 0.573\,\mathrm{m/s}$$

From equation (7.23)

$$h_f = \frac{4f}{d}\frac{u_0^2}{2g}\frac{l}{3} = \frac{4 \times 0.006}{0.1\,\mathrm{m}} \times \frac{(0.573\,\mathrm{m/s})^2}{2 \times 9.81\,\mathrm{m/s^2}} \times \frac{(4.5 \times 10^3)\mathrm{m}}{3}$$

$$= 6.024\,\mathrm{m}$$

Hence $\quad \Delta p^* = h_f \varrho g = 6.024 \,\text{m} \times 10^3 \,\text{kg/m}^3 \times 9.81 \,\text{m/s}^2$

$$= 59 \times 10^3 \,\text{N}\,\text{m}^{-2}.$$

(b) $\quad Q_0 = 0.00225 \,\text{m}^3 \text{s}^{-1} \quad$ and $\quad l = 2.25 \,\text{km}.$

The problem may now be solved as in part (a). Alternatively it can be solved by noting that $(Q_0)_b = (Q_0)_a/2$, $(u_0)_b = (u_0)_a/2$ and $l_b = l_a/2$. Hence

$$\left(\Delta p^*\right)_b = \frac{\left(u_0^2\right)_b}{\left(u_0^2\right)_a}\frac{l_b}{l_a}\left(\Delta p^*\right)_a = \frac{1}{4}\frac{1}{2}\left(59 \times 10^3\right)\text{N/m}^2$$

$$= 7.4 \times 10^3 \,\text{N}\,\text{m}^{-2} \qquad \qquad \square$$

7.9 CONDITIONS NEAR THE ENTRY TO THE PIPE

The formulae for the head loss to friction considered in the preceding sections are all applicable, strictly, only to the 'fully developed' flow found some distance downstream from the entry to the pipe or from other causes of disturbance to the flow. Near the entry to the pipe the variation of velocity across the section differs from the 'fully developed' pattern, and gradually changes until the final form is achieved.

Suppose that fluid from a large reservoir steadily enters a circular pipe through a smooth, bell-mouthed entry as indicated in Fig. 7.20. At first all the particles – except those in contact with the wall – flow with the same velocity. That is, the velocity profile is practically uniform across the diameter as shown at the left of the diagram. The effect of friction at the wall, however, is to slow down more and more of the fluid near the wall, so forming a boundary layer which increases in thickness until, ultimately, it extends to the axis of the pipe. Since the total flow rate past any section of the pipe is the same, the velocity of the fluid near the axis must increase to compensate for the retardation of fluid near the walls. The shape of the velocity profile thus changes until its final form – for laminar or turbulent flow according to the Reynolds number – is achieved.

Theoretically, an infinite distance is required for the final profile to be attained, but it is usual to regard the flow as fully developed when the velocity on the axis of the pipe is within 1% of its ultimate value. Figures given for the 'entry length' required to establish fully developed laminar flow vary somewhat, but a simple expression derived by H. L. Langhaar is $0.057(Re)d$, where $Re = \bar{u}d/\nu$ and $\bar{u}$ = mean velocity Q/A. So for $Re = 2000$, the highest value at which laminar flow can be counted on, the entry length is about 114 times the diameter. For turbulent flow the final state is reached sooner, the entry length is less dependent on Reynolds number and a value of about 50 times the diameter is common

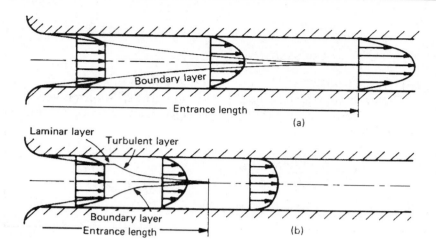

Fig. 7.20 Growth of boundary layer in a pipe (not to scale). (a) Laminar flow. (b) Turbulent flow.

for smooth pipes. If, however, the entry is sharp-edged or there are other factors producing turbulence at the inlet, the entry length is reduced.

For both laminar and turbulent flow the velocity gradient $\partial u/\partial r$ at the wall is higher in the entry length than for fully developed flow and so the shear stress at the wall is greater. The value of $\mathrm{d}h_f/\mathrm{d}l$ is also greater, so the total head lost is somewhat larger than if the flow were fully developed along the whole length of the pipe. If, however, the total length of the pipe is more than about 125 times its diameter the error is negligible in comparison with the uncertainty in the value of f.

7.10 QUASI-STEADY FLOW IN PIPES

Only steady flow in pipes has so far been considered. Chapter 12 will treat some problems of unsteady flow in which the acceleration (positive or negative) of the fluid is of considerable importance. There are, however, instances in which the rate of flow varies continuously with time – that is, the flow is strictly unsteady – and yet the acceleration of the fluid and the forces causing it are negligible. In these circumstances, the steady-flow energy equation applies with sufficient accuracy and the flow may be termed quasi-steady.

As an example, consider an open reservoir with a drain pipe of uniform diameter as illustrated in Fig. 7.21. The area A of the free surface in the reservoir is very large compared with the cross-sectional area a of the pipe, so the level in the reservoir and therefore the rate of flow through the pipe change only slowly. Moreover, the velocity in the reservoir is so small that friction there is negligible. The drain pipe discharges freely to atmosphere. At a time when the free surface in the reservoir is at a height h above the outlet of the pipe and conditions are 'quasi-steady'

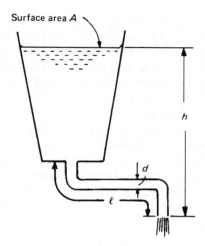

Surface area A

Fig. 7.21

$$h = \left(\frac{4fl}{d} + k\right)\frac{u^2}{2g} \qquad (7.24)$$

where u denotes the mean velocity in the pipe and k is a factor representing the sum of the 'minor' losses in the pipe. In a time interval δt the level of the free surface falls by an amount $-\delta h$. (Remember that δh means 'a very small *increase* of h' and so a decrease of h requires a minus sign.) The velocity of the free surface is therefore $-\mathrm{d}h/\mathrm{d}t$ and continuity then requires that

$$A(-\mathrm{d}h/\mathrm{d}t) = au \qquad (7.25)$$

From eqn 7.24 u may be expressed in terms of h, and substitution for u allows the integration of eqn 7.25:

$$t = \int_0^t \mathrm{d}t = \int_{h_1}^{h_2} \frac{-A\mathrm{d}h}{au} = \int_{h_1}^{h_2} \frac{-A\mathrm{d}h}{anh^{1/2}} \qquad (7.26)$$

where

$$n = \left\{\frac{2g}{(4fl/d) + k}\right\}^{1/2} \qquad (7.27)$$

Thus the time taken for the liquid level to fall from h_1 to h_2 (measured above the pipe outlet) may be determined. If A varies with h evaluation of the integral may not be possible by algebraic methods and recourse must be had to a graphical or numerical technique.

It is true that, as h decreases, u decreases and f changes. (Indeed, if the flow is laminar, the substitution for u in eqn 7.26 must be from the Hagen–Poiseuille formula, eqn 6.9.) If necessary, a succession of integrations could be performed, each over a small range of h and u, an appropriate mean value of f being used in each. Usually, however, it may be assumed that f is practically constant over the whole range of u involved. If eqn 7.26 is used to determine the time required to empty the reservoir completely, some inaccuracy also arises because conditions in the reservoir change significantly as the depth of liquid becomes very small. For example, if the pipe is connected to the base of the reservoir, a vortex may form at its entrance. The effective head is then altered and the swirling motion in the pipe results in a smaller mean axial velocity for a given loss of head. Unless the initial depth in the reservoir is small, however, these different conditions at the end of the emptying period occupy only a small proportion of the total emptying time t.

If flow into the reservoir (Q_{in}) occurs at the same time as the outflow, then the net volume flow rate out of the reservoir is $au - Q_{in}$, the continuity equation becomes

$$A(-dh/dt) = au - Q_{in}$$

and the expression for t is accordingly

$$t = \int_{h_1}^{h_2} \frac{-A dh}{anh^{1/2} - Q_{in}} \tag{7.28}$$

Quasi-steady flow from a reservoir to atmosphere via an orifice is similarly treated. Then the continuity equation is

$$A(-dh/dt) = C_d a \sqrt{(2gh)} - Q_{in}$$

where a represents the cross-sectional area of the orifice and h the head above it. Again A is much greater than a. The orifice size and the Reynolds number of the flow are both usually large enough for C_d to be assumed constant. For the special case of a reservoir with constant cross-section and with no inflow

$$t = \int_0^t dt = \int_{h_1}^{h_2} \frac{-A dh}{C_d a \sqrt{(2g)h}} = \frac{2A}{C_d a \sqrt{(2g)}}\left(h_1^{1/2} - h_2^{1/2}\right) \tag{7.29}$$

If two reservoirs are joined by a pipe, both ends of which are submerged (Fig. 7.22), the rate of flow between them is determined by the difference in the reservoir surface levels. As one surface falls, the other rises, so

$$A_1\left(-\frac{dz_1}{dt}\right) = Q = A_2 \frac{dz_2}{dt}$$

The rate at which the *difference* of levels changes is

$$\frac{dh}{dt} = \frac{d}{dt}(z_1 - z_2) = \frac{dz_1}{dt} - \frac{A_1}{A_2}\left(-\frac{dz_1}{dt}\right) = \frac{dz_1}{dt}\left(1 + \frac{A_1}{A_2}\right) \tag{7.30}$$

The continuity equation gives

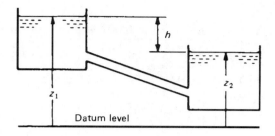

Fig. 7.22

$$A_1\left(-\frac{\mathrm{d}z_1}{\mathrm{d}t}\right) = au = anh^{1/2} \quad \left(\text{from eqn 7.27}\right).$$

Substitution from eqn 7.30 gives

$$\frac{-A_1}{1 + \left(A_1/A_2\right)}\frac{\mathrm{d}h}{\mathrm{d}t} = anh^{1/2}$$

The relation now has a *single* head variable h and may be integrated in the usual manner.

In any problem containing more than one head variable, care should be taken to distinguish between them, and integration should not be attempted until the principle of continuity has been used to express all heads (and differentials of heads) in terms of a single variable.

Example 7.6 Two vertical cylindrical water tanks, each open to atmosphere and of diameters 3 m and 2 m respectively, are connected by two pipes in parallel, each 50 mm diameter and 75 m long. Initially the water level in the larger tank is 1.8 m above that in the smaller. Assuming that entry and exit losses for each pipe total 1.5 times the velocity head, that the pipes are always full of water, and that f for each pipe has the constant value 0.007, determine the change in level in the larger tank in 15 minutes.

Solution
If the plan areas of the tanks are respectively A_1 and A_2, then from eqn 7.30,

$$\frac{\mathrm{d}h}{\mathrm{d}t} = \frac{\mathrm{d}z_1}{\mathrm{d}t}\left(1 + \frac{A_1}{A_2}\right) = \frac{\mathrm{d}z_1}{\mathrm{d}t}\left(1 + \frac{3^2}{2^2}\right) = \frac{13}{4}\frac{\mathrm{d}z_1}{\mathrm{d}t}$$

Therefore the total volume flow rate

$$Q = -A_1\frac{\mathrm{d}z_1}{\mathrm{d}t} = -\frac{4}{13}A_1\frac{\mathrm{d}h}{\mathrm{d}t} \qquad (7.31)$$

Although there are two pipes in parallel, the head lost is that in only one. So, at any instant,

$$h = \left(\frac{4fl}{d} + 1.5\right)\frac{u^2}{2g} = \left(\frac{4 \times 0.007 \times 75}{0.05} + 1.5\right)\frac{u^2}{2g}$$

$$= \frac{43.5}{19.62\,\text{m/s}^2}\frac{(Q/2)^2}{\left(\frac{\pi}{4} \times 0.05^2\,\text{m}^2\right)^2} = 1.438 \times 10^5 Q^2 \text{s}^2/\text{m}^5$$

Substituting for Q and A_1 in eqn 7.26 and rearranging, we get

$$dt = -\frac{4}{13}\left(\frac{\pi}{4}3^2\,\text{m}^2\right)\left(\frac{1.438 \times 10^5}{h}\frac{\text{s}^2}{\text{m}^5}\right)^{1/2} dh$$

$$= -\left(824.7\frac{\text{s}}{\text{m}^{1/2}}\right)h^{-1/2}dh$$

Integrating this equation from $t = 0$ to $t = 900\,\text{s}$ and from $h = 1.8\,\text{m}$ to $h = H$ gives

$$t = 900\,\text{s} = 2 \times 824.7\frac{\text{s}}{\text{m}^{1/2}}\left\{(1.8\,\text{m})^{1/2} - H^{1/2}\right\}$$

whence $H = 0.6336\,\text{m}$. The change in h is therefore $(1.8 - 0.6336)\,\text{m} = 1.1664\,\text{m}$ and the change in z_1 is $(4/13) \times 1.1664\,\text{m} = 0.359\,\text{m}$.

7.11   FLOW MEASUREMENT

Methods of measuring various quantities such as pressure and viscosity have been mentioned in other parts of this book in connection with the theory on which they are based. Here brief reference will be made to the principles of some other techniques; constructional details of instruments will, however, be omitted and no attempt will be made to mention every form of measurement.

7.11.1   Measurement of velocity

Among instruments for velocity measurement, other than the Pitot and Pitot-static tubes mentioned in Chapters 3 and 11, is the *hot-wire anemometer*, which depends on the facts that the electrical resistance of a wire is a function of its temperature; that this temperature depends on the heat transfer to the surroundings; and that the heat transfer is a function of the velocity of fluid past the wire. A fine platinum, nickel or tungsten wire, heated electrically, is held between the ends of two pointed prongs so as to be perpendicular to the direction of flow. The current through the wire may be adjusted to keep its temperature and therefore its resistance constant, the current being measured; or the current may be kept constant and the change in resistance determined by measuring the potential difference across the wire; or the potential difference may be held constant and the current measured. Whichever method of measurement is used, frequent calibration of the instrument against a known velocity is required. The rate of heat transfer depends also on the density and other properties of the fluid and so calibration should be carried out in the same fluid as that whose velocity is to be measured.

Because of its small size – the wire is seldom more than about 6 mm long or more than 0.15 mm diameter – the instrument may be used where velocity gradients are large, for example in boundary layers. For the same reason it responds very rapidly to fluctuations of velocity, and therefore, in conjunction with oscilloscopes and similar electronic apparatus, it has found wide use in measuring the intensity of turbulence. As a single wire responds principally to fluctuations parallel to the mean velocity, turbulence is usually investigated by using more than one wire.

The hot-wire anemometer is used mainly for measuring the velocity of gases. It has proved less successful in liquids because bubbles and small solid particles tend to collect on the wire and spoil the calibration. More recently a *hot-film anemometer* has been developed, in which a thin platinum film fused to a glass support takes the place of the hot wire. Although inferior in frequency response to the hot-wire type, it finds application where the use of a hot wire is precluded by its limited mechanical strength.

Mechanical devices embodying some form of rotating element may also be used for velocity measurement. For use in liquids these are called *current meters*; for use in gases, *anemometers*. In one type, hollow hemispheres or cones are mounted on spokes so as to rotate about a shaft perpendicular to the direction of flow (see Fig. 7.23). The drag on a hollow hemisphere or cone is greater when its open side faces a flow of given velocity, and so there is a net torque on the assembly when flow comes from any direction in the plane of rotation. The magnitude of the fluid velocity (but not its direction) determines the speed of rotation; on an anemometer this is usually indicated by a mechanical counter, and on a current meter by the number of once-per-revolution electrical contacts made in a known time interval. For higher velocities a propeller is used having its axis parallel to the fluid flow. This type is sensitive to the direction of flow particularly if the propeller is surrounded by a shielding cylinder. Each type requires calibration, usually either in a wind tunnel or by towing the instrument at known speed through still liquid.

An expensive but accurate instrument suitable for a very wide range of velocities is the *laser-Doppler anemometer*. This utilizes the Doppler effect by which waves from a given source reach an observer with an increased frequency if the source approaches him (or he approaches the source) and a reduced frequency if source and observer recede from one another. The phenomenon is most often noticed with sound waves: to a stationary listener the sound emitted by, for example, an express train appears to drop in pitch as the source of sound approaches, passes and then recedes from him. For light waves the effect is seldom directly observable because the velocity of light is so large compared with that of any moving object that the proportional change in frequency is barely detectable by ordinary spectroscopic techniques. However, if light that has undergone a Doppler frequency change is caused to beat with light at the original frequency, the change may be accurately measured and the velocity of the moving object thus deduced.

In the most common form of laser-Doppler anemometer the intense monochromatic light from a laser is split into two beams which are deflected by mirrors or lenses so as to intersect in a small region of the fluid. The 'reference' beam passes directly to a photo-detector; small particles in the fluid cause some of the light in the other beam to be scattered in the same direction as the reference beam, and thus to mix with it in the photo-detector. The beat frequency between the direct and the scattered light is determined by feeding the electrical output of the photo-detector into a wave analyser.

The expression for the beat frequency may be simply obtained by considering the interference fringes formed where the beams intersect: the distance d between successive fringes is $\lambda/\{2\sin(\theta/2)\}$ where λ represents the wavelength of the light and θ the acute angle between the two beams. A particle in the fluid – small compared with d – moving at velocity u in the plane of the two beams and perpendicular to the bisector of the angle θ will scatter light in moving through the bright regions between fringes and thus the scattered light will be modulated with a frequency $u/d - 2(u/\lambda)\sin(\theta/2)$. By measuring this frequency, u may be determined.

In practice there are often sufficient impurities in the fluid to make any special addition of particles unnecessary. Particular advantages of the technique are that no obstructions need be inserted into the fluid and no flow calibration is needed.

Among special techniques used in rivers and large water pipe-lines is the *salt-velocity method*, in which a quantity of a concentrated solution of salt is suddenly injected into the water. At each of two downstream cross-sections a pair of electrodes is inserted, the two pairs being connected in parallel. When the salt-enriched water passes between a pair

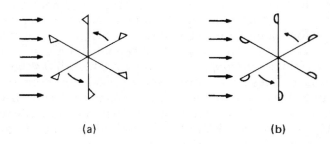

Fig. 7.23 (a) Current meter used in water. (b) Anemometer used in air.

(a)   (b)

of electrodes its higher conductivity causes the electric current to rise briefly to a peak. The mean velocity of the stream is determined from the distance between the two electrode assemblies and the time interval between the appearances of the two current peaks. Rather elaborate apparatus is needed for the automatic recording of data.

7.11.2 Measurement of discharge

Reference is also made to orifices, nozzles, venturi-meters and weirs in Chapters 3 and 10.

Since the drag on a submerged body depends on the velocity of flow past it, a number of meters have been devised in which the velocity (and thus, in suitable arrangements, the discharge) is indicated by a measurement of the drag force. For example, the deflection of a vane against a spring may be measured. In the *rotameter* (Fig. 7.24) flow occurs upwards through a transparent tube tapering outwards towards the top. A 'float' is carried up to the level where the velocity in the annular space between float and tube is such that the drag on the float just balances its weight minus its buoyancy. Slanting grooves cut into the circumference of the float cause it to rotate and thus to remain central in the tube. For a given homogeneous fluid the (steady) rate of flow may be read directly from a calibrated scale on the tube.

The *bend-meter* or *elbow-meter* depends on the fact that, since in curved flow pressure increases with radius, a pressure difference exists between the outer and inner walls of a pipe bend. Use of a suitable manometer allows the measurement of this pressure difference which is a function of the rate of flow. As the bend is usually already part of the piping system, this meter is simple and inexpensive, and after careful calibration *in situ* it is accurate.

Many patterns of discharge meter have been made that are in effect propeller-type current meters enclosed in a pipe of fixed cross-sectional area, the total discharge being indicated by a mechanical revolution counter.

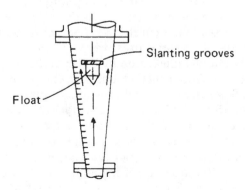

Slanting grooves

Float

Fig. 7.24 Rotameter.

Various forms of positive displacement meter are in use: some form of piston or partition is moved by the flow of fluid, and each movement corresponds to a certain volume. Such meters may be regarded as positive displacement pumps in reverse.

For rivers or large pipe-lines in which thorough mixing of the liquid occurs the *salt-dilution method* may be used. A concentrated solution of salt is fed into the stream at a known, steady, volume flow rate q, preferably from a number of points distributed over the cross-section. The concentration of salt is measured some distance downstream after the salt has become thoroughly diffused throughout the liquid. Let the mass of salt per unit volume of the original stream be s_0, the mass of salt per unit volume of the added solution be s_1 and the mass of salt per unit volume measured downstream be s_2. If the volume flow rate of the stream is Q then, when steady conditions have been attained, the mass of salt reaching the downstream sampling point in unit time is given by $Qs_0 + qs_1 = (Q + q)s_2$ whence

$$Q = q\left(\frac{s_1 - s_2}{s_2 - s_0}\right)$$

The concentrations may be determined by chemical titration or by electrical conductivity measurements.

Electro-magnetic meters have been developed in which fluid with a sufficiently high electrical conductivity flows through a non-conducting tube across a magnetic field. Electrodes embedded in the walls enable the induced e.m.f. to be measured. As in all electro-magnetic generators the e.m.f. is proportional to the number of magnetic flux lines cut by the conductor in unit time, and therefore to the mean velocity of the fluid. The instrument offers no obstruction whatever to the flow, and thus no additional head loss; the e.m.f. is independent of the density and viscosity of the fluid and is insensitive to velocity variations over the cross-section. The fluid, however, must be conductive.

7.11.3 Measurement of flow direction

An instrument to measure the direction of flow is termed a *yaw meter*. One simple type uses a pivoted vane and operates like the well-known weathercock; for a complete indication of direction two vanes are required, their pivot axes being (preferably) perpendicular.

Many designs of differential-pressure yaw meters exist but their common principle may be illustrated by the cylindrical type depicted in Fig. 7.25. A circular cylinder has its axis perpendicular to the plane of two-dimensional flow, and two small pressure tappings P_1 and P_2 normal to the surface of the cylinder are connected to the two sides of a manometer. The cylinder is rotated about its axis until the stagnation point S is mid-way between P_1 and P_2: the pressures at P_1 and P_2 are then equal as indicated by a null reading on the manometer. The flow direction is then the bisector of the angle P_1OP_2. For an ideal fluid the pressure recorded at P_1 or P_2 would be the static pressure of the

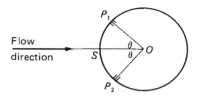

Fig. 7.25

undisturbed stream if P_1 and P_2 were each 30° from S (from eqn 9.27). Frictional effects, however, make the angle θ about $39\frac{1}{4}°$ in practice. Thus if the angle P_1OP_2 is $78\frac{1}{2}°$ the instrument may be used also to measure static pressure and, when turned through $39\frac{1}{4}°$, the stagnation pressure. The angles are larger when compressibility effects become significant. In other designs small Pitot tubes are used in place of pressure tappings on a cylinder.

For three-dimensional flow a sphere may be employed in place of the cylinder: three pressure tappings are then situated at the vertices of an equilateral triangle, and the pressures at all three points are equal when the flow is parallel to the radius to the centroid of the triangle.

The hot-wire anemometer can be adapted as a yaw meter for two-dimensional flow by incorporating two similar wires placed in the plane of flow at an angle to each other. A position is found in which the heat losses from the wires are equal; the flow direction is then that which bisects the angle between the wires. For three-dimensional flow three wires are required, arranged as adjoining edges of a regular tetrahedron.

REFERENCES

1. Haaland, S. E. 'Simple and explicit formulas for the friction factor in turbulent pipe flow', *J. Fluids Engng* **105**, 89–90 (1983).
2. 'Charts for the hydraulic design of channels and pipes', *Hydraulics Res. Pap. No. 2 of Hydraulics Res. Stn* (5th edn), HMSO, London (1983).
3. Jones, O. C. Jr, 'An improvement in the calculation of turbulent friction in rectangular ducts', *J. Fluids Engng* **98**, 173–81 (1976).
4. Jones, O. C. Jr, and Leung, J. C. M. 'An improvement in the calculation of turbulent friction in smooth concentric annuli', *J. Fluids Engng*, **103**, 615–23 (1981).
5. Ito, H. 'Friction factors for turbulent flow in curved pipes', *J. basic Engng* **81D**, 123–34 (1959).
6. Ito, H. 'Pressure losses in smooth pipe bends', *J. basic Engng* **82D**, 131–43 (1960).
7. Jeppson, R. W. *Analysis of Flow in Pipe Networks*, Ann Arbor Science, Ann Arbor, Michigan (1976).
8. Ward-Smith, A. J. *Internal Fluid Flow: The Fluid Dynamics of Flow in Pipes and Ducts*, pp. 492–5, Oxford University Press (1980).

FURTHER READING

Benedict, R. P. *Fundamentals of Temperature, Pressure and Flow Measurement* (3rd edn), Wiley, New York (1984).

Cheremisinoff, N. P. and Cheremisinoff, P. N. *Flow Measurement for Engineers and Scientists*, Dekker, New York (1988).

Clayton, C. G. (ed.) *Modern Developments in Flow Measurement*, Peter Peregrinus, London (1972).

Durst, F., Melling, A. and Whitelaw, J. H. *Principles and Practice of Laser–Doppler Anemometry* (2nd edn), Academic Press, London (1981).

Goldstein, R. J. *Fluid Mechanics Measurements* (2nd edn), Hemisphere Publishing Corporation, Washington (1996).

Hinze, J. O. 'Turbulent pipe-flow', in *The Mechanics of Turbulence, International Symposium* 1961, *Centre national de la Recherche Scientifique*, pp. 129–165, Gordon and Breach, New York (1964).

Lomas, C. G. *Fundamentals of Hot-Wire Anemometry*, Cambridge University Press (1986).

Moody, L. F. 'Friction factors for pipe flow', *Trans. Am. Soc. mech. Engrs* **66**, 671–84 (1944).

Nikuradse, J. 'Strömungsgesetze in rauhen Rohren', *VDI-Forschungsheft* **361** (1933).
English translations: *NACA Tech. Memo.* 1292 and *Petroleum Engineer* (1940) March, pp. 164–6; May, pp. 75, 78, 80, 82; June, pp. 124, 127, 128, 130; July, pp. 38, 40, 42; August, pp. 83, 84, 87.
(Nikuradse's work was discussed by Miller, B. in 'The laminar-film hypothesis', *Trans. Am. Soc. mech. Engrs* **71**, 357–67 (1949).)

Ower, E. and Pankhurst, R. C. *The Measurement of Air Flow* (5th edn), Pergamon, Oxford (1977).

Perry, A. E. *Hot-wire Anemometry*, Clarendon Press, Oxford (1982).

Scott, R. W. W. *Developments in Flow Measurement*, Applied Science, London (1982).

Fluid Flow Measurement in the mid-1980s (two volumes), HMSO, London (1986).

Ward-Smith, A. J. *Internal Fluid Flow: The Fluid Dynamics of Flow in Pipes and Ducts*, Oxford University Press (1980).

PROBLEMS

7.1 Calculate the power required to pump sulphuric acid (viscosity 0.04 Pa s, relative density 1.83) at 45 litres a second from a supply tank through a glass-lined 150 mm diameter pipe, 18 m long, into a storage tank. The liquid level in the storage tank is 6 m above that in the supply tank. For laminar flow $f = 16/Re$; for turbulent flow $f = 0.0014(1 + 100Re^{-1/3})$ if $Re < 10^7$. Take all losses into account.

7.2   In a heat exchanger there are 200 tubes each 3.65 m long and 30 mm outside diameter and 25 mm bore. They are arranged axially in a cylinder of 750 mm diameter and are equally spaced from one another. A liquid of relative density 0.9 flows at a mean velocity of 2.5 m/s through the tubes and water flows at 2.5 m/s *between* the tubes in the opposite direction. For all surfaces f may be taken as 0.01. Neglecting entry and exit losses, calculate (*a*) the *total* power required to overcome fluid friction in the exchanger and (*b*) the saving in power if the two liquids exchanged places but the system remained otherwise unaltered.

7.3   A hose pipe of 75 mm bore and length 450 m is supplied with water at 1.4 MPa. A nozzle at the outlet end of the pipe is 3 m above the level of the inlet end. If the jet from the nozzle is to reach a height of 35 m calculate the maximum diameter of the nozzle assuming that $f = 0.01$ and that losses at inlet and in the nozzle are negligible. If the efficiency of the supply pump is 70% determine the power required to drive it.

7.4   A straight smooth pipe 100 mm diameter and 60 m long is inclined at 10° to the horizontal. A liquid of relative density 0.9 and kinematic viscosity 120 mm²/s is to be pumped through it into a reservoir at the upper end where the gauge pressure is 120 kPa. The pipe friction factor f is given by $16/Re$ for laminar flow and by $0.08(Re)^{-1/4}$ for turbulent flow when $Re < 10^5$. Determine (*a*) the maximum pressure at the lower, inlet, end of the pipe if the mean shear stress at the pipe wall is not to exceed 200 Pa; (*b*) the corresponding rate of flow.

7.5   A trailer pump is to supply a hose 40 m long and fitted with a 50 mm diameter nozzle capable of throwing a jet of water to a height 40 m above the pump level. If the power lost in friction in the hose is not to exceed 15% of the available hydraulic power, determine the diameter of hose required. Friction in the nozzle may be neglected and f for the hose assumed to be in the range 0.007–0.01.

7.6   A pipe 900 m long and 200 mm diameter discharges water to atmosphere at a point 10 m below the level of the inlet. With a pressure at inlet of 40 kPa above atmospheric the steady discharge from the end of the pipe is 49 litres/s. At a point half way along the pipe a tapping is then made from which water is to be drawn off at a rate of 18 litres/s. If conditions are such that the pipe is always full, to what value must the inlet pressure be raised so as to provide an unaltered discharge from the end of the pipe? (The friction factor may be assumed unaltered.)

7.7   Two water reservoirs, the surface levels of which differ by 1.5 m, are connected by a pipe system consisting of a sloping pipe at each end, 7.5 m long and 75 mm diameter, joined by a horizontal pipe 300 mm diameter and 60 m long. Taking entry head losses as 0.5 $u^2/2g$ and $f = 0.005(1 + 25/d)$ where d mm is the pipe diameter, calculate the steady rate of flow through the pipe.

7.8 Kerosene of relative density 0.82 and kinematic viscosity 2.3 mm^2/s is to be pumped through 185 m of galvanized iron pipe ($k = 0.15$ mm) at 40 litres/s into a storage tank. The pressure at the inlet end of the pipe is 370 kPa and the liquid level in the storage tank is 20 m above that of the pump. Neglecting losses other than those due to pipe friction determine the size of pipe necessary.

7.9 Calculate the volume flow rate of the kerosene in Problem 7.8 if the pipe were 75 mm diameter.

7.10 A liquid of relative density 1.2 flows from a 50 mm diameter pipe A into a 100 mm diameter pipe B, the enlargement from A to B being abrupt. Some distance down-stream of the junction is a total-head tube facing the oncoming flow; this is connected to one limb of a U-tube manometer containing mercury (relative density 13.6). The other limb of the manometer is connected to a tapping in the side of pipe A. Calculate the mass flow rate of the liquid when the difference of mercury levels is 50 mm.

7.11 A single uniform pipe joins two reservoirs. Calculate the percentage increase of flow rate obtainable if, from the mid-point of this pipe, another of the same diameter is added in parallel to it. Neglect all losses except pipe friction and assume a constant and equal f for both pipes.

7.12 Two reservoirs are joined by a sharp-ended flexible pipe 100 mm diameter and 36 m long. The ends of the pipe differ in level by 4 m; the surface level in the upper reservoir is 1.8 m above the pipe inlet while that in the lower reservoir is 1.2 m above the pipe outlet. At a position 7.5 m horizontally from the upper reservoir the pipe is required to pass over a barrier. Assuming that the pipe is straight between its inlet and the barrier and that $f = 0.01$ determine the greatest height to which the pipe may rise at the barrier if the absolute pressure in the pipe is not to be less than 40 kPa. Additional losses at bends may be neglected. (Take atmospheric pressure = 101.3 kPa.)

7.13 Between the connecting flanges of two pipes A and B is bolted a plate containing a sharp-edged orifice C for which $C_c = 0.62$. The pipes and the orifice are coaxial and the diameters of A, B and C are respectively 150 mm, 200 mm and 100 mm. Water flows from A into B at the rate of 42.5 litres/s. Neglecting shear stresses at boundaries, determine (*a*) the difference of static head between sections in A and B at which the velocity is uniform, (*b*) the power dissipated.

7.14 Petrol of kinematic viscosity 0.6 mm²/s is to be pumped at the rate of 0.8 m³/s through a horizontal pipe 500 mm diameter. However, to reduce pumping costs a pipe of different diameter is suggested. Assuming that the absolute roughness of the walls would be the same for a pipe of slightly different diameter, and that, for $Re > 10^6$, f is approximately proportional to the cube root of the roughness, determine the diameter of pipe for which the pumping costs would be halved. Neglect all head losses other than pipe friction.

How are the running costs altered if n pipes of equal diameter are used in parallel to give the same total flow rate at the same Reynolds number as for a single pipe?

7.15 A pump delivers water through two pipes laid in parallel. One pipe is 100 mm diameter and 45 m long and discharges to atmosphere at a level 6 m above the pump outlet. The other pipe, 150 mm diameter and 60 m long, discharges to atmosphere at a level 8 m above the pump outlet. The two pipes are connected to a junction immediately adjacent to the pump and both have $f = 0.008$. The inlet to the pump is 600 mm below the level of the outlet. Taking the datum level as that of the pump inlet, determine the total head at the pump outlet if the flow rate through it is 0.037 m³/s. Losses at the pipe junction may be neglected.

7.16 A reservoir A, the free water surface of which is at an elevation of 275 m, supplies water to reservoirs B and C with water surfaces at 180 m and 150 m elevation respectively. From A to a junction D there is a common pipe 300 mm diameter and 16 km long. The pipe from D to B is 200 mm diameter and 9.5 km long while that from D to C is 150 mm diameter and 8 km long. The ends of all pipes are submerged. Calculate the rates of flow to B and C, neglecting losses other than pipe friction and taking $f = 0.01$ for all pipes.

7.17 A reservoir A feeds two lower reservoirs B and C through a single pipe 10 km long, 750 mm diameter, having a downward slope of 2.2×10^{-3}. This pipe then divides into two branch pipes, one 5.5 km long laid with a downward slope of 2.75×10^{-3} (going to B), the other 3 km long having a downward slope of 3.2×10^{-3} (going to C). Calculate the necessary diameters of the branch pipes so that the steady flow rate in each shall be 0.24 m³/s when the level in each reservoir is 3 m above the end of the corresponding pipe. Neglect all losses except pipe friction and take $f = 0.006$ throughout.

7.18 A pipe 600 mm diameter and 1 km long with $f = 0.008$ connects two reservoirs having a difference in water surface level of 30 m. Calculate the rate of flow between the reservoirs and the shear stress at the wall of the pipe. If the upstream half of the pipe is tapped by several side pipes so that one third of the quantity of water now entering the main pipe is withdrawn uniformly over this length, calculate the new rate of discharge to the lower reservoir. Neglect all losses other than those due to pipe friction.

7.19 A rectangular swimming bath 18 m long and 9 m wide has a depth uniformly increasing from 1 m at one end to 2 m at the other. Calculate the time required to empty the bath through two 150 mm diameter outlets for which $C_d = 0.9$, assuming that all conditions hold to the last.

7.20 A large tank with vertical sides is divided by a vertical partition into two sections A and B, with plan areas of $1.5 \, \text{m}^2$ and $7.5 \, \text{m}^2$ respectively. The partition contains a 25 mm diameter orifice ($C_d = 0.6$) at a height of 300 mm above the base. Initially section A contains water to a depth of 2.15 m and section B contains water to a depth of 950 mm. Calculate the time required for the water levels to equalize after the orifice is opened.

7.21 Two vertical cylindrical tanks, of diameters 2.5 m and 1.5 m respectively, are connected by a 50 mm diameter pipe, 75 m long for which f may be assumed constant at 0.01. Both tanks contain water and are open to atmosphere. Initially the level of water in the larger tank is 1 m above that in the smaller tank. Assuming that entry and exit losses for the pipe together amount to 1.5 times the velocity head, calculate the fall in level in the larger tank during 20 minutes. (The pipe is so placed that it is always full of water.)

7.22 A tank 1.5 m high is 1.2 m in diameter at its upper end and tapers uniformly to 900 mm diameter at its base (its axis being vertical). It is to be emptied through a pipe 36 m long connected to its base and the outlet of the pipe is to be 1.5 m below the bottom of the tank. Determine a suitable diameter for the pipe if the depth of water in the tank is to be reduced from 1.3 m to 200 mm in not more than 10 minutes. Losses at entry and exit may be neglected and f assumed constant at 0.008.

7.23 The diameter of an open tank, 1.5 m high, increases uniformly from 4.25 m at the base to 6 m at the top. Discharge takes place through 3 m of 75 mm diameter pipe which opens to atmosphere 1.5 m below the base of the tank. Initially the level in the tank is steady, water entering and leaving at a constant rate of 17 litres/s. If the rate of flow into the tank is suddenly doubled, calculate the time required to fill it completely. Assume that the pipe has a bell-mouthed entry with negligible loss and that f is constant at 0.01. A numerical or graphical integration is recommended.

7.24 A tank of plan area $5 \, \text{m}^2$, open to atmosphere at the top, is supplied through a pipe 50 mm diameter and 40 m long which enters the base of the tank. A pump, providing a constant gauge pressure of 500 kPa, feeds water to the inlet of the pipe which is at a level 3 m below that of the base of the tank. Taking *f* for the pipe as 0.008, calculate the time required to increase the depth of water in the tank from 0.2 m to 2.5 m. If the combined overall efficiency of pump and driving motor is 52%, determine (in kWh) the total amount of electricity used in this pumping operation.

7.25 A water main with a constant gauge pressure of 300 kPa is to supply water through a pipe 35 m long to a tank of uniform plan area 6 m^2, open to atmosphere at the top. The pipe is to enter the base of the tank at a level 2.9 m above that of the main. The depth of water in the tank is to be increased from 0.1 m to 2.7 m in not more than 15 min. Assuming that *f* has the constant value 0.007, and neglecting energy losses other than pipe friction, determine the diameter of pipe required.

7.26 A viscometer consists essentially of two reservoirs joined by a length of capillary tubing. The instrument is filled with liquid so that the free surfaces in the reservoirs are where the walls are in the form of vertical circular cylinders, each of diameter 20 mm. The capillary tube is 1 mm diameter and 400 mm long. Initially the difference in the free surface levels is 40 mm, and, for a particular liquid of relative density 0.84, the time taken for the higher level to fall by 15 mm is 478 s. Neglecting 'end effects' in the capillary tube and energy losses other than those due directly to viscosity, determine the viscosity of the liquid.

7.27 Water flows at a steady mean velocity of 1.5 m/s through a 50 mm diameter pipe sloping upwards at 45° to the horizontal. At a section some distance downstream of the inlet the pressure is 700 kPa and at a section 30 m further along the pipe the pressure is 462 kPa. Determine the average shear stress at the wall of the pipe and at a radius of 10 mm.

7.28 A fluid of constant density ϱ enters a horizontal pipe of radius R with uniform velocity V and pressure p_1. At a downstream section the pressure is p_2 and the velocity varies with radius r according to the equation $u = 2V\{1 - (r^2/R^2)\}$. Show that the friction force at the pipe walls from the inlet to the section considered is given by $\pi R^2(p_1 - p_2 - \varrho V^2/3)$.

8 Boundary layers and wakes

8.1 INTRODUCTION

The flow of a real fluid (except at extremely low pressures) has two fundamental characteristics. One is that there is no discontinuity of velocity; the second is that, at a solid surface, the velocity of the fluid relative to the surface is zero, the so-called no-slip condition. As a result there is, close to the surface, a region in which the velocity increases rapidly from zero and approaches the velocity of the main stream. This region is known as the *boundary layer*. It is usually very thin, but may sometimes be observed with the naked eye: close to the sides of a ship, for example, is a narrow band of water with a velocity relative to the ship clearly less than that of water further away.

Boundary layer

The increase of velocity with increasing distance from the solid surface involves relative movement between the particles in the boundary layer, and thus shear stresses are in evidence. Since the layer is usually very thin the velocity gradient – that is, the rate of increase of velocity with increasing distance from the surface – is high, and the shear stresses are therefore important. In 1904 the German engineer Ludwig Prandtl (1875–1953) suggested that the flow may be considered in two parts: (1) that in the boundary layer where the shear stresses are of prime importance, and (2) that beyond the boundary layer where (in general) velocity gradients are small and so the effect of viscosity is negligible. In this second part the flow is thus essentially that of an ideal fluid.

(The shape of the solid surface past which the fluid flows may be such as to deflect the streamlines, and therefore to produce velocity gradients even in an ideal fluid. The effect of the boundary layer, however, stems from the condition of zero velocity at the boundary itself, and so is additional to the effects of the deflection of streamlines.)

With increasing distance from the solid surface the velocity approaches that of the main stream asymptotically, and there is thus no sharp dividing line between the boundary layer and the rest of the flow. Nevertheless, Prandtl's concept of the boundary layer, where the influence of viscosity is concentrated, has bridged the gap between classical hydrodynamics (in which an inviscid fluid is postulated) and the observed behaviour of real fluids. The rapid advances in mechanics of fluids in the twentieth century are largely due to this important concept.

Our principal concern in this chapter will be to examine the flow in the boundary layer and its influence on the main flow. For simplicity, steady, two-dimensional flow of a constant-density fluid will be assumed.

8.2   DESCRIPTION OF THE BOUNDARY LAYER

The simplest boundary layer to study is that formed in the flow along one side of a thin, smooth, flat plate parallel to the direction of the oncoming fluid (Fig. 8.1). No other solid surface is near, and the pressure of the fluid is assumed uniform. If the fluid were ideal no velocity gradient would, in this instance, arise. The velocity gradients in a real fluid are therefore entirely due to viscous action near the surface.

The fluid, originally having velocity u_∞ in the direction of the plate, is retarded in the neighbourhood of the surface, and the boundary layer begins at the leading edge of the plate. As more and more of the fluid is slowed down the thickness of the layer increases. We have already noted that the boundary layer merges into the main flow with no sharp line of demarcation but, for convenience, the thickness of the layer may be taken as that distance from the surface at which the velocity reaches 99% of the velocity of the main stream. The flow in the first part of the boundary layer (close to the leading edge of the plate) is entirely laminar. With increasing thickness, however, the laminar layer becomes unstable, and the motion within it becomes disturbed. The irregularities of the flow develop into turbulence, and the thickness of the layer increases more rapidly. These changes take place over a short length known as the *transition region*. Downstream of the transition region the boundary layer is almost entirely turbulent, and its thickness increases further. For a plane surface over which the pressure is uniform the increase of thickness continues indefinitely. It should be noted that the y scale of Fig. 8.1 is greatly enlarged. At any distance x from the leading edge of the plate the boundary layer thickness δ (Greek 'delta') is very small compared with x.

The random secondary movements of turbulent flow must die out very close to the surface, and so beneath the turbulent boundary layer an even thinner *viscous sublayer* is formed in which the flow is basically laminar.

In a turbulent layer there is more intermingling of fluid particles and therefore a more nearly uniform velocity than in a laminar layer (Fig. 8.2). As a result the turbulent layer usually has a greater velocity gradient at the surface. Because of the essentially laminar flow immediately adjacent to the surface the shear stress there, τ_0, is given by

Transition region

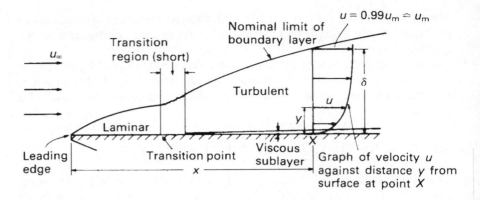

Fig. 8.1 Boundary layer on flat plate (*y* scale greatly enlarged).

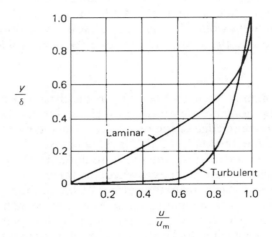

Fig. 8.2 Typical velocity distributions in laminar and turbulent boundary layers on a flat plate.

$$\tau_0 = \mu \left(\frac{\partial u}{\partial y} \right)_{y=0}$$

and thus the shear stress associated with a turbulent boundary layer is usually greater than that for a wholly laminar one. Moreover, from a turbulent layer there is a more ready interchange of particles with the main flow, and this explains the more rapid increase in thickness of a turbulent layer. Whereas the thickness of a laminar boundary layer increases as $x^{0.5}$ (when the pressure is uniform) a turbulent layer thickens approximately as $x^{0.8}$.

The point at which a laminar boundary layer becomes unstable depends on a number of factors. Roughness of the surface hastens the transition to turbulence, as does the intensity of turbulence in the main stream. The predominant factor, however, is the Reynolds number of the flow in the boundary layer. This is usually expressed as $u_\infty x/\nu$, where u_∞ represents the velocity of the oncoming flow far upstream, ν the kinematic viscosity of the fluid and x the distance from the leading edge (where the boundary layer starts) to the point in question. This

Reynolds number is given the symbol Re_x, the suffix x indicating that it is calculated with the distance x as the characteristic length. For values of $Re_x = u_\infty x/\nu$ below about 10^5 the laminar layer is very stable, but as the value increases transition is more and more easily induced, and when $Re_x > 2 \times 10^6$ it is difficult to prevent transition occurring even when the surface is smooth and there is no turbulence in the main stream. If the pressure p is not uniform over the surface other considerations arise, which are discussed in Section 8.8. In general, if $\partial p/\partial x$ is positive the critical value of Re_x at which transition occurs is lower; if $\partial p/\partial x$ is negative the critical Re_x is higher.

For long plates the boundary layer may be laminar over only a relatively short distance from the leading edge and then it is often assumed – with sufficient accuracy – that the layer is turbulent over its entire length.

8.3 THE THICKNESS OF THE BOUNDARY LAYER

Because the velocity within the boundary layer increases to the velocity of the main stream asymptotically, some arbitrary convention must be adopted to define the thickness of the layer. One possible definition of thickness is that distance from the solid surface in which the velocity reaches 99% of the velocity u_m of the main stream. The figure of 99% is an arbitrary choice: the thickness so defined is that distance from the surface beyond which in general we are prepared to neglect the viscous stresses. But if, for example, greater accuracy were desired a greater thickness would have to be specified. Other so-called 'thicknesses' are lengths, precisely defined by mathematical expressions, which are measures of the effect of the boundary layer on the flow.

One of these is the *displacement thickness*, δ^*. If u represents the velocity parallel to the surface and at a perpendicular distance y from it, as shown in Fig. 8.3, the volume flow rate per unit width through an element of thickness δy in two-dimensional flow is $u\,\delta y$. If, however, there had been no boundary layer the value would have been $u_m\,\delta y$. The

Displacement thickness

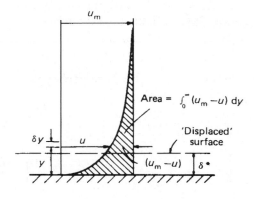

Fig. 8.3

total reduction of volume flow rate caused by the boundary layer is therefore $\int_0^\infty (u_m - u)dy$ (represented by the shaded area on the diagram) and this may be equated to $u_m \delta^*$. In other words, to reduce the total volume flow rate of a frictionless fluid by the same amount, the surface would have to be displaced outwards a distance δ^*.

$$\therefore \quad \delta^* = \frac{1}{u_m} \int_0^\infty (u_m - u)dy = \int_0^\infty \left(1 - \frac{u}{u_m}\right)dy \qquad (8.1)$$

The concept of displacement thickness often allows us to consider the main flow as that of a frictionless fluid past a 'displaced' surface instead of the actual flow past the actual surface.

Momentum thickness   Similarly a *momentum thickness* θ (Greek 'theta') may be defined. The fluid passing through an element of the boundary layer carries momentum at a rate $(\varrho u \, \delta y)u$ per unit width, whereas in frictionless flow the same amount of fluid would have momentum $\varrho u \, \delta y \, u_m$. For constant density, the total reduction in momentum flow rate $\int_0^\infty \varrho(u_m - u)u \, dy$ equals the momentum flow rate under frictionless conditions through a thickness θ.

$$\therefore \quad (\rho u_m \theta)u_m = \int_0^\infty \rho(u_m - u)u \, dy$$

whence

$$\theta = \int_0^\infty \frac{u}{u_m}\left(1 - \frac{u}{u_m}\right)dy \qquad (8.2)$$

■ **Example 8.1**   Under conditions of zero pressure gradient, the velocity profile in a laminar boundary may be represented by the approximate relation

$$\frac{u}{u_m} = 2\left(\frac{y}{\delta}\right) - \left(\frac{y}{\delta}\right)^2$$

where δ represents the thickness of the boundary layer. Calculate the displacement thickness, δ^*, and the momentum thickness, θ, when the boundary layer thickness is 0.6 mm.

Solution
We can use equations 8.1 and 8.2, replacing the upper limit of integration by δ. Hence

$$\delta^* = \int_0^\delta \left(1 - \frac{u}{u_m}\right)dy = \int_0^\delta \left(1 - 2\frac{y}{\delta} + \left(\frac{y}{\delta}\right)^2\right)dy$$

$$= \left[y - \frac{y^2}{\delta} + \frac{y^3}{3\delta^2}\right]_0^\delta = \delta\left[1 - 1 + \frac{1}{3}\right] = \frac{\delta}{3}$$

and

$$\theta = \int_0^\delta \frac{u}{u_m}\left(1 - \frac{u}{u_m}\right)dy = \int_0^\delta \left(2\frac{y}{\delta} - \left(\frac{y}{\delta}\right)^2\right)\left(1 - 2\frac{y}{\delta} + \left(\frac{y}{\delta}\right)^2\right)dy$$

$$= \int_0^\delta \left(2\frac{y}{\delta} - 5\left(\frac{y}{\delta}\right)^2 + 4\left(\frac{y}{\delta}\right)^3 - \left(\frac{y}{\delta}\right)^4\right)dy$$

$$= \left[\frac{y^2}{\delta} - \frac{5}{3}\frac{y^3}{\delta^2} + \frac{y^4}{\delta^3} - \frac{1}{5}\frac{y^5}{\delta^4}\right]_0^\delta = \delta\left[1 - \frac{5}{3} + 1 - \frac{1}{5}\right] = \frac{2}{15}\delta$$

Summarizing:

$$\delta^* = \frac{\delta}{3} \quad \text{and} \quad \theta = \frac{2}{15}\delta$$

so, substituting $\delta = 0.60$ mm, $\delta^* = 0.20$ mm and $\theta = 0.08$ mm. $\square$

8.4 THE MOMENTUM EQUATION APPLIED TO THE BOUNDARY LAYER

The mathematical methods required for an exact study of the flow in boundary layers are highly complex but the Hungarian–American engineer Theodore von Kármán (1881–1963) obtained very useful results by approximate methods based on the steady-flow momentum equation. As an example of his technique we shall apply the momentum equation to the steady flow in a boundary layer on a flat plate over which there may be a variation of pressure in the direction of flow.

Figure 8.4 shows a small length AE ($=\delta x$) of the plate. The width of the surface (perpendicular to the plane of the diagram) is assumed large so that 'edge effects' are negligible, and the flow is assumed wholly two-dimensional. The boundary layer is of thickness δ, and its outer edge is represented by BD. This line is not a streamline because with increasing distance x more fluid continually enters the boundary layer. Let C be the point on AB produced that is on the same streamline as D. No fluid of course crosses the streamline CD. We may take $ACDE$ as a suitable 'control volume'.

We suppose the (piezometric) pressure over the face AC to have the mean value p. (For simplicity we omit the asterisk from p^* in this section.) Then over the face ED the mean pressure is $p + (\partial p/\partial x)\delta x$. For unit width perpendicular to the diagram the differences of pressure therefore produce on the control volume a force in the x direction equal to:

$$p\,AC - \left(p + \frac{\partial p}{\partial x}\delta x\right)ED + \left(p + \frac{1}{2}\frac{\partial p}{\partial x}\delta x\right)(ED - AC) \tag{8.3}$$

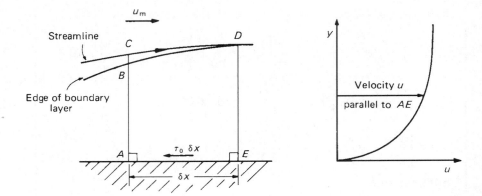

Fig. 8.4

where $p + \frac{1}{2}(\partial p/\partial x)\delta x$ is the mean value of the pressure over the surface CD. The expression 8.3 reduces to $-\frac{1}{2}(\partial p/\partial x)\delta x(ED + AC)$ and since, as $\delta x \to 0$, $AC \to ED$ in magnitude, the expression becomes $-(\partial p/\partial x)\delta x$ ED. The total x-force on the control volume is therefore

$$- \tau_0 \delta x - \frac{\partial p}{\partial x} \delta x\, ED \tag{8.4}$$

where τ_0 represents the shear stress at the boundary. By the steady-flow momentum equation this force equals the net rate of increase of x-momentum of the fluid passing through the control volume.

Through an elementary strip in the plane AB, distance y from the surface, of thickness δy and unit breadth perpendicular to the diagram, the mass flow rate is $\varrho u\, \delta y$. The rate at which x-momentum is carried through the strip is therefore $\varrho u^2\, \delta y$, and for the entire boundary layer the rate is $\int_0^\delta \varrho u^2\, \mathrm{d}y$. The corresponding value for the section ED is

$$\int_0^\delta \varrho u^2\, \mathrm{d}y + \frac{\partial}{\partial x}\left(\int_0^\delta \varrho u^2\, \mathrm{d}y\right)\delta x$$

Thus the net rate of increase of x-momentum of the fluid passing through the control volume $ACDE$ is

Flow rate of x-momentum through ED
  $-$ (flow rate of x-momentum through AB
  $+$ flow rate of x-momentum through BC)

$$= \left[\int_0^\delta \varrho u^2\, \mathrm{d}y + \frac{\partial}{\partial x}\left(\int_0^\delta \varrho u^2\, \mathrm{d}y\right)\delta x\right] - \left[\int_0^\delta \varrho u^2\, \mathrm{d}y + \varrho u_m^2(BC)\right]$$

$$= \frac{\partial}{\partial x}\left(\int_0^\delta \varrho u^2\, \mathrm{d}y\right)\delta x - \varrho u_m^2(BC) \tag{8.5}$$

where u_m represents the velocity (in the x direction) of the main stream outside the boundary layer at section AB. If p is a function of x, so is u_m, and thus u_m is not necessarily equal to the velocity far upstream.

The magnitude of BC is readily determined by the continuity principle. As the flow is steady:

Mass flow rate across AC = Mass flow rate across ED

i.e.

Mass flow rate across $BC = \varrho u_{\mathrm{m}}(BC)$

$\qquad$ = Mass flow rate across ED − Mass flow rate across AB

$$= \frac{\partial}{\partial x}\left(\int_0^\delta \varrho u\, \mathrm{d}y\right)\delta x$$

$$\therefore \quad \varrho u_{\mathrm{m}}^2(BC) = u_{\mathrm{m}}\frac{\partial}{\partial x}\left(\int_0^\delta \varrho u\, \mathrm{d}y\right)\delta x$$

and substitution into eqn 8.5 gives

Rate of increase of x-momentum

$$= \frac{\partial}{\partial x}\left(\int_0^\delta \varrho u^2\, \mathrm{d}y\right)\delta x - u_{\mathrm{m}}\frac{\partial}{\partial x}\left(\int_0^\delta \varrho u\, \mathrm{d}y\right)\delta x$$

Equating this to the total x-force on the control volume (eqn 8.4) and dividing by $-\delta x$ gives:

$$\tau_0 + \frac{\partial p}{\partial x}(ED) = u_{\mathrm{m}}\frac{\partial}{\partial x}\int_0^\delta \varrho u\, \mathrm{d}y - \frac{\partial}{\partial x}\int_0^\delta \varrho u^2\, \mathrm{d}y \qquad (8.6)$$

The acceleration of the fluid perpendicular to the surface is very small compared with that parallel to it, since δ is very small compared with x. (At the point where the boundary layer begins, i.e. at the leading edge of the plate, the acceleration perpendicular to the plate may be comparable in magnitude with that parallel to it. But this is so only for a short distance and, except for extremely low Reynolds numbers, the perpendicular acceleration may be assumed always negligible compared with that parallel to the surface.) Consequently, for a flat plate the variation of pressure with y is negligible, and the pressure in the boundary layer may be taken to be the same as that outside it. Outside the boundary layer the influence of viscosity is negligible, and the flow may therefore be assumed to correspond to that of an ideal fluid. Bernoulli's equation $p + \frac{1}{2}\varrho u_{\mathrm{m}}^2$ = constant (again, for simplicity, p represents piezometric pressure) may be applied to it, and differentiating with respect to x we obtain

$$\frac{\partial p}{\partial x} + \varrho u_{\mathrm{m}}\frac{\partial u_{\mathrm{m}}}{\partial x} = 0 \qquad (8.7)$$

From this, substitution for $\partial p/\partial x$ may be made in eqn 8.6. Then, noting that $ED = \delta = \int_0^\delta \mathrm{d}y$ and that ϱ is constant, we have

$$\tau_0 - \varrho u_{\mathrm{m}}\frac{\partial u_{\mathrm{m}}}{\partial x}\int_0^\delta \mathrm{d}y = \varrho u_{\mathrm{m}}\frac{\partial}{\partial x}\int_0^\delta u\, \mathrm{d}y - \varrho\frac{\partial}{\partial x}\int_0^\delta u^2\, \mathrm{d}y \qquad (8.8)$$

Since

$$u_{\mathrm{m}}\frac{\partial u}{\partial x} = \frac{\partial}{\partial x}(u_{\mathrm{m}}u) - u\frac{\partial u_{\mathrm{m}}}{\partial x}$$

eqn 8.8 becomes

$$\tau_0 = \varrho\frac{\partial}{\partial x}\int_0^\delta u_m\, u\, \mathrm{d}y - \varrho\frac{\partial u_m}{\partial x}\int_0^\delta u\, \mathrm{d}y - \varrho\frac{\partial}{\partial x}\int_0^\delta u^2\, \mathrm{d}y + \varrho\frac{\partial u_m}{\partial x}\int_0^\delta u_m\, \mathrm{d}y$$

$$= \varrho\frac{\partial}{\partial x}\int_0^\delta \left(u_m - u\right)u\, \mathrm{d}y + \varrho\frac{\partial u_m}{\partial x}\int_0^\delta \left(u_m - u\right)\mathrm{d}y \qquad (8.9)$$

Since $u_m - u$ becomes zero at the outer edge of the boundary layer the upper limit of the last two integrals may be changed to ∞. Then, from the definitions of displacement thickness and momentum thickness (eqns 8.1 and 8.2), eqn 8.9 simplifies to

$$\tau_0 = \varrho\frac{\mathrm{d}}{\mathrm{d}x}\left(u_m^2\theta\right) + \varrho\frac{\mathrm{d}u_m}{\mathrm{d}x}u_m\delta* \qquad (8.10)$$

The partial derivatives have given way to full derivatives since the quantities vary only with x.

Momentum integral equation

Equation 8.10 is the form in which the *momentum integral equation* of the boundary layer is usually expressed, and it forms the basis of many approximate solutions of boundary-layer problems. It is applicable to laminar, turbulent or transition flow in the boundary layer. Some assumption about the way in which u varies with y is necessary, however, in order that the displacement and momentum thickness may be evaluated.

It may be noted that if $\partial p/\partial x = 0$ then eqn 8.7 shows that $\partial u_m/\partial x = 0$ and eqn 8.10 reduces to

$$\frac{\tau_0}{\varrho u_m^2} = \frac{\mathrm{d}\theta}{\mathrm{d}x} \qquad (8.11)$$

8.5 THE LAMINAR BOUNDARY LAYER ON A FLAT PLATE WITH ZERO PRESSURE GRADIENT

In practice, the laminar part of the boundary layer is often so short that it may be neglected. Especially at low values of Reynolds number, however, there are several examples of flow for which the laminar boundary layer is important. The German engineer P.R.H. Blasius* (1883–1970) developed, analytically, equations for the flow in a laminar boundary layer on a flat plate with zero pressure gradient. It is therefore possible to compare results obtained by the approximate methods with those of Blasius. (His results are subject to the proviso that the Reynolds number Re_x is not extremely small because the assumption of negligible acceleration perpendicular to the boundary is then not valid. An improved solution is that by Kuo[1].)

* pronounced *Blah'-ze-ooce*.

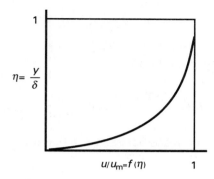

Fig. 8.5

In all laminar flow the shear stress $\tau = \mu \partial u / \partial y$ (where y is measured in a direction perpendicular to u). Thus, if y is measured outwards from the boundary, τ_0 is given by $\mu(\partial u / \partial y)_{y=0}$. Together with the assumption that the pressure gradient $\partial p / \partial x$ is zero, so that $\partial u_m / \partial x = 0$, substitution for τ_0 in eqn 8.9 gives

$$\mu \left(\frac{\partial u}{\partial y} \right)_{y=0} = \varrho \frac{\partial}{\partial x} \int_0^\delta (u_m - u) u \, \mathrm{d}y \qquad (8.12)$$

The momentum equation provides no information about the way in which u varies with y. It is reasonable, however, to assume that, although the boundary-layer thickness δ varies with x, the increase of u, from zero at the boundary to (very nearly) u_m at $y = \delta$, is the same at any value of x, apart from the scale of the y coordinates. That is, if $\eta = y/\delta$ so that η (Greek 'eta') always varies from 0 at the boundary to 1 at $y = \delta$, then u/u_m is always the same function of η – e.g. that depicted in Fig. 8.5 – so long as the boundary layer remains laminar. This assumption has been verified experimentally. Substituting $y = \eta \delta$ and $u = u_m f(\eta)$ in eqn 8.12 gives

$$\frac{\mu}{\delta} u_m \left[\frac{\partial f(\eta)}{\partial \eta} \right]_{\eta=0} = \varrho \frac{\partial}{\partial x} \left[u_m^2 \delta \int_0^1 \{ 1 - f(\eta) \} f(\eta) \mathrm{d}\eta \right]$$

Since $f(\eta)$ is assumed independent of x, the integral $\int_0^1 \{1 - f(\eta)\} f(\eta) \mathrm{d}\eta$ may be written as constant A and $[\partial f(\eta)/\partial \eta]_{\eta=0}$ as constant B. Then

$$\tau_0 = \frac{\mu}{\delta} u_m B = \varrho \frac{\partial}{\partial x} (u_m^2 A \delta) = \varrho u_m^2 A \frac{\mathrm{d}\delta}{\mathrm{d}x} \qquad (8.13)$$

Multiplying by δ / u_m and integrating with respect to x gives

$$\mu B x = \varrho u_m A \frac{\delta^2}{2} + \text{constant} \qquad (8.14)$$

If x is measured from the leading edge of the plate, $\delta = 0$ when $x = 0$, and so the constant in eqn 8.14 is zero.

$$\therefore \qquad \delta = \left(\frac{2\mu Bx}{\varrho u_m A}\right)^{1/2} \tag{8.15}$$

Also, from eqns 8.13 and 8.15

$$\tau_0 = \varrho u_m^2 A \frac{d\delta}{dx} = \varrho u_m^2 A \left(\frac{2\mu B}{\varrho u_m A}\right)^{1/2} \frac{1}{2} x^{-1/2} \tag{8.16}$$

and so the total friction force between $x = 0$ and $x = l$ for unit width *on one side* of the plate is

$$F = \int_0^l \tau_0 dx = \left[\varrho u_m^2 A\delta\right]_0^l = \varrho u_m^2 \theta_l = \left(2AB\mu\varrho u_m^3 l\right)^{1/2} \tag{8.17}$$

Skin friction coefficient   The dimensionless skin friction coefficient C_F defined by

$$\frac{\text{Mean Friction Stress}}{\frac{1}{2}\varrho u_m^2} = \frac{F}{\frac{1}{2}\varrho u_m^2 l}$$

is therefore given by

$$C_F = 2\left(\frac{2AB\mu}{\varrho u_m l}\right)^{1/2} \tag{8.18}$$

Equation 8.15 shows that the thickness of a laminar boundary layer on a flat plate with zero pressure gradient is proportional to the square root of the distance from the leading edge, and inversely proportional to the square root of the velocity of the main flow relative to the plate.

In order to evaluate δ from eqn 8.15, we must calculate A and B, and we therefore require the form of the function $f(\eta)$. Referring to Figure 8.5, the function $f(\eta)$ must satisfy the following conditions:

(I) $\dfrac{u}{u_m} = 0$ when $\eta = 0$

(II) $\dfrac{u}{u_m} = 1$ when $\eta = 1$

(III) $\dfrac{d(u/u_m)}{d\eta} = 0$ when $\eta = 1$

(IV) $\dfrac{d(u/u_m)}{d\eta} \neq 0$ when $\eta = 0$

The fourth condition follows because τ_0 is finite and

$$\tau_0 = \mu\left(\frac{\partial u}{\partial y}\right)_{y=0} = \frac{\mu u_m}{\delta}\left(\frac{d(u/u_m)}{d\eta}\right)_{y=0}$$

Example 8.2 Which of the following expressions describes better the velocity distribution for a laminar boundary layer on a flat plate in the absence of a streamwise pressure gradient?

(a) $\dfrac{u}{u_m} = \dfrac{3}{2}\left(\dfrac{y}{\delta}\right) - \dfrac{1}{2}\left(\dfrac{y}{\delta}\right)^3$

(b) $\dfrac{u}{u_m} = \dfrac{3}{2}\left(\dfrac{y}{\delta}\right) - \dfrac{1}{2}\left(\dfrac{y}{\delta}\right)^2$

Give the reasons for your decision.

Solution

Write $\eta = y/\delta$.

(a) $\dfrac{u}{u_m} = \dfrac{3}{2}\eta - \dfrac{1}{2}\eta^3$ (b) $\dfrac{u}{u_m} = \dfrac{3}{2}\eta - \dfrac{1}{2}\eta^2$

$\dfrac{u}{u_m} = 0$ when $\eta = 0$ $\dfrac{u}{u_m} = 0$ when $\eta = 0$

$\dfrac{u}{u_m} = 1$ when $\eta = 1$ $\dfrac{u}{u_m} = 1$ when $\eta = 1$

$\dfrac{d(u/u_m)}{d\eta} = \dfrac{3}{2} - \dfrac{3}{2}\eta^2$ $\dfrac{d(u/u_m)}{d\eta} = \dfrac{3}{2} - \eta$

$\dfrac{d(u/u_m)}{d\eta} = \dfrac{3}{2}$ when $\eta = 0$ $\dfrac{d(u/u_m)}{d\eta} = \dfrac{3}{2}$ when $\eta = 0$

$\dfrac{d(u/u_m)}{d\eta} = 0$ when $\eta = 1$ $\dfrac{d(u/u_m)}{d\eta} = \dfrac{1}{2}$ when $\eta = 1$

Velocity profile (a) is the better of the two, as it satisfies all four of the required boundary conditions, whereas profile (b) fails condition III.

Example 8.3 An approximate relation for the velocity profile in the laminar boundary layer subject to zero pressure gradient is

$$\frac{u}{u_m} = a_1\eta + a_2\eta^2$$

(a) Determine the values of the constants a_1 and a_2.
(b) Evaluate the constants A and B.
(c) Derive relations for the development of δ, δ^* and θ with x.

Solution

(a) Conditions I and II can be tested by substituting for η in the velocity profile equation

$$\frac{u}{u_m} = a_1 \eta + a_2 \eta^2$$

Condition I: When $\eta = 0$, $\dfrac{u}{u_m} = 0$

Thus the no-slip condition is satisfied automatically by the expression.

Condition II: When $\eta = 1$, $\dfrac{u}{u_m} = a_1 + a_2 = 1$

Differentiating

$$\frac{d(u/u_m)}{d\eta} = a_1 + 2a_2 \eta$$

Condition III: When $\eta = 1$, $\dfrac{d(u/u_m)}{d\eta} = a_1 + 2a_2 = 0$

Hence we require

$$a_1 + a_2 = 1$$

and

$$a_1 + 2a_2 = 0$$

These simultaneous equations are readily solved to yield

$$a_1 = 2 \quad \text{and} \quad a_2 = -1$$

so

$$f(\eta) = \frac{u}{u_m} = 2\eta - \eta^2$$

We can confirm that this relation also satisfies condition IV.

When $\eta = 0$, $\dfrac{d(u/u_m)}{d\eta} = a_1 = 2$ so $\dfrac{d(u/u_m)}{d\eta} \neq 0$ QED.

(b) $\qquad\qquad A = \int_0^1 \{1 - f(\eta)\} f(\eta) d\eta$

and

$$B = \left[\frac{\partial f(\eta)}{\partial \eta} \right]_{\eta=0}$$

$$A = \int_0^1 \{1 - (2\eta - \eta^2)\}(2\eta - \eta^2) d\eta = \int_0^1 (2\eta - 5\eta^2 + 4\eta^3 - \eta^4) d\eta$$

$$= \left[\eta^2 - \frac{5}{3}\eta^3 + \eta^4 - \frac{1}{5}\eta^5 \right]_0^1 = \frac{2}{15}$$

$$B = \left[\frac{\partial}{\partial \eta}(2\eta - \eta^2) \right]_{\eta=0} = [2 - 2\eta]_{\eta=0} = 2$$

(c) From eqn 8.15,

$$\delta = \left\{ \frac{4}{2/15} \left(\frac{\mu x}{\varrho u_{\mathrm{m}}} \right) \right\}^{1/2} = \frac{5.48\,x}{\sqrt{(Re_x)}} \qquad (8.19)$$

where $Re_x = xu_{\mathrm{m}}\varrho/\mu$ and may be termed the 'local' Reynolds number.

Equation 8.19 may be used to determine displacement and momentum thicknesses. Returning to eqn 8.1 and amending the upper limit of integration to suit our assumed velocity distribution, we have, with $y = \eta\delta$,

$$\delta^* = \delta \int_0^1 \{1 - f(\eta)\}\mathrm{d}\eta = \delta \int_0^1 (1 - 2\eta + \eta^2)\mathrm{d}\eta$$

$$= \delta \left[\eta - \eta^2 + \frac{\eta^3}{3} \right]_0^1 = \delta/3$$

Thus the displacement thickness is

$$\frac{1}{3} \times \frac{5.48\,x}{\sqrt{(Re_x)}} = \frac{1.826\,x}{\sqrt{(Re_x)}}$$

The momentum thickness can be determined from eqn 8.2:

$$\theta = \delta \int_0^1 f(\eta)\{1 - f(\eta)\}\mathrm{d}\eta = A\delta = \frac{2}{15} \times \frac{5.48\,x}{\sqrt{(Re_x)}} = \frac{0.730\,x}{\sqrt{(Re_x)}}$$

From eqn 8.16, the boundary shear stress is $0.365\,\varrho u_{\mathrm{m}}^2/\sqrt{(Re_x)}$ □

Other assumptions about the function $f(\eta)$ may be made, and these agree even better with Blasius's results (see Table 8.1). The success of this approximate method when applied to this problem suggests its use for problems where no exact solution is available – for example, when the boundary layer is turbulent. The method also may be extended to boundary layers on surfaces with mild curvature (that is, with a radius of curvature large compared with the boundary layer thickness) if a curvilinear coordinate system is used in which the x axis follows the curved boundary, and the y coordinates are everywhere perpendicular to it.

The thickness of such a laminar boundary layer is very small. For example, the maximum thickness of the layer (say, when $Re_x = 10^6$) is only about 0.75 mm in air at 100 m/s, and the layer then extends for about 150 mm from the leading edge of the plate. Measurement of boundary layer thickness is therefore difficult, and an experimental check on theoretical results is more easily obtained by measurements of the drag force exerted by the fluid on the plate. Table 8.1 shows, however, that results do not vary widely when different assumptions are

Table 8.1 Comparison of results for various approximate velocity distributions in a laminar boundary layer

$\dfrac{u}{u_m}$	$\dfrac{\delta\sqrt{(Re_x)}}{x}$	$\dfrac{\delta^*\sqrt{(Re_x)}}{x}$	$\dfrac{\theta\sqrt{(Re_x)}}{x}=c_f\sqrt{(Re_x)}$	$\left(\dfrac{\tau_0}{\varrho u_m^2}\right)\sqrt{(Re_x)}$
$2\eta - \eta^2$	5.48	1.826	0.730	0.365
$\frac{3}{2}\eta - \frac{1}{2}\eta^3$	4.64	1.740	0.646	0.323
$2\eta - 2\eta^3 + \eta^4$	5.84	1.751	0.685	0.343
$\sin\left(\dfrac{\pi}{2}\eta\right)$	4·80	1.743	0.655	0.328
Blasius's solution	–	1.721	0.664	0.332

used for the velocity distribution within the boundary layer. The usefulness of the approximate method is thus further demonstrated. Because of the asymptotic approach to the mainstream velocity, Blasius's solution does not give a finite value of the boundary layer thickness, but the value of y at which the velocity is $0.99\,u_m$ is about $4.91x/\sqrt{(Re_x)}$, and $u = 0.999\,u_m$ at $y \simeq 6.01x/\sqrt{(Re_x)}$.

Example 8.4 Air of density $1.21\,\mathrm{kg\,m^{-3}}$ and kinematic viscosity $1.5 \times 10^{-5}\,\mathrm{m^2\,s^{-1}}$ passes over a thin flat plate, of dimensions $1.2\,\mathrm{m} \times 1.2\,\mathrm{m}$, parallel to the airstream. If transition takes place at the trailing edge of the plate, determine

(a) the velocity of the airstream
(b) the frictional drag of the plate, D_F.

Assume that transition takes place at $Re_t = 5 \times 10^5$ and the velocity profile is given by

$$\frac{u}{u_m} = \frac{3}{2}\left(\frac{y}{\delta}\right) - \frac{1}{2}\left(\frac{y}{\delta}\right)^3$$

Solution
At the trailing edge $x = x_t = x_i$; $Re_x = Re_t = Re_i$; $\theta = \theta_l$.

(a) $$Re_t = \frac{u_m x_t}{v}$$

so

$$u_m = \frac{v\,Re_t}{x_t} = \frac{\left(1.5 \times 10^{-5}\right)\mathrm{m^2/s} \times \left(5 \times 10^{-5}\right)}{1.2\,\mathrm{m}} = 6.25\,\mathrm{m\,s^{-1}}$$

(b) From Table 8.1

$$\frac{\theta\sqrt{Re_x}}{x} = c_f\sqrt{Re_x} = 0.646$$

Hence

$$\theta = \frac{0.646\,x}{\sqrt{Re_x}} = \frac{0.646 \times 1.2\,\mathrm{m}}{\sqrt{(5)(10^5)}} = 1.096 \times 10^{-3}\,\mathrm{m}$$

From equation 8.17, the frictional drag per unit width on one side of the plate, F, is given by

$$F = \varrho u_\mathrm{m}^2 \theta_l = 1.21\,\mathrm{kg/m^3} \times \left(6.25\,\mathrm{m/s}\right)^2 \times \left(1.096 \times 10^{-3}\right)\mathrm{m}$$

$$= 5.18 \times 10^{-2}\,\mathrm{N\,m^{-1}}$$

Hence, for width $b = 1.2\,\mathrm{m}$, and summing the contributions on the two sides of the plate,

$$D_\mathrm{F} = 2Fb = 2 \times \left(5.18 \times 10^{-2}\right)\mathrm{N\,m^{-1}} \times 1.2\,\mathrm{m} = 0.124\,\mathrm{N}$$

□

8.6 THE TURBULENT BOUNDARY LAYER ON A SMOOTH FLAT PLATE WITH ZERO PRESSURE GRADIENT

Study of turbulent boundary layers is particularly important, because most of the boundary layers encountered in practice are turbulent for most of their length. The analysis of flow in a turbulent layer depends more heavily on experimental data than does that for a laminar layer. Unfortunately, however, reliable direct measurements on boundary layers are not always easy to obtain. Prandtl therefore suggested that, as much experimental information was available about turbulent flow in pipes, this could be used in the study of turbulent boundary layers on flat plates, on the grounds that the boundary layers in the two cases were not essentially different. In a pipe the thickness of the boundary layer in fully developed flow equals the radius, and the maximum velocity (along the axis) corresponds to the velocity u_m of the main stream past a flat plate.

For moderate values of Reynolds number a simple expression for the shear stress τ_0 at the boundary may be obtained from Blasius's formula for hydraulically smooth pipes (eqn 7.6) since, by eqn 7.4, $|\tau_0|/\varrho \bar{u}^2 = f/2$. Then, for a pipe of radius R,

$$|\tau_0| = \frac{1}{2}\varrho \bar{u}^2 0.079\left(\frac{\nu}{\bar{u}\,2R}\right)^{1/4} = \mathrm{const} \times \varrho u_\mathrm{max}^2\left(\frac{\nu}{u_\mathrm{max}R}\right)^{1/4} \quad (8.20)$$

since, for any particular velocity profile, the mean velocity $\bar{u}$ is a particular fraction of the maximum value u_max (on the pipe axis). The value of the constant on the right of eqn 8.20 need not concern us because the essential features of a turbulent boundary layer may be deduced without specifying the velocity profile. If R is now assumed equivalent to the boundary layer thickness δ on the flat plate, we have

$$\tau_0 = \text{const} \times \varrho u_{\text{m}}^2 \left(\frac{\nu}{u_{\text{m}} \delta} \right)^{1/4} \tag{8.21}$$

If $\partial p / \partial x = 0$ so that $\partial u_{\text{m}} / \partial x = 0$, substituting eqn 8.21 into eqn 8.9 gives

$$\text{const} \times \varrho u_{\text{m}}^2 \left(\frac{\nu}{u_{\text{m}} \delta} \right)^{1/4} = \varrho \frac{\partial}{\partial x} \int_0^\delta (u_{\text{m}} - u) u \, \mathrm{d}y$$

$$= \varrho u_{\text{m}}^2 \frac{\partial}{\partial x} \left\{ \delta \int_0^1 \left(1 - \frac{u}{u_{\text{m}}} \right) \frac{u}{u_{\text{m}}} \, \mathrm{d}\eta \right\} \tag{8.22}$$

where $\eta = y/\delta$. The integral in the final term of eqn 8.22 involves only dimensionless magnitudes and so its value is simply a number. Rearranging the equation then gives

$$\delta^{1/4} \frac{\mathrm{d}\delta}{\mathrm{d}x} = \text{const} \left(\frac{\nu}{u_{\text{m}}} \right)^{1/4}$$

the partial derivative giving place to the full derivative since δ varies only with x. Integrating with respect to x yields

$$\frac{4}{5} \delta^{5/4} = \text{const} \left(\frac{\nu}{u_{\text{m}}} \right)^{1/4} x + C \tag{8.23}$$

Determination of the integration constant C presents a problem. The turbulent boundary layer begins after the transition from the laminar layer, that is, at a non-zero value of x, and its initial thickness is unknown. Nevertheless Prandtl showed that reasonably good results are obtained if the boundary layer is supposed turbulent from $x = 0$. Although more refined assumptions are possible, Prandtl's has the advantage of simplicity, and so we set C equal to zero. For long plates the laminar part of the boundary layer is only a small proportion of the total length, and the error incurred by the assumption is small. Moreover, near the leading edge of the plate, turbulence is often induced in the boundary layer by a roughened surface or by 'trip wires' The length of the laminar part is then greatly reduced. It has to be admitted, however, that attempts to generate complete turbulence from $x = 0$ are apt to produce unwanted effects: the boundary layer is often thickened, and it may temporarily separate from the surface. In cases where the laminar portion is not negligible, results for the laminar and turbulent portions may be combined as shown in Section 8.7.

From eqn 8.23, then, we obtain

$$\delta = \text{const} \left(\frac{\nu}{u_{\text{m}}} \right)^{1/5} x^{4/5} \tag{8.24}$$

The total drag force F on one side of the plate (of length l) $= \int_0^l \tau_0 \, \mathrm{d}x$ for unit width, and substituting from eqn 8.24 into eqn 8.21 gives

$$F = \int_0^l \text{const} \times \varrho u_m^2 \left(\frac{v}{u_m}\right)^{1/4} \left\{\text{const}\left(\frac{v}{u_m}\right)^{1/5} x^{4/5}\right\}^{-1/4} dx$$

$$= \text{const} \times \varrho u_m^2 \left(\frac{v}{u_m}\right)^{1/5} \int_0^l x^{-1/5} dx = \text{const} \times \varrho u_m^2 \left(\frac{v}{u_m}\right)^{1/5} l^{4/5}$$

The corresponding skin-friction coefficient $C_F =$

$$\text{Mean friction stress} \div \frac{1}{2}\varrho u_m^2 = (F/l) \div \frac{1}{2}\varrho u_m^2$$

$$= \text{const}\left(\frac{v}{u_m l}\right)^{1/5} = \text{const}(Re_l)^{-1/5}$$

where $Re_l = u_m l/v$, the Reynolds number based on the total length l of the plate. Measurements of drag force indicate that the value of C_F is in fact

$$C_F = 0.074(Re_l)^{-1/5} \tag{8.25}$$

In Fig. 8.6 comparison may be made with the corresponding expression for a laminar boundary layer.

The relations just obtained are valid only for a limited range of Reynolds number since Blasius's relation (eqn 7.6) is valid only for values of $\bar{u}d/v$ (the pipe Reynolds number) less than 10^5. Moreover, the relation between u/u_m and η changes somewhat with Reynolds number. The expressions for the turbulent boundary layer on a flat plate are thus applicable only in the range $Re_x = 5 \times 10^5$ to 10^7. (Below $Re_x = 5 \times 10^5$ the boundary layer is normally laminar.)

Basically similar analyses may be made using equations for pipe flow with a greater range of validity than Blasius's, but the mathematical detail is then very tedious.

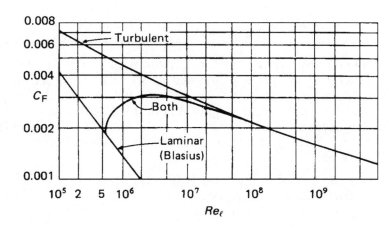

Fig. 8.6 Drag coefficients for a smooth flat plate.

For values of Re_x between 10^7 and 10^9, H. Schlichting assumed a velocity varying with the logarithm of the distance from the boundary (as in Section 8.12), and obtained the semi-empirical relation

$$C_F = \frac{0.455}{\left(\log_{10} Re_l\right)^{2.58}} = \frac{3.913}{\left(\ln Re_l\right)^{2.58}} \qquad (8.26)$$

It should be remembered that the whole of the analysis so far has applied exclusively to *smooth* plates. Appreciably different results are obtained for rough plates – as might be expected from results for flow in rough pipes. The relative roughness for a plate, however, varies with the boundary-layer thickness. The value of k/δ, where k represents the average height of the bumps on the surface, is very large at the front of the plate, and then decreases as δ increases. More complicated expressions for τ_0 (which may be deduced from those for the pipe friction factor f) are therefore required in the analysis of boundary layers on rough plates. The calculations are too lengthy and complex, however, to be reproduced here.

Also, we have considered only the case of zero pressure gradient along the plate. In many instances the pressure gradient is not zero, and if the pressure increases in the direction of flow, *separation* of the boundary layer is possible. This will be discussed in Section 8.8.

To sum up, the principal characteristics of the turbulent boundary layer on a flat plate with zero pressure gradient and moderate Reynolds number are these:

1. The thickness of the boundary layer increases approximately as the $\frac{4}{5}$ power of the distance from the leading edge (compared with the power $\frac{1}{2}$ for a laminar layer).
2. The shear stress τ_0 at the boundary is approximately inversely proportional to the fifth root of the local Reynolds number (compared with the square root for a laminar layer).
3. The total friction drag is approximately proportional to the $\frac{9}{5}$ power of the velocity of the main flow and the $\frac{4}{5}$ power of the length (compared with $\frac{3}{2}$ and $\frac{1}{2}$ powers respectively for the laminar layer).

■ **Example 8.5** A train is 100 m long, 2.8 m wide and 2.75 m high. The train travels at $180 \, \text{km} \, \text{h}^{-1}$ through air of density $1.2 \, \text{kg} \, \text{m}^{-3}$ and kinematic viscosity $1.5 \times 10^{-5} \, \text{m}^2 \text{s}^{-1}$. You may assume that the frictional drag of the train is equivalent to the drag of a turbulent boundary layer on one side of a flat plate of length $l = 100 \, \text{m}$ and breadth $b = 8.3 \, \text{m}$. Taking the constant in equation 8.24 equal to 0.37, calculate

(a) the boundary layer thickness at the rear of the train
(b) the frictional drag acting on the train, D_F
(c) the power required to overcome the frictional drag.

Solution

(a) $$u_m = \frac{180\,\text{km/h} \times 10^3\,\text{m/km}}{3600\,\text{s/h}} = 50\,\text{m}\,\text{s}^{-1}$$

Substituting in equation 8.24, with $x = l$

$$\delta = 0.37\left(\frac{\nu}{u_m}\right)^{1/5} l^{4/5} = (0.37)\left(\frac{(1.5 \times 10^{-5})\,\text{m}^2/\text{s}}{50\,\text{m/s}}\right)^{1/5} (100\,\text{m})^{4/5}$$

$$= 0.73\,\text{m} = 73\,\text{cm}$$

(b) $$Re_l = \frac{u_m l}{\nu} = \frac{50\,\text{m/s} \times 100\,\text{m}}{(1.5 \times 10^{-5})\,\text{m}^2/\text{s}} = 3.333 \times 10^8$$

From equation 8.25

$$C_F = \frac{2F}{\varrho u_m^2 l} = 0.074(Re_l)^{-1/5}$$

Hence

$$F = 0.037\varrho u_m^2 l(Re_l)^{-1/5}$$

$$= 0.037 \times 1.2\,\text{kg/m}^3 \times (50\,\text{m/s})^2 \times 100\,\text{m} \times (3.333 \times 10^8)^{-1/5}$$

$$= 219\,\text{N}\,\text{m}^{-1}$$

and

$$D_F = F \times b = 219\,\text{N/m} \times 8.3\,\text{m} = 1819\,\text{N}$$

(c) $$\text{Power} = D_F \times u_m = 1819\,\text{N} \times 50\,\text{m/s}$$

$$= 9.09 \times 10^4\,\text{W} = 90.9\,\text{kW.}$$ □

8.7 FRICTION DRAG FOR LAMINAR AND TURBULENT BOUNDARY LAYERS TOGETHER

In Sections 8.5 and 8.6 we discussed separately the properties of the laminar and turbulent boundary layers on a flat plate. When the plate is of such a length that the boundary layer changes from laminar to turbulent, and both portions make an appreciable contribution to the total frictional force, combined relations are required.

The transition from laminar to turbulent flow in the boundary layer depends, among other things, on the 'local' Reynolds number, Re_x, but the critical value at which the laminar layer becomes unstable is much less well-defined than the critical Reynolds number ($\bar{u}d/\nu$) in pipe flow. Usually, however, transition occurs at a value of Re_x between 3×10^5 and 5×10^5. At the transition point, there must be no discontinuity of the friction force (nor consequently of the momentum thickness θ) otherwise the shear stress there would be infinite. For zero pressure

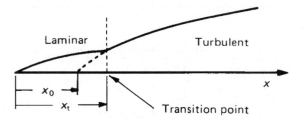

Fig. 8.7

gradient the value of θ at the transition point is therefore that given by Blasius's solution (Table 8.1) for the laminar boundary layer:

$$\theta_t = 0.664x_t\left(\frac{u_m x_t}{\nu}\right)^{-1/2} \tag{8.27}$$

where suffix t refers to the transition point.

As the transition region is extremely short we may suppose that the layer becomes fully turbulent at the point where the laminar layer breaks down. We suppose also that the shape of the turbulent layer is such that it could have started from a hypothetical leading edge at $x = x_0$ (Fig. 8.7). From eqn 8.11, for constant u_m and any type of boundary layer,

Total friction force per unit width over length x from leading edge

$$= \frac{1}{2}\varrho u_m^2 C_F x = \int_0^x \tau_0\, dx = \left[\varrho u_m^2 \theta\right]_0^x = \varrho u_m^2 \theta$$

and thus $\theta/x = C_F/2$.

So for the turbulent layer starting from $x = x_0$, and for $Re_x < 10^7$,

$$\frac{\theta}{x - x_0} = \frac{1}{2} \times 0.074\left\{\frac{(x - x_0)u_m}{\nu}\right\}^{-1/5} \tag{8.28}$$

(using eqn 8.25).

At $x = x_t$, $\theta = \theta_t$ and so

$$x_t - x_0 = \frac{\theta_t^{5/4} u_m^{1/4}}{0.037^{5/4} \nu^{1/4}}$$

Substitution for θ_t from eqn 8.27 gives

$$x_t - x_0 = \frac{0.664^{5/4} \nu^{3/8} x_t^{5/8}}{0.037^{5/4} u_m^{3/8}} = 36.9\left(\frac{\nu}{u_m x_t}\right)^{3/8} x_t \tag{8.29}$$

■ **Example 8.6** A boundary layer develops on a flat plate at zero pressure gradient. If the Reynolds number Re_l at the trailing edge of the plate is 5×10^6 and transition from laminar to turbulent flow occurs at the Reynolds number $Re_t = 5 \times 10^5$, determine

(a) the proportion of the plate occupied by the laminar boundary layer
(b) the skin friction coefficient C_F evaluated at the trailing edge.

Solution

(a) Since

$$Re_l = \frac{u_m l}{v} \quad \text{and} \quad Re_t = \frac{u_m x_t}{v}$$

it follows that

$$\frac{x_t}{l} = \frac{Re_t}{Re_l} = \frac{5 \times 10^5}{5 \times 10^6} = 0.1$$

From equation 8.29

$$\frac{x_t - x_0}{x_t} = 36.9 \left(\frac{1}{Re_t} \right)^{3/8} = \frac{36.9}{\left(5 \times 10^5 \right)^{3/8}} = 0.269$$

Hence

$$\frac{x_0}{x_t} = 1 - 0.269 = 0.731$$

and

$$\frac{x_0}{l} = \frac{x_0}{x_t} \frac{x_t}{l} = 0.731 \times 0.1 = 0.0731$$

At the trailing edge $x = l$, so 7.31% of the plate is occupied by a laminar boundary layer. The remaining 92.69% is turbulent.

(b) From equation 8.28

$$\frac{\theta_l}{l - x_0} = 0.037 \left\{ \frac{(l - x_0) u_m}{v} \right\}^{-1/5}$$

Since C_F and θ are related by the general expression $C_F = 2\theta/x$, we can determine C_F at the trailing edge substituting $x = l$ and $\theta = \theta_l$. Thus

$$C_F = \frac{2\theta_l}{l} = \frac{2 \times 0.037 \times (l - x_0)^{4/5} v^{1/5}}{l u_m^{1/5}}$$

$$= \frac{0.074}{Re_l^{1/5}} \left(1 - \frac{x_0}{l} \right)^{4/5} = 0.00318$$

This result compares with $C_F = 0.074(Re_l)^{-1/5} = 0.00338$ evaluated using equation 8.25 for an entirely turbulent boundary layer. □

All our discussion has so far been based on the assumption that the boundary surface is smooth, and for such conditions results may be represented graphically as in Fig. 8.6. Transition from a laminar boundary layer to a turbulent one, however, occurs sooner on a rougher surface, and the critical height of roughness elements causing a laminar layer to change prematurely to a turbulent one is given by

$$k_c = \frac{Nv}{u_k}$$

where N is a numerical factor practically independent of all variables except the shape of the roughness element, and u_k represents the velocity that would be found at $y = k$ if the roughness element were not there[2]. For a small roughness element $u_k \simeq k(\partial u/\partial y)_{y=0}$, so

$$k_c \simeq \frac{Nv}{k_c(\partial u/\partial y)_0} = \frac{Nv\mu}{k_c\tau_0}$$

whence, for a flat plate with zero pressure gradient,

$$k_c \simeq v\sqrt{\left(\frac{N}{\tau_0/\varrho}\right)} = \frac{N^{1/2}v}{\left(0.332u_m^2 \, Re_x^{-1/2}\right)^{1/2}} = \frac{1.74N^{1/2}v}{u_m} Re_x^{1/4} \quad (8.30)$$

Thus k_c increases as $x^{1/4}$. That is, as the distance from the leading edge increases and the boundary layer grows thicker, a greater roughness is needed to upset its stability. When the transition Reynolds number is reached, however, the laminar layer becomes turbulent whatever the roughness of the surface. Since the frictional force is less with a laminar boundary layer than with a turbulent one, the transition should be delayed as long as possible if minimum friction drag is to be achieved. Therefore the surface should be as smooth as possible near the leading edge, where the boundary layer is thinnest, although greater roughness may be tolerated further downstream.

8.8 EFFECT OF PRESSURE GRADIENT

8.8.1 Separation and flow over curved surfaces

We have so far considered flow in which the pressure outside the boundary layer is constant. If, however, the pressure varies in the direction of flow, the behaviour of the fluid may be greatly affected.

Favourable pressure gradient

Let us consider flow over a curved surface as illustrated in Fig. 8.8. (The radius of curvature is everywhere large compared with the boundary-layer thickness.) As the fluid is deflected round the surface it is accelerated over the left-hand section until at position C the velocity just outside the boundary layer is a maximum. Here the pressure is a minimum, as shown by the graph below the surface. Thus from A to C the pressure gradient $\partial p/\partial x$ is negative and the net pressure force on an

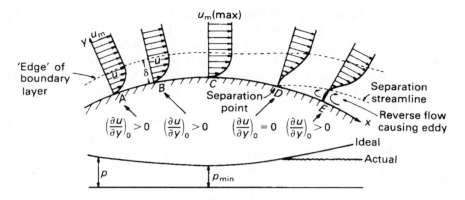

'Edge' of boundary layer

$\left(\dfrac{\partial u}{\partial y}\right)_0 > 0$ $\left(\dfrac{\partial u}{\partial y}\right)_0 > 0$ $\left(\dfrac{\partial u}{\partial y}\right)_0 = 0$ $\left(\dfrac{\partial u}{\partial y}\right)_0 > 0$

Separation point

Separation streamline

Reverse flow causing eddy

Ideal

Actual

Fig. 8.8

element in the boundary layer is in the forward direction. (For the curved surface, x is a curvilinear coordinate along the surface and y is perpendicular to it.) Such a pressure gradient is said to be *favourable*: it counteracts to some extent the 'slowing down' effect of the boundary on the fluid, and so the rate at which the boundary layer thickens is less than for a flat plate with zero pressure gradient (at a corresponding value of Re_x).

Separation

Beyond C, however, the pressure increases, and so the net pressure force on an element in the boundary layer opposes the forward flow. Although the pressure gradient $\partial p/\partial x$ has practically the same value throughout the cross-section of the boundary layer, its most significant effect is on the fluid closest to the surface. This is because the fluid there has less momentum than fluid further out, and so when its momentum is reduced still more by the net pressure force the fluid near the surface is soon brought to a standstill. The value of $\partial u/\partial y$ at the surface is then zero as at D. Further downstream, e.g. at E, the flow close to the surface has actually been reversed. The fluid, no longer able to follow the contour of the surface, breaks away from it. This breakaway before the end of the surface is reached is usually termed *separation* and it first occurs at the *separation point* where $(\partial u/\partial y)_{y=0}$ becomes zero.

Adverse pressure gradient

Separation is caused by the reduction of velocity in the boundary layer, combined with a positive pressure gradient (known as an *adverse* pressure gradient since it opposes the flow). Separation can therefore occur only when an adverse pressure gradient exists; flow over a flat plate with zero or negative pressure gradient will never separate before reaching the end of the plate, no matter how long the plate. (In an ideal fluid, separation from a continuous surface would never occur, even with a positive pressure gradient, because there would be no friction to produce a boundary layer along the surface.)

Separation streamline

The line of zero velocity dividing the forward and reverse flow leaves the surface at the separation point, and is known as the *separation streamline*. As a result of the reverse flow, large irregular eddies are formed in which much energy is dissipated as heat. The separated boundary layer tends to curl up in the reversed flow, and the region of disturbed fluid usually extends for some distance downstream. Since the energy of the eddies is dissipated as heat the pressure downstream remains approximately the same as at the separation point.

Separation occurs with both laminar and turbulent boundary layers, and for the same reasons, but laminar layers are much more prone to separation than turbulent ones. This is because in a laminar layer the increase of velocity with distance from the surface is less rapid, and the adverse pressure gradient can more readily halt the slow-moving fluid close to the surface. A turbulent layer can survive an adverse pressure gradient for some distance before separating. For any boundary layer, however, the greater the adverse pressure gradient, the sooner separation occurs. The layer thickens rapidly in an adverse pressure gradient, and the assumption that δ is small may no longer be valid.

A surface need not, of course, be curved to produce a pressure gradient. An adverse pressure gradient is, for example, found in a diffuser (see Section 7.6.3) and is the cause of the flow separation which occurs there unless the angle of divergence is very small.

Separation of the boundary layer greatly affects the flow as a whole. In particular the formation of a wake of disturbed fluid downstream, in which the pressure is approximately constant, radically alters the pattern of flow. The effective boundary of the flow is then not the solid surface but an unknown shape which includes the zone of separation. Because of the change in the pattern of flow the position of minimum pressure may be altered, and the separation point may move upstream from where the pressure was originally a minimum (e.g. point C in Fig. 8.8).

Once a laminar layer has separated from the boundary it may become turbulent. The mixing of fluid particles which then occurs may, in some circumstances, cause the layer to re-attach itself to the solid boundary so that the separation zone is an isolated bubble on the surface. Although not a common occurrence this does sometimes happen at the leading edge of a surface where excessive roughness causes separation of the laminar layer, which is followed by a turbulent layer downstream.

8.8.2 Predicting separation in a laminar boundary layer

Predicting the position at which separation may be expected is clearly important yet there is at present no exact theory by which this may readily be done. However, the momentum equation (8.10) allows some valuable approximate results to be obtained, especially for laminar boundary layers. One method, due to the English mathematician Sir Bryan Thwaites (1923–), is simple to use yet remarkably accurate.

Expanding the first term on the right of eqn 8.10 and then isolating the term containing $d\theta/dx$, we get

$$\varrho u_m^2 \frac{d\theta}{dx} = \tau_0 - 2\varrho u_m \frac{du_m}{dx}\theta - \varrho\frac{du_m}{dx}u_m\delta^*$$

From this, multiplication by $2\theta/\mu u_m$ gives

$$\frac{u_m}{v}\frac{d}{dx}(\theta^2) = \frac{2\theta\tau_0}{\mu u_m} - 2\left(2 + \frac{\delta^*}{\theta}\right)\frac{\theta^2}{v}\frac{du_m}{dx} \qquad (8.31)$$

To integrate eqn 8.31 we need to be able to correlate $\tau_0/\mu u_m$ and δ^* with the momentum thickness θ. For a laminar layer $\tau_0/\mu = (\partial u/\partial y)_{y=0}$ and so is obtainable from the velocity distribution. Thwaites examined many exact and approximate solutions for velocity distribution in laminar layers under various values of pressure gradient. His analysis showed that the right-hand side of eqn 8.31 is to a close approximation simply a function of the dimensionless quantity $(\theta^2/v)\,du_m/dx = \lambda$ say. Moreover the function is very nearly linear: $0.45 - 6\lambda$, the range of choice for each numerical coefficient being only about 2% of its value. Substituting this linear relation into eqn 8.31 we get

$$\frac{u_m}{v}\frac{d}{dx}(\theta^2) = 0.45 - \frac{6\theta^2}{v}\frac{du_m}{dx}$$

Multiplication by $v u_m^5$ and rearrangement then gives

$$u_m^6\frac{d}{dx}(\theta^2) + 6u_m^5\frac{du_m}{dx}\theta^2 = \frac{d}{dx}(u_m^6\theta^2) = 0.45 v u_m^5$$

which can be integrated between $x = 0$ and $x = x$ to yield

$$\theta^2 = \theta_0^2 + \frac{0.45v}{u_m^6}\int_0^x u_m^5 dx \qquad (8.32)$$

Here θ_0 denotes the momentum thickness at $x = 0$ but if, as is usual, x is measured from the upstream stagnation point, then $\theta_0 = 0$. Hence, for a given pressure gradient and therefore variation of u_m with x, θ can be determined.

For the special case of zero pressure gradient it is interesting to compare eqn 8.32 with Blasius's solution. When $u_m = $ constant eqn 8.32 integrates to $\theta^2 = 0.45vx/u_m = 0.45x^2/Re_x$, whence $\theta/x = 0.671/Re_x^{1/2}$ which is only 1% different from Blasius's exact result in Table 8.1.

From the results Thwaites analysed he also correlated (in tabular form) $\tau_0\theta/\mu u_m$ and δ^*/θ with λ, thus giving all the main parameters of a laminar boundary layer. In particular he found that separation (i.e. $\tau_0 = 0$) occurs when $\lambda \simeq -0.082$. Additional data that have since become available suggest, however, that a better average value for λ_{sep} might be nearer to -0.09.

Equation 8.32 always gives θ to within $\pm 3\%$, but the prediction of other parameters is less good, particularly in adverse pressure gradients. Although improvement in the results is possible with modifications to the correlations, this is unfortunately at the cost of greater complication.

For determining θ and the position of separation, however, Thwaites's original method gives satisfactory accuracy with very little calculation.

■

Example 8.7 If x denotes the distance along the boundary surface from the front stagnation point and the main-stream velocity is given by $u_m = a(1 + bx)^{-1}$, where a and b are positive constants, determine whether the boundary layer will separate and, if so, where.

Solution
Since

$$\frac{du_m}{dx} = \frac{-ab}{(1 + bx)^2}$$

is negative, dp/dx is positive (from eqn 8.7), i.e. adverse. Therefore the boundary layer will separate.

Let us assume that the Reynolds number is such that the separation occurs before the layer becomes turbulent. Then, taking $\theta_0 = 0$ since x is measured from the stagnation point, we have, from eqn 8.32,

$$\theta^2 = \frac{0.45v}{a^6(1 + bx)^{-6}} \int_0^x a^5(1 + bx)^{-5} dx$$

$$= \frac{-0.45v}{4a(1 + bx)^{-6}b}\left\{(1 + bx)^{-4} - 1\right\}$$

$$= \frac{0.45v}{4ab}\left\{(1 + bx)^6 - (1 + bx)^2\right\}$$

Also

$$\lambda = \frac{\theta^2}{v}\frac{du_m}{dx} = -\frac{\theta^2}{v}\frac{ab}{(1 + bx)^2}$$

Hence, at separation where $\lambda \simeq -0.09$ say,

$$\frac{0.09v(1 + bx)^2}{ab} = \theta^2 = \frac{0.45v}{4ab}\left\{(1 + bx)^6 - (1 + bx)^2\right\}$$

$$\therefore \quad (1 + bx)^4 = 1.8 \quad \text{and} \quad x = 0.1583/b$$

□ An exact computer solution for this example gives $x_{sep} = 0.159/b$.

8.8.3 Pressure drag

Pressure drag and form drag

When flow occurs past a plane surface parallel to it, the fluid exerts a drag force on the surface as a direct result of viscous action. The resultant frictional force in the downstream direction is usually known as the skin friction drag, and the factors on which its magnitude depends have been discussed in the preceding sections. But when flow occurs past a surface not everywhere parallel to the main stream there is in that direction an additional drag force resulting from differences of pressure over the surface. This force is known as the *pressure drag* or, since it depends on the form (i.e. the shape) of the boundary, the *form drag*. Thus, whereas the skin friction drag is the resultant of the forces tangential to the surface, the pressure drag is the resultant of the forces normal to the surface.

The sum of the skin friction drag and the pressure drag is termed the *profile drag*. (Vortex drag, which may occur on a body producing a lift force, is treated in Section 9.9.4.)

Profile drag

In almost all cases in which flow takes place round a solid body, the boundary layer separates from the surface at some point. (One exception is an infinitesimally thin flat plate parallel to the main stream.) Downstream of the separation position the flow is greatly disturbed by large-scale eddies, and this region of eddying motion is known usually as the *wake*. As a result of the energy dissipated by the highly turbulent motion in the wake the pressure there is reduced and the pressure drag on the body is thus increased. The magnitude of the pressure drag depends very much on the size of the wake and this, in turn, depends on the position of separation.

If the shape of a body is such that separation occurs only well towards the rear, and the wake is small, the pressure drag is also small, and it is skin friction that makes the major contribution to the total drag. Such a body is termed a *streamlined body*.

Streamlined body

For a *bluff body*, on the other hand, the flow is separated over much of the surface, the wake is large and the pressure drag is much greater than the skin friction.

Bluff body

It is usually the total drag that is of interest, and this is expressed in terms of a dimensionless *drag coefficient* C_D defined as

Drag coefficient

$$C_D = \frac{\text{Total drag force}}{\frac{1}{2}\varrho u_\infty^2 A}$$

where ϱ represents the density of the fluid and u_∞ the velocity (far upstream) with which the fluid approaches the body. Except for an aerofoil or hydrofoil, A usually represents the frontal area of the body (i.e. the projected area perpendicular to the oncoming flow). For a thin flat plate parallel to the oncoming flow, the drag is almost entirely due to skin friction and the area used is that of both sides of the plate (as for the skin-friction coefficient C_F). And in the case of an aerofoil, A represents the product of the span and the mean chord (see Section 9.9.1).

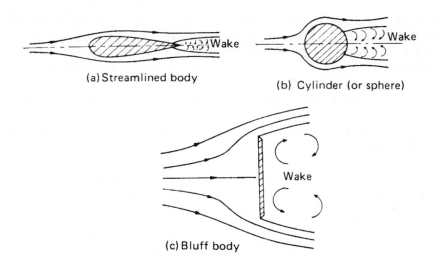

(a) Streamlined body   (b) Cylinder (or sphere)

(c) Bluff body

Fig. 8.9

The area used in the definition of C_D for a particular body should therefore be specified. The denominator of the coefficient is the product of the dynamic pressure of the undisturbed flow, $\frac{1}{2}\varrho u_\infty^2$, and the specified area. Being therefore a ratio of two forces, the coefficient is the same for two dynamically similar flows (see Section 5.2.3), so C_D is independent of the size of the body (but not of its shape) and is a function of Reynolds number. For velocities comparable with the speed of sound in the fluid it is a function also of the Mach number, and if an interface between two fluids is concerned (as in the motion of a ship for example) C_D depends also on the Froude number (see Section 5.3).

Examples of streamlined and bluff bodies are shown in Fig. 8.9. The body of maximum 'bluffness' is a thin flat plate normal to the flow (Fig. 8.9c). Skin friction has no component in the downstream direction (apart from an infinitesimal contribution across the very thin edges) and so the entire drag is the pressure drag. The importance of pressure drag may be judged from the fact that the drag on a square-shaped plate held perpendicular to a given stream may be 100 times the drag exerted when the plate is held parallel to that stream.

The flow pattern in the wake depends on the Reynolds number of the flow. Still restricting the discussion to two-dimensional flow, we may consider as an example the flow past an infinitely long, circular cylinder of diameter d, with its axis perpendicular to the flow. We suppose other solid surfaces to be far enough from the cylinder not to affect the flow near it. For very low values of $Re = u_\infty d/\nu$ (say $Re < 0.5$) the inertia forces are negligible, and the streamlines come together behind the cylinder as indicated in Fig. 8.10a. If Re is increased to the range 2–30 the boundary layer separates symmetrically from the two sides at the positions S, S (Fig. 8.10b) and two eddies are formed which rotate in opposite directions. At these Reynolds numbers the eddies remain unchanged in position, their energy being maintained by the flow from

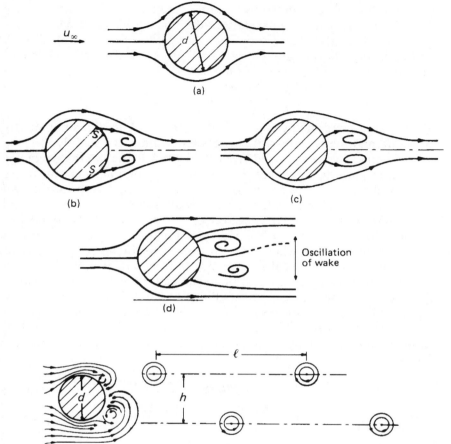

Fig. 8.10 Development of a wake behind a cylinder

Fig. 8.11 Vortex street.

the separated boundary layer. Behind the eddies, however, the main streamlines come together, and the length of the wake is limited. With increase of Re the eddies elongate as shown in (c) but the arrangement is unstable and at $Re \simeq 40$–70 (for a circular cylinder) a periodic oscillation of the wake is observed. Then, at a certain limiting value of Re (usually about 90 for a circular cylinder in unconfined flow), the eddies break off from each side of the cylinder alternately and are washed downstream. This limiting value of Re depends on the turbulence of the oncoming flow, on the shape of the cylinder (which, in general, may not be circular) and on the nearness of other solid surfaces.

Vortex street

In a certain range of Re above the limiting value, eddies are continuously shed alternately from the two sides of the cylinder and, as a result, they form two rows of vortices in its wake, the centre of a vortex in one row being opposite the point midway between the centres of consecutive vortices in the other row (Fig. 8.11). This arrangement of vortices is known as a *vortex street* or *vortex trail*. The energy of the vortices is, of course, ultimately consumed by viscosity, and beyond a certain distance from the cylinder the regular pattern disappears. Von Kármán considered the vortex street as a series of separate vortices in an ideal fluid, and deduced that the only pattern stable to small disturbances is that indicated in Fig. 8.11, and then only if

$$\frac{h}{l} = \frac{1}{\pi} \text{arcsinh} \, 1 = 0.281$$

a value later confirmed experimentally. (In a real fluid h is not strictly constant, and the two rows of vortices tend to diverge slightly.)

Each time a vortex is shed from the cylinder, the lateral symmetry of the flow pattern, and hence the pressure distribution round the cylinder, are disturbed. The shedding of vortices alternately from the two sides of the cylinder therefore produces alternating lateral forces and these may cause a forced vibration of the cylinder at the same frequency. This is the cause of the 'singing' of telephone or power wires in the wind, for example, when the frequency of vortex shedding is close to the natural frequency of the wires. For the same reason, wind passing between the slats of a Venetian blind can cause them to flutter. Vibrations set up in this way can affect chimneys and submarine periscopes, and they have also been responsible for the destruction of suspension bridges in high winds. The frequency, f, with which vortices are shed from an infinitely long circular cylinder, is given by the empirical formula

$$\frac{fd}{u_\infty} = 0.198\left(1 - \frac{19.7}{Re}\right) \tag{8.33}$$

for $250 < Re < 2 \times 10^5$.

Strouhal number

The dimensionless parameter fd/u_∞ is known as the *Strouhal number* after the Czech physicist Vincenz Strouhal (1850–1922) who, in 1878, first investigated the 'singing' of wires.

If the cylinder itself oscillates, the vortices are shed when, or nearly when, the points of maximum displacement are reached, and h is thus increased by twice the amplitude. Consequently both the width of the wake and the pressure drag are increased, the lateral force is increased and f decreased.

At high Reynolds numbers, the large angular velocities and rates of shear associated with the individual vortices cause these to disintegrate into random turbulence close to the cylinder, and a regular vortex street can no longer be observed. The undesirable vibration of a cylindrical body may be eliminated by attaching a longitudinal fin to the downstream side. If the length of the fin is not less than the diameter of the cylinder, interaction between the vortices is prevented; they thus remain in position and are not shed from the cylinder. For similar reasons, tall chimneys sometimes have helical projections, like large screw threads, which cause unsymmetrical three-dimensional flow and so discourage alternate vortex shedding.

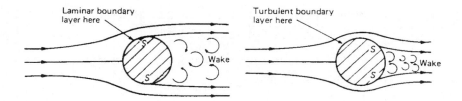

Laminar boundary layer here

Turbulent boundary layer here

Wake

Wake

Fig. 8.12

The width of the wake evidently depends on the positions of separation S, S (Fig. 8.12). We recall that a turbulent boundary layer is better able to withstand an adverse pressure gradient than is a laminar layer. For a turbulent layer, therefore, separation occurs further towards the rear of the cylinder, and the wake is thus narrower. As on a flat plate, the boundary layer becomes turbulent at a critical value of the Reynolds number (defined by $u_\infty s/\nu$, in which s represents the distance measured along the surface from the front stagnation point, where the boundary layer starts). That is, transition occurs where $s = \nu\, Re_t/u_\infty$. For low values of u_∞ the boundary layer is wholly laminar until it separates, but as u_∞ increases, the transition value of s decreases, and when it becomes less than the distance to the separation position the layer becomes turbulent. The separation position then moves further downstream, the wake becomes narrower and the drag coefficient less.

8.8.4 Profile drag of 'two-dimensional' bodies

We recall that the skin friction and pressure drag together constitute the profile drag and that this may be expressed in terms of the dimensionless drag coefficient $C_D = \text{Profile Drag} \div \frac{1}{2}\varrho u_\infty^2 A$, where A is – usually – the projected area of the body perpendicular to the oncoming flow. Both skin friction and pressure drag depend on Reynolds number, so C_D is a function of Re. (Its value is also affected to some extent by the roughness of the surface and the degree of turbulence in the oncoming fluid.)

Figure 8.13 shows the variation of C_D with Reynolds number for a smooth, infinitely long, circular cylinder with its axis perpendicular to the flow. At very low values of Re (say < 0.5) inertia forces are negligible compared with the viscous forces, and the drag is almost directly proportional to u_∞. In other words, C_D is approximately inversely proportional to Re, as indicated by the straight-line part of the graph. Skin friction here accounts for a large part of the total drag: in the limit, as $Re \rightarrow 0$, skin friction drag is two thirds of the total. When separation of the boundary layer occurs, pressure drag makes a proportionately larger contribution, and the slope of the (total) C_D curve becomes less steep. By $Re = 200$ the von Kármán vortex street is well established, and pressure drag then accounts for nearly 90% of the total. The drag coefficient reaches a minimum value of about 0.9 at $Re \simeq 2000$ and there is then a slight rise to 1.2 for $Re \simeq 3 \times 10^4$. One reason for this rise is the increasing turbulence in the wake; in addition, however, the position of

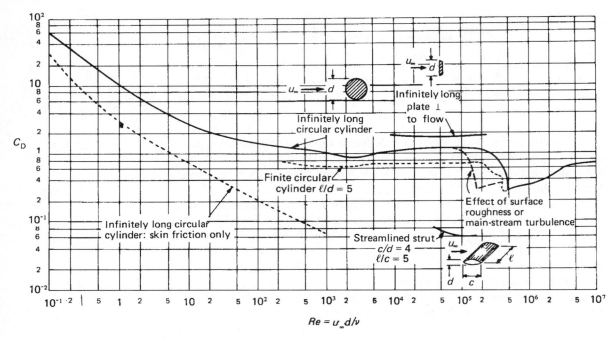

Fig. 8.13 Drag coefficient for two-dimensional bodies.

separation gradually advances upstream. This is because, once separation occurs, the pattern of flow outside the boundary layer changes, and the position of separation moves upstream somewhat. This shift in turn brings about a further change in the pattern of the surrounding flow, and yet another move of the separation position. A stable position is approached, however – which may be upstream from where the minimum pressure of an ideal fluid would be found (Fig. 8.8) – and the width of the wake is then constant so long as the boundary layer remains laminar. Pressure drag is now responsible for almost all the profile drag, the contribution of skin friction being insignificant.

At $Re \simeq 2 \times 10^5$ the boundary layer, hitherto entirely laminar, becomes turbulent before separation. It is then better able to withstand the adverse pressure gradient, and so the separation position moves further downstream and the wake narrows. In consequence, the drag falls markedly. Results for Reynolds numbers above 5×10^5 are limited, but it appears likely that after the sharp drop in C_D to about 0.3 there is a rise to about 0.7, occurring over the approximate range $5 \times 10^5 < Re < 4 \times 10^6$. Thereafter, since viscous effects are now relatively small, it is probable that C_D is practically independent of Re. It has been suggested[3] that the minimum C_D occurs when the laminar layer separates, becomes turbulent, re-attaches to the surface and then separates again, whereas at higher Reynolds numbers the separation is of the turbulent layer only. More experimental evidence is needed, however, to confirm this hypothesis.

The critical value of Re at which the large drop of C_D occurs is smaller both for a greater degree of turbulence in the main flow and for greater roughness of the surface upstream of the separation position. If a small roughness element such as a wire is placed on the surface of the cylinder upstream of the separation position, the transition from a laminar to a turbulent boundary layer occurs at a smaller Reynolds number. Paradoxically, therefore, the drag may be significantly reduced by increasing the surface roughness if the Reynolds number is such that a wholly laminar layer can by this means be made turbulent. The additional skin friction is of small importance.

For comparison, Fig. 8.13 also shows values of C_D for other 'two-dimensional' shapes. We clearly see the effect of 'streamlining' by which the separation position is moved as far towards the rear as possible. The shape of the rear of a body is much more important than that of the front in determining the position of separation. Lengthening the body, however, increases the skin friction, and the optimum amount of streamlining is that which makes the sum of skin friction and pressure drag a minimum.

For bluff bodies, such as a thin flat plate held perpendicular to the flow, the skin friction is negligible compared with the pressure drag, and so the effects of inertia forces become predominant at much lower values of Reynolds number than for a circular cylinder. Moreover, for the flat plate, except at low Reynolds numbers (say less than 100), separation of the boundary layer always occurs at the same place – the sharp edge of the plate – whatever the Reynolds number. The total drag is therefore proportional to the square of the velocity, and C_D is independent of Re.

It should be remembered that the results discussed in this section apply only to two-dimensional flow, that is, to a cylinder so long that 'end effects' are negligible. With reduction in the length, however, an increasing deviation from these results is found because – unless it is prevented – flow round the ends tends to reduce the pressure at the front of the body and increase it at the rear. The value of C_D therefore decreases as the length is reduced and these end effects become more significant. This is demonstrated by the drag coefficient of a thin, flat plate held perpendicular to the flow. The coefficient, although sensibly independent of Reynolds number for $Re > 1000$, varies markedly with the ratio of length to breadth of the plate:

Length/Breadth	1	2	4	10	18	∞
C_D	1.10	1.15	1.19	1.29	1.40	2.01

8.8.5 Profile drag of 'three-dimensional' bodies

The drag coefficient for a 'three-dimensional' body varies in a similar way to that for a 'two-dimensional' body. However, for a sphere or other body with axial symmetry, the vortex street and the associated alternating forces observed with a two-dimensional body do not occur.

Instead of a pair of vortices, a vortex ring is produced. This first forms at $Re \simeq 10$ (for a sphere) and moves further from the body as Re increases. For $200 < Re < 2000$, the vortex ring may be unstable and move downstream, its place immediately being taken by a new ring. Such movements, however, do not occur at a definite frequency, and the body does not vibrate.

Flow round bodies at very small Reynolds numbers (so-called *creeping motion*) is of interest in connection with the sedimentation of small particles. Separation does not occur and there is no disturbed wake. For a sphere past which fluid of infinite extent flows in entirely laminar conditions, Sir George G. Stokes (1819–1903) developed a mathematical solution in which the inertia forces were assumed negligible compared with the viscous forces. In his result, the total drag force is $3\pi d\mu u_\infty$, of which two-thirds is contributed by skin friction. Here d represents the diameter of the sphere, μ the absolute viscosity of the fluid and u_∞ the velocity of the undisturbed fluid relative to the sphere. Division by $\frac{1}{2}\varrho u_\infty^2 \times \pi d^2/4$ gives

$$C_D = \frac{24}{Re}$$

where $Re = u_\infty d/v$, a result that, because of the neglect of the inertia terms, agrees closely with experiment only for $Re < 0.1$. The Swedish physicist Carl Wilhelm Oseen (1879–1944) improved the solution by accounting in part for the inertia terms that Stokes had omitted. Oseen's solution, valid for $Re < 1$, is

$$C_D = \frac{24}{Re}\left(1 + \frac{3}{16}Re\right) \tag{8.34}$$

An empirical relation, acceptable up to about $Re = 100$, is

$$C_D = \frac{24}{Re}\left(1 + \frac{3}{16}Re\right)^{1/2} \tag{8.35}$$

It is interesting to note that in Stokes's solution the pressure at the upstream stagnation point exceeds the ambient pressure by $3\mu u_\infty/d$, that is, $6/Re$ times the value of $\frac{1}{2}\varrho u_\infty^2$ found in frictionless flow (Section 3.7.2). At the downstream stagnation point the pressure is $3\mu u_\infty/d$ *less* than the ambient pressure, whereas for an ideal fluid the pressures at the two stagnation points would be equal.

It will be recalled from Section 6.6.3 that, provided that Re is sufficiently small, Stokes's Law may be used to calculate the terminal velocity of a falling sphere, and it therefore forms the basis of a method for determining the viscosity of the fluid. However, if the diameter is so small that it is comparable to the mean free path of the fluid molecules, the sphere will be affected by Brownian movement, and the assumption of a fluid continuum is no longer valid.

At the Reynolds numbers for which Stokes's Law is valid the influence of viscosity extends far beyond the sphere. With increasing Reynolds numbers, however, viscosity's region of influence becomes sufficiently thin to merit the term 'boundary layer'. Separation of the layer begins at the downstream stagnation point and moves further forward as Re increases, until when $Re \simeq 1000$ a stable separation position is achieved at about 80° from the front stagnation point. Pressure drag increasingly takes precedence over skin friction, and C_D gradually becomes independent of Re, as shown in Fig. 8.14. At $Re \simeq 3 \times 10^5$, however, the boundary layer becomes turbulent before separation, and so the separation position is moved further downstream, the wake becomes smaller, and the value of C_D drops sharply from about 0.5 to below 0.1. Measurements made on small spheres towed through the atmosphere by aircraft have indicated a critical value of $Re = 3.9 \times 10^5$. Since atmospheric eddies are large in comparison with such a sphere these conditions correspond to a turbulence-free main stream. When the main stream is turbulent, however, the critical Reynolds number is reduced. Indeed, the sphere may be used as a turbulence indicator. The value of Re at which C_D for a smooth sphere is 0.3 (a value near the middle of the sudden drop) is quite a reliable measure of the intensity of turbulence.

The critical Reynolds number also depends on the surface roughness (see Fig. 8.14): the reduction of drag obtained when the boundary layer becomes turbulent is the reason why golf balls have a dimpled surface.

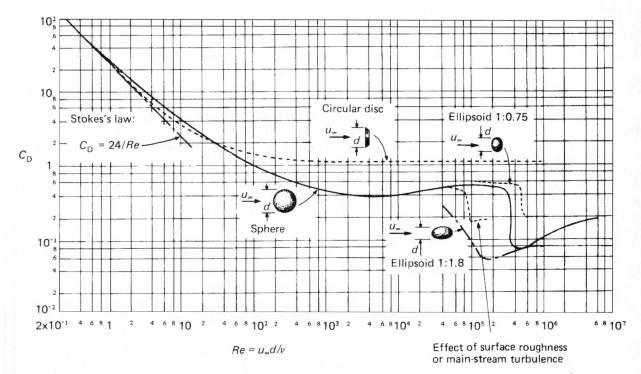

Fig. 8.14 Drag coefficients of smooth, axially symmetric bodies.

The study of flow past a sphere is the basis of the subject of *particle mechanics* which arises in problems of particle separation, sedimentation, filtration, elutriation, pneumatic transport and so on. Although in practice most particles are not spheres, results are usually related to those for a sphere since that has the simplest shape and is thus most easily examined both theoretically and experimentally.

For a sphere freely moving vertically through a fluid under steady (i.e. non-accelerating) conditions the drag force equals the difference between the weight and the buoyancy force, that is

$$C_D \frac{\pi}{4} d^2 \times \frac{1}{2} \varrho u_T^2 = \frac{\pi}{6} d^3 (\Delta \varrho) g$$

Here u_T represents the 'terminal velocity', that is, the velocity of the sphere relative to the undisturbed fluid when steady conditions have been achieved, and $\Delta \varrho$ the difference between the mean density of the sphere and the density of the fluid. Under these conditions, then,

$$C_D = 4d(\Delta \varrho) g / 3 \varrho u_T^2 \qquad (8.36)$$

Thus C_D, like Re, is a function of both diameter and terminal velocity. Consequently, if either d or u_T is initially unknown, neither C_D nor Re can be directly determined, and so the use of Fig. 8.14 would involve a trial-and-error technique.

(There is some experimental evidence, as yet inconclusive, that at moderate Reynolds numbers the apparent value of C_D for a falling sphere is slightly more than that for a fixed sphere. This might be because the falling sphere tends to oscillate from side to side instead of following an exactly vertical path. However, we shall here assume that a falling sphere has the same C_D as a fixed sphere for a given value of Re.)

For problems concerning terminal velocity the information about C_D is better presented as curves of $(4Re/3C_D)^{1/3}$ and $\{\frac{3}{4} C_D (Re)^2\}^{1/3}$ as in Fig. 8.15. If we need to determine the diameter for a given terminal velocity, we may calculate $(4Re/3C_D)^{1/3}$, which for these conditions equals $u_T \{\varrho^2 / g \mu (\Delta \varrho)\}^{1/3}$ and so is independent of the unknown d. Then from Fig. 8.15 the corresponding value of Re is taken and used to calculate d. On the other hand, if u_T is initially unknown we may determine $\{\frac{3}{4} C_D (Re)^2\}^{1/3} = d \{\varrho (\Delta \varrho) g / \mu^2\}^{1/3}$ and then obtain u_T from the corresponding value of Re.

It will be noted, however, that for a given value of $\{\frac{3}{4} C_D (Re)^2\}^{1/3}$ in the approximate range $2.3 \times 10^5 < Re < 3.8 \times 10^5$ there appear to be three possible values of Re, and therefore of velocity. The kink in the graph here is of course a consequence of the abrupt fall in C_D that occurs when the boundary layer becomes turbulent. The middle one of these three values of velocity represents an unstable condition because over the section AB of the curve a small increase of velocity leads to a reduction of drag and thus to further acceleration; therefore steady conditions cannot be maintained. Although both the outer values are stable, it is normally the lower one that is attained in practice, unless it is very close

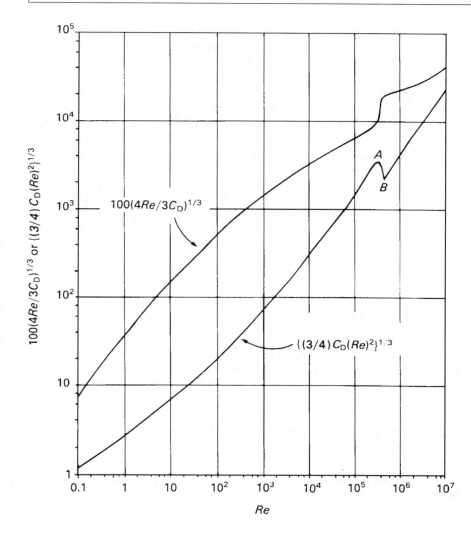

Fig. 8.15 Drag relations for isolated smooth spheres.

to the peak A or there is some pronounced unsteadiness in the fluid itself such as upward gusts of air if the sphere is falling through the atmosphere. This is of practical importance for some meteorological balloons and large hailstones.

As we have seen, for very small values of Re (say less than 0.1) Stokes's Law applies to the motion of a sphere. Then $C_D = 24/Re$ and eqn 8.36 yields

$$u_T = \frac{d^2 (\Delta \varrho) g}{18 \mu}$$

from which either u_T or d may be obtained algebraically.

It should be remembered that Fig. 8.15 applies only to spheres not near other particles or solid boundaries. The sphere also is assumed rigid: if the sphere itself consists of fluid (e.g. a raindrop in the atmosphere or a gas bubble in a liquid) its motion may be modified by flow inside it. The effect of such internal flow is, however, not usually significant for small spheres.

Results for some other bodies of revolution are shown in Fig. 8.14, and it will be noted that the decrease in C_D when the boundary layer becomes turbulent is somewhat more gradual for the more streamlined bodies than for the sphere. This is because of the smaller contribution made by pressure drag to the total drag of the streamlined bodies. For the circular disc, the body of maximum 'bluffness', there is no reduction in C_D because, except at low Reynolds numbers, C_D is independent of viscous effects, and the position of separation is fixed at the sharp edge of the disc. (It may be recalled that losses due to separation at the abrupt changes of section in pipe-lines (Section 7.6) are also practically independent of Reynolds number.)

Much detailed information about drag coefficients is given by Hoerner[4].

8.8.6 Separation from an aerofoil

If fluid flowing past a body produces a lift force on it (that is, a force perpendicular to the direction of the oncoming fluid), then any separation of the flow from the surface of the body not only increases the drag but also affects the lift. For the sake of definiteness we here consider a section of an aircraft wing. The 'angle of attack' is defined as the angle between the direction of the oncoming fluid and a longitudinal reference line in the cross-section. As we shall show in Section 9.9.2, an increase in the angle of attack causes the lift force on the aerofoil to increase – at least in the absence of boundary-layer effects. The adverse pressure gradient along the rear part of the upper (lower pressure) surface is thereby intensified until, at a particular value of the angle of attack, the boundary layer separates from the upper surface, and a turbulent wake is formed. At fairly small angles of attack the position of separation may be quite close to the trailing (rear) edge of the aerofoil. As the angle increases further, the position of separation moves forward, the wake becomes wider, and there is a marked increase of drag. In addition the main flow round the aerofoil is disrupted, and the lift no longer increases (Fig. 8.16). Figure 8.17 shows these variations of lift and drag – for a particular aerofoil at a given Reynolds number – in terms of the respective dimensionless coefficients C_L and C_D. Each coefficient is defined as the corresponding force divided by the product of $\frac{1}{2}\varrho u_\infty^2$ and the plan area of the aerofoil. (Note that they are plotted to different scales; C_L is much larger than C_D.) As we have mentioned, the angle of attack α is measured from some reference line in the aerofoil section, but, as this is chosen simply for geometrical convenience, the position $\alpha = 0$ has, in general, no particular significance. At high Reynolds numbers, the variation of C_L is seen to be nearly linear for a range of about $15°$. Because the ratio of lift to drag is a

Fig. 8.16 Stalled flow above an aerofoil.

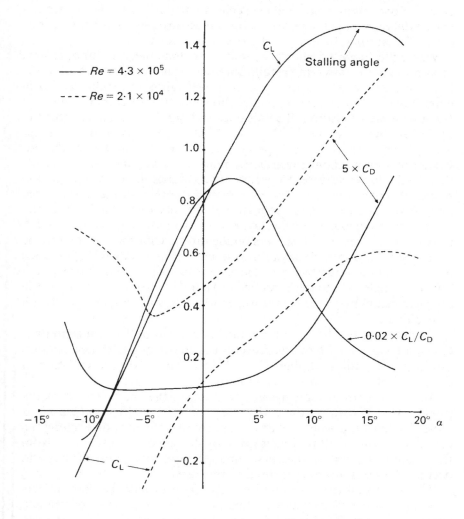

Fig. 8.17

measure of the efficiency of the aerofoil, the ratio C_L/C_D is usually plotted also.

The value of α at which C_L reaches its maximum value is known as the *stalling angle*, and the condition in which the flow separates from

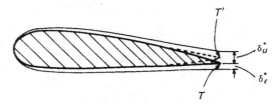

Fig. 8.18

practically the whole of the upper surface is known as *stall*. As in the case of flow round other bodies, such as circular cylinders, separation can be delayed if the boundary layer is made turbulent. Higher values of α and C_L can then be attained before stall occurs.

When the angle of attack is small – so that there is little, if any, separation of the flow – the results obtained from the Kutta – Joukowski law of ideal fluid flow theory (Section 9.8.6) agree reasonably well with experiment. With increasing angles, however, the discrepancy widens. Even without separation, the existence of boundary layers reduces the lift below the theoretical value. Since the lift results from a smaller average pressure over the upper surface than over the lower, the adverse pressure gradient towards the rear of the upper surface exceeds that on the lower surface. Consequently, towards the trailing edge, the boundary-layer thickness increases more rapidly on the upper surface and, in spite of the larger main-stream velocity over this surface, the displacement thickness of the layer at the trailing edge is, in general, greater on the upper surface than on the lower. Thus the effective shape of the aerofoil is modified, and the effective trailing edge T' (Fig. 8.18) is higher than the true trailing edge T by $\frac{1}{2}(\delta_u^* - \delta_l^*)$. In other words, the effective angle of attack is reduced. There is also a decrease in the effective camber (i.e. the curvature, concave downwards, of the aerofoil centre line).

As shown in Fig. 8.17, both C_L and C_D depend on Reynolds number (calculated as $u_\infty c/\nu$ where c represents the chord length of the aerofoil). For larger Reynolds numbers (e.g. of the order of 10^6) the variation of each coefficient with Re is negligible.

Another means of showing the variations of the coefficients is by the single curve known as the *polar diagram* (Fig. 8.19) in which C_L is plotted against C_D. Angles of attack are represented by points along the curve. The ratio of lift to drag is given by the slope of a line drawn from the origin to a point on the curve, and the maximum value of the ratio occurs when this line is tangential to the curve.

(If the aerofoil is not of infinite span, the flow past it is not entirely two-dimensional. Near the wing tips there are components of velocity perpendicular to the plane of the cross-section and these give rise, even with an ideal fluid, to vortex drag (see Section 9.9.4). This may be a major portion of the total drag.)

At high speeds the drag on an aircraft is very important, and much research has been directed towards increasing the ratio C_L/C_D for

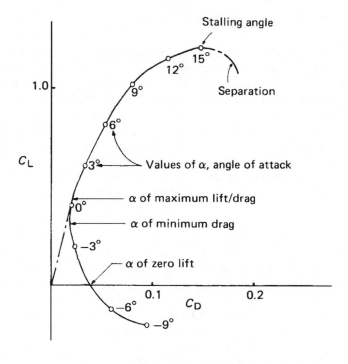

Fig. 8.19 Polar diagram for aerofoil.

Fig. 8.20 Flap increasing lift coefficient.

aircraft wings. For small angles of attack most of the drag results from skin friction, and so to reduce the drag it is desirable to maintain a laminar boundary layer over as much of the surface as possible. Since transition to turbulence is encouraged by an adverse pressure gradient, such as that downstream from the thickest part of the aerofoil, modern 'laminar-flow' aerofoils have their thickest part as far towards the rear as possible.

Under normal conditions the lift for high-speed aircraft may be obtained with only moderate values of C_L since lift $= \frac{1}{2}\varrho u_\infty^2$ (Plan Area)C_L and u_∞ is large. To achieve low landing and take-off speeds, however, a means of temporarily increasing C_L is desirable, even if C_D is also increased thereby. For this purpose flaps at the trailing edge are usually employed (Fig. 8.20). With the flap lowered, the effective curvature of the wing, and hence the lift, is increased. Many variations of this basic arrangement have been used, and by such means C_L may be approximately doubled.

8.9 BOUNDARY LAYER CONTROL

As we have seen, the drag on a body depends greatly on whether the boundary layer is laminar or turbulent, and especially on the position at which separation occurs. The reduction of drag is of the greatest importance in aircraft design. Much effort has been devoted both to reducing skin friction by delaying the transition from laminar to turbulent flow in the boundary layer, and also to delaying separation. Much may be done, by careful shaping of the body, to avoid small radii of curvature, particularly at its downstream end, but this does not prevent the continual thickening of the boundary layer, and where adverse pressure gradients are encountered it is difficult to prevent separation unless other means are adopted.

Separation may be prevented by accelerating the boundary layer in the direction of flow. Increased turbulence produced by artificial roughening of the surface will achieve this to some degree, but it is more effective either to inject fluid at high velocity from small backward-facing slots in the boundary surface or to extract slow-moving fluid by suction. A disadvantage of injecting extra fluid is that, if the layer is laminar, this process provokes turbulence, which itself increases the skin friction. The slotted wing (Fig. 8.21), which has been used for the control of separation on aircraft, rejuvenates the boundary layer on the upper surface with fast-moving air brought through a tapered slot. It is particularly effective at large angles of attack, for which separation would otherwise occur early in the boundary layer's journey, and therefore it also helps to increase lift (Section 8.8.6). A cowl (Fig. 8.22) can similarly reduce the drag of a blunt body (such as an aircraft engine).

One of the most successful methods of control is the removal of slow-moving fluid in the boundary layer by suction through slots or through a porous surface. Downstream of the suction position the boundary

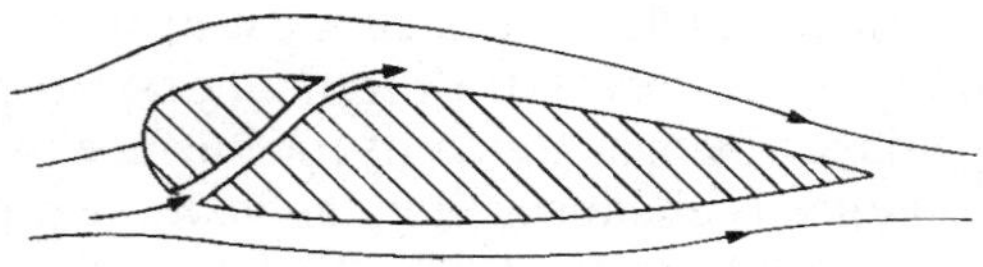

Fig. 8.21 Slotted wing.

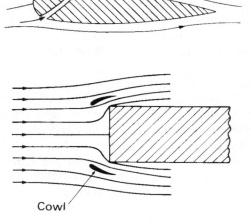

Fig. 8.22

Cowl

layer is thinner and faster and so better able to withstand an adverse pressure gradient. Suction also greatly delays the transition from laminar to turbulent flow in the boundary layer, and so reduces skin friction. It must be admitted, however, that the structural and mechanical complications are often intolerable.

On the upper surface of swept-back aircraft wings the boundary layer tends to move sideways towards the wing tips, thus thickening the layer there, and inducing early separation. To prevent the sideways movement, 'boundary-layer fences' – small flat plates parallel to the direction of flight – have been used with success.

More detailed discussion of boundary layer control will be found, for example, in the book by Thwaites[5].

8.10 EFFECT OF COMPRESSIBILITY ON DRAG

Our study of drag has so far been concerned only with flow of a single fluid in which the density is constant throughout. The relevant forces have been those of viscosity and inertia, so, for a body of given shape, the drag coefficient has been a function of Reynolds number only. However, when the velocity approaches that of sound in the fluid, elastic forces become significant, and the drag coefficient becomes a function also of Mach number, that is, the ratio of the fluid velocity (relative to the body) to the velocity of sound in the fluid.

Drag is always the result of shear forces and pressure differences, but when compressibility effects are significant the distribution of these quantities round a given body differs appreciably from that in flow at constant density. The abrupt rise of pressure that occurs across a shock wave (see Section 11.5) is particularly important. This is not simply because the difference of pressure between front and rear of the body is thereby affected. The adverse pressure gradient produced by the shock wave thickens the boundary layer on the surface, and encourages separation. The problem, however, is complicated by the interaction between shock waves and the boundary layer. Pressure changes cannot be propagated upstream in supersonic flow; in the boundary layer, however, velocities close to the surface are subsonic, so the pressure change across the shock wave can be conveyed upstream through the boundary layer. The result is to make the changes in quantities at the surface less abrupt. At high Mach numbers, heat dissipation as a result of skin friction causes serious rises of temperature and further complication of the problem.

At high values of Mach number the Reynolds number may be high enough for viscous effects to be relatively unimportant. But although a valuable simplification of problems is achieved if effects of either compressibility or viscosity may be neglected, it must be remembered that there is no velocity at which the effects of compressibility begin or those of viscosity cease, and in many situations the Reynolds number and Mach number are of comparable significance.

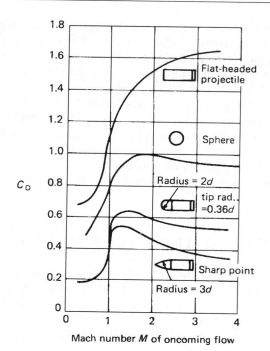

Fig. 8.23 Drag coefficients for bodies of revolution.

With a completely non-viscous fluid (which would produce no skin friction and no separation of the flow from the boundary) there would be zero drag in subsonic flow. In supersonic flow, however, the change of pressure across a shock wave would produce a drag even with a non-viscous fluid. This drag is known as *wave drag*.

The drag coefficient of a given body rises sharply as the Mach number M of the oncoming flow approaches 1.0 (Fig. 8.23). For a blunt body, for which the position of separation is fixed by its shape, the skin friction is small, and C_D continues to rise beyond $M = 1$, as a result of shock wave effects at or near the front of the body. These effects make the largest contribution to the total drag, so, for supersonic flow, streamlining the rear of such a body has little effect on the total drag. The greatest reduction of drag is achieved by making the nose of the body a sharp point (see Fig. 8.23) because wave drag is then confined to only a small region.

For other bodies the position of separation is closely associated with the shock phenomena. A shock wave first appears on the surface at the position of maximum velocity, and separation occurs close behind it. There is thus a sharp rise of C_D. An increase of Mach number, however, shifts the shock wave further downstream, the position of separation moves similarly and the width of the wake is thereby reduced. Thus, although wave drag is intensified with increasing velocity, C_D rises less than for a blunt body. With further increase of Mach number C_D decreases towards an asymptotic value. The contribution of the wake to the total drag is clearly limited because the absolute pressure in the wake cannot fall below zero.

For minimum drag in supersonic flow, the body should have a sharp forward edge, or conical nose, and the shape of the rear is of secondary importance. This requirements is the reverse of those at Mach numbers well below unity, for then the drag is least for a body well tapered at the rear and rounded at the front. A body well streamlined for subsonic velocities may thus be poorly shaped for supersonic velocities, and vice versa.

8.11 EDDY VISCOSITY AND THE MIXING LENGTH HYPOTHESIS

Turbulence involves entirely haphazard motions of small fluid particles in all directions, and it is impossible to follow the adventures of every individual particle. By considering average motions, however, attempts have been made to obtain mathematical relationships appropriate to turbulent flow, and in this section we shall briefly review some of these attempts.

It will be recalled from Section 1.6.1 that when adjacent layers of a fluid move at different velocities the interchange of molecules and the associated transfer of momentum between the layers tend to accelerate the slower one and to slow down the faster one. This action, resulting in a shear stress at the interface, is, at least in part, responsible for the phenomenon of viscosity. In turbulent flow, however, there is a continuous interchange not only of molecules over short distances but also of larger particles over larger distances. These particles may be regarded as taking increments of momentum (positive or negative) from one part of the fluid to another, and just as the interchange of molecules gives rise to a shear stress wherever relative motion exists, so does the interchange of these larger particles. If the general motion of the fluid is in straight and parallel lines, a 'turbulent shear stress' due to the interchange of particles may consequently be defined as

$$\bar{\tau}_{xz} = \eta \frac{\partial \bar{u}}{\partial y} \tag{8.37}$$

The bar over the u indicates that the velocity is the average velocity (in the x direction) at the point in question over an appreciable period of time. The symbol η (Greek 'eta') denotes what is usually now called the '*Eddy viscosity*'.

Eddy viscosity

Equation 8.37 was the suggestion (in 1877) of the French mathematician J. Boussinesq (1842–1929). Unlike the absolute viscosity μ, the eddy viscosity is not a property of the fluid but depends on the degree of turbulence of the flow and on the location of the point considered. When viscous action is also included the *total* shear stress in the xz plane is given by

$$\overline{\tau}_{xz} = \mu\frac{\partial\overline{u}}{\partial y} + \eta\frac{\partial\overline{u}}{\partial y} = (\mu + \eta)\frac{\partial\overline{u}}{\partial y} \qquad (8.38)$$

The magnitude of η may vary from zero (if the flow is laminar) to several thousand times that of μ. Its value depends on the momentum carried by the migrating particles and thus on the density of the fluid; it may be argued therefore that a *kinematic eddy viscosity* $\varepsilon = \eta/\varrho$ is entirely independent of the properties of the fluid and so characteristic only of the flow. (Many writers use the symbol ε (Greek 'epsilon') where we have used η. Moreover, not infrequently, the term eddy viscosity is used for the quantity $(\mu + \eta)$ in eqn 8.38, not for η alone.) Since values of η or ε cannot be predicted, eqn 8.37 is, however, of limited use.

Reynolds, in 1886, showed that expressions for the momentum interchange may be derived by considering that at a given point the instantaneous velocity u parallel to the x axis consists of the time-average value $\overline{u}$ plus a fluctuating component u'. In the y direction the fluctuating component is v'. Consider a surface perpendicular to the y direction and separating two adjacent fluid layers. Through a small element of this surface, of area δA, the mass transferred from one layer to the other is $\varrho(\delta A)v'$ in unit time, and the x-momentum carried in unit time is $\varrho(\delta A)v'u = \varrho(\delta A)v'(\overline{u} + u')$. Although the time-average value of v' is zero, that of $u'v'$ is not necessarily zero, so the time-average value of the x-momentum carried in unit time reduces to $\varrho(\delta A)\overline{u'v'}$, where $\overline{u'v'}$ represents the time-average value of the product $\overline{u'v'}$. An accelerating force on the lower layer of Fig. 8.24 would be produced, for example, by a positive value of u' and a negative value of v' and so, to agree with the sign convention mentioned in Section 1.6.1, a minus sign is introduced, thus giving the time-average of the turbulent shear stress on an xz-plane as

$$\overline{\tau}_{xz} = \frac{\text{Force}}{\text{Area}} = -\varrho\overline{u'v'} \qquad (8.39)$$

The stress represented by this expression is usually termed a Reynolds stress. Similar equations are of course obtained for Reynolds stresses on other planes.

Mixing length

In 1925 the German engineer Ludwig Prandtl (1875–1953) introduced the concept of a *mixing length*, that is, the average distance l, perpendicular to the mean flow direction, in which a small particle, moving towards slower layers, loses its extra momentum and takes on the mean velocity of its new surroundings. The idea is thus somewhat similar to the mean free path in molecular theory. In practice the particle does not move a distance l and then suddenly change velocity, but undergoes a gradual change. Nevertheless, on the assumption that the change in velocity $\Delta\overline{u}$ experienced by the particle in moving a distance l in the y direction is $l\,\partial\overline{u}/\partial y$, the average shear stress $\overline{\tau}_{xz}$ is $\varrho v'l\,\partial\overline{u}/\partial y$. Prandtl then supposed that v' is proportional to $\Delta\overline{u}$, i.e. to $l\,\partial\overline{u}/\partial y$, and hence $\overline{\tau}_{xz} \propto \varrho l^2(\partial\overline{u}/\partial y)^2$.

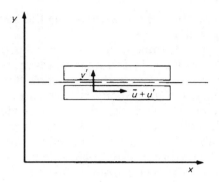

Fig. 8.24

The coefficient of proportionality, being merely a number, may be absorbed into the (unknown) l^2 and, accounting for the change of sign of $\bar{\tau}_{xz}$ with that of $\partial\bar{u}/\partial y$, one obtains:

$$\bar{\tau}_{xz} = \varrho l^2 \left|\frac{\partial\bar{u}}{\partial y}\right|\frac{\partial\bar{u}}{\partial y} \qquad (8.40)$$

The superiority of this equation over eqn 8.37 lies in plausible guesses being possible about the variation of l in certain parts of the flow. Close to a solid boundary, for example, Prandtl assumed l to be proportional to the distance from the boundary. From similarity considerations T. von Kármán (1881–1963) suggested that $l = k(\partial\bar{u}/\partial y)/(\partial^2\bar{u}/\partial y^2)$ where k was thought to be a universal constant equal to 0.40. Later experimental results, however, have shown variations of k, and among obvious loopholes in the Prandtl–von Kármán theory are that for flow in a circular pipe l is apparently independent of the pipe diameter and zero at the centre where $\partial\bar{u}/\partial y = 0$ (by symmetry) and yet mixing is there most intense! Even so the idea of mixing length permitted notable progress in the investigation of turbulent flow. Similar results are, however, obtainable by more rigorous methods, and the concept is now largely outmoded.

8.12 DISTRIBUTION OF VELOCITY IN TURBULENT FLOW

It was Prandtl who first deduced an acceptable expression for the variation of velocity in turbulent flow past a flat plate and in a circular pipe. He used his concept of 'mixing length' (Section 8.11) together with some intuitive assumptions about its variation with distance from the boundary. Using somewhat different assumptions, von Kármán and others obtained the same basic result. The expressions are all semi-empirical in that the values of the constant terms have to be determined by experiment but the forms of the expressions are derived theoretically.

However, the same results may be obtained from dimensional analysis without the hypothesis of mixing length and in fact with less far-reaching assumptions. We shall here use this more general method based on dimensional analysis.

For fully-developed turbulent flow in a circular pipe we require primarily to know the way in which the time-average value of the velocity varies with position in the cross-section. If the flow is steady – in the sense that the time-average velocity at a given point does not change with time – then considerations of symmetry indicate that this velocity u is the same at all points at the same distance from the pipe axis. The independent variables that may be supposed to affect the value of u are the density ϱ and viscosity μ of the fluid, the radius R of the pipe, the position of the point (distance y from the pipe wall, or radius $r = R - y$ from the axis), the roughness of the pipe wall – which may be represented by some characteristic height k of the surface 'bumps' (as in Section 7.3), and the shear stress τ_0 at the wall. (In place of τ_0, the pressure drop per unit length could be used as a variable, as it is simply related to τ_0 by eqn 7.3, but τ_0 – here taken as positive – is more convenient for our purpose.) Application of the principles of dimensional analysis suggests the following relation:

$$\frac{u}{(\tau_0/\varrho)^{1/2}} = \phi_1 \left\{ \frac{R}{v} \left(\frac{\tau_0}{\varrho} \right)^{1/2}, \frac{y}{R}, \frac{k}{R} \right\} \tag{8.41}$$

where $v = \mu/\varrho$ and $\phi\{\ \}$ means 'some function of'. (Suffixes on the ϕ symbols will be used to distinguish one function from another, but in general the form of the function will not be known.)

It is to be expected, however, that the effect of viscosity on the flow will be appreciable only near the walls, where the velocity gradient is large, and that its effect near the axis of the pipe will be slight. Similarly, although the roughness of the walls affects the value of $(\tau_0/\varrho)^{1/2}$ for a given rate of flow, it has little influence on the flow near the axis. Consequently, for the central part of the flow the 'velocity defect', that is, the difference $u_m - u$ between the maximum velocity u_m at the centre of the pipe and the velocity u elsewhere, may be supposed to depend on y/R only. That is

$$\frac{u_m - u}{(\tau_0/\varrho)^{1/2}} = \phi_2(\eta) \tag{8.42}$$

where $\eta = y/R$.

This hypothesis is well corroborated by experiment. It was first demonstrated by Sir Thomas E. Stanton in 1914; later Nikuradse and others obtained curves of $u(\tau_0/\varrho)^{-1/2}$ against η for a wide range of wall roughness, and when the points of maximum velocity were superposed the curves were coincident for much of the pipe cross-section. Only for a narrow region close to the wall did the curves differ. Equation 8.42 may therefore be regarded as a universal relation valid from $\eta = \Delta_1$, say, to the centre of the pipe where $\eta = 1$.

It may be noted that the smaller the value of the friction factor f and thus of τ_0, the smaller is $u_m - u$, and therefore the more nearly uniform is the velocity over the section.

8.12.1 Velocity distribution in smooth pipes and over smooth plates

For a smooth pipe, that is, one in which the projections on the wall do not affect the flow, eqn 8.41 becomes

$$\frac{u}{(\tau_0/\varrho)^{1/2}} = \phi_1(\xi, \eta) \text{ for all values of } \eta \qquad (8.43)$$

where $\xi = (R/v)(\tau_0/\varrho)^{1/2}$. In particular,

$$\frac{u_m}{(\tau_0/\varrho)^{1/2}} = \phi_1(\xi, 1) = \phi_3(\xi) \quad \text{for} \quad \eta = 1 \qquad (8.44)$$

Adding eqns 8.42 and 8.43 gives

$$\frac{u_m}{(\tau_0/\varrho)^{1/2}} = \phi_3(\xi) = \phi_1(\xi, \eta) + \phi_2(\eta) \quad \text{for} \quad \Delta_1 < \eta \leqslant 1 \quad (8.45)$$

We now consider a very thin layer of the flow close to the wall. Since the radius of the pipe is large compared with the thickness of this layer the flow in it will be hardly affected by the fact that the wall is curved. In other words, the radius of the pipe has negligible effect on the velocity near the wall. This supposition that the velocity near the wall depends on the distance from the wall but not on the pipe radius was put forward by Prandtl and it too is well supported by experimental evidence. For positions close to the wall, that is, for $0 < \eta < \Delta_2$, say, the function $\phi_1(\xi, \eta)$ in eqn 8.43 becomes $\phi_4(\xi\eta)$ since

$$\xi\eta = \frac{R}{v}\left(\frac{\tau_0}{\varrho}\right)^{1/2} \frac{y}{R} \text{ is independent of } R$$

$$\therefore \quad \frac{u}{(\tau_0/\varrho)^{1/2}} = \phi_4(\xi\eta) \quad \text{for} \quad 0 < \eta < \Delta_2 \qquad (8.46)$$

Experimental results suggest that $\Delta_2 > \Delta_1$; that is, that there is a region – even if only narrow one – where the ranges of validity of eqns 8.42 and 8.46 overlap. For this region where $\Delta_1 < \eta < \Delta_2$ both equations apply, and they may be added to give

$$\phi_2(\eta) + \phi_4(\xi\eta) = \frac{u_m}{(\tau_0/\varrho)^{1/2}} = \phi_3(\xi) \qquad (8.47)$$

(from eqn 8.44).

Differentiation of eqn 8.47 with respect to ξ gives

$$\eta\phi_4'(\xi\eta) = \phi_3'(\xi) \qquad (8.48)$$

where $\phi'(x)$ means $(\partial/\partial x)\phi(x)$. Differentiating eqn 8.48 with respect to η gives

$$\phi_4'(\xi\eta) + \xi\eta\phi_4''(\xi\eta) = 0 \qquad (8.49)$$

Equation 8.49 involves only the combined variable $\xi\eta$. Integrating with respect to $\xi\eta$ yields

$$\xi\eta\phi_4'(\xi\eta) = \text{constant} = A$$

i.e. $\phi_4'(\xi\eta) = A/\xi\eta$, and a further integration gives

$$\phi_4(\xi\eta) = A\ln(\xi\eta) + B$$

where B = constant. That is

$$\frac{u}{(\tau_0/\varrho)^{1/2}} = A\ln\left\{\frac{y}{v}\left(\frac{\tau_0}{\varrho}\right)^{1/2}\right\} + B \qquad (8.50)$$

Similar arguments may be used in considering turbulent flow between two parallel smooth plates separated by a distance $2h$. The expressions obtained differ from those above only in having h in place of R, and the final result (eqn 8.50) is independent of either h or R. Equation 8.50 applies also of course to turbulent flow over a single smooth flat plate ($h = \infty$).

Since the so-called 'logarithmic profile' represented by eqn 8.50 is of such general application, it is also termed the 'universal velocity distribution' for turbulent flow. It has been well confirmed by experiment. If measurements of u are made at various distances y from a flat plate, for example, a graph of u against $\ln\{y(\tau_0/\varrho)^{1/2}/v\}$ is indeed a straight line for a considerable range of values. We recall, however, from the derivation of the result, that it can be expected to apply over only a certain range of values of y. For one thing, the velocity does not in practice increase indefinitely with y as the equation suggests, but tends to the velocity of the main stream. And then, although at very small values of y it is difficult to obtain reliable experimental values of u, the equation clearly fails when $y \to 0$ since it predicts an infinite negative velocity at the boundary. This failure is hardly surprising in view of the existence of the viscous sub-layer: as turbulence is suppressed immediately next to the boundary the equation can apply only outside the sub-layer.

When experimental results for a circular pipe are plotted as a graph of $u(\tau_0/\varrho)^{-1/2}$ against $\ln\{y(\tau_0/\varrho)^{1/2}/v\}$ a very good straight line is obtained over a remarkably wide range (Fig. 8.25). Even so, $\phi_4(\xi\eta)$ in eqn 8.46 is only an approximation to $\phi_1(\xi, \eta)$ and it cannot be expected to be valid as far as the centre of the pipe because Prandtl's hypothesis that the velocity is independent of the pipe radius there breaks down. Moreover, symmetry requires $\partial u/\partial y [= R^{-1}(\tau_0/\varrho)^{1/2}\partial\phi_1/\partial\eta]$ to be zero at the axis and this condition cannot be met by $\phi_4(\xi\eta)$.

Figure 8.25 suggests that for large Reynolds numbers, and therefore large values of ξ, eqn 8.50 is valid except close to the pipe axis and close to the wall. The equation fails for very small values of η because of the presence of the sub-layer. From eqn 7.11 $\tau = \tau_0 r/R = \tau_0\{1 - (y/R)\}$. Within the viscous sub-layer, however, $y/R = \eta$ is so small that τ differs negligibly from τ_0. That is, the stress in the sub-layer may be considered constant. Integrating the laminar stress equation $\tau = \mu(\partial u/\partial y)$ and setting the integration constant to zero so that $u = 0$ when $y = 0$ we have $u = \tau_0 y/\mu$. Rearranged, this becomes

$$\frac{u}{(\tau_0/\varrho)^{1/2}} = \frac{y}{v}\left(\frac{\tau_0}{\varrho}\right)^{1/2} = \xi\eta$$

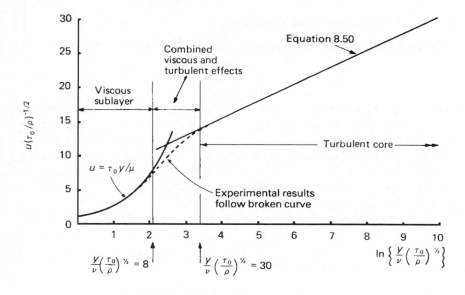

Fig. 8.25

This relation is plotted at the left-hand side of Fig. 8.25. The viscous sub-layer, however, has no definite edge where viscous effects end and appreciable turbulence begins, and so experimental points make a gradual transition between the two curves of Fig. 8.25.

Nikuradse made many detailed measurements of velocity distribution in turbulent flow for a wide range of Reynolds number, and his results suggest that $A = 2.5$ and $B = 5.5$. Substitution in eqn 8.50 gives

$$\frac{u}{\left(\tau_0/\varrho\right)^{1/2}} = 2.5 \ln\left\{\frac{y}{v}\left(\frac{\tau_0}{\varrho}\right)^{1/2}\right\} + 5.5 = 5.75 \log_{10}\left\{\frac{y}{v}\left(\frac{\tau_0}{\varrho}\right)^{1/2}\right\} + 5.5$$

This equation closely represents the velocity distribution in smooth pipes at fairly high Reynolds numbers. Points follow the 'viscous' line up to

$$\frac{y}{v}\left(\frac{\tau_0}{\varrho}\right)^{1/2} \simeq 8$$

and the 'log' law from a value of about 30. The 'thickness' δ_l of the viscous sub-layer is therefore given by

$$\frac{\delta_l}{v}\left(\frac{\tau_0}{\varrho}\right)^{1/2} \simeq 8$$

and since $f = |\tau_0|/\frac{1}{2}\varrho\bar{u}^2$ we have

$$\frac{\delta_l}{d} \simeq \frac{8v}{d\left(\tau_0/\varrho\right)^{1/2}} = \frac{8v}{d\bar{u}\left(f/2\right)^{1/2}} = \frac{8\sqrt{2}}{Re\sqrt{f}}$$

where d represents the pipe diameter, $\bar{u} = Q \div (\pi d^2/4)$ and $Re = \bar{u}d/v$.

8.12.2 Friction factor for smooth pipes

The parameter $(\tau_0/\varrho)^{1/2}$ is related to the friction factor f by eqn 7.4:

$$\frac{|\tau_0|}{\varrho \bar{u}^2} = \frac{f}{2}$$

We therefore use the relation for velocity distribution (eqn 8.50) to obtain an expression for the mean velocity $\bar{u}$. It is true that eqn 8.50 is not valid over the entire cross-section, but experimental results such as those represented in Fig. 8.25 show that Δ_1, the value of η at which the equation ceases to apply, is small, especially at high Reynolds numbers. As an approximation we therefore suppose that eqn 8.50 is valid for all values of η. Multiplying the equation by $2\pi r\,dr$, integrating between the limits $r = 0$ and $r = R$, and then dividing the result by πR^2 we obtain

$$
\begin{aligned}
\frac{\bar{u}}{(\tau_0/\varrho)^{1/2}} &= \frac{1}{\pi R^2} \int_0^R \left\{ A \ln(\xi\eta) + B \right\} 2\pi r\,dr \\
&= \int_1^0 \left\{ A \ln(\xi\eta) + B \right\} 2(1 - \eta)(-d\eta) \\
&= A \ln \xi - \frac{3}{2} A + B
\end{aligned}
\tag{8.51}
$$

This integration neglects the fact the eqn 8.50 does not correctly describe the velocity profile close to the axis, but as this inaccuracy is appreciable only as $\eta \to 1$ and $(1 - \eta)$ is then small, the effect on the value of the integral is not significant. Then, substituting for $\bar{u}(\tau_0/\varrho)^{-1/2}$ from eqn 7.4, we have

$$\left(\frac{2}{f}\right)^{1/2} = A \ln \xi - \frac{3}{2} A + B$$

Noting that

$$\xi = \frac{R}{v}\left(\frac{\tau_0}{\varrho}\right)^{1/2} = \frac{d}{2v}\bar{u}\left(\frac{f}{2}\right)^{1/2} = Re\left(\frac{f}{8}\right)^{1/2}$$

we have

$$f^{-1/2} = \frac{1}{\sqrt{2}}\left\{ A \ln(Re\,f^{1/2}) - A \ln \sqrt{8} - \frac{3}{2} A + B \right\}$$

Substituting Nikuradse's values of A and B and converting ln to $\log_{10}$ then gives

$$f^{-1/2} = 4.07 \log_{10}(Re\,f^{1/2}) - 0.6$$

In view of the approximations made it would be surprising if this equation exactly represented experimental results, especially as the last term is a difference between relatively large quantities. Good agreement with experiment is, however, obtained if the coefficients are adjusted to give

$$f^{-1/2} = 4\log_{10}\left(Re\,f^{1/2}\right) - 0.4 \qquad (8.52)$$

$$\left[\text{Or } \lambda^{-1/2} = 2\log_{10}\left(Re\,\lambda^{1/2}\right) - 0.8 \text{ where } \lambda = 4f.\right]$$

This expression has been verified by Nikuradse for Reynolds numbers from 5000 to 3×10^6. For Reynolds numbers up to 10^5, however, Blasius's equation (7.6) provides results of equal accuracy and is easier to use since eqn 8.52 has to be solved for f by trial.

8.12.3 Velocity distribution and friction factor for rough pipes

From Section 7.3 we recall that a pipe which is 'hydraulically smooth' is one in which the roughness projections on the wall are small enough to be submerged within the viscous sub-layer and to have no influence on the flow outside it. Since the sub-layer thickness is determined by the value of the parameter $y(\tau_0/\varrho)^{1/2}/\nu$, the maximum height k of roughness elements which will not affect the flow is governed by the value of $k(\tau_0/\varrho)^{1/2}/\nu$. Nikuradse's results for pipes artificially roughened with uniform grains of sand indicate that the roughness has no effect on the friction factor when $k(\tau_0/\varrho)^{1/2}/\nu < 4$ (approximately). On the other hand, if $k(\tau_0/\varrho)^{1/2}/\nu > 70$ (approximately) f becomes independent of Re, and this suggests that the effect of the roughness projections then quite overshadows viscous effects. For this 'rough zone of flow' the sub-layer is completely disrupted, and viscous effects are negligible. Equation 8.41 may then be written

$$\frac{u}{(\tau_0/\varrho)^{1/2}} = \phi\left(\frac{y}{R}, \frac{k}{R}\right) = \phi_5\left(\eta, \frac{R}{k}\right)$$

Here k may be regarded as some magnitude suitably characterizing the roughness size; it is not necessarily the average height of the projections.

By similar reasoning to that used for smooth pipes (except that R/k here takes the place of ξ) the equation for velocity distribution may be shown to take the form

$$\frac{u}{(\tau_0/\varrho)^{1/2}} = A\ln\left(\frac{y}{k}\right) + C \qquad (8.53)$$

A similar expression may be deduced for flow over rough plates; η then represents y/δ.

The constant A in eqn 8.53 is the same as that for smooth pipes. The constant C, however, differs from B. Equation 8.53 fails to give $\partial u/\partial y = 0$ on the pipe axis, but, over most of the cross-section, experimental results are well described by the equation

$$\frac{u}{(\tau_0/\varrho)^{1/2}} = 5.75 \log_{10}(y/k) + 8.48 \qquad (8.54)$$

It will be noted that the equation is independent of the relative roughness k/R: experimental points for a wide range of k/R all conform to a straight line when $u(\tau_0/\varrho)^{-1/2}$ is plotted against $\log_{10}(y/k)$.

Reasoning similar to that used in the previous section (8.12.2) leads to a relation for the friction factor:

$$f^{-1/2} = 4.07 \log_{10}(R/k) + 3.34$$

Slight adjustment of the coefficients to achieve better agreement with experiment gives

$$f^{-1/2} = 4 \log_{10}(R/k) + 3.48 = 4 \log_{10}(d/k) + 2.28 \qquad (8.55)$$

where d represents the diameter of the pipe.

All these relations apply only to fully-developed turbulent flow (and not, therefore, to the entrance length of a pipe) and to Reynolds numbers sufficiently high for f to be independent of Re (see Fig. 7.2).

Equation 8.55 is valid for values of $k(\tau_0/\varrho)^{1/2}/\nu > 70$, whereas the appropriate equation for smooth pipes (8.52) is valid for $k(\tau_0/\varrho)^{1/2}/\nu < 4$. For the range 4 to 70, f is dependent on both Re and the relative roughness. If $4 \log_{10}(d/k)$ is subtracted from both sides of eqns 8.52 and 8.55, we obtain

$$f^{-1/2} - 4 \log(d/k) = 4 \log\left(Re\, f^{1/2} k/d\right) - 0.4 \quad \text{for 'smooth' pipes}$$

and

$$f^{-1/2} - 4 \log(d/k) = 2.28 \quad \text{for 'fully rough' pipes.}$$

When $f^{-1/2} - 4\log(d/k)$ is plotted against $\log\left(Re f^{1/2} k/d\right)$, as shown in Fig. 8.26 two straight lines result – one for 'smooth' pipes and one for 'fully rough' pipes. For commercial pipes, in which the roughness is not uniform, an equivalent roughness size k may be deduced from the (constant) value of f at high Reynolds number. C. F. Colebrook[6] measured values of f for new commercial pipes and found that when the results were plotted in the form of Fig. 8.26, the points were closely clustered about a curve joining the two straight lines and having the equation

$$f^{-1/2} - 4 \log(d/k) = 2.28 - 4 \log\left\{1 + \frac{4.67d}{Re\, k f^{1/2}}\right\}$$

or, in rearranged form,

$$f^{-1/2} = -4 \log\left\{\frac{k}{3.71d} + \frac{1.26}{Re\, f^{1/2}}\right\} \qquad (8.56)$$

(It may readily be shown that as $k \to 0$ the expression approaches the 'smooth pipe' equation 8.52, and that when $Re \to \infty$ it approaches the 'fully rough' equation 8.55.)

The rather remarkable fact that the points for these commercial pipes all quite closely fit a single curve suggests that their random roughness may be adequately described by a single parameter k. The curve, however, differs significantly from that for uniform roughness (as the reasoning of Section 7.3 would lead us to expect).

Since f appears on both sides of Colebrook's equation it is awkward to use, so L. F. Moody constructed his chart for friction factor (Fig. 7.2).

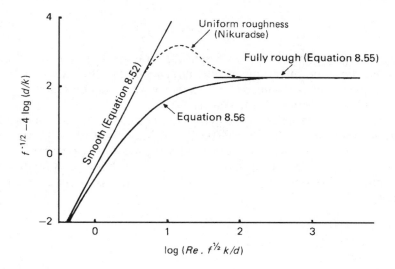

Fig. 8.26

Although eqn 8.56 and Moody's chart permit f to be determined when k and Re are known, there remains the problem of specifying the equivalent uniform roughness size k for a particular pipe. It is often not possible to conduct an experiment on the pipe at such a high Reynolds number that the constant value of f may be determined and k then deduced from eqn 8.55. Further research is being directed towards resolving such difficulties.

8.12.4 Universal features of the velocity distribution in turbulent flow

Finally, a few general points are worthy of note. In the expressions developed in the preceding analyses the factor A is common and so appears to be a universal constant. Although 2.5 is the usually accepted value, different workers have obtained somewhat different figures, and the value could well depend slightly on Reynolds number. For flow over flat plates the values $A = 2.4$ and $B = 5.84$ in eqn 8.50 have been found to give better agreement with experiment.

The 'velocity defect law' is independent of both Reynolds number and roughness. Thus if, as in the integration leading to eqn 8.51, this law is assumed to apply with sufficient accuracy over the entire cross-section, integration of the appropriate expressions based on the law enables values of the kinetic energy correction factor α (Section 3.5.3) and the momentum correction factor β (Section 4.2.1) to be evaluated. Hence we obtain

$$\alpha = 1 + \frac{15}{8}A^2 f - \frac{9}{4}A^3 \left(\frac{f}{2}\right)^{3/2} = 1 + 11.72 f - 12.43 f^{3/2} \quad (8.57)$$

and

$$\beta = 1 + \frac{5}{8}A^2f = 1 + 3.91f \qquad (8.58)$$

for all values of roughness and Reynolds number. For fully developed turbulent flow the maximum values in practice are thus about 1.1 and 1.04, respectively, but considerably higher values may be realized if the flow is subject to other disturbances.

If transfer of heat takes place across the boundary surface, variations of temperature and other properties, especially viscosity, with distance from the boundary may affect the results appreciably. Consideration of these matters is beyond the scope of the present book, and for further information reference should be made to works on convective heat transfer.

8.13 FREE TURBULENCE

Turbulence occurs not only when fluid flows past a solid boundary, but also when two fluid streams flow past each other with different velocities. Turbulent mixing then takes place between the streams so as to equalize their velocities. This 'free turbulence' – that is, turbulence not bounded by solid walls – occurs, for example, when a jet issues from a 'drowned' orifice into a large expanse of stationary fluid, or in the wake behind a body moving through a fluid otherwise at rest. The flow conditions are broadly similar to those in a turbulent boundary layer with a large velocity gradient in a direction perpendicular to the main flow. Because solid boundaries are absent, however, the effect of viscosity is negligible.

When a jet encounters stationary fluid it sets some of this in motion (see Fig. 8.27), a process known as *entrainment*. Unless the velocity of the fluid is comparable with the velocity of sound propagation, the pressure is substantially uniform through the jet and the surrounding fluid. Thus the net force on the jet is zero and the momentum in the axial direction remains constant, even though the amount of fluid in motion

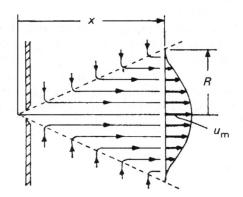

Fig. 8.27

increases. The average velocity therefore decreases. For a jet of circular cross-section the rate of momentum flow is

$$\int_0^R \varrho u^2 \, 2\pi r \, \mathrm{d}r = 2\pi \varrho u_m^2 R^2 \int_0^1 \left(\frac{u}{u_m}\right)^2 \eta \, \mathrm{d}\eta \qquad (8.59)$$

where u_m represents the maximum velocity (on the axis), $\eta = r/R$ and R represents the radius to the edge of the jet (beyond which the velocity is zero). The edge of the jet, however, like the edge of a boundary layer, is ill-defined. It is reasonable to assume that, except close to the orifice, the velocity 'profile' over the cross-section is similar for all axial positions. That is, u/u_m is a function of η only, so $\int_0^1 (u/u_m)^2 \eta \, \mathrm{d}\eta$ is a constant. Since the rate of momentum flow is constant it follows from eqn 8.59 that $u_m^2 R^2$ is constant.

Dimensional analysis suggests that the width of the jet increases linearly with the axial distance x, and so the velocity on the axis is given by $u_m = \text{constant}/x$ for a jet of circular cross-section. Experiment shows this rule to be closely followed. The angle of divergence is between $20°$ and $25°$ for a circular cross-section and about $5°$ greater for a jet issuing from a rectangular slit.

The methods used for studying the flow in turbulent boundary layers may be used also for free turbulence. For details reference should be made to the specialized books listed at the end of the chapter.

8.14 COMPUTATIONAL FLUID DYNAMICS

Our knowledge of classical fluid dynamics is based on experiment and theoretical analysis. In recent years a new branch of knowledge – known as computational fluid dynamics and denoted by the initials CFD – has emerged to provide engineers and applied mathematicians with a third and exceptionally powerful tool. Here we can do no more than set down a few words of introduction to this rapidly expanding subject, which has already developed a large literature in its own right.

The classical analysis of fluid flow problems involves the initial representation of the fluid motion by a physical model. This physical model is then interpreted as a mathematical model by setting down the conservation equations which the flow satisfies. These governing equations are then subject to analysis leading to closed-form solutions with the dependent variables, such as the velocity components and pressure, varying continuously throughout the flow field. Much of the content of this book consists of analytical solutions of this kind. This general approach and the closed-form solutions which result provide powerful insights into many important fluid flow problems, including the effects of changes in geometry and flow parameters such as Reynolds number. However, there are many circumstances where it is not possible to derive analytical solutions while at the same time retaining a mathematical model which displays all the essential features of the physics underlying the flow. Here the limitations of the analytical approach become evident. By way of contrast, CFD is capable of delivering solutions to the governing equations which fully describe all aspects of the physics of the flow (while still being subject to the limitations of our current understanding of turbulence).

The essential basis of CFD is as follows. The starting point is the specification of the governing equations of fluid dynamics – the equations of continuity, momentum and energy. These equations are then replaced by equivalent numerical descriptions which are solved by numerical techniques to yield information on the dependent variables – for example, the velocity components and pressure – at discrete locations in the flow field. The continuity, momentum and energy equations can be set down in a variety of different, but equivalent, forms and the choice of the appropriate form(s) is an important aspect of successfully obtaining numerical solutions.

8.14.1 The differential equations of fluid dynamics

The system of equations known as the Navier–Stokes equations takes account of the full three-dimensional, viscous nature of fluid motion. Derivations of these equations are given in many textbooks and we shall simply set down the more important results. It is worth noting that these equations for the instantaneous motion are equally valid in laminar and turbulent flows.

The equation of continuity expresses the principle of conservation of matter. The differential form of the equation is obtained by considering the flow into and out of an elementary control volume. For the rectangular Cartesian coordinate system, with coordinates x, y, z measured relative to a stationary frame of reference, and corresponding velocity components u, v, w, the continuity equation is

$$\frac{\partial \varrho}{\partial t} + \frac{\partial(\varrho u)}{\partial x} + \frac{\partial(\varrho v)}{\partial y} + \frac{\partial(\varrho w)}{\partial z} = 0 \qquad (8.60)$$

where ϱ is the fluid density.

The x, y and z components of the momentum equation are obtained by applying Newton's second law to an elementary volume of fluid. The resulting three equations for the rectangular Cartesian coordinate system are

$$\frac{\partial(\varrho u)}{\partial t} + \frac{\partial(\varrho u^2)}{\partial x} + \frac{\partial(\varrho u v)}{\partial y} + \frac{\partial(\varrho u w)}{\partial z} =$$

$$\varrho X - \frac{\partial p}{\partial x} + \frac{\partial}{\partial x}\left[\mu\left(2\frac{\partial u}{\partial x} - \frac{2}{3}\left(\frac{\partial u}{\partial x} + \frac{\partial v}{\partial y} + \frac{\partial w}{\partial z}\right)\right)\right]$$

$$+ \frac{\partial}{\partial y}\left[\mu\left(\frac{\partial u}{\partial y} + \frac{\partial v}{\partial x}\right)\right] + \frac{\partial}{\partial z}\left[\mu\left(\frac{\partial w}{\partial x} + \frac{\partial u}{\partial z}\right)\right] \qquad (8.61a)$$

$$\frac{\partial(\varrho v)}{\partial t} + \frac{\partial(\varrho u v)}{\partial x} + \frac{\partial(\varrho v^2)}{\partial y} + \frac{\partial(\varrho v w)}{\partial z} =$$

$$\varrho Y - \frac{\partial p}{\partial y} + \frac{\partial}{\partial y}\left[\mu\left(2\frac{\partial v}{\partial y} - \frac{2}{3}\left(\frac{\partial u}{\partial x} + \frac{\partial v}{\partial y} + \frac{\partial w}{\partial z}\right)\right)\right]$$

$$+ \frac{\partial}{\partial z}\left[\mu\left(\frac{\partial v}{\partial z} + \frac{\partial w}{\partial y}\right)\right] + \frac{\partial}{\partial x}\left[\mu\left(\frac{\partial u}{\partial y} + \frac{\partial v}{\partial x}\right)\right] \qquad (8.61b)$$

$$\frac{\partial(\varrho w)}{\partial t} + \frac{\partial(\varrho u w)}{\partial x} + \frac{\partial(\varrho v w)}{\partial y} + \frac{\partial(\varrho w^2)}{\partial z} =$$

$$\varrho Z - \frac{\partial p}{\partial z} + \frac{\partial}{\partial z}\left[\mu\left(2\frac{\partial v}{\partial z} - \frac{2}{3}\left(\frac{\partial u}{\partial x} + \frac{\partial v}{\partial y} + \frac{\partial w}{\partial z}\right)\right)\right]$$

$$+ \frac{\partial}{\partial x}\left[\mu\left(\frac{\partial w}{\partial x} + \frac{\partial u}{\partial z}\right)\right] + \frac{\partial}{\partial y}\left[\mu\left(\frac{\partial v}{\partial z} + \frac{\partial w}{\partial y}\right)\right] \qquad (8.61c)$$

Equations 8.61a–c, in which the symbols X, Y, Z represent the components of a body force, were first derived, independently, by Sir George Stokes and the French engineer, Louis Navier, and consequently the equations are known collectively as the Navier–Stokes equations.

The differential energy equation, which expresses the principle that energy is conserved, is

$$\frac{\partial(\varrho e)}{\partial t} + \frac{\partial(\varrho u e)}{\partial x} + \frac{\partial(\varrho v e)}{\partial y} + \frac{\partial(\varrho w e)}{\partial z} =$$

$$\varrho Q + \frac{\partial}{\partial x}\left(k\frac{\partial T}{\partial x}\right) + \frac{\partial}{\partial y}\left(k\frac{\partial T}{\partial y}\right) + \frac{\partial}{\partial z}\left(k\frac{\partial T}{\partial z}\right)$$

$$- p\left(\frac{\partial u}{\partial x} + \frac{\partial v}{\partial y} + \frac{\partial w}{\partial z}\right) - \lambda\left(\frac{\partial u}{\partial x} + \frac{\partial v}{\partial y} + \frac{\partial w}{\partial z}\right)^2$$

$$+ \mu\left\{2\left[\left(\frac{\partial u}{\partial x}\right)^2 + \left(\frac{\partial v}{\partial y}\right)^2 + \left(\frac{\partial w}{\partial z}\right)^2\right] + \left(\frac{\partial v}{\partial x} + \frac{\partial u}{\partial y}\right)^2\right.$$

$$+ \left.\left(\frac{\partial w}{\partial y} + \frac{\partial v}{\partial z}\right)^2 + \left(\frac{\partial u}{\partial z} + \frac{\partial w}{\partial x}\right)^2\right\} \qquad (8.62)$$

where e is the internal energy, λ is the bulk viscosity, Q is the heat addition per unit mass and k is the thermal conductivity.

Historically, only the momentum eqns 8.61a–c were referred to as the Navier–Stokes equations, but there appears to be an increasing tendency towards labelling the continuity, momentum and energy equations collectively as the Navier–Stokes equations. As noted previously, eqns 8.60–8.62 can be expressed in numerous alternative forms. The forms chosen here are the so-called conservation forms, which have particular advantages in the context of CFD.

8.14.2 Numerical procedures for solving the Navier–Stokes equations

There are two principal types of methods used for solving the governing flow equations. These are the finite-difference methods (FDM) and the finite-element methods (FEM). Two further approaches – the boundary-element method (BEM) and the finite-volume methods – are also available.

Finite-difference methods

The principle underlying FDM is the replacement of the partial derivatives in the differential eqns 8.60–8.62 by corresponding algebraic difference quotients. As a consequence the partial differential equations can be replaced by a system of algebraic equations, and these can be solved either by iterative methods or matrix inversion to yield the magnitude of the flow variables at discrete points in the flow field, known as grid points.

To illustrate some basic aspects of the FDM, refer to Fig. 8.28, which shows a regular grid pattern set out in the x–y plane. The grid points are identified by a suffix notation, the subscript i denoting the x-direction, and suffix j the y-direction. Suppose we wish to replace the pressure gradient term $\partial p/\partial x$ in eqn 8.61a by a finite difference quotient computed at the grid pint i,j; then the following options are available:

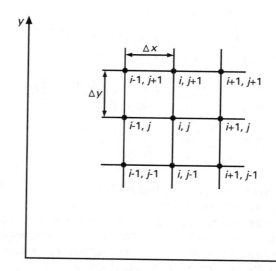

Fig. 8.28

$$\left(\frac{\partial p}{\partial x}\right)_{i,j} = \frac{p_{i,j+1} - p_{i,j}}{\Delta x} \quad \text{Forward difference}$$

$$\left(\frac{\partial p}{\partial x}\right)_{i,j} = \frac{p_{i,j} - p_{i,j-i}}{\Delta x} \quad \text{Backward difference}$$

$$\left(\frac{\partial p}{\partial x}\right)_{i,j} = \frac{p_{i,j+1} - p_{i,j-1}}{2\Delta x} \quad \text{Central difference}$$

Corresponding relations are easily derived for the other terms appearing in the partial differential equations. In establishing the algebraic equations, decisions have to be made from the choice of finite-difference formulations available. Although the process is straightforward in principle, in practice there are two particular aspects of the numerical procedures that require attention, namely the treatment of errors and stability. Errors can be of two kinds: round-off errors and discretization errors. Round-off errors arise during iteration as a consequence of the fact that calculations can only be performed to a finite number of significant figures. At any grid point, the discretization error is the difference between the exact analytical solution of the partial differential equations and the solution derived from the corresponding algebraic equations, assumed free from round-off errors. Discretization errors depend in part upon the fineness or coarseness of the grid constructed throughout the flow field, in part upon the rapidity with which flow variables change with respect to time and space, and also upon the sophistication of the relationships chosen to represent the difference quotients. The issue of stability arises in the following way. The ultimate goal of iterative numerical methods is to find the exact solution of the algebraic equations at all the grid points in the flow field. This goal is achieved in principle by moving to it, by a process of repetitive calculations, from a situation where the values of the flow variables at the grid points do not exactly satisfy the algebraic equations. Indeed to start the computation the assumed distribution of flow properties might only be a very rough approximation to the required final distribution. If the computational procedures give results which converge in a consistent way on the exact solution the numerical technique is described as stable. To achieve such desirable results certain stability criteria have to be satisfied. Otherwise the computation might diverge from the exact solution or may oscillate without convergence. In practice the goal of an exact solution to the finite-difference equations is an ideal that is rarely achievable, and a satisfactory solution is one that approximates to the exact solution within a specified level of accuracy.

The finite element method is another numerical technique for solving the governing partial differential equations, but, instead of working with the differential equations directly, it is based upon integral formulations, using weighting functions, of the differential equations. The flow field is divided into cells, known as elements, to form a grid. The elements can be of triangular or quadrilateral shape, and the sides of the

Finite element method

elements can be straight or curved. The objective of FEM is to determine the flow variables at selected points associated with each element. These selected points are known as nodes, and may be positioned at the corners, at the mid-side or at the centre of the element. Whereas the FDM requires an orderly, structured grid, with the FEM the grid need not be structured. The use of an unstructured grid allows complex geometrical shapes to be handled without undue complication, and in this regard provides FEM with a clear advantage over FDM. Another advantage of FEM over FDM is in regard to its robust and rigorous mathematical foundations, which embrace more precise definitions of accuracy than are inherent in FDM. Perhaps the main drawback of FEM is that it is intellectually much more demanding; in particular, with finite-difference methods the relationship between the derived algebraic equations and the partial differential equations they replace are much more self-evident than is the case for FEM.

Boundary element method

The defining feature of the BEM for solving numerically the partial differential equations is that all nodes are positioned on the boundary of the flow field, and there are no interior elements.

Finite-volume method

The starting point for the finite-volume method (FVM) is the discretization of the integral forms of the flow equations. This approach is particularly suited to the solution of flow fields containing discontinuities, such as a shock wave in a compressible flow field. The flow field is divided into cells, and the conservation equations are solved numerically to determine the magnitude of the flow variables at the nodes defined for each cell. The FVM shares the advantage of the FEM of accommodating an unstructured grid. Advocates of FVM claim that it combines the best feature of the FEM, namely its ability to handle complex geometries readily, with the virtue of the FDM, the simple and self-evident relationships between the finite-difference formulations and the partial differentials they replace.

REFERENCES

1. Kuo, Y. H. 'On the flow of an incompressible viscous fluid past a flat plate at moderate Reynolds numbers', *J. Math. & Phys.* **32**, 83–100 (1953).
2. Smith, A. M. O. and Clutter, D. W. 'The smallest height of roughness capable of affecting boundary-layer transition', *J. Aero/Space Sci.* **26**, 229–45, 256 (1959).
3. Roshko, A. 'Experiments on the flow past a circular cylinder at very high Reynolds numbers', *J. Fluid Mech.* **10**, 345–56 (1961).
4. Hoerner, S. F. *Fluid-Dynamic Drag*, S. F. Hoerner, Midland Park, New Jersey, U.S.A. (1958).
5. Thwaites, B. (Editor) *Incompressible Aerodynamics*, Chapter 6, Clarendon Press, Oxford (1987).

6. Colebrook, C. F. 'Turbulent flow in pipes, with particular reference to the transition region between the smooth and rough pipe laws', *J. Instn civ. Engrs*, **11,** 133–56 + tables (1939).

FURTHER READING

Anderson, J. D. *Computational Fluid Dynamics*, McGraw-Hill, London (1995).

Marris, A. W. 'A review on vortex streets, periodic wakes and induced vibration phenomena', *J. basic Engng*, **86D**, 815–96 (1964).

Rosenhead, L. (Editor) *Laminar Boundary Layers*, Clarendon Press, Oxford (1988).

Schlichting, H. *Boundary Layer Theory*, (7th edn), McGraw-Hill, New York (1979).

Steinman, D. B. 'Suspension bridges: the aerodynamic problem and its solution', *Am. Scient.* **42**, 397–438, 460 (1954).

Townsend, A. A. *The Structure of Turbulent Shear Flow*, (2nd edn), C.U.P., Cambridge (1975).

Wendt, J. F (Editor). *Computational Fluid Dynamics: An Introduction* (2nd edn), Springer, Berlin (1996).

On free turbulence:

Birkhoff, G. and Zarantonello, E. H. *Jets, Wakes and Cavities*, Academic Press, New York (1957).

FILMS

Hazen, D. C. *Boundary Layer Control*, Encyclopaedia Britannica Films or from Central Film Library.

Shapiro, A. H. *Fluid Dynamics of Drag*, Encyclopaedia Britannica Films or from Central Film Library.

PROBLEMS

8.1 Determine the ratios of displacement and momentum thickness to the boundary layer thickness when the velocity profile is represented by $u/u_m = \sin(\pi\eta/2)$ where $\eta = y/\delta$.

8.2 A smooth flat plate 2.4 m long and 900 mm wide moves lengthways at 6 m/s through still atmospheric air of density 1.21 kg/m^3 and kinematic viscosity 14.9 mm^2/s. Assuming the boundary layer to be entirely laminar, calculate the boundary layer thickness (i.e. the position at which the velocity is 0.99 times the free-stream velocity) at the trailing edge of the plate, the shear stress half-way along and the power required to move the plate. What power would be required if the boundary layer were made turbulent at the leading edge?

8.3 A smooth flat plate, 2.5 m long and 0.8 m wide, moves lengthways at 3 m/s through still water. The plate is assumed to be completely covered by a turbulent boundary layer in which the velocity distribution is given by $1 - (u/U) = (y/\delta)^{1/7}$ and the shear stress by $\tau = 0.023\varrho U^2(v/U\delta)^{1/4}$ where U denotes the steady velocity of the plate, u the velocity in the boundary layer at distance y from the plate, δ the thickness of the boundary layer, and ϱ and v the density and kinematic viscosity of the water respectively. If $v = 10^{-6}$ m²/s, calculate the total drag on the plate and the power required to move it.

8.4 Air (of kinematic viscosity 15 mm²/s) flows at 10.5 m/s past a smooth, rectangular, flat plate 300 mm × 3 m in size. Assuming that the turbulence level in the oncoming stream is low and that transition occurs at $Re = 5 \times 10^5$, calculate the ratio of the total drag force when the flow is parallel to the length of the plate to the value when the flow is parallel to the width.

8.5 A streamlined train is 110 m long, 2.75 m wide and with sides 2.75 m high. Assuming that the skin friction drag on sides and top equals that on one side of a flat plate 110 m long and 8.25 m wide, calculate the power required to overcome the skin friction when the train moves at 160 km/h through air of density 1.22 kg/m³ and viscosity 1.79×10^{-5} Pa s. How far is the laminar boundary layer likely to extend?

8.6 Air of constant density 1.2 kg/m³ and kinematic viscosity 14.5 mm²/s flows at 29 m/s past a flat plate 3 m long. At a distance 0.5 m from the leading edge a fine trip wire attached to the surface and set perpendicular to the flow induces abrupt transition in the boundary layer at that position and also an abrupt 15% increase in momentum thickness θ. Assuming steady two-dimensional flow at constant pressure, determine the drag coefficient for one side of the plate.

For the laminar boundary layer $\theta = 0.664x \, (Re_x)^{-1/2}$ and for the turbulent layer $\theta = 0.037x \, (Re_x)^{-1/5}$.

8.7 A honeycomb type of flow straightener is formed from perpendicular flat metal strips to give 25 mm square passages, 150 mm long. Water of kinematic viscosity 1.21 mm²/s approaches the straightener at 1.8 m/s. Neglecting the thickness of the metal, the effects of the small pressure gradient and of three-dimensional flow in the corners of the passages, calculate the displacement thickness of the boundary layer and the velocity of the main stream at the outlet end of the straightener. Applying Bernoulli's equation to the main stream deduce the pressure drop through the straightener.

8.8 What frequency of oscillation may be expected when air of kinematic viscosity 15 mm²/s flows at 22 m/s past a 3 mm diameter telephone wire which is perpendicular to the air stream?

8.9 The axis of a long circular cylinder is perpendicular to the undisturbed velocity of an unconfined air stream. For an ideal fluid the velocity adjacent to the surface of the cylinder would be $2U \sin \theta$, where U denotes the velocity far upstream and θ the angle between the radius and the initial flow direction. Neglecting changes of elevation and assuming that the pressure in the wake is sensibly uniform and that there are no discontinuities of pressure anywhere, estimate the position at which the boundary layer separates if the measured drag coefficient is 1.24. Why will this estimate be somewhat in error?

8.10 If, for the cylinder and air flow mentioned in Problem 8.9, the boundary layer is entirely laminar, use Thwaites's method (with $\lambda = -0.09$ at separation) to show that the position of separation is at $\theta \simeq 103.1°$.

8.11 Air of kinematic viscosity $15\,mm^2/s$ and density $1.21\,kg/m^3$ flows past a smooth 150 mm diameter sphere at 60 m/s. Determine the drag force. What would be the drag force on a 150 mm diameter circular disc held perpendicular to this air stream?

8.12 Calculate the diameter of a parachute (in the form of a hemispherical shell) to be used for dropping a small object of mass 90 kg so that it touches the earth at a velocity no greater than 6 m/s. The drag coefficient for a hemispherical shell with its concave side upstream is approximately 1.32 for $Re > 10^3$. (Air density $= 1.22\,kg/m^3$.)

8.13 Determine the diameter of a sphere of density $2800\,kg/m^3$ which would just be lifted by an air-stream flowing vertically upward at 10 m/s. What would be the terminal velocity of this sphere falling through an infinite expanse of water? (Densities: air $1.21\,kg/m^3$; water $1000\,kg/m^3$. Viscosities: air $18.0\,\mu Pa\,s$; water $1.0\,mPa\,s$.)

8.14 When water (kinematic viscosity $1.2\,mm^2/s$) flows steadily at the rate of 18.5 litres/s through 25 m of a 100 mm diameter pipe, the head loss is 1.89 m. Estimate the relative roughness, the velocity on the axis of the pipe and the shear stress at the wall.

8.15 Gas is pumped through a smooth pipe, 250 mm in diameter, with a pressure drop of 50 mm H_2O per kilometre of pipe. The gas density is $0.7\,kg/m^3$ and the kinematic viscosity is $18\,mm^2/s$. Use eqn 8.52 to determine the rate of flow and the shear stress at the wall of the pipe.

9 The flow of an ideal fluid

9.1 INTRODUCTION

The theory presented in this chapter stems from the 'classical hydrodynamics' by which eighteenth-century mathematicians sought to specify the motion of fluids by mathematical relations. These relations, however, could be developed only after certain simplifying assumptions had been made, the chief of which was that the fluid itself was 'ideal', in other words, that it had no viscosity and was incompressible. (An ideal fluid is also assumed to exhibit no surface tension effects and, if a liquid, not to vaporize.)

All real fluids do possess viscosity, and are, in some degree, compressible. Nevertheless there are many instances in which the behaviour of real fluids quite closely approaches that of the hypothetical ideal fluid. It has already been remarked (Section 8.1) that the flow of a real fluid may frequently be regarded as occurring in two regions: one, adjacent to the solid boundaries of the flow, is a thin layer in which viscosity has a considerable effect; in the other region, constituting the remainder of the flow, the viscous effects are negligible. In this latter region the flow is essentially similar to that of an inviscid fluid. As for compressibility, its effects are negligible, even for the flow of a gas, unless the velocity of flow is comparable with the speed with which sound is propagated through the fluid, or accelerations are very large. Consequently, relations describing the flow of an ideal fluid may frequently be used to indicate the behaviour of a real fluid away from the boundaries. The results so obtained may be only an approximation to the truth because of the simplifying assumptions made, although in certain instances the theoretical results are surprisingly close to the actual ones. In any event, they give valuable insight to the actual behaviour of the fluid.

In this chapter we shall attempt no more than an introduction to classical hydrodynamics and its application to a few simple examples of flow. Attention will be confined almost entirely to steady two-dimensional plane flow. In two-dimensional flow, we recall, two conditions are fulfilled: (a) a plane may be specified in which there is at no point any component of velocity perpendicular to the plane, and (b) the motion in that plane is exactly reproduced in all other planes parallel to it. Coordinate axes Ox, Oy, may be considered in the reference plane, and the velocity of any particle there may be specified completely by its

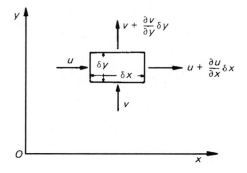

Fig. 9.1

components parallel to these axes. We shall use the following notation: u = component of velocity parallel to Ox; v = component of velocity parallel to Oy; $q^2 = u^2 + v^2$.

The condition for continuity of the flow may readily be obtained. Figure 9.1 shows a small element of volume, $\delta x \times \delta y$ in section, through which the fluid flows. The average velocities across each face of the element are as shown. For an incompressible fluid, volume flow rate into the element equals volume flow rate out; thus for unit thickness perpendicular to the diagram

$$u\delta y + v\delta x = \left(u + \frac{\partial u}{\partial x}\delta x \right)\delta y + \left(v + \frac{\partial v}{\partial y}\delta y \right)\delta x$$

whence

$$\frac{\partial u}{\partial x} + \frac{\partial v}{\partial y} = 0 \qquad (9.1)$$

9.2   THE STREAM FUNCTION

Figure 9.2 illustrates two-dimensional plane flow. It is useful to imagine a transparent plane, parallel to the paper and unit distance away from it; the lines in the diagram should then be regarded as surfaces seen edge on, and the points as lines seen end on. A is a fixed point but P is any point in the plane. The points A and P are joined by the arbitrary lines AQP, ARP. For an incompressible fluid flowing across the region shown, the volume rate of flow across AQP into the space $AQPRA$ must equal that out across ARP. Whatever the shape of the curve ARP, the rate of flow across it is the same as that across AQP; in other words, the rate of flow across the curve ARP depends only on the end points A and P. Since A is fixed, the rate of flow across ARP is a function only of the position of P. This function is known as the *stream function*, ψ (Greek letter 'psi'). The value of ψ at P therefore represents the volume flow rate across any line joining P to A at which point ψ is arbitrarily zero. (A may be at the origin of coordinates, but this is not necessary.)

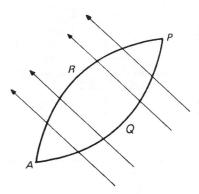

Fig. 9.2

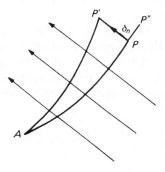

Fig. 9.3

If a curve joining A to P' is considered (Fig. 9.3), PP' being along a streamline, then the rate of flow across AP' must be the same as across AP since, by the definition of a streamline, there is no flow across PP'. The value of ψ is thus the same at P' as at P and, since P' was taken as *any* point on the streamline through P, it follows that ψ is constant along a streamline. Thus the flow may be represented by a series of streamlines at equal increments of ψ, like the contour lines on a map.

Now consider another point P'' in the plane, such that PP'' is a small distance δn perpendicular to the streamline through P, and $AP'' > AP$. The volume flow rate across the line AP'' is greater than that across AP by the increment $\delta \psi$ across PP''. If the average velocity perpendicular to PP'' (i.e. in the direction of the streamline at P) is q, then $\delta \psi = q \, \delta n$ and, as $\delta n \to 0$,

$$q = \partial \psi / \partial n \qquad (9.2)$$

Equation 9.2 indicates that the closer the streamlines for equal increments of ψ, the higher the velocity. It also shows that the position of A is immaterial. Any constant may be added to ψ without affecting the value of the velocity. Thus the equation $\psi = 0$ may be assigned to any convenient streamline. To determine in which direction the velocity is to be considered positive a sign convention is needed. The

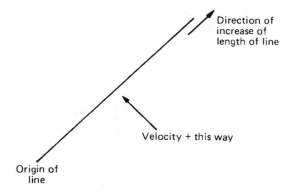

Direction of
increase of
length of line

Velocity + this way

Origin of
line

Fig. 9.4

usual convention is that a velocity is considered positive if, across the line drawn outwards from the fixed point A, the flow is from right to left. Thus the positive direction of a velocity normal to a line is obtained by turning 90° anticlockwise from the direction in which the line increases in length (Fig. 9.4).

The partial derivative of the stream function with respect to any direction gives, by eqn 9.2, the component of velocity at 90° anti-clockwise to that direction. Thus we may obtain expressions for the velocity components u and v:

$$u = -\frac{\partial \psi}{\partial y}; \quad v = +\frac{\partial \psi}{\partial x} \qquad (9.3)$$

An observer standing on the y axis and facing the direction of increase of y would see the component u, which is parallel to Ox, crossing the y axis from left to right. That is why the minus sign appears in the expression for u. On the other hand, an observer standing on the x axis and facing the direction of increase of x would see the component v crossing the x axis from right to left. Hence v has a plus sign. (Some writers have used the opposite convention.)

The above definition of ψ comes solely from kinematic considerations (together with the assumed incompressibility of the fluid). Whatever assumption may be made about the form of the function ψ, eqn 9.3 will truly describe the fluid motion. Whether such a motion is possible *under steady conditions*, however, is another matter and this depends on dynamic considerations (Section 9.7.4).

9.3 CIRCULATION AND VORTICITY

Across any line AP in the fluid the volume flow rate $= \int_A^P q_n \, ds$ where ds *Circulation*
represents the length of an infinitesimal element of the curve joining A and P, and q_n the component of the velocity perpendicular to it. Similarly we may consider the integral $\int_A^P q_s \, ds$ where q_s represents the component of velocity *along* an element of the curve AP wholly in the fluid.

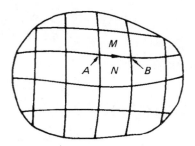

Fig. 9.5

If the integration is performed round a fixed closed circuit, i.e. $\oint q_s \mathrm{d}s$, the anticlockwise direction by convention being considered positive, the result is termed the *circulation* of that circuit. The circle on the integral sign indicates that the integration is performed once round the circuit in the anticlockwise direction. The word 'circulation' is here used in a special sense which is strictly defined by the integral $\oint q_s \mathrm{d}s$; it should not be thought that any particle necessarily 'circulates' round the circuit – after all, the velocity may be zero at some parts of the circuit.

The symbols Γ (Greek capital 'gamma') and K are both used for circulation: we shall use Γ.

The circulation round a large circuit equals the sum of the circulations round component small circuits (provided that the boundaries of all circuits are wholly in the fluid). This is readily demonstrated by Fig. 9.5. Suppose that a large circuit is subdivided into any number of smaller ones of which M and N are typical. Suppose that along the common boundary of circuits M and N the velocity is in the direction shown. Then the integral $\int_A^B q_s \, \mathrm{d}s$ along this common boundary makes an anticlockwise (i.e. positive) contribution to the circulation round circuit M but a clockwise (i.e. negative) contribution to that round circuit N. All such contributions from common boundaries therefore cancel in the total, which then consists only of the circulation round the periphery.

As an example of an elementary circuit arising from the subdivision of a larger one we consider the elementary rectangle, $\delta x \times \delta y$ in size, of Fig. 9.6. The velocities along the sides have the directions and average values shown. Starting at the lower left-hand corner we may add together the products of velocity and distance along each side, remembering that circulation is considered positive anticlockwise.

$$\text{Circulation, } \Gamma = u\delta x + \left(v + \frac{\partial v}{\partial x}\delta x \right)\delta y - \left(u + \frac{\partial u}{\partial y}\delta y \right)\delta x - v\delta y$$

$$= \frac{\partial v}{\partial x}\delta x \delta y - \frac{\partial u}{\partial y}\delta y \delta x \tag{9.4}$$

Vorticity

Now the *vorticity* at a point is defined as the ratio of the circulation round an infinitesimal circuit there to the area of that circuit (in the case of two-dimensional plane flow). Thus, from eqn 9.4,

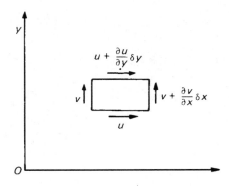

Fig. 9.6

$$\text{Vorticity, } \zeta \text{ (Greek 'zeta')} = \frac{\text{Circulation}}{\text{Area}} = \frac{\partial v}{\partial x} - \frac{\partial u}{\partial y} \qquad (9.5)$$

(In the more general case of three-dimensional flow the expression 9.5 represents only a component of the vorticity. Vorticity is a vector quantity whose direction is perpendicular to the plane of the small circuit round which the circulation is measured.)

Consider alternatively a small circular circuit of radius r (Fig. 9.7):

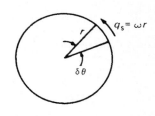

Fig. 9.7

$$\Gamma = \oint q_s \, ds = \oint \omega r \, r \, d\theta = r^2 \oint \omega \, d\theta = r^2 \, \overline{\omega} \, 2\pi$$

where $\overline{\omega}$ is the mean value of the angular velocity ω about the centre for all particles on the circle.

$$\zeta = \frac{\Gamma}{\text{Area}} = \frac{r^2 \overline{\omega} 2\pi}{\pi r^2} = 2\overline{\omega} \qquad (9.6)$$

That is, the vorticity at a point is twice the mean angular velocity of particles at that point.

If the vorticity is zero at all points in a region (except certain special points, called *singular* points, where the velocity or the acceleration is theoretically zero or infinite) then the flow in that region is said to be *irrotational*. Flow in regions where the vorticity is other than zero is said to be *rotational*. In practice there may be rotational motion in one part of a flow field and irrotational motion in another part.

The concept of irrotational flow lies behind much of what follows and we may here pause to consider its physical interpretation. Irrotational flow is flow in which each element of the moving fluid undergoes no *net* rotation (with respect to chosen coordinate axes) from one instant to another. A well-known example of irrotational motion is that of the carriages of the Big (Ferris) Wheel in a fairground: although each carriage follows a circular path as the wheel revolves, it does not rotate with respect to the earth – the passengers fortunately remain upright and continue to face in the same direction.

Irrotational flow

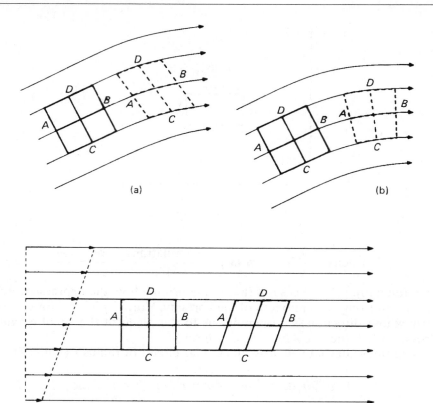

Fig. 9.8

Fig. 9.9

Two examples of fluid flow are depicted in Fig. 9.8. A small element of fluid is here represented by a quadrilateral with axes AB and CD. At (a) the axis AB rotates clockwise as the element moves along, but CD rotates an equal amount anticlockwise so that the *net* rotation is zero. Although undergoing distortion, the element is thus not rotating; the flow is irrotational. At (b), however, both axes rotate in the same direction: there is thus rotation but little distortion.

For an axis in the element originally parallel to Ox and of length δx, the angular velocity (anti-clockwise) is given by $(\partial v / \partial x)\partial x \div \partial x = \partial v / \partial x$. Similarly the angular velocity (anti-clockwise) of an axis originally parallel to Oy and of length δy is given by $-\partial u / \partial y$. The average angular velocity is therefore $\frac{1}{2}\{(\partial v / \partial x) - (\partial u / \partial y)\}$ which, from eqn 9.5, equals half the vorticity. The vorticity may in fact be alternatively defined as the algebraic sum of the angular velocities of two (momentarily) perpendicular line elements that move with the fluid and intersect at the point in question.

Rotation of a fluid particle can be caused only by a torque applied by shear forces on the sides of the particle. Such forces, however, are impossible in an ideal, inviscid, fluid and so the flow of an ideal fluid is necessarily irrotational. Where viscous forces are active, rotation of particles is, with one exception, inevitable. (The only possible instance of irrotational flow in the presence of viscous forces is the circulating flow round a circular cylinder rotating in an infinite expanse of fluid.) For example, in two-dimensional parallel shear flow, as illustrated in Fig. 9.9, the velocity over the cross-section is non-uniform. The movement of a fluid element then involves both its distortion and its rotation – although the fluid as a whole is not whirling about a fixed centre.

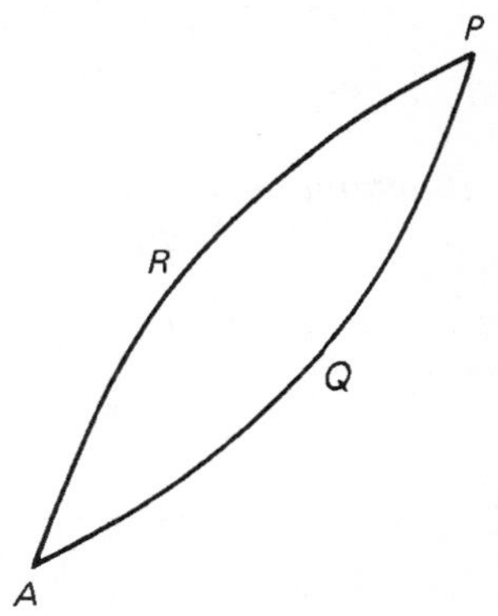

9.4 VELOCITY POTENTIAL

When the flow is irrotational the value of the integral $\int_A^P q_s \, ds$ (where q_s represents the component of velocity along an element ds of the curve AP) is independent of the path between A and P. For example, the circulation round the circuit $AQPRA$ of Fig. 9.10, wholly occupied by the fluid, equals

Fig. 9.10

$$\int_{A \atop Q}^{P} q_s \, ds + \int_{P \atop R}^{A} q_s \, ds = \int_{A \atop Q}^{P} q_s \, ds - \int_{A \atop R}^{P} q_s \, ds$$

For irrotational flow the circulation is zero and so

$$\int_{A \atop Q}^{P} q_s \, ds = \int_{A \atop R}^{P} q_s \, ds.$$

In other words, the value of the integral depends only on the position of P relative to A.

Putting

$$-\phi = \int_A^P q_s \, ds \qquad (9.7)$$

we have $\delta\phi = -q_s \, \delta s$ or $q_s = -\partial\phi/\partial s$. (The reason for the minus sign will appear in a moment.) The function ϕ is termed the *velocity potential*. If the line element, of length δs, is perpendicular to a streamline, then $q_s = 0$ and so $\delta\phi = 0$. Thus the velocity potential is constant along lines perpendicular to streamlines. These lines of constant velocity potential are known as *equipotential lines*. The velocity potential provides an alternative means of expressing the velocity components parallel to the coordinate axes in irrotational flow:

$$u = -\frac{\partial\phi}{\partial x}; \quad v = -\frac{\partial\phi}{\partial y} \qquad (9.8)$$

The minus signs in these equations arise from the convention that the velocity potential decreases in the direction of flow just as electrical potential decreases in the direction in which current flows. (Some writers adhere to the opposite convention by which $\delta\phi = q_s \, \delta s$.) The analogy with electrical potential should not, however, be pushed further than this. Velocity potential is simply a quantity defined mathematically by eqn 9.7. It is not a physical quantity that can be directly measured.

Its zero position, like that of the stream function, may be arbitrarily chosen.

Potential flow

However, whereas the stream function applies to both rotational and irrotational flow, it is only in irrotational flow that velocity potential has any meaning. This is because it is only in irrotational flow that the value of $\int_A^P q_s \, ds$ is independent of the path traversed from A to P. For this reason irrotational flow is often termed *potential flow*.

If the expressions for u and v from eqn 9.8 are substituted into the continuity relation 9.1 we obtain

$$\frac{\partial^2 \phi}{\partial x^2} + \frac{\partial^2 \phi}{\partial y^2} = 0 \qquad (9.9)$$

This is the two-dimensional form of Laplace's equation which finds application in many branches of science. All flows that conform with the principle of continuity therefore satisfy Laplace's equation if they are irrotational.

Similar substitutions in the expression for vorticity (9.5) yield

$$\zeta = \frac{\partial}{\partial x}\left(-\frac{\partial \phi}{\partial y}\right) - \frac{\partial}{\partial y}\left(-\frac{\partial \phi}{\partial x}\right) = -\frac{\partial^2 \phi}{\partial x \partial y} + \frac{\partial^2 \phi}{\partial y \partial x}$$

Since $\partial^2 \phi/\partial x \partial y = \partial^2 \phi/\partial y \partial x$ the vorticity must be zero when a velocity potential ϕ exists.

Any function ϕ that satisfies Laplace's equation (9.9) is a velocity potential of a possible flow. There is, however, an infinite number of solutions of Laplace's equation, and the actual ϕ for a particular flow is determined by the condition that at stationary boundaries the velocity *normal* to the surface must be zero. (For an ideal fluid there is no restriction on the velocity *tangential* to the surface.)

For irrotational flow the stream function ψ also satisfies Laplace's equation as may be seen by substituting the expressions 9.3 into the expression for vorticity (9.5) and equating to zero. Flows that satisfy Laplace's equation in ψ are irrotational ones; those that do not are rotational. The fact that for irrotational flow both ψ and ϕ satisfy Laplace's equation indicates that the ψ and ϕ of one pattern of flow could be interchanged to give ϕ and ψ of another pattern.

■

Example 9.1 In a two-dimensional flow field, the velocity components are given by

$$u = A\left(x^2 - y^2\right), \text{ and } \quad v = 4xy$$

where A is a constant.

(a) Determine the value of A.
(b) Find the stream function of the flow.

(c) Show that the flow is irrotational, and hence find the velocity potential.

Solution

(a) $\dfrac{\partial u}{\partial x} = \dfrac{\partial}{\partial x}\left[A\left(x^2 - y^2\right)\right] = 2Ax$, and $\dfrac{\partial v}{\partial y} = \dfrac{\partial}{\partial y}\left(4xy\right) = 4x$

The continuity condition must be satisfied. Hence

$$\frac{\partial u}{\partial x} + \frac{\partial v}{\partial y} = 0$$

Substituting:

$$2Ax + 4x = 0, \quad \therefore A = -2$$

(b) $$u = -\frac{\partial \psi}{\partial y} \quad \text{and} \quad v = \frac{\partial \psi}{\partial x}$$

Integrating each expression in turn:

$$\psi = -\int u\,dy + f_1(x) = -\int\left[-2\left(x^2 - y^2\right)\right]dy + f_1(x)$$
$$= 2\left(x^2 y - y^3/3\right) + f_1(x)$$

and $\quad \psi = \int v\,dx + f_2(y) = \int 4xy\,dx + f_2(y) = 2x^2 y + f_2(y)$

These two relations for ψ must be identical so we deduce that

$$\psi = 2x^2 y - \frac{2}{3}y^3 + \text{const.}$$

(c) $\dfrac{\partial v}{\partial x} = \dfrac{\partial}{\partial x}\left(4xy\right) = 4y \quad \text{and} \quad \dfrac{\partial u}{\partial y} = \dfrac{\partial}{\partial y}\left[-2\left(x^2 - y^2\right)\right] = 4y$

For irrotational flow

$$\frac{\partial v}{\partial x} - \frac{\partial u}{\partial y} = 0$$

Substituting:

$$4y - 4y = 0 \quad \text{QED.}$$

$$u = -\frac{\partial \phi}{\partial x}$$

Hence

$$\phi = -\int u\,dx + f_3(y) = 2x^3/3 - 2y^2 x + f_3(y)$$

and

$$v = -\frac{\partial \phi}{\partial y}$$

Hence

$$\phi = -\int v \mathrm{d}y + f_4(x) = -2xy^2 + f_4(x)$$

Comparing these two relations, we deduce

$$\phi = \frac{2}{3}x^3 - 2xy^2 + \text{const.}$$

☐

9.5 FLOW NETS

For any two-dimensional irrotational flow of an ideal fluid, two series of lines may be drawn: (1) lines along which ψ is constant, i.e. streamlines, and (2) lines along which ϕ is constant (equipotential lines). The latter, as we have seen, are perpendicular to the streamlines; thus lines of constant ψ and lines of constant ϕ together form a grid of quadrilaterals having 90° corners. This grid is known as a *flow net*, and it provides a simple yet valuable indication of the flow pattern.

It is customary to draw the streamlines at equal increments of ψ and the equipotential lines at equal increments of ϕ. Then, since $q = \partial \psi / \partial n$, the higher the velocity the closer the streamlines. Also, since $q = -\partial \phi / \partial s$, the higher the velocity the closer the equipotential lines. If Δs is made equal to Δn (Fig. 9.11) then in the limit as $\Delta s = \Delta n \to 0$ the quadrilaterals become perfect squares – except where q is zero or (theoretically) infinity. For increasing velocities the quadrilaterals become smaller; for decreasing velocities they become larger.

A flow net may be drawn for any region of two-dimensional flow. Fixed solid boundaries, since they have no flow across them, correspond to streamlines and, between these, other streamlines can be sketched in by guesswork. A set of smooth equipotential lines is then drawn so as to intersect the streamlines (including the fixed boundaries) perpendicularly and so spaced as to form approximate squares throughout the entire network.

The larger the intervals the more the quadrilaterals (in general) depart from perfect squares. In any event, it is usually found that equipotential lines which produce good 'squares' in one part of the pattern do not produce good 'squares' in another part when they are extended so as to intersect the streamlines perpendicularly. This failure arises from errors in the original, guessed, positions of the streamlines. Successive adjustments to both sets of lines are then made so as to improve the 'squareness' of the quadrilaterals. Particularly where the boundaries have abrupt changes of direction, good 'squares' may be difficult to obtain, but subdivision into smaller 'squares' may facilitate

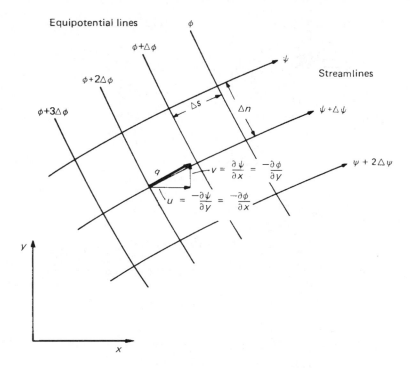

Fig. 9.11 Elements of a flow net.

the drawing of the net. The number of lines drawn depends on the accuracy required and the effort one is prepared to spend: the smaller the intervals between lines the more work is involved but the greater the accuracy.

For high accuracy a computer would be used, but the simple graphical technique is often valuable for obtaining approximate results quickly. Practice in the art of drawing flow nets enables good results to be achieved more rapidly, although even the most expert make much use of an eraser. Large scale diagrams on thick paper are clearly an advantage.

For a given set of boundary conditions there is only one possible pattern of the flow of an ideal fluid, and a correctly drawn flow net satisfying these conditions will represent this pattern. Velocities at any point may be deduced by the spacing of the streamlines; pressure variations may then be calculated from Bernoulli's equation. A few examples of flow nets are shown in Fig. 9.12.

9.5.1 Flow nets applied to real fluids

Although a flow net may be drawn for any arrangement of boundaries, it will not represent the actual pattern of flow unless the flow is everywhere irrotational. In practice, regions of rotational motion may be caused where the flow 'separates' from the boundary surface. Two examples of separation are indicated in Fig. 9.13. At point A the

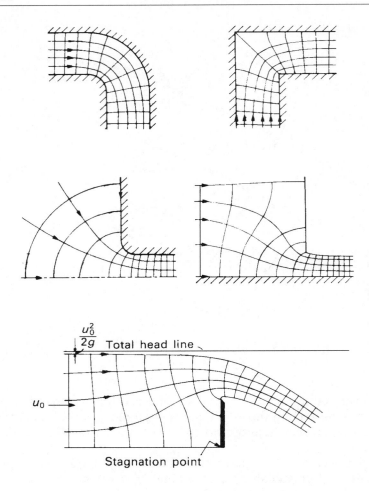

Fig. 9.12 Flow nets.

boundary abruptly turns towards the fluid, and the flow net indicates that the spacing of the streamlines would remain finite at this point even if the spacing elsewhere were infinitesimal. In other words, the velocity at A would be zero, and the point is therefore known as a *stagnation point*.

At point B, on the other hand, where the boundary abruptly turns away from the flow, the spacing of the streamlines would tend to zero while the spacing elsewhere was still finite. That is, the velocity would be theoretically infinite. Such a condition is physically impossible, and in fact the flow separates from the boundary so that a finite velocity may prevail.

An abrupt convex corner (that is, one with zero radius of curvature) would thus cause the flow of even an ideal fluid to leave the boundary. In a real fluid, as we saw in Section 8.8, the flow tends to leave a boundary whenever divergence of the streamlines is appreciable (as at the approach to a stagnation point or following a convex turn of the boundary). When separation occurs in practice, zones exist in which the fluid particles rotate and make no contribution to the main flow. Here,

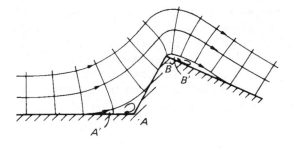

Fig. 9.13

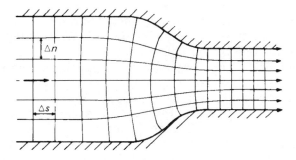

Fig. 9.14

then, the assumption of irrotational flow on which the flow net is based is no longer valid.

If a flow net is used to examine conditions in which the velocity adjacent to the boundary tends to increase and the streamlines therefore converge, it will provide a good approximation to the actual pattern of flow. But where a flow net suggests that the velocity should decrease along the boundary a real fluid is unable to obey, because the velocity at the boundary is already zero. The real fluid therefore separates from the boundary so that flow past that section may continue.

Once separation has occurred, a flow net drawn on the assumption that the solid boundaries are the outer streamlines no longer fits the facts. For example, the flow net of Fig. 9.14 describes the flow from left to right with reasonable accuracy. If, however, the flow were from right to left the same flow net pattern would be obtained but the streamlines would then diverge, not converge. As a result, considerable separation would in practice occur: the flow would continue past the enlargement with the streamlines still substantially parallel – even though the corners are well rounded – and thus quite fail to follow the boundaries. The more rapidly streamlines converge the better does a flow net represent the actual flow; if the streamlines diverge appreciably – that is, if the fluid undergoes an appreciable deceleration – then the flow net bears little relation to the actual flow.

If the edge of the separation zone can be defined, a flow net may still be used to indicate the flow outside that zone by treating the edge of the zone (instead of the solid boundary) as the outer streamline. In some instances it is possible to do this. Even when this cannot be done, however, the net, although not representing the actual flow, may use-fully indicate the regions in which separation may be expected. Thus it is valuable in indicating how boundaries may be 'streamlined' to reduce the chances of separation. If the boundary surface can be adjusted to follow the edges of the zones of separation, and thus to 'fill in' these zones, the separation can be eliminated, the flow pattern improved and the dissipation of energy reduced.

9.6   COMBINING FLOW PATTERNS

If two or more flow patterns are combined, the resultant flow pattern is described by a stream function that at any point is the algebraic sum of the stream functions of the constituent flows at that point. By this principle complicated motions may be regarded as combinations of simpler ones.

Figure 9.15 shows two sets of streamlines, ψ_1 etc. and ψ_2 etc., with intersections at P, Q, R, S etc. Consider the same arbitrary increment $\Delta\psi$ between successive streamlines of each set. Then, if A is some suitable reference point at which both stream functions are zero, the volume flow rate across any line AP is, by the definition of stream function, $\psi_1 + \Delta\psi + \psi_2$. Similarly, the volume flow rate across any line AQ is $\psi_1 + \psi_2 + \Delta\psi$. Consequently, both P and Q lie on the resultant streamline $\psi_1 + \psi_2 + \Delta\psi$ (here shown dotted). The points R and S likewise lie on the resultant line $\psi_1 + \psi_2 + 2\Delta\psi$.

The resultant flow pattern may therefore be constructed graphically simply by joining the points for which the total stream function has the same value. Care is necessary, however, in observing the sign conven-tion: flow is considered positive when it crosses a line such as AP from right to left (from the point of view of an observer looking from A towards P) and each $\Delta\psi$ must be an increment, not a decrement. Thus, if facing downstream, one would always see ψ increasing towards one's right.

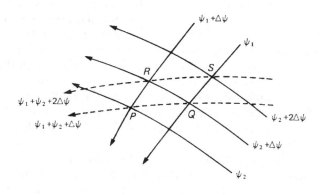

Fig. 9.15

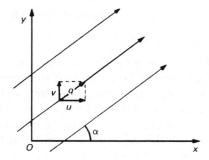

Fig. 9.16

This method was first described by W. J. M. Rankine (1820–72). The velocity components of the resultant motion are given by the algebraic sums of those for the constituent motions:

$$u = -\frac{\partial}{\partial y}(\psi_1 + \psi_2) = -\frac{\partial \psi_1}{\partial y} - \frac{\partial \psi_2}{\partial y} = u_1 + u_2 \quad \text{and similarly for } v.$$

For irrotational flows the velocity potentials are similarly additive: since Laplace's equation is linear in ϕ, then if ϕ_1 and ϕ_2 are each solutions of the equation, $(\phi_1 + \phi_2)$ is also a solution.

We shall now investigate a few examples of simple flow patterns and those combinations of them that lead to results of practical interest. For describing the flow patterns we shall mostly use the stream function ψ, but it should be remembered that for irrotational flows the velocity potential ϕ could be used for the same purpose.

9.7   BASIC PATTERNS OF FLOW

9.7.1   Rectilinear flow

For uniform flow with velocity q at an angle α to the x axis (as in Fig. 9.16), $u = q \cos \alpha$ and $v = q \sin \alpha$, the angle being considered positive in the anticlockwise direction from Ox.

$$\psi = \int \frac{\partial \psi}{\partial x} dx + \int \frac{\partial \psi}{\partial y} dy$$

$$= \int v \, dx + \int (-u) dy \quad \text{(from eqn 9.3)}$$

$$= vx - uy + C \quad \left(\text{since } u \text{ and } v \text{ are here constant}\right)$$

The integration constant C may take any convenient value. For uniform flow parallel to Ox, the streamline along Ox may be designated $\psi = 0$ and then, since $v = 0$ and $C = 0$, the flow is described by the equation

$$\psi = -uy \tag{9.10}$$

or, expressed in polar coordinates where $y = r\sin\theta$,

$$\psi = -ur\sin\theta \qquad (9.11)$$

Equation 9.10 is that of the streamline at a distance y above the x axis and $-uy$ is the volume rate of flow (for unit distance perpendicular to the x–y plane) between that streamline and the x axis where $\psi = 0$. If the integration constant C were not taken as zero this result would be unaffected: at distance y above Ox the value of ψ would be $-uy + C$ and along Ox ψ would be C; the difference would still be $-uy$. We shall therefore usually set integration constants equal to zero in subsequent integrations of $\mathrm{d}\psi$.

Similarly, if the flow is irrotational

$$\phi = \int \frac{\partial\phi}{\partial x}\,\mathrm{d}x + \int \frac{\partial\phi}{\partial y}\,\mathrm{d}y = -\int u\,\mathrm{d}x - \int v\,\mathrm{d}y = -ux - vy + C$$

and where $v = 0$ and $C = 0$, $\phi = -ux$.

9.7.2 Flow from a line source

A *source* is a point from which fluid issues uniformly in all directions. If, for two-dimensional flow, the flow pattern consists of streamlines uniformly spaced and directed radially outwards from one point in the reference plane (as in Fig. 9.17), the flow is said to emerge from a *line source*. (We remember that for two-dimensional flow what appears on the diagram as a point is to be regarded as a line seen end on.) The *strength m* of a source is the total volume rate of flow from it, the line source of two-dimensional flow being considered of unit length. The velocity q at radius r is given by

Volume rate of flow $\div$ Area perpendicular to velocity $= m/2\pi r$

for unit depth, since the velocity is entirely in the radial direction. As $r \to 0$, $q \to \infty$ and so no exact counterpart of a source is found in practice. However, except at the 'singular point' $r = 0$, a similar flow pattern would be achieved by the uniform expansion of a circular cylinder forcing fluid away from it, or, more approximately, by the uniform emission of fluid through the walls of a porous cylinder. Nevertheless, the concept of a source has its chief value in providing a basic mathematical pattern of flow which, as we shall see, can be combined with other simple patterns so as to describe flows closely resembling those found in practice.

In cases such as this, where circular symmetry is involved, polar coordinates are more suitable. The velocity q_r radially outwards $= \partial\psi/\partial n = -\partial\psi/r\,\partial\theta$. The minus sign arises from the 'right-to-left' convention and the fact that the angle θ is considered positive anti-clockwise; the length $\int r\,\mathrm{d}\theta$ of a circumferential line therefore increases in the anti-clockwise direction, and to an observer on this line and facing that direction, the radially outward flow would appear to move from left to

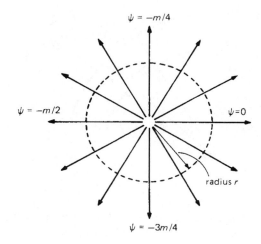

$\psi = -m/4$

$\psi = -m/2$

$\psi = 0$

radius r

$\psi = -3m/4$

Fig. 9.17

right, i.e. in the negative direction. A tangential velocity q_t is given by $\partial\psi/\partial r$; if one looked outwards along the radius one would see a positive (i.e. anticlockwise) tangential velocity coming from one's right and so the sign here is positive.

For flow from a source at the origin

$$q_t = \partial\psi/\partial r = 0 \text{ and } q_r = -(\partial\psi/r\partial\theta) = m/2\pi r$$

whence $\psi = -m\theta/(2\pi)$ where θ is in radian measure and taken in the range $0 \leqslant \theta < 2\pi$. [Also $-\partial\phi/\partial r = q_r = m/2\pi r$ and $-\partial\phi/r\,\partial\theta = q_t = 0$, so

$$\phi = -(m/2\pi)\ln(r/C)]$$

The streamlines are thus lines of constant θ, i.e. radii, and, for irrotational flow, the ϕ lines are concentric circles.

9.7.3 Flow to a line sink

A *sink*, the exact opposite of a source, is a point to which the fluid converges uniformly and from which fluid is continuously removed. The strength of a sink is considered negative, and the expressions for velocities and the functions ψ and ϕ are therefore the same as those for a source but with the signs reversed.

9.7.4 Irrotational vortex

A flow pattern in which the streamlines are concentric circles is known as a plane circular vortex. The particles of fluid moving in these concentric circles may, or may not, rotate on their own axes. If they do not, that is, if the flow is irrotational, the vortex is known as an irrotational or 'free' vortex, and it is this type which is our present concern. We shall therefore examine the conditions under which flow in a circular path is irrotational.

Free vortex

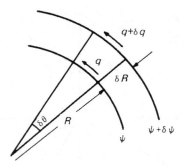

Fig. 9.18

Figure 9.18 shows a small element in the reference plane bounded by two streamlines and two radii. The velocities, considered positive anticlockwise, are q and $q + \delta q$ along the streamlines as shown, the velocity perpendicular to the streamlines of course being zero. The circulation Γ (positive anticlockwise) round this small element is therefore

$$(q + \delta q)(R + \delta R)\delta\theta - qR\delta\theta = (R\delta q + q\delta R)\delta\theta$$

higher orders of small magnitudes being neglected. Therefore the vorticity

$$\zeta = \frac{\text{Circulation}}{\text{Area}} = \frac{(R\delta q + q\delta R)\delta\theta}{R\delta\theta\,\delta R} = \frac{q}{R} + \frac{\delta q}{\delta R}$$

$$= \frac{q}{R} + \frac{\partial q}{\partial R} \quad \text{as } \delta R \to 0 \tag{9.12}$$

In this expression for vorticity, R represents the radius of curvature of the streamlines, not necessarily the polar coordinate.

For irrotational flow,

$$\frac{q}{R} + \frac{\partial q}{\partial R} = 0 \tag{9.13}$$

Now for a circular vortex, since the streamlines are concentric circles, the cross-sectional area of a stream-tube is invariant along its length. Continuity therefore requires the velocity q to be constant along each streamline. In other words, the velocity varies only with R. The partial derivative $\partial q/\partial R$ is then equal to the full derivative dq/dR and eqn 9.13 may then be written $dq/dR = -q/R$ which on integration gives

$$qR = \text{constant} \tag{9.14}$$

This equation describes the variation of velocity with radius in a 'free', i.e. irrotational, vortex. As $R \to 0$, $q \to \infty$ which in practice is impossible. In a real fluid, friction becomes dominant as $R = 0$ is approached, and so fluid in this central region tends to rotate like a solid body, that is, with velocity proportional to radius, and eqn 9.14 does not then apply. (Another possibility is that the centre may be occupied by a solid body or another fluid.) This discrepancy as $R \to 0$ does not render the theory of the irrotational vortex useless, however, for in most practical problems our concern is with conditions away from the central core.

The circulation round a circuit corresponding to a streamline of an irrotational vortex $= \Gamma = q \times 2\pi R$. Since $qR = $ constant (eqn 9.14) the circulation is also constant for the entire vortex. Thus the circulation round an infinitesimal circuit about the centre is this same non-zero constant, and the vorticity at the centre is therefore not zero. The 'free' vortex, then, although irrotational everywhere else, has a rotational core at the centre. The centre is a special, 'singular', point at which the velocity is theoretically infinite and so the equations do not necessarily apply there.

For a vortex centred at the origin of coordinates,

$$\psi = \int \frac{\partial \psi}{\partial r} dr + \int \frac{\partial \psi}{\partial \theta} d\theta = \int q \, dr + 0 = \int \frac{\Gamma}{2\pi r} dr = \frac{\Gamma}{2\pi} \ln\left(\frac{r}{r_0}\right) \quad (9.15)$$

where r_0 represents the radius at which (arbitrarily) $\psi = 0$. The constant Γ is known as the *vortex strength*.

We may note in passing that if the streamlines and equipotential lines for a source (Fig. 9.17) are interchanged, the flow pattern for an irrotational vortex is obtained, the ψ lines of one flow becoming the ϕ lines of the other.

At this point a further property of irrotational flow may be considered. The equation of motion will be derived for the small element of fluid between two streamlines shown in Fig. 9.19, the flow being assumed steady. At radius R from the centre of curvature the pressure is p, at radius $R + \delta R$ it is $p + \delta p$; consequently there is a net thrust on the element, towards the centre of curvature, equal to

$$\left(p + \delta p\right)\left(R + \delta R\right)\delta\theta - pR\delta\theta - 2\left(p + \frac{\delta p}{2}\right)\delta R \sin\frac{\delta\theta}{2}$$

(for unit thickness perpendicular to the diagram). As $\delta\theta \to 0$, $\sin(\delta\theta/2) \to \delta\theta/2$ and, with higher orders of small magnitudes neglected, the thrust reduces to $R\,\delta p\,\delta\theta$. (Shear forces, even if present, have no component perpendicular to the streamlines.)

The component of the weight acting radially inwards

$$= R\delta\theta\delta R\varrho g\left(\delta z/\delta R\right) = R\varrho g\,\delta\theta\,\delta z$$

where δz is the vertical projection of δR so that $\arccos(\delta z/\delta R)$ is the angle between the radius and the vertical. Thus the total inward force is

$$R\,\delta p\,\delta\theta + R\varrho g\,\delta\theta\,\delta z = \text{Mass} \times \text{Centripetal acceleration}$$

$$= \varrho R\,\delta\theta\,\delta R\left(q^2/R\right)$$

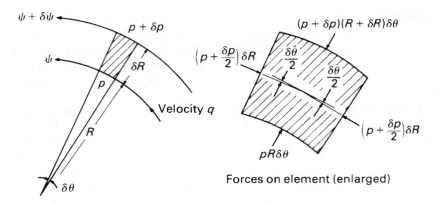

Fig. 9.19

Forces on element (enlarged)

Division by $R\varrho g\,\delta\theta$ gives

$$\frac{\delta p}{\varrho g} + \delta z = \frac{q^2}{R}\frac{\delta R}{g} \qquad (9.16)$$

Now, by Bernoulli's theorem for steady flow of a frictionless fluid:

$$\frac{p}{\varrho g} + \frac{q^2}{2g} + z = H$$

where H is a constant along a streamline (although it may differ from one streamline to another). Differentiation gives

$$\frac{\delta p}{\varrho g} + \frac{2q\,\delta q}{2g} + \delta z = \delta H \qquad (9.17)$$

Combining eqns 9.16 and 9.17 we obtain

$$\delta H = \frac{q\,\delta q}{g} + \frac{q^2}{Rg}\frac{\delta R}{} = \frac{q}{g}\left(\frac{\delta q}{\delta R} + \frac{q}{R}\right)\delta R$$

But $q\,\delta R = \delta\psi$ and from eqn 9.12

$$\frac{\delta q}{\delta R} + \frac{q}{R} = \zeta$$

$$\therefore \ \delta H = \zeta\frac{\delta\psi}{g} \qquad (9.18)$$

Since both H and ψ are constant along any streamline, the space between two adjacent streamlines corresponds to fixed values of δH and $\delta\psi$. By eqn 9.18 the vorticity ζ must also have a fixed value between the two streamlines and thus, in the limit, it is constant along a streamline. This, then, is the condition for *steady* motion: the particles retain their vorticity unchanged as they move along the streamline.

If in addition the motion is irrotational, then $\zeta = 0$ and hence $\delta H = 0$. That is, if steady flow is also irrotational, H is constant not only along a single streamline but over the whole region of flow.

Our deduction of the properties of an irrotational vortex has, it is true, been based on the supposition of an ideal fluid. Nevertheless, the motion of real fluids may closely approximate to an irrotational vortex. For example, fluid moving round a bend in a pipe or a channel tends to follow this pattern. If frictional effects are neglected and, upstream of the bend, the flow is along straight and parallel streamlines with steady, uniform velocity, then the Bernoulli constant H is the same for all streamlines. It of course remains so as the fluid moves into the bend, that is $\delta H = 0$; eqn 9.18 then shows that $\zeta = 0$ and so, by eqns 9.13 and 9.14, $qR = $ constant.

The net force on any element of fluid in a vortex towards the axis requires a decrease of piezometric pressure in that direction. This has consequences that can often be seen. There is, for example, the fall in the liquid surface when a vortex forms at the outlet of a bath. A similar effect occurs in whirlpools. The vortices shed from the wing tips of aircraft have a reduced temperature at the centre in addition to a lower pressure (since air is a compressible fluid), and, under favourable atmospheric conditions, water vapour condenses in sufficient quantity to form a visible 'vapour trail'. A large irrotational vortex in the atmosphere is known as a tornado: over land the low pressure at the centre causes the lifting of roofs of buildings and other damage; over water it produces a waterspout. In practice, the reduction of piezometric pressure at the centre usually causes some radial flow in addition. This is so, of course, for the 'drain-hole' vortex (see Section 9.8.8).

In an ideal fluid an irrotational vortex is permanent and indestructible. Tangential motion can be given to the fluid only by a tangential force, and this, in an ideal fluid, is impossible. Thus a complete irrotational vortex (i.e. one in which the streamlines are complete circles) could not be brought into being, although if it could once be established it could not then be stopped because that too would require a tangential force. In a real fluid, however, vortices are formed as a consequence of viscosity, and are eventually dissipated by viscosity. In an ideal fluid, a vortex cannot have a free end in the fluid because that would involve a discontinuity of pressure. (This is a consequence of Stokes's Theorem – see, for example, Duncan et al.[1], p. 78.) It must either terminate at a solid boundary or a free surface (like that of the 'drain-hole' vortex), or form a closed loop (like a smoke ring).

Example 9.2   A ventilating duct, of square section with 200 mm sides, includes a 90° bend in which the centre-line of the duct follows a circular arc of radius 300 mm. Assuming that frictional effects are negligible and that upstream of the bend the air flow is completely uniform, determine the way in which the velocity varies with radius in the bend. Taking the air density as constant at 1.22 kg/m³, determine the mass flow rate when a water U-tube manometer connected between the mid-points of the outer and inner walls of the bend reads 11.5 mm.

Solution

If upstream flow is uniform, all streamlines have the same Bernoulli constant and if friction is negligible all streamlines retain the same Bernoulli constant.

$$\therefore \quad \frac{\partial}{\partial R}\left(p^* + \frac{1}{2}\varrho q^2\right) = 0$$

$$\therefore \quad \varrho q \frac{\partial q}{\partial R} = -\frac{\partial p^*}{\partial R} = -\frac{\varrho q^2}{R} \left(\text{by eqn 9.16}\right)$$

$$\therefore \quad \frac{\mathrm{d}q}{q} = -\frac{\mathrm{d}R}{R}$$

which on integration gives $qR = \text{constant} = C$, the equation for a free vortex.

$$\frac{\partial p^*}{\partial R} = \frac{\varrho q^2}{R} = \frac{\varrho C^2}{R^3}$$

and integration gives

$$p_A^* - p_B^* = -\frac{1}{2}\varrho C^2 \left(\frac{1}{R_A^2} - \frac{1}{R_B^2}\right)$$

$$1000 \frac{\mathrm{kg}}{\mathrm{m}^3} \times 9.81 \frac{\mathrm{N}}{\mathrm{kg}} \times 0.0115\,\mathrm{m}$$

$$= -\frac{1}{2} \times 1.22 \frac{\mathrm{kg}}{\mathrm{m}^3} C^2 \left(\frac{1}{0.4^2} - \frac{1}{0.2^2}\right)\mathrm{m}^{-2}$$

whence $C = 3.141\,\mathrm{m}^2/\mathrm{s}$

$$\therefore \quad \text{Mass flow rate} = \varrho Q = \varrho(0.2\,\mathrm{m})\int_{0.2\,\mathrm{m}}^{0.4\,\mathrm{m}} q\,\mathrm{d}R$$

$$= \varrho C(0.2\,\mathrm{m})\int_{0.2\,\mathrm{m}}^{0.4\,\mathrm{m}} \frac{\mathrm{d}R}{R} = \varrho C(0.2\,\mathrm{m})\ln 2$$

$$= 1.22 \frac{\mathrm{kg}}{\mathrm{m}^3} 3.141 \frac{\mathrm{m}^2}{\mathrm{s}} 0.2\,\mathrm{m} \times \ln 2 = 0.531\,\mathrm{kg/s}$$

$\square$

9.7.5 Forced (rotational) vortex

This type of motion is obtained when all particles of the fluid have the same angular velocity about some fixed axis. That is, the fluid rotates about that axis like a solid body. Because an external torque is required to start the motion, the term 'forced vortex' has been used, although 'rigid-body rotation' might be preferable. Such motion may be produced by rotating about its axis a cylinder containing the fluid. Alternatively, rotation of a paddle in the fluid will produce forced vortex motion within its periphery (although beyond the periphery conditions are more nearly those of an irrotational vortex). Once steady conditions are established, there is no relative motion between the fluid particles and thus no shear forces exist, even in a real fluid. The velocity at radius R from the centre is given by ωR where ω represents the (uniform) angular velocity. Substituting $q = \omega R$ in the centripetal force equation (9.16) gives

$$\frac{\delta p}{\varrho g} + \delta z = \omega^2 R \frac{\delta R}{g}$$

from which integration yields

$$\frac{p}{\varrho g} + z = \frac{\omega^2 R^2}{2g} + \text{constant}$$

i.e. $\quad p* = \dfrac{\varrho \omega^2 R^2}{2} + \text{constant}$ \hfill (9.19)

where $p* = p + \varrho g z$.

Thus the piezometric pressure $p*$ increases with radius. Fluid may be supplied to the centre of a forced vortex and then ejected at the periphery at a much higher pressure. This principle is the basis of operation of the centrifugal pump, which will be considered in Chapter 13. When the discharge valve of the pump is closed, the blades of the impeller cause the fluid to rotate with substantially the same angular velocity ω as the pump shaft, and the increase of piezometric pressure from the inlet to outlet radius is given by

$$p_2^* - p_1^* = \frac{\varrho \omega^2}{2}\left(R_2^2 - R_1^2\right)$$

If a forced vortex is produced in a liquid in an open container, the pressure at the free surface of the liquid is atmospheric and therefore constant. Thus, for the free surface, $z = \omega^2 R^2/2g + \text{constant}$. If $z = z_0$ when $R = 0$ then

$$z - z_0 = \omega^2 R^2/2g$$

and if R is perpendicular to z (i.e. if the axis of rotation is vertical) the surface is a paraboloid of revolution (Fig. 9.20).

That the forced vortex is rotational is readily seen from the general expression (9.12) for vorticity where $q/R + (\partial q/\partial R) = \omega + \omega = 2\omega$ which is not zero. The fluid particles rotate about their own axes. Consequently, although the Bernoulli constant H is fixed for any particular streamline, it varies from one streamline to another. Bernoulli's equation could therefore be applied between points on the same streamline, but *not* between points on different streamlines.

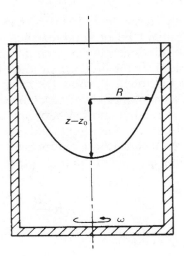

Fig. 9.20 Forced vortex about vertical axis formed in liquid in open container.

■ **Example 9.3** The axis of a closed cylindrical drum is vertical and its internal dimensions are: diameter 400 mm, height 560 mm. A small vertical filling tube (open to atmosphere) is connected to the centre of the top of the drum and the drum is filled entirely with an oil of density 900 kg/m³ to a point in the filling tube 40 mm above the inner surface of the top of the drum. Inside the drum and concentric with it is a set of paddles 200 mm diameter. What is the maximum speed at which the paddles may rotate about their vertical axis if the pressure in the oil is nowhere to exceed 150 kPa (gauge)? It may be assumed that all the oil within the central 200 mm diameter rotates as a forced vortex and that the remainder moves as a free vortex.

Solution
Since $\partial p^*/\partial r$ is always >0, maximum p^* is at greatest radius and maximum p is at greatest depth. Let ω = angular velocity of paddles. Then in the central forced vortex

$$p^* = \frac{1}{2}\varrho\omega^2 R^2 + C \qquad (\text{eqn } 9.19)$$

Let $z = 0$ on bottom of drum. Then, at $R = 0$ on bottom

$$p = \varrho g h = 900\frac{\text{kg}}{\text{m}^3}9.81\frac{\text{N}}{\text{kg}}(0.56 + 0.04)\text{m} = 5297\,\text{Pa} = C$$

$\therefore$ At $z = 0$ and $R = 0.1\,\text{m}$ $\left(\text{the outer edge of forced vortex}\right)$

$$p = \frac{1}{2} \times 900\frac{\text{kg}}{\text{m}^3}\omega^2(0.1\,\text{m})^2 + 5297\,\text{Pa} = \left(4.5\frac{\text{kg}}{\text{m}}\right)\omega^2 + 5297\,\text{Pa}$$

For free vortex $qR = $ constant $= K$ but there can be no discontinuity of velocity and so at $R = 0.1\,\text{m}$ velocities in forced and free vortices are the same.

$$\therefore\ K = \left(\omega 0.1\,\text{m}\right)0.1\,\text{m} = \left(0.01\,\text{m}^2\right)\omega$$

For any type of fluid motion in a circular path

$$\frac{\partial p^*}{\partial R} = \varrho\frac{q^2}{R} \qquad \left(\text{eqn } 9.16\right)$$

$\therefore$ In this free vortex

$$\frac{\partial p^*}{\partial R} = \varrho\frac{K^2}{R^3}$$

$$\text{whence } p^* = -\frac{\varrho K^2}{2R^2} + D \text{ where } D = \text{constant}$$

Where forced and free vortices join there can be no discontinuity of pressure. Hence at $z = 0$ and $R = 0.1\,\text{m}$

$$p = \left(4.5\frac{\text{kg}}{\text{m}}\right)\omega^2 + 5297\,\text{Pa} = -900\frac{\text{kg}}{\text{m}^3}\frac{\left(0.01\,\text{m}^2\right)^2\omega^2}{2\left(0.1\,\text{m}\right)^2} + D$$

$$\text{whence } D = \left(9.0\frac{\text{kg}}{\text{m}}\right)\omega^2 + 5297\,\text{Pa}$$

So, at $z = 0$ and $R = 0.2\,\text{m}$ (outer edge of drum)

$$p = D - \frac{\varrho K^2}{2R^2} = \left(9.0\frac{\text{kg}}{\text{m}}\right)\omega^2 + 5297\,\text{Pa} - 900\frac{\text{kg}}{\text{m}^3}\frac{\left(0.01\,\text{m}^2\right)^2\omega^2}{2\left(0.2\,\text{m}\right)^2}$$

$$= 150\,000\,\text{Pa}$$

$\therefore\ \omega = 135.6\,\text{rad/s}$, i.e. $21.57\,\text{rev/s}$

In practice the free vortex in the outer part of the motion would be modified because the velocity of a real fluid would have to be zero at the wall of the drum. $\qquad\square$

9.8 COMBINATIONS OF BASIC FLOW PATTERNS

9.8.1 Uniform rectilinear flow and line source

For simplicity we take a source (of strength m) at the origin of coordinates and combine its flow pattern with that of uniform flow at velocity U parallel to the line $\theta = 0$. The fluid is assumed to extend to infinity in all directions. The resulting streamline pattern is that of Fig. 9.21. From the source, the velocity radially outwards, $m/(2\pi r)$, decreases with increasing radius and so at some point to the left of O this velocity is exactly equal and opposite to that of the uniform stream on which it is superimposed. Hence the resultant velocity there is zero and the point (S) is consequently known as a *stagnation point*. At this point $m/(2\pi r) = U$, whence $r = m/(2\pi U)$. Fluid issuing from the source is unable to move to the left beyond S, and so it diverges from the axis $\theta = \pi$ and is carried to the right.

In two-dimensional flow a streamline represents a surface (viewed edge on) along which the velocity must everywhere be tangential; hence there can be no flow perpendicular to the surface. So, for steady two-dimensional flow of a frictionless fluid, any streamline may be supplanted by a thin, solid barrier. In particular, the resultant streamline diverging from S may be considered the barrier dividing the previously uniform stream from the source flow. Since, however, the flows on the two sides of this barrier do not interact, the pattern of streamlines outside the barrier is identical with that which would be obtained if the barrier were the contour of a solid body with no flow inside it. Thus the source is simply a hypothetical device for obtaining the form of the contour of the body deflecting the originally uniform flow.

Adding the stream functions for the uniform flow and the source we obtain, for the resultant flow,

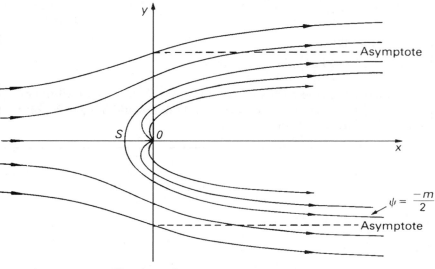

Fig. 9.21   At $r = \infty$, $u = U$ and $v = 0$.

$$\psi = -Uy + \left(-\frac{m\theta}{2\pi}\right) = -Ur \sin\theta - \frac{m\theta}{2\pi}$$

At the stagnation point, $\theta = \pi$; thus the value of ψ there is $-m/2$ and this value must be that everywhere along the streamline corresponding to the contour of the body. The contour is therefore defined by the equation

$$-Uy - \frac{m\theta}{2\pi} = -\frac{m}{2}$$

It extends to infinity towards the right, the asymptotic value of y being given by $m/2U$ (when $\theta \to 0$) or $-m/2U$ (when $\theta \to 2\pi$).

The velocity components at any point in the flow are given by

$$q_t = \partial\psi/\partial r = -U \sin\theta$$

and

$$q_r = -\partial\psi/r\ \partial\theta = +U \cos\theta + \left(m/2\pi r\right)$$

With $q^2 = q_t^2 + q_r^2$ pressures may then be calculated from Bernoulli's equation.

The body whose contour is formed by the combination of uniform rectilinear flow and a source is known as a *half body*, since it has a nose but no tail. It is a useful concept in studying the flow at the upstream end of symmetrical bodies long in comparison with their width – such as struts and bridge piers. The shape of the half body may be altered by adjusting the strength of the (imaginary) source in relation to U or, more generally, by the use of sources of various strengths at other positions along the axis to produce a contour with the desired equation. The upper part of the pattern could be regarded as representing the flow of liquid in a channel over a rise in the bed, the flow of wind over a hillside, or the flow past a side contraction in a wide channel.

Example 9.4 Air flowing past a wall encounters a step of height $h = 10\,\text{mm}$, the leading edge of which is profiled to avoid sharp changes of flow direction, as shown. The flow may be represented by the upper half-plane of a uniform flow of velocity U parallel to the x-axis and a line source of strength m. If the velocity of the airstream is $40\,\text{m}\,\text{s}^{-1}$ determine

(a) the strength of the line source
(b) the distance s the line source is located behind the leading edge of the step

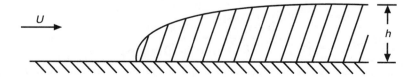

(c) the horizontal and vertical velocity components at a point on the step 5 mm above the initial wall surface.

Solution

By reference to Figure 9.21, the step may be modelled by the streamline passing through S, which is given by $\psi = -m/2$. From the text of Section 9.8.1, the stream function of the flow is

$$\psi = -Uy - \frac{m\theta}{2\pi} = -Ur\sin\theta - \frac{m\theta}{2\pi}$$

and the surface of the step is defined by

$$-Uy - \frac{m\theta}{2\pi} = -\frac{m}{2}$$

Also we note that the source is positioned at the origin of the coordinate system.

(a) The thickness of the step is defined at $\theta = 0$, where $y = h$. Hence

$$-Uh = -\frac{m}{2} \quad \text{or} \quad m = 2Uh = 2 \times 40\,\text{m/s} \times 0.01\,\text{m} = 0.8\,\text{m}^2\text{s}^{-1}$$

(b) $\quad s = \dfrac{m}{2\pi U} = \dfrac{0.8\,\text{m}^2/\text{s}}{2\pi \times 40\,\text{m/s}} = 3.18 \times 10^{-3}\,\text{m} = 3.18\,\text{mm}$

(c) The streamline defining the step is

$$\psi = -Uy - \frac{m\theta}{2\pi} = -Uh = -\frac{m}{2}$$

At $y = h/2$

$$-U\frac{h}{2} - \frac{m\theta}{2\pi} = -Uh \quad \text{or} \quad -U\frac{h}{2} - \frac{2Uh\theta}{2\pi} = -Uh$$

which simplifies to $\theta = \pi/2$, for which $x = 0$.

$$u = -\frac{\partial\psi}{\partial y} = -\frac{\partial}{\partial y}\left(-Uy - \frac{m\theta}{2\pi}\right) = -\frac{\partial}{\partial y}\left(-Uy - \frac{m\arctan(y/x)}{2\pi}\right)$$

$$= U + \frac{m}{2\pi}\frac{x}{x^2 + y^2} = 40\,\text{m/s} + \frac{0.8\,\text{m}^2/\text{s}}{2\pi}\frac{0\,\text{m}}{0\,\text{m}^2 + (0.005\,\text{m})^2}$$

$$= 40\,\text{m s}^{-1}$$

$$v = \frac{\partial\psi}{\partial x} = \frac{\partial}{\partial x}\left(-Uy - \frac{m\theta}{2\pi}\right) = \frac{\partial}{\partial x}\left(-Uy - \frac{m\arctan(y/x)}{2\pi}\right)$$

$$= \frac{m}{2\pi}\frac{y}{x^2 + y^2} = \frac{0.8\,\text{m}^2/\text{s}}{2\pi}\frac{0.005\,\text{m}}{0\,\text{m}^2 + (0.005\,\text{m})^2} = 25.5\,\text{m s}^{-1}$$

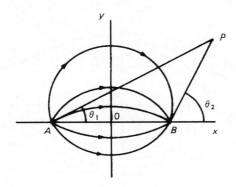

Fig. 9.22

9.8.2   Source and sink of numerically equal strength

The streamline pattern formed by this combination is shown in Fig. 9.22, the assumption again being made that the fluid extends to infinity in all directions. If the strength of the source (at A) is m and that of the sink (at B) is $-m$, then the stream function of the combined flow is

$$\psi = -\frac{m\theta_1}{2\pi} + \frac{m\theta_2}{2\pi} = \frac{m}{2\pi}(\theta_2 - \theta_1) \tag{9.20}$$

For any point P in the flow $|\theta_2 - \theta_1| = \angle APB$. Lines of constant ψ (i.e. streamlines) are therefore curves along which $\angle APB$ is constant, in other words, circular arcs of which AB is the base chord.

If A is at $(-b, 0)$ and B at $(b, 0)$ then

$$\tan\theta_1 = y/(x + b) \text{ and } \tan\theta_2 = y/(x - b)$$

$$\therefore\ \tan(\theta_2 - \theta_1) = \frac{\tan\theta_2 - \tan\theta_1}{1 + \tan\theta_2 \tan\theta_1} = \frac{y/(x - b) - y/(x + b)}{1 + \left\{y^2/(x^2 - b^2)\right\}}$$

$$= \frac{2by}{x^2 - b^2 + y^2}$$

and hence (from eqn 9.20)

$$\psi = \frac{m}{2\pi}\arctan\frac{2by}{x^2 - b^2 + y^2} \tag{9.21}$$

The angle is between 0 and π if $y > 0$, or between 0 and $-\pi$ if $y < 0$.

9.8.3   Source and sink of numerically equal strength, combined with uniform rectilinear flow

Uniform rectilinear flow with velocity U parallel to the line $\theta = 0$ may now be added to the source and sink combination of Section 9.8.2. The resultant stream function is

$$\psi = -Uy + \frac{m}{2\pi}(\theta_2 - \theta_1) = -Uy + \frac{m}{2\pi}\arctan\frac{2by}{x^2 - b^2 + y^2}$$

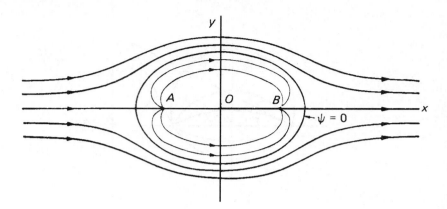

Fig. 9.23

With the source to the left of the origin a stagnation point is expected upstream of the source (as for the half body, Section 9.8.1) and a second stagnation point downstream of the sink. If a stagnation point is at a distance s from O along the x axis the combined velocity there is

$$U - \frac{m}{2\pi(s - b)} + \frac{m}{2\pi(s + b)} = 0$$

whence

$$s = \pm b \sqrt{\left(1 + \frac{m}{\pi U b}\right)}$$

At the stagnation points, $y = 0$ and $\theta_2 - \theta_1 = 0$ and so there $\psi = 0$. They therefore lie on the line $\psi = 0$ which is symmetrical about both axes and, as shown in Fig. 9.23, encloses all the streamlines running from the source to the sink. The line $\psi = 0$ is usually known as the *Rankine oval* after W. J. M. Rankine (1820–72) who first developed the technique of combining flow patterns. (N.B. It is *not* an ellipse.) Although $\psi = 0$ along this line, the velocity is not zero throughout its length. However, for frictionless flow, the contour of a solid body may be put in place of this oval streamline; the flow pattern outside the oval is therefore that of an originally uniform stream deflected by a solid body of that oval shape.

The shape of the solid boundary may be altered by varying the distance between source and sink or, as in the case of the half body, by varying the value of m relative to U – the source and sink are, after all, quite hypothetical. Other shapes may be obtained by the introduction of additional sources and sinks along the x axis, although the total strength of these must remain zero. Rankine developed ship contours in this way.

Example 9.5 A Rankine oval of length L and breadth B is produced in an otherwise uniform stream of velocity $5\,\text{m s}^{-1}$, by a source and sink 75 mm apart. Determine L if $B = 125$ mm.

Solution

The surface of the Rankine oval is defined by $\psi = 0$. As the oval is symmetrically disposed about the y-axis, its maximum thickness occurs when $x = 0$. Write $t = B/2$. Hence

$$0 = -Ut + \frac{m}{2\pi}\arctan\frac{2tb}{-b^2 + t^2}$$

which can be written

$$\frac{2\pi Ut}{m} = \arctan\frac{2tb}{t^2 - b^2}$$

Substituting, $b = 37.5$ mm, $t = 62.5$ mm, $U = 5\,\text{m s}^{-1}$. Hence

$$m = 2\pi Ut / \arctan\frac{2tb}{t^2 - b^2} = \frac{2\pi \times 5\,\text{m/s} \times 0.0625\,\text{m}}{1.081}$$

$$= 1.816\,\text{m}^2\text{s}^{-1}$$

$$L = 2s = 2b\left(1 + \frac{m}{\pi U b}\right)^{1/2}$$

$$= 2 \times 37.5\,\text{mm} \times \left(1 + \frac{1.816\,\text{m}^2/\text{s}}{\pi \times 5\,\text{m/s} \times \left(3.75 \times 10^{-2}\,\text{m}\right)}\right)^{1/2}$$

$$= 152\,\text{mm}$$

9.8.4 The doublet

If, in the pattern illustrated in Fig. 9.22, the source and sink are moved indefinitely closer together but the product $m \times 2b$ is maintained finite and constant, the resulting pattern is said to correspond to a *doublet* or *dipole*. The angle APB becomes zero and the streamlines become circles tangent to the x axis. (This line, joining source and sink, is known as the axis of the doublet and is considered positive in the direction sink to source.)

From eqn 9.21, as $2b \to 0$,

$$\psi \to \frac{m}{2\pi}\left(\frac{2by}{x^2 - b^2 + y^2}\right) \to \frac{Cy}{x^2 + y^2} = \frac{Cr\sin\theta}{r^2} = \frac{C\sin\theta}{r} \quad (9.22)$$

where $C = \text{constant} = mb/\pi$ and r, θ are polar coordinates.

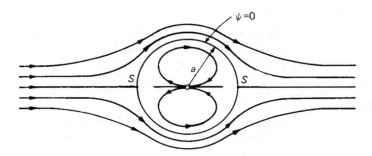

Fig. 9.24

9.8.5 Doublet and uniform rectilinear flow

If a doublet at the origin with its axis in the $-x$ direction is combined with uniform rectilinear flow in the $+x$ direction, the resultant stream function is

$$\psi = -Uy + \frac{C \sin \theta}{r} = -Ur \sin \theta + \frac{C \sin \theta}{r} \qquad (9.23)$$

This is a limiting case of the combination discussed in Section 9.8.3: when the source and sink merge to form a doublet the Rankine oval becomes a circle. Equation 9.23 shows that the streamline $\psi = 0$ is found when $\theta = 0$, $\theta = \pi$ or $C = Ur^2$. In other words, $\psi = 0$ along the x axis and where

$$r = \sqrt{(C/U)} = \text{constant.}$$

With the substitution

$$C/U = a^2 \qquad (9.24)$$

eqn 9.23 becomes

$$\psi = -U\left(r - \frac{a^2}{r}\right) \sin \theta \qquad (9.25)$$

This equation represents the pattern formed when, in an ideal fluid of infinite extent, an originally uniform steady flow, parallel to the x axis, is deflected by a circular cylinder of radius a with its axis at the origin (Fig. 9.24). The velocity at any point in the flow may be expressed in terms of its radial and tangential components:

$$q_r = -\frac{1}{r}\frac{\partial \psi}{\partial \theta} = U\left(1 - \frac{a^2}{r^2}\right) \cos \theta$$

$$q_t = \frac{\partial \psi}{\partial r} = -U\left(1 + \frac{a^2}{r^2}\right) \sin \theta$$

In particular, at the surface of the cylinder $r = a$, so $q_r = 0$ (of course) and

$$q_t = -2U \sin \theta \qquad (9.26)$$

Stagnation points (S, S) occur at $\theta = 0$ and $\theta = \pi$, and the velocity at the surface has a maximum magnitude at $\theta = \pi/2$ and $\theta = 3\pi/2$.

The distribution of pressure round the cylinder may also be determined. For the frictionless, irrotational flow considered, Bernoulli's equation gives

$$p_\infty^* + \frac{1}{2}\varrho U^2 = p^* + \frac{1}{2}\varrho q^2$$

Here p_∞^* represents the piezometric pressure far upstream where the flow is unaffected by the presence of the cylinder. So, substituting from eqn 9.26, we obtain

$$p^* = p_\infty^* + \frac{1}{2}\varrho U^2 - 2\varrho U^2 \sin^2 \theta \qquad (9.27)$$

This expression is independent of the sign of $\sin\theta$ and so the variation of p^* with θ is symmetrical about both x and y axes. Consequently, the net force exerted by the fluid on the cylinder in any direction is zero (apart from a possible buoyancy force corresponding to the ϱgz part of p^*). This result is to be expected from the symmetry of the flow pattern. Although the fluid exerts a thrust on the upstream half of the cylinder, it exerts an equal and opposite force on the downstream half. As the fluid is supposed ideal no tangential forces can be exerted, and so the net force on the cylinder is zero. Indeed, by more complicated mathematics, it may be shown that an ideal fluid exerts zero net force (apart from buoyancy) on a body of any shape wholly immersed in it.

d'Alembert's Paradox

This result conflicts with practical experience, and the contradiction is known as *d'Alembert's Paradox* after J. R. d'Alembert* (1717–83) who first obtained the mathematical result. It constituted, for a long time, a great stumbling-block in the development of classical hydrodynamics. As we have seen (Section 8.8.2), in a real fluid flowing past a circular cylinder, viscous action causes the flow to separate from the downstream surface of the cylinder and form a wake in which the pressure variation differs from the theoretical, as shown in Fig. 9.25. It should not be concluded, however, that the study of the flow of an ideal fluid is entirely useless. Using the concept of the boundary layer we may combine the results for an ideal fluid with boundary-layer theory in a very useful manner.

Virtual mass

The flow pattern shown in Fig. 9.24 is the steady one which would be seen by an observer at rest relative to the cylinder. A cylinder moving through fluid otherwise at rest would present the same steady pattern to an observer moving with the cylinder, but a different, unsteady, one to an observer at rest. This latter pattern may be deduced by

* Pronounced *Dal-arm-bair'*.

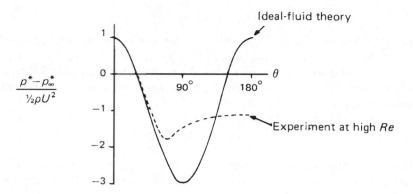

Fig. 9.25

superimposing on the former a uniform velocity of $-U$. The relative velocities and the pressures, of course, remain unchanged. The uniform flow ingredient of the steady pattern is thereby nullified, and there remains simply that part of the doublet pattern outside the cylinder (i.e. for $r > a$).

Now for this doublet pattern

$$\psi = \frac{C \sin \theta}{r} = \frac{Ua^2}{r} \sin \theta \quad \left(\text{from eqns 9.22 and 9.24}\right)$$

$$q_r = -\frac{1}{r}\frac{\partial \psi}{\partial \theta} = -\frac{Ua^2}{r^2} \cos \theta \quad \text{and} \quad q_t = \frac{\partial \psi}{\partial r} = -\frac{Ua^2}{r^2} \sin \theta$$

$$\therefore \; q^2 = q_r^2 + q_t^2 = U^2 a^4 / r^4$$

Thus the magnitude of the velocity varies only with r. In an annular element of radius r and width δr, the kinetic energy is $\frac{1}{2}(2\pi r \, \delta \varrho) U^2 a^4 / r^4$ per unit axial length of the cylinder. The total kinetic energy per unit length in the whole pattern is therefore

$$\int_a^\infty \pi \varrho U^2 a^4 \frac{dr}{r^3} = \frac{1}{2}\pi \varrho U^2 a^2 = \frac{1}{2}M'U^2 \qquad (9.28)$$

where $M' = \varrho \pi a^2$. When the cylinder is set in motion it has to be given kinetic energy but, in addition, the fluid that it displaces by its movement has to be given an amount of kinetic energy $\frac{1}{2}M'U^2$. The work done in accelerating the cylinder is therefore greater than if only the cylinder itself had to be moved, and the result is the same as if the mass of the cylinder were greater by an amount M'. This amount M' is known as the *added mass* or *induced mass* or *hydrodynamic mass* of the cylinder, and the sum of the actual mass and the added mass is known as the *virtual mass*. For the circular cylinder in unrestricted two-dimensional flow the added mass equals the mass of the fluid displaced by the cylinder, but this is not a general result applicable to bodies of any shape. For a sphere in unrestricted three-dimensional flow the added mass is half the mass of the fluid displaced by the sphere. For bodies not completely symmetrical the added mass depends on the direction of motion. Moreover, as the added mass depends on the flow pattern, it is affected by the presence of other boundaries.

Even where the flow pattern for a real fluid closely approximates to that of irrotational motion, this pattern is not attained immediately when the velocity of the body changes. Owing to viscosity, bodies accelerated in real fluids experience other effects, and measured values of virtual mass differ somewhat from those predicted by theory that assumes an ideal fluid. Nevertheless, added mass is frequently apparent in practice. For aircraft its effects are small because of the relatively small mass of the air displaced, but in the docking and mooring of balloons or ships it is important. And in walking through water one notices the greater force needed to accelerate one's legs.

9.8.6 Doublet, uniform rectilinear flow and irrotational vortex

A particularly useful combination of flow patterns is obtained when, to the pattern of Section 9.8.5, is added that of an irrotational vortex with its centre at the doublet. Since the streamlines of the vortex are concentric circles about the doublet they do not cut the cylindrical surface, and so their superposition on the flow pattern round the cylinder is valid. Adding the stream function for the irrotational vortex (eqn 9.15) to eqn 9.25 we obtain

$$\psi = -U\left(r - \frac{a^2}{r}\right)\sin\theta + \frac{\Gamma}{2\pi}\ln\left(\frac{r}{r_0}\right) \qquad (9.29)$$

as the stream function of the combined flow. (We recall that the value of r_0 is quite arbitrary: it simply determines the position of the line $\psi = 0$ for the vortex.)

Whereas the flow past the cylinder without the vortex motion is symmetrical (Fig. 9.24), the addition of the vortex increases the magnitude of the velocity on one side of the cylinder and reduces it on the other. The pattern is no longer completely symmetrical; in consequence, the distribution of pressure round the cylinder is not symmetrical, and there is a net transverse force. We shall now determine the velocity distribution, and from that the pressure distribution and the net force.

The tangential velocity $q_t = \partial\psi/\partial r = -U\{1 + (a^2/r^2)\}\sin\theta + \Gamma/(2\pi r)$ and at the surface of the cylinder, where the radial component is necessarily zero,

$$\left(q_t\right)_{r=a} = -2U\sin\theta + \frac{\Gamma}{2\pi a}$$

At the stagnation point, $q_t = 0$ and therefore $\sin\theta = \Gamma/(4\pi aU)$. For anticlockwise vortex motion, Γ is positive and, provided that $\Gamma/(4\pi aU) < 1$, the two stagnation points are moved above the x axis, although symmetry about the y axis is maintained as shown in Fig. 9.26a. When $\Gamma/(4\pi aU) = 1$ the two stagnation points merge at $\theta = \pi/2$ (Fig. 9.26b).

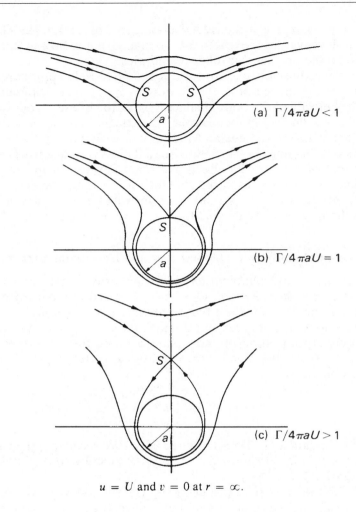

(a) $\Gamma/4\pi aU < 1$

(b) $\Gamma/4\pi aU = 1$

(c) $\Gamma/4\pi aU > 1$

Fig. 9.26

$u = U$ and $v = 0$ at $r = \infty$.

If $\Gamma/(4\pi aU) > 1$, q_t cannot be zero anywhere on the cylinder, and the stagnation point moves along the y axis out into the flow (Fig. 9.26c).

On the cylinder surface the velocity must be wholly tangential. The pressure is then given by Bernoulli's equation:

$$p = \text{constant} - \frac{1}{2}\varrho\left(q_t\right)^2_{r=a}$$

(If buoyancy is disregarded the z terms may be ignored.) Therefore

$$p = \text{constant} - \frac{1}{2}\varrho\left(4U^2 \sin^2\theta - \frac{2U\Gamma \sin\theta}{\pi a}\right)$$

the term independent of θ being incorporated in the constant. The first term in the bracket has the same value for θ as for $(\pi - \theta)$ and only the second term can contribute to the net force on the cylinder.

The force radially inwards on a small element of surface area is $pa\,\delta\theta$ per unit length of the cylinder. The components of the total force are:

$$F_x = -\int_0^{2\pi} pa \cos\theta \, d\theta$$

$$F_y = -\int_0^{2\pi} pa \sin\theta \, d\theta$$

The x-component is zero, as the symmetry of the flow pattern about the y-axis suggests. Remembering that only the last term in the expression for p will be effective in the integration, we obtain

$$F_y = -\frac{\varrho U\Gamma}{\pi} \int_0^{2\pi} \sin^2\theta \, d\theta = -\varrho U\Gamma \text{ per unit length of cylinder} \quad (9.30)$$

The minus sign is a consequence of our sign conventions for U and Γ. For instance, if the main flow is from left to right in a vertical plane (i.e. U is positive) and Γ is positive (i.e. anticlockwise), then F_y is negative, i.e. downwards. This is to be expected: the flow patterns combine to give a higher velocity below the cylinder than above it and consequently a lower pressure below than above, and a net downward force.

This transverse force is known as the *Magnus effect* after the German physicist H. G. Magnus (1802–70) who investigated it experimentally in 1852. For a given value of the circulation Γ, the force is independent of the radius a. Indeed, it was later shown by the German, M. Wilhelm Kutta (1867–1944) and the Russian, Nikolai E. Joukowski* (1847–1921), independently, that for a body of any shape in two-dimensional flow, the transverse force per unit length is $-\varrho U\Gamma$ in the plane of flow and is perpendicular to the direction of U. The result is therefore known as the Kutta–Joukowski law. It is one of the most useful results of ideal fluid flow theory.

Magnus effect

The Magnus effect is in no way dependent on viscosity. However, with a real fluid, a circulatory flow near the cylinder surface can be produced by rotation of the cylinder, since the latter drags a layer of fluid round with it. This kind of effect is largely responsible for the deflection of golf or tennis balls that are 'cut' or 'sliced' and therefore have a spin about a vertical axis – a phenomenon remarked on by Newton in 1672. (For spheres the phenomenon is now sometimes known as the Robins effect, after Benjamin Robins (1707–51) who in 1742 demonstrated the deflection of spinning musket balls.) A cricket ball given 'top spin' has a smaller velocity relative to the undisturbed air at the top than at the bottom, so there is a net downward force on the ball that keeps its trajectory low. The flow of air round balls is three-dimensional, so the expression 9.30 does not apply; moreover, viscous action to some degree invalidates Bernoulli's equation used in the derivation of the formula. The transverse force may also be modified by an unsymmetrical wake

Robins effect

* Pronounced *Zhoo-koff'-skee.*

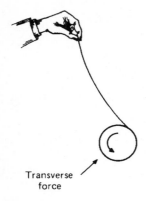

Transverse
force

Fig. 9.27

behind the ball. There is, however, no doubt of the reality of the Magnus effect. It may be simply demonstrated by holding the free end of a length of cotton coiled round a long cylinder of paper; when the cylinder is allowed to fall the resulting rotation produces a Magnus effect which causes the cylinder to be deflected from a vertical path, as shown in Fig. 9.27.

Experimental results for the transverse force on a rotating cylinder are markedly different from that given by eqn 9.30. Especially is this so if the cylinder is relatively short (with a length less than, say, 10 times the diameter). Then the flow across the ends from the high-pressure side to the low-pressure side has a considerable effect. This end flow may be largely eliminated by fitting flat discs, of about twice the cylinder diameter, to the ends. This does not completely solve the problem because, in a real fluid, the discs introduce other 'end effects'. In any event, the viscosity of a real fluid alters the flow pattern from that of an ideal fluid, so the effective circulation produced by the surface drag of a rotating cylinder is rarely more than half the theoretical value. The practical difficulties of reproducing the theoretical conditions are discussed in detail by W. M. Swanson[2].

The transverse force on a rotating cylinder has been used as a form of ship propulsion. The rotor-ship designed by Anton Flettner (1885–1961) about 1924 had large vertical cylinders on the deck. These were rapidly rotated and, being acted on by the natural wind, took the place of sails. However, although technically successful, the rotor-ship proved uneconomic. By far the most important application of the Kutta–Joukowski law is in the theory of the lift force produced by aircraft wings and the blades of propellers, turbines and so on. The law shows that, for a transverse force to be produced, a circulation round the body is always required, in addition to the velocity of translation. Some aspects of the application to aerofoils will be considered in Section 9.9.

9.8.7 Vortex pair

Two irrotational vortices with strengths equal in magnitude but opposite in sign (i.e. turning in opposite directions) constitute a *vortex pair*. The stream function of the combined flow is given by:

$$\psi = \frac{(-\Gamma)}{2\pi}\ln\frac{r_1}{r_0} + \frac{\Gamma}{2\pi}\ln\frac{r_2}{r_0} = \frac{\Gamma}{2\pi}\ln\frac{r_2}{r_1}$$

This equation yields the symmetrical pattern shown in Fig. 9.28a.

Each vortex is affected by the movement of the fluid due to the other, and each therefore moves in a direction perpendicular to the line joining their centres. From eqn 9.15 the velocity of each centre is given by $\Gamma/(2\pi\,2b)$. Figure 9.28a represents the pattern seen instantaneously by a stationary observer. However, an observer moving with the velocity of the vortex centres ($\Gamma/4\pi b$) would see the pattern of Fig. 9.28b.

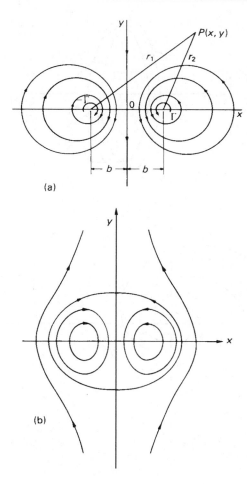

(a)

(b)

Fig. 9.28

To bring the vortex centres to rest relative to the observer, a uniform velocity of magnitude $\Gamma/4\pi b$ may be superimposed on the pattern in the opposite direction to the observer's motion. More generally, uniform rectilinear flow, at velocity V parallel to the y axis, combined with the vortex pair gives the stream function

$$\psi = \frac{\Gamma}{2\pi}\ln\frac{r_2}{r_1} + Vx = \frac{\Gamma}{2\pi}\ln\left\{\frac{(x-b)^2 + y^2}{(x+b)^2 + y^2}\right\}^{1/2} + Vx$$

It may readily be shown that, provided $V < \Gamma/\pi b$, two stagnation points are formed on the y axis, and a closed curve analogous to a Rankine oval (Section 9.8.3) is obtained. At those stagnation points, $x = 0$ and hence $\psi = 0$; thus the oval contour has the equation $\psi = 0$. This curve always encloses the same particles of fluid which move with the vortex pair.

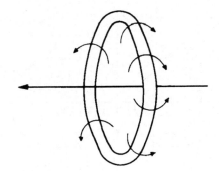

Fig. 9.29

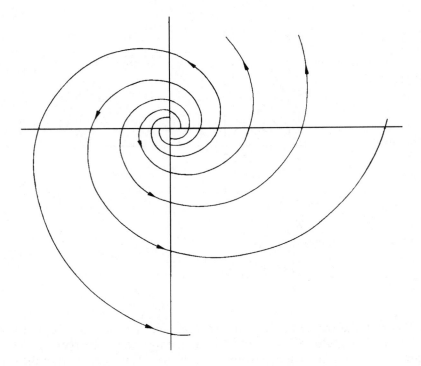

Fig. 9.30

A vortex pair may be seen when a flat blade such as a knife is held vertically in water and moved for a short distance perpendicular to its breadth. If it is then rapidly withdrawn from the water, a vortex pair, produced by friction at the blade edges, will advance as described above. A small dimple in the water surface marks the position of each vortex.

In Fig. 9.28b the central streamline along the y axis may be regarded as a solid boundary without the pattern on either side being affected. Therefore a single irrotational vortex close to a plane wall moves along it just as though another vortex were mirrored in the surface.

The action of one vortex on another is seen in a smoke ring: each element of the ring is affected by the velocity field of the other elements, and so the ring advances with uniform velocity as shown in Fig. 9.29.

9.8.8 Irrotational vortex and line source (spiral vortex)

This combination gives a resultant pattern in which flow moves outwards in a spiral path. The stream function is

$$\psi = \frac{\Gamma}{2\pi}\ln\frac{r}{r_0} - \frac{m\theta}{2\pi}$$

For any streamline, ψ = constant (say $C/2\pi$), whence $\ln (r/r_0) = (C + m\theta)/\Gamma$ which is the equation of a logarithmic spiral. This pattern of flow (Fig. 9.30) is of considerable practical importance in fluid machinery – for example, in the volutes of centrifugal pumps.

The converse combination of irrotational vortex and line sink describes the flow approaching the outlet in the base of a container (the 'drain hole' vortex) or, more approximately, that between the guide vanes and the runner of a Francis turbine.

9.9   ELEMENTARY AEROFOIL THEORY

9.9.1 Definitions

It will be useful at this point to define a few terms commonly used in reference to aerofoils.

Chord line: A straight line in the plane of the aerofoil cross-section, which serves as a datum. It is commonly taken as the line joining the centres of curvature of the leading (i.e. front) edge and trailing (i.e. rear) edge. (Although other definitions are sometimes used, this one is precise enough for our present purpose.) It is not necessarily an axis of symmetry.

Chord, c: The length of the chord line produced to meet the leading and trailing edges.

Span, b: The overall length of the aerofoil (in the direction perpendicular to the cross-section).

Plan area, S: The area of the projection of the aerofoil on a plane perpendicular to the section (or *profile*) and containing the chord line. For an aerofoil with a cross-section constant along the span, Plan area = Chord × Span.

Mean chord, $\bar{c} = S/b$

Aspect ratio, $A\!\!R$ or $A = Span/Mean\ chord = b/\bar{c} = b^2/S$

Lift, L: That component of the total aerodynamic force on the aerofoil, which is perpendicular to the direction of the oncoming fluid. Lift is not necessarily vertical.

Drag, D: That component of the total aerodynamic force on the aerofoil, which is parallel to the direction of the oncoming fluid.

Lift coefficient, $C_L = L/(\frac{1}{2}\varrho U^2 S)$

Drag coefficient, $C_D = D/(\frac{1}{2}\varrho U^2 S)$. In these expressions, which are both dimensionless, U represents the velocity (relative to the aerofoil) of the fluid far upstream.

Angle of attack or *Angle of incidence*, α: (Angle of attack is the preferable name, as 'angle of incidence' is sometimes used with other meanings.) The angle between the chord line and the direction of the oncoming fluid. More significantly, zero angle of attack is sometimes defined as that for which the lift is zero.

9.9.2 Aerofoils of infinite span

We have already noted that the Kutta–Joukowski law (eqn 9.30) is applicable to the two-dimensional flow of an ideal fluid round a body of any shape. Joukowski also showed that the pattern of flow round a circular cylinder could be used to deduce the pattern for a body of different (but mathematically related) shape. The mathematical process by which this may be done is known as *conformal transformation*. Details of the technique are outside the scope of this book; it may be said here, however, that geometrical shapes and patterns are transformed into other shapes and patterns, but the fact that a pattern represents the solution of a particular problem is preserved. Although straight lines are in general transformed into curves, angles at intersections are unchanged, and so are quantities such as circulation. The importance of the results for the circular cylinder, then, is that they form the starting point for transformations to flow patterns for other bodies.

Since the same basic principle governs the production of a transverse force on any body – aerofoil, hydrofoil, propeller blade, boat sail etc. – it is sufficient to consider as an example the wing of an aircraft in level flight. We assume for the moment that the wing is of infinite span; thus the flow at the section considered is two-dimensional, being unaffected by the transverse velocity components at the ends of the span. Now, for the flow past a circular cylinder without circulation, the stagnation points are at $\theta = 0$ and $\theta = \pi$ (Fig. 9.24). The corresponding transformation of this pattern to that for an aerofoil is shown in Fig. 9.31a. The exact positions of the stagnation points S_1 and S_2 on the aerofoil depend on the angle of attack with respect to the oncoming flow. To an ideal fluid, these positions would present no difficulty (provided that the radius of curvature of the trailing edge was not actually zero). But when a real fluid, having flowed along the underside of the aerofoil, is called

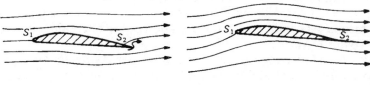

Fig. 9.31   (a)   (b)

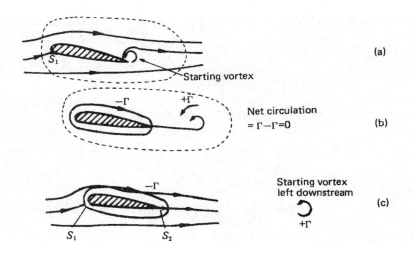

Net circulation
$= \Gamma - \Gamma = 0$ (b)

Starting vortex
left downstream   (c)

Fig. 9.32

upon to turn the sharp corner at the trailing edge in order to reach the stagnation point S_2, it cannot do so. This is because the adverse pressure gradient from the trailing edge to S_2 would cause the boundary layer to separate at the corner (as we saw in Section 8.8.1).

In fact, for a real fluid the only stable position for the stagnation point S_2 is at the trailing edge. As the motion begins, the condition illustrated in Fig. 9.31a lasts only for an instant. A shift of the point S_2 to the trailing edge corresponds to a shift of the downstream stagnation point in Fig. 9.24 from $\theta = 0$ to $\theta < 0$. To achieve this a negative (clockwise) circulation is required. As Fig. 9.26a shows, the forward stagnation point moves similarly, and the transformation of this revised pattern to that for the aerofoil shape is shown in Fig. 9.31b. Thus, for stable conditions, a circulation has to be established round the aerofoil, its magnitude being determined by the shift required to bring the stagnation point S_2 to the trailing edge.

This is Joukowski's 'stagnation hypothesis'. It tells us that a clockwise circulation is required, but not how it is generated. In an ideal fluid there is no process by which circulation can be generated – nor, for that matter, any process by which circulation, once established, can be changed. Frederick W. Lanchester (1878–1946) and Ludwig Prandtl (1875–1953) first explained the production of circulation round an aerofoil as follows.

Starting vortex

The initial separation of a real fluid at the trailing edge causes fluid on the upper surface to move from the stagnation point S_2 towards the edge. This flow is in the opposite direction to that of the ideal fluid, and consequently an eddy, known as the *starting vortex*, is formed, as shown in Fig. 9.32. This starting vortex is rapidly washed away from the edge but, in leaving the aerofoil, it generates an equal and opposite circulation round the aerofoil. The vortex which remains with the aerofoil is known as the bound vortex. In this way, the net circulation round the

Fig. 9.33 (Crown Copyright. Reproduced with permission).

dotted curve in Fig. 9.32 remains zero. This must be so to satisfy a theorem by William Thomson, later Lord Kelvin (1824–1907): in a frictionless fluid the circulation, around a closed curve that moves with the fluid so as always to enclose the same particles, does not change with time. (We recall that the viscosity of a fluid is in evidence only where velocity gradients are appreciable, and that elsewhere the mean motions of the fluid closely resemble the behaviour of an ideal fluid. The dotted curve lies in a region of essentially ideal flow.)

Kutta–Joukowski condition

The circulation round the aerofoil, produced as a reaction to the starting vortex, brings S_2 nearer to the trailing edge. Similar starting vortices produce increased circulation round the aerofoil until S_2 reaches the stable position at the edge (Fig. 9.32c). The condition for the circulation in ideal flow to bring S_2 to the trailing edge is known as the *Kutta–Joukowski condition*. (In practice, the circulation required to give stable conditions is slightly less than that for the Kutta–Joukowski condition because the boundary layer slightly alters the effective shape of the aerofoil.)

Experimental observations have amply confirmed the existence of starting vortices (Fig. 9.33). Whenever conditions are changed, either by an alteration in the upstream velocity U or in the angle of attack α, fresh vortices are formed, and the circulation round the aerofoil takes on a new value. Once they have left the aerofoil, the starting vortices have no further effect on the flow round it, and they are ultimately dissipated by viscous action.

Although the viscosity of a real fluid causes the formation of a starting vortex and thus the generation of circulation round the aerofoil, the transverse force, lift, is not greatly affected by the magnitude of the viscosity. If the angle of attack is small, so that flow does not separate appreciably from the upper surface, and if two-dimensional conditions are fulfilled, the measured lift forces for thin aerofoils agree remarkably well with the Kutta–Joukowski law, lift $= -\varrho U\Gamma$ per unit length.

9.9.3 Aerofoils of finite span

We have so far restricted our discussion to two-dimensional flow. This can occur only if the aerofoil has an infinite span or if it extends between parallel frictionless end walls. If, however, the span of the aerofoil is finite and the ends do not meet walls, then motion of the fluid takes place in the span-wise direction, and the effect of this additional motion is very important.

When a lift force is produced on the aerofoil the pressure on the underside is greater than that on the upper. (For the sake of definiteness we again consider the wing of an aircraft in level flight, the lift force then being upwards.) Consequently fluid escapes round the ends, and on the underside there is a flow outwards from the centre to the ends, while on the upper side flow occurs from the ends towards the centre. These movements, superimposed on the main flow, distort the overall pattern (Fig. 9.34). At the trailing edge, the fluid from the upper surface forms a surface of discontinuity with that from the underside. Vortices are set up, and the surface of discontinuity is a sheet of vortices – each vortex acting like a roller bearing between the upper and lower layers. Since the vortex sheet is unstable, however, the individual vortices entwine to form two concentrated vortices trailing from the aerofoil close to the tips (Fig. 9.34). On an aircraft wing, the reduction of pressure at the core of these tip vortices is accompanied by a reduction of temperature, and, under certain conditions of humidity and ambient temperature, atmospheric moisture condenses to form visible 'vapour trails' extending several kilometres across the sky. With ships' propellers the reduction of pressure at the tips of the blades may cause bubbles to form which are seen following spiral paths downstream.

The formation of tip vortices does not violate Thomson's theorem (Section 9.9.2) because these are of equal but opposite magnitude and

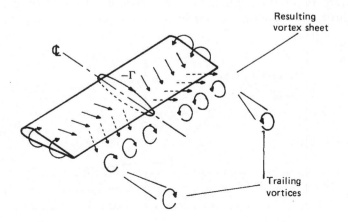

Resulting vortex sheet

$-\Gamma$

Trailing vortices

Fig. 9.34

the net circulation remains zero. Their existence in fact maintains the property of a vortex that it cannot terminate in the fluid but only at a solid boundary. The circulation round the aerofoil, which produces the lift, derives from the bound vortex whose axis is along the span. There is no solid boundary at the ends of the aerofoil and so the circulation cannot stop there. It continues in the tip vortices. These, in turn, connect with the starting vortex downstream so that there is a complete vortex ring. In a real fluid, of course, the starting vortex and the downstream ends of the tip vortices are extinguished by viscous action, and only the bound vortex and the forward ends of the tip vortices persist, forming a so-called horseshoe vortex.

The pressure difference between top and bottom of an unbounded aerofoil must decrease to zero at the ends; consequently the lift per unit length of span also decreases to zero there. In practice, then, neither the lift nor the circulation is uniformly distributed along the span, and the variation approximates to a semi-ellipse (Fig. 9.35). The philosophical difficulty of having a non-constant circulation along the aerofoil was resolved by Prandtl. He supposed the circulation to be the result of adding separate vortices of different lengths, each having its own pair of trailing vortices in the vortex sheet.

9.9.4 Vortex drag

The tip vortices induce a downward component of velocity, known as the *downwash velocity*, in the fluid passing over the aerofoil. When the downwash velocity v_i is combined with the velocity of the approaching fluid the *effective* angle of attack is altered to α_e as shown in Fig. 9.36. If the aerofoil is now treated as one of infinite span but set at the effective angle of attack α_e, the lift force is given by $L_e = -\varrho U_e \Gamma$ per unit length of span, and is perpendicular to U_e. The force L_e may be resolved into two perpendicular components: the useful lift L, normal to U, and a component D_i, parallel to U in the rearward direction, called the *vortex drag*, formerly called the *induced drag*.

In a real fluid an aerofoil is subject to a drag force even if the flow is completely two-dimensional. One contribution to this drag force is made directly by viscous action in the boundary layer; another arises

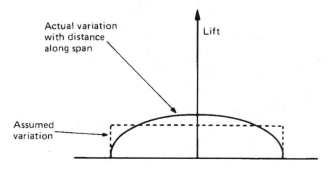

Fig. 9.35

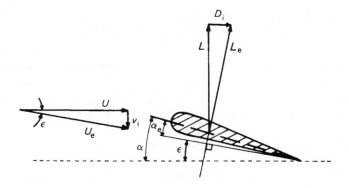

Fig. 9.36

because the flow separates to form a turbulent wake at the rear, and the flow pattern then differs from that for an ideal fluid. Vortex drag, however, is additional to these and, as it depends entirely on the downwash velocity induced by the tip vortices, it would be obtained with an ideal fluid. Drag in an ideal fluid may seem paradoxical: the work done against the vortex drag appears as the kinetic energy of the fluid in the tip vortices which are continuously left behind.

From similar triangles in Fig. 9.36

$$D_i/L_e = v_i/U_e \tag{9.31}$$

Now Prandtl showed that for an aerofoil producing an elliptical distribution of lift (to a close approximation this is true for aircraft wings) (Fig. 9.35), the downwash velocity v_i is constant along the span and equal in magnitude to

Elliptical lift distribution

$$-\Gamma_0/2b \tag{9.32}$$

where Γ_0 represents the circulation in the centre of the aerofoil of span b. At a distance x from the centre the circulation $\Gamma = \Gamma_0 \{1 - (x/\frac{1}{2}b)^2\}^{1/2}$.

$$\therefore L_e = \int_{-b/2}^{b/2} -\varrho U_e \Gamma \, dx = -\varrho U_e \Gamma_0 \int_{-b/2}^{b/2} \left\{ 1 - \left(x / \frac{1}{2}b \right)^2 \right\}^{1/2} \, dx$$

$$= -\varrho U_e \Gamma_0 b\pi/4 \tag{9.33}$$

Combining eqns 9.31, 9.32 and 9.33 we obtain

$$D_i = \frac{v_i L_e}{U_e} = \frac{-\Gamma_0 L_e}{2bU_e} = \frac{4L_e}{\varrho U_e b\pi} \frac{L_e}{2bU_e} = \frac{2}{\varrho \pi b^2} \left(\frac{L_e}{U_e} \right)^2 = \frac{2}{\varrho \pi b^2} \left(\frac{L}{U} \right)^2$$

Division by $\frac{1}{2}\varrho U^2 S$ gives the result in terms of the dimensionless coefficients:

$$C_{Di} = C_L^2 \bigg/ \left(\frac{\pi b^2}{S} \right) = C_L^2 / (\pi A\!R) \qquad (9.34)$$

To minimize vortex drag on subsonic aircraft the aspect ratio $A\!R$ of the wings is therefore usually as large as structural limitations allow.

It may also be shown that the elliptical distribution of lift here assumed is the condition for the vortex drag to be a minimum for a given value of the lift.

Equation 9.34 enables the effect of vortex drag to be separated out, and thus data obtained at one aspect ratio can be converted for aerofoils with the same profile but a different aspect ratio. The change in the effective angle of attack is

$$\arctan(v_i/U) = \arctan(D_i/L) = \arctan(C_{Di}/C_L) = \arctan\left(\frac{C_L}{\pi A\!R} \right)$$

■ **Example 9.6** An aerofoil of span 10 m and mean chord 2 m has a lift coefficient of 0.914 and a drag coefficient of 0.0588 for an angle of attack of 6.5°. If the distribution of lift over the span is elliptical, what are the corresponding lift and drag coefficients for an aerofoil of the same profile and effective angle of attack, but aspect ratio 8.0?

Solution
Aspect ratio = 10 m/2 m = 5

$$C_{Di} = \frac{C_L^2}{\pi A\!R} = 0.914^2/5\pi = 0.0532$$

The effective angle of attack is less than the nominal value by

$$\arctan \frac{C_L}{\pi A\!R} = \arctan \frac{0.914}{5\pi} = 3.33°$$

i.e. effective angle is 6.5° − 3.33° = 3.17°. For $A\!R = \infty$ and so no vortex drag, $C_D = 0.0588 − 0.0532 = 0.0056$. (This represents the drag due to skin friction and the wake.)

For $A\!R = 8.0$, $L = L_e \cos\varepsilon \simeq L_e$ since ε is small. Therefore if the effective angle of attack is unchanged, the lift is not appreciably different and $C_L = 0.914$ again.

$$C_{Di} = \frac{0.914^2}{8\pi} = 0.0332 \quad \text{and} \quad C_D = 0.0056 + 0.0332 = 0.0388$$

The value of ε is now $\arctan(C_L/\pi A\!R) = \arctan(0.914/8\pi) = 2.08°$, and so, if the effective angle of attack is to be unchanged, the nominal angle must be 3.17° + 2.08° = 5.25°.

Phenomena associated with the flow of real fluids round aerofoils are considered in Section 8.8.6 and the effect of compressibility is briefly treated in Section 11.9.

REFERENCES

1. Duncan, W. J., Thom, A. S. and Young, A. D. *An Elementary Treatise on the Mechanics of Fluids* (2nd edn), Arnold, London (1970).
2. Swanson, W. M. 'The Magnus effect: A summary of investigations to date', *J. Basic Engng*, **83** (*Trans. ASME, Series D*), 461–70 (1961).

FURTHER READING

Lamb, H. *Hydrodynamics* (6th edn), C.U.P., Cambridge (1932).
Milne-Thomson, L. M. *Theoretical Hydrodynamics* (5th edn), Macmillan, London (1968).
Robertson, J. M. *Hydrodynamics in Theory and Application*, Prentice-Hall, Englewood Cliffs, N.J. (1965).
Vallentine, H. R. *Applied Hydrodynamics* (S.I. edn), Butterworth, London (1969).

FILM

Shapiro, A. H. *Vorticity*, Encyclopaedia Britannica Films or from Central Film Library.

PROBLEMS

9.1 Is a flow for which $u = 3$ m/s, $v = 8$ m/s possible in an incompressible fluid? Can a potential function exist?

9.2 Which of the following functions could represent the velocity potential for the two-dimensional flow of an ideal fluid? (*a*) $x + 5y$; (*b*) $3x^2 - 4y^2$; (*c*) $\cos(x - y)$; (*d*) $\ln(x + y)$; (*e*) arctan (x/y); (*f*) arccosec (x/y).

9.3 Show that the two-dimensional flow described (in metre-second units) by the equation $\psi = x + 2x^2 - 2y^2$ is irrotational. What is the velocity potential of the flow? If the density of the fluid is 1.12 kg/m^3 and the piezometric pressure at the point $(1,-2)$ is 4.8 kPa, what is the piezometric pressure at the point $(9,6)$?

9.4 Determine the two-dimensional stream function corresponding to $\phi = A \ln(r/r_0)$ where A is a constant. What is the flow pattern?

9.5 A two-dimensional source at the origin has a strength $(3\pi/2)$ m³/s per metre. If the density of the fluid is $800\,\mathrm{kg/m^3}$ calculate the velocity, the piezometric pressure gradient and the acceleration at the point (1.5 m, 2 m).

9.6 An enclosed square duct of side s has a horizontal axis and vertical sides. It runs full of water and at one position there is a curved right-angled bend where the axis of the duct has radius r. If the flow in the bend is assumed frictionless so that the velocity distribution is that of a free vortex, show that the volume rate of flow is related to Δh, the difference of static head between the inner and outer sides of the duct, by the expression

$$Q = \left(r^2 - \frac{s^2}{4}\right)(sg\Delta h/r)^{1/2}\ln\left(\frac{2r+s}{2r-s}\right)$$

9.7 An open cylindrical vessel, having its axis vertical, is 100 mm diameter and 150 mm deep and is exactly two-thirds full of water. If the vessel is rotated about its axis, determine at what steady angular velocity the water would just reach the rim of the vessel.

9.8 A hollow cylindrical drum of internal diameter 250 mm is completely filled with an oil of relative density 0.9. At the centre of the upper face is a small hole open to atmosphere. The drum is rotated is 15 rev/s about its axis (which is vertical). Determine the pressure of the oil at the circumference of the drum and the net thrust on the upper circular face when steady conditions have been attained.

9.9 Two radii r_1 and r_2, $(r_2 > r_1)$, in the same horizontal plane have the same values in a free vortex and in a forced vortex. The tangential velocity at radius r_1 is the same in both vortices. Determine, in terms of r_1, the radius r_2 at which the pressure difference between r_1 and r_2 in the forced vortex is twice that in the free vortex.

9.10 A set of paddles of radius R is rotated with angular velocity ω about a vertical axis in a liquid having an unlimited free surface. Assuming that the paddles are close to the free surface and that the fluid at radii greater than R moves as a free vortex, determine the difference in elevation between the surface at infinity and that at the axis of rotation.

9.11 A closed cylindrical drum of diameter 500 mm has its axis vertical and is completely full of water. In the drum and concentric with it is a set of paddles 200 mm diameter which are rotated at a steady speed of 15 rev/s. Assuming that all the water within the central 200 mm diameter rotates as a forced vortex and that the remainder moves as a free vortex, determine the difference of piezometric pressure between the two radii where the linear velocity is 6 m/s.

9.12 A hollow cylindrical drum has an internal diameter of 600 mm and is full of oil of relative density 0.9. At the centre of the upper face is a small hole open to atmosphere. Concentric with the axis of the drum (which is vertical) is a set of paddles 300 mm in diameter. Assuming that all the oil in the central 300 mm diameter rotates as a forced vortex with the paddles and that the oil outside this diameter moves as a free vortex, calculate the additional force exerted by the oil on the top of the drum when the paddles are steadily rotated at 8 rev/s.

9.13 In an infinite two-dimensional flow field a sink of strength $-3 \text{ m}^3/\text{s}$ per metre is located at the origin and another of strength $-4 \text{ m}^3/\text{s}$ per metre at $(2 \text{ m}, 0)$. What is the magnitude and direction of the velocity at $(0, 2 \text{ m})$? Where is the stagnation point?

9.14 A tall cylindrical body having an oval cross-section with major and minor dimensions $2X$ and $2Y$ respectively is to be placed in an otherwise uniform, infinite, two-dimensional air stream of velocity U parallel to the major axis. Assuming irrotational flow and a constant density, show that an appropriate flow pattern round the body may be deduced by postulating a source and sink each of strength $|m|$ given by the simultaneous solution of the equations

$$m/\pi U = \left(X^2 - b^2\right)/b \quad \text{and} \quad b/Y = \tan\left(\pi U Y/m\right).$$

Determine the maximum difference of pressure between points on the surface.

9.15 To produce a Rankine oval of length 200 mm and breadth 100 mm in an otherwise uniform infinite two-dimensional stream of velocity 3 m/s (parallel to the length) what strength and positions of source and sink are necessary? What is the maximum velocity outside the oval?

9.16 The nose of a solid strut 100 mm wide is to be placed in an infinite two-dimensional air stream of velocity 15 m/s and density 1.23 kg/m^3 and is to be made in the shape of a half-body. Determine the strength of the corresponding source, the distance between the stagnation point and the source, the equation of the surface in rectangular coordinates based on the source as origin, and the difference in pressure between the stagnation point and the point on the strut where it is 50 mm wide.

9.17 To the two-dimensional infinite flow given by $\psi = -Uy$ are added two sources, each of strength m, placed at $(0, a)$ and $(0, -a)$ respectively. If $m > |2\pi Ua|$, determine the stream function of the combined flow and the position of any stagnation points. Sketch the resulting body contour and determine the velocity at the point where the contour cuts the y axis.

9.18 Estimate the hydrodynamic force exerted on the upstream half of a vertical, cylindrical bridge pier 1.8 m diameter in a wide river 3 m deep which flows at a mean velocity of 1.2 m/s.

9.19 An empty cylinder with plane ends, 300 mm in external diameter and 4 m long, is made entirely from sheet steel (relative density 7.8) 6 mm thick. While completely submerged in water it is accelerated from rest in a horizontal direction perpendicular to its axis. Neglecting end effects and effects due to viscosity, calculate the ratio of the accelerating force required to the force needed to give the cylinder the same acceleration from rest in air.

9.20 On a long circular cylinder with its axis perpendicular to an otherwise uniform, infinite, two-dimensional stream, the stagnation points are at $\theta = 60°$ and $\theta = 120°$. What is the value of the lift coefficient?

9.21 At a speed of 6 m/s the resistance to motion of a rotor-ship is 80 kN. It is propelled by two vertical cylindrical rotors, each 3 m diameter and 9 m high. If the actual circulation generated by the rotors is 50% of that calculated when viscosity and end effects are ignored, determine the magnitude and direction of the rotor speed necessary when the ship travels steadily south-east at 6 m/s in a 14 m/s north-east wind. For these conditions determine the positions of the stagnation points and the difference between the theoretical maximum and minimum pressures. (Assume an air density of $1.225\,\text{kg/m}^3$.)

9.22 Show that flow from a two-dimensional source of strength m at $(a, 0)$ deflected by an impervious wall along the y axis is described by

$$\psi = (-m/2\pi)\arctan\{2xy/(x^2 - y^2 - a^2)\}$$

9.23 Water leaves the guide passages of an inward-flow turbine at a radius of 1.2 m. Its velocity is then 20 m/s at an angle of 70° to the radius. It enters the runner at a radius of 900 mm. Neglecting friction and assuming that the flow is entirely two-dimensional, calculate the drop in piezometric pressure between the guide passages and the entry to the runner.

9.24 A kite may be regarded as equivalent to a rectangular aerofoil of 900 mm chord and 1.8 m span. When it faces a horizontal wind of 13.5 m/s at 12° to the horizontal the tension in the guide rope is 102 N and the rope is at 7° to the vertical. Calculate the lift and drag coefficients, assuming an air density of $1.23\,\text{kg/m}^3$.

9.25 A rectangular aerofoil of 100 mm chord and 750 mm span is tested in a wind-tunnel. When the air velocity is 30 m/s and the angle of attack 7° the lift and drag are 32.8 N and 1.68 N respectively. Assuming an air density of 1.23 kg/m^3 and an elliptical distribution of lift, calculate the coefficients of lift, drag and vortex drag, the corresponding angle of attack for an aerofoil of the same profile but aspect ratio 5.0, and the lift and drag coefficients at this aspect ratio.

10 Flow with a free surface

10.1 INTRODUCTION

In previous chapters, a flowing fluid has usually been assumed to be bounded on all sides by solid surfaces. For liquids, however, flow may take place when the uppermost boundary is the free surface of the liquid itself. The cross-section of the flow is not then determined entirely by the solid boundaries, but is free to change. As a result, the conditions controlling the flow are different from those governing flow that is entirely enclosed. Indeed, the flow of a liquid with a free surface is, in general, much more complicated than flow in pipes and other closed conduits.

If the liquid is bounded by side walls – such as the banks of a river or canal – the flow is said to take place in an *open channel*. The free surface is subjected (usually) only to atmospheric pressure and, since this pressure is constant, the flow is caused by the weight of the fluid – or, more precisely, a component of the weight. As in pipes, uniform flow is accompanied by a drop in piezometric pressure, $p + \varrho gz$, but for an open channel it is only the second term, ϱgz, that is significant, and uniform flow in an open channel is always accompanied by a fall in the level of the surface.

Open channels are frequently encountered. Natural streams and rivers, artificial canals, irrigation ditches and flumes are obvious examples; but pipe-lines or tunnels that are not completely full of liquid also have the essential features of open channels. Water is the liquid usually involved, and practically all the experimental data for open channels relate to water at ordinary temperatures.

Even when the flow is assumed to be steady and uniform, complete solutions of problems of open channel flow are usually more difficult to obtain than those for flow in pipes. For one thing there is a much wider range of conditions than for pipes. Whereas most pipes are of circular cross-section, open channels may have cross-sections ranging from simple geometrical shapes to the quite irregular sections of natural streams. The state of the boundary surfaces, too, varies much more widely – from smooth timber, for instance, to the rough and uneven beds of rivers. The choice of a suitable friction factor for an open channel is thus likely to be much more uncertain than a similar choice for a pipe. Also the fact that the surface is free allows many other phenomena to occur which can markedly affect the behaviour of the fluid.

10.2 TYPES OF FLOW IN OPEN CHANNELS

The flow in an open channel may be uniform or non-uniform, steady or unsteady. It is said to be *uniform* if the velocity of the liquid does not change – either in magnitude or direction – from one section to another in the part of the channel under consideration. This condition is achieved only if the cross-section of the flow does not change along the length of the channel, and thus the depth of the liquid must be unchanged. Consequently, uniform flow is characterized by the liquid surface being parallel to the base of the channel. Constancy of the velocity across any one section of the stream is not, however, required for uniformity in the sense just defined; it is sufficient for the velocity to be the same at corresponding points of all cross-sections.

Flow in which the liquid surface is *not* parallel to the base of the channel is said to be non-uniform, or, more usually, *varied* since the depth of the liquid continuously varies from one section to another. The change in depth may be rapid or gradual, and so it is common to speak of rapidly varied flow and gradually varied flow. (These terms refer only to variations from section to section along the channel – not to variations with time.) Uniform flow may of course exist in one part of a channel while varied flow exists in another part.

Flow is termed *steady* or *unsteady* according to whether the velocity, and hence the depth, at a particular point in the channel varies with *time*. In most problems concerned with open channels the flow is steady – at least approximately. Problems of unsteady flow do arise, however; if there is a surge wave, for example, the depth at a particular point changes suddenly as the wave passes by.

The type of flow most easily treated analytically is *steady uniform flow*, in which the depth of the liquid changes neither with distance along the channel nor with time. The various types of flow are depicted in Fig. 10.1. In these diagrams, as in most others in this chapter, the slope of the channels is much exaggerated: most open channels have a very small slope, of the order perhaps of 1 in 1000.

In practice, non-uniform, or varied, flow is found more frequently than strictly uniform flow. Especially is this so in short channels because a certain length of channel is required for the establishment of uniform flow. Nevertheless, much of the theory of flow in open channels is necessarily based on the behaviour of the liquid in uniform flow.

In addition, the flow in an open channel, like that in a pipe, may be either laminar or turbulent. Which of these types of flow exists depends on the relative magnitude of viscous and inertia forces, the Reynolds number, ul/v, again being used as the criterion. For the characteristic length l it is customary to use the hydraulic mean depth, m (see Section

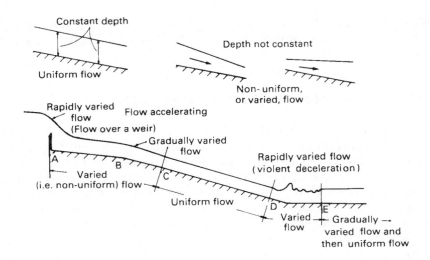

Fig. 10.1

10.4), and the lower critical value of Reynolds number is then about 600. However, laminar flow in open channels seldom occurs in cases of practical interest; it is perhaps most commonly observed in the small grooves in domestic draining boards set at a small slope. In channels of engineering interest, completely turbulent flow may invariably be assumed: the fact that the surface of a flowing liquid occasionally appears smooth and glassy is no indication that turbulent flow does not exist underneath. The inertia forces usually far outweigh the viscous forces. Thus it is not ordinarily necessary to consider in detail the effect of Reynolds number on the flow in a channel.

A further important classification of open channel flow is derived from the magnitude of the Froude number of the flow. The relation of this quantity to the flow in open channels is discussed in later sections. It may, however, be said here that when the velocity of the liquid is small it is possible for a small disturbance in the flow to travel against the flow and thus affect the conditions upstream. The Froude number (as defined in Section 10.9) is then less than 1.0, and the flow is described as *tranquil*. If, on the other hand, the velocity of the stream is so high that a small disturbance cannot be propagated upstream but is washed downstream, then the Froude number is greater than 1.0 and the flow is said to be *rapid*. When the Froude number is exactly equal to 1.0, the flow is said to be *critical*.

To sum up, then, a complete description of the flow thus always consists of four characteristics. The flow will be

1. Either uniform or non-uniform (varied)
2. Either steady or unsteady
3. Either laminar or turbulent
4. Either tranquil or rapid.

10.3 THE STEADY-FLOW ENERGY EQUATION FOR OPEN CHANNELS

In open channels we are concerned only with fluids of constant density and temperature changes are negligible. Therefore at any particular point the mechanical energy per unit weight is represented by the sum of three terms:

$$\frac{p}{\varrho g} + \frac{u^2}{2g} + z$$

Now if the streamlines are sensibly straight and parallel – and even in gradually varied flow the curvature of streamlines is usually very slight – there is a hydrostatic variation of pressure over the cross-section. In other words, the pressure at any point in the stream is governed only by its depth below the free surface. (Where there is appreciable curvature of the streamlines – as in rapidly varied flow – there are accelerations perpendicular to them, and consequently differences of pressure additional to the hydrostatic variation. Also, if the slope of the channel is exceptionally large, say greater than 1 in 10, there is a modification of the hydrostatic pressure variation even when the streamlines are straight and parallel. This is because lines perpendicular to the streamlines – along which the piezometric pressure is constant (Section 3.6) – cannot then be considered vertical.)

When the pressure variation is hydrostatic, a point at which the (gauge) pressure is p is at a depth $p/\varrho g$ below the surface, and so the sum $(p/\varrho g) + z$ (see Fig. 10.2) represents the height of *the surface* above datum level. The expression for the mechanical energy per unit weight is thus simplified to

$$\text{Height of surface above datum} + u^2/2g \qquad (10.1)$$

We see that the height of the individual streamline above datum has no place in the expression. If it be further assumed that at the section considered the velocity is the same along all streamlines, then the expression 10.1 has the same value for the entire stream.

In practice, however, a uniform distribution of velocity over a section is never achieved. The actual velocity distribution in an open channel is

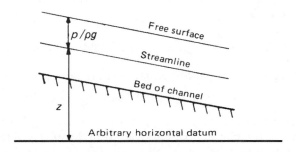

Fig. 10.2

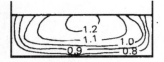

Fig. 10.3 Contours of constant velocity in a rectangular channel (figures are proportions of mean velocity).

influenced both by the solid boundaries (as in closed conduits such as pipes) and by the free surface. Bends in the channel and irregularities in the boundaries also have an effect. The irregularities in the boundaries of open channels are usually so large, and occur in such a random manner, that each channel has its own peculiar pattern of velocity distribution. Nevertheless, it may in general be said that the maximum velocity usually occurs at a point slightly below the free surface (at from 0.05 to 0.25 times the full depth) and that the average velocity, which is usually of the order of 85% of the velocity at the surface, occurs at about 0.6 of the full depth below the surface. A typical pattern for a channel of rectangular section is shown in Fig. 10.3.

As a result of this lack of uniformity of velocity over a cross-section, the velocity head, $u^2/2g$, representing the kinetic energy of the fluid per unit weight, has too low a value if calculated from the average velocity $\bar{u}$. To compensate for the error $\alpha \bar{u}^2/2g$ may be used in place of $\bar{u}^2/2g$, where α is the *kinetic energy correction factor* (Section 3.5.3). Experiments show that the value of α varies from 1.03 to as much as 1.6 in irregular natural streams, the higher values generally being found in small channels.

The calculation of the momentum of the stream is also affected by a non-uniform distribution of velocity. The momentum carried by the fluid past a particular cross-section in unit time is given by $\beta Q \varrho \bar{u}$ where Q represents the volume flow rate, ϱ the density of the liquid, and β the *momentum correction factor* (Section 4.2.1). The value of β varies from 1.01 to about 1.2.

In straight channels of regular cross-section, however, the effects of a non-uniform velocity distribution on the calculated velocity head and momentum flow rate are not normally of importance. Indeed, other uncertainties in the numerical data are usually of greater consequence. Unless accurate calculations are justified, therefore, it is usual to assume that the factors α and β are insufficiently different from unity to warrant their inclusion in formulae.

10.3.1 Energy gradient

In practice, as the liquid flows from one section to another, friction causes mechanical energy to be converted into heat, and thus lost. If the loss per unit weight is denoted by h_f, then for steady flow between two sections (1) and (2)

$$\left(\text{Height of surface}\right)_1 + \frac{u_1^2}{2g} - h_f = \left(\text{Height of surface}\right)_2 + \frac{u_2^2}{2g}$$

With reference to Fig. 10.4, the equation may be written

$$h_1 + z_1 + u_1^2/2g - h_f = h_2 + z_2 + u_2^2/2g \qquad (10.2)$$

where the hs represent vertical depths of liquid in the channel, and the zs the heights of the channel bed above datum level. To take account of non-uniformity of velocity over the cross-section, we may write

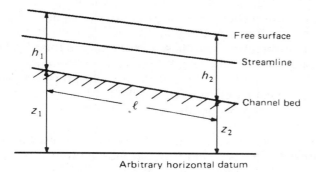

Fig. 10.4

$$h_1 + z_1 + \alpha_1 \bar{u}_1^2/2g - h_f = h_2 + z_2 + \alpha_2 \bar{u}_2^2/2g \qquad (10.3)$$

The rate at which mechanical energy is lost to friction may be expressed by h_f/l, where l represents the length of channel over which the head loss h_f takes place. This quantity h_f/l may be termed the *energy gradient* since it corresponds to the slope of a graph of the total mechanical energy per unit weight plotted against distance along the channel. In the special case of uniform flow, $\bar{u}_1 = \bar{u}_2$, $\alpha_1 = \alpha_2$ and $h_1 = h_2$ in eqn 10.3. Therefore $h_f = z_1 - z_2$. The energy gradient is thus the same as the actual, geometrical, gradient of the channel bed and of the liquid surface. This, it must be emphasized, is necessarily true only for uniform flow in open channels. In discussing non-uniform flow it is important to distinguish carefully from one another the energy gradient, the slope of the free surface and the slope of the bed.

10.4 STEADY UNIFORM FLOW – THE CHÉZY EQUATION

Steady uniform flow is the simplest type of open channel flow to analyse, although in practice it is not of such frequent occurrence as might at first be supposed. Uniform conditions over a length of the channel are achieved only if there are no influences to cause a change of depth, there is no alteration of the cross-section of the stream, and there is no variation in the roughness of the solid boundaries. Indeed, strictly uniform flow is scarcely ever achieved in practice, and even approximately uniform conditions are more the exception than the rule. Nevertheless, when uniform flow is obtained the free surface is parallel to the bed of the channel (sometimes termed the *invert*) and the depth from the surface to the bed is then termed the *normal* depth.

The basic formula describing uniform flow is due to the French engineer Antoine de Chézy* (1718–98). He deduced the equation from the results of experiments conducted on canals and on the River Seine in 1769. Here, however, we shall derive the expression analytically.

* pronounced *Shay'-zee.*

In steady uniform (or *normal*) flow there is no change of momentum, and thus the net force on the liquid is zero. Figure 10.5 represents a stretch of a channel in which these conditions are found. The slope of the channel is constant, the length of channel between the planes 1 and 2 is l and the (constant) cross-sectional area is A. It is assumed that the stretch of the channel considered is sufficiently far from the inlet (or from a change of slope or of other conditions) for the flow pattern to be fully developed.

Now the 'control volume' of liquid between sections 1 and 2 is acted on by hydrostatic forces F_1 and F_2 at the ends. However, since the cross-sections at 1 and 2 are identical, F_1 and F_2 are equal in magnitude and have the same line of action; they thus balance and have no effect on the motion of the liquid. Hydrostatic forces acting on the sides and bottom of the control volume are perpendicular to the motion, and so they too have no effect. The only forces we need consider are those due to gravity and the resistance exerted by the bottom and sides of the channel. If the average stress at the boundaries is τ_0, the total resistance force is given by the product of τ_0 and the area over which it acts, that is, by $\tau_0 Pl$ where P represents the 'wetted perimeter' (Fig. 10.6).

It is important to notice that P does *not* represent the total perimeter of the cross-section since the free surface is not included. Only that part of the perimeter where the liquid is in contact with the solid boundary is relevant here, for that is the only part where resistance to flow can be exerted. (The effect of the air at the free surface on the resistance is negligible compared with that of the sides and bottom of the channel).

For zero net force in the direction of motion, the total resistance must exactly balance the component of the weight W. That is

$$\tau_0 Pl = W \sin \alpha = Al\varrho g \sin \alpha$$

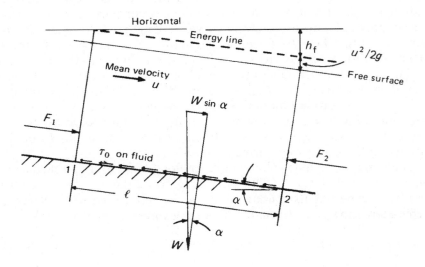

Fig. 10.5

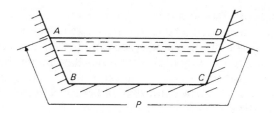

Fig. 10.6

whence

$$\tau_0 = \frac{A}{P}\varrho g \sin \alpha \qquad (10.4)$$

For uniform flow, however, $\sin \alpha = h_f/l$, the energy gradient defined in Section 10.3.1. Denoting this by i we may therefore write $\tau_0 = (A/P)\varrho g i$.

We now require an expression to substitute for the average stress at the boundary, τ_0. In almost all cases of practical interest, the Reynolds number of the flow in an open channel is sufficiently high for conditions to correspond to the 'rough zone of flow' (see Fig. 7.2) in which simple viscous effects are negligible and the stress at the boundary is proportional to the square of the mean velocity. By analogy with eqn 7.4, we may therefore take $\tau_0 = \frac{1}{2}\varrho u^2 f$, where f is sensibly independent of u. Substituting for τ_0 in eqn 10.4 gives

$$\frac{1}{2}\varrho u^2 f = (A/P)\varrho g i$$

whence

$$u^2 = \frac{2g}{f}\frac{A}{P}i = \frac{2g}{f}mi \qquad (10.5)$$

where $m = A/P$.

The quantity m is termed the *hydraulic mean depth* or *hydraulic radius*. For the channel depicted in Fig. 10.6, for example, m would be calculated by dividing the cross-sectional area $ABCD$ by the wetted perimeter, i.e. the length $ABCD$ *only*.

Taking square roots in eqn 10.5 and putting

$$C = \sqrt{(2g/f)} \qquad (10.6)$$

we arrive at Chézy's equation

$$u = C\sqrt{(mi)} \qquad (10.7)$$

Since u is the average velocity of flow over the cross-section, the discharge through the channel is given by

$$Q = Au = AC\sqrt{(mi)} \qquad (10.8)$$

The factor C is usually known as Chézy's coefficient. Its dimensional formula is $[g^{1/2}] = [L^{1/2}T^{-1}]$ since f is a dimensionless magnitude. Consequently the expression for the magnitude of C depends on the system of units adopted. It used to be thought that C was a constant for all sizes of channel, but it is now recognized that its value depends to some extent on the size and shape of the channel section, as well as on the roughness of the boundaries. The study of flow in pipes has shown that the friction factor f depends both on Reynolds number and on the relative roughness k/d. Thus Chézy's coefficient C may be expected to depend on Re and k/m (the hydraulic mean depth, m, is used here as the most significant characteristic length of the system), although for the fully turbulent flow usually encountered in open channels the dependence on Re is slight, and k/m is by far the more important factor. Although open channels vary widely in the shape of their cross-sections, the use of the hydraulic mean depth m largely accounts for differences of shape. Experience suggests that the shape of the cross-section has little effect on the flow if the shear stress τ_0 does not vary much round the wetted perimeter. For the simpler cross-sectional shapes, therefore, the hydraulic mean depth by itself may be regarded as adequate in describing the influence of the cross-sectional form on the flow – at least as a first approximation. For channels of unusual shape, however, the hydraulic mean depth should be used with considerable caution.

Many attempts have been made to correlate the large amount of available experimental data and so enable the value of C for a particular channel to be predicted. We shall here do no more than mention a few such formulae. All are based on analyses of experimental results.

The simplest expression, and one that is very widely used, is that ascribed to the Irish engineer Robert Manning (1816–97). This formula gives

$$C = m^{1/6}/n$$

in other words, when combined with Chézy's equation (10.7), Manning's expression becomes

$$u = m^{2/3}i^{1/2}/n \left(\text{for metre-second units}\right) \qquad (10.9)$$

(In Central Europe this is known as Strickler's formula and $1/n$ as the Strickler coefficient.)

The n in eqn 10.9 is often known as Manning's roughness coefficient. For the equation to be dimensionally homogeneous it appears that n should have the dimensional formula $[\mathrm{TL}^{-1/3}]$. It is, however, illogical that an expression for the roughness of a surface should involve dimensions in respect to time, and it may be seen from comparison with eqn 10.6 that the formula would be more logically written

$$u = \left(Ng^{1/2}/n\right)m^{2/3}i^{1/2}$$

If N is regarded as a numeric, then n takes the dimensional formula $[\mathrm{L}^{1/6}]$. However, because g is not explicitly included eqn 10.9 must be regarded as a numerical formula, suitable only for a particular set of units. With the figures usually quoted for n, the units are those based on the metre and the second.

The expression may be adapted for use with foot-second units by changing the numeric from 1.0 to 1.49. Then

Table 10.1 Approximate values of Manning's roughness coefficient n for straight, uniform channels

Smooth cement, planed timber	0.010
Rough timber, canvas	0.012
Cast iron, good ashlar masonry, brickwork	0.013
Vitrified clay, asphalt, good concrete	0.015
Rubble masonry	0.018
Firm gravel	0.020
Canals and rivers in good condition	0.025
Canals and rivers in bad condition	0.035

(For use with metre-second units in eqn 10.9 or foot-second units in eqn 10.9a).

$$u = (1.49/n)m^{2/3}i^{1/2}\left(\text{for foot-second units}\right) \qquad (10.9a)$$

This change allows the same *numbers* for n to be used in either eqn 10.9 or eqn 10.9a. Table 10.1 gives a few representative values of n, but it should be realized that they are subject to considerable variation. The selection of an appropriate value requires much judgement and experience, and for discussion of the factors affecting n, more specialist works must be consulted.

Other empirically-based formulae that have been used include those due to Kutter, Bazin and Thijsse, but Manning's remains much the most popular. It should never be forgotten that the data available for the construction of such formulae are the results of experiments in which water at ordinary temperatures was used under conditions producing fully-developed turbulent flow at high values of Reynolds number. To apply formulae of this kind to conditions not closely similar to those on which the formulae are based is, to say the least, hazardous. It should be realized that flow in channels of small size or at an unusually small velocity may well have a Reynolds number lower than that for which the formulae are truly applicable.

The formulae, such as Chézy's and Manning's, that account for the friction in an open channel have no connection with the Froude number and are thus applicable to tranquil or rapid flow. It is, however, emphasized again that they apply only to steady uniform flow.

Flow in open channels, like flow in pipes, is also subject to 'minor' losses resulting from additional turbulence at abrupt changes of section, bends, or other disturbances to the flow. These additional energy losses are, however, normally negligible compared with the friction in the channel as a whole.

10.5 THE BOUNDARY LAYER IN OPEN CHANNELS

We saw in Chapter 7 that the friction factor, f, in pipe flow is closely related to the relative roughness k/d. It is natural to expect that Chézy's coefficient C is closely related to the roughness of the boundaries of open channels; indeed, the results of Nikuradse and others who have studied flow in pipes may be expected to shed some light on the way in which friction influences flow in open channels.

Whereas for a circular pipe the diameter is usually regarded as the significant linear measurement, the hydraulic mean depth (with less justification) serves the same purpose for open channels. Since $C = \sqrt{(2g/f)}$ (eqn 10.6), it is possible to re-plot Nikuradse's data (Fig. 7.1) in the manner of Fig. 10.7. Since the hydraulic mean depth of the circular section – with which Nikuradse was concerned – is $d/4$, $4m$ has been substituted for d.

Figure 10.7 shows the way in which C may be expected to depend on Reynolds number and the relative roughness k/m. In the 'rough zone' of the graph, C is constant for a particular value of k/m, and it is clearly to these conditions that Manning's roughness coefficient n, for example, applies.

Even the simplest shapes of cross-section for open channels lack the axial symmetry of a circular pipe, and Fig. 10.7 can be expected to represent open channel flow only in a qualitative manner. Nevertheless, for the rough zone, it is of interest to combine eqn 10.6 with eqn 8.55 which gives the friction factor for turbulent flow in rough pipes:

$$\frac{C}{(2g)^{1/2}} = f^{-1/2} = 4 \log_{10}(d/k) + 2.28$$

Substituting $m^{1/6}/n$ for C (as in eqn 10.9), and $4m$ for d, then gives (with metre-second units)

$$\frac{m^{1/6}}{(19.62)^{1/2} n} = 4 \log_{10}(4m/k) + 2.28$$

whence

$$n = \frac{0.0564 \, m^{1/6}}{\log_{10}(14.86 \, m/k)}$$

Because an open channel lacks axial symmetry, if for no other reason, the numerical factors in this expression are hardly trustworthy. Nevertheless the expression does suggest that, with the logarithmic type of

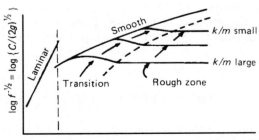

Fig. 10.7

velocity profile to be expected in turbulent flow, n is not very sensitive to changes in k, and even less so to changes in m. Many attempts have been made to apply theories developed for turbulent flow in pipes to that in open channels. The difficulties, however, are great; in addition, the effects of the free surface and of non-uniform shear stress round the wetted perimeter are uncertain; and as yet no conclusive theory has emerged (although a good review, with 200 references, is given by Carter et al.[1]). Even so, the qualitative conclusions drawn from Fig. 10.7 about the relation of C to the Reynolds number and to the roughness size are valid – at least for rigid channel boundaries.

In alluvial channels, however, in which the surfaces are composed of movable sand or gravel, the roughness elements are not permanent, but depend on the flow. Ripples and dunes may form in the boundary material, and the spacing of these humps may be much greater than the spacing of the irregularities on the walls of pipes or on rigid boundaries of open channels. Under such conditions as Nikuradse investigated – where the roughness projections are close together – the wake behind one projection interferes with the flow around those immediately downstream. The larger irregularities formed in alluvial channels, however, have a different kind of effect, which, in turn, results in a much larger friction loss than the size of the individual particles alone would suggest. (Flow in alluvial channels is also discussed in detail by Carter et al.[1].)

10.6   OPTIMUM SHAPE OF CROSS-SECTION

The Chézy formula and the Manning formula (or any of the others describing uniform flow in an open channel) show that, for any given value of slope, surface roughness and cross-sectional area, the discharge Q increases with increase in the hydraulic mean depth m. Therefore the discharge is a maximum when m is a maximum, that is, when, for a given area, the wetted perimeter is a minimum (since $m = A/P$ by definition). A cross-section having such a shape that the wetted perimeter is a minimum is thus, from a hydraulic point of view, the most efficient. Not only is it desirable to use such a section for the sake of obtaining the maximum discharge for a given cross-sectional area, but a minimum wetted perimeter requires a minimum of lining material, and so the most efficient section tends also to be the least expensive.

It may be shown that, of all sections whose sides do not slope inwards towards the top, the semicircle has the maximum hydraulic mean depth. This mathematical result, however, is not usually the only consideration. Although semicircular channels are in fact built from prefabricated sections, for other forms of construction the semicircular shape is impractical. Trapezoidal sections are very popular, but when the sides are made of a loose granular material its 'angle of repose' may limit the angle of the sides.

Another point is this. The most efficient section will give the maximum discharge for a given area and, conversely, the minimum area for a given discharge. This does not, however, necessarily imply that such a channel, if constructed below ground level, requires the minimum excavation. After all, the surface of the liquid will not normally be exactly level with the tops of the sides. Nevertheless the minimum excavation may, in certain instances, be an overriding requirement. Factors other than the hydraulic efficiency may thus determine the best cross-section to be used for an open channel.

However, when the hydraulic efficiency is the chief concern, determining the most efficient shape of section for a given area is simply a matter of obtaining an expression for the hydraulic mean depth, differentiating it and equating to zero to obtain the condition for the maximum. For example, for a channel section in the form of a symmetrical trapezium with horizontal base (Fig. 10.8),

$$\text{Area } A = bh + h^2 \cot \alpha$$

$$\text{Wetted perimeter } P = b + 2h \operatorname{cosec} \alpha$$

Since $b = (A/h) - h \cot \alpha$,

$$m = \frac{A}{P} = \frac{A}{(A/h) - h \cot \alpha + 2h \operatorname{cosec} \alpha}$$

For a given value of A, this expression is a maximum when its denominator is a minimum, that is, when $(-A/h^2) - \cot \alpha + 2 \operatorname{cosec} \alpha = 0$. (The second derivative, $2A/h^3$, is clearly positive and so the condition is indeed that for a minimum.) Thus

$$A = h^2 (2 \operatorname{cosec} \alpha - \cot \alpha) \tag{10.10}$$

Substituting this value in the expression for m gives $m_{\max} = h/2$. In other words, for maximum efficiency a trapezoidal channel should be so proportioned that its hydraulic mean depth is half the central depth of flow. Since a rectangle is a special case of a trapezium (with $\alpha = 90°$) the optimum proportions for a rectangular section are again given by $m = h/2$; taking $A = bh = 2h^2$ (from eqn 10.10) we get $b = 2h$.

A further exercise in differential calculus shows that, if α may be varied, a minimum perimeter and therefore maximum m is obtained when $\alpha = 60°$. This condition, taken in conjunction with the first, shows that the most efficient of all trapezoidal sections is half a regular hexagon.

The concept of the most efficient section as considered here applies only to channels with rigid boundaries. For channels with erodible boundaries, e.g. of sand, the design must take account of the maximum

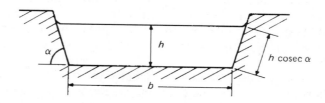

Fig. 10.8

shear stress, τ_0, on the boundary. Such considerations as this, however, are outside the scope of this book, and reference must be made to more specialist works, e.g. Chow[2].

10.7 FLOW IN CLOSED CONDUITS ONLY PARTLY FULL

Closed conduits only partly full are frequently encountered in civil engineering practice, particularly as drains and sewers. Because the liquid has a free surface its flow is governed by the same principles as if it were in a channel completely open at the top. There are, however, special features resulting from the convergence of the boundaries towards the top. For conduits of circular section (see Fig. 10.9) the area of the cross-section *of the liquid* is

$$r^2\theta - 2\left(\frac{1}{2}r\sin\theta\, r\cos\theta\right) = r^2\left(\theta - \frac{1}{2}\sin 2\theta\right)$$

and the wetted perimeter $= 2r\theta$. From an equation such as Manning's (eqn 10.9) the mean velocity and the discharge may then be calculated for any value of θ and hence of h. Figure 10.10 shows these variations, the variables u, Q and h being expressed as proportions of their values when the conduit is full. It will be seen that the maximum discharge and maximum mean velocity are both greater than the values for the full conduit. Differentiation of the appropriate expressions shows that the

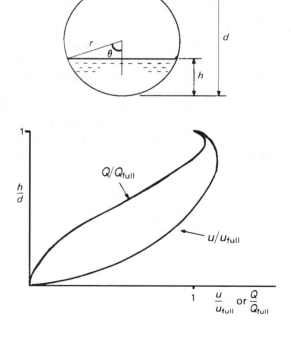

Fig. 10.9

Fig. 10.10

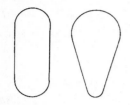

Fig. 10.11

maximum discharge occurs when h/d is about 0.94, and the maximum velocity when h/d is about 0.81. These figures are based on the assumption that Manning's roughness coefficient n is independent of the depth of flow. In fact, n may increase by as much as 25% as the depth is reduced from d to about $d/4$, so depths for maximum discharge and maximum mean velocity are slightly underestimated by the simple analysis. Although it might seem desirable to design such a conduit to operate under the conditions giving maximum discharge, the corresponding value of h/d is so near unity that in practice the slightest obstruction or increase in frictional resistance beyond the design figure would cause the conduit to flow completely full.

The circular shape is frequently modified in practice. For example, when large fluctuations in discharge are encountered oval or egg-shaped sections (Fig. 10.11) are commonly used. Thus at low discharges a velocity high enough to prevent the deposition of sediment is maintained. On the other hand, too large a velocity at full discharge is undesirable as this could lead to excessive scouring of the lining material.

10.8 SIMPLE WAVES AND SURGES IN OPEN CHANNELS

The flow in open channels may be modified by waves and surges of various kinds which produce unsteady conditions. Any temporary disturbance of the free surface produces waves: for example, a stone dropped into a pond causes a series of small surface waves to travel radially outwards from the point of the disturbance. If the flow along a channel is increased or decreased by the removal or insertion of an obstruction – for example, by the sudden opening or closing of a sluice gate – surge waves are formed and propagated upstream and downstream of the obstruction. In certain circumstances, tidal action may cause a surge, known as a *bore*, in large estuaries and rivers, e.g. the River Severn. A positive wave is one that results in an increase in the depth of the stream; a negative one causes a decrease in depth.

Let us consider the simple positive surge illustrated in Fig. 10.12. To avoid too much algebraic complication we assume a straight channel of uniform width whose cross-section is a rectangle with horizontal base. We suppose also that the slope of the bed is zero (or so nearly zero that the weight of the liquid has a negligible component in the direction of

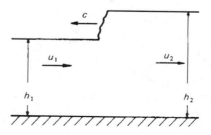

Fig. 10.12

flow). Uniform flow at velocity u_1 and depth h_1, as depicted at the left of the diagram, is disturbed by, for example, the closing of a gate down-stream so that a positive surge travels upstream, with (constant) velocity c (relative to the bed of the channel). A short distance downstream of the wave the flow has again become uniform with velocity u_2 and depth h_2.

The change of velocity from u_1 to u_2 caused by the passage of the wave is the result of a net force on the fluid, the magnitude of which is given by the momentum equation. To apply the *steady-flow* momentum equa-tion, however, coordinate axes must be chosen that move with the wave. The wave then appears stationary, conditions at any point fixed with respect to those axes do not change with time, and the velocities are as shown in Fig. 10.13. The net force acting on the fluid in the control volume indicated is the difference between the horizontal thrusts at sections 1 and 2. These sections are sufficiently near each other for friction at the boundaries of the fluid to be negligible. If the streamlines at these two sections are substantially straight and parallel then the variation of pressure is hydrostatic and the total thrust on a vertical plane is therefore $\varrho g(h/2)h = \varrho gh^2/2$ for unit width. (As a rectangular section is assumed, in which the velocities u_1 and u_2 are uniform across the width, it is sufficient to consider only a unit width of the channel.) With the further assumption that the velocity is sufficiently uniform over the depth at sections 1 and 2 for the momentum correction factor to differ negligibly from unity, the steady-flow momentum equation yields:

$$\overrightarrow{\text{Net force on fluid in control volume}} = \frac{\varrho gh_1^2}{2} - \frac{\varrho gh_2^2}{2}$$

$$= \overrightarrow{\text{Rate of increase of momentum}} = \varrho Q(u_2 - u_1) \qquad (10.11)$$

By continuity, $Q = (u_1 + c)h_1 = (u_2 + c)h_2$ whence

$$u_2 = (u_1 + c)\frac{h_1}{h_2} - c \qquad (10.12)$$

Substituting for Q and u_2 in eqn 10.11 gives

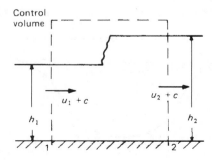

Fig. 10.13

$$\frac{\varrho g}{2}\left(h_1^2 - h_2^2\right) = \varrho(u_1 + c)h_1\left\{(u_1 + c)\frac{h_1}{h_2} - c - u_1\right\}$$

$$= \varrho(u_1 + c)^2 \frac{h_1}{h_2}(h_1 - h_2)$$

whence

$$u_1 + c = \left(gh_2\right)^{1/2}\left(\frac{1 + h_2/h_1}{2}\right)^{1/2} \qquad (10.13)$$

If the wave is of small height, that is $h_2 \simeq h_1 \simeq h$, then eqn 10.13 reduces to $u_1 + c = (gh)^{1/2}$. In other words, the velocity of the wave *relative to the undisturbed liquid ahead of it* is $(gh)^{1/2}$. No restriction, it should be noted, was placed on the values of u_1 and u_2: either may be zero or even negative, but the analysis is still valid.

This derivation applies only to waves propagated in rectangular channels. For a channel of any shape it may be shown that the velocity of propagation of a small surface wave is $(g\bar{h})^{1/2}$ relative to the undisturbed liquid where $\bar{h}$ represents the *mean* depth as calculated from

$$\frac{\text{Area of cross-section}}{\text{Width of liquid surface}} = \frac{A}{B} \quad \text{(see Fig. 10.14)}$$

Although $(g\bar{h})^{1/2}$ represents the velocity of propagation of a very small surge wave, it should not be forgotten that a larger positive wave will be propagated with a higher velocity – as shown by eqn 10.13. Moreover, the height of the wave does not remain constant over appreciable distances; frictional effects, which in the above analysis were justifiably assumed negligible in the short distance between sections 1 and 2, gradually reduce the height of the wave.

In Fig. 10.12 the wave was shown with a more or less vertical front. For a positive wave this is the stable form. If the wave originally had a sloping front (as in Fig. 10.15) it could be regarded as the superposition of a number of waves of smaller height. As the velocity of propagation of such smaller waves is given by $(g\bar{h})^{1/2}$, the velocity of the uppermost elements would be somewhat greater than that of the lower ones. Thus the sloping front of the waves would tend to become vertical.

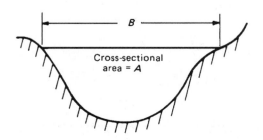

Cross-sectional
area = A

Fig. 10.14

The opposite, however, is true for a negative wave (Fig. 10.16). Since the upper elements of the wave move more rapidly, the wave flattens and the wave front soon degenerates into a train of tiny wavelets.

Small disturbances in a liquid may often cause waves of the shape shown in Fig. 10.17. This sort of wave may simply be regarded as a positive wave followed by a negative one and, provided that it is only of small height, its velocity of propagation is $(g\bar{h})^{1/2}$ as for the small surge wave.

Where the scale is small, as for example in small laboratory models, another type of wave may be of significance. This is a 'capillary wave' which results from the influence of surface tension. In fact, the velocity of propagation of any surface wave is governed by both gravity forces and surface tension forces but for the single surge wave in a channel of 'civil engineering' size the effect of surface tension is negligible.

For waves on the surface of deep water, for example on the surface of the sea, different considerations apply. This is largely because the assumption of a uniform distribution of velocity over the entire depth is not valid for large depths. Moreover, ocean waves appear as a succession or *train* of waves in which the waves follow one another closely, so an individual wave cannot be considered separately from those next to it. We shall give some consideration to waves of this type in Section 10.13. For the moment our concern is with the propagation of small disturbances in open channels of small depth.

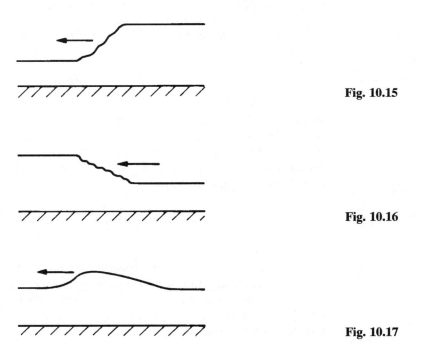

Fig. 10.15

Fig. 10.16

Fig. 10.17

10.9 SPECIFIC ENERGY AND ALTERNATIVE DEPTHS OF FLOW

As we saw in Section 10.3, the total head (that is, the total mechanical energy per unit weight of liquid) is given by $p/\varrho g + u^2/2g + z$ in which z represents the height of the point in question above some arbitrary horizontal datum plane. If the channel slope is small and the streamlines are straight and parallel, so that the variation of pressure with depth is hydrostatic – and this is a most important proviso – then the sum $(p/\varrho g) + z$ is equivalent to the height of the free surface above the same datum. If, at a particular position, the datum level coincides with the bed of the channel, then the local value of the energy per unit weight is given by $h + u^2/2g$, where h represents the depth of flow at that position. (We consider channels in which the velocity distribution over the cross-section is sensibly uniform.) This quantity $h + u^2/2g$ is usually termed the *specific energy*, E (or occasionally 'specific head'). The name, though in very wide use, is in many respects unsatisfactory, particularly as 'specific energy' has other meanings in other contexts. It should here be regarded as no more than a useful 'shorthand' label for $h + u^2/2g$.

The fact that the bed of the channel may not be precisely horizontal does not matter: specific energy is essentially a *local* parameter, applied over a short length of the channel in which any change of bed level is negligible.

From the form of the expression $E = h + u^2/2g$ we see that a particular value of E could be composed of a small value of h and a large value of $u^2/2g$, or of a large value of h and a small value of $u^2/2g$, even when the volume flow rate Q remains unaltered. Since $u =$ the mean velocity $= Q/A$, where A represents the cross-sectional area of the stream at the position considered, we may write

$$E = h + \frac{1}{2g}\left(\frac{Q}{A}\right)^2 \tag{10.14}$$

Although, as we shall see later, the principles may be applied to channels with any shape of cross-section, we may for the moment consider a wide channel of rectangular cross-section so as to illustrate the fundamentals with the minimum of mathematical complication. If the width of such a rectangular section is b, then from eqn 10.14

$$E = h + \frac{1}{2g}\left(\frac{Q}{bh}\right)^2 = h + \frac{1}{2g}\left(\frac{q}{h}\right)^2 \tag{10.15}$$

where $q = Q/b$.

Equation 10.15 relates the specific energy E, the depth h and the discharge per unit width q. In the most general case, of course, all three quantities vary, but particular interest attaches to those instances in which q is constant while h and E vary, and to those in which E is constant while h and q vary.

If q is kept constant E may be plotted against h as in Fig. 10.18 (it is more suitable to plot E along a horizontal than a vertical axis); alternatively, if E is kept constant h may be plotted against q as in Fig. 10.19.

Let us first consider Fig. 10.18. This is usually known as the specific-energy diagram. With q constant, a small value of h corresponds to a high velocity and thus, as h tends to zero, $u^2/2g$ tends to infinity and so also does E. Hence the specific energy curve is asymptotic to the E axis. Conversely, as h increases, the velocity becomes smaller, the $u^2/2g$ term becomes insignificant compared with h, and E tends to h. The upper part of the specific-energy curve is thus asymptotic to the line $E = h$, which, if identical scales are used on the two axes, has slope unity on the diagram. There is clearly a minimum value of E between these two extremes. This minimum occurs at a value of h known as the *critical depth*, h_c.

For each value of E other than the minimum it is seen that there are two possible values of h, one greater and one less than h_c. (Although eqn 10.15 is a cubic in h, the third root is always negative and therefore physically meaningless.) These two values are known as *alternative*

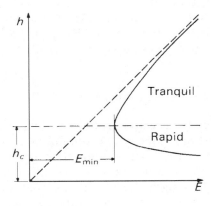

Fig. 10.18

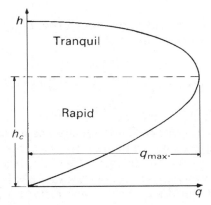

Fig. 10.19

depths. We shall examine the significance of these a little later, but first we study the conditions under which the critical depth is found.

The conditions for the critical depth are those for minimum E, which, for a channel of rectangular section, may be found by differentiating eqn 10.15 with respect to h and equating the result to zero.

$$\frac{\partial E}{\partial h} = 1 + \frac{q^2}{2g}\left(-\frac{2}{h^3}\right)$$

This expression is zero when $q^2/gh^3 = 1$, i.e. when $h = (q^2/g)^{1/3}$. This value of h is the critical depth h_c and so we may write

$$h_c = \left(q^2/g\right)^{1/3} \tag{10.16}$$

The corresponding minimum value of E is obtained by substituting the value of q from eqn 10.16 in eqn 10.15:

$$E_{min} = h_c + gh_c^3/2gh_c^2 = \frac{3}{2}h_c \tag{10.17}$$

These relations, it must be emphasized, refer *only* to channels of rectangular cross-section.

It is also of interest to examine the situation in which the specific energy E is kept constant while h and q vary, as in Fig. 10.19. This curve shows that q reaches a maximum value for a particular value of h. Equation 10.15 may be rearranged to give $q^2 = 2gh^2(E - h)$. Differentiating with respect to h gives $2q(\partial q/\partial h) = 2g(2Eh - 3h^2)$ and so $\partial q/\partial h = 0$ when

$$h = \frac{2}{3}E \tag{10.18}$$

This, however, is identical with eqn 10.17 and so it may be said that at the critical depth the discharge is a maximum for a given specific energy, or that the specific energy is a minimum for a given discharge. Thus, if in a particular channel the discharge is the maximum obtainable then somewhere along its length the conditions must be critical. It is not surprising that the conditions for maximum q and minimum E correspond: the curves of Figs. 10.18 and 10.19 are plotted from the same equation.

Since $u = Q/bh = q/h$, the velocity corresponding to the critical depth may be determined from eqn 10.16.

$$u_c = \frac{q}{h_c} = \frac{\left(gh_c^3\right)^{1/2}}{h_c} = \left(gh_c\right)^{1/2} \tag{10.19}$$

The velocity u_c, which occurs when the depth is at its critical value h_c, is known as the *critical velocity*. This velocity has no connection with the critical velocity at which turbulent flow becomes laminar, and it is perhaps unfortunate that the term has been duplicated.

Equation 10.19 is an expression for the critical velocity in a channel of rectangular section but the corresponding result for a channel with any shape of section may be obtained simply. As before (eqn 10.14), for uniform velocity the specific energy at a particular section is given by

$$E = h + Q^2/2gA^2$$

Differentiation with respect to h gives

$$\frac{\partial E}{\partial h} = 1 + \frac{Q^2}{2g}\left(-\frac{2}{A^3}\right)\frac{\partial A}{\partial h}$$

Now, from Fig. 10.20, the small increase of area δA, which corresponds to a small increase δh in the depth, is given by $B\delta h$ where B is the breadth of the *surface*. Hence, as $\delta h \to 0$, $\delta A/\delta h \to \partial A/\partial h = B$.

$$\therefore \quad \frac{\partial E}{\partial h} = 1 - \frac{Q^2}{gA^3}B$$

which is zero when $Q^2 = gA^3/B$, i.e. when $gA/B = Q^2/A^2 = u^2$. If A/B is regarded as the mean depth of the section and is represented by $\bar{h}$ then

$$u_c = \left(gA/B\right)^{1/2} = \left(g\bar{h}\right)^{1/2} \tag{10.20}$$

(This mean depth $\bar{h}$ must not be confused with the hydraulic mean depth A/P.)

The great importance of the 'critical' conditions is that they separate two distinct types of flow: that in which the velocity is less than the critical value and that in which the velocity exceeds the critical value. As we have seen in Section 10.8, the critical velocity $u_c = (g\bar{h})^{1/2}$ corresponds to the velocity of propagation (relative to the undisturbed liquid) of a small surface wave in shallow liquid. Thus, when the velocity of flow is less than the critical velocity, it is possible for a small surface wave to be propagated upstream as well as downstream. Any small disturbance to the flow can cause a small surface wave to be formed, and this wave may be regarded as carrying, to the liquid further away, information about the disturbance. If the wave, as messenger, can be propagated against the flow then the liquid upstream will be informed of the disturbance and its behaviour will be influenced accordingly. When the flow velocity is less than $(g\bar{h})^{1/2}$, then, the behaviour of the liquid upstream can be influenced by events downstream.

If, on the other hand, the flow velocity is greater than the critical, the liquid travels downstream faster than a small wave can be propagated upstream. Information about events downstream cannot therefore be transmitted to the liquid upstream, and so the behaviour of the liquid is

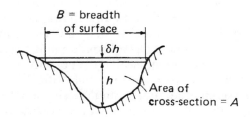

B = breadth of surface

δh

h

Area of cross-section = A

Fig. 10.20

not controlled by downstream conditions. In these circumstances small waves may be propagated only in the downstream direction.

When the flow is just at the critical velocity a relatively large change of depth causes only a small change of specific energy (as is shown by Fig. 10.18). Consequently small undulations on the surface are easily formed under these conditions. A small wave that attempts to travel upstream, however, makes no progress, and so is known as a *standing wave*. The appearance of stationary waves on the surface of flowing liquid is therefore an indication of critical flow conditions.

Thus we see that the behaviour of liquid flowing in an open channel depends very largely on whether the velocity of flow is greater or less than the critical velocity.

The ratio of the mean velocity of flow to the velocity of propagation of a small disturbance is $u/(g\bar{h})^{1/2}$ and this is seen to be of the same form as the Froude number, the dimensionless parameter we met in Section 5.3.2. If the mean depth $\bar{h}$, defined as the cross-sectional area divided by the surface width, is used as the characteristic length in the expression for Froude number, and the mean velocity of flow for the characteristic velocity, then $u/(g\bar{h})^{1/2} = Fr$. Thus for critical conditions $u = (g\bar{h})^{1/2}$ and $Fr = 1$. Alternatively it may be said that the critical Froude number is unity.

Uniform flow at the critical velocity may be produced in a long open channel if the slope is suitable. The value of the slope for which critical uniform flow is achieved is known as the *critical slope*. Using the Chézy formula, we obtain $u_c = (g\bar{h})^{1/2} = C(mi)^{1/2}$. In uniform flow the energy gradient i and the slope of the bed s are equal and so the critical slope s_c is defined by $(g\bar{h})^{1/2} = C(ms_c)^{1/2}$. A slope less than the critical slope is known as *mild*; a slope greater than the critical is *steep*. Since, however, the critical slope is a function of the depth and therefore of the rate of flow, a given channel may have a mild slope for one rate of flow but a steep slope for another rate.

Flow in which the velocity is less than the critical velocity is referred to as *tranquil*; flow in which the velocity is greater than the critical is known as *rapid* or *shooting*. Other pairs of names are often used, the most common alternatives probably being sub-critical and super-critical. These last-mentioned terms, it should be noted, refer to the velocity; sub-critical velocity, however, corresponds to a depth greater than the critical depth, and super-critical velocity corresponds to a depth smaller than the critical depth. The terms sub- and super-critical are thus apt to give rise to confusion. The same criticism may be levelled at the terms subundal and superundal (from Latin *unda* = a wave). In this book we shall use the terms tranquil and rapid. Tranquil flow, then, has a velocity less than the critical velocity and consequently a Froude number less than 1; rapid flow has a Froude number greater than 1.

It will be noticed that – provided that the velocity is uniform over the cross-section – the relations governing the critical conditions of flow involve only the volume rate of flow and the shape of the cross-section. They do not depend on the roughness of the boundaries.

Example 10.1 The cross-section of a river is rectangular and the width is 25 m. At a point where the river bed is horizontal the piers of a bridge restrict the width to 20 m. A flood of $400\,m^3\,s^{-1}$ is to pass under the bridge with the upstream depth a minimum. Given that this occurs when the flow under the bridge is critical, determine:

(a) the depth of the water under the bridge
(b) the depth of water upstream.

Solution
Denote upstream conditions by suffix 1 and those under the bridge by suffix 2. The flow under the bridge is critical. Hence, from Fig. 10.19, $h_1 > h_2$. Also $u_2 = \sqrt{gh_2}$. From continuity

$$Q = b_2 h_2 u_2 = b_2 g^{1/2} h_2^{3/2}$$

Hence

$$h_2 = \left(\frac{Q}{b_2 g^{1/2}}\right)^{2/3} = \left(\frac{400\,m^3/s}{20\,m \times \sqrt{9.81\,m/s^2}}\right)^{2/3} = 3.44\,m$$

Since energy is conserved

$$h_1 + \frac{u_1^2}{2g} = h_2 + \frac{u_2^2}{2g} = h_2 + \frac{h_2}{2} = \frac{3h_2}{2}$$

or

$$h_1 + \frac{1}{2g}\left(\frac{Q}{b_1 h_1}\right)^2 = \frac{3h_2}{2}$$

Substituting:

$$h_1 + \frac{1}{2 \times 9.81\,m/s^2}\left(\frac{400\,m^3/s}{25\,m \times h_1\,m}\right)^2 = \frac{3 \times 3.44\,m}{2}$$

which on multiplying throughout by h_1^2 becomes

$$h_1^3 - 5.16\,h_1^2 + 13.05 = 0$$

This cubic equation is solved by trial-and-error to yield $h_1 = 4.52\,m$. □

10.9.1 The use of the specific-energy curve in dimensionless form

The specific-energy curve shown in Fig. 10.18 shows the relation between the specific energy E and the depth h for a particular rate of flow. For a different, yet still unvarying, rate of flow, the curve relating E and h would be of similar shape but the value of E corresponding to each value of h would of course be different. A more general curve, applicable to any value of q, may be obtained by reducing eqn 10.15 to a dimensionless form. (A channel of rectangular section is again taken

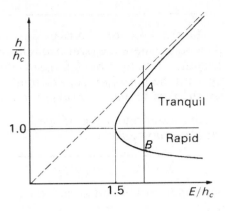

Fig. 10.21

as an example). Dividing eqn 10.15 by the critical depth h_c and then substituting gh_c^3 for q^2 from eqn 10.16 we obtain

$$\frac{E}{h_c} = \frac{h}{h_c} + \frac{1}{2}\left(\frac{h_c}{h}\right)^2 \qquad (10.21)$$

in which every term involves only a ratio of two lengths and is therefore dimensionless. The discharge per unit width q no longer appears explicitly in the equation, although q determines the value of h_c. Figure 10.21 shows the curve representing eqn 10.21. From such a diagram as this we may determine the two possible values of h/h_c (one greater than unity and one less than unity), as indicated for example by the points A and B, corresponding to a particular value of E/h_c. The depths corresponding to the points A and B are the alternative depths.

Similarly Fig. 10.19, showing the relation between rate of flow and the depth for a given specific energy, may be re-plotted in dimensionless form. If eqn 10.15 is divided by $q^2_{max} = gh_c^3$, a relation is obtained between q/q_{max} and h/h_c.

$$\frac{E}{gh_c^3} = \frac{h}{gh_c^3} + \frac{1}{2gh^2}\left(\frac{q}{q_{max}}\right)^2$$

$$\therefore \frac{2E}{h_c}\left(\frac{h}{h_c}\right)^2 = 2\left(\frac{h}{h_c}\right)^3 + \left(\frac{q}{q_{max}}\right)^2$$

and, using eqn 10.18 for constant E,

$$\left(\frac{q}{q_{max}}\right)^2 = 3\left(\frac{h}{h_c}\right)^2 - 2\left(\frac{h}{h_c}\right)^3 \qquad (10.22)$$

10.10   THE HYDRAULIC JUMP

In Section 10.8, attention was turned to the propagation of a surge wave in an open channel. For a positive wave travelling upstream in a horizontal channel of rectangular cross-section, eqn 10.13 relates the velocity of flow, the velocity of propagation of the wave, and the depths of flow before and after the wave. It is clear that for a particular set of conditions the wave velocity c may be zero, that is, the wave may be stationary relative to the bed of the channel. Such a stationary surge wave, through which the depth of flow increases, is known as a hydraulic jump. (The term 'standing wave' has also been used, but in modern usage this is reserved for a stationary wave of very small height, whereas the depths before and after a hydraulic jump are appreciably different.)

Equation 10.13 points at once to the essential features of a hydraulic jump. Putting $c = 0$ in that equation we obtain

$$u_1 = (gh_2)^{1/2}\left(\frac{1 + h_2/h_1}{2}\right)^{1/2} = (gh_1)^{1/2}\left(\frac{h_2}{h_1}\right)^{1/2}\left(\frac{1 + h_2/h_1}{2}\right)^{1/2} \tag{10.23}$$

$$\therefore\ u_1\big/(gh_1)^{1/2} = Fr_1 = \left\{\left(\frac{h_2}{h_1}\right)\left(\frac{1 + h_2/h_1}{2}\right)\right\}^{1/2}$$

Since $h_2 > h_1$ this expression is greater than unity. In other words, before a hydraulic jump the Froude number is always greater than unity and the flow is rapid.

What of the flow after the jump? From the continuity relation $u_1bh_1 = u_2bh_2$ we have $u_2 = u_1h_1/h_2$.

$$\therefore\ u_2 = (gh_2)^{1/2}\left(\frac{1 + h_2/h_1}{2}\right)^{1/2}\left(\frac{h_1}{h_2}\right)$$

and

$$Fr_2 = \frac{u_2}{(gh_2)^{1/2}} = \left\{\frac{(h_1/h_2)^2 + (h_1/h_2)}{2}\right\}^{1/2}$$

which is less than 1. Hence the flow after a hydraulic jump is always tranquil. (Although for mathematical simplicity we have here considered a channel of rectangular section, the same conditions apply to sections of any shape.)

A hydraulic jump, then, is an abrupt change from rapid to tranquil flow: the depth of the liquid is less than h_c before the jump and greater than h_c after it. The rapid flow before the jump may arise in a number of ways. The liquid may, for example, be released into the channel at high velocity from under a sluice gate, or enter via a steep spillway. However, the rapid flow thus produced cannot persist indefinitely in a channel where the slope of the bed is insufficient to sustain it. For the particular rate of flow concerned, the depth corresponding to uniform flow in the channel is determined by the roughness of the boundaries and the slope of the bed. For a mild slope this depth is greater than the critical depth, i.e. uniform flow would be tranquil.

A gradual transition from rapid to tranquil flow is not possible. As the depth of rapid flow increases, the lower limb of the specific-energy diagram (Fig. 10.18) is followed from right to left, that is, the specific energy decreases. If the increase of depth were to continue as far as the critical value, an increase of specific energy would be required for a further increase of depth to the value corresponding to uniform flow downstream. However, in such circumstances an increase in specific energy is impossible. In uniform flow the specific energy remains constant and the *total* energy decreases at a rate exactly corresponding to the slope of the bed. For any depth less than that for uniform flow the velocity is higher, and friction therefore consumes energy at a greater rate than the loss of gravitational energy. That is, the energy gradient is greater than the slope of the channel bed, and so, as shown in Fig. 10.22, the specific energy must decrease.

Consequently, a hydraulic jump forms before the critical depth is reached so that uniform flow can take place immediately after the jump. It represents a discontinuity in which the simple specific-energy relation is temporarily invalid (because the streamlines are then by no means straight and parallel). The depth of the tranquil flow after the jump is determined by the resistance offered to the flow, either by some obstruction such as a weir or by the friction forces in a long channel. The jump causes much eddy formation and turbulence. There is thus an appreciable loss of mechanical energy, and both the total energy and the specific energy after the jump are less than before.

Determining the change in depth that occurs at the hydraulic jump is frequently important. Although a hydraulic jump may be formed in a channel with any shape of cross-section, we shall again consider here only a channel of uniform rectangular cross-section because the analysis for other sections is mathematically complicated.

A rearrangement of eqn 10.23 gives

$$h_2^2 + h_1 h_2 - 2u_1^2 h_1/g = 0$$

Putting $u_1 = q/h_1$, where q represents the discharge per unit width, we get

$$h_1 h_2^2 + h_1^2 h_2 - 2q^2/g = 0 \tag{10.24}$$

whence

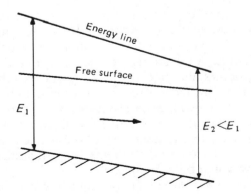

Fig. 10.22

$$h_2 = -\frac{h_1}{2} {}^{+}_{(-)} \sqrt{\left(\frac{h_1^2}{4} + \frac{2q^2}{gh_1}\right)} = -\frac{h_1}{2} + \sqrt{\left(\frac{h_1^2}{4} + \frac{2h_1 u_1^2}{g}\right)} \qquad (10.25)$$

(The negative sign for the radical is rejected because h_2 cannot be negative.) Equation 10.24 is symmetrical in respect of h_1 and h_2 and so a similar solution for h_1 in terms of h_2 may be obtained by interchanging the suffixes. The depths of flow on both sides of a hydraulic jump are termed the *conjugate depths* for the jump.

We recall that the following assumptions have been made:

1. The bed is horizontal (or so nearly so that the component of weight in the direction of flow may be neglected) and the rectangular cross-section uniform (i.e. the channel is not tapered). (For the hydraulic jump in channels of non-zero slope see Chow[2], p. 425.)
2. The velocity over each of the cross-sections considered is so nearly uniform that mean velocities may be used without significant error.
3. The depth is uniform across the width.
4. Friction at the boundaries is negligible. This assumption is justifiable because the jump occupies only a short length of the channel.
5. Surface tension effects are negligible.

The loss of mechanical energy that takes place in the jump may be readily determined from the energy equation. If the head lost in the jump is h_j then

$$h_j = \left(h_1 + \frac{u_1^2}{2g}\right) - \left(h_2 + \frac{u_2^2}{2g}\right)$$

$$= h_1 - h_2 + \frac{q^2}{2g}\left(\frac{1}{h_1^2} - \frac{1}{h_2^2}\right)$$

$$= h_1 - h_2 + \left(\frac{h_1 h_2^2 + h_1^2 h_2}{4}\right)\left(\frac{1}{h_1^2} - \frac{1}{h_2^2}\right)$$

(from eqn 10.24)

$$= \left(h_2 - h_1\right)^3 \big/ 4 h_1 h_2 \qquad (10.26)$$

This amount is represented by the distance h_j on Fig. 10.23b. This dissipation of energy is a direct result of the considerable turbulence in the wave: friction at the boundaries makes a negligible contribution to it. The 'frictional' forces in the wave are in the form of innumerable pairs of action and reaction and so, by Newton's Third Law, annul each other in the net force on the control volume considered in deriving eqns 10.13 and 10.23. Fortunately then, the exact form of the jump between sections 1 and 2 is of no consequence in our analysis.

The dissipation of energy is by no means negligible. Indeed, the hydraulic jump is a very effective means of reducing unwanted energy in a stream. For example, if water from a steep spillway is fed into a

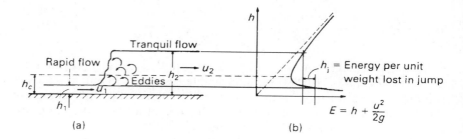

Fig. 10.23

(a)   (b)

channel, severe scouring of the bed may well occur if the rapid flow is allowed to continue. A hydraulic jump arranged to occur at the foot of the spillway, however, dissipates much of the surplus energy, and the stream may then be safely discharged as tranquil flow. The final result of the dissipation of energy is, of course, that the temperature of the liquid is raised somewhat.

The position at which the jump occurs is always such that the momentum relation is satisfied: the value of h_2 is determined by conditions downstream of the jump, and the rapid flow continues until h_1 has reached the value which fits eqn 10.25. The method of calculating the position will be indicated in Section 10.12.1.

The turbulence in a jump may be sufficient to induce large quantities of air into the liquid. The presence of bubbles of air reduces the effective density of the liquid, and, as a result, the depth immediately after the jump may be greater than that predicted by eqn 10.25. The great turbulence in a hydraulic jump is sometimes put to advantage in mixing two liquids thoroughly together.

Equation 10.26 tells us that a hydraulic jump is possible only from rapid to tranquil flow and not vice versa. If h_2 were less than h_1 then h_j would be negative, that is, there would be a *gain* of energy. This would contravene the Second Law of Thermodynamics. A hydraulic jump, then, is an irreversible process.

It is instructive to write eqn 10.25 in the following form:

$$\frac{h_2}{h_1} = -\frac{1}{2} + \sqrt{\left(\frac{1}{4} + \frac{2u_1^2}{gh_1}\right)} = -\frac{1}{2} + \sqrt{\left\{\frac{1}{4} + 2(Fr_1)^2\right\}} \quad (10.27)$$

Equation 10.27 emphasizes the importance of the Froude number as a parameter describing flow in open channels. We see that the ratio of the conjugate depths, h_2/h_1, is a function of the initial Froude number only and that the larger the initial Froude number the larger the ratio of the depths. When $Fr_1 = 1$, $h_2/h_1 = 1$ and the jump becomes a standing wave of infinitesimal height.

For small jumps, that is, those in which h_2/h_1 is not greater than 2.0 (and consequently Fr_1 for rectangular sections is not greater than $\sqrt{3}$), the surface does not rise abruptly but passes through a series of undulations gradually diminishing in size. Such a jump is known as an *undular jump* (see Fig. 10.24). For larger values of h_2/h_1 and Fr_1, however, the

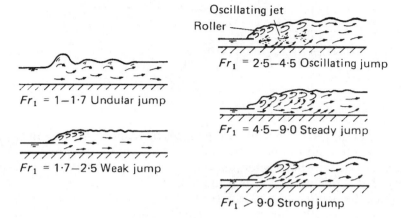

Fig. 10.24 Various types of hydraulic jump in rectangular channels.

jump is *direct*, that is, the surface does rise fairly abruptly (as shown in Fig. 10.23a).

At the wave front there is a 'roller' (rather like an ocean wave about to break on the shore). This results from the upper layers of the wave tending to spread over the oncoming rapid stream. The frictional drag of the rapid stream penetrating underneath, and the transfer of momentum from the lower layers by eddies, however, prevent the upper layers moving upstream.

When h_2/h_1 is between about 3.0 and 5.5, oscillations may be caused which result in irregular waves being transmitted downstream. For values of h_2/h_1 between 5.5 and 12 the jump is stable and a good dissipator of energy. The length of the jump (that is, the horizontal distance between the front of the jump and a point just downstream of the roller) is usually of the order of five times its height.

These figures refer only to channels of rectangular section. In other channels the shape of the jump is often complicated additionally by cross-currents.

10.10.1 The force applied to obstacles in a stream

The techniques used to study a hydraulic jump may also be used to calculate the force experienced by an obstacle placed in a moving stream. Such an obstacle may be a pier of a bridge, a large block placed on the bed of the stream, a buoy, or an anchored boat. If the obstacle illustrated in Fig. 10.25 exerts a force F on the liquid (in the upstream direction) then, with the assumptions of a horizontal bed, uniform velocity over the cross-section, uniform depth across the width and negligible boundary friction, the steady-flow momentum equation is:

$$\frac{\varrho g h_1}{2} h_1 - F - \frac{\varrho g h_2}{2} h_2 = \varrho q (u_2 - u_1) \qquad (10.28)$$

for unit width of a uniform rectangular channel.

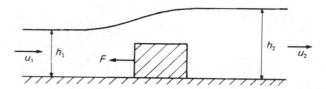

Fig. 10.25

This equation is true whatever the values of h_1 and h_2 in relation to the critical depth: both h_1 and h_2 may be greater than h_c, both may be less than h_c or one may be greater and one less than h_c. It may, however, be shown that if the initial flow is tranquil the effect of the applied force is to reduce the depth of the stream (so that $h_2 < h_1$), although a limit is set at the critical depth h_c because the specific energy is then a minimum. A further increase in the obstructing force F beyond the value giving $h_2 = h_c$ merely raises the upstream level. If, on the other hand, the initial flow is rapid, the depth is increased by the application of the force (i.e. $h_2 > h_1$). Indeed, if the force is large enough – because, for example, the obstacle is large compared with the cross-sectional area of the channel – the depth may be increased beyond the critical value via a hydraulic jump. In all these cases eqn 10.28 holds and the force F may thus be calculated.

10.11   THE OCCURRENCE OF CRITICAL CONDITIONS

In analysing problems of flow in open channels it is important to know at the outset whether critical flow occurs anywhere and, if so, at which section it is to be found, especially because these conditions impose a limitation on the discharge (as indicated by Fig. 10.19).

Critical conditions are of course to be expected at a section where tranquil flow changes to rapid flow. Such a situation is illustrated in Fig. 10.26. A long channel of mild slope (i.e. $s < s_c$) is connected to a long channel of steep slope (i.e. $s > s_c$). (The slopes are greatly exaggerated in the diagram.) At a sufficiently large distance from the junction the depth in each channel is the *normal* depth corresponding to the particular slope and rate of flow; that is, in the channel of mild slope there is uniform tranquil flow, and in the other channel there is uniform rapid flow. Between these two stretches of uniform flow the flow is non-uniform, and at one position the depth must pass through the critical value as defined for example by eqn 10.19. This position is close to the junction of the slopes. If both channels have the same constant shape of cross-section the critical depth remains unchanged throughout, as shown by the dashed line on the diagram (the effect of the difference in slopes being negligible). If the change of slope is abrupt (as in the diagram) there is appreciable curvature of the streamlines near the junction. The assumption of a hydrostatic variation of pressure with depth – on which the concept of specific energy is based – is then not

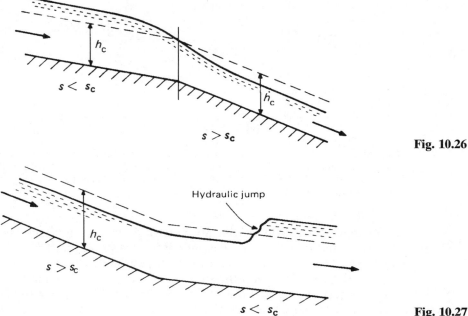

Fig. 10.26

Fig. 10.27

justified and the specific-energy equation is only approximately valid. In these circumstances the section at which the velocity is given by $(g\bar{h})^{1/2}$ is not exactly at the junction of the two slopes, but slightly upstream from it.

The discharge of liquid from a long channel of steep slope to a long channel of mild slope requires the flow to change from rapid to tranquil (Fig. 10.27). The rapid flow may persist for some distance downstream of the junction, and the change to tranquil flow then takes place abruptly at a hydraulic jump.

Critical flow, however, may be brought about without a change in the slope of the channel bed. A raised part of the bed, or a contraction in width, may in certain circumstances cause critical flow to occur. We shall now study these conditions in Sections 10.11.1–10.11.4.

10.11.1 The broad-crested weir

We consider first the flow over a raised portion of the bed, which extends across the full width of the stream. Such an obstruction is usually known as a weir. Figure 10.28a depicts a weir with a broad horizontal crest, and at a sufficient height above the channel bed for the cross-sectional area of the flow approaching the weir to be large compared with the cross-sectional area of the flow over the top of it. We assume that the approaching flow is tranquil. The upstream edge of the weir is well rounded so as to avoid undue eddy formation and thus loss of mechanical energy. We suppose that downstream of the weir there is

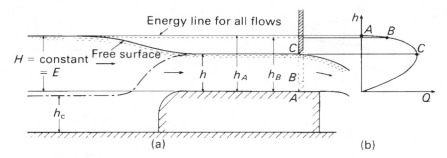

Fig. 10.28 Notice that, for a channel of rectangular section and constant width, the critical depth as shown by —·—·— is the same for a given rate of flow at any point along the bed (apart from the slight deviations at the corners because the streamlines are not parallel there).

no further obstruction (so that the rate of flow is not limited except by the weir) and that the supply comes from a reservoir sufficiently large for the surface level upstream of the weir to be constant.

Over the top of the weir the surface level falls to give a depth h there. Moreover, provided that the channel width is constant, that the crest is sufficiently broad (in the flow direction) and friction is negligible, the liquid surface becomes parallel to the crest. To determine the value of the depth h we may imagine that downstream of the weir – or perhaps near the end of it as indicated by the dotted lines in Fig. 10.28a – the flow is controlled by a movable sluice gate. If, initially, the gate is completely closed, the liquid is stationary and the surface level above the crest of the weir is the same as the level in the reservoir. This level corresponds to the energy available and H represents the specific energy for the liquid on the crest of the weir. To correspond to these conditions we have the point A on the curve of h against Q for the given specific energy (Fig. 10.28b). If the gate is raised slightly to a position B, a small rate of flow Q_B takes place, as represented by point B on the h–Q curve, and the depth is then h_B. Further raising of the gate brings the flow rate to Q_C, the maximum, and since the gate is then just clear of the surface of the liquid, no additional raising has any effect on the flow. With no other obstruction downstream the discharge is the maximum possible for the specific energy available. That is, the flow is critical over the crest of the weir and $h = h_c$. Even if it were possible to reduce h below the critical depth, the h–Q curve shows that the rate of flow would be less than Q_C, not greater.

If the channel is of rectangular cross-section, $h_c = (q^2/g)^{1/3}$, where q represents the discharge per unit width, Q/b. Thus $q = g^{1/2}h_c^{3/2}$. From this equation it is evident that the rate of flow could be calculated simply from a measurement of h_c, and consequently the broad-crested weir is a useful device for gauging the rate of flow. In practice, however, it is not easy to obtain an accurate measurement of h_c directly. When critical flow occurs there are usually many ripples on the surface and so the depth seldom has a steady value. It is thus more satisfactory to express the rate of flow in terms of the specific energy. From eqn 10.18 (again for a channel of rectangular section) $h_c = 2E/3$, so $q = g^{1/2}(2E/3)^{3/2}$. Therefore

$$Q = g^{1/2}b\left(\frac{2E}{3}\right)^{3/2} = \left(1.705\frac{m^{1/2}}{s}\right)bE^{3/2} = \left(3.09\frac{ft^{1/2}}{s}\right)bE^{3/2} \quad (10.29)$$

The specific energy E in eqn 10.29 is that over the crest of the weir, the crest itself being the datum. If, however, the velocity u of the liquid upstream is so small that $u^2/2g$ is negligible, and if friction at the approach to the weir is also negligible, the specific energy over the weir is given simply by the term H. It should be particularly noted that H must be measured above the crest as datum. It is *not* the full depth in the channel upstream. For a negligible 'velocity of approach', therefore, eqn 10.29 becomes

$$Q = \left(1.705\frac{m^{1/2}}{s}\right)bH^{3/2} = \left(3.09\frac{ft^{1/2}}{s}\right)bH^{3/2} \quad (10.30)$$

In the above analysis no account has been taken of friction which, in practice, is always present in some degree. The effect of friction over the crest, and the curvature of the streamlines, is to reduce the value of the discharge somewhat below the value given by eqn 10.30. If the crest of the weir is insufficiently broad in comparison with the head H, the liquid surface may not become parallel to the crest before the downstream edge is reached (see Fig. 10.29). Although critical flow may still be obtained (in the sense that the specific energy is a minimum for the particular discharge occurring) it takes place at a section where the streamlines have appreciable curvature and so the specific energy relation $E = h + u^2/2g$ does not strictly apply. In these circumstances eqn 10.29 underestimates the true rate of flow. Even when the crest is very broad, so that sensibly parallel flow is obtained, the truly critical conditions, in which the specific energy is a minimum for the particular discharge, must, because of friction, occur at the downstream edge, where the streamlines are curved. Appreciable variations in the numerical factors in eqn 10.30 are thus possible and especially for weirs having cross-sections different from the more or less rectangular one shown in Fig. 10.28a. If a broad-crested weir is to be used for reliable

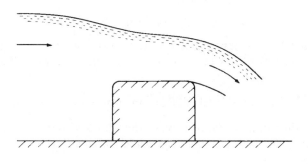

Fig. 10.29

determination of the rate of flow, calibration is therefore necessary. Since the development of the venturi flume (Section 10.11.4) the use of broad-crested weirs for measurement has in fact greatly declined.

A weir is sometimes used in a situation where the velocity of the stream approaching it is not negligible. The height of the upstream surface level above the crest of the weir is thus somewhat less than the specific energy ($H = E - u^2/2g$). The measured value of H may, however, be used as a first approximation to E, and a value of Q calculated using this figure. The 'velocity of approach' may then be estimated from this value of Q and the cross-sectional area of the channel at the section where the measurement of H is made. From this value of velocity (say u_1) the term $u_1^2/2g$ may be calculated, and then $H + u_1^2/2g$ may be taken as a closer approximation to the specific energy so that another, more accurate, value of Q may be determined.

■

Example 10.2   A block 40 mm thick is placed across the bed of a rectangular horizontal channel 400 mm wide to form a broad-crested weir. The depth of liquid just upstream of the weir is 70 mm. Calculate the rate of flow on the assumption that conditions over the block are critical, and make one correction for the velocity of approach. The effects of friction and of curvature of the streamlines may be assumed negligible.

Solution
As a first approximation the velocity of approach is assumed negligible. Then the specific energy relative to the crest of the weir is $(70 - 40)\,\text{mm} = 30\,\text{mm}$.

Therefore the first approximation to Q is $1.705\,bH^{3/2}\,\text{m}^{1/2}/\text{s}$, the effects of friction and curvilinear motion being neglected. That is

$$Q = 1.705 \times 0.4 \times (0.03)^{3/2}\,\text{m}^3/\text{s} = 3.544 \times 10^{-3}\,\text{m}^3/\text{s}$$

For this rate of flow the velocity of approach would be

$$\frac{3.544 \times 10^{-3}\,\text{m}^3/\text{s}}{0.070 \times 0.4\,\text{m}^2} = 0.1266\,\text{m/s}$$

$$\therefore\ u_1^2/2g = 0.000816\,\text{m} = 0.816\,\text{mm}$$

Consequently a closer approximation to the specific energy relative to the crest of the weir would be $(30 + 0.816)\,\text{mm} = 30.82\,\text{mm}$. Then

$$Q = 1.705 \times 0.4 \times (0.03082)^{3/2}\,\text{m}^3/\text{s} = 3.689 \times 10^{-3}\,\text{m}^3/\text{s}.$$

If a further approximation were desired another, better, value of the upstream velocity (say u_2) could be calculated

$$\left(\frac{3.689 \times 10^{-3}\,\mathrm{m}^3/\mathrm{s}}{0.070 \times 0.4\,\mathrm{m}^2} \right)$$

from which $u_2^2/2g$ could be determined and added on to $H = 30\,\mathrm{mm}$ to give the next approximation to the specific energy. One correction for velocity of approach, however, gives sufficient accuracy. (A rigorous solution of the problem is possible but the resulting expression is too complicated for practical use. The method of successive approximation given here is much simpler and entirely adequate: the final value of Q in this problem is $3.703 \times 10^{-3}\,\mathrm{m}^3/\mathrm{s}$.)

The velocity of approach here is necessarily calculated as the mean velocity over the whole cross-section of the approach channel. The distribution of velocity over this cross-section is never perfectly uniform and so the extra term to be added to the measured value of H should, strictly, be slightly greater than the calculated $u^2/2g$. However, the effect on the final result is so small that such refinements are of little consequence. ☐

As a measuring device the broad-crested weir has the advantages that it is simple to construct and has no sharp edge which can wear and thus alter the coefficient. Moreover, it does not cause an appreciable raising of the surface level upstream, and the results are little affected by conditions downstream provided that critical flow occurs over the crest. On the other hand, the limitations of the theoretical analysis make it unwise to use, for any but approximate measurements, a weir that has not been calibrated.

10.11.2 Drowned weir and free outfall

Critical flow does not always occur over a rise in the bed of a channel. The maximum discharge takes place only in the absence of greater restrictions downstream. If the depth downstream of the weir is sufficiently increased, the flow over the weir may not be critical, and the weir is then said to be 'drowned'. For example, another obstruction placed downstream may serve to increase the depth between itself and the weir, as shown in Fig. 10.30. In this case the flow may be critical over the top of the downstream obstacle. Or the channel may discharge into a reservoir in which the level is high enough to maintain a depth greater than the critical over the weir (Fig. 10.31). In some instances the friction in a long channel downstream of the weir may be sufficient to prevent critical conditions occurring over the weir.

When a broad-crested weir is 'drowned' a depression of the surface over the crest still takes place, but it is not sufficient for the critical depth to be reached. In no circumstances, indeed, can a depth less than the critical be achieved: if a weir over which critical flow exists is raised still further above the channel bed, the liquid surface rises by

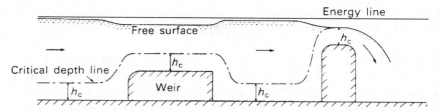

Fig. 10.30 Broad-crested weir 'drowned' by obstruction downstream (over which flow is critical). For a channel of rectangular section and constant width, $h_c = (q^2/g)^{1/3}$ is constant, except where streamlines are not parallel.

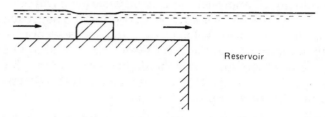

Fig. 10.31 Broad-crested weir 'drowned' by high level in downstream reservoir.

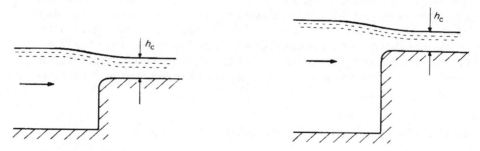

Fig. 10.32 If the depth of flow over a weir is already critical, raising the weir increases the depth upstream, and flow over the weir remains critical.

a corresponding amount so that the depth over the crest is still the critical depth (Fig. 10.32).

Arguments similar to those used in discussing the broad-crested weir may be applied to a *free outfall* (Fig. 10.33) from a long channel of mild slope. The discharge is a maximum for the specific energy available, and the flow must pass through the critical conditions in the vicinity of the brink. At the brink the streamlines have a pronounced curvature and so the usual specific-energy relation $E = h + u^2/2g$ is invalid. The depth immediately over the brink is in fact less than the value given by the usual expression for critical depth. For example, in a channel of uniform rectangular section, the depth $(q^2/g)^{1/3}$ (eqn 10.16) is found a short distance upstream (between three and four times $(q^2/g)^{1/3}$) and the depth at the brink itself is about 71% of this depth (Rouse[3]).

10.11.3 Rapid flow approaching a weir or other obstruction

In Sections 10.11.1 and 10.11.2, the flow approaching the weir has been considered tranquil. Even in a channel of mild slope, however, rapid flow may be produced. This may occur when the flow enters the channel down a spillway or from under a sluice gate. For a sufficient head upstream of the sluice, the discharge through the aperture may be great enough for the flow to be rapid. In a channel of rectangular cross-section, for example, if the discharge per unit width is q, the critical depth is given by $(q^2/g)^{1/3}$, and if this exceeds the height of the aperture the flow is necessarily rapid. The velocity head upstream of the sluice is normally negligible, and so the discharge is determined by the difference in head $h_0 - h_1$ across the sluice opening (see Fig. 10.34). Thus

$$q = C_d h_1 \sqrt{\left\{ 2g \left(h_0 - h_1 \right) \right\}} \tag{10.31}$$

(The coefficient of discharge C_d is close to unity for a well-rounded aperture.)

If friction is negligible the total energy of the liquid remains unchanged. But when the rapid flow reaches the raised part of the bed the *specific* energy is reduced by an amount equal to the height of the weir, z. (We recall that specific energy must always be calculated above the base of the flow at the point in question.) Thus, for a moderate value of z, the appropriate point on the specific-energy diagram (Fig. 10.35) moves from position 1 to position 2. In such a case as that illustrated the flow remains rapid but the depth comes closer to the critical depth h_c.

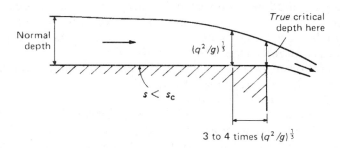

Fig. **10.33** (Vertical scale exaggerated.)

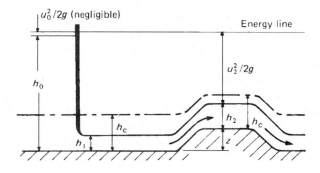

Fig. **10.34**

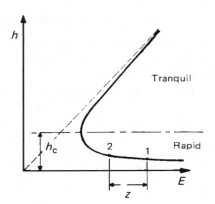

Fig. 10.35

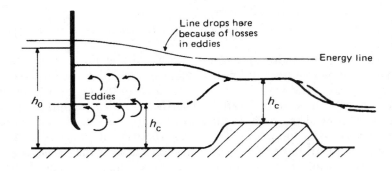

Fig. 10.36

For a larger value of z the specific energy may be reduced to the minimum value and the conditions over the weir are then critical. The flow is not, however, controlled by the weir as it was when the approaching flow was tranquil (Section 10.11.1). Rapid flow can never be controlled by conditions downstream and in the present example the flow will already have been determined by the difference of head across the sluice, and therefore by eqn 10.31.

If z is greater still, no further reduction in the specific energy is possible and so the depth upstream of the weir is increased. The flow is therefore tranquil instead of rapid. The critical depth h_c still exists on the crest of the weir and the conditions are closely similar to those investigated in Section 10.11.1. The sluice is now *drowned*, that is, the liquid from it discharges into slower moving liquid at a depth greater than the height of the aperture, as shown in Fig. 10.36. The weir now does exercise control over the discharge, since the approaching flow is at a depth greater than the critical. Much turbulent, eddying motion develops where the liquid discharged from the sluice encounters the slower moving liquid ahead of it. There is thus an appreciable reduction in the specific energy over this part of the flow. Over the crest of the weir, however, there is little production of eddies since the flow here is converging before becoming (approximately) parallel. (In this example there would be no point in using the broad-crested weir as a metering device, since the sluice itself, being simply a kind of orifice, could be so used.)

10.11.4 The venturi flume

Tranquil flow in an open channel may become critical not only by passing over a raised portion of the bed (a broad-crested weir) but also by undergoing a contraction in width. Such an arrangement, shown (in plan view) in Fig. 10.37, is usually known as a venturi flume, although its relation to a venturi in a pipe is slight. As a device for measuring the rate of flow the venturi flume has certain advantages over a broad-crested weir. The principal ones are that the loss of head experienced by the liquid in passing through the flume is much less than that in flowing over a broad-crested weir, and that the operation of the flume is not affected by the deposition of silt. If the liquid surface downstream is not maintained at too high a level, maximum discharge is achieved through the narrowest section (known as the *throat*) for the same reasons that the discharge over a broad-crested weir reaches a maximum in the absence of appreciable restrictions downstream. The flow at the throat is therefore critical, and the flume is said to be under *free discharge* conditions. For a rectangular section in which the streamlines are straight and parallel the velocity at the throat is therefore given by $\sqrt{(gh_2)}$, and the discharge by

$$Q = b_2 h_2 \sqrt{(gh_2)} \tag{10.32}$$

where b_2 represents the width of the throat, and h_2 the corresponding depth.

Measurements of b_2 and h_2 would therefore be sufficient for the calculation of the discharge. The exact position at which the critical conditions exist, however, is not easy to determine, and measurement of h_2 is thus impracticable. If, however, there is negligible friction in the upstream, converging, part of the flume and the slope of the bed is also negligible over this distance, then

$$h_1 + \frac{u_1^2}{2g} = h_2 + \frac{u_2^2}{2g} = h_2 + \frac{h_2}{2} = \frac{3}{2} h_2 \tag{10.33}$$

where suffix 1 refers to quantities upstream of the contraction.

Moreover

$$u_1 = \frac{Q}{b_1 h_1} = \frac{b_2 h_2 \sqrt{(gh_2)}}{b_1 h_1}$$

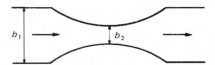

Fig. 10.37 Plan view of a venturi flume.

and this expression substituted for u_1 in eqn 10.33 gives

$$h_1 + \frac{1}{2g}\left(\frac{b_2}{b_1}\right)^2 \frac{h_2^3}{h_1^2} g = \frac{3}{2}h_2$$

i.e.

$$\left(h_1/h_2\right)^3 + \frac{1}{2}\left(b_2/b_1\right)^2 = \frac{3}{2}\left(h_1/h_2\right)^2 \tag{10.34}$$

Equation 10.34 shows that, since the ratio b_2/b_1 is fixed for a particular flume, the ratio of depths h_1/h_2 is constant whatever the rate of flow, provided that the discharge is 'free' (that is, the velocity at the throat is critical). Therefore eqn 10.32 may be re-written

$$Q = b_2 g^{1/2} h_2^{3/2} = \frac{b_2 g^{1/2} h_1^{3/2}}{r^{3/2}} \tag{10.35}$$

where $r = h_1/h_2$, and the rate of flow through a given flume may be determined by measuring the *upstream* depth h_1.

Since $b_1 > b_2$, the discharge per unit width $q_1 < q_2$, and thus, as Fig. 10.19 shows, $h_1 > h_2$ for tranquil flow upstream, i.e. $r > 1$.

The value of r is, of course, given by the solution of eqn 10.34. Of the three roots the only one meeting the requirement $r > 1$ is

$$r = 0.5 + \cos\left(2\theta/3\right)$$

where $0 \leqslant \theta \leqslant 90°$ and $\sin\theta = b_2/b_1$. If b_1 is large compared with b_2, r becomes equal to 1.5 and eqn 10.35 may then be written

$$Q = \frac{b_2 g^{1/2} h_1^{3/2}}{\left(1.5\right)^{3/2}} = \left(1.705\,\frac{m^{1/2}}{s}\right)b_2 h_1^{3/2} = \left(3.09\,\frac{ft^{1/2}}{s}\right)b_2 h_1^{3/2} \tag{10.36}$$

This equation may be derived alternatively by substituting for h_2 from eqn 10.33 into eqn 10.32:

$$Q = b_2 g^{1/2}\left(\frac{2}{3}\right)^{3/2}\left(h_1 + \frac{u_1^2}{2g}\right)^{3/2} \tag{10.37}$$

When b_1 is very large, u_1 is small and $u_1^2/2g$ may be neglected. Even for $b_2/b_1 = 1/3$, $r = 1.474$ and so the error involved in neglecting the velocity of approach and using eqn 10.36 would be only 1.7% in r and 2.54% in Q.

Corrections for velocity of approach may be applied in the same manner as for the broad-crested weir (Section 10.11.1). If u_1 is at first neglected in eqn 10.37 an approximate value of Q may be calculated from which $u_1 = Q/b_1 h_1$ can be estimated for use in a second approximation. For a flume that has already been calibrated, however, this arithmetical process is not necessary as the cubic equation for r will have been solved once for all.

A coefficient of discharge C_d is required to account for the small amount of friction between inlet and throat and for the effects of curvature of the streamlines which the above theory ignores. Values of C_d between 0.95 and 0.99 are obtainable in practice.

All the foregoing applies to a flume under conditions of 'free discharge', that is, where the liquid surface downstream is not maintained at too high a level. The level in the outlet from the flume continues to fall and thus rapid flow exists where the width of the passage again increases. If conditions downstream of the flume are such that the velocity is greater than the critical, the surface of the liquid issuing from the flume gradually merges into the normal depth of flow in the downstream channel, and any excess energy possessed by the liquid emerging from the flume is appropriated by friction.

On the other hand, if conditions downstream are such as to demand a velocity less than the critical, the flow has to change from rapid to tranquil. This change normally takes place through a hydraulic jump (Fig. 10.38). The jump occurs at a point where the depth of the rapid flow is such as to give the correct depth of the subsequent tranquil flow (according to eqn 10.25 if the section is rectangular). A limiting position is that in which the jump – then shrunk to zero height – occurs at the throat itself. If the downstream level is raised further (usually beyond about 80% of the depth upstream of the flume) the velocity at the throat no longer reaches the critical and the flume is said to be 'drowned'. In these circumstances the ratio h_1/h_2 is no longer constant and so the rate of flow cannot be determined from a measurement of the upstream depth only. Determination of the discharge is still possible, but only if the depth h_2 at the throat is also measured. Flumes are usually designed, however, to 'run free' (i.e. not to be drowned) for all expected conditions.

A number of modern flumes, intended to deal with a wide range of discharge, have not only a contraction in width at the throat but also a hump on the bed. The rise in the bed enables such flumes to 'run free' with a higher downstream level and yet avoids the use of an excessively small throat width which would give rise to undue dissipation of energy at the large rates of flow. (As for the broad-crested weir, the upstream head must then be measured relative to the top of the hump.) Flumes of more complex shapes are also constructed but the fundamental principles of operation are identical to those for the simple type considered here.

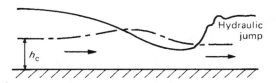

Fig. 10.38 Flow through venturi flume. The critical depth at the throat is greater because the narrowing of the channel increases the discharge per unit width q and for a rectangular section $h_c = (q^2/g)^{1/3}$.

As a measuring device the venturi flume has the advantage that under ideal conditions the loss of mechanical energy may be kept as low as 10% in a good design. The hydraulic jump that is usually present downstream is the cause of much of the dissipation of energy. Yet even if there is no hydraulic jump the losses principally occur downstream of the throat since eddies readily form where the section expands again. Like the broad-crested weir, the flume is suitable for measurement only when the upstream flow is tranquil.

Example 10.3   A venturi flume installed in a rectangular horizontal channel 800 mm wide has a throat of width 300 mm. The depths of water upstream and at the throat are 450 mm and 350 mm, respectively. Calculate:

(a) the flow rate
(b) the Froude number at the throat.

Conditions downstream of the flume are now altered and a hydraulic jump occurs, while the upstream depth is maintained at 450 mm. Calculate:

(c) the depth of water at the throat
(d) the new flow rate.

Solution
For parts (a) and (b), from the way the question is posed it is not known whether the flow through the flume is critical or not. In these circumstances, it is important to use the general flow relations, and not the more restricted relations which are specific to critical flow.
   (a) The continuity equation is

$$Q = u_1 h_1 b_1 = u_2 h_2 b_2$$

The general energy equation is

$$h_1 + \frac{u_1^2}{2g} = h_2 + \frac{u_2^2}{2g}$$

or

$$h_1 + \frac{1}{2g} \frac{Q^2}{\left(h_1 b_1\right)^2} = h_2 + \frac{1}{2g} \frac{Q^2}{\left(h_2 b_2\right)^2}$$

Substituting

$$0.45\,\text{m} + \frac{1}{2 \times 9.81\,\text{m/s}^2} \frac{Q^2\left(\text{m}^3/\text{s}\right)^2}{\left(0.45\,\text{m} \times 0.8\,\text{m}\right)^2}$$

$$= 0.35\,\text{m} + \frac{1}{2 \times 9.81\,\text{m/s}^2} \frac{Q^2\left(\text{m}^3/\text{s}\right)^2}{\left(0.35\,\text{m} \times 0.3\,\text{m}\right)^2}$$

which is solved to give $Q = 0.154\,\mathrm{m^3\,s^{-1}}$.

(b) $Fr_2 = \dfrac{u_2}{(gh_2)^{1/2}} = \dfrac{Q}{g^{1/2}b_2h_2^{3/2}}$

$\qquad = \dfrac{0.154\,\mathrm{m^3/s}}{(9.81\,\mathrm{m/s^2})^{1/2} \times 0.3\,\mathrm{m} \times (0.35\,\mathrm{m})^{3/2}} = 0.79$

This calculation shows that the flow at the throat is sub-critical.

For parts (c) and (d) the fact that a hydraulic jump occurs downstream indicates that the flow in the throat is critical. Hence the relations for critical flow may now be used.

(c) From equation 10.34

$$\left(\frac{h_1}{h_2}\right)^3 + \frac{1}{2}\left(\frac{b_2}{b_1}\right)^2 = \frac{3}{2}\left(\frac{h_1}{h_2}\right)^2$$

or

$$\frac{3}{2}\left(\frac{h_1}{h_2}\right)^2 - \left(\frac{h_1}{h_2}\right)^3 = \frac{1}{2}\left(\frac{b_2}{b_1}\right)^2 = \frac{1}{2}\left(\frac{0.3\,\mathrm{m}}{0.8\,\mathrm{m}}\right)^2 = 0.0703$$

which has the solution

$$\frac{h_1}{h_2} = 0.5 + \cos\frac{2\arcsin(b_2/b_1)}{3} = 0.5 + \cos\frac{2\arcsin(0.3/0.8)}{3}$$

$$\qquad = 1.467$$

Hence

$$h_2 = \frac{0.45\,\mathrm{m}}{1.467} = 0.307\,\mathrm{m}$$

(d) $Q = g^{1/2}b_2h_2^{3/2} = (9.81\,\mathrm{m/s^2})^{1/2} \times 0.3\,\mathrm{m} \times (0.307\,\mathrm{m})^{3/2} = 0.160\,\mathrm{m^3\,s^{-1}}$.

$\qquad\qquad\qquad\qquad\qquad\qquad\qquad\qquad\qquad\qquad\qquad\qquad\square$

10.12 GRADUALLY VARIED FLOW

Uniform flow, which we studied in Sections 10.4–10.7, is generally to be found only in artificial channels because the condition requires a cross-section constant in shape and area. Consequently the liquid surface must be parallel to the bed of the channel, and this in turn demands that the slope of the bed be constant. With a natural stream, such as a river, the shape and size of cross-section and also the slope of the bed usually vary appreciably, and true uniform flow is extremely rare. Indeed, even for artificial channels uniform flow is a condition that is approached asymptotically and so, strictly speaking, is never attained at all. The equations for uniform flow therefore give results that are only approximations to the truth when applied to flow in natural channels and, even so, care should be taken that they are not applied to long lengths of the channels over which the conditions are not even approximately constant.

For a particular shape of channel and for a given discharge and bed slope, there is only one depth at which uniform flow can take place. This depth is known as the *normal depth*. There are, however, innumerable ways in which the same steady rate of flow can pass along the same channel in non-uniform flow. The liquid surface is then not parallel to the bed, and takes the form of a curve.

There are, broadly speaking, two kinds of steady, non-uniform flow. In one the changes of depth and velocity take place over a long distance. Such flow is termed *gradually varied flow*. In the other type of non-uniform flow the changes of depth and velocity take place in only a short distance and may, in fact, be quite abrupt (as in a hydraulic jump). This local non-uniform flow is termed *rapidly varied flow*. There is in practice no rigid dividing line between these two types, but for the purposes of analysis gradually varied flow is regarded as that in which the changes occur slowly enough for the effects of the acceleration of the liquid to be negligible. It is important to realize this limitation of the analysis: formulae based on the assumption of *gradually* varied flow should not be applied to flow in which the changes take place more rapidly.

Gradually varied flow may result from a change in the geometry of the channel – for example, a change in the shape of the cross-section, a change of slope, or an obstruction – or from a change in the frictional forces at the boundaries. It can occur when the flow is either tranquil or rapid but, as we have already seen, a change from tranquil to rapid flow or vice versa usually occurs abruptly.

When, in tranquil flow, the depth is increased upstream of an obstruction the resulting curve of the liquid surface is usually known as a 'backwater curve'. The converse effect – a fall in the surface as the liquid approaches a free outfall from the end of the channel, for example – is termed a 'downdrop' or 'drawdown curve'. Both curves are asymptotic to the surface of uniform flow (Fig. 10.39). (Both are sometimes loosely referred to as backwater curves.)

10.12.1 The equations of gradually varied flow

It is frequently important in practice to be able to estimate the depth of a stream at a particular point or to determine the distance over which the effects of an obstruction such as a weir are transmitted upstream. Such information is given by the equation representing the surface 'profile'. The depth at a particular section determines the area of the cross-section and hence the mean velocity; consequently the surface profile defines the flow. We need, then, to investigate the way in which the depth and velocity change with distance along the channel. Since the changes in gradually varied flow occupy a considerable distance, the effects of boundary friction are very important. (This is in distinction to

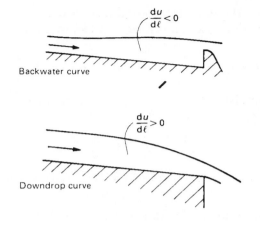

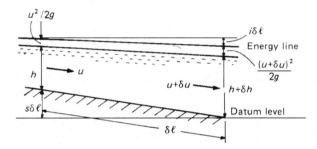

Fig. 10.39

Fig. 10.40

rapidly varied flow, where the boundary friction is usually neglected in comparison with the other forces involved.)

In developing the equations it is essential to distinguish between the slope i of the energy line and the slope s of the channel bed: in non-uniform flow they are not identical as in uniform flow.

Figure 10.40 depicts a short length δl of the channel in which the flow is steady. Over this length the bed falls by an amount $s\,\delta l$, the depth of flow increases from h to $h + \delta h$ and the mean velocity increases from u to $u + \delta u$. We assume that the increase in level δh is constant across the width of the channel. (It is true that if the breadth of the channel is unaltered an increase of u implies a reduction of h; but we recall that 'δh' means 'a very small *increase* of h'. If, in fact, h decreases, then δh is negative and will be seen to be so in the final result.)

If it is assumed that the streamlines in the flow are sensibly straight and parallel and that the slope of the bed is small so that the variation of pressure with depth is hydrostatic, then the total head above the datum level is $h + s\,\delta l + \alpha u^2/2g$ at the first section. As in Section 10.3, α is the kinetic energy correction factor accounting for non-uniformity of velocity over the cross-section. Its value is not normally known: for simplicity we here assume that it differs only slightly from unity so that, without appreciable error, it may be omitted as a factor. Similarly, the total head above datum at the second section is $h + \delta h + (u + \delta u)^2/2g$. If the loss of head in friction is i for unit length of the channel then

$$h + s\,\delta l + \frac{u^2}{2g} - i\,\delta l = h + \delta h + \frac{(u + \delta u)^2}{2g} \qquad (10.38)$$

Rearrangement gives $\delta h = (s - i)\delta l - u\,\delta u/g$, the term in $(\delta u)^2$ being neglected. Therefore in the limit as $\delta l \to 0$

$$\frac{\mathrm{d}h}{\mathrm{d}l} = (s - i) - \frac{u}{g}\frac{\mathrm{d}u}{\mathrm{d}l} \qquad (10.39)$$

To eliminate $\mathrm{d}u/\mathrm{d}l$ we make use of the continuity equation

$$Au = \text{constant} \qquad (10.40)$$

which, when differentiated with respect to l, yields

$$A\,\mathrm{d}u/\mathrm{d}l + u\,\mathrm{d}A/\mathrm{d}l = 0$$

so

$$\frac{\mathrm{d}u}{\mathrm{d}l} = -\frac{u}{A}\frac{\mathrm{d}A}{\mathrm{d}l} \qquad (10.41)$$

To evaluate $\mathrm{d}A/\mathrm{d}l$ we assume that, although the cross-section may be of any shape whatever, the channel is 'prismatic', that is, its shape and alignment do not vary with l. Thus A changes only as a result of a change in h. Figure 10.20 shows that δA then equals $B\,\delta h$ where B denotes the *surface* width of the cross-section. So putting $\mathrm{d}A/\mathrm{d}l = B\,\mathrm{d}h/\mathrm{d}l$ in eqn 10.41 and then substituting into eqn 10.39 we obtain

$$\frac{\mathrm{d}h}{\mathrm{d}l} = (s - i) + \frac{u^2 B}{gA}\frac{\mathrm{d}h}{\mathrm{d}l}$$

whence

$$\frac{\mathrm{d}h}{\mathrm{d}l} = \frac{s - i}{1 - (u^2 B/gA)} = \frac{s - i}{1 - (u^2/g\bar{h})} \qquad (10.42)$$

where $\bar{h}$ = the mean depth A/B. This equation represents the slope of the free surface with respect not to the horizontal but to the bed of the channel.

Examination of the equation shows that when $i = s$, $\mathrm{d}h/\mathrm{d}l = 0$ and there is uniform flow. The term $u^2/g\bar{h}$ will be recognized as the square of the Froude number; therefore when the flow is critical the denominator of the right-hand side of eqn 10.42 is zero. We must not, however, jump to the conclusion that $\mathrm{d}h/\mathrm{d}l$ is therefore infinite and the liquid surface is perpendicular to the bed. Such a conclusion is at odds with the initial assumption of *gradually* varied flow and in such circumstances the equation is not valid. The equation may nevertheless apply to critical flow if the numerator of the expression is also zero, for there is then no mathematical necessity for $\mathrm{d}h/\mathrm{d}l$ to be infinite.

In tranquil flow $u < \sqrt{(g\bar{h})}$ and so the denominator of the expression is positive. In rapid flow, on the other hand, the denominator is negative. Thus if the slope of the bed is less than that corresponding to the rate of dissipation of energy by friction, that is, if $s < i$, then $\mathrm{d}h/\mathrm{d}l$ is negative for tranquil flow and the depth decreases in the direction of flow. For rapid flow $\mathrm{d}h/\mathrm{d}l$ is positive in these circumstances. At critical conditions $\mathrm{d}h/\mathrm{d}l$ changes sign; this is why liquid flowing at the critical depth has an unstable, wavy surface and exactly critical conditions are not maintained over a finite length of a channel. The minor irregularities that are always present on the boundaries cause small variations of h, and so $\mathrm{d}h/\mathrm{d}l$ is then alternately positive and negative, thus causing the wavy appearance.

Equation 10.42 can represent a number of different types of surface profile and these will be considered briefly in Section 10.12.2. Our present concern, however, is the integration of the equation. As i is a function of u (and of other things) the equation cannot be integrated directly. To proceed further the assumption is made that the value of i at a particular section is the same as it would be for uniform flow having the same velocity and hydraulic mean depth. That is to say, a formula for uniform flow, such as Chézy's or Manning's, may be used to evaluate i, and the corresponding roughness coefficient is also assumed applicable to the varied flow. The assumption is not unreasonable, and it does give reliable results. Any error it introduces is no doubt small compared with the uncertainties in selecting suitable values of Chézy's coefficient or Manning's roughness coefficient. If Chézy's formula is used $i = u^2/(C^2 m)$ and if Manning's formula is used $i = n^2 u^2/m^{4/3}$ for metre-second units or $n^2 u^2/(1.49^2 m^{4/3}) \simeq n^2 u^2/(2.22 m^{4/3})$ for foot-second units. For a given steady discharge Q, u is a function of h only and so the integration of eqn 10.42 is possible. Only in a few special cases, however, is an algebraic integration possible, and even then the result is complicated. The solution is therefore normally obtained by either a numerical or a graphical integration.

The problem under investigation is more often that of determining the position in the channel at which a particular depth is reached, i.e. l is required for a particular value of h, so eqn 10.42 may be inverted. Then

$$l = \int_{h_1}^{h_2} \frac{1 - u^2/g\overline{h}}{s - i} \, \mathrm{d}h \qquad (10.43)$$

The following example illustrates the general method of solution.

Example 10.4 A dam is built across a channel of rectangular cross-section which carries water at the rate of $8.75\,\mathrm{m}^3/\mathrm{s}$. As a result the depth just upstream of the dam is increased to $2.5\,\mathrm{m}$. The channel is $5\,\mathrm{m}$ wide and the slope of the bed is 1 in 5000. The channel is lined with concrete (Manning's $n = 0.015$). How far upstream is the depth within $100\,\mathrm{mm}$ of the normal depth?

Solution

To determine the normal depth h_0 Manning's formula is applied (with metre-second units):

$$\frac{8.75\,\mathrm{m}^3/\mathrm{s}}{(5 \times h_0)\mathrm{m}^2} = u = \frac{m^{2/3}i^{1/2}}{n} = \frac{1}{0.015}\left(\frac{5h_0}{5 + 2h_0}\right)^{2/3}\left(\frac{1}{5000}\right)^{1/2}$$

Solution by trial gives $h_0 = 1.800\,\mathrm{m}$. Since the channel is rectangular in section the critical depth $h_\mathrm{c} = (q^2/g)^{1/3} = (1.75^2/9.81)^{1/3}\,\mathrm{m} = 0.678\,\mathrm{m}$. This value is considerably less than the actual depth and the flow is therefore tranquil.

We require a solution of eqn 10.43 between the limits $2.5\,\mathrm{m}$ and $(1.800 + 0.100)\,\mathrm{m} = 1.9\,\mathrm{m}$. To illustrate a simple technique of numerical integration it will be sufficient to divide the range of depth, $2.5\,\mathrm{m}$ to $1.9\,\mathrm{m}$, into three equal parts. Then, starting at the dam (the position where the depth is known) and proceeding upstream, the calculations may be set out in tabular form (Table 10.2; Fig. 10.41).

This method is well suited to computer programming and it may be applied to channels of variable cross-section, roughness and slope by adopting suitable average values of these parameters over each increment of length Δl. Moreover, although the above example concerns a 'backwater' curve in tranquil flow, the same technique may be used for any type of gradually varied flow.

For example, it may be used in calculating the position of a hydraulic jump. As we saw in Section 10.10 the depth h_2 downstream of the jump

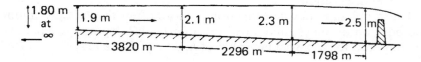

Fig. 10.41

Table 10.2

h (m)	Average h (m)	A (m²)	P (m)	m (m)	u (m/s)	u^2/gh	$1 - u^2/gh$	$i \times 10^{4\dagger}$	$(s - i)10^4$	dl/dh	Δh (m)	Δl (m)
2.5–2.3	2.4	12	9.8	1.224	0.729	0.0226	0.9774	0.913	1.087	8990	0.2	1798
2.3–2.1	2.2	11	9.4	1.170	0.795	0.0293	0.9707	1.155	0.845	11480	0.2	2296
2.1–1.9	2.0	10	9.0	1.111	0.875	0.0390	0.9610	1.497	0.503	19100	0.2	3820

$$7914$$

Total length about 7·9 km

† From Manning's formula $i = n^2u^2/m^{4/3} = 2\cdot25 \times 10^{-4}u^2/m^{4/3}$ for metre-second units.

is determined by the conditions there (e.g. the normal depth appropriate to the rate of flow and the slope of the bed). The momentum relation (eqn 10.25 if the section is rectangular and uniform) then fixes the upstream depth h_1. The rapid flow upstream of the jump is of gradually increasing depth as energy is quickly dissipated by friction, and the distance of the jump from the upstream control section (e.g. a sluice gate), where the depth is known, is given by eqn 10.43. The step-by-step integration proceeds from that upstream control section towards the jump.

In all examples of gradually varied flow the accuracy of the results depends largely on the magnitude of the steps in h because the actual surface curve is in effect being approximated by a series of straight lines. However, there is usually little point in seeking great accuracy because the value of n can only be approximate; moreover, especially in natural channels, the discharge may not be exactly constant and the roughness, the shape of the cross-section and the slope may vary continuously from place to place.

10.12.2   Classification of surface profiles

Equation 10.42 applies to any type of gradually varied flow, subject to the assumptions involved in its derivation. The surface profile may have a variety of forms, depending on how the flow is controlled by weirs or other obstructions, changes in bed slope and so on. These different circumstances affect the relative magnitudes of the quantities appearing in the equation and $\mathrm{d}h/\mathrm{d}l$ is accordingly positive or negative. It is helpful, therefore, to have a logical system of classification of the different types of possible surface profile. All problems of varied flow, no matter how complex, may be studied by considering separate lengths of the channel, in each of which the flow corresponds to one of the various types of surface profile. The separate lengths may be studied one by one until the entire problem has been covered.

The primary classification refers to the slope of the bed. This may be: adverse (A), i.e. 'uphill' or negative slope; zero, i.e. the bed is horizontal (H) and $s = 0$; mild (M), i.e. $s < s_c$; critical (C), i.e. $s = s_c$; or steep (S), i.e. $s > s_c$. The profiles are further classified according to the depth of the stream: the depth may be greater or less than the normal depth, and greater or less than the critical depth. If the depth is greater than both the normal depth (h_0) and the critical depth (h_c), the profile is of type 1; if the actual depth is between h_0 and h_c the profile is of type 2; and if it is less than both the profile is of type 3.

The twelve possible types are illustrated in Fig. 10.42 (where, for the sake of clarity, the slopes are greatly exaggerated). Even a steep slope differs from the horizontal by only a few degrees, so it is immaterial whether the depths are measured vertically (as shown) or perpendicular to the bed. In practice, the longitudinal distances are usually so great that it is impossible to distinguish between uniform and gradually varied flow with the naked eye.

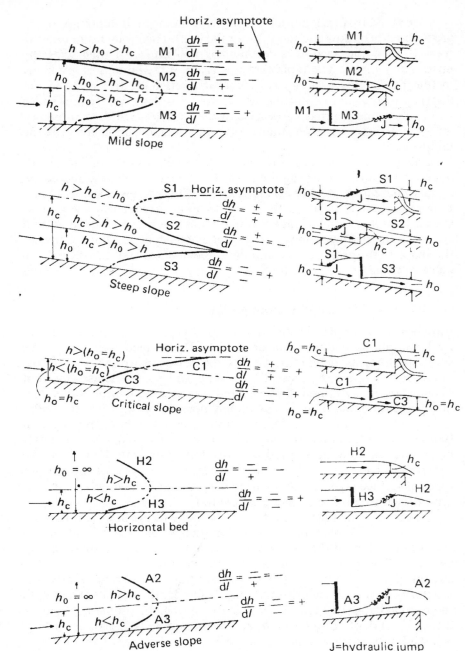

Fig. 10.42 Types of non-uniform flow (*N.B.* all slopes exaggerated).

With five letter categories (A, H, M, C, S) and three number categories (1, 2, 3) it might appear that 15 different types of profile can occur. However, normal flow is impossible on either an adverse or a zero slope, so A1 and H1 curves do not exist. Also, when the slope is critical, h_0 and h_c are the same and thus a C2 curve cannot occur.

Certain general features are evident. All the type 1 curves approach a horizontal asymptote as the velocity is progressively reduced with increasing depth. Moreover, all curves (except the C curves) that approach the normal depth line $h = h_0$ do so asymptotically: this is because uniform flow takes place at sections remote from any disturbances. The curves that cross the critical depth line $h = h_c$ do so perpendicularly because at critical conditions the denominator of eqn 10.42 is zero and dh/dl is thus (mathematically) infinite. But, as we have seen, the assumptions underlying the theory are then invalid, and so those parts of the curves are shown dotted. The C curves are exceptions to these statements for then h_0 is identical with h_c and a curve cannot approach a line asymptotically and perpendicularly at the same time.

By way of example, the M curves may be briefly considered. On a mild slope the normal, uniform flow is tranquil and so $h_0 > h_c$. For the M1 curve $h > h_0$, $i < s$ and since the flow is tranquil the denominator of eqn 10.42 is positive. Therefore dh/dl is positive and the depth increases in the direction of flow. The M1 curve is the usual 'backwater' curve caused by a dam or weir. As h approaches h_0, i approaches s and dh/dl $\rightarrow 0$; thus the normal depth is an asymptote at the upstream end of the curve. For the M2 curve, $h_0 > h > h_c$. The numerator of the expression is now negative, although the denominator is still positive. Therefore dh/dl is negative. This curve represents accelerated flow and is the downdrop curve approaching a free outfall at the end of a channel of mild slope. For the M3 curve $h < h_c$, the flow is rapid, the denominator becomes negative and so dh/dl is again positive; thus the depth increases in the downstream direction as shown. This curve results from an upstream control such as a sluice gate. Since the bed slope is insufficient to sustain the rapid flow a hydraulic jump will form at a point where the equation for the jump can be satisfied.

The slope of a given channel may, of course, be classified as mild for one rate of flow, critical for another, and steep for a third. The qualitative analysis of surface profiles is applicable to channels of any shape and roughness, provided that local variations of slope, shape, roughness and so on are properly taken into account.

10.13   OSCILLATORY WAVES

We now consider disturbances moving over the free surface of a liquid in a periodic manner – ocean waves for example. At any one position some liquid rises above the mean level and then subsides below it, and in doing so appears to travel over the surface. But what actually travels over the surface is simply the *form* of the disturbance. There is practically no net movement of the liquid itself: an object floating on the surface moves forward with the crest of the wave but, in the succeeding trough, returns to almost its original position. As ocean waves move into shallow water close to the shore, the regular wave motion is modified, but even so there is no net flow of water towards the shore (apart from the small amount produced by a rise or fall of the tide).

An analogous phenomenon is seen when wind passes over a field of grain: 'waves' travel across the upper surface although evidently the stalks are not moved across the field but merely oscillate about a mean position.

The study of oscillatory waves is a vast one and has led to a good deal of complex mathematical theory. Here we shall attempt only an outline of those parts of the subject with wide practical application. First we shall determine the velocity c with which a wave moves over the free surface.

10.13.1 The basic equations of motion

Observations show that ocean waves are transmitted practically unchanged in size and shape over great distances and this indicates that effects of viscosity are very small. Accordingly, we assume that the liquid is essentially an inviscid fluid, and if we suppose that the motion was originally generated from rest it can be regarded as irrotational (see Section 9.3).

Figure 10.43 depicts two-dimensional flow in the plane of the diagram, the fluid being unlimited in the x (horizontal) direction. We assume a rigid horizontal bed; any other fixed boundaries are vertical planes in the x direction. The effect of any boundary layers is usually negligible. Shear stresses between the liquid and the atmosphere above it are negligible too, and, as the density of air is so much less than that of the liquid, we also disregard the effect of air being set in motion by the waves. The wavelength λ is defined as the distance between corresponding points on consecutive waves, e.g. crest-to-crest or trough-to-trough.

If the wave shape is constant then so is its velocity c in the x direction; hence to an observer also travelling in that direction at velocity c the flow pattern would appear steady. The simplest assumption about this steady flow pattern is that variations with x are sinusoidal. (This

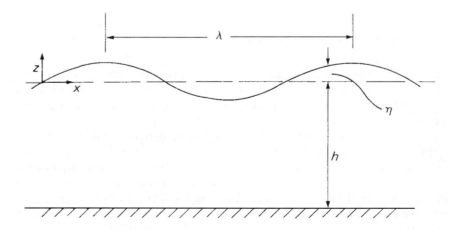

Fig. 10.43

assumption is not, in fact, restrictive since, by Fourier's theorem, more complicated motions can be described by combining sinusoidal variations.) For any point in the liquid the stream function ψ (see Section 9.2) may thus be represented by

$$\psi = cz + f(z) \sin mx \qquad (10.44)$$

where z denotes the vertically upwards coordinate, $f(z)$ is a function of z to be determined, and $m = 2\pi/\lambda$. The cz term accounts for the uniform velocity $-c$ applied to make the flow appear steady to the moving observer.

For irrotational motion, ψ_x must obey Laplace's equation

$$\frac{\partial^2 \psi}{\partial x^2} + \frac{\partial^2 \psi}{\partial z^2} = 0$$

(see Section 9.4), so, substituting from eqn 10.44, we get

$$-m^2 f + \frac{d^2 f}{dz^2} = 0$$

the solution of which is

$$f = A \sinh(B + mz)$$

where A and B are constants. Thus eqn 10.44 becomes

$$\psi = cz + A \sinh(B + mz)\sin mx$$

The free surface, where $z = \eta$ say, must be composed of streamlines and so there

$$\psi = \text{constant} = c\eta + A \sinh(B + m\eta)\sin mx \qquad (10.45)$$

So that $d\eta/dx = 0$ at a crest or a trough the position $x = 0$ must be midway between a trough and a crest. We may conveniently designate the free-surface streamline $\psi = 0$; eqn 10.45 then shows that this line must cross $x = 0$ where $z = 0$ as shown in Fig. 10.43. Moreover, because the bed, where $z = -h$, also consists of streamlines, ψ at the bed is independent of x, and so $\sinh\{B + m(-h)\} = 0$, i.e. $B = mh$. Therefore, in general,

$$\psi = cz + A \sinh\{m(h + z)\}\sin mx \qquad (10.46)$$

In particular, at the free surface,

$$0 = c\eta + A \sinh\{m(h + \eta)\}\sin mx \qquad (10.47)$$

Hence, if, as is usual for ocean waves (except when close to the shore), η is small compared with both h and λ, the free surface has a sinusoidal form given by

$$\eta \simeq \left(-\frac{A}{c}\sinh mh\right)\sin mx \qquad \text{A (10.48)}$$

(In what follows equations marked A are subject to the restrictions $\eta \ll h, \eta \ll \lambda$.)

Since we are considering steady flow, the elevation η of the free surface can be related to the velocity there by Bernoulli's equation. We therefore require expressions for velocity and pressure.

Using values for components of velocity from eqn 9.3 (with z in place of y) and making use of eqn 10.46, we have

$$\left(\text{Velocity}\right)^2 = \left(\frac{\partial \psi}{\partial x}\right)^2 + \left(-\frac{\partial \psi}{\partial z}\right)^2$$
$$= A^2 m^2 \sinh^2\{m(h + z)\}\cos^2 mx$$
$$+ \left[-c - Am \cosh\{m(h + z)\}\sin mx\right]^2 \quad (10.49)$$

Putting $z = \eta$ gives the velocity at the free surface. Then substituting for A from eqn 10.47 and neglecting terms in $m^2\eta^2$ (on the assumption that $\eta \ll \lambda$) we obtain

$$\left(\text{Surface velocity}\right)^2 = c^2\left[1 - 2m\eta \coth\{m(h + \eta)\}\right]$$
$$\simeq c^2\left(1 - 2m\eta \coth mh\right) \qquad \text{A} \quad (10.50)$$

The pressure above the surface is atmospheric but, because the surface is not plane, the pressure in the liquid is, in general, modified by surface tension (γ). For unit distance perpendicular to the plane of Fig. 10.44, the surface tension force is γ and at the position P its vertical component is $\gamma \sin\theta$. At Q the vertical component is

$$\gamma \sin\theta + \frac{\mathrm{d}}{\mathrm{d}x}(\gamma \sin\theta)\delta x$$

Hence the net upwards surface tension force on PQ is

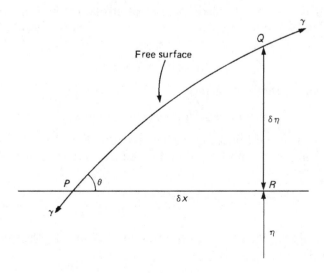

Fig. 10.44

$$\frac{\mathrm{d}}{\mathrm{d}x}(\gamma \sin \theta)\delta x$$

If a mean gauge pressure p acts over PR, the total upwards force on the fluid in the control volume PQR

$$= \frac{\mathrm{d}}{\mathrm{d}x}(\gamma \sin \theta)\delta x + p\,\delta x$$

= Rate of increase of vertical momentum of the fluid through PQR.

This momentum term, however, is proportional both to δx and to $\delta \eta$; so, after dividing the equation by δx and letting $\delta \eta \to 0$, the momentum term vanishes and we obtain

Pressure immediately below the free surface

$$= p = -\frac{\mathrm{d}}{\mathrm{d}x}(\gamma \sin \theta) = -\gamma \frac{\mathrm{d}}{\mathrm{d}x}\left[\frac{\mathrm{d}\eta/\mathrm{d}x}{\left\{1 + (\mathrm{d}\eta/\mathrm{d}x)^2\right\}^{1/2}}\right] \qquad (10.51)$$

For a wave in which η is small compared with both h and λ, we have, from eqn 10.48,

$$\frac{\mathrm{d}\eta}{\mathrm{d}x} = -\frac{A}{c}m \sinh mh \cos mx = m\eta \cot mx \qquad \text{A (10.52)}$$

and, since this is everywhere small compared with unity, eqn 10.51 reduces to $p = -\gamma \mathrm{d}^2\eta/\mathrm{d}x^2$. Substitution from eqn 10.52 then gives $p = \gamma m^2 \eta$.

Applying Bernoulli's equation between the points $\eta = \eta$ and $\eta = 0$ on the free-surface streamline (or, strictly, a streamline in the liquid infinitesimally close to the surface), and using eqn 10.50, we obtain

$$\gamma m^2 \eta + \frac{1}{2}\varrho c^2(1 - 2m\eta \coth mh) + \varrho g\eta = \frac{1}{2}\varrho c^2 \qquad \text{A (10.53)}$$

whence

$$c^2 = \left(\frac{\gamma m}{\varrho} + \frac{g}{m}\right)\tanh mh = \left(\frac{2\pi\gamma}{\varrho\lambda} + \frac{g\lambda}{2\pi}\right)\tanh\left(2\pi h/\lambda\right) \qquad \text{A (10.54)}$$

Waves for which this result is true, that is, those for which η is everywhere small compared with both h and λ, are known as Airy waves after Sir George B. Airy (1801–92), who first analysed them mathematically.

For waves not meeting these conditions the analysis is much more complicated, and in what follows we shall therefore restrict ourselves almost entirely to Airy waves. (The corresponding equations are those marked A.) Although application of this theory may lead to some error in practice, this is often tolerable in comparison with uncertainties in measurements.

The velocity c is known as the *phase velocity* because points of the same phase (i.e. of equal η) move at this velocity, regardless of the shape of the wave. In general, c depends on λ, so the wave is then said to be *dispersive.* This is because if a wave of general shape were split into components of different wavelengths the components would move at different velocities and thus become separated, i.e. dispersed.

Equation 10.54 shows that the effect of surface tension on the phase velocity is negligible if

$$\frac{2\pi\gamma}{\varrho\lambda} \ll \frac{g\lambda}{2\pi} \quad \text{i.e. if } \lambda \gg 2\pi\left(\frac{\gamma}{\varrho g}\right)^{1/2}$$

For water this last quantity is about 17 mm, and so we can safely disregard the effect of surface tension on ocean waves, for example. On the other hand, for waves of very short length, usually termed *ripples*, the term in eqn 10.54 involving g becomes negligible. For ripples on water with $\lambda = 3$ mm, say, gravity affects c by only 1.5%. When λ is small compared with h then $\tanh(2\pi h/\lambda) \to 1$ and the velocity of capillary waves (that is, those governed principally by surface tension) $\to (2\pi\gamma/\varrho\lambda)^{1/2}$. The frequency, that is, the number of wave crests passing a given fixed point in unit time interval, is c/λ which, for $\lambda = 3$ mm on deep water, is 131 Hz. Thus capillary waves can be generated by a tuning fork held in a liquid – although they rapidly decay and cannot be seen for more than a few centimetres. Nevertheless, the measurement of the length of waves produced by a tuning fork of known frequency is the basis of one method of determining surface tension. (In practice, non-uniformity of the value of γ caused by contamination of the surface can affect the results somewhat for short waves.)

'Deep water' waves are often regarded as those for which $h > \lambda/2$. Then $\tanh(2\pi h/\lambda)$ differs from unity by less than 0.004 and we can take

$$c^2 = \frac{2\pi\gamma}{\varrho\lambda} + \frac{g\lambda}{2\pi}$$

This has a minimum value when $\lambda = 2\pi(\gamma/\varrho g)^{1/2}$ ($\simeq 17$ mm for water); then $c_{min} = (4g\gamma/\varrho)^{1/4}$ ($\simeq 0.23$ m/s for water).

For 'long' waves, in which $\lambda \gg h$, $\tanh(2\pi h/\lambda) \to 2\pi h/\lambda$. If the effect of γ is negligible we then obtain $c = (gh)^{1/2}$ as for the single surge wave (Section 10.8).

■ **Example 10.5** Gravity waves on water with a mean depth of 4 m have a period of 5 s. What is the wavelength?

Solution
From eqn 10.54

$$c^2 = \frac{g\lambda}{2\pi} \tanh \frac{2\pi h}{\lambda}$$

surface tension being neglected. But $c = \lambda/T$ where T denotes the period, i.e. (frequency)$^{-1}$

$$\therefore \lambda = \frac{gT^2}{2\pi} \tanh \frac{2\pi h}{\lambda}$$

This equation will have to be solved for λ by trial. Since $\tanh x$ is zero when $x = 0$ and then fairly rapidly approaches unity as x increases, taking $\tanh(2\pi h/\lambda) = 0.75$ say gives a reasonable first approximation. Then $\lambda = (9.81 \times 5^2/2\pi)0.75\,\text{m} = 29.27\,\text{m}$. The next value is $(9.81 \times 5^2/2\pi)\tanh(2\pi \times 4/29.27)\,\text{m} = 27.15\,\text{m}$ and similarly the following one is $28.44\,\text{m}$. We could continue refining the result in this way but as the approach to the solution evidently oscillates we can speed up the process by using the *average* of the last two results for the next trial. So here we take $\frac{1}{2}(27.15 + 28.44)\,\text{m} = 27.80\,\text{m}$ to calculate the next value of λ as $(9.81 \times 5^2/2\pi) \times \tanh(8\pi/27.80)\,\text{m} = 28.04\,\text{m}$. The final result is $\lambda = 27.95\,\text{m}$ □

Example 10.6 Waves run up on a shore with a period of 12 s. Determine their phase velocity and wavelength in deep water well away from the shore. ■

Solution

For ocean waves, surface tension effects can be ignored. From equation 10.54, with $\gamma = 0$:

$$c^2 = \frac{g\lambda}{2\pi}\tanh\left(\frac{2\pi h}{\lambda}\right)$$

For deep water

$$\tanh\left(\frac{2\pi h}{\lambda}\right) \to 1$$

Hence, at the limit

$$c = \left(\frac{g\lambda}{2\pi}\right)^{1/2}$$

which combines with $c = \lambda/T$ to yield

$$c = \frac{gT}{2\pi} = \frac{9.81\,\text{m/s}^2 \times 12\,\text{s}}{2\pi} = 18.74\,\text{m s}^{-1}$$

and

$$\lambda = cT = 18.74\,\text{m/s} \times 12\,\text{s} = 225\,\text{m} \qquad \square$$

10.13.2 Movement of individual particles

To describe the motion with respect to stationary axes we must remove the velocity $-c$ which was imposed to make the flow appear steady. As the conditions at a particular point will then depend also on time t, we must use $x - ct$ in place of x: this is because conditions existing at time

t and position x are reproduced unchanged at time $t + \Delta t$ and position $x + c\Delta t$. Using eqn 10.46 we obtain, for any position (x, z),

Absolute horizontal velocity of particle =

$$c - \frac{\partial \psi}{\partial z} = -Am \cosh\{m(h + z)\} \sin\{m(x - ct)\} \qquad (10.55)$$

Absolute vertical velocity of particle =

$$\frac{\partial \psi}{\partial x} = Am \sinh\{m(h + z)\} \cos\{m(x - ct)\} \qquad (10.56)$$

Integrating these expressions with respect to t gives displacements X, Z, respectively, from the mean position. If the displacements are small (as they are for waves of small amplitude) we may approximate x and z by their mean values $\bar{x}$ and $\bar{z}$, and thus obtain

$$X = -\frac{A}{c} \cosh\{m(h + \bar{z})\} \cos\{m(\bar{x} - ct)\}$$

$$Z = -\frac{A}{c} \sinh\{m(h + \bar{z})\} \sin\{m(\bar{x} - ct)\}$$

Hence

$$\frac{X^2}{(A^2/c^2)\cosh^2\{m(h + \bar{z})\}} + \frac{Z^2}{(A^2/c^2)\sinh^2\{m(h + \bar{z})\}} = 1 \quad \text{A} \ (10.57)$$

which is the equation of an ellipse with semi-axes

$$\left|\frac{A}{c} \cosh\{m(h + \bar{z})\}\right| \quad \text{horizontally and} \quad \left|\frac{A}{c} \sinh\{m(h + \bar{z})\}\right| \quad \text{vertically}$$

Both axes decrease as the mean depth of the particle $(-\bar{z})$ increases, the vertical axis necessarily becoming zero when $\bar{z} = -h$. Figure 10.45 illustrates the elliptical orbits of typical particles in shallow liquid; however, when $h \gg \lambda$, $\tanh mh \to 1$, i.e. $\cosh mh \to \sinh mh$ and the orbits

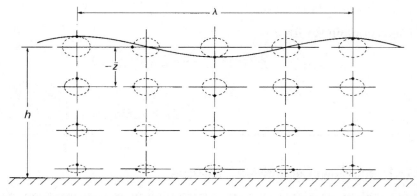

Fig. 10.45 The diagram shows the instantaneous positions of particles on their elliptical path lines. For a wave travelling from left to right the particles move clockwise.

near the free surface (where $\bar{z} \to 0$) become circles. If the depth h is much less than λ, then $m(h + \bar{z})$ is always much less than 1, thus $\cosh\{m(h + \bar{z})\} \to 1$ and the horizontal axis of an orbit is practically independent of $\bar{z}$. The horizontal component of velocity is also practically independent of $\bar{z}$; the vertical velocity, and therefore acceleration, is small and thus – in this particular case – a hydrostatic variation of pressure may be assumed. The vertical displacement Z increases practically linearly with $\bar{z}$, i.e. from the bed to the free surface.

If the amplitude of a wave is not infinitesimal, as here assumed, the paths followed by individual particles are not quite closed curves and the particles in fact slowly advance with the wave.

10.13.3 Wave energy

The energy contained in a wave is the sum of contributions from gravitational energy, kinetic energy and free surface energy.

For unit width perpendicular to the x direction, an element of fluid $\delta x \times \delta z$ in size has gravitational energy $(\varrho g\, \delta x\, \delta z)z$ relative to the equilibrium level $z = 0$. For a complete wavelength the total gravitational energy in a sinusoidal wave is therefore   *Gravitational energy*

$$\int_0^\lambda \left(\int_0^\eta \varrho g z \,\mathrm{d}z \right) \mathrm{d}x = \frac{1}{2}\varrho g \int_0^\lambda \eta^2 \,\mathrm{d}x$$

$$= \frac{1}{2}\varrho g a^2 \int_0^\lambda \sin^2\{m(x - ct)\} \,\mathrm{d}x$$

$$= \frac{1}{4}\varrho g a^2 \lambda$$

where a represents the amplitude of the wave, i.e. the maximum value of η, and, as before, $m = 2\pi/\lambda$.

The kinetic energy (per unit width) for a complete wavelength can be calculated from the velocity components for a particle (eqns 10.55 and 10.56):   *Kinetic energy*

$$\mathrm{KE} = \frac{1}{2}\varrho \int_0^\lambda \left(\int_{-h}^\eta \left[A^2 m^2 \cosh^2\{m(h + z)\} \sin^2\{m(x - ct)\} \right. \right.$$

$$\left. \left. + A^2 m^2 \sinh^2\{m(h + z)\} \cos^2\{m(x - ct)\} \right] \mathrm{d}z \right) \mathrm{d}x$$

$$= \frac{1}{2}\varrho A^2 m^2 \int_0^\lambda \left(\int_{-h}^\eta \left[\cosh^2\{m(h + z)\} - \cos^2\{m(x - ct)\} \right] \mathrm{d}z \right) \mathrm{d}x$$

$$= \frac{1}{4}\varrho A^2 m^2 \int_0^\lambda \left[\frac{1}{2m} \sinh\{2m(h + \eta)\} - (h + \eta)\cos\{2m(x - ct)\} \right] \mathrm{d}x$$

Again assuming that $\eta \ll h$ and $m\eta \ll 1$, we obtain

$$\mathrm{KE} = \frac{1}{8}\varrho A^2 m\lambda \sinh 2mh$$

Substituting $A = -ca\, \mathrm{cosech}\, mh$ (eqn 10.48), and then for c^2 from eqn 10.54 gives

$$\mathrm{KE} = \frac{1}{4}a^2\lambda\left(\varrho g + \gamma m^2\right)$$

Free surface energy

If surface tension has a significant effect, we should also add γ(length of surface $-\lambda$), which is the work done in stretching the surface in one wavelength when the wave was formed. The length of the surface is

$$\int_0^\lambda \left\{1 + \left(\mathrm{d}\eta/\mathrm{d}x\right)^2\right\}^{1/2}\mathrm{d}x$$

which for small amplitudes is

$$\int_0^\lambda \left[1 + a^2 m^2 \cos^2\left\{m(x - ct)\right\}\right]^{1/2}\mathrm{d}x$$
$$\approx \int_0^\lambda \left[1 + \frac{1}{2}a^2 m^2 \cos^2\left\{m(x - ct)\right\}\right]\mathrm{d}x = \lambda + \frac{1}{4}a^2 m^2 \lambda$$

so the free surface energy over one wavelength is $\frac{1}{4}\gamma a^2 m^2 \lambda$.

$\therefore$ The total energy for one wavelength

= gravitational energy + kinetic energy + free surface energy

$$= \frac{1}{2}a^2\lambda\left(\varrho g + \gamma m^2\right) \hspace{3cm} \text{A (10.58)}$$

Although we assumed that the bed is horizontal (h uniform) variations of h have no appreciable effect on the result provided that h exceeds (say) $\lambda/2$.

10.13.4 Rate of energy transmission

Although the wave *shape* moves with velocity c this is not necessarily the velocity with which energy is transmitted through the liquid. The energy is carried by the particles of liquid and, as we have seen, these do not all move at velocity c.

Any small element of liquid carries an amount of energy $p^* + \frac{1}{2}\varrho(u^2 + v^2)$ per unit volume where p^* denotes the piezometric pressure and u, v the horizontal and vertical components of velocity. In any fixed vertical plane perpendicular to the x direction the volume flow rate through a small element of height δz is $u\,\delta z$ per unit breadth and so the total amount of energy transferred across that plane per unit time interval is

$$\int_{-h}^{\eta} \left\{ p* + \frac{1}{2}\varrho\left(u^2 + v^2\right) \right\} u\,dz \qquad (10.59)$$

Since the flow is assumed irrotational Bernoulli's equation may be applied between any two points in steady flow. One of these points may be taken on the free surface at $\eta = 0$; then, for the *steady* conditions and waves of small amplitude,

$$p* + \frac{1}{2}\varrho\left\{ (u - c)^2 + v^2 \right\} = \frac{1}{2}\varrho c^2 \qquad \text{A (10.60)}$$

(as in eqn 10.53). Hence $p* + \frac{1}{2}\varrho(u^2 + v^2) = \varrho uc$, so the expression 10.59 becomes

$$\varrho c \int_{-h}^{\eta} u^2\,dz.$$

Substitution from eqn 10.55 for u allows the rate of transfer of energy across the plane to be evaluated as

$$\frac{1}{2}\varrho c A^2 m^2 \left(h + \frac{1}{2m} \sinh 2mh \right) \sin^2\left\{ m(x - ct) \right\} \quad \text{A (10.61)}$$

where η has again been neglected in comparison with h.

The *mean* rate of energy transfer per unit breadth is obtained by integrating the expression 10.61 with respect to t for the passage of one wavelength through the fixed plane, and then dividing by the corresponding period λ/c. Substituting for A the value $-ca\,\text{cosech}\,mh$ (from eqn 10.48) and then for c^2 from eqn 10.54 we obtain

Mean rate of energy transfer per unit breadth

$$= \frac{1}{4}ca^2\left(1 + 2mh\,\text{cosech}\,2mh\right)\left(\varrho g + \gamma m^2\right)$$

Dividing this by the mean total energy per unit plan area $\frac{1}{2}a^2(\varrho g + \gamma m^2)$ (from eqn 10.58) gives the velocity of energy transmission

$$\frac{1}{2}c\left(1 + 2mh\,\text{cosech}\,2mh\right) \qquad \text{A (10.62)}$$

This, it will be seen, is always less than c.

10.13.5 Group velocity

If an otherwise still liquid is disturbed at a particular position – for example, when a stone is thrown into a pond or when a boat moves through calm water – a group of waves is produced that moves away from the point of disturbance. It will often be noticed that the group advances at a velocity less than that of the individual waves within it. Individual waves appear to move forward through the group, grow to a maximum amplitude, and then diminish before vanishing entirely at the front of the group.

This happens because the group has components that in general are of slightly different wavelength (and therefore slightly different veloc-

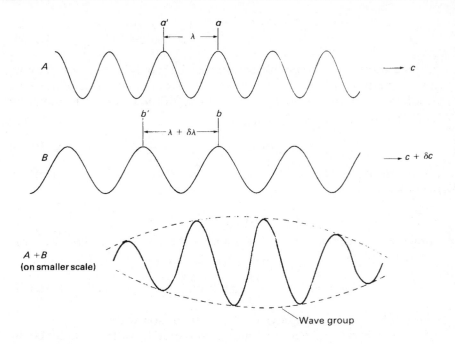

Fig. 10.46

ity). As a simple example, let us consider two trains of waves A and B that combine together as shown in Fig. 10.46. If trains A and B differ slightly in wavelength, the 'total' wave form resulting from their addition has a slowly varying amplitude, giving the appearance of groups of waves alternating with intervals of almost still liquid. (A similar phenomenon can occur in any kind of wave motion, e.g. the 'beats' produced by the conjunction of two trains of sound waves of nearly equal wavelength, or the amplitude modulation of radio waves.)

At a particular instant the crests a and b coincide, so the maximum combined amplitude for the group is at this position. A little later the faster waves (B, say) will have gained a distance $\delta\lambda$ relative to the slower waves (A) and then crests a' and b' will coincide. Since the relative velocity between trains A and B is δc, this takes a time $\delta\lambda/\delta c$. However, while the point where crests coincide has moved back a distance λ (relative to A), A itself has in this time interval $\delta\lambda/\delta c$ moved forward a distance $c(\delta\lambda/\delta c)$. Thus the maximum combined amplitude has moved forward a *net* distance $c\,\delta\lambda/\delta c - \lambda$. The velocity with which it does so is therefore

$$\frac{c\,\delta\lambda/\delta c - \lambda}{\delta\lambda/\delta c} = c - \lambda\frac{\delta c}{\delta\lambda}$$

which, as δc and $\delta\lambda$ both tend to zero, becomes

$$c - \lambda\frac{\mathrm{d}c}{\mathrm{d}\lambda} = c + m\frac{\mathrm{d}c}{\mathrm{d}m} \qquad (10.63)$$

This is known as the group velocity c_g, and clearly, unless the individual wave velocity c is independent of the wavelength, the group velocity differs from c.

Provided that there are only slight variations of c and λ, the result is true for any number of waves of any type. But, in particular, eqn 10.63 shows that for small surface waves

$$c_g = c + m \frac{\mathrm{d}c}{\mathrm{d}m} = \frac{c}{2}\left\{2 + \frac{m}{c^2}\frac{\mathrm{d}(c^2)}{\mathrm{d}m}\right\}$$

$$= \frac{c}{2}\left(\frac{3\gamma m^2 + \varrho g}{\gamma m^2 + \varrho g} + \frac{2mh}{\sinh 2mh}\right) \quad \text{A (10.64)}$$

If surface tension has a negligible effect, the first term in the final bracket is unity; c_g is then (by coincidence) the same as the velocity of energy transmission (eqn 10.62) and it varies between $c/2$ (when $mh \to \infty$) and c (when $mh \to 0$). (This last condition is found close to a sloping beach, and this is why we do not there see groups of waves moving more slowly than individual waves.) However, for small waves in which surface tension is significant the group velocity differs from the velocity of energy transmission and may exceed the velocity of individual waves.

Example 10.7 Waves run up on a shore with a period of 12 s; see Example 10.6.

(a) Estimate the time elapsed since the waves were generated in a storm occurring 800 km out to sea.
(b) Estimate the depth at which the waves begin to be significantly influenced by the sea bed as they approach the shore .

Solution
From Example 10.6, $c = 18.74\,\text{m/s}$ and $\lambda = 225\,\text{m}$.

(a) Waves travel over long distances at the group velocity c_g. Denote the distance and time by x and t respectively. Hence $c_g = x/t$.

Noting that, for large h, $c_g = c/2$, we obtain

$$t = \frac{x}{c_g} = \frac{2x}{c} = \frac{2 \times 800\,\text{km} \times 10^3\,\text{m/km}}{18.74\,\text{m/s}} = 85.38 \times 10^3\,\text{s}$$

$$= 23.7\,\text{hours}$$

(b) We can examine two alternative criteria.

$$\text{Criterion I:} \quad h = \frac{\lambda}{2} = \frac{225\,\text{m}}{2} = 112.5\,\text{m}$$

Criterion II: The influence of the sea bed is felt when the $\tanh(2\pi h/\lambda)$ term gives a 1% change in λ:

$$\tanh\left(\frac{2\pi h}{\lambda}\right) = 0.99 \quad \text{or} \quad \frac{2\pi h}{\lambda} = 2.65$$

Hence

$$h = \frac{2.65\lambda}{2\pi} = \frac{2.65 \times 225\,\text{m}}{2\pi} = 94.9\,\text{m}$$

The answers show that h lies in the range between about 95 m and 113 m, say 100 m.

□

10.13.6 Waves moving into shallower liquid

We now consider what happens when waves move, for example, towards a gradually sloping beach. Although the depth of liquid decreases, the wave period T cannot change. This is because the number of waves passing a fixed position in unit time interval is $1/T$, and if this varied from one position to another then the number of waves entering a given region would differ from the number leaving, so the number of waves in the region would increase or decrease indefinitely. By using the fundamental relation

$$\lambda = cT \tag{10.65}$$

λ can be eliminated from eqn 10.54 and, with T constant, it is then readily shown that dc/dh is always positive. Consequently a decrease of depth h entails a reduction of velocity c. (It is true that eqn 10.54 is based on the assumption $h = $ constant, but the slope of the sea bed is usually so small that the change of h is negligible over a single wavelength. This also means that there will be negligible reflection of a wave from the sloping bed. We disregard too the effects of any additional currents there may be – for example, close to the outlet of a river.)

This reduction of wave velocity may produce refraction effects similar to those in optics. For example, a uniform train of waves approaching a beach obliquely is deflected so that the wave crests become more nearly parallel to the contour lines of the bed, as in Fig. 10.47.

If refraction causes horizontal lines perpendicular to the wave crests to converge, the wave energy is constricted to a passage of decreasing width and thus the wave amplitude increases – sometimes quite dramatically. This is why large waves are often found at headlands at the sides of a bay; and it is an important consideration in the design of harbours.

When the depth becomes little more than the amplitude, the Airy theory ceases to hold and the wave profile is increasingly distorted. The crests become sharper and the troughs flatter. Moreover, the velocity of propagation of the upper part of the profile is greater than that of the lower part; consequently the crests curl forwards and finally break. Breaking usually occurs when $a \simeq \frac{3}{4}h$.

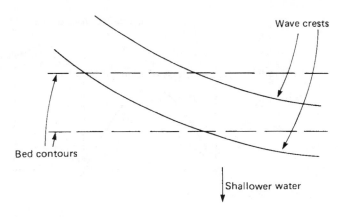

Wave crests

Bed contours

Shallower water

Fig. 10.47

10.13.7 Standing waves

If two trains of waves of the same amplitude, wavelength and period, but travelling in opposite directions, are combined, the result is a set of standing (i.e. stationary) waves. This can happen when a series of waves is reflected by a fixed solid boundary perpendicular to the direction of propagation: the two individual wave trains are then formed of the incident waves travelling with velocity c and the reflected waves with velocity $-c$. For example, if the individual waves are small-amplitude sine waves, the equation of the free surface for the resulting standing wave is

$$\eta = a \sin\{m(x - ct)\} + a \sin\{m(x + ct)\} = 2a \sin mx \cos mct$$

$$\text{A (10.66)}$$

That is, at any instant the free surface is a sine curve but its amplitude $2a \cos mct$ varies continuously with time. The values of x that give $\eta = 0$ are independent of t and thus the wave profile does not travel over the surface; instead it simply rises and falls as indicated by the dotted lines in Fig. 10.48. The positions at which η is always zero are called *nodes* and those of maximum vertical motion are called *antinodes*. Equation 10.66 shows that the wavelength $\lambda = 2\pi/m$ is the same as for the original waves; so too is the period λ/c.

From eqn 10.47 $A = -ca \operatorname{cosech}\{m(h + a)\}$ where a again denotes the maximum value of η. Substituting this into eqn 10.55 we see that the horizontal velocity of a particle in a single moving wave (not necessarily of small amplitude) is

$$mca \operatorname{cosech}\{m(h + a)\} \cosh\{m(h + z)\} \sin\{m(x - ct)\}$$

If we add to this the value for a wave of equal amplitude travelling with velocity $-c$ we obtain the horizontal velocity of a particle in a standing wave:

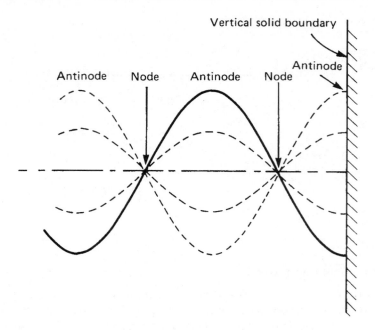

Fig. 10.48

$$mca \; \text{cosech}\{m(h+a)\} \cosh\{m(h+z)\}$$
$$\times \left[\sin\{m(x-ct)\} - \sin\{m(x+ct)\}\right]$$
$$= -2mca \; \text{cosech}\{m(h+a)\} \cosh\{m(h+z)\}\cos mx \sin mct$$

Similarly the vertical velocity may be shown to be

$$-2mca \; \text{cosech}\{m(h+a)\} \sinh\{m(h+z)\} \sin mx \sin mct$$

As the ratio of these velocity components is independent of t, each particle – if it does not move far from its mean position (x, z) – moves to and fro in a straight line, the direction of which varies from vertical beneath the antinodes (where $\cos mx = 0$) to horizontal beneath the nodes (where $\sin mx = 0$).

If the liquid is confined in a channel with completely closed vertical ends, then only certain values of wavelength are possible. Because the horizontal velocity at the closed ends must be zero, each end must coincide with an antinode and so the length of the channel must be an integral number of half-wavelengths. If, however, one end of the channel is closed but the other is open, i.e. connected with an infinite amount of liquid, then that end must be a node (no vertical movement) and the length of the channel must be an odd number of quarter-wavelengths. These standing wave oscillations are thus seen to be analogous to the longitudinal vibrations of the air in an organ pipe or other wind instrument. Once the motion is established it can continue, even with large amplitudes, with very little dissipation of energy.

A bay may behave like a channel open at one end if waves or tides from the open sea arrive with a frequency equal to that at which standing waves oscillate. The water in the bay is then set into resonance and very high amplitudes may occur at the inner shore even though there is only moderate vertical movement at the mouth of the bay.

Lakes or harbours may act like channels with closed ends. For example, wind action may move water towards one end, and, after the wind drops, the water may oscillate for a considerable time. Such standing waves are usually termed *seiches* (pronounced 'saishes'). On the surface the horizontal movement at the nodes may be several times the vertical movement at the antinodes, and boats may thus be wrenched from their moorings.

10.14 CONCLUSION

In this chapter we have restricted our discussion of phenomena occurring in flow with a free surface to the case of a liquid in contact with the atmosphere, the density of the latter being negligible in comparison with that of the liquid. Similar phenomena can, however, take place when the interface is between two fluids whose densities are much more nearly equal. Consideration of these further examples of free surface flow is beyond the scope of this book, but it is worth remarking that the phenomena discussed in this chapter have their counterparts in the atmosphere and in lakes and oceans whenever there is relative movement of one layer of fluid over another layer of slightly greater density. Such a situation occurs in the atmosphere when a stream of cold air flows under warmer air. Similarly, in a large bulk of liquid a current of cold liquid may flow under warmer liquid; or salt water or sediment-laden water may flow under fresh water. The effect of gravity on the situation is then represented, not by the weight per unit mass g, but by $\{1 - (\varrho_1/\varrho_2)\}g$ where ϱ_1, ϱ_2 represent the densities of the two fluids, and $\varrho_1/\varrho_2 < 1$.

REFERENCES

1. Carter, R. W., Einstein, H. A., Hinds, J., Powell, R. W. and Silberman, E. 'Friction factors in open channels', *Proc. Am. Soc. civ. Engrs*, **89** HY2, 97–143 (1963).
   See also supplementary bibliography, *ibid.* **90** HY4, 226–7 (1964).
2. Chow, V. T. *Open-Channel Hydraulics*, McGraw-Hill, New York (1959).
3. Rouse, H. 'Discharge characteristics of the free overfall', *Civil Engng*, **6**, 257–60 (1936).

FURTHER READING

Ackers, P., White, W. R., Perkins, J. A. and Harrison, A. J. M. *Weirs and Flumes for Flow Measurement*, Wiley, Chichester (1978).

Chow, V. T. *Open-Channel Hydraulics*, McGraw-Hill, New York (1959).

Crapper, G. D. *Introduction to Water Waves*, Ellis Horwood, Chichester (1984).

French, R. H. *Open-Channel Hydraulics*, McGraw-Hill, New York (1985).

Henderson, F. M. *Open-Channel Flow*, Macmillan, New York (1966).

Kinsman, Blair *Wind Waves: Their Generation and Propagation on the Ocean Surface*. Prentice-Hall, Englewood Cliffs, N. J. (1965) and Dover, New York (1984).

FILM

Bryson, A. E. *Waves in Fluids*, Encyclopaedia Britannica Films or from Central Film Library.

PROBLEMS

10.1   A channel of symmetrical trapezoidal section, 900 mm deep and with top and bottom widths 1.8 m and 600 mm respectively, carries water at a depth of 600 mm. If the channel slopes uniformly at 1 in 2600 and Chézy's coefficient is $60 \, \mathrm{m}^{1/2}/\mathrm{s}$, calculate the steady rate of flow in the channel.

10.2   An open channel of trapezoidal section, 2.5 m wide at the base and having sides inclined at 60° to the horizontal, has a bed slope of 1 in 500. It is found that when the rate of flow is $1.24 \, \mathrm{m}^3/\mathrm{s}$ the depth of water in the channel is 350 mm. Assuming the validity of Manning's formula, calculate the rate of flow when the depth is 500 mm.

10.3   A long channel of trapezoidal section is constructed from rubble masonry at a bed slope of 1 in 7000. The sides slope at arctan 1.5 to the horizontal and the required flow rate is $2.8 \, \mathrm{m}^3/\mathrm{s}$. Determine the base width of the channel if the maximum depth is 1 m. (Use Table 10.1.)

10.4   A long concrete channel of trapezoidal section with sides that slope at 60° to the horizontal is to carry $0.3 \, \mathrm{m}^3$ of water a second. Determine the optimum dimensions if the bed slope is 1 in 1800. (Use Table 10.1.)

10.5   A conduit 1 m diameter and 3.6 km long is laid at a uniform slope of 1 in 1500 and connects two reservoirs. When the reservoir levels are low the conduit runs partly full and when the depth is 700 mm the steady rate of flow is 0.325 m³/s. The Chézy coefficient is given by $Km^{1/6}$, where K is a constant and m represents the hydraulic mean depth. Neglecting losses of head at entry and exit, calculate K and the rate of flow when the conduit is full and the difference between reservoir levels is 4.5 m.

10.6   A circular conduit is to satisfy the following conditions: capacity when flowing full, 0.13 m³/s; velocity when the depth is one quarter the diameter, not less than 0.6 m/s. Assuming uniform flow, determine the diameter and the slope if Chézy's coefficient is 58 m¹/²/s.

10.7   Show that the surface of the liquid in a circular conduit under conditions of maximum discharge subtends an angle of 302.4° at the centre. Determine the diameter of such a conduit at a slope of 1 in 10⁴ and carrying a maximum discharge of 2.8 m³/s. Assume that $u = (80 \, \text{m}^{1/3}/\text{s}) \, m^{2/3} i^{1/2}$.

10.8   A long horizontal channel has a base width of 1 m and sides at 60° to the horizontal. When the flow in the channel is 0.85 m³/s the depth is 500 mm. The discharge is suddenly reduced so that a surge wave of amplitude 150 mm is propagated upstream. Determine the new rate of flow, the velocity of the wave and the Froude numbers before and after the wave.

10.9   In a long rectangular channel 3 m wide the specific energy is 1.8 m and the rate of flow is 12 m³/s. Calculate two possible depths of flow and the corresponding Froude numbers. If Manning's $n = 0.014$ what is the critical slope for this discharge?

10.10 The cross-section of a channel is a parabola with a vertical axis. Determine the critical velocity and critical depth in terms of the specific energy E.

10.11 The cross-section of a river 30 m wide is rectangular. At a point where the bed is approximately horizontal the width is restricted to 25 m by the piers of a bridge. If a flood of 450 m³/s is to pass the bridge with the minimum upstream depth, describe the flow past the piers and calculate the upstream depth.

10.12 A long, straight open channel has a base width of 1 m and sides that slope outwards at 60° to the horizontal. The bed has a uniform slope of 1 in 250 and the channel ends in a free outfall. A small surface wave travelling slowly upstream becomes stationary at a section where the boundary surface is slightly smoother. If the normal depth of flow is 150 mm and the Chézy coefficient is given by $Km^{1/6}$, where m represents the hydraulic mean depth, estimate the value of K for the smoother surface and the mean shear stress on this surface.

10.13 Water flows at $5.4\,\text{m}^3/\text{s}$ under a wide sluice gate into a rectangular prismatic channel 3.5 m wide. A hydraulic jump is formed just downstream of a section where the depth is 380 mm. Calculate the depth downstream of the jump and the power dissipated in it.

10.14 Water discharges at the rate of $8.5\,\text{m}^3/\text{s}$ from under a sluice gate into a long rectangular channel 2.5 m wide which has a slope of 0.002. A hydraulic jump is formed in which the ratio of conjugate depths is 2.5. Estimate the value of Manning's n for the channel.

10.15 A sluice across a rectangular prismatic channel 6 m wide discharges a stream 1.2 m deep. What is the flow rate when the upstream depth is 6 m? The conditions downstream cause a hydraulic jump to occur at a place where concrete blocks have been placed on the bed. What is the force on the blocks if the depth after the jump is 3.1 m?

10.16 Water flows at a depth of 1.2 m in a rectangular prismatic channel 2.7 m wide. Over a smooth hump 200 mm high on the channel bed a drop of 150 mm in the water surface is observed. Neglecting frictional effects, calculate the rate of flow.

10.17 Uniform flow occurs with a depth of 900 mm in a rectangular prismatic channel 2.5 m wide. If Manning's $n = 0.015$ and the bed slope is 1 in 1200 what is the minimum height of hump in the bed over which critical flow will be produced?

10.18 A venturi flume of rectangular section, 1.2 m wide at inlet and 600 mm wide at the throat, has a horizontal base. Neglecting frictional effects in the flume calculate the rate of flow if the depths at inlet and throat are 600 mm and 560 mm respectively. A hump of 200 mm is now installed at the throat so that a standing wave is formed beyond the throat. Assuming the same rate of flow as before, show that the increase in upstream depth is about 67.4 mm.

10.19 A venturi flume installed in a horizontal rectangular channel 700 mm wide has a uniform throat width of 280 mm. When water flows through the channel at $0.140\,\text{m}^3/\text{s}$, the depth at a section upstream of the flume is 430 mm. Neglecting friction, calculate the depth of flow at the throat, the depth at a section just downstream of the flume where the width is again 700 mm, and the force exerted on the stream in passing through the flume.

10.20 A rectangular prismatic channel 1.5 m wide has a slope of 1 in 1600 and ends in a free outfall. If Manning's n is 0.015 how far from the outlet is the depth 750 mm when the flow rate is $1.25\,\text{m}^3/\text{s}$? (Use a tabular integration with three equal steps.)

10.21 Water runs down a 50 m wide spillway at $280\,m^3/s$ on to a long concrete apron ($n = 0.015$) having a uniform downward slope of 1 in 2500. At the foot of the spillway the depth of flow is 600 mm. How far from the spillway will a hydraulic jump occur? (For this very wide channel taking $m = h$ gives acceptable accuracy).

10.22 A pressure gauge fixed 1 m above the sea bed at a position where the mean water depth is 15 m records an average maximum (gauge) pressure of 145 kPa and period 9 s. Determine the wavelength and height. (Density of sea-water $= 1025\,kg/m^3$.)

10.23 A ripple tank contains liquid of density $875\,kg/m^3$ to a depth of 4 mm. Waves of length 8.5 mm are produced by a reed vibrating at 25 Hz. Determine the surface tension of the liquid, the mean rate of energy transfer and the group velocity of the waves.

10.24 In a rectangular wave-tank 4.5 m wide containing fresh water, waves of total height (trough to crest) 0.5 m and period 5 s are generated over a still-water depth of 4 m. Verifying the assumptions made, determine the wave speed, wavelength, group velocity and total power. For a position midway between a trough and a crest and at half the still-water depth determine the velocity of a particle and the static pressure. What are the semi-axes of a particle orbit at this position?

10.25 Ocean waves, with a period of 8 s and amplitude 0.6 m in deep water, approach the shore in the normal direction. A device 80 m long for extracting power from the waves is installed parallel to the shore in water 5 m deep. If there is negligible dissipation of energy before the waves reach the device and its efficiency is 50%, what power is produced? What is the amplitude of the waves immediately before this position? (Density of sea-water $= 1025\,kg/m^3$.)

11 | Compressible flow of gases

11.1 INTRODUCTION

Although all fluids are to some extent compressible, only gases show a marked change of density with a change of pressure or temperature. Even so, there are many examples of the flow of gases in which the density does not change appreciably, and theory relating to constant-density fluids may then adequately describe the phenomena of flow. In this chapter, however, we turn attention to flow in which changes of pressure and velocity are associated with significant changes of density.

In general, significant changes of density are those greater than a few per cent, although there is no sharp dividing line between flows in which the density changes are important and those in which they are unimportant. Significant density changes in a gas may be expected if the velocity (either of the gas itself or of a body moving through it) approaches or exceeds the speed of propagation of sound through the gas, if the gas is subject to sudden accelerations or if there are very large changes in elevation. This last condition is rarely encountered except in meteorology and so (apart from the references in Chapter 2 to the equilibrium of the atmosphere) is not considered in this book.

Because the density of a gas is related to both the pressure and the temperature all changes of density involve thermodynamic effects. Account therefore must be taken of changes in internal energy of the gas, and thermodynamic relations must be satisfied in addition to the laws of motion and continuity. Furthermore, new physical phenomena are encountered. The study of flow in which density varies is thus a good deal more complex than that with constant density, and in this chapter we shall attempt only an introduction to it. It must be assumed that, at this stage, the reader knows something of thermodynamics, but in the next section will be found a few brief reminders of essential points required in studying flow with variable density.

11.2 THERMODYNAMIC CONCEPTS

The density ϱ of a particular gas is related to its absolute pressure p and absolute temperature T by the *equation of state*. For a perfect gas this takes the form

$$p = \varrho RT \tag{1.5}$$

The conditions under which a gas may be assumed 'perfect' are discussed in Section 1.4 and attention is there drawn to the precise definition of the constant R and to its dimensional formula.

From Section 3.5.1 we recall the First Law of Thermodynamics, from which was derived the Steady-Flow Energy Equation (3.13). Since thermal energy and mechanical energy are interchangeable, amounts of either may be expressed in terms of the same units. The inclusion of the 'mechanical equivalent of heat', J, in algebraic equations is therefore unnecessary. It is in any case simply a 'conversion factor' equal to unity and is required only when data about heat and mechanical energy are expressed with different units.

When the physical properties of a gas (e.g. its pressure, density and temperature) are changed, it is said to undergo a *process*. The process is said to be *reversible* if the gas and its surroundings could subsequently be completely restored to their initial conditions by transferring back to (or from) the gas exactly the amounts of heat and work transferred from (or to) it during the process. A reversible process is an ideal never achieved in practice. Viscous effects and friction dissipate mechanical energy as heat which cannot be converted back to mechanical energy without further changes occurring. Also, heat passes by conduction from hotter to cooler parts of the system considered, and heat flow in the reverse direction is not possible. In practice, then, all processes are, in various degrees, irreversible. A process may be considered reversible, however, if velocity gradients and temperature gradients are small, so that the effects of viscosity and heat conduction are negligible.

If, in the course of a process, no heat passes from the gas to its surroundings or from the surroundings to the gas, then that process is said to be *adiabatic*. This is not to say that the internal energy of the gas does not change: the gas may be in a thermally insulated container, and yet some of the kinetic energy of particles may be converted to internal energy by viscous action. An adiabatic process therefore is not necessarily reversible.

Work done on a fluid by its surroundings may compress it or increase its kinetic energy (including the energy of eddies in turbulence) and potential energy. During an infinitesimal reversible (i.e. frictionless) process the work done *on* the fluid is therefore

$$p(-\delta V) + \delta(\text{KE}) + \delta(\text{PE})$$

where p represents its absolute pressure and V its volume. (The minus sign appears because δ means 'a very small *increase* of'.) From the First Law of Thermodynamics

Heat transferred *to* fluid + Work done *on* fluid
= Increase in internal energy
+ Increase in Kinetic and Potential Energies

$$\therefore \delta Q_{rev} + p(-\delta V) = \delta(\text{Internal Energy})$$

or, with reference to unit mass of the fluid,

$$\delta q_{rev} = \delta e + p\delta(1/\varrho) \qquad \text{R*} \quad (11.1)$$

At constant volume ($\delta V = 0$), the specific heat capacity c_v is equal to $(\delta q_{rev}/\delta T)_v = (\partial e/\partial T)_v$ as $\delta T \to 0$. (The suffix v indicates that the volume v per unit mass is held constant.)

At constant pressure $\delta q_{rev} = \delta e + p\delta(1/\varrho) = \delta e + \delta(p/\varrho)$ and for a thermally perfect gas this becomes $\delta q_{rev} = \delta e + \delta(RT)$. Therefore the specific heat capacity at constant pressure $c_p = (\delta q_{rev}/\delta T)_p = (\partial e/\partial T)_p + R$. But for a thermally perfect gas the internal energy consists only of the kinetic energy of the molecules and so is not dependent on their spacing. That is, e is a function of T only and so

$$\left(\frac{\partial e}{\partial T}\right)_v = \frac{de}{dT} = \left(\frac{\partial e}{\partial T}\right)_p$$

$$\therefore c_p - c_v = R \qquad \text{TPG} \quad (11.2)$$

Entropy

For a given system, that is, a given collection of matter, specific entropy s (i.e. entropy per unit mass) is defined by the relation $\delta s = \delta q_{rev}/T$. For the purposes of definition a reversible process is supposed, because between two states (say 1 and 2) of the system the change of specific entropy

$$s_2 - s_1 = \int_1^2 ds$$

is then the same whatever the details of the process, that is, whatever the 'path' of the integral between states 1 and 2. The zero of the s scale is arbitrary, but, once it has been fixed, a certain value of s corresponds to each state of the system. Consequently, a difference of entropy between two states is a definite amount in no way depending on whether the *actual* process is reversible.

Isentropic process

A process in which the entropy does not change ($\delta s = 0$) is termed *isentropic*.

Constant entropy requires $\delta q_{rev} = 0$, a condition achieved if no heat passes between the system and its surroundings and no mechanical energy is converted to thermal energy by friction. A frictionless, adiabatic process is therefore isentropic. In practice such a process is

---

* Here and elsewhere in this chapter the main assumptions involved in each equation are noted alongside: A = adiabatic conditions; B = Boyle's Law; I = isentropic; NF = negligible friction; PG = perfect gas; R = reversible process; SF = steady flow; TPG = thermally perfect gas (i.e. R = constant.)

closely approximated if there is little friction and the changes occur rapidly enough for little transfer of heat to take place across the boundaries.

In the absence of friction, the system considered gains as much thermal energy as its surroundings lose, and the *total* change of entropy is zero. When friction acts, the system gains more thermal energy than its surroundings lose, so the total entropy increases. There is no process by which the total entropy can decrease: for a reversible process the total entropy is constant; for all others it increases.

An important relation may be deduced for a frictionless adiabatic process in a gas. Referring throughout to unit mass of gas, we may write for a frictionless (i.e. reversible) process: $\delta q_{rev} = T\delta s$ and so specific heat capacity $= T\delta s/\delta T$. Hence $c_p = T(\partial s/\partial T)_p$ and $c_v = T(\partial s/\partial T)_v$. Now since entropy is a function of state (i.e. of absolute pressure p and volume v per unit mass) we may write

$$\delta s = \left(\frac{\partial s}{\partial v}\right)_p \delta v + \left(\frac{\partial s}{\partial p}\right)_v \delta p$$

where a suffix indicates the quantity held constant. Temperature does not appear as a separate variable since it is uniquely related to p and v by the equation of state. Since the assumed process is isentropic, however, $\delta s = 0$ and so, as δp and δv both tend to zero,

$$\left(\frac{\partial p}{\partial v}\right)_s = -\frac{(\partial s/\partial v)_p}{(\partial s/\partial p)_v} = -\frac{(\partial s/\partial T)_p}{(\partial v/\partial T)_p} \Big/ \frac{(\partial s/\partial T)_v}{(\partial p/\partial T)_v}$$

$$= -\gamma \frac{(\partial p/\partial T)_v}{(\partial v/\partial T)_p} \qquad \text{A, R} \quad (11.3)$$

where $\gamma = c_p/c_v$.

If the gas obeys Boyle's Law $pv = f(T)$, where $f(T)$ represents any function of T, then

$$\left(\frac{\partial p}{\partial T}\right)_v = \frac{1}{v}\frac{df}{dT} \quad \text{and} \quad \left(\frac{\partial v}{\partial T}\right)_p = \frac{1}{p}\frac{df}{dT}$$

These expressions substituted in eqn 11.3 give

$$\left(\frac{\partial p}{\partial v}\right)_s = -\frac{\gamma p}{v} \qquad \text{A, R, B} \quad (11.4)$$

For a constant value of γ, integration then yields $\ln p = -\gamma \ln v +$ constant. Thus

$$pv^\gamma = \text{constant} = p/\varrho^\gamma \quad \text{A, R, B, } \gamma \text{ const.} \quad (11.5)$$

For a perfect gas $p/\varrho T = $ constant and the isentropic relation 11.5 may be alternatively expressed

$$\frac{p^{(\gamma-1)/\gamma}}{T} = \text{constant} \qquad \text{A, R, PG} \quad (11.6)$$
$$\text{i.e. I, PG}$$

The equation of state, which relates p, ϱ and T, strictly refers to conditions of thermodynamic equilibrium. Experience shows, however, that this equation, and others derived from it, may in practice be used for non-equilibrium conditions except when the departure from equilibrium is extreme (as in explosions).

Enthalpy

Certain combinations of fluid properties occur so frequently in thermodynamic problems that they may usefully be given symbols of their own. One such combination is $E + pV = H$, known as *enthalpy*. The specific enthalpy (i.e. enthalpy per unit mass) is given by $h = e + pv = e + p/\varrho$. For many vapours, values of h (as a function of temperature and pressure) may be obtained from appropriate tables and charts. However, if c_p is constant, h is linearly related to temperature. This is because, for constant pressure, $\delta q_{\text{rev}} = \delta e + \delta(p/\varrho) = \delta h$ from eqn 11.1. Consequently $c_p = (\partial q_{\text{rev}}/\partial T)_p = (\partial h/\partial T)_p$ and so, with c_p constant and the zero of h taken at the absolute zero of temperature,

$$h = c_p T \qquad c_p \text{ const} \quad (11.7)$$

11.3 ENERGY EQUATION WITH VARIABLE DENSITY: STATIC AND STAGNATION TEMPERATURE

The general energy equation for the steady flow of any fluid was presented in Section 3.5.2. This equation may be applied to any two points along a streamline. If no heat is transferred to or from the fluid between these points and no 'machine' work is done, we may set $q = 0$ and $w = 0$ and so obtain

$$0 = \left(\frac{p_2}{\varrho_2} + \frac{u_2^2}{2} + gz_2 \right) - \left(\frac{p_1}{\varrho_1} + \frac{u_1^2}{2} + gz_1 \right) + e_2 - e_1 \quad \text{SF, A} \quad (11.8)$$

Since $e + p/\varrho$ may be written as specific enthalpy h, and as points 1 and 2 were arbitrarily chosen, eqn 11.8 may be expressed as

$$h + \frac{1}{2}u^2 + gz = \text{constant along a streamline} \quad \text{SF, A} \quad (11.9)$$

This is the general form of the equation for steady, adiabatic flow in which the fluid (gas, liquid or vapour) neither does work on its surroundings nor has work done on itself. If, however, c_p is constant then, from eqn 11.7,

$$c_p T + \frac{1}{2}u^2 + gz = \text{constant} \quad \text{SF, A, } c_p \text{ const} \quad (11.10)$$

Also, since for a perfect gas

$$c_p T = c_p \frac{p}{R\varrho} = \left(\frac{c_p}{c_p - c_v} \right) \frac{p}{\varrho} = \left(\frac{\gamma}{\gamma - 1} \right) \frac{p}{\varrho}$$

another useful form is

$$\left(\frac{\gamma}{\gamma - 1} \right) \frac{p}{\varrho} + \frac{1}{2}u^2 + gz = \text{constant} \qquad \text{SF, A, PG} \quad (11.11)$$

It will be noted that the assumptions implicit in these equations do not include that of no friction. This is because the equations include internal as well as mechanical energy. Friction merely involves the conversion of energy of one kind to an equivalent amount of another kind which, for adiabatic conditions, remains in the fluid, and so the total energy is unchanged.

For gases the gz term is usually regarded as negligible compared with, for example, the first term of eqn 11.11 because ϱ is small and changes of z are usually small also. (The concept of piezometric pressure $p + \varrho gz$ cannot of course be used where the density is variable.) Equation 11.9 may then be reduced to

$$h + \frac{1}{2}u^2 = \text{constant along a streamline} \qquad \text{SF, A} \quad (11.12)$$

It is clear that in steady adiabatic flow an increase of velocity must be accompanied by a decrease of enthalpy and a decrease of velocity by an increase of enthalpy. For a given streamline the specific enthalpy is a maximum when the velocity is zero (at a stagnation point), and this maximum value is termed the *stagnation enthalpy*, h_0. From eqn 11.7 the corresponding *stagnation temperature* T_0 is h_0/c_p and so the energy equation may be written

$$c_p T + \frac{1}{2}u^2 = c_p T_0 \qquad \text{SF, A, } c_p \text{ const} \quad (11.13)$$

If an attempt is made to measure the temperature of a flowing gas by placing a thermometer or similar device in the stream, the temperature recorded will be greater than T. Equation 11.13 shows that the stagnation temperature exceeds T by $u^2/2c_p$. For air $c_p = 1005 \, \text{J}/(\text{kg K})$ and the excess is therefore

$$\frac{u^2}{2010 \text{J}/(\text{kg K})} = \frac{u^2}{2010} \frac{\text{s}^2 \text{K}}{\text{m}^2}$$

Thus the stagnation temperature for an air stream at, say, 200 m/s exceeds the ordinary 'static' temperature by about 20 K. And the nose cone of a rocket travelling through air at, say, 2 km/s must withstand a temperature rise approaching 2000 K! For these extreme conditions the use of eqn 11.13 involves some inaccuracy because of the variation of c_p during the large increases of temperature and pressure. However, although the stagnation temperature would be reached at the stagnation point on a thermometer bulb, the temperature would rise less at other points on it, so the mean temperature recorded by an ordinary thermometer would be somewhat less then the stagnation temperature. The static temperature cannot be directly measured by any stationary instrument. (It could be measured only by a thermometer or other instrument moving at the same velocity as the gas.)

■

Example 11.1   Air flows adiabatically through a pipe. At a plane, denoted by suffix 1, the temperature is $-2\,°C$, its pressure is $1.50 \times 10^5\,N\,m^{-2}$, and the air moves at a speed of $270\,m/s$. At plane 2 the pressure is $1.20 \times 10^5\,N\,m^{-2}$ and the speed of the air is $320\,m/s$. Calculate the following properties of the air:

(a) the density at plane 1
(b) the stagnation temperature
(c) the temperature and density at plane 2.

Solution

(a) From the equation of state

$$\varrho_1 = \frac{p_1}{RT_1} = \frac{\left(1.5 \times 10^5\right)N/m^2}{287\,J/(kg\,K) \times 271\,K} = 1.93\,kg\,m^{-3}$$

(b) From eqn 11.13, and using the value for c_p from Appendix 2,

$$T_0 = T_1 + \frac{u_1^2}{2c_p} = 271\,K + \frac{\left(270\,m/s\right)^2}{2 \times 1005\,J/(kg\,K)} = 307.3\,K$$

(c) Since T_0 is constant in adiabatic flow:

$$T_2 = T_0 - \frac{u_2^2}{2c_p} = 307.3\,K - \frac{\left(320\,m/s\right)^2}{2 \times 1005\,J/(kg\,K)} = 256.4\,K$$

and

$$\varrho_2 = \frac{p_2}{RT_2} = \frac{\left(1.2 \times 10^5\right)N/m^2}{287\,J/(kg\,K) \times 256.4\,K} = 1.63\,kg\,m^{-3}$$

□

11.4   ELASTIC WAVES

If the pressure at a point in a fluid is altered, the density is also altered – even if only slightly – and in consequence individual particles undergo small changes in position. To maintain a continuum, adjacent particles also change position and thus the new pressure is progressively, yet rapidly, transmitted throughout the rest of the fluid. Indeed, in a completely incompressible fluid any disturbances would be propagated with infinite velocity because all particles would have to change position simultaneously. Even in an actual fluid, changes of pressure are transmitted so rapidly that the time necessary for them to be spread throughout the fluid may often be negligible compared with the time taken for the original change. Thus, in previous chapters of this book we have assumed that pressure adjustments occur simultaneously throughout the fluid. But if the pressure at a point is suddenly altered, or the fluid is moving with high velocity relative to some solid body, then the exact speed with which pressure changes are transmitted is of great importance, This speed is determined by the relation between changes of pressure and changes of density, that is, by the elastic properties of the fluid.

Consider an instant after a small change of pressure has been caused at some point in the fluid. The change may have resulted from the movement of a solid body, such as a piston, the breaking of a thin membrane across which a pressure difference existed, or an electrical discharge such as lightning. Not all the fluid has yet experienced the pressure change and so at a certain distance from the place where the change originated there is a more or less abrupt discontinuity of pressure. This discontinuity is known as a pressure wave: ahead of it is the original pressure p; behind it is the new pressure $p + \delta p$. Figure 11.1 shows a part of the wave W, small enough to be considered plane. (A wave produced by the movement of a plane piston in a duct of constant cross-section would, in the absence of friction, remain entirely plane but, even if the wave spreads as a spherical surface radially outwards in all directions, a sufficiently small part may be considered plane.) The wave is being propagated towards the right at velocity c.

The existence of the pressure difference δp across the wave indicates that the layer of fluid immediately in front of the wave has an unbalanced force acting on it and is consequently accelerated. As the wave proceeds other layers are similarly accelerated. Once the full increase of pressure has been established, however, no further acceleration of the fluid occurs. Thus the component of the fluid's velocity perpendicular to

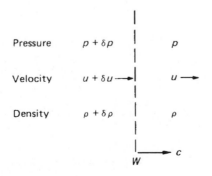

Pressure	$p + \delta p$	p
Velocity	$u + \delta u \rightarrow$	$u \rightarrow$
Density	$\rho + \delta \rho$	ρ

W $\xrightarrow{\ c\ }$

Fig. 11.1

the wave changes from u (which may be positive, negative or zero) to $u + \delta u$.

Since the velocity of the fluid at a point changes as the wave passes that point, the flow is not steady and so cannot be analysed by equations developed for steady flow. However, from a set of coordinate axes moving with velocity c the wave will appear stationary, and velocities measured with respect to those axes will not change with time. So Fig. 11.2 shows the situation as seen by someone moving with the new axes, that is, as steady flow. Continuity requires that, across an area ΔA of the wave,

$$(\varrho + \delta\varrho)(u - c + \delta u)\Delta A = \varrho(u - c)\Delta A$$

whence

$$(c - u)\delta\varrho = (\varrho + \delta\varrho)\delta u \qquad (11.14)$$

In view of the thinness of the wave frictional effects may be neglected. Then, for a control volume enclosing the area ΔA of the wave, the momentum equation gives

Force towards right $= (p + \delta p)\Delta A - p\Delta A$

$\qquad\qquad = $ Rate of increase of momentum towards right

$\qquad\qquad = \varrho(u - c)\Delta A(-\delta u)$

whence

$$\delta p = \varrho(c - u)\delta u \qquad \text{NF} \quad (11.15)$$

Elimination of δu from eqns 11.14 and 11.15 gives

$$(c - u)^2 = \left(\frac{\varrho + \delta\varrho}{\varrho}\right)\frac{\delta p}{\delta\varrho} \qquad \text{NF} \quad (11.16)$$

For a weak pressure wave, for which $\delta p \to 0$ and $\delta\varrho \to 0$, eqn 11.16 becomes

$$c - u = \sqrt{(\partial p/\partial\varrho)} \qquad \text{NF} \quad (11.17)$$

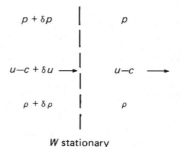

Fig. 11.2

Since the left-hand side of this equation represents the velocity of the pressure wave relative to the fluid ahead of it, a small pressure change is propagated at a velocity $\sqrt{(\partial p/\partial \varrho)}$ relative to the fluid.

Now for a weak pressure wave the changes in pressure, density and temperature are exceedingly small. Not only is the friction resulting from the very small change of velocity negligible, but the extreme smallness of the temperature differences and the rapidity of the propagation together indicate that transfer of heat across the wave is also extremely small. Consequently the passage of the wave is a process which, to a close approximation, may be considered both adiabatic and frictionless. That is, the process is isentropic and we may write

$$a = \sqrt{\left(\frac{\partial p}{\partial \varrho}\right)_s} \qquad \text{I} \quad (11.18)$$

where $a = c - u$. (It may be observed in passing that for an entirely incompressible fluid $\delta\varrho = 0$ whatever the value of δp and so a would be infinite, that is, changes of pressure would be transmitted instantaneously through the fluid.) For any fluid the bulk modulus of elasticity K is defined by $\varrho\, \partial p/\partial \varrho$ (eqn 1.8), so $a = \sqrt{(K_s/\varrho)}$. (For a mixture of a liquid and a gas, a is less than for either the liquid or the gas separately. When a liquid contains gas bubbles it is more compressible, so K is reduced, but ϱ is little affected. When a gas contains small drops of liquid K is little affected but ϱ is increased. In each case, therefore, a is reduced.)

If the fluid is a gas obeying Boyle's Law, then from eqn 11.4

$$\left(\frac{\partial p}{\partial \varrho}\right)_s = \frac{\mathrm{d}v}{\mathrm{d}\varrho}\left(\frac{\partial p}{\partial v}\right)_s = \frac{p\gamma}{\varrho} \quad \text{since } \varrho = \frac{1}{v} \qquad \text{B, I}$$

The speed of propagation of a very small pressure wave (*relative to the fluid*) is therefore

$$a = \sqrt{(p\gamma/\varrho)} \qquad \text{B, I} \quad (11.19)$$

Alternatively, from the equation of state for a thermally perfect gas,

$$a = \sqrt{(\gamma RT)} \qquad \text{TPG, I} \quad (11.20)$$

Sonic velocity

Sound is propagated by means of a succession of very small pressure waves in which δp is alternately positive and negative. (The faintest sound that the human ear can detect unaided corresponds to a pressure fluctuation of about 3×10^{-5} Pa; the loudest that can be tolerated without physical pain corresponds to a fluctuation of about 100 Pa.) The velocity represented by eqns 11.13–11.20 is therefore known as the *speed of sound* or *velocity of sound* or *sonic velocity* or *acoustic velocity* in the gas. As it is a function of temperature it varies in general from point to point in the fluid. The validity of the assumptions made in deriving the expressions is indicated by the excellent agreement found with experimental determinations of the velocity of sound. For air of moderate humidity $\gamma = 1.4$ and $R = 287$ J/(kg K); so at 15 °C

$$a = \sqrt{\left\{1.4 \times 287 \frac{J}{kg\,K}(273 + 15)K\right\}} = 340\,m/s$$

The velocity of sound is appreciably less at high altitudes because of the lower temperature there.

It is important to note that the preceding expressions refer only to waves in which the change of pressure is very small compared with the pressure itself. Waves in which a comparatively large pressure change occurs will be considered in Section 11.5. The assumption of constant entropy is not justified for these larger waves, and they move at velocities greater than that of sound.

Mach number

When the velocity of the fluid at a particular point is less than the velocity of sound there, small pressure waves can be propagated both upstream and downstream. When, however, the velocity of the fluid exceeds the local sonic velocity a, a small pressure wave cannot be propagated upstream. A velocity equal to a thus sharply divides two essentially different types of flow. It is useful to express the velocity of the fluid in terms of the sonic velocity. The ratio (fluid velocity) ÷ (sonic velocity) is known as the Mach number (see Section 5.3.4). The symbol commonly used is M: this is not consistent with the double-letter notation usual for dimensionless parameters but is more widely used than Ma and is simpler in algebraic work. Fluid velocities less than the sonic velocity are known as *subsonic* ($M < 1$), those greater than the sonic velocity as *supersonic* ($M > 1$). For an entirely incompressible fluid a would be infinite and M therefore always zero.

11.4.1 The Mach cone

Small elastic waves are the means by which 'messages' are transmitted from one point in the fluid to another. Messages can be sent in a particular direction, however, only if the velocity of the fluid in the opposite direction is less than the sonic velocity. It of course does not matter whether the origin of the wave is at a fixed point in space and the fluid is moving, or the fluid is stationary and the source of the wave is moving through the fluid.

To see how flow patterns are affected by such waves let us consider a small solid body, such as a tiny projectile, moving in a straight line through fluid which – except as it is disturbed by the passage of the body – is stationary. We suppose first that the (steady) velocity u of the body is less than the local sonic velocity a. The movement of the body generates pressure waves in the fluid which are transmitted with velocity a radially in all directions. If the body is at point A (Fig. 11.3) at time $t = 0$, then at a short time t later the waves that originated from the body when it was at A will have grown into the surface of a sphere of radius at. During that time the body itself moves a distance $AB = ut$. Since $u < a$, the pressure waves in the forward direction are able to travel

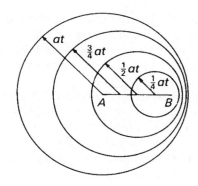

Fig. 11.3

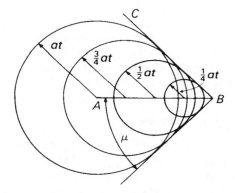

Fig. 11.4

ahead of the body and B is inside the sphere of radius at (and also inside other spheres formed by pressure waves started at intermediate times.) The waves travelling ahead of the body inform the fluid of the body's approach, so the fluid has an opportunity to prepare for its arrival.

The picture is very different when $u > a$. After a time t the body has travelled a distance $ut > at$ and is therefore outside the sphere formed by the pressure waves sent out at time $t = 0$ (Fig. 11.4). That is, the body travels faster than the message, and thus arrives at point B unannounced. The unprepared fluid then has to move suddenly, thus producing the sharp discontinuities known as shock waves (Section 11.5). The line CB is a common tangent to all the spheres formed by the waves, and so is the generator of a cone having its axis in the direction of motion and its vertex at B, the instantaneous position of the body. The cone therefore advances with the body into the undisturbed fluid.

The semi-vertex angle μ of the cone is given by *Mach angle*

$$\sin \mu = \frac{at}{ut} = \frac{1}{M} \qquad (11.21)$$

where M represents the Mach number u/a. The angle μ is known as the *Mach angle*, the cone as the *Mach cone* and a line such as CB as a *Mach line*.

Zone of silence

Outside the cone the fluid is completely unaffected by the waves, and this region is often known as the *zone of silence*, whereas the space inside the cone is termed the *zone of action* or *region of influence*. Across the surface of the Mach cone there are abrupt changes of pressure and density.

It should be remembered that the discussion here has been restricted to *small* waves and also to a body small enough to be considered as a single point. The behaviour of larger bodies, such as aircraft, travelling at supersonic velocities is similar to that discussed here, although close to the body itself the pressure increases are more than infinitesimal and are thus propagated with a velocity greater than *a*. Consequently the Mach cone of an actual body travelling at supersonic velocity has a rounded apex (see Section 11.5.2).

Changes of density in a gas give rise to small changes in its refractive index and so, by suitable optical arrangements (see Section 11.12), abrupt changes of density, as across the surface of a Mach cone, may be made visible. Measurements of the Mach angle are then possible and the Mach number may be deduced from eqn 11.21.

11.4.2 Propagation of finite waves

A finite change of pressure across a wave may be regarded as the sum of a series of infinitesimal changes. Suppose that a compression wave (i.e. one producing an increase of density) is caused by the motion of a piston, for example. At the start of the motion an infinitesimal wave travels with sonic velocity into the undisturbed fluid. A second tiny wave follows immediately, but the fluid into which this second wave moves has already been traversed by the first wave. That fluid is therefore at a slightly higher pressure, density and temperature than formerly. If for simplicity a perfect gas is assumed so that $a = \sqrt{(\gamma R T)}$ (although the conclusion may be shown to apply to any fluid) it is clear that the second wave travels slightly faster than the first. Similarly the third wave is propagated with a slightly higher velocity than the second and so on. The pressure distribution at a certain time might be as shown in Fig. 11.5a, and a little while later would become like that shown at b. In practice, separate small waves would not be distinguishable but, instead, a gradual rise of pressure which becomes steeper as it advances. Before long the wave front becomes infinitely steep and a sharp discontinuity of pressure results. This is known as a *shock wave*, and it will be studied in the next section.

A similar argument applied to a rarefaction wave shows that this becomes *less* steep as shown in Figs. 11.5c and d. The foremost edge of the wave proceeds into the undisturbed fluid with the sonic velocity in that fluid, but all other parts of the wave have a lower velocity. Consequently no effect of a pressure decrease, no matter how large or how sudden it is initially, is propagated at more than the sonic velocity in the undisturbed fluid. A pressure increase, on the other hand, may be propagated at a velocity greater than that of sound.

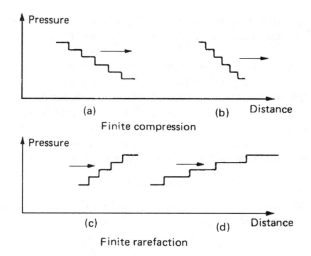

Fig. 11.5

11.5   SHOCK WAVES

Whereas in Section 11.4 we considered a pressure change of infinitesimal size, we now turn attention to an abrupt finite pressure change known as a *shock*. The possibility of such an abrupt change in a compressible fluid was envisaged by the German mathematician G. F. Bernhard Riemann (1826–66) and the theory was developed in detail by W. J. M. Rankine (1820–72) and the French physicist Henri Hugoniot (1851–87).

In practice a shock is not absolutely abrupt, but the distance over which it occurs is of the order of only a few times the mean free path of the molecules (about $0.3\,\mu m$ in atmospheric air). For most purposes, therefore, the changes in flow properties (pressure, density, velocity and so on) may be supposed abrupt and discontinuous and to take place across a surface termed the *shock wave*. (In photographs a thickness apparently greater than $0.3\,\mu m$ may be observed because the wave is seldom exactly plane or exactly parallel to the camera axis.) We shall not concern ourselves here with what happens within the very narrow region of the shock itself, for such a study is very complex and involves non-equilibrium thermodynamics. Moreover, the analysis that follows will be restricted to a perfect gas because a general solution is algebraically complicated and explicit results cannot usually be obtained. Qualitatively, however, the phenomena discussed apply to any gas.

11.5.1   Normal shock waves

We consider first a *normal shock wave*, that is, one perpendicular to the direction of flow. Such shocks may occur in the diverging section of a convergent-divergent nozzle or in front of a blunt-nosed body. We shall see that in every case the flow upstream of the shock is supersonic, while

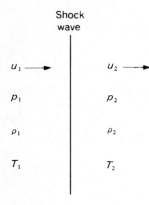

Shock wave

$u_1 \longrightarrow$ $u_2 \longrightarrow$

p_1 p_2

ρ_1 ρ_2

T_1 T_2

Fig. 11.6 Normal shock.

that downstream is subsonic and at a higher pressure. We shall see too that the changes occurring in a shock are not reversible and so not isentropic.

Our first objective is to determine the relations between quantities upstream and downstream of the shock. As shown in Fig. 11.6, quantities upstream are denoted by suffix 1 and those downstream by suffix 2. To obtain steady flow we consider the shock stationary and so the velocities u_1, u_2 are relative to it. As the shock region is so thin, any change in the cross-sectional area of a stream-tube from one side of the shock to the other is negligible, and so the continuity relation is simply

$$\varrho_1 u_1 = \varrho_2 u_2 \tag{11.22}$$

If effects of boundary friction are negligible, the momentum equation for a stream-tube of cross-sectional area A is

$$\left(p_1 - p_2\right)A = \left(\text{Mass flow rate}\right)\!\left(u_2 - u_1\right)$$

i.e.

$$p_1 - p_2 = \varrho_2 u_2^2 - \varrho_1 u_1^2 \tag{11.23}$$

Since $a^2 = p\gamma/\varrho$ and Mach number $M = u/a$, this may be written

$$p_1 - p_2 = p_2\gamma M_2^2 - p_1\gamma M_1^2$$

whence

$$\frac{p_2}{p_1} = \frac{1 + \gamma M_1^2}{1 + \gamma M_2^2} \tag{11.24}$$

If there is no net heat transfer to or from the stream-tube considered, the adiabatic energy equation (11.10) may be used:

$$c_p T_1 + \frac{1}{2}u_1^2 = c_p T_2 + \frac{1}{2}u_2^2 \tag{11.25}$$

Thus, if c_p remains constant, the stagnation temperature (defined by eqn 11.13) does not change across the shock. Putting $u = aM = M\sqrt{(\gamma RT)}$ for a perfect gas we obtain

$$c_p T_1 + \frac{1}{2}M_1^2\gamma R T_1 = c_p T_2 + \frac{1}{2}M_2^2\gamma R T_2$$

whence

$$\frac{T_2}{T_1} = \frac{c_p + \frac{1}{2}M_1^2\gamma R}{c_p + \frac{1}{2}M_2^2\gamma R} = \frac{1 + \frac{1}{2}(\gamma - 1)M_1^2}{1 + \frac{1}{2}(\gamma - 1)M_2^2} \tag{11.26}$$

From the equation of state and from eqn 11.22

$$\frac{T_2}{T_1} = \frac{p_2\varrho_1}{p_1\varrho_2} = \frac{p_2 u_2}{p_1 u_1} = \frac{p_2 M_2}{p_1 M_1}\sqrt{\left(\frac{\gamma R T_2}{\gamma R T_1}\right)} \tag{11.27}$$

whence

$$\frac{T_2}{T_1} = \left(\frac{p_2 M_2}{p_1 M_1}\right)^2$$

Substitution from eqns 11.24 and 11.26 now gives

$$\frac{1 + \frac{1}{2}(\gamma - 1)M_1^2}{1 + \frac{1}{2}(\gamma - 1)M_2^2} = \left(\frac{1 + \gamma M_1^2}{1 + \gamma M_2^2}\right)^2 \frac{M_2^2}{M_1^2} \qquad (11.28)$$

An obvious and trivial solution of this equation is $M_1^2 = M_2^2$; that is, conditions upstream and downstream are identical and no shock exists. Simplification in which $M_1^2 - M_2^2$ is factored out, however, gives

$$M_2^2 = \frac{1 + \left(\frac{\gamma - 1}{2}\right)M_1^2}{\gamma M_1^2 - \frac{\gamma - 1}{2}} \qquad (11.29)$$

From this result the downstream Mach number may be calculated and the ratios of pressure, temperature, density and velocity then obtained from eqns 11.24, 11.26 and 11.27. Equation 11.29 shows that if $M_1 = 1$ then $M_2 = 1$, and that if $M_1 > 1$, $M_2 < 1$.

From the equations obtained above ϱ_2/ϱ_1 may be expressed in terms of p_2/p_1:

$$\frac{\varrho_2}{\varrho_1} = \frac{\gamma - 1 + (\gamma + 1)p_2/p_1}{\gamma + 1 + (\gamma - 1)p_2/p_1} \qquad (11.30)$$

The fact that eqn 11.30, the *Rankine–Hugoniot relation*, is not the same as that for an adiabatic *reversible* process (p/ϱ^γ = constant) indicates that the changes occurring in a shock are not reversible and so not isentropic. The greater the value of p_2/p_1 the more does the Rankine–Hugoniot relation diverge from the isentropic: in air ($\gamma = 1.4$), for example, when $p_2/p_1 = 10$ (i.e. $M_1 = 2.95$) the density ratio across a normal shock is 3.81, whereas by an isentropic compression the value 5.18 would be obtained.

Rankine–Hugoniot relation

In the equations so far deduced no restriction has been placed on the value of M_1, and an abrupt rarefaction would seem just as possible as an abrupt compression. Calculation of the entropy change across the shock shows, however, that in steady flow a rarefaction shock – in which $M_2 > M_1$ and $p_2 < p_1$ – is impossible. By the definition of entropy and eqn 11.1

$$s_2 - s_1 = \int_1^2 \frac{\mathrm{d}q_{\mathrm{rev}}}{T} = c_v \int_1^2 \frac{\mathrm{d}T}{T} + \int_1^2 \frac{p\,\mathrm{d}(1/\varrho)}{T}$$

$$= c_v \ln(T_2/T_1) + \int_1^2 \varrho R\,\mathrm{d}(1/\varrho) = c_v \ln(T_2/T_1) - R\ln(\varrho_2/\varrho_1)$$

Substitution for T_2/T_1, ϱ_2/ϱ_1 and then for M_2 gives

$$s_2 - s_1 = c_v \ln\left[\frac{2\gamma M_1^2 - \gamma + 1}{\gamma + 1}\left\{\frac{2 + (\gamma - 1)M_1^2}{(\gamma + 1)M_1^2}\right\}^\gamma\right] \quad (11.31)$$

This expression is positive if $M_1 > 1$ but negative if $M_1 < 1$. Since a decrease of entropy in an adiabatic process is impossible, a shock can exist in steady flow only when the upstream velocity is supersonic. This is in accord with the conclusions drawn in Section 11.4.2: if a sudden rarefaction wave is formed (e.g. by the collapse of an evacuated vessel) it is not stable and rapidly decays into a gradual pressure change. The increase of entropy experienced by the gas passing through a shock wave results principally from conduction of heat across the wave, from the gas already compressed to that not yet compressed.

Expression 11.31 is plotted in Fig. 11.7 and it is seen that the slope of the curve is zero at $M_1 = 1$. Thus the flow through very weak shocks (such as the sound waves considered in Section 11.4) may be regarded as isentropic to a close approximation, and an *infinitesimal* rarefaction wave is just possible.

We have already noted that (for adiabatic conditions) the stagnation temperature remains unchanged across a shock. However, because of the dissipation of mechanical energy in the shock, the stagnation pressure is reduced. From eqn 11.13, we have $c_p T + \frac{1}{2}u^2 = c_p T_0$ and so with the aid of eqns 11.2 and 11.20

$$\frac{T_0}{T} = 1 + \frac{u^2}{2c_p T} = 1 + \frac{M^2 \gamma RT}{2c_p T} = 1 + \left(\frac{\gamma - 1}{2}\right)M^2$$

Since stagnation pressure is defined as that pressure which would be reached if the fluid were brought to rest by a *reversible* adiabatic, that is, an isentropic, process, we have, from eqn 11.6,

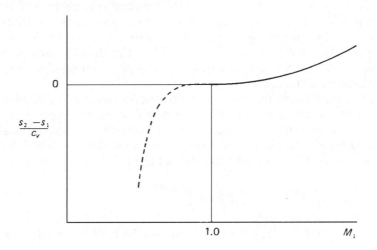

Fig. 11.7

$$\frac{p_0}{p} = \left(\frac{T_0}{T}\right)^{\gamma/(\gamma - 1)} = \left\{1 + \left(\frac{\gamma - 1}{2}\right)M^2\right\}^{\gamma/(\gamma - 1)} \qquad (11.32)$$

The reduction of stagnation pressure across a shock is indicated by the ratio

$$\frac{(p_0)_2}{(p_0)_1} = \frac{(p_0)_2}{p_2}\frac{p_2}{p_1}\frac{p_1}{(p_0)_1}$$

which, on substitution from eqns 11.32, 11.24 and 11.29, becomes

$$\frac{(p_0)_2}{(p_0)_1} = \left\{\frac{(\gamma + 1)M_1^2}{2 + (\gamma - 1)M_1^2}\right\}^{\gamma/(\gamma - 1)}\left\{\frac{\gamma + 1}{2\gamma M_1^2 - \gamma + 1}\right\}^{1/(\gamma - 1)} \qquad (11.33)$$

The greater the departure from isentropic conditions the greater the loss of stagnation pressure. This is illustrated in Fig. 11.8 where the expression 11.33 is plotted – together with M_2 and the ratios of pressure, temperature and density – against values of M_1 for air ($\gamma = 1.4$). These functions are also tabulated in Appendix 3, Table A3.1. It may be noted from eqn 11.29 that, as $M_1 \to \infty$, $M_2 \to \{(\gamma - 1)/2\gamma\}^{1/2} = 0.378$ for air; the density ratio tends to $(\gamma + 1)/(\gamma - 1)$ ($= 6$ for air) and the velocity ratio to $(\gamma - 1)/(\gamma + 1)$ ($= 1/6$ for air).

The equations may also be solved for u_1 in terms of the pressure ratio:

$$u_1 = \left[\frac{p_1}{2\varrho_1}\left\{\gamma - 1 + (\gamma + 1)\frac{p_2}{p_1}\right\}\right]^{1/2}$$

Since u_1 is the upstream velocity relative to the shock wave this result expresses the velocity with which a moving shock advances into stationary fluid. For an infinitesimal pressure change $p_2/p_1 = 1$ and the expression reduces to the velocity of sound $\sqrt{(p_1\gamma/\varrho_1)}$. For values of p_2/p_1 greater than unity, the velocity of propagation is always greater than the velocity of sound. Thus the shock waves produced by explosions, for example, are propagated with velocities in excess of sonic velocity.

In practice, friction at boundaries and conduction of heat through the gas cause the changes in flow properties across a shock to be slightly less than those predicted by the foregoing analysis. Moreover, for strong shocks (with $M_1 > 5$, say) a variation of specific heat capacities is noticeable, and if the temperature rises greatly (say above 250°C) dissociation and ionization phenomena may occur.

11.5.2 Oblique shock waves

An oblique shock wave is one that is not perpendicular to the flow. It arises, for example, when supersonic flow is caused to change direction by a boundary surface converging towards the flow. Figure 11.9 illustrates a (stationary) oblique shock wave in two-dimensional flow. The portion of the wave considered is assumed plane and the velocities of

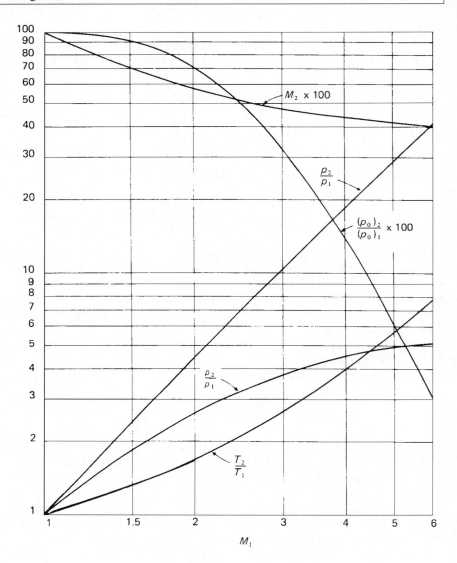

Fig. 11.8 Curves for normal shock ($\gamma = 1.4$).

the fluid on each side of the shock may be split into components u_n and u_t, respectively normal and tangential to the wave. The pressure change across the shock causes a reduction in the normal velocity component but leaves the tangential component unaltered. Consequently the flow is deflected away from the normal to the wave. Although the normal component u_{2n} downstream of the shock must be subsonic, the *total* velocity $\sqrt{(u_{2n}^2 + u_t^2)}$ may still be supersonic if u_t is sufficiently large. In other words, M_2 is always less than M_1, but M_2 may be greater or less than unity.

Changes of flow properties across the shock may be determined in the same way as for the normal shock. The continuity and momentum

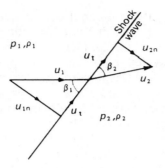

Fig. 11.9

relations (corresponding to eqns 11.22 and 11.23 for the normal shock) are:

$$\varrho_1 u_{1n} = \varrho_2 u_{2n} \qquad (11.34)$$

and

$$p_1 - p_2 = \varrho_2 u_{2n}^2 - \varrho_1 u_{1n}^2 \qquad (11.35)$$

The energy equation for adiabatic conditions (11.10) may be written

$$\text{constant} = c_p T + \frac{1}{2}u^2 = c_p T + \frac{1}{2}\left(u_n^2 + u_t^2\right)$$

But since u_t is unchanged across the shock the energy equation reduces to

$$c_p T_1 + \frac{1}{2}u_{1n}^2 = c_p T_2 + \frac{1}{2}u_{2n}^2 \qquad (11.36)$$

Equations 11.34–11.36 differ from those for a normal shock only in using the normal component of flow velocity in place of the full velocity. Therefore the subsequent equations developed for the normal shock are applicable to the oblique shock if $u\sin\beta$ is substituted for u, and $M\sin\beta$ for M.

The angles β_1 and β_2 of Fig. 11.9 are related by

$$\frac{\tan\beta_2}{\tan\beta_1} = \frac{u_{2n}}{u_{1n}} = \frac{\varrho_1}{\varrho_2} = \frac{p_1 T_2}{p_2 T_1}$$

With the aid of eqns 11.24, 11.26 and 11.29 this expression becomes

$$\frac{\tan\beta_2}{\tan\beta_1} = \frac{2 + (\gamma - 1)M_1^2 \sin^2\beta_1}{(\gamma + 1)M_1^2 \sin^2\beta_1}$$

and elimination of β_2 in favour of the angle of deflection $(\beta_1 - \beta_2)$ then gives

$$\tan(\beta_1 - \beta_2) = \frac{2\cot\beta_1\left(M_1^2 \sin^2\beta_1 - 1\right)}{M_1^2(\gamma + \cos 2\beta_1) + 2} \qquad (11.37)$$

Equation 11.37 shows that $\beta_1 - \beta_2 = 0$ when $\beta_1 = 90°$ (a normal shock wave) or when $M_1 \sin\beta_1 = 1$. This second condition is the limiting case when the normal velocity component $= a_1$ and the pressure rise is infinitesimal; β_1 is then the Mach angle (Section 11.4.1).

Equation 11.37 is plotted in Fig. 11.10 from which it is seen that, for a given value of M_1, the deflection has a maximum value, and that a particular deflection below this maximum is given by two values of β_1.

As an example we may consider a wedge-shaped solid body of semi-vertex angle $\beta_1 - \beta_2$ placed symmetrically in a uniform supersonic flow of Mach number M_1 (Fig. 11.11). On reaching the wedge the flow is deflected through the angle $\beta_1 - \beta_2$ and if this is less than the maximum for the given value of M_1 an oblique shock wave is formed at the nose of the wedge as shown in Fig. 11.11a. Now Fig. 11.10 shows that two values of β_1 will satisfy the oblique shock equations, and the question arises: which one occurs in practice?

For the smaller value of β_1 the normal component $M_1 \sin \beta_1$ would be smaller and so (as shown by Figs 11.7 and 11.8) the corresponding pressure rise and entropy increase would also be smaller. The wave that

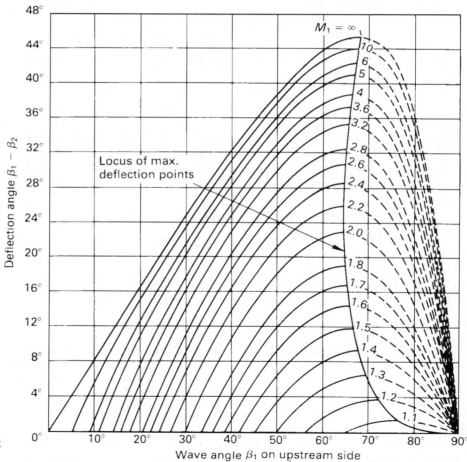

Fig. 11.10 Oblique shock relations for $\gamma = 1.4$.

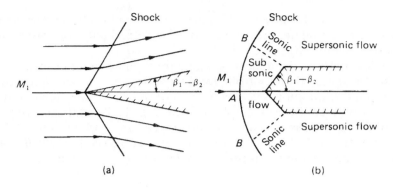

(a)   (b)   **Fig. 11.11**

has this smaller value of β_1, and is known as a weak, or ordinary, oblique shock wave, is the one that usually appears. The strong, or extraordinary, wave, corresponding to the larger value of β_1, occurs only if the boundary conditions of the flow are such as to require the greater pressure rise across the strong wave.

Figure 11.10 also shows that for a given value of M_1 there is a maximum value of $\beta_1 - \beta_2$ for which an oblique shock wave is possible. This maximum increases from zero when $M_1 = 1$ to arccosec γ ($= 45.6°$ for air) when $M_1 = \infty$. (This is a mathematical rather than a physical result since for $M_1 = \infty$ there would be infinite rises of pressure and temperature which would invalidate the assumption of a perfect gas.) Conversely, it may be said that for a specified deflection angle (less than arccosec γ) an oblique shock wave is possible only if M_1 exceeds a certain minimum value. If the semi-vertex angle of the wedge is greater than the maximum value of $\beta_1 - \beta_2$ for the given Mach number the shock wave cannot be attached to the wedge because that would require the flow to turn through an angle greater than the maximum. In these circumstances the required deflection can be achieved only if the flow first becomes subsonic and for this the shock wave must be detached from the nose of the wedge as shown in Fig. 11.11b. (The shock wave actually leaves the wedge when M_2 just becomes unity, but this occurs when the semi-vertex angle is only a fraction of a degree less than $(\beta_1 - \beta_2)_{max}$.) Similar detached waves are formed in front of blunt objects.

At a large distance from the wedge (or other solid body) the size of the body may be considered negligible in comparison with the distance, and thus the situation is similar to that discussed in Section 11.4.1. In other words, with increasing distance from the body the shock wave must degenerate into a *Mach wave*, that is, a wave of infinitesimal strength, at an angle arccosec M_1 to the oncoming flow. Consequently, whatever the shape of the body a detached shock wave in front of it is always curved. At A the wave is normal to the flow, and so subsonic conditions are necessarily produced in the region between A and the nose of the body. With increasing distance from the axis, however, the shock becomes weaker and its angle decreases towards that of a Mach wave; thus the region of subsonic flow behind the wave extends only to some position B. In the subsonic part of the flow the streamlines can of course accommodate themselves to whatever change of direction is imposed by the shape of the body.

Values of M_1 or the deflection angle through an oblique shock wave can be calculated directly from eqn 11.37. However, when β_1 has to be determined from known values of M_1 and $(\beta_1 - \beta_2)$ solutions are most readily obtained from a graph such as Fig. 11.10. Alternatively, an algebraic solution is available if both numerator and denominator on the right of eqn 11.37 are multiplied by $\mathrm{cosec}^2\beta_1 \ (= 1 + \cot^2\beta_1)$. The resulting cubic equation in $\cot\beta_1$ may be solved by calculating

$$A = \frac{1}{3}\left(1 + \frac{\gamma + 1}{2}M_1^2\right)\tan(\beta_1 - \beta_2)$$

$$B = A^2 + \frac{1}{3}\left(M_1^2 - 1\right)$$

$$C = A^3 + \frac{1}{3}\left(\frac{\gamma + 1}{4}M_1^4 + \frac{\gamma - 1}{2}M_1^2 + 1\right)\tan(\beta_1 - \beta_2)$$

$$\theta = \arccos\left(C/B^{3/2}\right)$$

Then $\cot\beta_1 = 2B^{1/2}\sin(30° \pm \theta/3) - A$ where the positive alternative sign corresponds to the ordinary wave and the negative sign to the extraordinary wave.

The equations and results referred to here apply only to two-dimensional flow; similar effects, however, are obtained in three-dimensional flow, although the values of limiting deflection angles and so on are somewhat different.

Since the shock wave ahead of a solid body is curved, different streamlines undergo different changes of direction and the flow properties downstream are not uniform. In particular, the velocity is not uniform and so viscous forces come into play. The analysis of flow behind a detached shock wave is in fact very difficult, and a complete solution has yet to be obtained. It is clear, however, that for a given upstream Mach number the pressure rise through the central, nearly normal, part of a detached wave is greater than that through an oblique wave. The drag force on a body moving at supersonic velocity is therefore greater if the wave is detached from its nose, and that is why small nose angles are desirable on supersonic aircraft and rockets.

Example 11.2 An airstream, at a temperature of 238 K, is moving at a speed of 773 m/s when it encounters an oblique shock wave. The shock angle β_1 is 38°. Find

(a) the angle through which the airstream is deflected
(b) the final Mach number
(c) the pressure ratio across the wave.

Solution
(a) Upstream of the shock, the speed of sound is given by

$$a_1 = \left(\gamma R T_1\right)^{1/2} = \left(1.4 \times 287\,\text{J}/(\text{kg\,K}) \times 238\,\text{K}\right)^{1/2} = 309.2\,\text{m/s}$$

$$M_1 = u_1/a_1 = (773\,\text{m/s})/(309.2\,\text{m/s}) = 2.5$$

$$\tan\beta_2 = \tan\beta_1 \left[\frac{2 + (\gamma - 1)M_1^2 \sin^2\beta_1}{(\gamma + 1)M_1^2 \sin^2\beta_1}\right]$$

$$= (0.7813)\left[\frac{2 + (0.4 \times 6.25 \times 0.379)}{2.4 \times 6.25 \times 0.379}\right] = 0.405$$

Hence

$$\beta_2 = 22.05°$$

The deflection angle is $(\beta_1 - \beta_2) = 15.95°$.
  (b) The tangential velocity component is given by

$$u_t = u_1\cos\beta_1 = u_2\cos\beta_2$$

Hence

$$u_2 = u_1\frac{\cos\beta_1}{\cos\beta_2} = 773\,\text{m/s} \times \frac{0.788}{0.927} = 657\,\text{m/s}$$

Since $\quad c_p T_1 + \dfrac{1}{2}u_1^2 = c_p T_2 + \dfrac{1}{2}u_2^2$

$$T_2 = T_1 + \frac{1}{2c_p}\left(u_1^2 - u_2^2\right)$$

$$= 238\,\text{K} + \frac{1}{2 \times 1005\,\text{J}/(\text{kg\,K})}\left[(773\,\text{m/s})^2 - (657\,\text{m/s})^2\right]$$

$$= 320.5\,\text{K}$$

Hence

$$M_2 = u_2/a_2 = u_2 / \left(\gamma R T_2\right)^{1/2}$$

$$= (657\,\text{m/s})/\left(1.4 \times 287\,\text{J}/(\text{kg\,K}) \times 320.5\,\text{K}\right)^{1/2}$$

$$= 1.83$$

  (c) Since

$$\frac{\tan \beta_2}{\tan \beta_1} = \frac{p_1}{p_2} \frac{T_2}{T_1}$$

$$\frac{p_2}{p_1} = \frac{T_2}{T_1} \frac{\tan \beta_1}{\tan \beta_2} = \frac{320.5}{238} \times \frac{0.7813}{0.405} = 2.60$$

11.5.3 Reflection and intersection of oblique shock waves

When an oblique wave meets a solid boundary a reflection wave may be formed. In the example shown in Fig. 11.12 the original flow is parallel to the boundary; an oblique shock wave is produced, by the wedge W for example, and the flow downstream of this wave is therefore deflected towards the boundary by the angle $\beta_1 - \beta_2$. If the flow downstream of the wave is still supersonic, the flow can again become parallel to the boundary only through another shock wave which counteracts the original deflection. The second wave BC may thus be regarded as the 'reflection' of the original wave AB at the boundary.

Shock waves are not like light waves, and the angles of 'incidence' and 'reflection' are in general not equal. For example, if the original flow has Mach number 2.5 and the original wave makes an angle of 30° with the boundary (see Fig. 11.13) then, for $\gamma = 1.4$, the deflection is 8.0° (from Fig. 11.10 or eqn 11.37) and $M_2 = 2.17$ (eqn 11.29). At the 'reflected' wave the deflection must be 8.0° in the opposite direction and this requires the wave to be at 26.3° to the boundary – not 30°. (As noted in Section 11.5.2, of the two mathematically possible solutions only that corresponding to the weaker shock is normally observed.)

The deflection through the first wave may be so large that it exceeds the value obtainable from the intermediate Mach number M_2. If, in the example just discussed, the initial wave angle were 45° instead of 30°, the deflection would be 21.6° and M_2 would be 1.569. But the maximum deflection possible from an upstream Mach number 1.569 is only 13.9° and this is insufficient to bring the flow again parallel to the boundary. The pattern of waves then takes the form illustrated in Fig. 11.14 in which BB' is an approximately normal shock wave, so that downstream

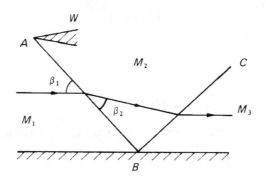

Fig. 11.12

of it subsonic flow is produced adjacent to the boundary. The pressures and directions of the fluid streams passing through those parts of BC and BB' close to B must be the same, but their difference in velocity causes them to be separated by a vortex sheet or 'slip surface' BV (which, however, is soon diffused by turbulence). This sort of reflection was first observed by Ernst Mach, and is therefore known as a Mach reflection. (The analysis here is somewhat idealized: in practice the reflection of an oblique shock wave from a wall is influenced by the boundary layer there.)

If two oblique shock waves meet, and the flow downstream of each remains supersonic, the usual result is the formation of another pair of waves springing from the intersection, on the downstream side. There is

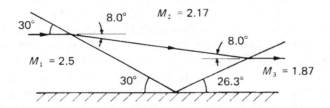

Fig. 11.13

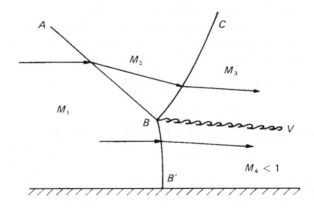

Fig. 11.14

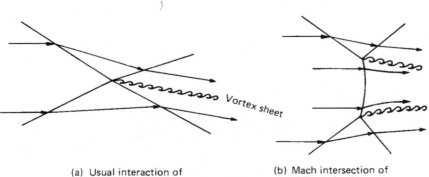

(a) Usual interaction of oblique shock waves

(b) Mach intersection of oblique shock waves

Fig. 11.15

also a vortex sheet if the velocities after the second pair of waves are not equal (Fig. 11.15a). As with the reflection of waves, however, the combination of intermediate Mach number and wave angle may not be suitable, and then a 'Mach intersection' is formed (Fig. 11.15b).

11.6 SUPERSONIC FLOW ROUND A CORNER

Figure 11.16 illustrates steady two-dimensional supersonic flow past a boundary consisting of two plane surfaces which make an angle $\delta\theta$ with each other. This angle, although greatly exaggerated for clarity in the diagram, is infinitesimal. The flow approaching the corner is supposed uniform and so far from other boundary surfaces as to be unaffected by them. We shall see that if the corner is convex, as in the diagram, the gas undergoes an expansion through an infinitesimal Mach wave which makes an acute angle μ with the original direction of flow. It will be recalled from the consideration of entropy change in Section 11.5.1 that only an infinitesimal expansion wave is possible; this is why the present analysis must be restricted to an infinitesimal angle of turn, $\delta\theta$. The corner constitutes an infinitesimal disturbance to the flow, so μ is the Mach angle of the flow. Since the upstream conditions are uniform, the value of μ is the same for all streamlines: that is, the wave is straight and the downstream conditions must all be uniform too.

Across the wave the velocity changes from u to $u + \delta u$, but because there is no pressure gradient along the wave the component of velocity parallel to the wave remains unaltered. That is

$$u\cos\mu = \left(u + \delta u\right)\cos\left(\mu + \delta\theta\right)$$
$$= \left(u + \delta u\right)\left(\cos\mu\cos\delta\theta - \sin\mu\sin\delta\theta\right)$$

As $\delta\theta$ is very small, $\cos\delta\theta \to 1$ and $\sin\delta\theta \to \delta\theta$. Hence

$$\left(u + \delta u\right)\delta\theta\sin\mu = \left(u + \delta u - u\right)\cos\mu$$

giving

$$\delta\theta = \frac{\delta u}{u + \delta u}\cot\mu \to \frac{\delta u}{u}\cot\mu = \frac{\delta u}{u}\left(M^2 - 1\right)^{1/2} \qquad (11.38)$$

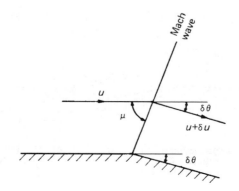

Fig. 11.16

since

$$\sin \mu = 1/M \qquad (11.21)$$

Equation 11.38 shows that, with positive $\delta\theta$ measured from the original direction in Fig. 11.16, δu is positive. In other words, the velocity increases round a convex corner. As the component parallel to the Mach wave remains unchanged the normal component must increase and, to satisfy the continuity relation $\varrho u_n = $ constant, the density must decrease.

If an initially uniform flow makes a succession of small turns, as in Fig. 11.17, there will be a number of regions of uniform flow separated by Mach waves emanating from the corners. If the straight portions between the corners, and also the angles $\delta\theta_1$, $\delta\theta_2$ etc., are indefinitely decreased a continuously curved surface is obtained, from which an infinite number of Mach waves is generated.

Since opportunity for heat transfer at these high velocities is so slight the process across each infinitesimal wave may be considered adiabatic. Then, since changes of elevation are negligible, the energy equation (11.11) may be written

$$\left(\frac{\gamma}{\gamma-1}\right)\frac{p}{\varrho} + \frac{1}{2}u^2 = \frac{a^2}{\gamma-1} + \frac{1}{2}u^2 = \text{constant } C$$

whence

$$\frac{1}{(\gamma-1)M^2} + \frac{1}{2} = \frac{C}{u^2} \qquad \text{SF, A, PG} \quad (11.39)$$

Differentiating and then substituting for C/u^2 we obtain

$$-\frac{2\,dM}{(\gamma-1)M^3} = -\frac{2C\,du}{u^3}$$

$$= -\frac{2\,du}{u}\left\{\frac{1}{(\gamma-1)M^2} + \frac{1}{2}\right\} \qquad \text{SF, A, PG} \quad (11.40)$$

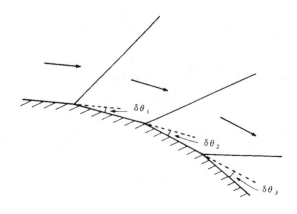

Fig. 11.17

Eliminating du/u from eqns 11.38 and 11.40 enables $d\theta$ to be expressed in terms of the single parameter M:

$$d\theta = \frac{\left(M^2 - 1\right)^{1/2} dM}{\left(\gamma - 1\right)M^3\left\{\dfrac{1}{\left(\gamma - 1\right)M^2} + \dfrac{1}{2}\right\}} \qquad \text{SF, A, PG} \quad (11.41)$$

Since $M > 1$, dM is positive when $d\theta$ is positive and so M increases round a convex corner. Thus $\sin\mu$ $(= 1/M)$ decreases; that is, the Mach waves make successively smaller angles with the oncoming streamlines, as shown in Fig. 11.17. A particular case is that in which the individual Mach waves all intersect at a common point. Such flow is known as a *centred expansion*. An important instance is that in which the common centre for the waves is on the boundary itself, that is, at a corner of finite angle (Fig. 11.18). The flow is unaffected up to the Mach wave A; in the fan-shaped region between waves A and B the gas expands gradually and isentropically with a gradual change in flow direction; beyond wave B uniform conditions again prevail. Although this results in a discontinuous change of properties from state 1 to state 2 at the corner itself (like a rarefaction 'shock'), the discontinuity is infinitesimal in extent and so the Second Law of Thermodynamics is not violated.

Whether the expansion is centred or not, the total change in conditions through the series of Mach waves may be determined by integrating eqn 11.41. Putting $M^2 - 1 = $ (say) x^2 yields

$$d\theta = \frac{2x^2 dx}{\left(x^2 + 1\right)\left\{\gamma + 1 + \left(\gamma - 1\right)x^2\right\}}$$

$$= \frac{\left(\gamma + 1\right)dx}{\gamma + 1 + \left(\gamma - 1\right)x^2} - \frac{dx}{x^2 + 1}$$

which on integration gives

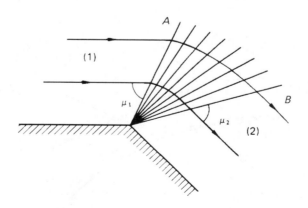

Fig. 11.18

$$\theta = \left(\frac{\gamma + 1}{\gamma - 1}\right)^{1/2} \arctan\left\{\left(\frac{\gamma - 1}{\gamma + 1}\right)^{1/2} x\right\} - \arctan x$$

$$= \left(\frac{\gamma + 1}{\gamma - 1}\right)^{1/2} \arctan\left[\left\{\left(\frac{\gamma - 1}{\gamma + 1}\right)(M^2 - 1)\right\}^{1/2}\right] - \operatorname{arcsec} M$$

$$\text{SF, A, I, PG} \quad (11.42)$$

Since θ may be measured from an arbitrary datum direction the integration constant has been set at zero (and thus $\theta = 0$ when $M = 1$). The result – tabulated in Appendix 3, Table A3.2 – was first obtained in 1907 by L. Prandtl and Theodor Meyer*; such flow round a convex corner is usually known as a Prandtl–Meyer expansion and θ is known as the Prandtl–Meyer angle. It represents the angle through which a stream, initially at sonic velocity, must be turned to reach the given Mach number M. The change in direction needed for the flow to expand from M_1 to M_2 may be determined as the change of the Prandtl–Meyer angle between the limits M_1 and M_2.

Example 11.3   A uniform stream of air ($\gamma = 1.4$) at Mach number 1.8 and static pressure 50 kPa is expanded round a 10° convex bend. What are the conditions after the bend?

Solution
From eqn 11.42 or Table A3.2, since $M_1 = 1.8$, $\theta_1 = 20.73°$

$$\therefore \ \theta_2 = 30.73° \text{ and } M_2 = 2.162$$

For an isentropic process in a perfect gas

$$\frac{p_2}{p_1} = \left(\frac{T_2}{T_1}\right)^{\gamma/(\gamma-1)} = \left[\frac{1 + \left(\frac{\gamma - 1}{2}\right)M_1^2}{1 + \left(\frac{\gamma - 1}{2}\right)M_2^2}\right]^{\gamma/(\gamma-1)} \quad \text{(from eqns 11.6 and 11.26)}$$

$$\therefore \ p_2 = 50\,\text{kPa}\left(\frac{1 + 0.2 \times 1.8^2}{1 + 0.2 \times 2.162^2}\right)^{3.5} = 28.5\,\text{kPa}$$

The maximum deflection theoretically possible would be that corresponding to acceleration from $M = 1$ to $M = \infty$, that is

---

* pronounced *My'-er*.

$$\theta_{max} = \left\{ \left(\frac{\gamma + 1}{\gamma - 1} \right)^{1/2} \frac{\pi}{2} - \frac{\pi}{2} \right\} \text{ (in radian measure), i.e. } 130.5° \text{ for } \gamma = 1.4$$

The gas would then have expanded to zero pressure and temperature. If the boundary turned away from the flow by more than θ_{max} a void would form next to the boundary. In practice, however, the maximum deflection would not be achieved because the temperature cannot fall to absolute zero without liquefaction of the gas. Moreover, as the pressure approaches absolute zero the assumption of a fluid continuum is no longer tenable.

11.6.1 Supersonic flow over a concave boundary

Uniform supersonic flow approaching an infinitesimal concave corner undergoes a change of direction through a Mach wave; this change is described by eqn 11.38 except that the sign of $\delta\theta$ is changed. Thus a compression results instead of an expansion and the Mach angle μ increases. As a result, a series of small concave corners produces Mach waves which converge and form an envelope building up into an oblique shock wave (Fig. 11.19). Flow through the shock wave, however, is not isentropic and so the Prandtl–Meyer relation does not hold there. For a sharp bend the shock wave forms at (or very near to) the surface itself (as described in Section 11.5.2).

11.6.2 Supersonic flow between two boundaries

We have so far considered initially uniform flow past single boundaries sufficiently far from other surfaces to be uninfluenced by them. A characteristic of such flow is that the Mach waves are straight and flow conditions along each wave are constant. If, however, supersonic flow occurs between two boundaries, both of which are curved, each generates a family of Mach waves. Flow conditions along a Mach wave of one family are affected by those of the other family, and so where two

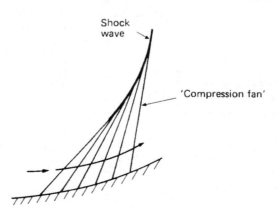

Fig. 11.19

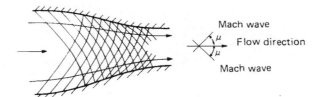

Fig. 11.20 Families of Mach waves between curved boundaries.

families intersect the waves are no longer straight (Fig. 11.20). A number of special graphical and tabular techniques have been developed for dealing with problems of flow between curved boundaries, but these are beyond the scope of this book.

11.7   THE PITOT TUBE IN FLOW WITH VARIABLE DENSITY

When a Pitot-static tube is used to determine the velocity of a constant-density fluid the stagnation pressure and static pressure need not be separately measured: it is sufficient to measure their difference. A high-velocity gas stream, however, may undergo an appreciable change of density in being brought to rest at the front of the Pitot-static tube, and in these circumstances stagnation and static pressures must be separately measured. Moreover, if the flow is initially supersonic, a shock wave is formed ahead of the tube, and so results for supersonic flow differ essentially from those for subsonic flow. We first consider the Pitot-static tube in uniform subsonic flow.

The process by which the fluid is brought to rest at the nose of the tube is assumed to be frictionless and adiabatic. From the energy equation (11.13) and from eqn 11.6 we therefore obtain

$$\frac{u^2}{2} = c_p(T_0 - T) = c_p T_0 \left\{ 1 - \left(\frac{p}{p_0} \right)^{(\gamma-1)/\gamma} \right\} \qquad \text{SF, I, PG} \quad (11.43)$$

where suffix 0 again refers to stagnation conditions (see Fig. 11.21). For measuring T_0 it is usual to incorporate in the instrument a small thermocouple surrounded by an open-ended jacket. If T_0 and the ratio of static to stagnation pressure are known the velocity of the stream may then be determined from eqn 11.43.

The influence of compressibility is best illustrated, however, by using the Mach number. Rearranging eqn 11.13 and noting that $a^2 = \gamma R T = (\gamma - 1)c_p T$, we obtain

$$\frac{T_0}{T} = 1 + \frac{u^2}{2c_p T} = 1 + \left(\frac{\gamma - 1}{2} \right) M^2 \qquad \text{SF, A, PG} \quad (11.44)$$

and then from eqn 11.6,

$$\frac{p_0}{p} = \left\{ 1 + \left(\frac{\gamma - 1}{2} \right) M^2 \right\}^{\gamma/(\gamma-1)} \qquad \text{SF, I, PG} \quad (11.45)$$

Fig. 11.21

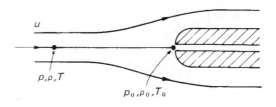

For subsonic flow $(\gamma - 1) M^2/2 < 1$ and so the right-hand side of eqn 11.45 may be expanded by the binomial theorem to give

$$\frac{p_0}{p} = 1 + \frac{\gamma}{2}M^2 + \frac{\gamma}{8}M^4 + \frac{\gamma(2 - \gamma)}{48}M^6 + \ldots$$

whence

$$p_0 - p = \frac{p\gamma M^2}{2}\left\{1 + \frac{M^2}{4} + \left(\frac{2 - \gamma}{24}\right)M^4 + \ldots\right\}$$

$$= \frac{1}{2}\varrho u^2\left\{1 + \frac{M^2}{4} + \left(\frac{2 - \gamma}{24}\right)M^4 + \ldots\right\} \quad \text{SF, I, PG} \quad (11.46)$$

Comparing eqn 11.46 with the result for a fluid of constant density (Section 3.7.2) we see that the bracketed quantity (sometimes known as the *compressibility factor*) represents the effect of compressibility. Table 11.1 indicates the variation of the compressibility factor with M when $\gamma = 1.4$.

It is seen that for $M < 0.2$ compressibility affects the pressure difference by less than 1%, and the simple formula for flow at constant density is then sufficiently accurate. For larger values of M, however, the compressibility must be taken into account.

For supersonic flow eqn 11.45 is not valid because a shock wave forms ahead of the Pitot tube (Fig. 11.22), so the fluid is not brought to rest entirely isentropically. The nose of the tube is always so shaped (i.e. the semi-angle is greater than the maximum deflection obtainable through an oblique shock – see Section 11.5.2) that the shock wave is detached. If the axis of the tube is parallel to the oncoming flow the wave may be assumed normal to the streamline leading to the stagnation point. The pressure rise across the shock is therefore given by eqn 11.24:

$$\frac{p_2}{p_1} = \frac{1 + \gamma M_1^2}{1 + \gamma M_2^2}$$

In the subsonic region downstream of the shock there is a gradual isentropic pressure rise according to eqn 11.32 and so

$$\frac{(p_0)_2}{p_1} = \frac{(p_0)_2}{p_2}\frac{p_2}{p_1} = \left\{1 + \left(\frac{\gamma - 1}{2}\right)M_2^2\right\}^{\gamma/(\gamma-1)}\left(\frac{1 + \gamma M_1^2}{1 + \gamma M_2^2}\right)$$

Table 11.1 Variation of 'compressibility factor' for air

M	$\dfrac{p_0 - p}{\frac{1}{2}\varrho u^2}$
0·1	1.003
0·2	1.010
0·3	1.023
0·4	1.041
0·5	1.064
0·6	1.093
0·7	1.129
0·8	1.170
0·9	1.219
1·0	1.276

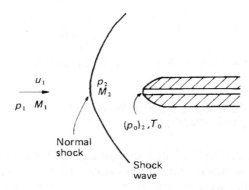

Fig. 11.22

Using eqn 11.29 to express M_2 in terms of M_1 finally yields Rayleigh's formula:

$$\frac{(p_0)_2}{p_1} = \left\{ \frac{(\gamma + 1)^{\gamma+1}}{2\gamma M_1^2 - \gamma + 1} \left(\frac{M_1^2}{2} \right)^{\gamma} \right\}^{1/(\gamma-1)}$$

$$= \frac{166.9 M_1^7}{\left(7 M_1^2 - 1\right)^{2.5}} \quad \text{when } \gamma = 1.4 \qquad (11.47)$$

(Values are tabulated in Appendix 3, Table A3.1.)

Although a conventional Pitot-static tube gives satisfactory results at Mach numbers low enough for no shock waves to form, it is unsuitable in supersonic flow because its 'static' holes, being in the region downstream of the shock, do not then register p_1. Nor do they register p_2 since this is found only on the central streamline, immediately behind the normal part of the shock wave. Consequently p_1 is best determined independently – for example, through an orifice in a boundary wall well upstream of the shock. (Such measurements of static pressure *close* to the shock are unreliable because the pressure rise through the shock can be transmitted upstream through the subsonic flow in the boundary layer.) Where independent measurement of p_1 is not possible, a special Pitot-static tube may be used in which the static holes are much further back (about 10 times the outside diameter of the tube) from the nose. The oblique shock wave on each side of the tube has by then degenerated into a Mach wave across which the pressure rise is very small.

When $M_1 = 1$ the pressure rise across the shock is infinitesimal, so eqns 11.45 and 11.47 both give $p_0/p_1 = \{(\gamma + 1)/2\}^{\gamma/(\gamma-1)}$ ($= 1.893$ for $\gamma = 1.4$). A smaller value of p_0/p therefore indicates subsonic flow, a larger value supersonic flow.

Equation 11.47 enables the upstream Mach number to be calculated from the ratio of stagnation to static pressure. Since (for a perfect gas) the *stagnation* temperature does not change across a shock wave

$$c_p T_0 = c_p T_1 + \frac{u_1^2}{2} = c_p \frac{u_1^2}{\gamma R M_1^2} + \frac{u_1^2}{2}$$

Thus u_1 also may be calculated if T_0 is determined.

11.8 ONE-DIMENSIONAL FLOW WITH NEGLIGIBLE FRICTION

We recall from Section 1.8.3 that the concept of 'one-dimensional' flow in any form of conduit is that all relevant quantities (velocity, pressure, density and so on) are uniform over any cross-section of that conduit. Thus the flow can be described in terms of only one coordinate (distance along the axis – which is not necessarily straight) and time. The flow of a real fluid is never strictly one-dimensional because of the presence of boundary layers but the assumption provides satisfactory solutions of many problems in which the boundary layer is not very thick and there are no abrupt changes of cross-section. In this section we shall use the one-dimensional idealization to examine instances of flow in which the effects of compressibility are of particular importance.

For a cross-section of area A over which ϱ and u are uniform the continuity equation for steady flow is

$$\varrho A u = \text{constant} \qquad\qquad \text{SF} \quad (11.48)$$

Differentiating and then dividing by $\varrho A u$ gives

$$\frac{d\varrho}{\varrho} + \frac{dA}{A} + \frac{du}{u} = 0 \qquad\qquad \text{SF} \quad (11.49)$$

If significant changes of cross-sectional area, and therefore of ϱ and u, occur over only a short length of the conduit, for example in a nozzle, frictional effects may be neglected in comparison with these changes.

Then Euler's equation for steady, frictionless flow (3.8) may be used (with dz neglected):

$$\frac{dp}{\varrho} + udu = 0 \qquad\qquad \text{NF, SF} \quad (11.50)$$

If adiabatic conditions hold, then, in the absence of friction, the flow is isentropic. So we may set

$$\frac{dp}{\varrho} = \left(\frac{\partial p}{\partial \varrho}\right)_s \frac{d\varrho}{\varrho} = a^2 \frac{d\varrho}{\varrho}$$

(from eqn 11.18) where a represents the sonic velocity. Equation 11.50 becomes

$$a^2 \frac{d\varrho}{\varrho} + udu = 0 \qquad\qquad \text{A, NF, SF} \quad (11.51)$$

and substitution for $d\varrho/\varrho$ from eqn 11.49 gives

$$\frac{dA}{A} = \frac{du}{u}\left(\frac{u^2}{a^2} - 1\right)$$

or, since u/a = the Mach number M,

$$\frac{dA}{A} = \frac{du}{u}\left(M^2 - 1\right) \qquad\qquad \text{A, NF, SF} \quad (11.52)$$

Several important conclusions may be drawn from eqn 11.52. For subsonic velocities ($M < 1$), dA and du must be opposite in sign. That is, an increase of cross-sectional area causes a reduction of velocity and vice versa. This result is familiar from studies of constant-density flow. For supersonic velocities, however, $M^2 - 1$ is positive and so dA and du are of the same sign. An increase of cross-sectional area then causes an increase of velocity and a reduction of cross-sectional area a reduction of velocity.

When $u = a$ so that $M = 1$, dA must be zero and (since the second derivative is positive) A must be a minimum. If the velocity of flow equals the sonic velocity anywhere it must therefore do so where the cross-section is of minimum area. However, dA could also be zero when $du = 0$. At a position of minimum cross-section the velocity is therefore either equal to the sonic velocity, or it is a maximum (for subsonic flow) or a minimum (for supersonic flow).

To obtain the relation between pressure and velocity, eqn 11.50 must be integrated. For a gas obeying Boyle's Law under adiabatic frictionless conditions p/ϱ^γ = constant (eqn 11.5), and using this to substitute for ϱ enables us to integrate eqn 11.50 to

$$\left(\frac{\gamma}{\gamma - 1}\right)\frac{p}{\varrho} + \frac{1}{2}u^2 = \text{constant}$$

$$= \left(\frac{\gamma}{\gamma - 1}\right)\frac{p_0}{\varrho_0} \qquad \text{B, } \gamma \text{ const, A, NF, SF} \quad (11.53)$$

where suffix '0' denotes the stagnation conditions, i.e. at zero velocity (as in a large storage reservoir, for example). When the velocity u equals $\sqrt{(p\gamma/\varrho)}$ we obtain

$$\left(\frac{\gamma}{\gamma-1}\right)\frac{p_c}{\varrho_c} + \frac{1}{2}\frac{p_c\gamma}{\varrho_c} = \left(\frac{\gamma}{\gamma-1}\right)\frac{p_0}{\varrho_0} \qquad \text{B, } \gamma \text{ const, A, NF, SF} \quad (11.54)$$

Here suffix 'c' denotes the 'critical' conditions at which the flow velocity equals the local sonic velocity. (These critical conditions have no connection with those referring to the liquefaction of gases under pressure.) Therefore

$$\frac{p_c}{\varrho_c} = \left(\frac{2}{\gamma+1}\right)\frac{p_0}{\varrho_0} \qquad \text{B, } \gamma \text{ const, A, NF, SF} \quad (11.55)$$

Application of these principles is found in the *de Laval nozzle*, developed by the Swedish engineer Carl Gustaf Patrik de Laval (1845–1913) for accelerating steam to supersonic velocity in his high-speed steam turbine. Similar nozzles are used nowadays in jet propulsion units and rocket motors. Initially the velocity is subsonic and acceleration occurs in the converging portion (Fig. 11.23). The 'critical' conditions described by eqn 11.55 occur at the 'throat' of minimum cross-section and the further acceleration in supersonic flow occurs in the diverging portion. (In practice, owing to viscous effects, $u = a$ slightly downstream of the throat.)

Critical pressure ratio   By means of eqn 11.55, the *critical pressure* p_c found at the throat may be expressed by the *critical pressure ratio*

$$\frac{p_c}{p_0} = \frac{2}{\gamma+1}\left(\frac{\varrho_c}{\varrho_0}\right) = \frac{2}{\gamma+1}\left(\frac{p_c}{p_0}\right)^{1/\gamma}$$

whence

$$\frac{p_c}{p_0} = \left(\frac{2}{\gamma+1}\right)^{\gamma/(\gamma-1)} \qquad \text{B, } \gamma \text{ const, I, SF} \quad (11.56)$$

For air ($\gamma = 1.4$) this critical pressure ratio is 0.528, whereas for *superheated* steam ($\gamma \simeq 1.3$) $p_c/p_0 \simeq 0.546$. (Although superheated steam does not behave as a perfect gas ($p/\varrho T$ is not independent of temperature) it does obey Boyle's Law fairly well over moderate ranges of pressure.) If the pressure of a vapour such as steam falls below the saturation pressure, however, liquid droplets may condense from it. When the drop in pressure occurs rapidly the vapour may expand be-

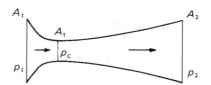

Fig. 11.23

yond the saturation pressure as a supersaturated vapour before condensing. Nevertheless, if condensation does occur the vapour ceases to obey Boyle's Law and γ also changes. Equation 11.56 is then no longer valid.

In the diverging part of the de Laval nozzle the velocity, now supersonic, continues to increase and the pressure therefore drops further below the critical value p_c.

For a nozzle of given throat area A_t (or for any duct with minimum cross-sectional area A_t) the mass rate of flow $\varrho A u$ is a maximum when ϱu at the throat is a maximum, that is, when

$$\frac{d(\varrho u)}{dp} = u\frac{d\varrho}{dp} + \varrho\frac{du}{dp} = 0$$

i.e. when

$$\frac{dp}{d\varrho} = -\frac{u}{\varrho}\frac{dp}{du}$$

From Euler's equation (11.50), however, $dp/du = -\varrho u$ and so the maximum flow occurs when $dp/d\varrho = u^2$; that is, for the assumed isentropic conditions, when $u = a$. By means of eqns 11.19, 11.5 and 11.56, this maximum mass flow rate may be written

$$m_{max} = \varrho_c A_t u_c = \varrho_c A_t \left(\frac{p_c \gamma}{\varrho_c}\right)^{1/2}$$

$$= A_t\left\{p_c\gamma\varrho_0\left(\frac{p_c}{p_0}\right)^{1/\gamma}\right\}^{1/2}$$

$$= A_t\left\{p_0\varrho_0\gamma\left(\frac{2}{\gamma+1}\right)^{(\gamma+1)/(\gamma-1)}\right\}^{1/2} \qquad \text{B, } \gamma \text{ const, I, SF} \quad (11.57)$$

The maximum mass flow rate is thus a function only of the stagnation conditions and the minimum cross-sectional area A_t. No matter how much the pressure p_2 at the downstream end is reduced, or how the shape of the duct may be changed upstream or downstream of the minimum cross-section, this maximum flow rate cannot be exceeded. Under such conditions the duct or nozzle is said to be *choked*. A physical explanation of the phenomenon is that the 'news' of any reduction of pressure downstream of the throat has to be transmitted in the form of a rarefaction wave. But, as we have seen (Section 11.4.2), the fastest portion of such a wave has a velocity that only just equals sonic velocity. Thus once the velocity at the throat reaches the sonic velocity no 'messages' can be transmitted upstream and so the fluid there is quite 'unaware' of any further reduction of pressure at the downstream end of the duct.

Example 11.4 Air flows through a convergent-divergent nozzle from a reservoir in which the temperature is 291 K. At the nozzle exit the pressure is $28 \times 10^3 \text{Nm}^{-2}$ and the Mach number is 2.4. With $\gamma = 1.4$ and $R = 287 \text{J kg}^{-1}\text{K}^{-1}$ and assuming isentropic flow conditions, calculate:

(a) the pressures in the reservoir and at the nozzle throat
(b) the temperature and velocity of the air at the exit

Solution

(a) From equation 11.45:

$$p_0 = p\left\{1 + \left(\frac{\gamma - 1}{2}\right)M^2\right\}^{\gamma/(\gamma-1)}$$

$$= \left(28 \times 10^3\right)\text{N/m}^2 \times \left\{1 + 0.2 \times \left(2.4\right)^2\right\}^{3.5}$$

$$= 409 \times 10^3 \, \text{N m}^{-2}$$

and

$$p_c = p_0\left\{1 + \left(\frac{\gamma - 1}{2}\right)M^2\right\}^{-\gamma/(\gamma-1)}$$

$$= \left(409 \times 10^3\right)\text{N/m}^2 \times \left\{1 + 0.2 \times \left(1\right)^2\right\}^{-3.5}$$

$$= 216 \times 10^3 \, \text{N m}^{-2}$$

(b) From equation 11.44 at exit:

$$T_0 = T\left\{1 + \left(\frac{\gamma - 1}{2}\right)M^2\right\}$$

Hence

$$T = T_0\left\{1 + \left(\frac{\gamma - 1}{2}\right)M^2\right\}^{-1} = 291\text{K} \times \left\{1 + 0.2 \times \left(2.4\right)^2\right\}^{-1}$$

$$= 135.2\,\text{K}$$

Also, $M = u/a$ and $a = \left(\gamma RT\right)^{1/2}$.

Hence

$$u = M\left(\gamma RT\right)^{1/2} = 2.4 \times \left(1.4 \times 287 \,\text{J}/\left(\text{kg}\,\text{K}\right) \times 135.2\,\text{K}\right)^{1/2}$$

$$= 559.4\,\text{m/s}$$

Figure 11.24 shows how the pressure varies with distance along a given nozzle, for various values of the pressure p_2 beyond the nozzle. We may consider the flow into the nozzle to come from a large reservoir in which the velocity is always negligible so that the pressure there is the stagnation pressure p_0. Adiabatic, frictionless flow is assumed through-out. Combination of the equations.

$$\left(\frac{\gamma}{\gamma-1}\right)\frac{p}{\varrho}+\frac{u^2}{2}=\left(\frac{\gamma}{\gamma-1}\right)\frac{p_0}{\varrho_0} \qquad \text{B, } \gamma \text{ const, I, SF.}$$

$$p/\varrho^\gamma = \text{constant} \qquad \text{B, } \gamma \text{ const, I}$$

and mass flow rate $m = \varrho A u$ gives

$$\left(\frac{p}{p_0}\right)^{2/\gamma}-\left(\frac{p}{p_0}\right)^{(\gamma+1)/\gamma}=\left(\frac{\gamma-1}{2\gamma}\right)\frac{m^2}{A^2 p_0 \varrho_0} \qquad \text{B, } \gamma \text{ const, I, SF} \quad (11.58)$$

For a particular mass flow rate and stagnation conditions eqn 11.58 thus provides a single relation between p and A.

We now suppose that p_0 is fixed but that the external pressure p_2 may be varied at will. If p_2 equals p_0 there is no flow and the pressure is p_0 throughout the nozzle, as represented by the line OB in Fig. 11.24. Reduction of p_2, however, causes a reduction in the pressure at the end of the nozzle; as the velocity has nowhere yet reached the sonic velocity this reduction can be 'telegraphed' upstream and a pressure graph such as ODE is obtained. This is seen to be similar in shape to that for the flow of a liquid in a venturi-meter.

Further decrease of the external pressure increases the velocity at the throat of the nozzle and reduces the pressure there until p_2 corresponds to the point F. The velocity at the throat is then the critical value and the pressure distribution is represented by OCF. If p_2 is further reduced the conditions in the convergent part of the nozzle remain unchanged and the upstream pressure distribution follows the single curve OC. The nozzle is choked and eqn 11.57 substituted in eqn 11.58 then gives

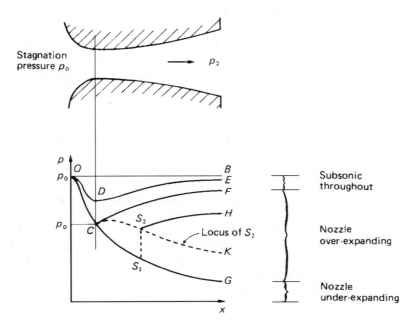

Fig. 11.24

$$\left(\frac{A_t}{A}\right)^2\left(\frac{2}{\gamma+1}\right)^{(\gamma+1)/(\gamma-1)} = \frac{2}{\gamma-1}\left\{\left(\frac{p}{p_0}\right)^{2/\gamma} - \left(\frac{p}{p_0}\right)^{(\gamma+1)/\gamma}\right\}$$

B, γ const, I, SF　(11.59)

If the external pressure corresponds exactly to point F there is an adiabatic compression according to the curve CF and the velocity downstream of the throat is entirely subsonic. Only this value of the external pressure, however, allows a compression from the point C. On the other hand, a continuous isentropic expansion at supersonic velocity is possible only according to the curve CG (which represents the relation between p and A of eqn 11.59). In a nozzle designed to produce supersonic flow the smooth expansion OCG is the ideal and the ratio p_0/p_G is known as the *design pressure ratio* of the nozzle.

It remains to consider what happens when p_2 corresponds to neither F nor G. For external pressures between F and G flow cannot take place without the formation of a shock wave and consequent dissipation of energy. In such circumstances the nozzle is said to be *over-expanding*. If p_2 is reduced slightly below F a normal shock wave is formed downstream of the throat. Since eqn 11.59 holds only for isentropic flow it is valid only as far as the shock, and so the curve CG is followed only as far as S_1, say: there is an abrupt rise of pressure through the shock (S_1S_2) and then subsonic deceleration of the flow with rise of pressure to p_H. The location of the shock is exactly determined by the need for the subsequent compression (S_2H) to lead to the exit pressure p_H. Downstream of the shock, the integrated Euler equation, the isentropic relation $p/\varrho^\gamma = $ constant, and the continuity equation together provide a relation between pressure and cross-sectional area in terms of the exit conditions (H). Upstream of the shock eqn 11.59 holds, and the shock forms where the value of A is such as to give the appropriate pressure ratio across the shock.

As the external pressure is lowered the shock moves further from the throat until, when $p_2 = p_K$, the point S_1 has moved to G and S_2 to K. For exit pressures less than p_K the flow within the entire diverging part of the nozzle is supersonic and follows the line CG; if p_2 lies between K and G a compression must occur outside the nozzle to raise the pressure from p_G to the external pressure. This compression involves oblique shock waves and the subsequent events cannot be described in 'one-dimensional' terms.

At the design conditions $p_2 = p_G$ and the pressure in the exit plane of the nozzle is the same as the external pressure. When the external pressure is below p_G the nozzle is said to be *under-expanding* and the expansion to pressure p_2 must be completed outside the nozzle. The curve CG is followed and the exit velocity is that corresponding to p_G; the additional expansion then takes place through oblique expansion (Mach) waves. Here, too, a 'one-dimensional' description of events is not possible. An extreme case of an under-expanding nozzle is one having no diverging portion at all.

(a) External pressure greater than pressure for isentropic flow throughout nozzle (nozzle over-expanding). Normal shock wave just inside nozzle.

(b) $p_G/p_2 = 0.66$. Nozzle still over-expanding but isentropic flow throughout nozzle; compression outside it through oblique shocks, formed at edges of nozzle exit. Simple intersection of oblique waves not possible, so Mach intersection results.

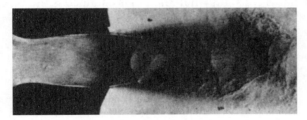

(c) $p_G/p_2 = 0.85$. Nozzle still over-expanding; external pressure lower than in (b). Oblique shocks now weaker; they cross jet and are reflected from opposite surface of jet as fans of expansion waves. This is because boundary condition at edge of jet is not unchanged direction (as at a solid boundary) but unchanged pressure. Pressure rise through incident shock wave can be cancelled only by an expansion. Hence diamond pattern in which alternate compression and expansion continue until damped out by viscous action.

(d) $p_G/p_2 = 1.37$. External pressure less than pressure inside nozzle (nozzle under-expanding). Fans of expansion waves formed at edges of nozzle exit, through which pressure is lowered to ambient value. Jet diverges and expansion waves are reflected from opposite surface as compression waves. Hence diamond pattern in which alternate expansion and compression continue until damped out by viscous action.

Fig. 11.25 (Crown Copyright. Reproduced with permission).

The foregoing is illustrated by the photographs of Fig. 11.25 which were obtained by the use of the schlieren technique (see Section 11.12). The theory agrees reasonably well with measurements of pressure within the nozzle if the external pressure is only a little lower than p_F. However, separation of the boundary layer from the walls is greatly encouraged by the abrupt pressure increase across a shock wave and, with lower values of the external pressure, the normal shock wave may not fill the entire cross-section. Also, oblique shock waves may be formed close to the wall.

■
> **Example 11.5** A convergent-divergent nozzle designed to give an exit Mach number of 1.8 when used with helium ($\gamma = 5/3$) is used with air ($\gamma = 1.4$) under conditions that produce a normal shock just inside the nozzle. Determine the Mach number just before the shock and thus the stagnation pressure at inlet if the absolute pressure beyond the exit is $30\,\mathrm{kPa}$.

Solution
What remains constant when the gas is changed is the geometry of the nozzle. From eqn 11.57

$$\text{Mass flow rate} = \varrho A u = M A (p \gamma \varrho)^{1/2}$$

$$= A_t \left\{ p_0 \varrho_0 \gamma \left(\frac{2}{\gamma + 1} \right)^{(\gamma+1)/(\gamma-1)} \right\}^{1/2}$$

$$\therefore \left(\frac{A}{A_t} \right)^2 = \frac{p_0 \varrho_0 \gamma}{M^2 p \gamma \varrho} \left(\frac{2}{\gamma + 1} \right)^{(\gamma+1)/(\gamma-1)}$$

$$= \frac{1}{M^2} \left(\frac{p_0}{p} \right)^{(\gamma+1)/\gamma} \left(\frac{2}{\gamma + 1} \right)^{(\gamma+1)/(\gamma-1)}$$

$$= \frac{\left(1 + \dfrac{\gamma - 1}{2} M^2 \right)^{(\gamma+1)/(\gamma-1)}}{M^2} \left(\frac{2}{\gamma + 1} \right)^{(\gamma+1)/(\gamma-1)}$$

$$= \frac{\left(1 + \dfrac{1}{3} \times 1.8^2 \right)^4}{1.8^2} \left(\frac{3}{4} \right)^4 = 1.828 \text{ for helium}$$

$$= \frac{\left(1 + 0.2 M^2 \right)^6}{M^2} \frac{1}{1.2^6} \text{ for air, whence (by trial)}$$

$$M = 1.715$$

Across the shock

$$\frac{p_2}{p_1} = \frac{2\gamma M_1^2 - (\gamma - 1)}{\gamma + 1} \quad \text{(from eqns 11.24 and 11.29)}$$

$$= \frac{2.8 \times 1.715^2 - 0.4}{2.4} = 3.265$$

$$\therefore p_1 = 30\,\text{kPa}/3.265 = 9.19\,\text{kPa}$$

From eqn 11.32,

$$\frac{p_{0,1}}{p_1} = \left(1 + \frac{\gamma - 1}{2} M_1^2\right)^{\gamma/(\gamma-1)}$$

$$= \left(1 + 0.2 \times 1.715^2\right)^{3.5} = 5.05$$

$$\therefore p_{0,1} = 46.4\,\text{kPa}$$

11.9   HIGH-SPEED FLOW PAST AN AEROFOIL

Compressibility effects of great importance arise when an aircraft flies at or near the speed of sound. A full treatment of the subject would require a book to itself, and in this section we shall merely consider in general terms what happens when the velocity of flow past an aerofoil is gradually increased from subsonic to supersonic. We shall consider only an aerofoil such as a thin wing of large span and a section where the flow is uninfluenced by fuselage, wing-tips or engine nacelles.

When M_∞, the Mach number of the oncoming flow, is small the flow pattern closely resembles that of an incompressible fluid (as discussed in Section 9.9.2). As M_∞ is increased, however, compressibility affects to some extent the pressure p at points on the surface of the aerofoil, even when the flow everywhere is subsonic. The difference between p and the upstream pressure p_∞ is usefully expressed as a dimensionless pressure coefficient $C_p = (p - p_\infty)/\frac{1}{2}\varrho_\infty u_\infty^2$ where ϱ_∞ and u_∞ represent respectively the density and velocity upstream of the aerofoil. In particular C_p at the stagnation point increases according to eqn 11.46. More generally, as was shown by Prandtl and the English mathematician Hermann Glauert (1892–1934), C_p is approximately proportional to $(1 - M_\infty^2)^{-1/2}$ (except near the leading and trailing edges). As a result, the lift coefficient

$$C_L = L\bigg/\frac{1}{2}\varrho_\infty u_\infty^2 S$$

(Section 9.9.1) for an unchanged angle of attack is affected similarly.

The foregoing holds so long as the flow everywhere is subsonic. However, at the point in the flow pattern where the velocity is greatest the temperature (by eqn 11.13) is least. Therefore at this point (usually above the aerofoil, close to the section of maximum thickness) $a = \sqrt{(\gamma RT)}$ is least and $M = u/a$ is greatest. Here, then, the flow may become supersonic well before the free-stream Mach number M_∞ reaches unity. This phenomenon characterizes the important 'transonic' range in which both subsonic and supersonic flow occur in the flow pattern. The value of M_∞ at which the transonic range begins is generally known as the lower critical Mach number M_c. Supersonic flow first appears, therefore, in a small region over the upper aerofoil surface, where the velocity is a maximum (and the pressure a minimum). A return to subsonic flow at a higher pressure must take place towards the trailing edge and, as in a de Laval nozzle, this can happen only through a shock wave as shown in Fig. 11.26a. This shock wave extends only a limited distance from the aerofoil surface because the main flow is still subsonic. Also, of course, the wave stops short of the surface because the velocity decreases to zero in the boundary layer. Similar changes occur in the flow past the underside of the aerofoil although, in order to produce lift, the reduction of pressure and increase of velocity on this side are in general smaller, and the formation of shock waves is thus somewhat delayed (Fig. 11.26b).

Now it will be recalled from Sections 8.8.1 and 8.8.6 that an adverse pressure gradient promotes separation of the boundary layer, and that

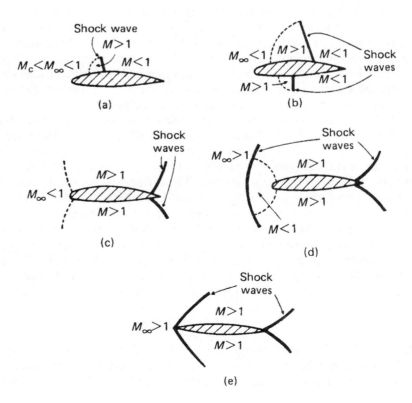

Fig. 11.26

such separation from the surface of an aerofoil markedly affects the lift and drag. The abrupt pressure rise through the shock has a similar effect even though the angle of attack of the aerofoil is much below that ordinarily causing stall. The boundary-layer separation induced by the shock, and the consequent loss of lift and increase of drag are together termed *shock stall*. The Prandtl–Glauert rule, which implies a rise of lift coefficient to infinity as $M_\infty \to 1$, now of course breaks down.

In practice, matters are considerably complicated by interaction between the shock wave and the boundary layer. For one thing, the subsonic flow in the boundary layer permits the pressure rise to be transmitted upstream, so the region of the shock is spread out near the surface. Also the separation of the boundary layer affects the formation of the shock wave. For our broad outline, however, we may leave aside these complications and note that as M_∞ is further increased (although still less than unity) the supersonic region spreads both fore and aft, and the shock wave moves towards the trailing edge (Fig. 11.26c). The divergence of the streamlines beyond the thickest part of the aerofoil causes an *increase* of velocity in supersonic flow; there is thus a further reduction of pressure, and the shock becomes stronger so as to provide the return to the required downstream pressure.

While M_∞ is less than unity there is only a single shock wave on each side of the aerofoil, but as soon as M_∞ exceeds unity a new shock wave forms upstream of the leading edge. The flow is then supersonic except for a small region just behind the front wave (Fig. 11.26d). Two oblique shock waves spring from points close to the trailing edge, their obliquity increasing as M_∞ increases. An increase of M_∞ also reduces the size of the subsonic region, and if the nose of the aerofoil is sufficiently pointed (see Section 11.5.2) the frontal wave may attach itself to the aerofoil at high Mach numbers and an entirely supersonic régime ensue (Fig. 11.26e).

Between the front and rear oblique shock waves, the supersonic flow over the convex surfaces produces rarefaction waves through which the pressure is successively reduced. These diverge from each other (see Section 11.6); they thus meet the shock waves and so, with increasing distance from the aerofoil, the latter are gradually reduced in strength and made more oblique until they are entirely dissipated.

The effects of these phenomena on the lift and drag coefficients of a thin aerofoil are illustrated in Fig. 11.27. (Reference to the effect of compressibility on drag is also made in Section 8.10.) For completely supersonic flow the variation of C_L is given by an approximate theory by J. Ackeret as proportional to $(M_\infty^2 - 1)^{-1/2}$ for moderate values of M_∞. In the transonic range, however, the force on the aerofoil depends on the size and position of the various regions of subsonic and supersonic flow, and therefore markedly on the shape of the section and the angle of attack. Theoretical solutions are not possible except for certain special cases, and even these are very complicated.

The formation of shock waves above and below the aerofoil alters the position of its centre of pressure, and this, together with the

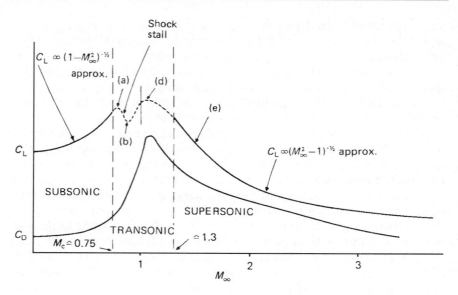

Fig. 11.27 Effect of Mach number on C_L and C_D for thin aerofoil at constant angle of attack. Points (a) (b) (d) (e) correspond approximately to similarly lettered parts of Fig. 11.26.

phenomenon of shock stall, presents considerable problems in the control of aircraft flying in the transonic range.

11.10 FLOW WITH VARIABLE DENSITY IN PIPES OF CONSTANT CROSS-SECTION

We have so far considered flow of a gas where friction may, at least to a first approximation, be neglected. Friction must be accounted for, however, when flow takes place in pipe-lines and similar conduits. Here we shall discuss steady flow in a pipe of constant cross-section and (except in Section 11.10.3) we shall assume the velocity to be sufficiently uniform over the section for the flow to be adequately described in 'one-dimensional' terms. Since the change of pressure resulting from friction gives rise to a change of density, and thus of velocity, matters are more complicated than for an incompressible fluid. The properties of the fluid are also affected by heat transferred through the walls of the pipe. If the pipe is well insulated the heat transfer may be negligible and the changes therefore adiabatic (but not, of course, isentropic). In short pipes where no specific provision is made for heat transfer the conditions may approximate to adiabatic. On the other hand, for flow at low velocities in long, uninsulated pipes an appreciable amount of heat may be transferred through the pipe walls, and if the temperatures inside and outside the pipe are similar the flow may be approximately isothermal (i.e. at constant temperature). This is so, for example, in long compressed-air pipe-lines and in low-velocity flows generally.

To account for the changes of fluid properties we must consider initially an infinitesimal length of the pipe. For adiabatic flow the relation involving the friction force may be integrated over a finite length of

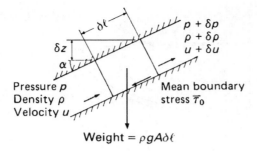

Weight $= \rho g A \delta \ell$

Fig. 11.28

the pipe by introducing the continuity equation and the energy equation. For flow with heat transfer the energy equation is not applicable, but for isothermal conditions integration is made possible by inserting the condition $T =$ constant.

Applying the steady-flow momentum equation to an element with cross-sectional area A, perimeter P and length δl (Fig. 11.28), we have

$$pA - \left(p + \delta p\right)A - \bar{\tau}_0 P\delta l - \varrho g A \delta l \sin\alpha = \varrho A u \delta u$$

Substituting $\bar{\tau}_0 = \frac{1}{2}f\varrho u^2$ (eqn 7.4), $\delta l \sin\alpha = \delta z$ and dividing by $\varrho g A$ we obtain

$$-\delta z = \frac{\delta p}{\varrho g} + \frac{u\delta u}{g} + \frac{f\delta l}{A/P}\frac{u^2}{2g} \qquad (11.60)$$

The final term corresponds to Darcy's formula for head lost to friction. Equation 11.60 applies to any steady one-dimensional flow where A is constant, but, for integration, the relation between density and pressure must be known and this depends on the degree of heat transfer. We consider first the case of zero heat transfer.

11.10.1 Adiabatic flow in a pipe

Before embarking on the integration of eqn 11.60 for adiabatic conditions it will be instructive to look at the problem in general thermodynamic terms. When no external work is done and changes of elevation may be neglected, the steady-flow energy equation (11.12) and the continuity equation $m = \varrho A u$ together give

$$\text{Constant } h_0 = h + \frac{1}{2}u^2 = h + \frac{1}{2}\left(\frac{m}{\varrho A}\right)^2 \qquad (11.61)$$

For given values of m, A and the stagnation enthalpy h_0, curves of h against ϱ could be plotted from eqn 11.61. A more significant relation, however, is that between h and specific entropy s. For any pure substance, s, like h and ϱ, is a function of state and so may be determined from values of h and ϱ. In particular, for a perfect gas $s - s_1 = c_v \ln\{(h/h_1)(\varrho_1/\varrho)^{\gamma-1}\}$. Starting from a specified state (point 1 on Fig. 11.29) the curve of h against s traces the states through which the

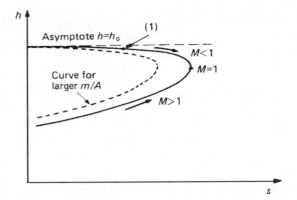

Fig. 11.29

substance must pass in an adiabatic process. Such a curve is termed a Fanno curve in honour of Gino Fanno, the Italian engineer who first studied its properties. All Fanno curves show a maximum value of s. At this maximum

$$\mathrm{d}s = 0 = \frac{1}{T}\mathrm{d}q_{\mathrm{rev}} = \frac{1}{T}\left\{\mathrm{d}e + p\mathrm{d}\left(\frac{1}{\varrho}\right)\right\} \tag{11.62}$$

Also, from the definition of specific enthalpy,

$$\mathrm{d}h = \mathrm{d}\left(e + p/\varrho\right) = \mathrm{d}e + p\mathrm{d}\left(\frac{1}{\varrho}\right) + \frac{1}{\varrho}\mathrm{d}p \tag{11.63}$$

Equations 11.62 and 11.63 together give

$$\mathrm{d}h = \frac{1}{\varrho}\mathrm{d}p \tag{11.64}$$

Differentiation of eqn 11.61 gives $0 = \mathrm{d}h - (m^2/A^2)(\mathrm{d}\varrho/\varrho^3)$ and substitution for $\mathrm{d}h$ from eqn 11.64 then shows that when s is a maximum

$$u^2 = \frac{m^2}{A^2\varrho^2} = \left(\frac{\partial p}{\partial \varrho}\right)_s = a^2 \qquad \text{(from eqn 11.18)}$$

That is, the specific entropy is a maximum when the Mach number is unity. The upper branch of the curve in Fig. 11.29, which approaches the stagnation enthalpy h_0, thus corresponds to subsonic flow, and the lower branch to supersonic flow. Since for adiabatic conditions the entropy cannot decrease, friction acts to increase the Mach number in subsonic flow and to reduce the Mach number in supersonic flow. Changes in other properties may be deduced as shown in Table 11.2. Moreover, as friction involves a continual increase of entropy, sonic velocity can be reached only at the exit end of the pipe, if at all.

If sonic velocity is to be reached in a particular pipe then, for given inlet conditions and exit pressure, a certain length is necessary. If the actual length is less than this 'limiting' value sonic conditions are not reached. If the length of the pipe is increased beyond the limiting value an initially subsonic flow will be choked; that is, the rate of flow will be

Table 11.2 Property changes during adiabatic flow in pipe of constant cross section

Property		Subsonic flow	Supersonic flow
Mach number	M	Increases	Decreases
Specific enthalpy	h	Decreases	Increases
Velocity	u	Increases (!)	Decreases
Density	ϱ	Decreases	Increases
Temperature	T	Decreases	Increases
Pressure	p	Decreases	Increases (!)
$Re = \dfrac{\varrho u d}{\mu} = \dfrac{\text{const.}}{\mu}$		Increases	Decreases
(μ increases with T)			
Stagnation temperature	T_0	Constant	Constant

reduced so as again to give sonic conditions at the outlet. An initially supersonic flow will also be adjusted to give sonic conditions at exit: a normal shock will form near the end of the pipe and the resulting subsonic flow will accelerate to sonic conditions at the exit. Further increase of length would cause the shock to move towards the inlet of the pipe and then into the nozzle producing the supersonic flow so that the flow would become entirely subsonic in the pipe.

These conclusions are valid for all gases. To obtain explicit relations from the integration of eqn 11.60, however, we assume the gas to be perfect. For simplicity we also neglect the gravity term δz. Then in the limit as $\delta l \to 0$

$$\frac{\mathrm{d}p}{\varrho} + u\mathrm{d}u + \frac{f\mathrm{d}l}{A/P}\frac{u^2}{2} = 0 \tag{11.65}$$

This equation can be integrated only when the number of variables is reduced, and a solution is most conveniently obtained in terms of Mach number. As the flow is not isentropic the temptation to use the relation $p/\varrho^\gamma = $ constant must, however, be resisted.

Differentiating the energy equation (11.13) gives $c_p\mathrm{d}T + u\,\mathrm{d}u = 0$ and division by $u^2 = M^2\gamma RT$ then yields

$$\frac{c_p}{\gamma RM^2}\frac{\mathrm{d}T}{T} + \frac{\mathrm{d}u}{u} = 0 \tag{11.66}$$

Also, since $u^2 = M^2\gamma RT$,

$$2\frac{\mathrm{d}u}{u} = 2\frac{\mathrm{d}M}{M} + \frac{\mathrm{d}T}{T} \tag{11.67}$$

Eliminating $\mathrm{d}T/T$ from eqns 11.66 and 11.67 and noting that $c_p/R = \gamma/(\gamma - 1)$ we then obtain

$$\frac{\mathrm{d}u}{u} = \frac{\mathrm{d}M/M}{\frac{1}{2}(\gamma - 1)M^2 + 1} \tag{11.68}$$

Successively using $p = \varrho RT$, $\varrho u =$ constant (by continuity) and eqns 11.67 and 11.68, we have

$$\frac{dp}{p} = \frac{d\varrho}{\varrho} + \frac{dT}{T} = -\frac{du}{u} + \frac{dT}{T} = \frac{du}{u} - 2\frac{dM}{M}$$

$$= -\left\{\frac{(\gamma - 1)M^2 + 1}{\frac{1}{2}(\gamma - 1)M^2 + 1}\right\}\frac{dM}{M} \qquad (11.69)$$

If we divide the momentum equation 11.65 by $u^2 = M^2\gamma p/\varrho$ and substitute for du/u and dp/p from eqns 11.68 and 11.69, we obtain

$$\left\{\frac{M^2 - 1}{\frac{1}{2}(\gamma - 1)M^2 + 1}\right\}\frac{dM}{\gamma M^3} + \frac{fdl}{2A/P} = 0 \qquad (11.70)$$

We now have only two main variables, M and l, and these are separated. But the friction factor f is a function of Reynolds number and this is not constant (see Table 11.2). Changes of temperature are appreciable only at high velocities, and therefore in general only at high Reynolds numbers. Fortunately, the higher the Reynolds number the less f depends on it, especially for rough pipes. For fully developed flow (that is, at a distance greater than say 50 times the diameter from the pipe entrance) the value of f is apparently uninfluenced by Mach number. The boundary layer, however, is greatly affected by the oblique shock waves that in supersonic flow form near the entrance, and so the apparent value of f is then notably reduced below the value for subsonic flow at the same Reynolds number. However, if we consider a mean value

$$\bar{f} = \frac{1}{l}\int_0^l fdl$$

we may integrate eqn 11.70 between points 1 and 2 a distance l apart:

$$\frac{1}{2\gamma}\left(\frac{1}{M_2^2} - \frac{1}{M_1^2}\right) + \frac{\gamma + 1}{4\gamma}\ln\left\{\left(\frac{M_2}{M_1}\right)^2\frac{(\gamma - 1)M_1^2 + 2}{(\gamma - 1)M_2^2 + 2}\right\} + \frac{\bar{f}l}{2A/P} = 0$$

$$(11.71)$$

We may note in passing that if dM/M is eliminated between eqns 11.69 and 11.70 and $\varrho u^2/\gamma M^2$ substituted for p we obtain

$$\frac{dp}{dl} = -\frac{f}{2A/P}\varrho u^2\left\{\frac{1 + (\gamma - 1)M^2}{1 - M^2}\right\}$$

This expression reduces to Darcy's formula (eqn 7.2) as $M \to 0$, so, if the Mach number remains low, the fluid may, with small error, be treated as incompressible. For air at $M = 0.1$, for example, the error in dp/dl is 1.41%. When M is not small, however, solutions to problems require the use of eqn 11.71 or its equivalent.

Changes of Mach number along the pipe are readily related to changes of other parameters. Substituting $u = M\sqrt{(\gamma RT)}$ in the energy equation we obtain

$$c_p T_1 + \frac{1}{2} M_1^2 \gamma RT_1 = c_p T_2 + \frac{1}{2} M_2^2 \gamma RT_2$$

whence

$$\frac{T_1}{T_2} = \frac{1 + \frac{1}{2}(\gamma - 1)M_2^2}{1 + \frac{1}{2}(\gamma - 1)M_1^2} \tag{11.72}$$

Moreover

$$\frac{u_1}{u_2} = \frac{M_1}{M_2}\sqrt{\frac{T_1}{T_2}} = \frac{M_1}{M_2}\left\{ \frac{1 + \frac{1}{2}(\gamma - 1)M_2^2}{1 + \frac{1}{2}(\gamma - 1)M_1^2} \right\}^{1/2} \tag{11.73}$$

and, from the gas law and continuity,

$$\frac{p_1}{p_2} = \frac{\varrho_1 T_1}{\varrho_2 T_2} = \frac{u_2}{u_1}\frac{T_1}{T_2} = \frac{M_2}{M_1}\left\{ \frac{1 + \frac{1}{2}(\gamma - 1)M_2^2}{1 + \frac{1}{2}(\gamma - 1)M_1^2} \right\}^{1/2} \tag{11.74}$$

The limiting conditions at which an initially subsonic flow is choked are given by putting $M_2 = 1$. Then

$$\left.\begin{aligned}
\frac{T_1}{T_c} &= \frac{\frac{1}{2}(\gamma + 1)}{1 + \frac{1}{2}(\gamma - 1)M_1^2} \\[2em]
\frac{u_1}{u_c} &= M_1\left\{ \frac{\frac{1}{2}(\gamma + 1)}{1 + \frac{1}{2}(\gamma - 1)M_1^2} \right\}^{1/2} \\[2em]
\frac{p_1}{p_c} &= \frac{1}{M_1}\left\{ \frac{\frac{1}{2}(\gamma + 1)}{1 + \frac{1}{2}(\gamma - 1)M_1^2} \right\}^{1/2}
\end{aligned}\right\} \tag{11.75}$$

Since T_c, u_c, p_c are constants for a given flow they may be regarded as convenient reference values for temperature, velocity and pressure. Values of the ratios given by the relations 11.75 for various values of M_1 have been tabulated (e.g. Appendix 3, Table A3.3) and because $T_2/T_1 = (T_2/T_c) \div (T_1/T_c)$, for instance, they may be found useful also for non-choked flows. (Notice, however, that Table A3.3 gives values of the pressure ratio in the reciprocal form p_c/p because that yields much greater accuracy when linear interpolation is used.)

From eqn 11.71 the limiting length (that is the length necessary for the Mach number to change from M_1 to unity) is given by

$$\frac{\bar{f}\gamma l_{\max}}{A/P} = \frac{1}{M_1^2} - 1 - \frac{\gamma+1}{2}\ln\left\{\frac{\gamma-1}{\gamma+1} + \frac{2}{M_1^2(\gamma+1)}\right\} \quad (11.76)$$

Values of $\bar{f}l_{\max}P/A$ are tabulated (e.g. Table A3.3). For choked flow M_1 can therefore be calculated (a value of $\bar{f}$ being assumed) and, from other known data for position 1, u_1 and the mass flow rate determined. If necessary, a revised estimate of f can be obtained from the value of Reynolds number at position 1 and a new calculation made. For subsonic flow the variation of f along the pipe is not usually large, and the use of the Reynolds number for position 1 is sufficiently accurate. The tabulated values of $\bar{f}l_{\max}P/A$ may also be used for non-choked subsonic flow since a hypothetical extension of the pipe can be assumed which would increase the Mach number from M_2 to unity. The actual length thus corresponds to the difference between values obtained for M_1 and for M_2; i.e.

$$l = \left(l_{\max}\right)_{M_1} - \left(l_{\max}\right)_{M_2}.$$

For a non-choked subsonic flow the pressure of the fluid at exit must be that of the surroundings, that is, a known quantity p_2, say.

Now, since $\varrho = m/Au$ (where m represents the mass flow rate), we have for any point in the pipe

$$p = \frac{m}{Au}RT = \frac{m}{AM\sqrt{(\gamma RT)}}RT = \frac{m}{AM}\left(\frac{RT}{\gamma}\right)^{1/2}$$

$$= \frac{m}{AM}\left[\frac{RT_0}{\gamma\left\{1 + \frac{1}{2}(\gamma-1)M^2\right\}}\right]^{1/2} \quad (11.77)$$

(use being made of the energy equation 11.44).

For a given pipe, mass flow rate and stagnation conditions, calculation of either the upstream or the downstream pressure is quite straightforward. Suppose that p_2 is required, p_1 already being known. Equation 11.77 yields

$$M^2 = \frac{1}{\gamma - 1}\left[-1 + \sqrt{\left\{1 + 2\left(\frac{\gamma - 1}{\gamma}\right)\frac{m^2 R T_0}{p^2 A^2}\right\}}\right] \qquad (11.78)$$

so M_1^2 may be calculated. From eqn 11.71 M_2 may then be determined by iteration (a reasonable first approximation is often obtained by neglecting the log term); alternatively, Table A3.3 may be used to obtain the value of $fl_{max}P/A$ corresponding to M_1; deducting flP/A, where l represents the actual pipe length, gives the value of $fl_{max}P/A$ corresponding to M_2; and M_2 may be read from the table. Equation 11.77 then gives p_2. Or values from the table may be used to give $p_2 = p_1 \times (p_c/p_1) \div (p_c/p_2)$.

Determining the mass flow rate m is rather more troublesome. The known values of p_1 and p_2 give the ratio $p_1/p_2 = (p_c/p_2) \div (p_c/p_1)$. Data for the pipe provide $flP/A = f(P/A)\{(l_{max})_1 - (l_{max})_2\}$. From inspection of Table A3.3 we find the pair of values M_1 and M_2 that give the required pressure ratio and difference of length parameters. Then inserting either inlet or outlet conditions in eqn 11.77 yields m. For example, with the data $l = 15$ m, pipe diameter 50 mm, $p_1 = 100$ kPa, $p_2 = 50$ kPa, $T_0 = 300$ K, $f = 0.008$, $\gamma = 1.4$, and $R = 287$ J/kg K we obtain $p_1/p_2 = 2$ and

$$\frac{f}{A/P}\left\{\left(l_{max}\right)_1 - \left(l_{max}\right)_2\right\} = \frac{0.008}{\left(0.05\,\mathrm{m}/4\right)}\{15\,\mathrm{m}\} = 9.6.$$

Since $p_1 > p_2$ the flow is subsonic (see Table 11.2).

Table A3.3 then shows that $M_1 \simeq 0.22$ and $M_2 \simeq 0.42$. In view of the uncertainty in the value of f, it is hardly worth refining the values of M_1 and M_2 by interpolation. Using $p_1 = 100$ kPa and $M_1 = 0.22$ in eqn 11.77 we obtain $m = 0.175$ kg/s.

If the pipe size has to be determined, obtaining a precise result will be very tedious. Fortunately, however, only an approximate result is normally needed and adequate accuracy may be obtained by assuming isothermal conditions (see Section 11.10.2).

If the flow is from a large reservoir or from the atmosphere, the initial stagnation pressure $(p_0)_1$ may be known but not the initial pressure p_1 in the pipe. It is then necessary to assume that there is only a small drop in pressure as the fluid enters the pipe; in other words, that to a first approximation $(p_0)_1$ and p_1 are identical. When a value of M_1 has been determined (from eqn 11.78) the corresponding value of $(p/p_0)_1$ may be obtained from Table A3.2 on the assumption that the drop in pressure at the pipe inlet arises solely because the fluid acquires velocity, i.e. that there is no 'entry loss' giving rise to additional turbulence and thus increase of entropy. Multiplying this value of p/p_0 by p_0 provides a new value of p_1 and the calculations are then repeated for a second approximation.

Supersonic flow can be maintained only for short distances in pipes of constant cross-section. Even if the initial Mach number were infinity, eqn 11.76 shows that the limiting length would be

$$\frac{A/P}{\bar{f}\gamma}\left\{\frac{\gamma+1}{2}\ln\left(\frac{\gamma+1}{\gamma-1}\right)-1\right\}$$

For $\gamma = 1.4$ and a circular pipe ($A/P = d/4$) with $\bar{f}$ as low as 0.0025 this length is only 82 times the diameter d. (As we have already remarked, however, for supersonic flow the value of f is uncertain for lengths less than about 50 times the diameter.) At these high velocities, the rate at which friction dissipates mechanical energy is large, and supersonic flow in a pipe is generally better avoided. If supersonic flow is subsequently required the gas may be expanded in a convergent-divergent nozzle.

■

Example 11.6   A length of pipe of diameter 20 mm is connected to a reservoir containing air, as shown in the diagram.

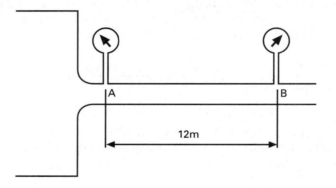

Two pressure gauges are connected to wall tappings at A and B, which are 12 m apart. Tapping A is at the entrance plane to the pipe.

When the gauge pressure in the reservoir is 410 kPa, the gauge pressures recorded at A and B are 400 kPa and 180 kPa, respectively. Calculate:

(a) the value of the friction factor for the pipe
(b) the overall length of the pipe, L, if the flow exhausts to atmosphere
(c) the mass flow rate if the reservoir temperature is 294 K.

Assume atmospheric pressure is 100 kPa and treat the rounded entry as isentropic.

Solution
Convert the gauge pressures to absolute pressures. Thus

$$p_0 = 510\,\text{kPa} \quad p_A = 500\,\text{kPa} \quad p_B = 280\,\text{kPa}$$

(a) At A, $p_A/p_0 = 500/510 = 0.980$. From the Isentropic Flow Tables (Appendix 3), $M_A = 0.17$. From the Fanno Flow Tables (Appendix 3) for $M_A = 0.17$ and $\gamma = 1.4$, $p_c/p_A = 0.1556$ and $(fl_{max}P/A)_A = 21.37$.

Hence

$$p_c = p_A \times (p_c/p_A) = 500\,\text{kPa} \times 0.1556 = 778\,\text{kPa}$$

and

$$p_c/p_B = 77.8/280 = 0.278$$

From the Fanno Tables at $p_c/p_B = 0.278$, $M_B = 0.302$ and $(fl_{max}P/A)_B = 5.21$. For a circular pipe $P/A = 4/d$, so

$$\left(fl_{max}P/A\right)_A - \left(fl_{max}P/A\right)_B = \frac{f \times 12\,\text{m} \times 4}{0.02\,\text{m}} = 21.37 - 5.21$$

yielding $f = 0.00673$

(b) At exit, $p_c/p = 77.8/100 = 0.778$. From the Fanno Tables, $(fl_{max}P/A) = 0.07$. Hence

$$\frac{Lm}{12\,\text{m}} = \frac{21.37 - 0.07}{21.37 - 5.21}$$

yielding $L = 15.82\,\text{m}$

(c) Rewriting eqn 11.77, and evaluating it at B:

$$m = ApM\left[\frac{\gamma\{1 + (\gamma - 1)M^2/2\}}{RT_0}\right]^{1/2}$$

$$= \frac{\pi}{4}(0.02\,\text{m})^2 \times (2.8 \times 10^5)\text{Pa} \times 0.302$$

$$\times \left[\frac{1.4 \times \{1 + 0.4 \times (0.302)^2/2\}}{287\,\text{J}/(\text{kg}\,\text{K}) \times 294\,\text{K}}\right]^{1/2}$$

$$= 0.109\,\text{kg}\,\text{s}^{-1} \qquad \qquad \square$$

11.10.2 Isothermal flow in a pipe

In Section 11.10.1 we assumed that there was no heat transfer across the walls of the pipe. In general, of course, any quantity of heat may be transferred to or from the fluid. This general case is too complicated to be considered in this book, but a particular example of practical interest is that in which the heat transfer is such as to keep the temperature of the fluid constant, that is, in which the flow is *isothermal*. For gases such flow is usually achieved at low velocities in long pipes not thermally insulated because there is then opportunity for sufficient heat transfer through the pipe walls to maintain the gas at (or near) the temperature of the surroundings. Although Mach numbers are usually low the assumption of constant density is untenable because of the significant changes of pressure.

We again seek the integral of the momentum equation (11.65):

$$\frac{\mathrm{d}p}{\varrho} + u\mathrm{d}u + \frac{fu^2}{2A/P}\mathrm{d}l = 0$$

The heat transfer through the walls invalidates the energy equation (11.9) but we introduce the condition $T = $ constant. For a perfect gas the pressure is then proportional to the density, so from the continuity relation $\varrho u = $ constant we obtain

$$\frac{\mathrm{d}u}{u} = -\frac{\mathrm{d}\varrho}{\varrho} = -\frac{\mathrm{d}p}{p} \qquad (11.79)$$

(As the temperature is constant the restriction is to a gas obeying Boyle's Law. The equation of state for a perfect gas is here used for algebraic simplicity, but $R = p/\varrho T$ need not be independent of temperature.)

Dividing eqn 11.65 by u^2, substituting for $\mathrm{d}u/u$ from eqn 11.79 and rearranging we obtain

$$\frac{f\mathrm{d}l}{2A/P} = -\frac{\mathrm{d}p}{\varrho u^2} + \frac{\mathrm{d}p}{p} \qquad (11.80)$$

and then, putting $u = m/\varrho A$ and $\varrho = p/RT$,

$$\frac{f\mathrm{d}l}{2A/P} = -\frac{A^2}{m^2}\frac{p}{RT}\mathrm{d}p + \frac{\mathrm{d}p}{p} \qquad (11.81)$$

Now Reynolds number $= u\varrho d/\mu = $ constant$/\mu$ and since μ (except at extreme pressures) is a function of temperature alone the Reynolds number is constant in isothermal flow. For uniform roughness along the length of the pipe f is therefore constant. Equation 11.81 may consequently be integrated directly between points 1 and 2 a distance l apart:

$$\frac{fl}{2A/P} = \frac{A^2}{2m^2RT}\left(p_1^2 - p_2^2\right) - \ln\frac{p_1}{p_2} = \frac{2fl}{d} \text{ for a circular pipe} \quad (11.82)$$

Alternatively, since

$$p = \varrho RT = \frac{m}{Au}RT = \frac{m}{AM}\sqrt{\left(\frac{RT}{\gamma}\right)}$$

eqn 11.82 may be written

$$\frac{fl}{2A/P} = \frac{1}{2\gamma}\left(\frac{1}{M_1^2} - \frac{1}{M_2^2}\right) - \ln\frac{M_2}{M_1} \qquad (11.83)$$

To obtain p_1 or p_2 from eqn 11.82 a solution by trial is necessary, but the log term is often small compared with the others and so may be neglected in a first approximation. The approximate result may then be used to calculate $\ln(p_1/p_2)$ and a more accurate solution thus obtained.

It is clear from eqn 11.82 that not only is m zero when $p_2 = p_1$, but $m \to 0$ as $p_2 \to 0$. Consequently, for some value of p_2 between p_1 and zero, m must reach a maximum. Differentiating eqn 11.82 with respect to p_2 while p_1 is held constant yields

$$0 = -\frac{A^2}{m^3 RT}\frac{\mathrm{d}m}{\mathrm{d}p_2}\left(p_1^2 - p_2^2\right) - \frac{A^2 p_2}{m^2 RT} + \frac{1}{p_2}$$

and the maximum value of m is given by setting $\mathrm{d}m/\mathrm{d}p_2 = 0$. Then $m^2 = A^2 p_2^2/RT = A^2 \varrho_2^2 RT$. Hence $u_2^2 = m^2/A^2\varrho_2^2 = RT$, so $M_2 = u_2/\sqrt{(\gamma RT)} = \sqrt{(1/\gamma)}$.

The reason why an outlet Mach number of $\sqrt{(1/\gamma)}$ corresponds to the maximum mass flow rate is seen from eqn 11.80. This gives

$$\frac{\mathrm{d}p}{\mathrm{d}l} = \frac{f\varrho u^2}{2A/P}\bigg/\left(\frac{\varrho u^2}{p} - 1\right) = \frac{f\varrho u^2}{2A/P}\bigg/\left(\gamma M^2 - 1\right) \qquad (11.84)$$

Thus when $M = \sqrt{(1/\gamma)}\ (= 0.845$ for air) the pressure gradient is infinite; that is, there is a discontinuity of pressure (and also, by eqn 11.79, of density and velocity).

For a given initial Mach number M_1, there is thus a limiting length for continuous isothermal flow, and this is given by setting $M_2 = \sqrt{(1/\gamma)}$ in eqn 11.83:

$$\frac{fl_{\max}}{A/P} = \frac{1}{\gamma M_1^2} - 1 + \ln\left(\gamma M_1^2\right)$$

If the actual length were made greater than $l_{\max}$ the rate of flow would adjust itself so that $M = \sqrt{(1/\gamma)}$ was not reached until the end of the pipe. Thus it is seen that the phenomenon of choking may occur in isothermal flow but that the limiting value of the Mach number is (theoretically) $\sqrt{(1/\gamma)}$ instead of unity as in adiabatic flow.

In practice, however, isothermal flow at Mach numbers close to $\sqrt{(1/\gamma)}$ cannot be obtained. For a perfect gas under isothermal conditions both p/ϱ and the internal energy per unit mass e are constant, and so the steady-flow energy equation (3.13) reduces to

$$q = \frac{1}{2}u_2^2 - \frac{1}{2}u_1^2 \qquad (11.85)$$

the gravity terms again being neglected. Hence if points 1 and 2 are separated by a distance δl the rate of heat transfer per unit length $\mathrm{d}q/\mathrm{d}l = u\,\mathrm{d}u/\mathrm{d}l$. Substitution from eqns 11.79 and 11.84 then gives

$$\frac{\mathrm{d}q}{\mathrm{d}l} = u\frac{\mathrm{d}u}{\mathrm{d}l} = -\frac{u^2}{p}\frac{\mathrm{d}p}{\mathrm{d}l} = \frac{f\gamma M^2 u^2}{2\left(1 - \gamma M^2\right)A/P}$$

For conditions near to $M = \sqrt{(1/\gamma)}$, and for large velocities generally, the required high values of $\mathrm{d}q/\mathrm{d}l$ are difficult to achieve and the flow becomes more nearly adiabatic. Indeed, the limiting condition $M = \sqrt{(1/\gamma)}$ cannot be achieved at all because dq/dl would then have to be infinite.

Except for high Mach numbers results for isothermal and adiabatic flow do not in fact differ widely. This is because in adiabatic flow at low Mach numbers there is little variation of temperature (see Table A3.3). In isothermal flow

$$\frac{p_2}{p_1} = \frac{\varrho_2}{\varrho_1} = \frac{u_1}{u_2} = \frac{M_1}{M_2} \tag{11.86}$$

and comparing eqns 11.86 and 11.74 (for adiabatic flow) we obtain, for the same values of p_1, M_1 and M_2,

$$\frac{(p_2)_{\text{isoth.}}}{(p_2)_{\text{adiab.}}} = \left\{ \frac{1 + \frac{1}{2}(\gamma - 1)M_2^2}{1 + \frac{1}{2}(\gamma - 1)M_1^2} \right\}^{1/2}$$

Even in the extreme case where $M_1 = 0$ and $M_2 = \sqrt{(1/\gamma)}$ this ratio is only 1.069 (for $\gamma = 1.4$).

Example 11.7 For flow at low Mach numbers, the general eqn 11.82 for flow through a pipe of diameter d simplifies to

$$\frac{A^2}{2m^2RT}(p_1^2 - p_2^2) = \frac{2fl}{d} \quad \text{or} \quad p_1^2 - p_2^2 = \frac{4fl}{d}\left(\frac{m}{A}\right)^2 RT$$

(a) Use this equation to calculate the diameter of pipe 145 m long required to transmit 0.32 kg of air per second, if the inlet pressure and temperature are, respectively, $800 \times 10^3\,\mathrm{N\,m^{-2}}$ and 288 K, and the pressure drop is not to exceed $300 \times 10^3\,\mathrm{N\,m^{-2}}$. Take $f = 0.006$.
(b) Calculate the entry and exit Mach numbers.
(c) Determine the pressure halfway along the pipe.

Solution

(a) Since $A = \pi d^2/4$,

$$\left(800 \times 10^3\,\mathrm{N/m^2}\right)^2 - \left(500 \times 10^3\,\mathrm{N/m^2}\right)^2$$

$$= \frac{4 \times 0.006 \times 145\,\mathrm{m}}{d(\mathrm{m})} \times \left(\frac{0.32\,\mathrm{kg/s} \times 4}{\pi d^2 (\mathrm{m^2})}\right)^2 \times 287\,\mathrm{J/(kg\,K)}$$

$$\times\, 288\,\mathrm{K}$$

which yields $d^5 = 1.224 \times 10^{-7}\,\mathrm{m^5}$ or $d = 0.0415\,\mathrm{m} = 41.5\,\mathrm{cm}$.

(b) From the continuity condition, $m = \varrho A u$. At entry

$$\varrho = \frac{p}{RT} = \frac{\left(800 \times 10^3\right) \text{N/m}^2}{287 \, \text{J/}\!\left(\text{kg K}\right) \times 288 \, \text{K}} = 9.679 \, \text{kg m}^{-3}$$

and

$$A = \frac{\pi\left(0.0415\,\text{m}\right)^2}{4} = 0.00135 \, \text{m}^2$$

Hence

$$u = \frac{m}{\varrho A} = \frac{0.32 \, \text{kg/s}}{9.679 \, \text{kg/m}^3 \times 0.00135 \, \text{m}^2} = 24.5 \, \text{m/s}$$

The speed of sound is given by

$$a = \left(\gamma R T\right)^{1/2} = \left(1.4 \times 287 \, \text{J/}\!\left(\text{kg K}\right) \times 288 \, \text{K}\right)^{1/2} = 340 \, \text{m/s}$$

Therefore

$$M_1 = \frac{u_1}{a} = \frac{24.5 \, \text{m/s}}{340 \, \text{m/s}} = 0.072$$

and

$$M_2 = \frac{p_1}{p_2} M_1 = \frac{800}{500} \times 0.072 = 0.115$$

(c) The pressure distribution is of the form $p_1^2 - p_2^2 = kl$, where k is a constant. Denoting the pressure at $x = l/2$ by p_x, it follows that $p_1^2 - p_x^2 = kl/2$. Hence

$$p_1^2 - p_2^2 = 2\left(p_1^2 - p_x^2\right)$$

or

$$
\begin{aligned}
p_x &= \left[\left(p_1^2 + p_2^2\right)\!/2\right]^{1/2} \\
&= \left[\left(\left(8\,000 \times 10^3 \, \text{N/m}^2\right)^2 + \left(500 \times 10^3 \, \text{N/m}^2\right)^2\right)\!/2\right]^{1/2} \\
&= 667 \times 10^3 \, \text{N m}^{-2}
\end{aligned}
$$

□

11.10.3   Laminar flow in a circular pipe

The parabolic distribution of velocity occurring with laminar flow in a circular pipe invalidates the 'one-dimensional' assumption on which the momentum equation (11.60) is based. Since the velocity varies both with radius and with axial distance the problem strictly involves a partial differential equation, but no general solution of this is obtainable. However, laminar flow is to be expected only at low velocities; isothermal conditions are therefore likely and it may be shown that the change of velocity in the axial direction has a negligible effect on the velocity distribution over the cross-section. So, with pressure, temperature and density uniform over the cross-section, we may apply Poiseuille's formula (eqn 6.8) to a short length δl of the pipe. For a thermally perfect gas

$$m = \varrho Q = -\frac{\pi d^4 \varrho}{128\mu}\frac{\mathrm{d}p}{\mathrm{d}l} = -\frac{\pi d^4}{128\mu}\frac{p}{RT}\frac{\mathrm{d}p}{\mathrm{d}l} \quad \text{TPG} \quad (11.87)$$

the gravity term again being neglected. Integration with T and μ constant gives

$$m = \frac{\pi d^4\left(p_1^2 - p_2^2\right)}{256\mu RTl} \quad \text{TPG} \quad (11.88)$$

This is seen to correspond with Poiseuille's formula, eqn 6.8, provided that the mean density $= \frac{1}{2}(p_1 + p_2)/RT$ is used.

(With T constant, R need not be independent of temperature. Equation 11.88 is thus restricted to a gas obeying Boyle's law.)

11.11 ANALOGY BETWEEN FLOW WITH VARIABLE DENSITY AND FLOW WITH A FREE SURFACE

In several respects the flow of gases is similar to the flow of liquids in open channels. This is because similar equations of energy, momentum and continuity apply to the two types of flow. The behaviour of a gas depends significantly on whether a small change of pressure can be propagated upstream, that is, on whether the flow is subsonic or supersonic. As will be recalled from Chapter 10, the behaviour of a liquid in an open channel depends notably on whether a small surface wave can be propagated upstream, that is, on whether the flow is tranquil or rapid. For gas flows the Mach number $u/\sqrt{(p\gamma/\varrho)}$ relates the velocity of flow to the velocity of propagation of a small pressure wave through the fluid; for flow in an open channel the corresponding ratio is the Froude number $u/\sqrt{(g\bar{h})}$ where $\bar{h}$ represents the mean depth given by cross-sectional area ÷ surface width, and $\sqrt{(g\bar{h})}$ represents the velocity of propagation of a small surface wave relative to the liquid. Sonic conditions in a gas ($M = 1$) correspond to critical conditions in open-channel flow ($Fr = 1$). The choking phenomenon in gas flow in a convergent-divergent nozzle, for example, corresponds to the attainment of critical flow in the throat of a venturi flume or over the crest of a spillway.

Thermodynamic considerations have shown that, for steady conditions, a change from supersonic to subsonic flow can take place only through the discontinuity known as a shock. Likewise, in steady open-channel flow a change from rapid to tranquil flow is possible only through a hydraulic jump. A hydraulic jump, then, is the open-channel analogue of the shock.

Comparison of the continuity equations $\varrho u = $ constant for gas flow of constant cross-section and $hu = $ constant for a rectangular open channel, indicates that the gas density ϱ and the liquid depth h are analogous, that is

$$\frac{\varrho}{\varrho_0} \equiv \frac{h}{h_0} \qquad (11.89)$$

where ϱ_0 and h_0, the stagnation conditions, are used as suitable reference values. For frictionless open-channel flow in which vertical accelerations are negligible compared with g (so that the pressure at a point depends only on the depth below the free surface) the energy equation (10.1) may be written $u^2 = 2g(h_0 - h)$. The maximum possible velocity u_{max} would occur when $h = 0$ and so $(u/u_{max})^2 = (h_0 - h)/h_0$. The corresponding equations for flow of a perfect gas are $u^2 = 2c_p(T_0 - T)$ and $(u/u_{max})^2 = (T_0 - T)/T_0$ with the result that

$$\frac{T}{T_0} \equiv \frac{h}{h_0} \qquad (11.90)$$

For a perfect gas $p/p_0 = (\varrho/\varrho_0)(T/T_0)$ and thus from eqns 11.89 and 11.90

$$\frac{p}{p_0} \equiv \left(\frac{h}{h_0}\right)^2 \qquad (11.91)$$

For isentropic gas flow, however, $p/p_0 = (\varrho/\varrho_0)^\gamma$ and, in the light of eqns 11.89 and 11.91, the analogy can be exact only if $\gamma = 2$. Quantitative results from the analogy, for any real gas (for which $1 < \gamma \leq \frac{5}{3}$), cannot therefore be expected, although, fortunately, the effect of a change of γ is seldom large. The analogy is thus chiefly useful in providing a simple means of investigating phenomena qualitatively by visual observation.

Since flow through a shock wave is non-isentropic, quantitative agreement with measurements on a hydraulic jump is even less close, but useful qualitative results may still be obtained if the changes across the discontinuity are not too great. For example, if a 'two-dimensional' body is placed in an open channel where the flow is rapid, waves are produced on the liquid surface, and these are similar to the shock and expansion waves formed about a similar body in two-dimensional supersonic gas flow. Surface tension effects can cause difficulties in this technique, however, particularly in the formation of capillary waves which may be mistaken for the analogues of the shock and expansion waves. For this reason, the depth of liquid is usually between about 3 mm and 10 mm.

The subject is reviewed, and 126 references given, by Hoyt[1]. For a fuller treatment of the theory see, for example, Black and Mediratta[2].

11.12   FLOW VISUALIZATION

In a gas local changes of density, especially the abrupt changes occurring across a shock wave, may be made visible because of an accompanying change of refractive index n. (For values of n close to unity $(n - 1)/\varrho =$ constant.) Of the three principal techniques utilizing this property the *shadowgraph* method is the simplest. Light from a point source, or in a parallel beam, passes through the gas flow and on to a screen (Fig. 11.30). If the gas density is uniform the light rays, even if refracted, have an unchanged spacing and so illuminate the screen uniformly. If, however, there is a sharp density gradient (e.g. across a shock wave) in a direction normal to the light beam, the rays either diverge or converge according to whether the density gradient increases or decreases. Thus the position of a shock wave is indicated on the screen by a brighter band (on the high density side where the rays converge) next to a darker band (on the low density side where the rays diverge). Since the method depends on the relative deflection of the rays, that is, on

$$\frac{\partial^2 n}{\partial y^2} + \frac{\partial^2 n}{\partial z^2} \simeq \text{constant}\left(\frac{\partial^2 \varrho}{\partial y^2} + \frac{\partial^2 \varrho}{\partial z^2}\right)$$

it is best suited to the study of sharp rather than gradual changes of density.

The second method is the *schlieren* or Töpler system, invented by August J. I. Töpler (1836–1912) in 1867. In the simplest version, monochromatic light from a narrow, uniformly illuminated rectangular slit AB (Fig. 11.31) is collimated by a lens L_1, passes through the gas flow, is brought to a focus by lens L_2 and projected on to a screen (or photographic plate). At the focal plane of L_2 – where an image $A'B'$ of the slit AB is formed – a knife-edge is introduced to cut off part of the light. If the gas density is uniform the illumination on the screen, although reduced by the knife-edge, is also uniform. If, however, in any part of the flow a density gradient exists in a direction perpendicular

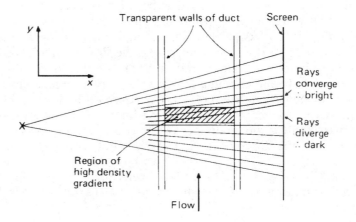

Fig. 11.30 Shadowgraph.

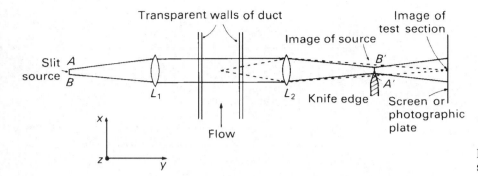

Fig. 11.31 Schlieren
system.

both to the light beam and to the knife-edge, light rays are refracted
and, depending on the sign of the density gradient, more or less light is
intercepted by the knife-edge. Corresponding parts of the image of the
test section on the screen are therefore darker or brighter. The change
in brightness is proportional to the density gradient. Only density gradients normal to the knife-edge are indicated, however, so two perpendicular knife-edges may be required.

In modern practice concave mirrors, being less expensive, are usually
employed instead of lenses. If a white source is used with strips of
coloured glass in place of the knife-edge, the changes of density gradient
may be represented by changes of colour rather than of brightness. The
method is widely used to obtain qualitative results: although it is theoretically possible to obtain quantitative estimates of density from
schlieren photographs the practical difficulties are considerable.

The third technique uses the *interferometer*, and the Mach–Zehnder
arrangement is that most widely employed (Fig. 11.32). A parallel
monochromatic light beam meets a half-silvered plate P_1 so that part of
the beam is reflected, and part transmitted. By means of fully-silvered
mirrors M_1, M_2 the two part beams, only one of which passes through the
flow being investigated, are brought together again and recombined by
a second half-silvered plate P_2. The apparatus may be so adjusted that,
with no flow in the test section, the two beams joining at P_2 are in phase
and so reinforce each other, making the screen uniformly bright. The
beam through the test section may be retarded by an increase of density
there; if this is sufficient to put the beams out of phase by half a
wavelength (or an odd multiple of half a wavelength), so that the crests
of waves in one beam coincide with troughs in the other, the screen is
uniformly dark. Lines of maximum and minimum brightness therefore
represent contours of constant density, and the density increment
between successive contours corresponds to a phase shift of half a
wavelength. Alternatively, the 'splitter plates' P_1 and P_2 may be slightly
rotated from the position that gives uniform brightness of the screen at
no-flow conditions to one that gives a series of interference fringes
parallel to the axes of rotation. Non-uniform density in the test section
then causes local distortion of these otherwise equally-spaced fringes,

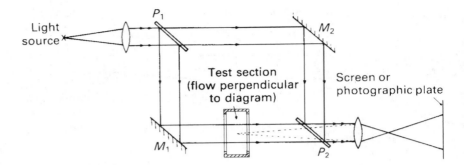

Fig. 11.32 Mach–
Zehnder interferometer.

and measurement of the fringe displacement permits calculation of the density at that point relative to the density in an undisturbed part of the flow.

The interferometer is costly and demands very high accuracy of the optical components and vibration-absorbing supports. It does, however, yield quantitative information about the density throughout the flow, and not just where the density changes rapidly.

In association with a heated wire shedding a filament of warmer, less dense, gas these optical methods may also be used to indicate a filament line in gas flow otherwise at constant density.

REFERENCES

1. Hoyt, J. W. 'The hydraulic analogy for compressible gas flow', *Appl. Mech. Rev.* **15**, 419–25 (1962).
2. Black, J. and Mediratta, O. P. 'Supersonic flow investigations with a "hydraulic analogy" water channel', *Aeronaut. Q.* **2**, 227–53 (1951).

FURTHER READING

Anderson, J. D. *Modern Compressible Flow*. McGraw-Hill, New York (1982).

Benedict, R. P. *Fundamentals of Gas Dynamics*, Wiley, New York (1983).

John, J. E. A. *Gas Dynamics*, Allyn and Bacon, Boston (1969).

Merzkirch, W. *Flow Visualization*, Academic Press, New York (1974).

'Schlieren methods', *N. P. L. Notes on Applied Science, No.* 31, HMSO, London (1963).

Shapiro, A. H. *The Dynamics and Thermodynamics of Compressible Fluid Flow*, Ronald Press, New York (1953, reprinted 1983).

Zucrow, M. J. and Hoffman, J. D. *Gas Dynamics*, Wiley, New York (1976).

FILMS

Shell Film Unit. *High Speed Flight*: Part 1 'Approaching the speed of sound'; Part 2 'Transonic flight'; Part 3 'Supersonic flight'.

Bryson, A. E *Waves in Fluids*, Encyclopaedia Britannica Films or from Central Film Library.

Shell Film Unit *Schlieren*.

Kline, S. J. *Flow Visualization*, Encyclopaedia Britannica Films or from Central Film Library.

PROBLEMS

(For air take $\gamma = 1.4$, $R = 287\,\text{J/(kg K)}$, $c_p = 1005\,\text{J/(kg K)}$. All pressures quoted are absolute.)

11.1   Atmospheric air at 101.3 kPa and 15 °C is accelerated isentropically. What are its velocity and density when the Mach number becomes 1.0 and what is the maximum velocity theoretically obtainable? Why could this maximum not be achieved in practice?

11.2   Air flows isentropically from atmosphere (pressure 101.5 kPa and temperature 15 °C) to a 600 mm square duct where the Mach number is 1.6. Calculate the static pressure, the velocity and the mass flow rate in the duct. What is the minimum cross-sectional area upstream of this section?

11.3   A schlieren photograph of a bullet shows a Mach angle of 40°. If the pressure and density of the undisturbed air are respectively 101.3 kPa and 10 °C what is the approximate temperature at the nose of the bullet?

11.4   A normal shock wave forms in front of a 'two-dimensional' blunt-nosed obstacle in a supersonic air stream. The pressure at the nose of the obstacle is three times the static pressure upstream of the shock wave. Determine the upstream Mach number, the density ratio across the shock and the velocity immediately after the shock if the upstream static temperature is 10 °C. If the air were subsequently expanded isentropically to its original pressure what would its temperature then be?

11.5   A supersonic air stream at 35 kPa is deflected by a wedge-shaped 'two-dimensional' obstacle of total angle 20° mounted with its axis parallel to the oncoming flow. Shock waves are observed coming from the apex at 40° to the original flow direction. What is the initial Mach number and the pressure rise through the shocks? If the stagnation temperature is 30 °C what is the velocity immediately downstream of the shock waves?

11.6   If the shocks in Problem 11.5 meet solid boundaries parallel to the initial flow, determine the angle at which they are reflected and the Mach number downstream of the reflected shocks.

11.7   Through what angle must a uniform air stream with Mach number 1.5 be turned so that its static pressure is halved?

11.8   A Pitot-static tube in a wind-tunnel gives a static pressure reading of 40.7 kPa and a stagnation pressure 98.0 kPa. The stagnation temperature is 90 °C. Calculate the air velocity upstream of the Pitot-static tube.

11.9   An aircraft flies at 8000 m altitude where the atmospheric pressure and temperature are respectively 35.5 kPa and -37 °C. An air-speed indicator (similar to a Pitot-static tube) reads 740 km/h, but the instrument has been calibrated for variable-density flow at sea-level conditions (101.3 kPa and 15 °C). Calculate the true air speed and the stagnation temperature.

11.10 Air from a large reservoir at 700 kPa and 40 °C flows through a converging nozzle, the exit area of which is 650 mm². Assuming that frictional effects are negligible, determine the pressure and temperature in the exit plane of the nozzle and the mass flow rate when the ambient pressure is (a) 400 kPa, (b) 100 kPa.

11.11 Air is to flow through a convergent-divergent nozzle at 1.2 kg/s from a large reservoir in which the temperature is 20 °C. At the nozzle exit the pressure is to be 14 kPa and the Mach number 2.8. Assuming isentropic flow, determine the throat and exit areas of the nozzle, the pressures in the reservoir and the nozzle throat and the temperature and velocity of the air at exit.

11.12 A convergent-divergent nozzle originally designed to give an exit Mach number of 1.8 with air is used with argon ($\gamma = 5/3$). What is the ratio of the entry stagnation pressure to the exit pressure when a normal shock is formed just inside the nozzle exit?

11.13 The mass flow rate of superheated steam ($\gamma = 1.3$) is to be measured by a venturi meter having inlet and throat diameters 100 mm and 50 mm respectively. If the upstream stagnation pressure and stagnation temperature are respectively 200 kPa (absolute) and 150 °C, what is the maximum mass flow rate that can be measured reliably in this way? Effects of friction and heat transfer may be neglected. (Thermodynamic tables will be needed.)

11.14 Superheated steam from a large reservoir in which the pressure is 1 MPa flows adiabatically through a convergent-divergent nozzle for which the cross-sectional area at exit is twice that at the throat. The pressure beyond the exit is 700 kPa. Determine the Mach number of the flow in the exit plane, the cross-sectional area at the plane where a normal shock may be expected in the nozzle, and the Mach number immediately upstream of the shock. What ambient pressure at exit would be necessary to produce isentropic supersonic flow without shocks? What exit pressure would give the same mass flow rate but with subsonic conditions throughout? Assume that frictional effects are negligible, that γ for superheated steam $= 1.3$, and that the steam remains superheated and with constant specific heat capacities throughout. (*Hint:* From eqn 11.57 substitute for m in eqn 11.58 to obtain p_0 downstream of the shock.)

11.15   Air flows adiabatically at the rate of 2.7 kg/s through a horizontal 100 mm diameter pipe for which a mean value f = 0.006 may be assumed. If the inlet pressure and temperature are 180 kPa and 50 °C what is the maximum length of the pipe for which choking will not occur? What are then the temperature and pressure at the exit end and half way along the pipe?

11.16   Air enters a 150 mm diameter pipe (mean f = 0.006) at 730 kPa and 30 °C. For a flow rate of 2.3 kg/s what is the pressure 2 km from the inlet when the flow is (a) adiabatic, (b) isothermal?

11.17   Calculate the diameter of a pipe 140 m long required to transmit 0.32 kg of air per second under isothermal conditions if the inlet pressure and temperature are respectively 800 kPa and 15 °C and the pressure drop is not to exceed 200 kPa. Take f = 0.006.

11.18   Air flows isothermally at 15 °C through 45 m of 75 mm diameter pipe (f = 0.008). Calculate the rate at which heat is transferred to the surroundings if the pressures at inlet and outlet are respectively 600 kPa and 240 kPa.

11.19   Determine the mass flow rate of air under adiabatic conditions through a 50 mm diameter pipe 85 m long from a large reservoir in which the pressure and temperature are 300 kPa and 15 °C. The pressure at the outlet of the pipe is 120 kPa. Assume a negligible pressure drop at the pipe inlet and a mean value of 0.006 for f. To what value would the inlet pressure need to be raised to increase the mass flow rate by 50%?

12 Unsteady flow

12.1 INTRODUCTION

Previous chapters of this book have been concerned almost exclusively with steady flow – that is, flow in which the velocity, pressure, density and so on at a particular point do not change with time. Admittedly, flow is rarely steady in the strictest sense of the term: in turbulent flow, for example, countless small variations of velocity are superimposed on the main velocity. But if the values of velocity and of other quantities at any particular point are averaged over a period of time and the resulting mean values are unchanging, then steadiness of flow may be assumed.

In unsteady, or non-steady flow, however, the mean values at a particular point do vary with time. Such variations add considerably to the difficulties of solving problems that involve unsteady flow. Indeed, the majority of such problems are too complex for normal algebraic methods to yield a solution. Nevertheless, certain problems of unsteady flow are amenable to analytical solution, and two or three will be briefly considered in this chapter.

Problems of unsteady flow may be put into one of three broad categories, according to the rate at which the change occurs. In the first group are problems in which the changes of mean velocity, although significant, take place slowly enough for the forces causing the temporal acceleration to be negligible compared with other forces involved. An example of this sort of problem is the continuous filling or emptying of a reservoir, discussed in Section 7.10. The second category embraces problems in which the flow changes rapidly enough for the forces producing temporal acceleration to be important: this happens in reciprocating machinery, such as positive displacement pumps, and in hydraulic and pneumatic servo-mechanisms. In the third group may be placed those instances in which the flow is changed so quickly, as for example by the sudden opening or closing of a valve, that elastic forces become significant.

Oscillatory motions, in which certain cycles of events are repeated, are also classified as unsteady. Examples of such motion are tidal movements, the oscillation of liquids in U-tubes and other vessels, and the vibrations encountered in acoustics.

12.2 INERTIA PRESSURE

Any volume of fluid undergoing an acceleration, either positive or negative, must be acted upon by a net external force. This force, distributed over the boundaries of the volume, corresponds to a difference of piezometric pressure, and this difference is commonly known as the *inertia pressure*. If circumstances are such that not all elements of the fluid concerned undergo the same acceleration, the inertia pressure may be difficult, if not impossible, to calculate. Here, however, we shall restrict consideration to flow in a pipe, of uniform cross-section and full of the fluid being accelerated. Changes of density ϱ will be assumed negligible.

For a stream-tube of negligible curvature and of constant cross-section, the acceleration force is given by the product of the mass of fluid in the tube and the acceleration. Thus, if the tube has a length l and cross-sectional area A, the mass of the fluid concerned is $\varrho A l$ and the accelerating force is $\varrho A l(\partial u/\partial t)$ where u denotes the instantaneous velocity of the fluid. (There is no change of velocity along the length of the tube.) If this force arises because the piezometric pressure at the upstream end exceeds that at the downstream end by an amount p_i, then

$$p_i A = \varrho A l \frac{\partial u}{\partial t}$$

The 'inertia pressure', p_i, is thus equal to $\varrho l(\partial u/\partial t)$, and the corresponding head, $h_i = p_i/\varrho g = (l/g)(\partial u/\partial t)$. For the bundle of stream-tubes filling a pipe of uniform cross-section, the mean inertia pressure is $\varrho l(\partial \bar{u}/\partial t)$ and the corresponding mean inertia head

$$h_i = \frac{l}{g} \frac{\partial \bar{u}}{\partial t} \qquad (12.1)$$

where $\bar{u}$ represents the mean instantaneous velocity over the cross-section. These expressions are of course positive or negative according to whether the velocity is increasing or decreasing with time.

In some types of hydraulic machine the inertia pressure required to accelerate a column of liquid may be so large that the pressure at the downstream end of the column falls to the vapour pressure of the liquid. As a result the column of liquid breaks and leaves a pocket of vapour. Not only is the efficiency of the machine thereby reduced but the subsequent collapse of this vapour pocket may produce dangerously high impact pressures. This is very important when a reciprocating pump has a long suction pipe: the speed of the pump is usually limited by the need to keep the inertia pressure at the beginning of the suction stroke to a moderate value.

The initiation of flow in a pipe-line is governed by inertia pressure. Let us suppose that a pipe of uniform cross-section is to convey fluid from a reservoir in which the piezometric pressure is constant. The pipe has at its downstream end a valve which is initially closed, and the

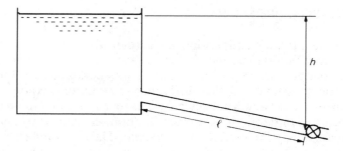

Fig. 12.1

pressure beyond the valve is constant (see Fig. 12.1). When the valve is opened, the difference of piezometric pressure between the ends of the pipe is applied to the fluid within it, and the rate of flow increases from zero to the steady value determined by the frictional and other losses in the pipe. Even if the valve could be opened instantaneously the fluid would not reach its full velocity instantaneously: its acceleration can never be greater than that corresponding to the available difference of piezometric pressure.

The acceleration is a maximum immediately after the valve is opened because the entire difference of piezometric pressure is then available for accelerating the fluid. As the velocity increases, however, energy is dissipated by friction; thus the piezometric pressure difference available for acceleration decreases and the acceleration itself is reduced.

Let the loss of head to friction be given by $ku^2/2g$, where u represents the mean velocity over the cross-section of the pipe and k has such a value that the term includes appropriate 'minor' losses in addition to $(4fl/d)(u^2/2g)$. (Friction and other losses in unsteady flow are assumed – with the support of experimental evidence – to be the same as in steady flow at the same instantaneous velocity.) Then at any instant the head available for accelerating the fluid is $h - ku^2/2g$. Thus, from eqn 12.1,

$$h - k\frac{u^2}{2g} = \frac{l}{g}\frac{\partial u}{\partial t}$$

The velocity u here varies only with the time and so the partial derivative may be written as the full derivative du/dt. Rearrangement gives the differential equation

$$dt = \frac{l}{g}\left(\frac{du}{h - ku^2/2g}\right) \tag{12.2}$$

Although the friction factor f, and consequently k, varies somewhat with u, an approximate result may be obtained by assuming k to be constant. When the maximum velocity u_{max} has been attained $h = ku_{max}^2/2g$. The integration of eqn 12.2 therefore yields

$$t = \frac{l}{g}\frac{2g}{k}\int_0^u \frac{\mathrm{d}u}{u_{max}^2 - u^2}$$

$$= \frac{l}{ku_{max}}\ln\left(\frac{u_{max} + u}{u_{max} - u}\right) \tag{12.3}$$

Equation 12.3 indicates that u becomes precisely equal to u_{max} only after an infinite time, but the velocity reaches, say, 99% of the maximum value within quite a short period. In any event we have here idealized the problem by the assumption that the fluid is incompressible. In practice even the slight compressibility of a liquid permits the propagation of elastic waves (see Section 12.3) and the subsequent damping of these waves brings about the equilibrium state sooner than eqn 12.3 suggests.

In the system depicted in Fig. 12.1 the rate of flow, once established, would slowly decline because the reservoir would gradually empty. Under these conditions, however, the change in velocity resulting from a change in overall head h would take place so slowly that the corresponding inertia head would be negligible. This class of problem has been considered in Section 7.10. For situations in which the temporal acceleration (either positive or negative) is appreciable, a term for the inertia head must be included. Its introduction, however, frequently renders the solution of the problem much more difficult.

N. R. Gibson's *inertia-pressure method*, used for determining the mean velocity of a liquid in a long pipe-line, involves rapidly closing a valve at the downstream end of the pipe and recording a diagram of pressure p against time t for a point immediately upstream of the valve. For a pipe of uniform cross-section and length l,

$$\int p \, \mathrm{d}t = l\varrho u$$

The value of the integral is determined graphically, and the original mean velocity u may then be calculated.

Example 12.1 A pump draws water from a reservoir and delivers it through a pipe 150 mm diameter, 90 m long, to a tank in which the free surface level is 8 m higher than that in the reservoir. The flow rate is steady at 0.05 m³/s until a power failure causes the pump to stop. Neglecting 'minor' losses in the pipe and in the pump, and assuming that the pump stops instantaneously, determine for how long flow into the tank continues after the power failure. The friction factor f may be taken as constant at 0.007 and elastic effects in the water or pipe material may be disregarded.

Solution
Before power failure, steady velocity in pipe

$$u_1 = \frac{0.05\,\text{m}^3/\text{s}}{\frac{\pi}{4}(0.15\,\text{m})^2} = 2.829\,\text{m/s}$$

After the power failure the static head of 8 m and the friction head both oppose the flow.

$$\therefore -(8\,\text{m}) - \frac{4fl}{d}\frac{u^2}{2g} = \frac{l}{g}\frac{du}{dt}$$

whence

$$\frac{du}{dt} = -(8\,\text{m})\frac{g}{l} - \frac{2f}{d}u^2$$

$$\therefore t = -\int_{u_1}^{0} \frac{du}{(8\,\text{m})\frac{g}{l} + \frac{2f}{d}u^2} = -\frac{d}{2f}\int_{u_1}^{0} \frac{du}{\frac{gd}{fl}(4\,\text{m}) + u^2}$$

$$= \frac{d}{2f}\sqrt{\left(\frac{fl}{gd(4\,\text{m})}\right)}\arctan\left\{u_1\sqrt{\left(\frac{fl}{gd(4\,\text{m})}\right)}\right\}$$

Substituting $u_1 = 2.829\,\text{m/s}$ and the values of d, f, l and g we get $t = 2.618\,\text{s}$.

12.3 PRESSURE TRANSIENTS

We now consider the third category of unsteady flow phenomena: those in which the changes of velocity occur so rapidly that elastic forces are important. As a result of the elasticity of the fluid – and also the lack of perfect rigidity of solid boundaries – changes of pressure do not take place instantaneously throughout the fluid, but are propagated by pressure waves. A change of velocity at a particular point in a fluid always gives rise to a change of pressure, and an important instance of such pressure changes is the phenomenon commonly known as 'water hammer' in pipe-lines. The name is perhaps a little unfortunate because not only water but any fluid – liquid or gas – may be involved.

It is a common experience that when a domestic water tap is turned off very quickly a heavy knocking sound is heard and the entire pipe vibrates. These effects follow from the rise in pressure brought about by the rapid deceleration of the water in the pipe when the tap is turned off. A similar phenomenon may occur in a pumping station owing to the slamming shut of non-return valves when a pumping set is shut down.

Not infrequently the increases of pressure caused by water hammer are sufficient to fracture the pipes, and for this reason alone the study of the phenomenon is of considerable practical importance.

To see why it is necessary to account for the elasticity of the fluid we may first consider the simple case of a fluid, originally flowing with a certain velocity in a pipe, being brought to rest by the closing of a valve at the downstream end of the pipe. If the fluid were entirely incompressible and the walls of the pipe perfectly rigid, then all the particles in the entire column of fluid would have to decelerate together. From Newton's Second Law, the more rapid the deceleration the greater would be the corresponding force, and with an instantaneous closure of the valve all the fluid would be stopped instantaneously and the force would be infinite. In fact, however, even a liquid is to some extent compressible, so its constituent particles do not decelerate uniformly. An instantaneous closure of the valve would not bring the entire column of fluid to a halt instantaneously: only those particles of fluid in contact with the valve would be stopped at once, and the others would come to rest later. Although an instantaneous closure of a valve is not possible in practice, an extremely rapid closure may be made, and the concept of instantaneous valve closure is valuable as an introduction to the study of what actually happens.

When a domestic water tap is suddenly turned off, the knocking sound produced can be heard not only at the tap but also – and often just as strongly – elsewhere in the house; it is evident, then, that the disturbance caused by the sudden closing of the tap must travel along the pipe to other parts of the system. To understand how the disturbance is transmitted along the pipe we may consider the instantaneous closing of a valve – the water tap, for example – at the end of a pipe. Just before the closure the pipe is full of fluid moving with a certain velocity (Fig. 12.2a). If the valve is suddenly closed, the fluid immediately next to the valve is stopped (Fig. 12.2b). For the time being, however, the fluid farther upstream continues to move as though nothing had happened. Consequently the fluid next to the valve is compressed slightly; its pressure is increased and the pipe (no longer assumed perfectly rigid) expands slightly as a result of the rise in pressure. The next element of fluid now finds an increased pressure in front of it; therefore it too comes to rest, is itself compressed and expands the pipe slightly. Each element of the fluid column thus stops the element following it until all the fluid in the pipe has been brought to rest. At any instant after the closing of the valve, but before all the fluid has stopped, there is a discontinuity of conditions in the pipe – represented by the line XX in Fig. 12.2c. On the valve side of XX the fluid has stopped and has been compressed; also, unless it is perfectly rigid, the pipe is slightly expanded.

On the other side of XX, however, the fluid is still moving with its original velocity, and the pressure and the pipe diameter still have their original values. As each successive element of the fluid is halted, the discontinuity XX moves farther away from the valve. In this way the change of velocity is transmitted along the pipe: the closing of the valve

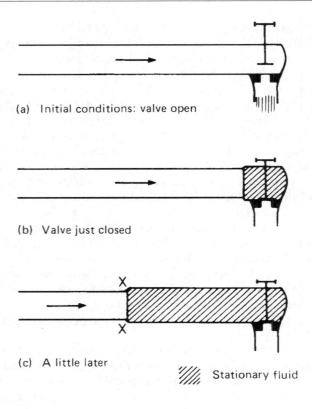

(a) Initial conditions: valve open

(b) Valve just closed

(c) A little later

Stationary fluid

Fig. 12.2

directly stops only the fluid that comes in contact with it; the remainder has to be brought to rest by a 'message' passed along the pipe from one fluid particle to another, each 'telling' the next that it must stop. The travelling discontinuity, known as a *pressure wave* or *pressure transient*, is in fact the 'message'.

(Pressure waves are also transmitted through the material of the pipe walls; their effect on pressure changes in the fluid, however, is almost always negligible, so we may disregard them here.)

We now see that neglecting the compressibility of the fluid – in other words assuming that all the fluid particles change velocity together – is legitimate only if the time of travel of the pressure wave is negligibly small compared with the time during which the change of velocity takes place.

When a fluid is suddenly stopped its behaviour is closely similar to that of a train of loosely coupled railway wagons. If the locomotive suddenly stops, the wagon immediately behind it compresses the buffer springs between itself and the locomotive. The force in the buffer springs increases to a value sufficient to stop the wagon. The second wagon then behaves likewise; it is stopped by the compressive force in the springs between itself and the first wagon. This process takes place successively along the whole length of the train. The compression of the buffer springs in the train is analogous to the compression of the fluid in the pipe-line. When a wagon has stopped, the force in the buffer springs at the front must equal the force in the springs at the rear (otherwise there would be a net force on the wagon and it would move). Similarly, a pressure wave in a fluid suddenly alters the pressure as it passes a particular point; but after that the pressure there stays at its new value.

There are, incidentally, many other examples of the transmission of a wave through an elastic material. Shock waves (considered in Chapter 11) are transmitted through gases. A sharp blow applied to one end of a long, weak coil spring causes a deflection of the coils that may be seen to travel to the opposite end of the spring. A stretched rubber tube may behave similarly. And if a series of similar, more gradual, disturbances occur regularly several times a second, musical sounds may be produced as, for example, when a succession of waves travels to and fro along the length of an organ pipe or other wind instrument.

12.3.1 The velocity and magnitude of pressure waves

We now need to consider the rate at which a change of pressure is transmitted through the fluid. Sections 11.4 and 11.5 dealt with the transmission of a pressure wave in a gas; here we restrict the analysis to a liquid, and consider also the effect of non-rigid boundaries.

Figure 12.3a illustrates a pipe in which liquid flowing from left to right at velocity u is brought to rest by a pressure wave XX moving from right to left. For the undisturbed liquid to the left of the diagram the pressure is p, the density is ϱ, and the cross-sectional area of the pipe is A. After the wave has passed, these quantities have become respectively $p + \Delta p$, $\varrho + \delta\varrho$, $A + \delta A$. (The change of density is small and so is the change of area – unless the pipe is made of exceptionally distensible material, such as very thin rubber or plastic – but the change of pressure Δp is not

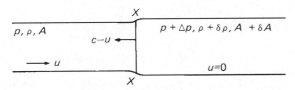

(a) Coordinate axes fixed relative to pipe

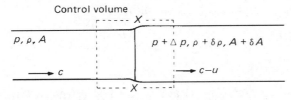

(b) Coordinate axes moving with wave

Fig. 12.3

necessarily small compared with p.) For the time being we shall disregard frictional effects because the friction head is usually small compared with the change of head caused by the pressure wave. The velocity is therefore considered uniform over the cross-section.

Let the pressure wave travel towards the left with a speed c relative to the oncoming liquid. The speed of the wave *relative to the pipe* is therefore $c - u$. The conditions will appear steady if we refer to coordinate axes moving with the wave (as in Fig. 12.3b). The continuity relation across the wave is then

$$\varrho A c = (\varrho + \delta\varrho)(A + \delta A)(c - u) \tag{12.4}$$

The change of density is directly related to the change of pressure by the bulk modulus of the liquid, K:

$$\frac{\delta\varrho}{\varrho} = \frac{\Delta p}{K} \quad \left(\text{from eqn 1.8}\right)$$

The difference in pressure across the wave gives rise to a force $-A\Delta p$ towards the right (if δA is negligible compared with A) and so, for the steady flow conditions of Fig. 12.3b,

$$-A\Delta p = \left(\text{Mass flow rate}\right)\left\{(c - u) - c\right\} = -\varrho A c u$$

whence

$$u = \Delta p / \varrho c \tag{12.5}$$

Dividing eqn 12.4 by ϱA and substituting for $\delta\varrho/\varrho$ and for u we obtain

$$c = \left(1 + \frac{\Delta p}{K}\right)\left(1 + \frac{\delta A}{A}\right)\left(c - \frac{\Delta p}{\varrho c}\right)$$

whence

$$\varrho c^2 = \frac{\Delta p\left(1 + \dfrac{\delta A}{A}\right)\left(1 + \dfrac{\Delta p}{K}\right)}{\dfrac{\Delta p}{K}\left(1 + \dfrac{\delta A}{A}\right) + \dfrac{\delta A}{A}} \tag{12.6}$$

Compared with Δp the bulk modulus K of a liquid is always very large; for example, the value for cold water at moderate pressures is about $2\,\text{GPa}$. Therefore $\Delta p/K$ may be neglected in comparison with 1, and so may $\delta A/A$. Equation 12.6 then simplifies to

$$\varrho c^2 = \frac{\Delta p}{\dfrac{\Delta p}{K} + \dfrac{\delta A}{A}} \tag{12.7}$$

(The assumptions $(\Delta p/K) \ll 1$ and $(\delta A/A) \ll 1$ together imply that $u \ll c$, as is evident from eqn 12.4. Hence it does not matter whether c in eqn

12.7 is regarded as representing a velocity relative to the liquid or to the pipe.)

For a perfectly rigid pipe $\delta A = 0$ and eqn 12.7 then gives $c = \sqrt{(K/\varrho)}$. In other words, the velocity with which a wave is propagated relative to the liquid in a rigid pipe is the same as the velocity of sound in an infinite expanse of the liquid. (For pipes of very small diameter, frictional effects reduce this value slightly.)

If the pipe is not perfectly rigid, however, the value of $\delta A/A$ must be determined. This depends on the material of the pipe and also on its freedom of movement. We shall here consider only the case of a pipe so mounted that there is no restriction of its longitudinal movement (it may be assumed to have numerous frictionless expansion joints). An increase of pressure Δp in the pipe causes a hoop stress tending to burst the pipe. For a circular pipe in which the thickness t of the wall is small compared with the diameter d, the hoop stress is given by $(\Delta p)d/2t$ and for elastic deformation the corresponding hoop strain is thus $(\Delta p)d/2tE$, where E represents the elastic modulus of the material of the pipe.

This hoop strain gives rise to a longitudinal strain (which may be calculated from the Poisson ratio). The longitudinal strain, however, does not enter the present problem – the pipe merely 'slides over' the column of liquid inside it and the pipe has no longitudinal *stress*. For a circular pipe, A is proportional to d^2, so $\delta A/A = 2\delta d/d = 2$(hoop strain) $= (\Delta p)d/tE$. Substitution of this value in eqn 12.7 gives

$$c = \sqrt{\left\{\left[\frac{1}{\varrho\left(\dfrac{1}{K} + \dfrac{d}{tE}\right)}\right]\right\}}$$

i.e.

$$c = \sqrt{\left(\frac{K'}{\varrho}\right)}$$

where

$$\frac{1}{K'} = \frac{1}{K} + \frac{d}{tE} \tag{12.8}$$

Here K' may be regarded as the effective bulk modulus of the liquid when in the pipe. (As mentioned earlier, the transmission of elastic waves also occurs through the material of the pipe walls, but the effect of these waves is normally negligible.)

In cases where the longitudinal movement of the pipe is restrained, as for example in the fuel-injection system of a diesel engine (where pressure transients have particular significance), the effective bulk modulus of the liquid is slightly different. It may be calculated by again determining the appropriate value of $\delta A/A$. However, even if all longitudinal movement of the pipe is prevented, the resulting wave velocity is little affected, and, for a thin-walled pipe, eqn 12.8 is sufficiently accurate for most purposes.

A few figures will show the orders of magnitude. A representative value of K for water is 2.05 GPa and so the wave velocity in a rigid pipe is

$$\sqrt{\left(\frac{K}{\varrho}\right)} = \sqrt{\left(\frac{2.05 \times 10^9\,\text{N/m}^2}{10^3\,\text{kg/m}^3}\right)} = \sqrt{\left(2.05 \times 10^6\,\text{m}^2/\text{s}^2\right)} = 1432\,\text{m/s}$$

Other liquids give values of the same order.

For a steel pipe ($E = 200\,\text{GPa}$) of, say, 75 mm diameter and 6 mm thickness

$$\frac{1}{K'} = \frac{1}{2.05 \times 10^9\,\text{Pa}} + \frac{75\,\text{mm}}{6\,\text{mm} \times 200 \times 10^9\,\text{Pa}}$$

whence $K' = 1.817 \times 10^9\,\text{Pa}$ and so

$$c = \sqrt{\left(\frac{1.817 \times 10^9\,\text{N/m}^2}{10^3\,\text{kg/m}^3}\right)} = 1348\,\text{m/s}$$

Although c by its dependence on K varies slightly with pressure and temperature, sufficient accuracy is usually obtained by regarding it as constant for a particular pipe line. (Gas bubbles in the liquid, however, appreciably reduce the value of K and therefore of c.)

It is important to realize that the velocity of wave propagation is not a velocity with which particles of matter are moving. For this reason the word *celerity* is sometimes used rather than velocity. It represents the rate at which a message can be 'telegraphed' through the fluid. As another example of a 'message' being sent through the fluid we may consider a pump delivering fluid through a long pipe-line into a reservoir. If the power to the pump suddenly fails, fluid continues to be discharged from the outlet end of the pipe until the pressure wave produced by the stopping of the pump reaches there: the pressure wave – this time a negative one – provides the only means by which the fluid at the outlet end 'knows' that the input has failed.

The magnitude of the rise in pressure caused by reducing the velocity from u to zero is given by eqn 12.5 as $\varrho c u$. In a more general case the velocity would be changed from u_1 to u_2. The analysis, however, would be modified only in so far as the zero of the velocity scale would be altered; the corresponding rise in pressure would be given by $\varrho c(u_1 - u_2)$. This may be expressed as a change of head by dividing by ϱg (the change of ϱ, we recall, is small):

$$h_2 - h_1 = \frac{c}{g}(u_1 - u_2), \quad \text{i.e.} \quad \Delta h = -\frac{c}{g}\Delta u \qquad (12.9)$$

Substituting in eqn 12.9 values appropriate for water shows that a reduction of velocity of 3 m/s corresponds to an increase of head of about 440 m (about 4.3 MPa). Such an increase is too large to be neglected in the design of a pipe system. The equation shows too that, just as a sudden reduction of velocity gives rise to an increase of pressure, so a sudden increase of velocity causes a reduction of pressure. It will be noticed also that the change of head is independent of the length of the pipe – in distinction to eqn 12.1 for the inertia head in an incompressible fluid.

The kinetic energy lost by the liquid when its velocity is reduced is converted entirely into strain energy by the compression of the liquid itself and the stretching of the pipe (if there is no friction loss). No mechanical energy has been lost: it is merely stored in the compressed material and may be released when the strain is removed.

Not all pressure waves result from sudden changes of velocity or pressure. A gradual change, however, may be regarded as the sum of a series of small instantaneous changes. Thus the transmission of such a 'gradual' pressure wave along a pipe is similar to that of a 'sudden' wave. Consequently, for a pipe-line of unlimited length, any particular pattern of change at one point will be reproduced at another point; there is, however, a time delay corresponding to the time that the pressure wave takes to travel the distance between the two points.

12.3.2   The reflection of waves

The problem is somewhat complicated if the pipe-line is not of unlimited length. This is because a pressure wave, on reaching the end of the pipe, is 'reflected'; in other words, another wave is produced which returns along the pipe to the starting point of the first wave. This reflected wave may be larger or smaller than the first, according to the conditions at the end where the 'reflection' takes place. The simplest case is that for the horizontal pipe illustrated in Fig. 12.4. Let us assume that flow towards the valve is stopped by the instantaneous closing of the valve. Frictional effects are for the moment disregarded. An increase of pressure is caused just upstream of the valve, and this is propagated with celerity c as a 'positive' pressure wave along the pipe towards the reservoir. At a time l/c later the wave reaches the reservoir. All the fluid in the pipe-line has now been brought to rest; it has been compressed and the pipe has been expanded slightly. (These conditions are indicated on the third pressure diagram in Fig. 12.4.)

Although the fluid is at rest it is not in equilibrium. Let us assume that the reservoir is very large in relation to the cross-sectional area of the pipe, so that the velocity in the reservoir is always negligible and the pressure therefore constant. There is now a discontinuity between the pipe and the reservoir: the fluid in the pipe is at the increased pressure, whereas that in the reservoir is at the original pressure and its velocity is already zero. As a result of the discontinuity of pressure, fluid begins to flow from the pipe back into the reservoir so as to equalize the

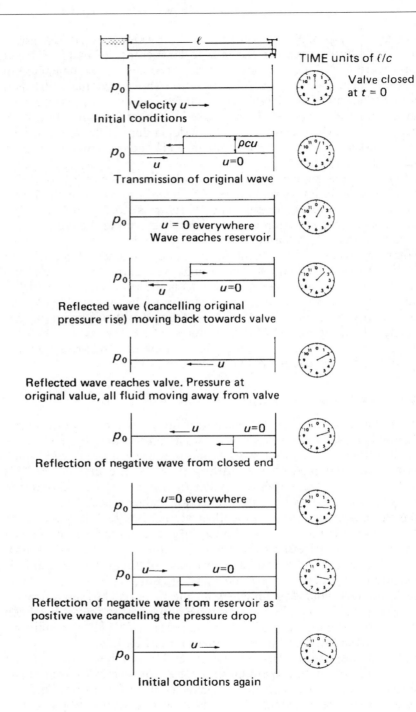

TIME units of ℓ/c

Valve closed at $t = 0$

Fig. 12.4 Pressure diagrams showing transmission and reflection of individual waves. (Friction and velocity heads assumed negligible compared with the changes of static head.)

pressures at that end of the pipe. In other words, the discontinuity that constitutes the pressure wave now moves back towards the valve. Because of the relation between the changes of pressure and velocity (eqn 12.5), the velocity with which the fluid now moves towards the reservoir is of magnitude u. The result is equivalent to the superposition of a 'negative' or 'unloading' wave on the original 'positive' one so as to nullify it. In other words, the reflection of a wave at a completely 'open' end (that is, an end connected to a reservoir of infinite extent) gives a second wave, equal in magnitude to the first but opposite in sign. (See fourth pressure diagram in Fig. 12.4.)

The reflected wave travels the length of the pipe in a time l/c, so it reaches the valve at a time $2l/c$ after the closing of the valve. The pressure has now been reduced everywhere to its original value, the fluid density and the pipe diameter are back to their original values – but all the fluid is moving back from the valve. This reverse movement decompresses the fluid immediately next to the valve, with the result that there is a fall in pressure. The magnitude of this pressure change is – ideally – the same as the magnitude of the previous changes because the corresponding change of velocity is again u. Therefore a negative wave is now propagated from the valve to the reservoir. This illustrates the nature of a reflection from a completely closed end: the magnitude of the reflected wave equals that of the incident wave, and the sign remains unchanged. When, after the next time interval of l/c, the negative wave reaches the open, reservoir, end there is an unbalanced state, with a higher pressure in the reservoir than in the pipe. Therefore fluid flows from the reservoir into the pipe; in other words, reflection of the wave takes place with a change of sign so that the existing negative wave is nullified by a positive wave. This positive wave is propagated towards the valve and reaches it at the end of the fourth time interval of l/c.

Conditions have now been reached that are identical with those existing at the moment the valve was closed: all the fluid in the pipe is moving towards the valve with velocity u, and the pressure is back to the original value. The complete cycle of events (as illustrated in Fig. 12.4) is therefore repeated and, in the absence of friction, would be repeated indefinitely, each cycle occupying a period of time $4l/c$. In practice, energy is gradually dissipated by friction and imperfect elasticity, so the waves diminish in intensity and die away.

The time needed for a pressure wave to make the round trip from valve to reservoir and back again is $2l/c$. Thus, for an instantaneous closing of the valve, the excess pressure at a point immediately before the valve remains constant for a time interval $2l/c$; the pressure is then altered by the arrival of the negative, unloading, wave. Similarly, the subsequent drop of pressure at this point remains constant for the same length of time. A graph of pressure against time for a point adjacent to the valve is therefore as shown in Fig. 12.5a. At a distance x from the reservoir the time required for the round trip to the reservoir and back is only $2x/c$ and so the duration of any rise or fall of pressure from the original value there is $2x/c$. Figure 12.5b shows the pressure–time graph

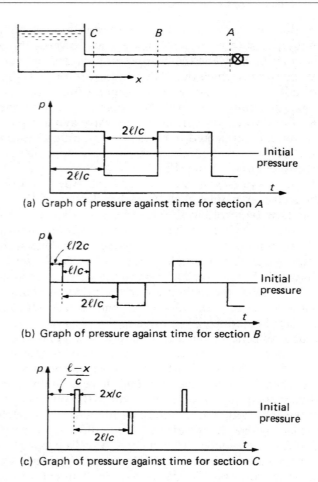

(a) Graph of pressure against time for section A

(b) Graph of pressure against time for section B

Fig. 12.5

(c) Graph of pressure against time for section C

for a point half-way along the pipe ($x = l/2$). When $x \to 0$, near the reservoir, the rise or fall of pressure persists for a very short time only (Fig. 12.5c).

If the negative wave has an amplitude greater than the original absolute pressure of the liquid then the conditions represented by Fig. 12.5 are modified, since, in practice, the absolute pressure cannot fall below zero. In fact, before the pressure reaches zero the liquid boils, and a cavity filled with the vapour of the liquid is formed. (Air or other gases may also come out of solution.) In such cases the lower part of the pressure diagrams is cut short at the vapour pressure of the liquid, and the changes in pressure may then follow a very complicated pattern.

Moreover, when a vapour cavity subsequently collapses, liquid may rush into it at a velocity greater than that of the original flow. The sudden stopping of this motion may then produce a positive pressure greater than that caused by the initial valve closure. It is therefore important to avoid, if possible, the vaporization of the liquid (or the release of substantial amounts of air or other gases from solution).

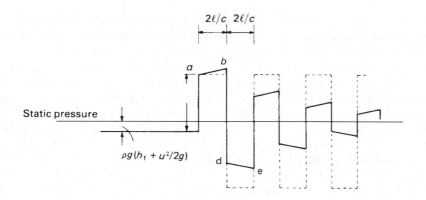

Fig. 12.6 Effect of friction. Variation of pressure next to the valve with time after complete valve closure. Theoretical (no friction) line shown dotted.

The effects of friction losses are indicated in Fig. 12.6, which shows the perhaps surprising fact that a greater rise of pressure may occur with friction than without it. When the velocity of the fluid is reduced, so is the head lost to friction; the head available at the downstream end of the pipe consequently rises somewhat as layer after layer of the fluid is slowed down. This secondary effect is transmitted back from each layer in turn with celerity c, and so the full effect is not felt at the valve until a time $2l/c$ after its closure. In Fig. 12.6 this effect is indicated by the upward slope of the line ab. During the second time interval of $2l/c$ velocities and pressure amplitudes have reversed signs, and thus the line de slopes slightly downwards. However, energy is also dissipated by viscous forces during the small movements of individual particles as the fluid is compressed and expanded. This dissipation of energy, known as damping, always tends to reduce the amplitude of the pressure waves and so bring ab and de nearer to the horizontal. Indeed, the effect of the damping may sometimes exceed that of the pipe friction, and so the lines ab and de may even converge towards the equilibrium pressure line (which, for complete closure of the valve, corresponds to the static pressure line). Friction forces also oppose the flow of the fluid back towards the reservoir, so the velocity of this flow is somewhat less in magnitude than the original. Consequently the amplitude of a wave is reduced at each reflection until the final equilibrium pressure is reached. Except in pipes of very small diameter, friction does not appreciably affect the celerity with which waves are propagated.

It is difficult in calculations to account accurately for friction. The neglect of friction is often justified because the friction head is small compared with the head produced by the 'water hammer'. In all cases, however, values 'on the safe side' are obtained by assuming that the initial head at the valve is the same as the head in the reservoir, and neglecting subsequent frictional effects.

In addition to the reflection of waves that takes place at open or closed ends, partial reflections occur at changes of section in the pipe-line, and at junctions. These partial reflections are considered in the more specialized works mentioned at the end of this chapter.

Our discussion so far has all been based on the concept of instantaneous closure of the valve. An absolutely instantaneous closure is physically impossible, but if the valve movement is completed in a time less than $2l/c$ the results are not essentially different. The pressure at the valve is gradually built up as the valve is closed; the maximum pressure reached, however, is the same as with instantaneous closure, because the conversion of kinetic energy of the fluid to strain energy is completed before any reflected waves have had time to reach the valve. If, on the other hand, the valve movement is not finished within a time $2l/c$, then not all of the kinetic energy has been converted to strain energy before a reflected wave arrives to reduce the pressure again. The maximum pressure rise thus depends on whether the time during which valve movement occurs is greater or less than $2l/c$. Movements taking less than $2l/c$ are said to be *rapid*; those taking longer than $2l/c$ are said to be *slow*. (When pressure changes are considered for a point between the valve and the reservoir – at a distance x, say, from the reservoir – the relevant time interval is $2x/c$.)

Similar effects follow from a sudden opening of the valve, although the primary result is a reduction of pressure behind the valve instead of a rise.

12.3.3   Slow closure of the valve

If the valve is closed in a time longer than $2l/c$ account must be taken of waves returning to the valve from the reservoir. The subsequent changes of pressure, as the waves travel to and fro in the pipe, may be very complex, and determining the pressure fluctuations requires a detailed step-by-step analysis of the situation. In this section we shall outline a simple, arithmetical, method of solution so as to focus attention on the physical phenomena without the distraction of too much mathematics. In Section 12.3.4 we shall look at the mathematical basis of a more general technique.

The essential feature of the arithmetical method is the assumption that the movement of the valve takes place, not continuously, but in a series of steps occurring at equal intervals of $2l/c$ or a sub-multiple of $2l/c$. The partial closure represented by each step is equal in magnitude to that achieved by the actual, continuous, movement during the time interval following the step, and the valve movement for each step is assumed to be instantaneous. Between the steps the valve is assumed stationary. Each of the instantaneous partial closures generates its own particular wave – similar in form to those depicted in Fig. 12.5 – which travels to and fro in the pipe until it is completely dissipated by friction. The total effect of all these individual waves up to a particular moment approximates fairly closely to the effect produced by the actual, continuous, valve movement. (During any interval while the valve is being closed, the pressure at the valve builds up to a maximum which occurs at the end of the interval. If each interval is so chosen that a wave sent off at the beginning does not return before the end of the interval, then a knowledge of the precise way in which the continuous valve move-ment occurs is unnecessary: eqn 12.9 gives $h_2 - h_1$ for a reduction of velocity from u_1 to u_2, regardless of the manner in which u changes with time.)

Let us suppose that the steady flow conditions before the valve closure begins are represented by a head h_0 at the valve and a velocity u_0 in the pipe. If at the end of the first chosen interval the head and velocity are respectively h_1 and u_1 then these must be related by eqn 12.9:

$$h_1 - h_0 = \frac{c}{g}(u_0 - u_1) \tag{12.10}$$

Another relation between h_1 and u_1 is required if either is to be calcu-lated. For the simple case in which the valve discharges to atmosphere, the valve may be regarded as similar to an orifice, and so

$$Au = Q = C_d A_v \sqrt{(2gh)} \tag{12.11}$$

where A represents the cross-sectional area of the pipe in which the velocity is u, A_v the area of the valve opening and C_d the corre-sponding coefficient of discharge. Putting $C_d(A_v/A)\sqrt{(2g)} = B$ in eqn 12.11 we have

$$u = B\sqrt{h} \tag{12.12}$$

The factor B is usually known as the 'valve opening factor' or 'area coefficient'. It should be noted that C_d is not necessarily constant, and the variation of B with the valve setting usually has to be determined by experiment for each design of valve. From eqn 12.12 $u_1 = B_1\sqrt{h_1}$ and simultaneous solution of this equation with eqn 12.10 gives correspond-ing values of h_1 and u_1.

Such a closure in which B varies linearly with time t, i.e. $B = B_0(1 - t/T)$, is frequently termed 'straight-line closure'. It does not imply uniform motion of the valve. With a gate valve (Fig. 12.7), for example, in which the gate is moved uniformly, the rate of reduction of area is initially small, but very much greater towards the end of the closure. It is therefore common to arrange for such valves on long pipe-lines to close slowly during the final stages of the movement.

A wave returning from the reservoir is negative in sign and it thus offsets – partially at least – rises in pressure caused by further reductions of velocity. A numerical example will show how allowance is made for the reflected waves.

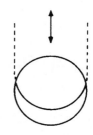

Fig. 12.7 Gate valve.

Example 12.2 A pipe carries water at a (mean) velocity of 2 m/s, and discharges to atmosphere through a valve. The head at the valve under steady flow conditions is 100 m (pipe friction being neglected). Length of pipe = 2400 m; wave celerity c = 1200 m/s.

The valve is closed so that the area coefficient B is reduced uniformly in 8 s. Determine the maximum and minimum heads developed.

Solution
An interval $2l/c$ here is $2 \times 2400\,\text{m}/(1200\,\text{m/s}) = 4\,\text{s}$, so the complete closure occupies two intervals of $2l/c$. We shall consider the valve movement to take place in 10 steps, that is, each interval of $2l/c$ will be subdivided into five parts, and we shall neglect friction.

The rise in head corresponding to any reduction of velocity is given by

$$\Delta h = -\frac{c}{g}\Delta u = -\frac{1200\,\text{m/s}}{9.81\,\text{m/s}^2}\Delta u = -(122.3\,\text{s})\Delta u \quad (12.13)$$

Initially $2\,\text{m/s} = B_0\sqrt{(100\,\text{m})}$, so $B_0 = 0.2\,\text{m}^{1/2}/\text{s}$. If the valve is completely closed in ten steps the numerical values of B are

$$0.2, 0.18, 0.16, \ldots 0.02, 0$$

It is helpful to draw a diagram of the pressure changes *adjacent to the valve* which are caused by the separate instantaneous partial closures. These diagrams (Fig. 12.8) show at a glance the sign of the change corresponding to any wave at any moment.

Although the simultaneous solution of eqns 12.12 and 12.13 is possible algebraically, solution by trial is often quicker. The calculations may be set out in tabular form as shown here.

B (m$^{1/2}$/s)	Head at valve, h (m)	u (m/s)	Δh (m)
0.2	100	2.00	–

These are the initial conditions. At the first step B is reduced from 0.2 to $0.18\,\text{m}^{1/2}/\text{s}$. Then, with metre-second units,

$$\Delta h = 122.3(2.00 - u_1)\,\text{m} \text{ and } u_1 = 0.18\sqrt{(100 + \Delta h)}\,\text{m/s}$$

These equations are satisfied by $u_1 = 1.903\,\text{m/s}$ and $\Delta h = 11.82\,\text{m}$.

| 0.18 | 100+11.82=111.82 | 1.903 | 11.82 |

Similarly for the next step:

0.16	100+11.82+13.61=125.43	1.792	13.61
0.14	100+11.82+13.61+15.73=141.16	1.663	15.73
0.12	100+11.82+13.61+15.73+18.18=159.34	1.515	18.18
0.10	100+11.82+13.61+15.73+18.18+21.01=180.35	1.343	21.01

The pressure diagrams, Fig. 12.8, show that for the next step – the reduction of B from 0.10 to 0.08 – matters are slightly complicated by the return of the first wave, that of magnitude 11.82 m. This rise of pressure is cancelled as the wave returns (its sign was changed on reflection from the reservoir) and the wave is then reflected from the valve *with the same (i.e. negative) sign*. Consequently the +11.82 m is dropped from the *h* column and a −11.82 m takes its place. The next line of the table is therefore

B (m$^{1/2}$/s)	Head at valve, h (m)	u (m/s)	Δh (m)
0.08	100 − 11.82 + 13.61 + 15.73 + 18.18 + 21.01 + 30.41 = 187.12	1.094	30.41

Since the closure of the valve is not yet completed, the returning 11.82 m wave does not meet an absolutely closed end. Its reflection is therefore not complete; that is, its amplitude after reflection is somewhat less than 11.82 m. But the degree of completeness of the reflection happily does not concern us. The amplitude of the new wave (here 30.41 m) must be such as to give a total head (187.12 m) which satisfies the equation $u = 0.08\sqrt{h}$. The new wave thus effectively incorporates the loss by reflection of the wave that has just arrived and reversed. (If the 11.82 m wave had not returned to the valve, Δh would have been 24.29 m. The extra 6.12 m accounts for the fact that only 5.70 m of the 11.82 m is actually reflected. The total h is the same.) This is why the interval between successive steps must be either $2l/c$ or a sub-multiple of $2l/c$: an 'old' wave is reflected from the valve at the same moment as a new one sets off, and so the new wave can incorporate the reflection loss of the old.

(Metre-second units throughout)   Head change at valve

B changed from	Δh (calculated as below)
0.20 to 0.18	11.82
0·18 to 0·16	13.61
0.16 to 0.14	15.73
0.14 to 0.12	18.18
0.12 to 0.10	21.01
0.10 to 0.08	30.41

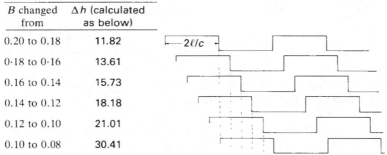

Fig. 12.8

The pressure diagrams show that for the next step (B changed from 0.08 to 0.06) the 13.61 m wave returns and the corresponding pressure rise at the valve changes sign. The table then continues:

B (m$^{1/2}$/s)	Head at valve, h (m)	u (m/s)	Δh (m)
0.06	$100 - 11.82 - 13.61 + 15.73 + 18.18$ $+\ 21.01 + 30.41 + 32.15 = 192.05$	0.831	32.15
0.04	$100 - 11.82 - 13.61 - 15.73 + 18.18$ $+\ 21.01 + 30.41 + 32.15 + 33.53$ $=\ 194.12$	0.557	33.53
0.02	$100 - 11.82 - 13.61 - 15.73 - 18.18$ $+\ 21.01 + 30.41 + 32.15 + 33.53$ $+\ 34.26 = 192.02$	0.277	34.26
0	$100 - 11.82 - 13.61 - 15.73 - 18.18$ $-\ 21.01 + 30.41 + 32.15 + 33.53$ $+\ 34.26 + 33.90 = 183.90$	0	33.90

Although the valve is now closed the waves continue their movement, and the pressure increments at the valve change sign according to Fig. 12.8. However, as the velocity at the valve is zero and must remain zero, no new waves can be generated.

B (m$^{1/2}$/s)	Head at valve, h (m)	u (m/s)	Δh (m)
0	$100 + 11.82 - 13.61 - 15.73 - 18.18$ $-\ 21.01 - 30.41 + 32.15 + 33.53$ $+\ 34.26 + 33.90 = 146.72$	0	–
0	$100 + 11.82 + 13.61 - 15.73 - 18.18$ $-\ 21.01 - 30.41 - 32.15 + 33.53$ $+\ 34.26 + 33.90 = 109.64$	0	–
0	$100 + 11.82 + 13.61 + 15.73 - 18.18$ $-\ 21.01 - 30.41 - 32.15 - 33.53$ $+\ 34.26 + 33.90 = 74.04$	0	–
0	$\ldots = 41.88$	0	–
0	$\ldots = 16.10$	0	–
0	$\ldots = 53.28$	0	–

The table shows that a maximum head of about 194 m is reached (almost twice the static head), and a minimum of about 16 m is reached after the valve is closed. In this example, if the valve remained closed, then, in the absence of friction, the waves would continue to travel to and fro indefinitely. In practice, of course, friction would fairly soon damp out the waves. In some circumstances (a high initial velocity or a low initial head at the valve) the minimum head reached might – apparently – be substantially less than zero. The column of liquid would then break, vaporization would occur, and the subsequent behaviour would be unpredictable.

The technique indicated here is applicable to the partial closure of a valve and also to the opening of a valve. In the latter case, the opening causes a drop of pressure at the valve, and thus an increase of velocity. The pressure waves sent upstream from the valve are negative, and the velocity and head at the valve usually approach their final values asymptotically. The pressure drop is limited to the difference between the initial values on the two sides of the valve, and there are no dangerous pressure changes as for valve closure. The velocity of flow changes gradually, and the assumptions that the fluid is incompressible and that the pipe is rigid usually give results (by the simpler analysis of Section 12.2) to a satisfactory degree of accuracy.

Uniform opening or closing of a valve is seldom achieved in practice, and so results based on the assumption of uniform movement can only be approximations to the truth. It must be emphasized that we have here considered only the simple case of a valve discharging to atmosphere. In other instances – for example, where a further length of pipe-line is connected to the downstream side of the valve – the movement of the valve may cause changes of pressure on the downstream side also, and these would affect the rate of flow through the valve at any given moment. If the pipe-line leads to a hydraulic turbine, and the changes of velocity of the fluid are brought about by the governor mechanism, then the flow may depend on the speed of the machine, the setting of the entry vanes and other factors. In pumping systems the pressure rise resulting from 'water hammer' may cause the pump to operate as a turbine and to run in reverse at high speeds. Usually, therefore, these systems incorporate devices to circumvent such trouble.

The arithmetical integration illustrated here shows the physical basis of the 'water hammer' phenomenon, but it neglects friction and any changes in elevation of the pipe (which would cause changes of static pressure additional to the 'water hammer' effects). Various graphical methods have been developed from it, yielding results more quickly (though less accurately). Some account can be taken of friction by assuming that the corresponding head loss is concentrated at individual points rather than distributed along the whole length of the pipe, but this adds considerably to the complexity.

In the next section, therefore, we derive equations not subject to these limitations and which are suitable for solution by digital computer.

12.3.4 The method of characteristics

Since the velocity and pressure in a pipe subject to pressure transients are continuous functions of position and time, they are described essentially by partial differential equations. The method of characteristics is a general technique in which such partial differential equations are converted into particular simultaneous total differential equations which, after being expressed in finite-difference form, can be solved by computer. The effects of friction and differences of elevation can be retained in the equations and complex pipe systems can be dealt with. This method is not the only one possible for solving the equations but it is increasingly used for the study of problems involving pressure transients, so we shall here derive the finite-difference equations and outline the method to the point where a computer program can be written.

The equation of motion   Figure 12.9 shows a short length δx of the pipe through which a pressure wave travels upstream. The velocity of the fluid is assumed uniform over each cross-section and so the fluid in the length δx may be regarded as a single particle to which Newton's Second Law can be directly applied. Then the force in the direction of flow

$$pA - \left(p + \frac{\partial p}{\partial x}\delta x \right)\left(A + \frac{\partial A}{\partial x}\delta x \right) + \left(p + \frac{1}{2}\frac{\partial p}{\partial x}\delta x \right)\frac{\partial A}{\partial x}\delta x$$

$$- \tau_0 P\,\delta x - mg\sin\alpha$$

$$= \text{Mass} \times \text{Acceleration} = m\left(u\frac{\partial u}{\partial x} + \frac{\partial u}{\partial t} \right) \tag{12.14}$$

Here the third term on the left-hand side consists of the average pressure at the sides of the element multiplied by the projected area of the sides perpendicular to the flow direction. The term therefore represents the component in the flow direction of the force exerted by the pipe walls on the fluid particle. In the fourth term τ_0 represents the mean frictional shear stress at the pipe wall and P the mean perimeter. We assume that τ_0 has the same value as in steady flow at the same velocity and, from eqn 7.4, we can substitute $\tau_0 = \frac{1}{2}f\varrho u^2$. However, the friction force always opposes the motion, so, to ensure that the term changes sign as u changes sign, we put $u|u|$ in place of u^2 (where $|u|$ means the magnitude of u regardless of its sign). A variation of f with u can be incorporated into the equation if desired and so can 'minor' losses.

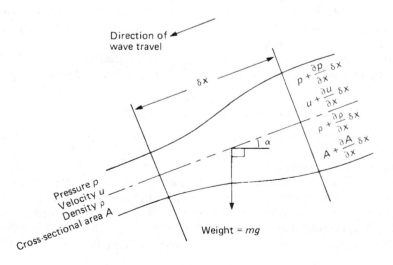

Fig. 12.9

On the right-hand side of the equation, the expression for acceleration is taken from eqn 3.1. Neglecting higher orders of small quantities, we write the mass m as $\varrho A\,\delta x$, and so, after division by $-\varrho A\,\delta x$, eqn 12.14 reduces to

$$\frac{1}{\varrho}\frac{\partial p}{\partial x} + \frac{1}{2}f\frac{u|u|}{A/P} + g\sin\alpha + u\frac{\partial u}{\partial x} + \frac{\partial u}{\partial t} = 0 \tag{12.15}$$

For the unsteady conditions being studied   *The continuity equation*

> Rate at which mass enters the volume of length δx
> = Rate at which mass leaves the volume
> + Rate of increase of mass within the volume

i.e.

$$\varrho Au = \left\{\varrho Au + \frac{\partial}{\partial x}(\varrho Au)\delta x\right\} + \frac{\partial}{\partial t}(\varrho A\,\delta x)$$

$$\therefore\ 0 = \frac{\partial}{\partial x}(\varrho Au) + \frac{\partial}{\partial t}(\varrho A)$$

$$= u\frac{\partial}{\partial x}(\varrho A) + \varrho A\frac{\partial u}{\partial x} + \frac{\partial}{\partial t}(\varrho A)$$

$$= u\frac{\partial p}{\partial x}\frac{\mathrm{d}}{\mathrm{d}p}(\varrho A) + \varrho A\frac{\partial u}{\partial x} + \frac{\partial p}{\partial t}\frac{\mathrm{d}}{\mathrm{d}p}(\varrho A) \tag{12.16}$$

Dividing eqn 12.16 by $\mathrm{d}(\varrho A)/\mathrm{d}p$ and putting

$$\frac{A}{\dfrac{\mathrm{d}}{\mathrm{d}p}(\varrho A)} = c^2 \tag{12.17}$$

we obtain

$$u\frac{\partial p}{\partial x} + \varrho c^2\frac{\partial u}{\partial x} + \frac{\partial p}{\partial t} = 0 \tag{12.18}$$

The left-hand side of eqn 12.17 equals

$$\frac{A}{A\dfrac{\mathrm{d}\varrho}{\mathrm{d}p} + \varrho\dfrac{\mathrm{d}A}{\mathrm{d}p}} = \frac{A}{\dfrac{A\varrho}{K} + \varrho\dfrac{\mathrm{d}A}{\mathrm{d}p}}$$

(from the definition of bulk modulus, eqn 1.8), so the result corresponds to eqn 12.7. That is, c represents the celerity of a small wave for which $\Delta p/K$ and $\delta A/A$ are both small compared with unity (as assumed in the derivation of eqn 12.7). However, eqn 12.17 is no more than a convenient mathematical substitution and the method is not restricted to waves of small amplitude.

The characteristic equations

Multiplying the continuity equation (12.18) by a factor λ and adding the result to the equation of motion (12.15) yields

$$\frac{1}{\varrho}\frac{\partial p}{\partial x} + \frac{1}{2}f\frac{u|u|}{A/P} + g\sin\alpha + u\frac{\partial u}{\partial x} + \frac{\partial u}{\partial t}$$

$$+ \lambda\left(u\frac{\partial p}{\partial x} + \varrho c^2\frac{\partial u}{\partial x} + \frac{\partial p}{\partial t}\right) = 0$$

i.e.

$$\left\{(u + \lambda\varrho c^2)\frac{\partial u}{\partial x} + \frac{\partial u}{\partial t}\right\} + \lambda\left\{\left(u + \frac{1}{\lambda\varrho}\right)\frac{\partial p}{\partial x} + \frac{\partial p}{\partial t}\right\}$$

$$+ \frac{1}{2}f\frac{u|u|}{A/P} + g\sin\alpha = 0 \qquad (12.19)$$

For movement in the x direction the full derivative of velocity u is (as we saw in Section 3.2)

$$\frac{\mathrm{d}u}{\mathrm{d}t} = \frac{\partial u}{\partial x}\frac{\mathrm{d}x}{\mathrm{d}t} + \frac{\partial u}{\partial t}$$

Consequently the first main bracket { } in eqn 12.19 may be written $\mathrm{d}u/\mathrm{d}t$ provided that

$$u + \lambda\varrho c^2 = \frac{\mathrm{d}x}{\mathrm{d}t} \qquad (12.20)$$

Similarly the second main bracket may be written $\mathrm{d}p/\mathrm{d}t$ if

$$u + \frac{1}{\lambda\varrho} = \frac{\mathrm{d}x}{\mathrm{d}t} \qquad (12.21)$$

The conditions 12.20 and 12.21 can be simultaneously satisfied only if

$$\lambda = \pm\frac{1}{\varrho c}, \quad \text{that is, if} \quad \frac{\mathrm{d}x}{\mathrm{d}t} = u \pm c \qquad (12.22)$$

Hence, for the positive alternative sign,

$$\frac{\mathrm{d}u}{\mathrm{d}t} + \frac{1}{\varrho c}\frac{\mathrm{d}p}{\mathrm{d}t} + \frac{1}{2}f\frac{u|u|}{A/P} + g\sin\alpha = 0 \qquad (12.23)$$

provided that

$$\frac{\mathrm{d}x}{\mathrm{d}t} = u + c \qquad (12.24)$$

and, for the negative sign,

$$\frac{\mathrm{d}u}{\mathrm{d}t} - \frac{1}{\varrho c}\frac{\mathrm{d}p}{\mathrm{d}t} + \frac{1}{2}f\frac{u|u|}{A/P} + g\sin\alpha = 0 \qquad (12.25)$$

provided that

$$\frac{\mathrm{d}x}{\mathrm{d}t} = u - c \qquad (12.26)$$

The original partial differential equations (12.15 and 12.18) have thus been converted into total differential equations known as the characteristic equations. It must be emphasized that these equations are not simultaneous. However, the conditions under which they are separately applicable can be visualized, as in Fig. 12.10, on a map where the independent variables x and t are the coordinates. Equation 12.23 holds along a line such as AP for which the slope $\mathrm{d}t/\mathrm{d}x$ has the value $(u + c)^{-1}$ given by eqn 12.24. Similarly, eqn 12.25 holds along a line such as BP with slope $(u - c)^{-1}$ (eqn 12.26). (The slope of BP is negative since c always exceeds u.) Lines such as AP, BP are known as *characteristics*.

Suppose that the conditions for the point $A(x_A, t_A, u_A, p_A)$ are all known. If Δx and Δt are both small, conditions at point P will be only slightly different and can be expressed by a finite-difference form of eqn 12.23.

$$\frac{u_P - u_A}{\Delta t} + \frac{1}{\varrho c}\frac{(p_P - p_A)}{\Delta t} + \frac{1}{2}f\frac{u_A|u_A|}{A/P} + g\sin\alpha_A = 0 \quad (12.27)$$

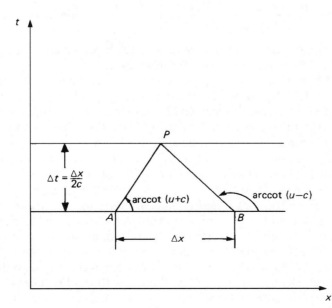

Fig. 12.10

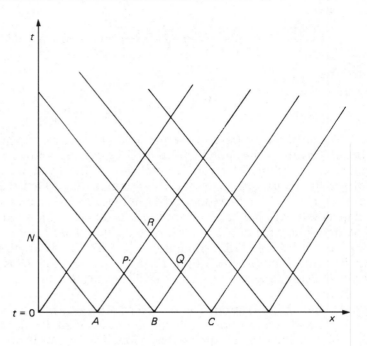

Fig. 12.11

The friction term is evaluated by taking u equal to the known value at A, and the value of α is assumed constant over the small distance $(u + c)\Delta t$ along the pipe. Similarly, if conditions at B are also known we can use eqn 12.25 in finite-difference form:

$$\frac{u_P - u_B}{\Delta t} - \frac{1}{\varrho c}\frac{(p_P - p_B)}{\Delta t} + \frac{1}{2}f\frac{u_B|u_B|}{A/P} + g\sin\alpha_B = 0 \quad (12.28)$$

Hence the particular values u_P and p_P can be determined by solving these finite-difference equations simultaneously.

The technique can clearly be extended to cover a complete x–t 'map'. We suppose that the total length of a pipe is divided into an integral number of short lengths, each Δx, as shown in Fig. 12.11. If, at time $t = 0$, all pressures and velocities are known, then, for a time interval Δt later, the conditions at points such as P and Q in the next row of intersections of characteristics can be calculated from the initial values respectively at A and B and at B and C. (The value of Δt must be $\Delta x/2c$, as shown by the geometry of Fig. 12.10.) Values for point R can then be established from those at P and Q, and similarly for the whole x–t map.

If c is constant and large compared with u, the slopes of the characteristics become $1/c$ and $-1/c$, respectively, and their intersections occur at regular intervals of x and t over the whole map. Strictly, the velocity u changes with both x and t (and c may change too if the liquid contains air or other gases because K may then change with p). The characteristics are then curved rather than straight and interpolation becomes necessary to obtain results for the intersections[1].

At the upstream end of the pipe ($x = 0$) no positive-slope characteristic   *Boundary conditions*
is available for calculating conditions at a point such as N (Fig. 12.11).
Therefore the appropriate upstream boundary condition has to be used
instead. This may be, for example, a constant value of either u or p, or
a specified variation of one of them as a function of t, or an algebraically
specified connection between u and p as for flow through a valve.
Similarly, the boundary condition at the downstream end of the pipe
($x = l$) must be used there in place of a negative-slope characteristic.
(If u is not negligible compared with c or if c is not constant, the
characteristics will not necessarily intersect on the boundary lines $x = 0$
and $x = l$. Interpolation then becomes necessary.)

At a junction of two or more pipes the equation of continuity must be
satisfied at every instant and (if any 'minor' losses are disregarded) the
pressures where each pipe meets the junction must at every instant be
the same. In each of the various finite-difference equations the same
time interval must of course be used; since $\Delta t = \Delta x/2c$ and $\Delta x = l/n$,
where n is an integer, the values of n for each pair of connecting pipes
should ideally be so chosen that

$$l_1/n_1 c_1 = l_2/n_2 c_2 \qquad (12.29)$$

If eqn 12.29 cannot be satisfied by fairly small values of n_1 and n_2, one is
faced either with excessive computing time (because large values of n
give large numbers of intervals Δx and Δt) or with using an interpolation
technique for one or more of the pipes because for a given value of Δt
the intersections of characteristics do not exactly correspond to the
intervals Δx. A compromise thus has to be reached between too much
computing time and too much complexity and loss of accuracy caused
by interpolation.

Example 12.3   A centrifugal pump running at constant speed
takes kerosene from a large reservoir and feeds it through a non-
return valve to a horizontal 150 mm diameter steel pipe, 1035 m
long, connected in series to a 200 mm diameter steel pipe, 850 m
long, laid at an upwards slope of $\alpha = $ arcsin 0.1 (Fig. 12.12).
The wall thickness of each pipe is 10 mm. The 200 mm pipe discharges through a valve to a large reservoir in which the pressure

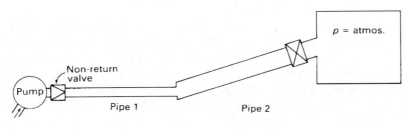

Fig. 12.12

is atmospheric. Derive the equations to determine the behaviour of the system when this valve is closed uniformly in 7.5 s.

Solution

For each pipe, the celerity of a small wave is given by eqn 12.8. Taking K and ϱ for kerosene as 1.36 GPa and 814 kg/m^3, respectively, and E for steel as 200 GPa we get, for pipe 1

$$K' = \left(\frac{1}{1.36} + \frac{150}{10 \times 200} \right)^{-1} \text{GPa} = 1.234 \text{ GPa}$$

and hence

$$c_1 = \sqrt{\left(\frac{1.234 \times 10^9}{814} \right)} \text{ m/s} = 1231 \text{ m/s}$$

For pipe 2,

$$K' = \left(\frac{1}{1.36} + \frac{200}{10 \times 200} \right)^{-1} \text{GPa} = 1.197 \text{ GPa}$$

and hence

$$c_2 = \sqrt{\left(\frac{1.197 \times 10^9}{814} \right)} \text{ m/s} = 1213 \text{ m/s}$$

In order to have the same finite time difference Δt for each pipe l/nc should be the same (eqn 12.29). i.e.

$$\frac{n_1}{n_2} = \frac{l_1 c_2}{l_2 c_1} = \frac{1035 \times 1213}{850 \times 1231} = 1.199$$

so we can take $n_1 = 6$, $n_2 = 5$. (These values will give adequate accuracy; higher values would be too expensive in computing time.)

$$\therefore \ \Delta x_1 = 1035 \text{ m}/6 = 172.5 \text{ m}; \quad \Delta x_2 = 850 \text{ m}/5 = 170 \text{ m};$$

and

$$\Delta t = \frac{172.5 \text{ m}}{2 \times 1231 \text{ m/s}} = \frac{170 \text{ m}}{2 \times 1213 \text{ m/s}} = 0.0701 \text{ s}$$

We now need values for the initial steady flow. By using Darcy's formula with an appropriate value of f for each pipe and data for pressure drop through the valve, the total steady-flow head losses can be calculated in terms of the volume flow rate Q. The sum of these losses together with the gain in elevation (850 m $\times$ 0.1 = 85 m) must equal the head provided by the pump under steady conditions. If this pump head can be expressed as a polynomial, for example

$$H = C_1 + C_2Q + C_3Q^2 \qquad (12.30)$$

then the resulting equation can be solved to give the initial value of Q. From this the initial head H at the pump can be determined; subtraction of the head losses between the pump and any point in the pipe system then enables us to calculate the head and therefore the pressure at that point. In this way we obtain the initial values of velocity and pressure at each subdivision point in the two pipes.

As an example, consider the second pipe. For interior points of this pipe (that is, points not at $x = 0$ or $x = 850\text{m}$) the finite-difference equations 12.27 and 12.28 become

$$\frac{u_P - u_A}{0.0701\,\text{s}} + \frac{p_P - p_A}{814\,\frac{\text{kg}}{\text{m}^3}\,1213\,\frac{\text{m}}{\text{s}}\,0.0701\,\text{s}} + \frac{1}{2}f\,\frac{u_A|u_A|}{\frac{1}{4} \times 0.2\,\text{m}}$$

$$+ 9.81 \times 0.1\,\frac{\text{m}}{\text{s}^2} = 0 \qquad (12.31)$$

and

$$\frac{u_P - u_B}{0.0701\,\text{s}} - \frac{p_P - p_B}{814\,\frac{\text{kg}}{\text{m}^3}\,1213\,\frac{\text{m}}{\text{s}}\,0.0701\,\text{s}} + \frac{1}{2}f\,\frac{u_B|u_B|}{\frac{1}{4} \times 0.2\,\text{m}}$$

$$+ 9.81 \times 0.1\,\frac{\text{m}}{\text{s}^2} = 0 \qquad (12.32)$$

Initially the velocities are positive and $u_A = u_B$. Simultaneous solution of these two equations gives the values u_P and p_P for an 'interior' point P after the first time interval of 0.0701 s.

For the outlet end of the pipe ($x = 850\text{m}$) there is no negative-slope characteristic bringing information about conditions at a time Δt earlier. Thus only the first of these two equations (12.31) is available, and in place of the second we use the boundary condition given by the valve 'area coefficient', for example:

$$u_P = B\sqrt{p_P} = B_0\left(1 - \frac{t}{7.5\,\text{s}}\right)\sqrt{p_P}$$

for the 'straight-line' closure assumed that $t \leqslant 7.5\text{s}$. Beyond $t = 7.5\text{s}$ (i.e. after the 107th interval Δt) the boundary condition changes to that of a completely closed end, i.e. $u_2 = 0$.

At the upstream boundary of pipe 2, the junction with pipe 1, only eqn 12.32 is available, and this must be solved using the conditions at the junction: (a) the pressures in the two pipes must be the same (the 'minor' head loss corresponding to the diameter change being neglected); (b) continuity: $u_1 = (200/150)^2u_2$. In other words, at the junction a negative-slope characteristic for pipe 2 intersects a positive-slope characteristic for pipe 1. If point

A is in pipe 1, B in pipe 2 and P at the junction, the simultaneous finite-difference equations are eqn 12.32, in which u_P is the value for pipe 2, and

$$\frac{(u_1)_P - (u_1)_A}{0.0701\,\text{s}} + \frac{p_P - p_A}{814 \times 1231 \times 0.0701\,\text{kg/m}^2}$$

$$+ \frac{1}{2}f\frac{(u_1)_A|(u_1)_A|}{\frac{1}{4} \times 0.15\,\text{m}} = 0 \qquad (12.33)$$

Also $(u_1)_P = 16(u_2)_P/9$ from continuity.

At the upstream, pump, end of pipe 1 the boundary condition that must be used together with the negative-slope characteristic for pipe 1 is that given by the pump equation (12.30):

$$p = \varrho g H = \varrho g\left(C_1 + C_2 A u + C_3 A^2 u^2\right)$$

provided that the non-return valve remains open. If at some stage the flow in the pipe reverses, then the non-return valve will shut (instantaneously, we will assume) and the boundary condition there becomes $u_1 = 0$ until the liquid again moves in the $+x$ direction.

Values of the friction factor f can be adjusted at each stage of the calculation since f is a function of Reynolds number and therefore of u.

Details of the computer programming will of course depend on the computer language used.

Pressure transients are important not only in long pipe-lines. The transmission of elastic waves is of great significance in the pipes of diesel-engine injection systems: here the compression of the fuel may exceed the volume to be injected on each stroke. Analogous phenomena with a gas may be encountered in ventilating ducts and in the exhaust systems of reciprocating engines.

12.4   SURGE TANKS

Many of the problems associated with pressure transients may be circumvented by the use of a surge tank. In hydro-electric installations, for example, the turbines must frequently be supplied with water via a long pipe-line or a tunnel cut through rock. If the electric power taken from the generators is suddenly altered, the turbines tend to change speed. The governors must counteract this tendency to change speed, by altering the flow rate to the machines. The consequent acceleration or deceleration of water in the pipe-line may give rise to 'water hammer'. The

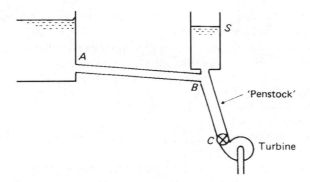

Fig. 12.13

minimizing of water hammer is doubly desirable: not only may water hammer produce dangerously high pressures within the pipe-line or tunnel, but the pressure changes impede the governing.

Figure 12.13 shows the essential features of the arrangement. The simplest type of surge tank is an open vertical cylinder of large diameter, as shown at S. It may be constructed of steel, or tunnelled in rock. Owing to the lie of the land it is seldom practicable for the entry to the surge tank to be immediately next to the turbines, but it should be as close to them as possible. The upstream pipe-line AB is of small slope, and the top of the surge tank S is higher than the water level in the reservoir. If the load on the turbines is suddenly reduced, the governing mechanism acts to decrease the rate of flow of water to them. The rate of flow in the line AB cannot at once drop to the required new value, and the temporary surplus of water goes into the surge tank S. The rise of water here then provides a hydrostatic head that decelerates the water in AB. If the required deceleration is exceptionally great the water may be allowed to overflow from the top of the surge tank; the maximum pressure in the pipe-line is then limited. The shorter length of pipe BC is still subject to water-hammer effects and so must be strongly enough constructed to withstand the increased pressures.

A no less important feature of a surge tank is that it provides a reserve supply of water to make up a temporary deficiency of flow down AB when the demand at the turbines is increased. In the absence of a surge tank the drop in pressure at the turbines could be excessive when a sudden demand required the acceleration of the water column in the supply pipe. As the water level in the surge tank is drawn down, the difference of head along AB is increased, and so the water there is accelerated until the rate of flow in AB equals that required by the turbines.

Apart from the changes of sign, the analysis of what happens when the flow rate is increased is essentially the same as that for reduction of flow rate, and it will be sufficient here to consider the reduction of flow rate in relation to a simple cylindrical surge tank. The section AB has in effect an open reservoir at each end, and so water-hammer effects there are slight. Consequently the compressibility of the liquid in AB may be neglected and the deceleration of the flow treated as a simple inertia problem similar to those discussed in Section 12.2. In *BC*, however, the flow is subject to water hammer, and so the deceleration there must be analysed separately by the methods already discussed.

We shall use the following notation (see also Fig. 12.14):

A = cross-sectional area of surge tank
a = cross-sectional area of upstream pipe-line
  (the cross-sectional area of the section *BC* may be different: that is of no consequence)
u = mean velocity in upstream pipe-line at any instant
y = depth of level in surge tank at that instant *below* datum
h_f = head lost to friction in the *upstream* pipe-line at that instant.

The continuity relation is

$$au = A\left(-\frac{dy}{dt}\right) + Q \tag{12.34}$$

where Q represents the volume rate of flow continuing along the pipe *BC* to the turbines.

A second relation is given by considering the deceleration of liquid in the upstream pipe-line. We shall assume that A is so large compared with a that the head required to decelerate the liquid in the surge tank is negligible compared with that required to decelerate the liquid in the upstream pipe-line. Velocity head, friction in the surge tank and entrance losses will also be assumed negligible. The head at B is thus equal to the vertical distance below the instantaneous water level in the surge tank.

Under steady conditions the level in the surge tank would be constant and y would exactly equal h_f. But, at any instant while the surge is taking place, the level in the tank is higher than that corresponding to steady conditions. An additional head of $(h_f - y)$ is thus available at the base of

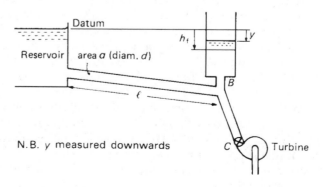

N.B. *y* measured downwards

Fig. 12.14

the tank to decelerate the liquid in the upstream pipe. From the expression for inertia head (eqn 12.1)

$$h_f - y = \frac{l}{g}(\text{deceleration}) = \frac{l}{g}\left(-\frac{du}{dt}\right) \qquad (12.35)$$

Even when $Q = 0$ (for complete shut-down of the turbines), an accurate analytical integration of eqns 12.34 and 12.35 is impossible because of the variation of the friction factor f with Reynolds number. However, the substitution of finite increments for the differentials makes numerical integration possible, and in the course of this, allowance may be made for the variation of f (and also, if need be, of Q).

Usually the problem is to determine the maximum height of the surge and the time interval required for this maximum to be reached. If the initial, steady, conditions are known, eqn 12.34 may be used to obtain dy/dt at the beginning of the surge, and hence the value of y at the end of a small time interval Δt. This value of y, substituted in eqn 12.35, gives the change of velocity Δu during the time interval Δt and hence the value of u at the end of that interval. These results are used as the starting point for similar calculations for a subsequent time interval, and the process is repeated as many times as necessary until the maximum height in the surge tank (i.e. minimum value of y) is obtained.

(For the special case $Q = 0$, the assumption of a constant value of f yields the solution

$$u^2 = \frac{2gd}{4fl}\left(y + \frac{ad}{4fA}\right) + C \exp\left(\frac{4fAy}{ad}\right) \qquad (12.36)$$

in which the constant C may be determined from the initial conditions. This equation may be applied over ranges in which f varies little, and it is useful in preliminary calculations for the entire range because it overestimates the total change of y.)

Steady conditions require the excess head $(h_f - y)$ to be zero, so the level in the tank falls immediately after the maximum height has been reached. The level then oscillates about the steady position where $y = h_f$ until the movements are damped out by friction. The oscillations of the system may be very important, particularly if the governing mechanism of the turbines is sufficiently sensitive to operate in step with them. It is of course essential that the maximum positive value of y should not be such as to empty the surge tank and so allow air to enter the penstock.

The simple cylindrical surge tank considered here has certain disadvantages. A surge tank has two principal functions – first, to minimize water-hammer effects and, second, to act as a reservoir either taking in surplus water when the demand is reduced, or meeting an increased demand while the water in the upstream pipe-line is accelerating. These two functions are in no way separated in the simple cylindrical tank, and consequently it is somewhat sluggish in operation. A number of different types of tank have therefore been devised in attempts to improve the operating characteristics for particular installations. The more complex tanks may have a cross-section varying with height, have over-flow devices or have damping arrangements such as a restriction in the entrance. Compound tanks are sometimes used, and occasionally – where a great difference of level between the ends of the upstream pipe-line makes an open tank impossible – closed tanks with compressed air above the water level are employed.

For further information on surge tanks and their operation, more specialized works should be consulted.

■ **Example 12.4** A reservoir supplies water at a steady mean velocity u to the turbine of a power plant through a long pipe in which the friction loss may be assumed to be proportional to u^2. The system is protected against high-pressure transients by means of a surge tank.

(a) If the flow to the turbine is stopped instantaneously, show that at any time t the level y in the surge tank is related to u by an equation of the form

$$2KY\left[u\frac{d^2u}{dy^2} + \left(\frac{du}{dy}\right)^2\right] - 2Ku\frac{du}{dy} + 1 = 0$$

which has the solution

$$u^2 = \frac{1}{K}\left(y + Y\right) + C\exp\left(\frac{y}{Y}\right)$$

where C, K and Y are constants.

(b) Water from a reservoir is supplied to a power plant through a pipe of diameter 0.75 m and length 1500 m at a steady flow rate of 1.2 m³/s. A surge tank of diameter 3 m is connected 100 m upstream of the turbine. If the base of the surge tank is 20 m below the free surface of the reservoir, estimate the height of tank required to accommodate instantaneous complete shut-down of the system without overflowing. The friction factor may be assumed constant and equal to 0.006.

Solution

(a) For $Q = 0$, the continuity equation is

$$au = A\left(-\frac{dy}{dt}\right)$$

The inertia head relation is

$$h_f - y = \frac{l}{g}\left(-\frac{du}{dt}\right)$$

The head loss is given by

$$h_f = \frac{4fl}{2gd}u^2 = Ku^2$$

where

$$K = \frac{4fl}{2gd}$$

Differentiation of the expression for inertia head with respect to t gives

$$\frac{dh_f}{dt} - \frac{dy}{dt} = -\frac{l}{g}\frac{d^2u}{dt^2} \quad \text{or} \quad \frac{dh_f}{dt} + \frac{a}{A}u = -\frac{l}{g}\frac{d^2u}{dt^2}$$

Differentiation of the head loss equation with respect to t yields

$$\frac{dh_f}{dt} = 2Ku\frac{du}{dt}$$

which can be combined with the previous equation to give

$$2Ku\frac{du}{dt} + \frac{a}{A}u + \frac{l}{g}\frac{d^2u}{dt^2} = 0$$

Since

$$\frac{d}{dt} = \frac{d}{dy}\frac{dy}{dt} = -\frac{a}{A}u\frac{d}{dy}$$

we can operate on the previous relation to finally obtain

$$2KY\left[u\frac{d^2u}{dy^2} + \left(\frac{du}{dy}\right)^2\right] - 2Ku\frac{du}{dy} + 1 = 0$$

where

$$2KY = \frac{la}{gA}$$

(b) $K = \dfrac{4fl}{2gd} = \dfrac{4 \times 0.006 \times (1500 - 100)\,\text{m}}{2 \times 9.81\,\text{m/s}^2 \times 0.75\,\text{m}} = 2.283\,\text{s}^2\,\text{m}^{-1}$

$2KY = \dfrac{la}{gA} = \dfrac{1400\,\text{m} \times (0.75\,\text{m})^2}{9.81\,\text{m/s}^2 \times (3\,\text{m})^2} = 8.919\,\text{s}^2$

Hence

$$Y = \frac{2KY}{2K} = \frac{8.919\,\text{s}^2}{2 \times 2.283\,\text{s}^2/\text{m}} = 1.953\,\text{m}$$

When $t = 0$

$$y_0 = h_{f0} = Ku_0^2 = 2.283\,\text{s}^2/\text{m} \times \left(\frac{1.2\,\text{m}^3/\text{s} \times 4}{\pi(0.75\,\text{m})^2}\right)^2 = 16.844\,\text{m}$$

At $t = 0$

$$u_0^2 = \frac{1}{K}(y_0 + Y) + C\exp\left(\frac{y_0}{Y}\right)$$

which, since $y_0 = Ku_0^2$, can be written as

$$\frac{Y}{K} = -C\exp\left(\frac{y_0}{Y}\right)$$

Substituting

$$\frac{1.953\,\text{m}}{2.283\,\text{s}^2/\text{m}} = -C(\text{m}^2/\text{s}^2)\exp\left(\frac{16.844\,\text{m}}{1.953\,\text{m}}\right)$$

from which $C = -1.537 \times 10^{-4}\,\text{m}^2/\text{s}^2$

To determine the height of the surge tank, we consider the condition $y = y_{max}$ when $u = 0$. Thus

$$0 = \frac{1}{K}(y_{max} + Y) + C\exp\left(\frac{y_{max}}{Y}\right)$$

or

$$\frac{1}{2.283\,\text{s}^2/\text{m}}(y_{max} + 1.953)\text{m} = 0.0001537\,\text{m}^2/\text{s}^2\,\exp\left(\frac{y_{max}}{1.953}\right)$$

The value of y_{max} is found by trial-and-error. A good first approximation is $y_{max} = -Y$. Solution: $y_{max} = -1.95\,\text{m}$.

Hence the minimum height of the surge tank $= (20 + 1.95)\,\text{m} = 21.95\,\text{m}$. The actual design height should exceed the minimum required, say 23 m.

REFERENCE

1. Fox, J. A. *Hydraulic Analysis of Unsteady Flow in Pipe Networks* (Chapter 4), Macmillan, London (1977).

FURTHER READING

Borg, J. E. Hydraulic transients, *Davis' Handbook of Applied Hydraulics*, Section 22 (edited by V. J. Zippano and H. Masen) 4th edn, McGraw-Hill, New York (1993).

Fox, J. A. *Hydraulic Analysis of Unsteady Flow in Pipe Networks*, Macmillan, London (1977).

Jaeger, C. *Fluid Transients in Hydro-electric Engineering Practice*, Blackie, Glasgow and London (1977).

Rich, G. R. *Hydraulic Transients*, (2nd edn), Dover, New York (1963).

Wylie, E. B. and Streeter, V. L. *Fluid Transients*, McGraw-Hill, New York (1978).

PROBLEMS

12.1 Verify that the inertia head was justifiably neglected in Problems 7.20 and 7.21.

12.2 A turbine which normally operates under a net head of 300 m is supplied with $2.5\,m^3$ of water per second through a pipe 1 m diameter and 1.6 km long for which $f = 0.005$. During a test on the turbine governor the flow to the turbine is gradually stopped over an interval of 8 s, the retardation of the water being proportional to $t^{5/4}$, where t represents the time measured from the beginning of the shut-down. Neglecting 'minor losses' and assuming an incompressible fluid in a rigid pipe with f independent of Reynolds number, determine the head at the turbine inlet and the velocity in the pipe at $t = 6\,s$.

12.3 A valve at the outlet end of a pipe 1 m diameter and 600 m long is rapidly opened. The pipe discharges to atmosphere and the piezometric head at the inlet end of the pipe is 23 m (relative to the outlet level). The head loss through the open valve is 10 times the velocity head in the pipe, other 'minor' losses amount to twice the velocity head, and f is assumed constant at 0.005. What velocity is reached after 12 s?

12.4 A pump draws water from a reservoir and delivers it at a steady rate of 115 litres/s to a tank in which the free surface level is 12 m higher than that in the reservoir. The pipe system consists of 30 m of 225 mm diameter pipe ($f = 0.007$) and 100 m of 150 mm diameter pipe ($f = 0.008$) arranged in series. Determine the flow rate 2 s after a failure of the power supply to the pump, assuming that the pump stops instantaneously. Neglect 'minor' losses in the pipes and in the pump, and assume an incompressible fluid in rigid pipes with f independent of Reynolds number.

12.5 A hydraulic lift cage of mass 225 kg, carrying a load of 1 Mg, is fixed on the vertical plunger of a hydraulic ram. It is counterbalanced by a 180 kg weight on a cable over a pulley above the plunger. Water from a mains supply at 2.75 MPa operates the lift through a horizontal pipe 60 m long, 40 mm diameter, $f = 0.006$. The gland friction at the plunger is

1.13 kN and the initial upward acceleration of the cage is to be 1.5 m/s^2. Calculate the plunger diameter and the maximum steady lifting speed with the full load. Neglect 'minor' losses in the pipe.

12.6 Determine the maximum time for rapid valve closure on a pipe-line 600 mm diameter, 450 m long made of steel ($E = 207$ GPa) with a wall thickness of 12.5 mm. The pipe contains benzene (relative density 0.88, $K = 1.035$ GPa) flowing at 0.85 m^3/s. It is not restricted longitudinally.

12.7 In a pipe of length 500 m and uniform circular cross-section, water flows at a steady velocity of 2 m/s and discharges to atmosphere through a valve. Under steady conditions the static head just before the valve is 300 m. Calculate the ratio of internal diameter to wall thickness of the pipe so that, when the valve is completely and instantaneously closed, the increase in circumferential stress is limited to 20 MPa, and determine the maximum time for which the closure could be described as rapid. The bulk modulus of water = 2 GPa, and the elastic modulus of the pipe material = 200 GPa.

12.8 Oil of relative density 0.85 flows at 3 m/s through a horizontal pipe 3 km long. The normal pressure at a valve at the outlet end of the pipe is 700 kPa and the effective bulk modulus of the oil is 1.24 GPa. If the amplitude of any pressure wave is not to exceed 300 kPa what is the maximum percentage increase in the area of flow tolerable during a 'rapid' opening of the valve? What is the meaning of 'rapid' in this context?

12.9 A valve at the end of a horizontal pipe 750 m long is closed in 10 equal steps each of $2l/c$ where $c = 1200$ m/s. The initial head at the valve, which discharges to atmosphere, is 144 m and the initial velocity in the pipe 3.6 m/s. Neglecting frictional effects, determine the head at the valve after 1.25, 2.50 and 3.75 s.

12.10 Show that, if the friction loss in a pipe-line is proportional to the square of the velocity, the oscillatory motion of the level in a simple, open, cylindrical surge tank following complete shut-down of the turbines in a hydro-electric plant is given by an equation of the form

$$\frac{d^2H}{dt^2} + \alpha\left(\frac{dH}{dt}\right)^2 + \beta H = 0$$

in which H represents the instantaneous height of the surge tank level above the reservoir level, and α and β are constants, α being positive or negative according as the flow along the pipe-line is towards or away from the surge tank. Determine α and β for a surge tank diameter 30 m, a pipe-line diameter 4.5 m and a length of pipe-line from reservoir to surge tank 730 m. Immediately before the turbines are shut down the rate of flow along the pipe-line is 42.5 m³/s and the level in the surge tank is stationary and 1 m below the water level in the reservoir.

12.11 An 800 mm diameter pipe 1200 m long supplies water at 1.1 m³/s to a power plant. A simple, open, cylindrical surge tank 2.5 m in diameter is connected 120 m upstream of the turbines. The base of the surge tank is 15 m below the water level in the reservoir. Neglecting entrance and other losses in the surge tank and taking f constant at 0.007, estimate the height of tank required to cope with instantaneous complete shut-down of the turbines without overflowing.

13 The principles of fluid machines

13.1 INTRODUCTION

Turbine

A fluid machine is a device either for converting the energy held by a fluid into mechanical energy or vice versa. The mechanical energy is usually transmitted by a rotating shaft: a machine in which energy from the fluid is converted directly to the mechanical energy of a rotating member is known as a *turbine* (from the Latin *turbo*, a circular motion); if, however, the initial mechanical movement is a reciprocating one the term *engine* or *motor* is used. A machine in which the converse process – the transfer of energy from moving parts to the fluid – takes place is given the general title of *pump*. When the fluid concerned is a gas other terms may be used.

Compressor

If the primary object is to increase the pressure of the gas, the machine is termed a *compressor*.

Fan

On the other hand, a machine used primarily for causing the movement of a gas is known as a *fan* or *blower*. In this case the change in static pressure is quite small – usually sufficient only to overcome the resistance to the motion – so the variation of density is negligible and the fluid may be regarded as incompressible.

We shall not consider here every kind of machine that has been devised, nor describe constructional details or the practical operation of machines. Our concern is simply with the basic principles of mechanics of fluids that are brought into play. Books giving further information on fluid machinery are listed at the end of the chapter.

Positive-displacement machines

Although a great variety of fluid machines is to be found, any machine may be placed in one of two categories: the *positive-displacement* group or the *rotodynamic* group. The functioning of a positive-displacement machine derives essentially from changes of the volume occupied by the fluid within the machine. This type is most commonly exemplified by those machines, such as reciprocating pumps and engines, in which a piston moves to and fro in a cylinder (a suitable arrangement of valves ensures that the fluid always moves in the direction appropriate to either a pump or an engine). Also in this category are diaphragm pumps, in which the change of volume is brought about by the deformation of flexible boundary surfaces (an animal heart is an example of this form of pump), and gear pumps in which two rotors similar to gear wheels mesh together within a close-fitting housing. Although hydrodynamic effects may be associated with a positive-displacement machine, the operation of the machine itself depends only on mechanical and hydrostatic principles. This is not to say that such a machine is easy to design, but since few principles of the mechanics of fluids are involved our consideration of positive-displacement machines in this book will be very brief.

Rotodynamic machines, on the other hand, do present hydrodynamic problems. All these machines have a rotor, that is, a rotating part through which the fluid passes. In a turbine this rotor is called the *runner*, for a pump the term *impeller* is more often used. The fluid has a component of velocity and therefore of momentum in a direction tangential to the rotor, and the rate at which this tangential momentum is changed corresponds to a tangential force on the rotor. In a turbine there is a reduction of the tangential momentum of the fluid in the direction of movement of the rotor; thus energy is transferred from the fluid to the rotor and hence to the output shaft. In a pump, energy from the rotor is used to increase the tangential momentum of the fluid; subsequent deceleration of the fluid produces a rise in pressure.

Rotodynamic machines

Rotodynamic machines have several advantages over the positive-displacement type. The flow from most positive-displacement machines is unsteady whereas, for normal conditions of operation, that from a rotodynamic machine is essentially steady. Most positive-displacement machines require small clearances between moving and stationary parts, and so are unsuited to handling fluids containing solid particles; in general, rotodynamic machines are not restricted in this way. If discharge from a positive-displacement pump is prevented – for example, by the closing of a valve – the pressure within the pump rises and either the pump stops or some part of the casing bursts; if the discharge valve of a rotodynamic pump is closed, however, the rotating impeller merely churns the fluid round, and the energy consumed is converted to heat. Moreover, for dealing with a given overall rate of flow a rotodynamic machine is usually less bulky than one of positive-displacement type.

13.2 RECIPROCATING PUMPS

From the point of view of mechanics of fluids a positive-displacement machine holds interest principally because of the unsteady nature of the flow. By way of illustration we shall consider briefly a reciprocating pump handling a liquid. The motion of the piston outwards (i.e. towards the right in Fig. 13.1) causes a reduction of pressure in the cylinder, and thus fluid flows into the cylinder through the inlet valve. The reverse movement of the piston causes an increase of pressure in the cylinder; the inlet valve then closes and the outlet valve opens so that fluid is discharged into the delivery pipe. Usually the operation of the valves is automatic, being controlled by the pressure in the cylinder. In other designs valves may give place to ports (i.e. apertures) in the walls of the cylinder, and these ports are covered and uncovered by the movement of the piston.

If p represents the pressure in the cylinder, and A the cross-sectional area of the piston, then the axial force exerted by the fluid on the piston is pA. If the piston moves through a small distance δx, the work done by the force is therefore $pA\,\delta x = p\delta V$ where δV represents the volume 'swept' by the piston movement. The total net work done by the pump is thus given by $\int p\,dV$, calculated round the complete cycle, that is, by the area enclosed by a graph of pressure against volume. For an

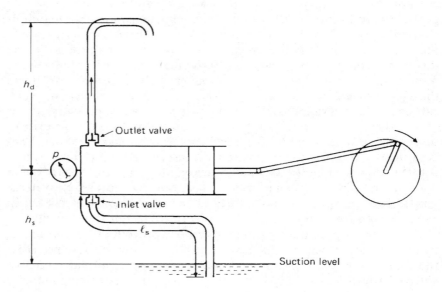

Fig. 13.1

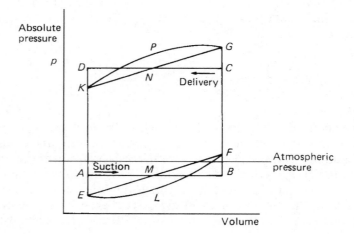

Fig. 13.2

incompressible fluid the ideal form of the diagram would be a simple rectangle (as $ABCD$ in Fig. 13.2). In practice, however, the acceleration and deceleration of the piston give rise to corresponding accelerations and decelerations of the liquid in the associated pipe-lines. At the beginning of the suction (i.e. outward) stroke of the piston, for example, the fluid in the suction (i.e. inlet) pipe has to be accelerated, and an additional difference of pressure is required. As a result the pressure in the cylinder falls by an amount corresponding to AE in Fig. 13.2, the magnitude of AE being given by eqn 12.1:

$$\text{Inertia head} = \frac{l_s}{g}(\text{Acceleration})$$

$$\text{or Inertia pressure} = \varrho l_s (\text{Acceleration})$$

where l_s represents the length of the suction pipe (not necessarily the same as h_s). Moreover, by continuity:

$$\text{Acceleration of liquid in pipe} = \text{Acceleration of piston} \times A/a_s$$

where A represents the cross-sectional area of the piston and a_s the cross-sectional area of the suction pipe. To decelerate the liquid at the end of the suction stroke, a corresponding rise of pressure in the cylinder is needed, so, as a consequence of the inertia of the liquid in the suction pipe, the base of the pressure-volume diagram is modified from AB to EMF. (For simple harmonic motion of the piston, EMF is a straight line.) A further modification of the diagram results from the effect of friction and other losses in the suction pipe. These are zero at the ends of the stroke when the velocity is zero, and a maximum at mid-stroke (again for simple harmonic motion of the piston) when the velocity is at its maximum. The base of the diagram therefore becomes ELF. Inertia and friction in the delivery pipe cause similar modifications to the upper part of the diagram, and so the actual shape is $ELFGPK$. The effects of inertia and friction in the cylinder itself are normally negligible. It should be noted that maximum frictional losses occur when the inertia head is zero (mid-stroke), and the maximum inertia effect occurs when the frictional losses are zero (ends of stroke).

The speed of such a pump is usually restricted by the pressure corresponding to the point E of the diagram. The higher the speed the greater the accelerations during each stroke and the lower the pressure in the cylinder at the beginning of the suction stroke. The pressure must not be allowed to fall to the value at which dissolved gases are liberated from the liquid. Under these conditions a cavity would form in the suction pipe and pumping would temporarily cease. A more serious matter would be the collapse of the cavity later in the stroke; the liquid would then rush towards the cylinder with destructive violence. 'Water-hammer' effects resulting from the sudden movement of the valves may sometimes be of importance too.

The pulsations of the flow in either the suction or delivery pipe may be largely eliminated by connecting a large, closed air-vessel to the pipe, at a point close to the pump. Its action is analogous to that of the surge-tank discussed in Section 12.4. Such a vessel fitted to the suction pipe (see Fig. 13.3) would obviate the changes of pressure in the cylinder due to the inertia of the liquid in the suction pipe. With a practically steady velocity in the pipe the friction loss would be constant and less than the previous maximum value. The lower part of the pressure–volume diagram would therefore be modified as shown in Fig. 13.4, *AE'* corresponding to the 'steady' friction loss based on the steady mean velocity. Since the point *E'* corresponds to a higher pressure than *E* on Fig. 13.2, the restriction on speed is made much less severe by the fitting of an air-vessel on the suction pipe.

In multi-cylinder pumps a number of cylinders are connected in parallel, their cranks being equally spaced over 360°. The fluctuating discharges from the individual cylinders are thus added together as indicated in Fig. 13.5. The (approximately) simple harmonic variations of flow during the delivery strokes of the individual cylinders here combine to produce only a 'ripple' on an otherwise steady total discharge. In the example illustrated, there are three cylinders with cranks at 120° to one another. For 60° of rotation (between 90° and 150°, for example) one cylinder gives nearly its maximum delivery while the other two are open to suction. Then for the next 60° the falling output

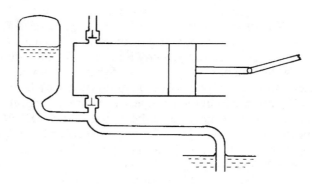

Fig. 13.3 Air vessel fitted to suction pipe.

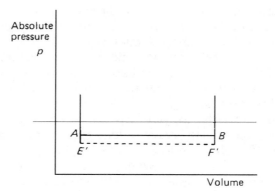

Fig. 13.4

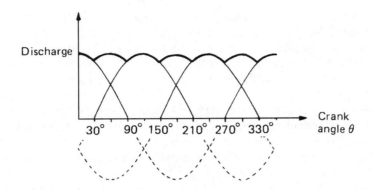

Fig. 13.5 Discharge from a three-cylinder reciprocating pump.

of the first cylinder is augmented by the increasing output of one of the others.

The benefits resulting from the much smaller fluctuations of velocity in both delivery and suction pipes of a multi-cylinder pump are thus similar to those derived from the use of a large air vessel, and so this is then unnecessary.

In practice the discharge for each working stroke of a reciprocating pump differs slightly from the volume displaced by the piston movement. This is a consequence of leakage and the imperfect operation of the valves. A 'coefficient of discharge' is thus introduced (equal to the ratio of the actual discharge per working stroke to the swept volume). The discrepancy is alternatively expressed as a 'percentage slip' where

$$\% \text{ slip} = (1 - C_d) \times 100$$

Reciprocating pumps, and positive-displacement pumps generally, are most suitable for low rates of flow and particularly for high pressures. For greater rates of flow and lower pressures rotodynamic machines are usually more satisfactory.

13.3 ROTODYNAMIC MACHINES

We now turn attention to those machines that depend for their functioning on principles of fluid dynamics. They are distinguished from positive-displacement machines in requiring relative motion between the fluid and the moving part of the machine. The latter consists of a rotor having a number of vanes or blades, and there is a transfer of energy between the fluid and the rotor. Whether the fluid does work on the rotor (as in a turbine) or the rotor does work on the fluid (as in a pump), the machine may be classified in the first instance according to the main direction of the fluid's path in the rotor. In a *radial-flow* machine the path is wholly or mainly in the plane of rotation; the fluid enters the rotor at one radius and leaves it at a different radius. Examples of this type of machine are the Francis turbine and the centrifugal pump. If, however, the main flow direction is parallel to the axis of rotation, so that any fluid particle passes through the rotor at a practically constant radius, then the machine is said to be an *axial-flow* machine. The Kaplan turbine and the 'propeller' or axial-flow pump are examples of this type. If the flow is partly radial and partly axial the term *mixed-flow* machine is used.

There are considerable similarities between turbines and pumps, and several of the formulae we shall derive are applicable to both types of machine. Indeed, a pump, for example, may be operated 'in reverse' as a turbine; dual-purpose machines having this facility are used in 'hydraulic pumped storage' schemes. (These are arrangements whereby, during periods of small demand for electric power – for example, at night – a dual-purpose machine driven by an electric motor pumps water to a high-level reservoir. At periods of peak demand the machine runs as a turbine and the electric motor as an alternator so that power is fed back to the electricity supply.) In efficiency, however, such dual-purpose machines are somewhat inferior to those intended only for one-way conversion of energy. We shall refer to the similarities between turbines and pumps again but, for the sake of explicitness, we shall fix attention first on turbines.

13.3.1 Types of turbine

As we have seen, one classification of turbines is based on the predominant direction of the fluid flow through the runner. In addition, turbines may be placed in one of two general categories: (a) impulse and (b) reaction. (These names have little justification except long usage: they should be regarded as no more than useful labels.) In both types the fluid passes through a runner having blades. The momentum of the fluid in the tangential direction is changed and so a tangential force on the runner is produced. The runner therefore rotates and performs useful work, while the fluid leaves it with reduced energy. The important feature of the impulse machine is that there is no change of static pressure across the runner. In the reaction machine, on the other hand, the static pressure decreases as the fluid passes through the runner.

For any turbine the energy held by the fluid is initially in the form of pressure. For a turbine in a hydro-electric scheme, water comes from a high-level reservoir: in a mountainous region several hundred metres head may thus be available, although water turbines are in operation in other situations where the available head is as low as three metres or less. For a steam turbine, the pressure of the working fluid is produced by the addition of heat in a boiler; in a gas turbine pressure is produced by the chemical reaction of fuel and air in a combustion chamber.

The impulse turbine has one or more fixed nozzles, in each of which this pressure is converted to the kinetic energy of an unconfined jet. The jets of fluid then impinge on the moving blades of the runner where they lose practically all their kinetic energy and, ideally, the velocity of the fluid at discharge is only just sufficient to enable it to move clear of the runner. As already mentioned, the term 'impulse' has little justification: 'constant pressure' would perhaps be better. In a reaction machine the change from pressure to kinetic energy takes place gradually as the fluid moves through the runner, and for this gradual change of pressure to be possible the runner must be completely enclosed and the passages in it entirely full of the working fluid.

Machines in which the fluid undergoes an appreciable change of density involve thermodynamic principles also, but in this book we shall confine our attention to those using constant-density fluids and operating under steady conditions. We shall deal first with impulse turbines since they are sufficiently different from reaction machines to justify separate consideration, and are in many ways simpler.

13.3.2 The Pelton wheel

This is the only hydraulic turbine of the impulse type now in common use and is named after Lester A. Pelton (1829–1908), the American engineer who contributed much to its development in about 1880. It is an efficient machine, particularly well suited to high heads.

The rotor consists of a circular disc with several (seldom less than 15) spoon-shaped 'buckets' evenly spaced round its periphery (see Figs. 13.6 and 13.7). One or more nozzles are mounted so that each directs a jet along a tangent to the circle through the centres of the buckets. Down the centre of each bucket is a 'splitter' ridge, which divides the oncoming jet into two equal portions and, after flowing round the smooth inner surface of the bucket, the fluid leaves it with a relative velocity almost opposite in direction to the original jet. The notch in the outer rim of each bucket (see Fig. 13.7) prevents the jet to the preceding bucket being intercepted too soon; it also avoids the deflection of the

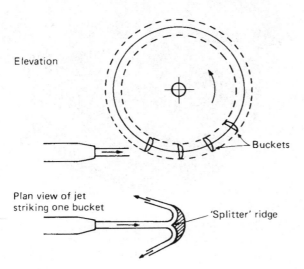

Fig. 13.6 Single-jet Pelton wheel.

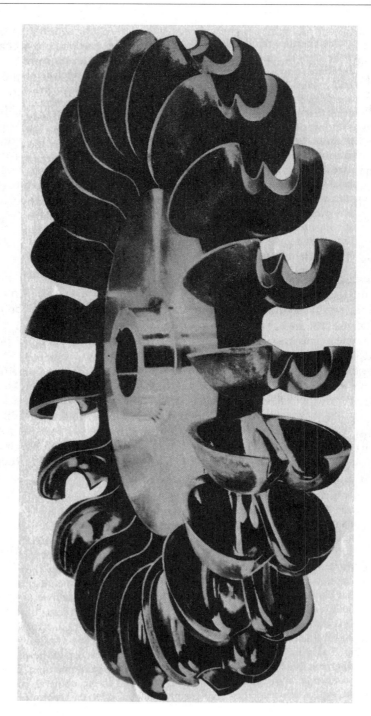

Fig. 13.7 Runner of Pelton wheel. (By courtesy of Biwater Industries Ltd.)

fluid towards the centre of the wheel as the bucket first meets the jet. The maximum change of momentum of the fluid – and hence the maximum force driving the wheel round – would be obtained if the bucket could deflect the fluid through 180°. In practice, however, the deflection is limited to about 165° if the fluid leaving one bucket is not to strike the back of the following one.

It should be noted that the flow only partly fills the buckets, and the fluid remains in contact with the atmosphere. Thus, once the jet has been produced by the nozzle, the static pressure of the fluid is atmospheric throughout the machine.

As for all rotodynamic machinery, it is important to distinguish clearly between so-called absolute velocities (i.e. velocities relative to the earth) and the velocities of the fluid relative to the moving blades. Here, although the fluid leaves the buckets with a high relative velocity, its *absolute* velocity is usually small so that little kinetic energy is wasted.

Usually the shaft of a Pelton wheel is horizontal, and then not more than two jets are used. If the wheel is mounted on a vertical shaft a larger number of jets (up to six) is possible. The nozzles, however, must never be spaced so closely that the spent fluid from one jet interferes with another jet. Because of the symmetry of the buckets, the side thrusts produced by the fluid in each half should balance – although it is usual for small thrust bearings to be fitted on the shaft to cope with lapses from this ideal.

The transfer of work from the fluid to the buckets takes place according to the Momentum Equation, and so it is necessary to examine the rate at which the momentum of the fluid is changed. Since momentum is a vector quantity, the process may be indicated geometrically, the vector quantity being represented by a line (the vector) whose length corresponds to the magnitude of the quantity.

In the study of rotodynamic machinery the construction and interpretation of vector diagrams are so important that a brief reminder of the principles on which such diagrams are based will not be out of place. The sense of a vector's direction is suitably indicated by an arrow-head on the vector. Addition of quantities is then represented by placing the corresponding vectors together 'nose to tail' (see Fig. 13.8a). The

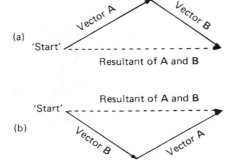

Fig. 13.8 The resultant in (b) is the same as in (a) although vectors **A** and **B** are used in different order.

resultant vector is equivalent to the sum of the others because, from the same starting point, the same destination is reached. It will be noticed (Fig. 13.8b) that the order in which the individual vectors are placed is immaterial. The subtraction of vector quantities may be represented by placing the vectors together 'nose to nose' or 'tail to tail'. In Fig. 13.8, for example, the quantity **B** is the difference between the resultant and the quantity **A**; diagram (a) shows the resultant vector and vector **A** placed 'tail to tail', and diagram (b) shows them placed 'nose to nose'. It is only the arrow-heads that indicate whether vector quantities are being added or subtracted, so it is essential to include the arrow-heads in every vector diagram. To investigate the changes of momentum that concern us here, we could draw diagrams in which the vectors represent amounts of momentum. But under steady conditions the mass of fluid flowing through the machine in unit time is constant, and if the velocities are sensibly uniform over the appropriate cross-sections then we may simplify the analysis by letting the vectors represent velocities. When a change of velocity has been determined, multiplication by the rate of mass flow through the machine gives the corresponding rate of change of momentum.

The resultant force corresponding to the change of momentum of the fluid may be in any direction, but only that component of it in the direction of movement of the rotor causes the rotation, and only this component of the total force does any work. We are principally concerned, then, with changes of momentum in a direction tangential to the periphery of the rotor. This direction is usually known as the direction of whirl, and the component of the absolute velocity of the fluid in this direction is known as the *velocity of whirl*.

To return to the Pelton wheel: Fig. 13.9 shows a section through a bucket that is being acted on by a jet. The plane of section is parallel to the axis of the wheel and contains the axis of the jet. The absolute velocity of the jet is determined by the head available at the nozzle, that is, the gross head H_{gr} minus the head loss h_f due to friction in the pipeline. The other symbols we shall use are these:

v_1 = absolute velocity of jet just before striking bucket
v_2 = absolute velocity of fluid leaving bucket

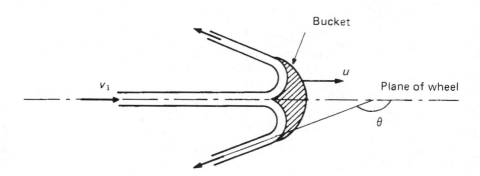

Fig. 13.9

ω = angular velocity of wheel

r = radius from axis of wheel to axis of jet striking bucket

u = absolute velocity of bucket at this radius = ωr

R_1 = velocity of oncoming jet *relative to bucket*

R_2 = velocity of fluid leaving bucket *relative to bucket*

θ = angle through which fluid is deflected by bucket

C_v = coefficient of velocity for the nozzle – usually between 0.97 and 0.99

Q = volume rate of flow from nozzle

ϱ = density of fluid

The jet velocity v_1 is given by $C_v\sqrt{(2gH)}$ where H represents the net head, $H_{gr} - h_f$. The velocity head of the fluid in the pipe-line is normally negligible compared with H.

During the time that any one bucket is being acted on by the jet the wheel turns through a few degrees and so the direction of motion of the bucket changes slightly. The effect of this change, however, is small, and it is sufficient here to regard the direction of the bucket velocity u as the same as that of v_1. Since the radius of the jet is small compared with that of the wheel, all the fluid may be assumed to strike the bucket at radius r. It is also assumed that all the fluid leaves the bucket at radius r and that the velocity of the fluid is steady and uniform over sections 1 and 2 where the values v_1 and v_2 are considered.

The relative velocity R_1 at the moment when the fluid meets the bucket is given by $R_1 = v_1 - u$. (Since v_1 and u are collinear, the diagram of velocity vectors is simply a straight line as in Fig. 13.10). The relative velocity R_2 with which the fluid leaves the bucket is somewhat less than the initial relative velocity R_1. There are two reasons for this. First, although the inner surfaces of the buckets are polished so as to minimize frictional losses as the fluid flows over them, such losses cannot be entirely eliminated. Second, some additional loss is inevitable as the fluid strikes the splitter ridge, because the ridge cannot have zero thickness. These losses of mechanical energy reduce the relative velocity between fluid and bucket. We therefore write $R_2 = kR_1$ where k is a fraction slightly less than unity.

(a) Inlet to bucket

(The vectors are actually colinear, but are shown slightly separated here for clarity.)

(b) Outlet from bucket

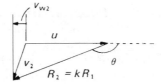

Fig. 13.10

As the bucket is symmetrical it is sufficient to consider only that part of the flow which traverses one side of it. The diagram of velocity vectors at outlet is therefore the triangle in Fig. 13.10. To obtain the absolute velocity of the fluid at discharge from the bucket, we must add (vectorially) the relative velocity R_2 to the bucket velocity u. The velocity of whirl at outlet, v_{w2}, is the component of v_2 in the direction of the bucket movement. The direction of u being taken as positive, $v_{w2} = u - R_2\cos(\pi - \theta)$. At inlet the velocity of whirl is v_1, so the change in the whirl component is

$$\Delta v_w = v_1 - \{u - R_2\cos(\pi - \theta)\} = R_1 + R_2\cos(\pi - \theta)$$
$$= R_1(1 - k\cos\theta) \tag{13.1}$$

The mass flow rate in the jet $= Q\varrho$ and so the rate of change of momentum in the whirl direction $= Q\varrho(\Delta v_w)$. This corresponds to the force driving the wheel round. The torque on the wheel is therefore $Q\varrho(\Delta v_w)r$ and the power output is $Q\varrho(\Delta v_w)r\omega = Q\varrho(\Delta v_w)u$.

The energy arriving at the wheel is in the form of kinetic energy of the jet, and is given by $\frac{1}{2}Q\varrho v_1^2$ per unit time. Therefore, the wheel efficiency,

$$\eta_w = \frac{Q\varrho(\Delta v_w)u}{\frac{1}{2}Q\varrho v_1^2} = \frac{2u(\Delta v_w)}{v_1^2}$$

Substituting for Δv_w from eqn 13.1 and putting $R_1 = v_1 - u$ gives

$$\eta_w = \frac{2u(v_1 - u)(1 - k\cos\theta)}{v_1^2} \tag{13.2}$$

which, if k is assumed constant, is a maximum when $u/v_1 = \frac{1}{2}$.

The wheel efficiency represents the effectiveness of the wheel in converting the kinetic energy of the jet into mechanical energy of rotation. Not all this energy of rotation is available at the output shaft of the machine, because some is consumed in overcoming friction in the bearings and some in overcoming the 'windage', that is the friction between the wheel and the atmosphere in which it rotates. In addition to these losses there is a loss in the nozzle (which is why C_v is less than unity). The overall efficiency is therefore less than the wheel efficiency. Even so, an overall efficiency of 85–90% may usually be achieved in large machines. Moreover, as the losses due to bearing friction and windage increase rapidly with speed, the peak of overall efficiency occurs when the ratio u/v_1 (often termed the *speed ratio*) is slightly less than the theoretical value of 0.5; the figure usually obtained in practice is about 0.46.

Equation 13.2 indicates that a graph of efficiency against bucket velocity is parabolic in form, as illustrated in Fig. 13.11.

The foregoing analysis has been idealized in many respects. The behaviour of the fluid is not strictly steady since, as the buckets successively come into the jet, conditions on any one bucket are varying

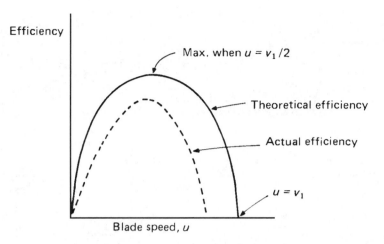

Fig. 13.11

with time. Also there is often considerable scattering and splashing of the fluid on leaving the buckets. This simplified analysis, however, does show the essential features of the functioning of this type of turbine.

A Pelton wheel is almost invariably used to drive an electrical generator mounted on the same shaft. It is designed to operate at the conditions of maximum efficiency, and the governing of the machine must be such as to allow the efficiency to be maintained even when the power demand at the shaft varies. No variation of the angular velocity, and hence of bucket velocity u, can normally be permitted (for this would alter the frequency of the electrical output). The control must therefore be in the volume rate of flow Q, and yet there must be no change in the jet velocity because that would alter the speed ratio u/v_1 from its optimum value of about 0.46. Since $Q = Av_1$ it follows that the control must be effected by a variation of the cross-sectional area A of the jet. This is usually achieved by a spear valve in the nozzle (Fig. 13.12a). Movement of the spear along the axis of the nozzle increases or decreases the annular area between the spear and the housing. The spear is so shaped, however, that the fluid coalesces into a circular jet and the effect of the spear movement is to vary the diameter of the jet. Sudden reduction of the rate of flow could result in serious water-hammer problems, and so deflectors (Fig. 13.12b) are often used in association with the spear valve. These plates temporarily deflect the jet so that not all of the fluid reaches the buckets; the spear valve may then be moved slowly to its new position and the rate of flow in the pipe-line reduced gradually. Diffusing plates in the surface of the spear are sometimes used for the same purpose.

In the design of a Pelton wheel, two parameters are of particular importance: the ratio of the bucket width to the jet diameter, and the ratio of the wheel diameter to the jet diameter. If the bucket width is too small in relation to the jet diameter, the fluid is not smoothly deflected

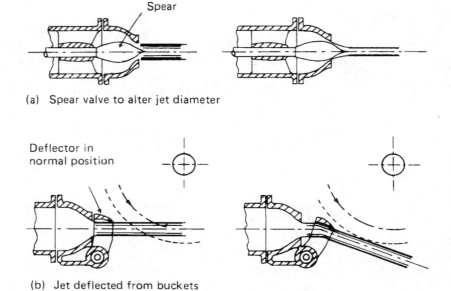

(a) Spear valve to alter jet diameter

(b) Jet deflected from buckets

Fig. 13.12

by the buckets and, in consequence, much energy is dissipated in turbulence and the efficiency drops considerably. On the other hand, if the buckets are unduly large, friction on the surfaces is unnecessarily high. The optimum value of the ratio of bucket width to jet diameter has been found to be between 4 and 5. The ratio of wheel diameter to jet diameter has in practice a minimum value of about 10; smaller values involve either too close a spacing of the buckets or too few buckets for the whole jet to be used. There is no upper limit to the ratio, but the larger its value the more bulky is the entire installation; the smaller ratios are thus usually desirable.

Since $v_1 = C_v\sqrt{(2gH)}$ the jet velocity is determined by the head available at the nozzle, and the bucket velocity u for maximum efficiency is given as approximately $0.46\,v_1$. Then, since $u = \omega r$, the radius of the pitch circle of the buckets may be calculated for a given shaft speed. The required power output P determines the volume rate of flow Q since

$$P = Q\varrho gH\eta_o$$

where η_o represents the overall efficiency of the turbine. Then, with v_1 already determined, the total cross-sectional area of the jets is given by Q/v_1. It is worth re-emphasizing here that H represents the head available *at the nozzles*, i.e. the gross head of the reservoir less the head lost to friction in the supply pipe. The overall efficiency quoted for a machine always refers to the ability of the machine itself to convert fluid energy to useful mechanical energy: the efficiency thus accounts for losses in the machine but not for losses that occur before the fluid reaches it.

Although the Pelton wheel is efficient and reliable when operating under large heads, it is less suited to smaller heads. To develop a given output power under a smaller head the rate of flow would need to be greater, with a consequent increase in jet diameter. (True, a greater rate of flow can be achieved by the use of more jets, but their number is usually limited to four – occasionally six.) The increase of jet diameter in turn requires an increase of wheel diameter. Since, moreover, the jet and bucket velocities are reduced as the head is reduced, the machine becomes very bulky and slow-running. In fact, for lower heads, turbines of the reaction type are more suitable.

13.3.3   Reaction turbines

The principal distinguishing features of a reaction turbine, we recall, are that only part of the overall head is converted to velocity head before the runner is reached, and that the working fluid, instead of engaging only one or two blades at a time (as in an impulse machine), completely fills all the passages in the runner. Thus the pressure of the fluid changes gradually as it passes through the runner. Figures 13.13 and 13.14 illustrate a *Francis* turbine, a radial-flow machine of the kind developed by the American engineer James B. Francis (1815–92).

Although some smaller machines of this type have horizontal shafts, the majority have vertical shafts as shown in the figure. The fluid enters a spiral casing (called a *volute* or *scroll case*) which completely surrounds the runner. The cross-sectional area of the volute decreases along the fluid path in such a way as to keep the fluid velocity constant

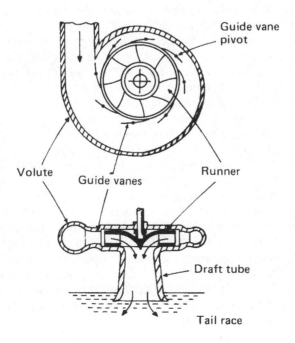

Fig. 13.13 Radial-flow (Francis) turbine. Large turbines may also have a 'stay ring' of fixed vanes outside the ring of guide vanes. The main function of the stay vanes is to act as columns helping to support the weight of the electrical generator above the turbine. They are so shaped as to conform to the streamlines of the flow approaching the guide vanes.

Fig. 13.14 Runner of Francis turbine being lowered into the volute casing during works assembly. The guide vanes and their pivots can also be seen. (By courtesy of Biwater Industries Ltd.)

in magnitude. From the volute the fluid passes between stationary guide vanes mounted all round the periphery of the runner. The function of these guide vanes is to direct the fluid on to the runner at the angle appropriate to the design. Each vane is pivoted and, by a suitable mechanism, all may be turned in synchronism so as to alter the flow rate through the machine, and hence the power output, as required by the governing gear. These vanes are also known as *wicket gates*. In its passage through the runner the fluid is deflected by the runner blades so that its angular momentum is changed.

From the centre of the runner the fluid is turned into the axial direction and flows to waste via the *draft tube*. The lower end of the draft tube must, under all conditions of operation, be submerged below the level of the water in the *tail race*, that is, the channel carrying the used water away. Only in this way can it be ensured that a hydraulic turbine is full of water. A carefully designed draft tube is also of value in gradually reducing the velocity of the discharged water so that the kinetic energy lost at the outlet is minimized.

The properties of the machine are such that, at the 'design point', the absolute velocity of the fluid leaving the runner has little, if any, whirl component. The Francis turbine is particularly suitable for medium heads (that is, from about 15 to 300 m) and overall efficiencies exceeding 90% have been achieved for large machines.

An inward-flow turbine such as this has the valuable feature of being to some extent self-governing. This is because a 'centrifugal head' like that in a forced vortex is developed in the fluid rotating with the runner. The centrifugal head balances part of the supply head. If for any reason the rotational speed of the runner falls, the centrifugal head also falls, with the result that a higher rate of flow through the machine is possible and the speed of the runner rises again. The converse action results from an increase of speed.

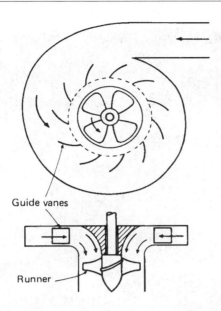

Guide vanes

Runner

Fig. 13.15 Axial-flow (propeller) turbine.

The runner of a reaction turbine is always full of the working fluid, whereas in an impulse machine only a few of the runner blades are in use at any one moment. The reaction turbine is therefore able to deal with a larger quantity of fluid for a given size of runner. For a runner of given diameter the greatest rate of flow possible is achieved when the flow is parallel to the axis. Such a machine is known as an *axial-flow* reaction turbine or *propeller* turbine.

From Fig. 13.15 it will be seen that the arrangement of guide vanes for a propeller turbine is (usually) similar to that for a Francis machine. The function of the vanes is also similar: to give the fluid an initial motion in the direction of whirl. Between the guide vanes and the runner, the fluid in a propeller turbine turns through a right-angle into the axial direction and then passes through the runner. The latter usually has four or six blades and closely resembles a ship's propeller. Apart from frictional effects, the flow approaching the runner blades is that of a free vortex (whirl velocity inversely proportional to radius) whereas the velocity of the blades themselves is directly proportional to radius. To cater for the varying relation between the fluid velocity and the blade velocity as the radius increases, the blades are twisted as shown in Fig. 13.16, the angle with the axis being greater at the tip than at the hub. The blade angles may be fixed if the available head and the load are both fairly constant, but where these quantities may vary a runner is used on which the blades may be turned about their own axes while the machine is running. When both guide-vane angle and runner-blade angle may thus be varied, a high efficiency can be maintained over a wide range of operating conditions. Such a turbine is known as a *Kaplan* turbine after its inventor, the Austrian engineer Viktor Kaplan (1876–1934) (see Fig. 13.16).

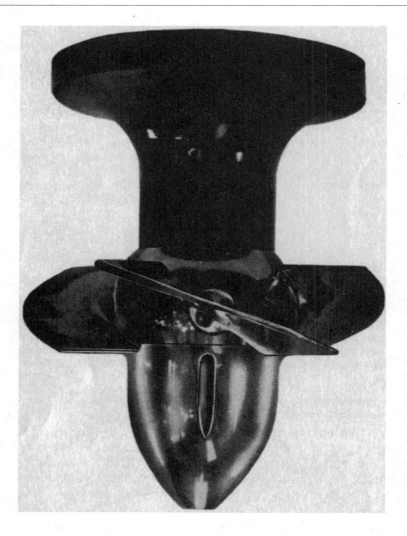

Fig. 13.16 Runner of Kaplan turbine. (By courtesy of Biwater Industries Ltd.)

In addition to the Francis (radial-flow) turbine and the axial-flow type are so-called mixed-flow machines in which the fluid enters the runner in a radial direction and leaves it with a substantial axial component of velocity. In such machines, radial and axial flow are combined in various degrees and there is a continuous transition of runner design from radial-flow only to axial-flow only.

Net head across a reaction turbine

The effective head across any turbine is the difference between the head at inlet to the machine and the head at outlet from it. As a reaction turbine must operate 'drowned', that is, completely full of the working fluid, a draft tube is fitted, and so important is the function of the draft tube that it is usually considered as part of the turbine. The kinetic energy of the fluid finally discharged into the tail race is wasted: the draft tube is therefore made divergent so as to reduce the velocity at outlet to a minimum. The angle between the walls of the tube and the axis is limited, however, to about 8° so that the flow does not separate from the walls and thereby defeat the purpose of the increase in cross-sectional area. Flow conditions within the draft tube may be studied by applying the energy equation between any point in the tube and a point in the tail race. The pressure at outlet from the runner is usually less than the atmospheric pressure of the final discharge; the turbine should therefore not be set so high above the tail water that the pressure at outlet from the runner falls to such a low value that cavitation (see Section 13.3.6) occurs.

The net head across the machine corresponds, then, to the total difference in level between the supply reservoir and the tail water, minus the losses external to the machine (that is, those due to pipe friction and the kinetic head at outlet from the draft tube). Figure 13.17 indicates that the net head across the turbine is

$$H = \text{total head at inlet to machine } (C)$$
$$- \text{ total head at discharge to tail race}$$

$$= \frac{p_C}{\varrho g} + \frac{v_C^2}{2g} + z_C - \frac{v_E^2}{2g} \qquad (13.3)$$

It will be noticed that, for a given difference between the levels of the supply reservoir and the tail water, the net head across a reaction turbine is greater than that for an impulse machine. The discharge from the runner of an impulse machine is necessarily at atmospheric pressure, and so the portion z_C of the total difference of levels is not then available for use by the turbine. For the high heads normally used by impulse turbines, however, this loss is not of great importance.

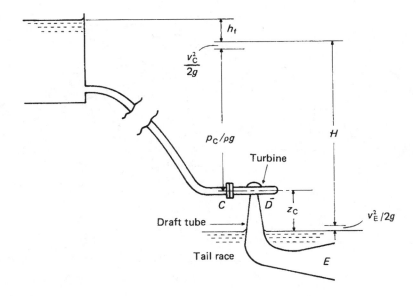

Fig. 13.17

13.3.4 Basic equations for rotodynamic machinery

No fundamentally new relations are required to study the flow through rotodynamic machines. The equation of continuity, the momentum equation and the general energy equation are used, but certain special forms of these relations may be developed. In particular an expression is required for the transfer of energy between the fluid and the rotor, and consideration is here restricted to steady conditions. The relation we shall obtain applies in principle to any rotor whatever. For the sake of explicitness, however, Fig. 13.18 represents the runner of a Francis turbine in which the fluid velocities are entirely in the plane of rotation. We use the following symbols:

v = absolute velocity of fluid
u = peripheral velocity of blade at point considered
R = relative velocity between fluid and blade
v_w = velocity of whirl, i.e. component of absolute velocity of fluid in direction tangential to runner circumference
r = radius from axis of runner
ω = angular velocity of runner
Suffix 1 refers to conditions at inlet to runner
Suffix 2 refers to conditions at outlet from runner.

The only movement of the runner blades is in the circumferential direction, and so only force components in this direction perform work. Our present concern, therefore, is with changes of momentum of the fluid in the circumferential direction: there may be changes of momentum in other directions also but the corresponding forces have no moments about the axis of rotation of the rotor. From the relation

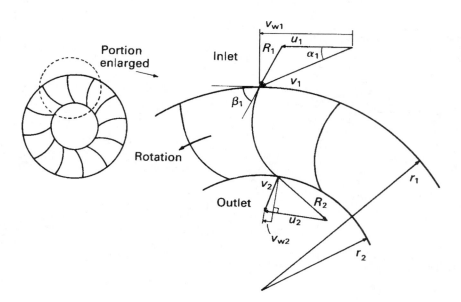

Fig. 13.18 Portion of a Francis turbine runner.

Torque about a given fixed axis = Rate of increase of angular
momentum about that axis

the torque on the fluid must be equal to the angular momentum of the
fluid leaving the rotor per unit time *minus* the angular momentum of the
fluid entering the rotor per unit time.

At inlet, any small particle of fluid, of mass δm, has momentum $\delta m\, v_{w1}$
in the direction tangential to the rotor. Its angular momentum (that is,
moment of momentum) is therefore $\delta m\, v_{w1} r_1$. Suppose that, of the total
(constant) mass flow rate $\dot{m}$, a part $\delta \dot{m}$ passes through a small element
of the inlet cross-section across which values of v_{w1} and r_1 are uniform.
Then the rate at which angular momentum passes through that small
element of inlet cross-section is $\delta \dot{m}\, v_{w1} r_1$, and the total rate at which
angular momentum enters the rotor is $\int v_{w1} r_1\, \mathrm{d}\dot{m}$, the integral being
taken over the entire inlet cross-section. Similarly, the total rate at
which angular momentum leaves the rotor is $\int v_{w2} r_2\, \mathrm{d}\dot{m}$, this integral
being evaluated for the entire outlet cross-section. The rate of increase
of angular momentum of the fluid is therefore

$$\int v_{w2} r_2\, \mathrm{d}\dot{m} - \int v_{w1} r_1\, \mathrm{d}\dot{m}$$

and this equals the torque exerted on the fluid. If there are no shear
forces at either inlet or outlet cross-section that have a moment about
the axis of the rotor, then this torque on the fluid must be exerted by the
rotor. By Newton's Third Law, a change of sign gives the torque exerted
on the rotor by the fluid:

$$T = \int v_{w1} r_1\, \mathrm{d}\dot{m} - \int v_{w2} r_2\, \mathrm{d}\dot{m} \tag{13.4}$$

Equation 13.4 was, in principle, first given by Leonhard Euler* (1707–
83) and is sometimes known as Euler's equation. It is a fundamental
relation for all forms of rotodynamic machinery – turbines, pumps, fans
or compressors. Although the equation has here been developed for a
rotor, it applies also to a stationary member (stator) through which the
angular momentum of the fluid is changed. A torque equal and opposite
to T has to be applied to a stator – usually by fixing bolts – to prevent its
rotation.

It is worth emphasizing that eqn 13.4 is applicable regardless of
changes of density or components of velocity in other directions.
Moreover, the shape of the path taken by the fluid in moving from inlet
to outlet is of no consequence: the expression involves only conditions
at inlet and outlet. In particular, it is independent of losses by turbulence, friction between the fluid and the blades, and changes of temperature. True, these factors may affect the velocity of whirl at outlet: they
do not, however, undermine the truth of eqn 13.4.

The torque *available* from the shaft of a turbine is somewhat less than
that given by eqn 13.4 because of friction in bearings and friction
between the runner and the fluid outside it.

For a rotor, the shaft work done in unit time interval is

* pronounced *Oy'-ler*.

$$T\omega = \int v_{w1}\omega r_1 \, \mathrm{d}\dot{m} - \int v_{w2}\omega r_2 \, \mathrm{d}\dot{m}$$

$$= \int u_1 v_{w1} \, \mathrm{d}\dot{m} - \int u_2 v_{w2} \, \mathrm{d}\dot{m} \qquad (13.5)$$

since $u = \omega r$.

The integrals in eqns 13.4 and 13.5 can in general be evaluated only if it is known in what way the velocity varies over the inlet and outlet cross-sections of the rotor. However, a simple result is obtained if the product $v_w r$ is constant at each cross-section concerned. This may be so if there is no significant variation of r at inlet or outlet (as in the rotor illustrated in Fig. 13.18) and v_w may be assumed uniform at each section. This assumption would be realistic only if the number of vanes guiding the fluid on to the rotor and also the number of blades in the rotor were large so that there would be no significant variation of either inlet or outlet values of v_w with angular position. In such a case, eqn 13.5 becomes

$$T\omega = u_1 v_{w1} \int \mathrm{d}\dot{m} - u_2 v_{w2} \int \mathrm{d}\dot{m} = \dot{m}(u_1 v_{w1} - u_2 v_{w2}) \qquad (13.6)$$

Equation 13.6 is also obtained if the products $v_w r$ are constant both at inlet and outlet, even though v_w and r are not individually constant. Since the relation $v_w r = $ constant is that describing the velocity distribution in a 'free vortex' (Section 9.7.4) fluid machines designed according to this condition are frequently termed 'free-vortex machines'. Although this design criterion is widely used for axial-flow machines, it often has to be abandoned because of space limitations, especially close to the hub.

The shaft work done by the fluid per unit mass is obtained by dividing eqn 13.5 by the total mass flow rate $\dot{m}$. Then

Work done per unit mass of fluid

$$= \frac{1}{\dot{m}}\left(\int u_1 v_{w1} \, \mathrm{d}\dot{m} - \int u_2 v_{w2} \, \mathrm{d}\dot{m}\right)$$

$$= u_1 v_{w1} - u_2 v_{w2} \text{ if the products } uv_w \text{ are individually constant} \quad (13.7)$$

Dividing this expression by g gives work per unit weight, i.e. head, and that form is sometimes known as the 'Euler head' (or 'runner head').

To a turbine the energy available per unit mass of the fluid is gH where $H = $ the net head. The hydraulic efficiency of the turbine is thus $(u_1 v_{w1} - u_2 v_{w2})/gH$ (if the products uv_w are uniform). This represents the effectiveness with which energy is transferred from the fluid to the runner. It should be distinguished from the overall efficiency of the machine because, owing to such losses as friction in the bearings and elsewhere, not all the energy received by the runner is available at the output shaft.

Similar relations apply to a pump. There the transfer of energy is from rotor to fluid instead of from fluid to rotor and so the expressions 13.4–13.7 have reversed signs.

Many machines are so constructed that uniformity of conditions at inlet and outlet is impossible to achieve. In an axial-flow machine the blade velocity u and the blade angle β both vary along the blade; any velocity vector diagram will therefore apply, in general, only to one radius. In mixed-flow machines the fluid leaves the rotor at various radii. Even Francis turbines nowadays usually have some 'mixed flow' at the outlet; moreover, the inlet and outlet edges of the blades are not always parallel to the axis of rotation. Thus the expressions in which the products uv_w are assumed uniform do not apply exactly to the flow considered as a whole.

In any case, the assumption that the velocities at inlet and outlet are uniform with respect to angular position is not fulfilled even for a rotor in which all the flow is in the plane of rotation. Individual particles of fluid may have different velocities. Since guide vanes and rotor blades are both limited in number, the directions taken by individual particles may differ appreciably from that indicated by the velocity diagram. Even the average direction of the relative velocity may differ from that of the blade it is supposed to follow. Thus the velocity diagrams and the expressions based on them should be regarded as only a first approximation to the truth. In spite of these defects, however, the simple theory is useful in indicating how the performance of a machine varies with changes in the operating conditions, and in what way the design of a machine should be altered in order to modify its characteristics.

With the limitations of the theory in mind we may examine the velocity diagrams further. Figure 13.18 shows the relative velocity of the fluid at inlet in line with the inlet edge of the blade. This is the ideal condition, in which the fluid enters the rotor smoothly. (A small angle of attack (see Section 9.9.1) is generally desirable, but this rarely exceeds a few degrees.) If there is an appreciable discrepancy between the direction of R_1 and that of the blade inlet, the fluid is forced to change direction suddenly on entering the rotor. Violent eddies form, a good deal of energy is dissipated as useless heat, and the efficiency of the machine is consequently lowered. For all rotodynamic machines the correct alignment of blades with the velocities relative to them is very important. For the inlet diagram of a turbine (as in Fig. 13.18) the angle α_1, defining the direction of the absolute velocity of the fluid, is determined by the setting of guide vanes. Smooth entry conditions can be achieved for a wide range of blade velocities and rates of flow by adjustment of the guide vanes and therefore of the angle α_1. For each value of the angle α_1 there is, however, only one shape of inlet velocity diagram that gives the ideal conditions. The angle of R_1 is then determined by the geometry of the vector diagram. At outlet the direction of the relative velocity R_2 is determined by the outlet angle of the blade (β_2) and the geometry of the outlet diagram then determines the magnitude and direction of the absolute velocity v_2.

Not all the energy of the fluid is used by a turbine runner. That remaining unused is principally in the form of kinetic energy and so, for high efficiency, the kinetic energy of the fluid at outlet should be small. For a given rate of flow the minimum value of v_2 occurs when v_2 is perpendicular to u_2 in the outlet vector diagram. The whirl component v_{w2} is then zero, the expression 13.7 for example becomes simply $u_1 v_{w1}$ and the hydraulic efficiency $u_1 v_{w1}/gH$. Other losses in the machine do not necessarily reach their minimum values at the same conditions, however, and a small whirl component is therefore sometimes allowed in practice. The ideal outlet vector diagram is, in any case, not achieved under all conditions of operation. Nevertheless, a zero, or nearly zero, whirl component at outlet is taken as a basic requirement in turbine design.

It is instructive to derive an alternative form of eqn 13.7. Assuming uniform velocities at inlet and outlet and referring again to the inlet vector diagram of Fig. 13.18 we have

$$R_1^2 = u_1^2 + v_1^2 - 2u_1 v_1 \cos \alpha_1 = u_1^2 + v_1^2 - 2u_1 v_{w1}$$

$$\therefore \; u_1 v_{w1} = \frac{1}{2}\left(u_1^2 + v_1^2 - R_1^2\right) \tag{13.8}$$

Similarly

$$u_2 v_{w2} = \frac{1}{2}\left(u_2^2 + v_2^2 - R_2^2\right) \tag{13.9}$$

Substituting eqns 13.8 and 13.9 in eqn 13.7, we obtain

Work done by the fluid per unit mass

$$\frac{1}{2}\left\{\left(v_1^2 - v_2^2\right) + \left(u_1^2 - u_2^2\right) - \left(R_1^2 - R_2^2\right)\right\} \tag{13.10}$$

In an axial-flow machine the fluid does not move radially, and so for a particular radius $u_1 = u_2$ and the term $(u_1^2 - u_2^2)$ is zero. In a radial or mixed-flow machine, however, each of the terms in the expression is effective. For a turbine, that is, a machine in which work is done *by* the fluid, the expression 13.10 must be positive. This is most easily achieved by the inward-flow arrangement. Then $u_1 > u_2$ and, since the flow passages decrease rather than increase in cross-sectional area, R_2 usually exceeds R_1. The contributions of the second and third brackets to the work done by the fluid are thus positive. By appropriate design, however, an outward-flow turbine, although seldom desirable, is possible. An inward-flow machine has a number of advantages, an important one being, as already mentioned, that it is to some extent self-governing.

Conversely, in a pump work is done *on* the fluid and so the expression 13.10 then needs to be negative. Outward flow is thus more suitable for a pump.

13.3.5 Similarity laws and specific speed

We now consider the application of the principles of dynamic similarity – discussed in Chapter 5 – to rotodynamic machines. By the use of these principles it becomes possible to predict the performance of one machine from the results of tests on a geometrically similar machine, and also to predict the performance of the same machine under conditions different from the test conditions. Geometric similarity, it will be recalled, is a prerequisite of dynamic similarity. For fluid machines the geometric similarity must apply to all significant parts of the system: the rotor, the entrance and discharge passages and, in the case of a reaction turbine or a pump, the conditions in the tail race or sump. Machines that are geometrically similar in these respects form a *homologous series*; members of such a series are therefore simply enlargements or reductions of one another.

The performance of any machine depends not only on its size but on its shape – which is common to all machines in the same homologous series. The shape of machine required in a particular application may therefore be determined by reference to the known characteristics of various homologous series.

Dynamic similarity also requires kinematic similarity, that is, corresponding velocities in a constant ratio. If two machines are to be dynamically similar, then, in particular, those velocities represented in the vector diagrams for inlet and outlet of the rotor of one machine must be similar to the corresponding velocities in the other machine. Geometric similarity of the inlet diagrams and of the outlet diagrams is therefore a necessary condition for dynamic similarity. Now, as we saw in Section 5.2, the most succinct statement of the conditions for kinematic similarity (fixed ratio of velocities) and dynamic similarity (fixed ratio of forces) is that certain dimensionless parameters, representing these ratios, are the same for each of the systems being compared. To determine these dimensionless parameters we may use dimensional analysis.

For a machine of a given shape, and handling a homogeneous fluid of constant density, the relevant variables are shown in Table 13.1.

The average height of the boundary roughness, k, may be regarded as included in the definition of the 'shape' of the machine, although similarity of surface roughness is difficult to achieve between machines of

Table 13.1

		Dimensional formula
D	rotor diameter, here chosen as a suitable measure of the size of the machine	$[L]$
Q	volume rate of flow through the machine	$[L^3T^{-1}]$
N	rotational speed	$[T^{-1}]$
H	difference of head across machine, i.e. energy per unit weight	$[L]$
g	weight per unit mass	$[LT^{-2}]$
ϱ	density of fluid	$[ML^{-3}]$
μ	viscosity of fluid	$[ML^{-1}T^{-1}]$
P	power transferred between fluid and rotor	$[ML^2T^{-3}]$

different size. Cavitation (see Section 13.3.6) is, for the moment, assumed absent.

The weight per unit mass, g, enters as a variable because we are concerned with transformations of static head to velocity head or vice versa, and g appears in the expression for velocity head. This, however, is the only way in which g comes into the problem: in most machines flow with a free surface does not occur, and even in an impulse turbine the velocity of the jet is so high that the effect of gravitational forces is negligible. The analysis may therefore be simplified by considering energy per unit mass, gH, rather than H alone. This manoeuvre also has the advantage of enabling the dimensional analysis to discriminate between H and D, both of which here have the dimensional formula [L] although they do not have the same physical significance. (The same simplification would be achieved by considering pressure difference instead of gH.) The number of separate variables is thus seven: D, Q, N, (gH), ϱ, μ, P; as there are three reference magnitudes, four dimensionless Πs may be expected.

(The equation connecting the four Πs is not the only one involving these variables. The rate of flow Q is determined by the head H, the size of the machine and the resistance to flow through it; thus, for a given machine with given blade angles and a given guide vane – or spear valve – setting, Q cannot be altered independently of H and N. Incorporating this other relation for Q would reduce the number of truly independent variables by one. An explicit algebraic expression relating the variables cannot be hoped for, however, and four Πs are here retained so as to establish the important dimensionless parameters readily.)

If D, N and ϱ are the variables used to eliminate the reference magnitudes, the following dimensionless parameters may be obtained:

$$\Pi_1 = \frac{Q}{ND^3} \quad \Pi_2 = \frac{gH}{N^2 D^2} \quad \Pi_3 = \frac{\varrho N D^2}{\mu} \quad \Pi_4 = \frac{P}{\varrho N^3 D^5} \quad (13.11)$$

Let us now consider the significance of these parameters. All lengths of the machine are proportional to D, and thus all areas to D^2. The average velocity of the fluid at a particular section is given by $Q \div$ (the cross-sectional area perpendicular to the velocity) which is proportional to Q/D^2. In fact, Q/D^2 may be regarded as the typical velocity of the fluid, all other fluid velocities – relative or absolute – in the machine being proportional to it. The peripheral velocity of the rotor (and therefore the blade velocity u) is proportional to the product ND. Thus

$$\Pi_1 = \frac{Q/D^2}{ND} \propto \frac{|\text{Fluid velocity}|}{|\text{Blade velocity}|}$$

Now if Π_1 is the same for two geometrically similar machines the ratio of |Fluid velocity| to |Blade velocity| is the same, and this is the condition for similarity of the vector diagrams. In other words, kinematic similarity of two geometrically similar machines is achieved if Q/ND^3

(sometimes known as the 'discharge number' or 'flow coefficient') is the same for each.

The product of Π_1 and Π_3 is $(Q/D^2)(D\varrho/\mu)$ and, if Q/D^2 is taken as the typical velocity of the fluid, the result represents the Reynolds number. If Π_1 is held the same to obtain kinematic similarity, then Π_3 is proportional to the Reynolds number. In the normal range of speed and size, the flow in fluid machines is highly turbulent. The influence of viscosity is therefore small and, for most purposes, Π_3 may be disregarded as a significant parameter.

Combining the power parameter Π_4 with Π_1 and Π_2 in the form $\Pi_4/\Pi_1\Pi_2$ gives $P/\varrho QgH$. For a turbine ϱQgH represents the energy given up by the fluid in unit time in passing through the machine. Therefore, since P has been defined as the power transferred between fluid and rotor, $P/\varrho QgH$ represents the hydraulic efficiency η_h of the turbine. The output power of a turbine is somewhat less than this because of windage and other mechanical losses. If the mechanical efficiency is assumed constant, however, then the shaft power may be used in place of P in the dimensionless parameters and the overall efficiency $\eta_o = \eta_{mech}\eta_h$ in place of ϱ_h. (For a pump, the hydraulic efficiency is given by the reciprocal form $\varrho QgH/P$.)

With the provisos that the effects of differences of Reynolds number are negligible so that Π_3 may be omitted, that the lapse from true geometric similarity caused by differences of relative surface roughness is also of small effect, and that the mechanical efficiency is constant, the relation connecting the variables may be written

$$\phi_1\left(\frac{Q}{ND^3}, \frac{gH}{N^2D^2}, \frac{P}{\varrho N^3 D^5}\right) = 0$$

or, with another arrangement of the Πs,

$$\phi_2\left(\eta, \frac{gH}{N^2D^2}, \frac{P}{\varrho N^3 D^5}\right) = 0 \qquad (13.12)$$

If data obtained from tests on a model turbine, for example, are plotted so as to show the variation of the dimensionless parameters Q/ND^3, gH/N^2D^2, $P/\varrho N^3 D^5$, η with one another, then, subject to the provisos already mentioned, the graphs are applicable to any machine in the same homologous series. The curves for other homologous series would naturally be different, but one set of curves would be sufficient to describe the performance of all the members of one series.

Particularly useful in showing the characteristics of turbines are results obtained under conditions of constant speed and head. For a particular machine and a particular incompressible fluid, D and ϱ are constant. Then gH/N^2D^2 is constant. From eqn 13.12 η is then simply a function of P, and the results may be presented in the form shown in Fig. 13.19.

For a turbine using a particular fluid the operating conditions are expressed by values of N, P and H. It is important to know the range of

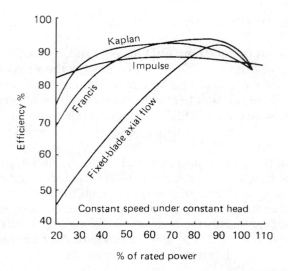

Fig. 13.19 Typical efficiency curves.

these conditions which can be covered by a particular design (i.e. shape) of machine. Such information enables us to select the type of machine best suited to a particular application, and serves as a starting point in its design. We require, therefore, a parameter characteristic of all the machines of a homologous series and independent of the size represented by D. A parameter involving N, P and H but not D is obtained by dividing $(\Pi_4)^{1/2}$ by $(\Pi_2)^{5/4}$. This yields the dimensionless parameter

$$\frac{NP^{1/2}}{\varrho^{1/2}(gH)^{5/4}} \tag{13.13}$$

For complete similarity of flow in machines of a homologous series, each of the dimensionless parameters must be unchanged throughout the series. Consequently the expression 13.13 must be unchanged. A particular value of this ratio therefore relates all the combinations of N, P and H for which the flow conditions are similar in the machines of that homologous series.

Interest naturally centres on the conditions for which the efficiency is a maximum, so, in calculating the value of the expression 13.13, it is customary to use values of N, P and H that correspond to maximum efficiency. There is in general only one pair of values of $P/\varrho N^3 D^5$ and $gH/N^2 D^2$ for which the efficiency is a maximum (see eqn 13.12). For a given homologous series, therefore, we are concerned with a unique set of flow conditions, and thus a unique value of $NP^{1/2}\varrho^{-1/2}(gH)^{-5/4}$ is obtained which may be designated K_n. Whatever the conditions of operation, provided only that they correspond to maximum efficiency, the machines of a particular homologous series, that is, of a particular shape, have a particular value of K_n. Machines of a different shape have, in general, a different value of K_n. Consequently, the value of K_n obtained from a set of values of N, P and H indicates the shape of machine that meets those conditions.

Hitherto the only incompressible fluid used in practical turbines has been cold water, and they have been operated only on the surface of the earth. The values of ϱ and g have therefore never been appreciably different from $1000\,\text{kg/m}^3$ and $9.81\,\text{m/s}^2$ respectively. Consequently engineers have often omitted from K_n the constant ϱ and g terms, and worked with values of $NP^{1/2}/H^{5/4}$. The last expression has been termed the *specific speed*, N_s. The rather unfortunate name arises from the concept of a hypothetical *specific* turbine – a member of the same homologous series as the actual turbine but so reduced in size as to generate one unit of power (e.g. 1 hp) under one unit of head (e.g. 1 ft) at maximum efficiency. The numeric, *but not the units*, obtained on substituting values in $NP^{1/2}/H^{5/4}$ then corresponds simply to the speed at which the specific turbine would run.

The 'specific speed' is not dimensionless, since the ϱ and g terms have been dropped, nor it is a speed, either linear or angular. It is better regarded as a 'shape parameter' which conveniently identifies the homologous series and thus classifies the geometrical shape of the machine. As it is not dimensionless, its numerical value depends on the units with which the constituent magnitudes are expressed. The usual British and American practice has been for N to be expressed in rev/min, P the *shaft* power in hp and H the difference of head across the turbine in feet. In most of Europe N has been measured in rev/min, P in metric horsepower and H in metres. No agreement has, it seems, yet been reached on future usage with SI units.

The dimensionless parameter K_n is often known as the dimensionless specific speed to distinguish it from N_s. However, K_n is truly dimensionless only if N involves the radian measure of angle. Yet N is invariably measured as number of revolutions per unit time interval and, since it would be pedantic to insist that this figure be multiplied by 2π, K_n is usually calculated with N as number of revolutions per unit time interval. Consequently 'rev' remains as the unit of K_n. (Similar remarks of course apply to other 'dimensionless' parameters in which N appears.)

When the site of the installation and the output required from a turbine are known, the value of K_n (or of N_s) may be calculated and the type of machine best suited to these conditions selected. For the principal types of turbine, experience has shown the values in Table 13.2 most suitable.

Figure 13.20 indicates the variation of specific speed with the shape of the turbine runner. (The values quoted should be regarded as approximate because the shape of other parts of the turbine – for example, the volute, the guide passages and the draft tube – affects the specific speed to some extent.) For given values of H and P, N increases with N_s and K_n. With the same peripheral runner velocity, a larger value of N implies

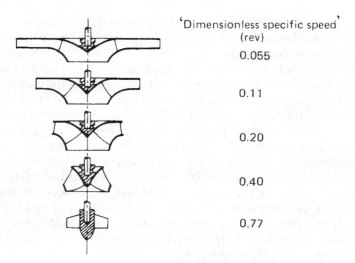

'Dimensionless specific speed'
(rev)

0.055

0.11

0.20

0.40

0.77

Fig. 13.20

Table 13.2

Type of turbine	Approximate range of 'dimensionless specific speed' $K_n = NP^{1/2}\varrho^{-1/2}(gH)^{-5/4}$ (rev)[†]
Pelton (if there is more than one jet, P is taken as total power ÷ number of jets)	0.015–0.024
Francis (not for heads above about 370 m)	0.055–0.37
Kaplan (not for heads above about 60 m)	0.3–0.8
(The upper limits of K_n decrease somewhat with increase of head.)	

[†] Multiplying these figures by 273 gives N_s in 'British units'; multiplying by 1214 gives N_s in 'metric units'.

a smaller value of D and so, in general, lower cost. For this reason, where a choice lies between two machines of different specific speeds, the designer usually prefers that with the higher value. Machines of high specific speeds, however, are limited to low heads because of cavitation – an important factor that we shall discuss in the next section.

■ **Example 13.1** A vertical-shaft Francis turbine of 'dimensionless specific speed' 0.09 rev rotates at 11 rev/s and the internal pressure loss through the machine is given by $p_1 = (1.19Q^2 - 1.43Q + 0.47)$ MPa when the flow rate is Q m³/s. The output power at maximum efficiency is 200 kW. At inlet the rotor has a diameter of 0.5 m, a height of 60 mm, and the blades occupy 5% of the circumference. The mean diameter at outlet from the rotor is 0.325 m and the radial velocities may be assumed equal at inlet and outlet. If the mechanical efficiency is constant, calculate the hydraulic and overall efficiencies of the turbine, the outlet angle of the guide vanes, and the rotor blade angles at inlet and outlet if there is no whirl at outlet.

Solution

Maximum hydraulic efficiency occurs for minimum pressure loss, i.e. when

$$\frac{dp_1}{dQ} = 2.38Q - 1.43 = 0$$

$$\therefore Q_{opt} = 1.43/2.38 = 0.601 \, \left(m^3/s\right)$$

and minimum $p_1 = \left\{1.19(0.601)^2 - (1.43 \times 0.601) + 0.47\right\}$ MPa

$$= 40.4 \, \text{kPa} \equiv \frac{40.4 \times 10^3 \, \text{Pa}}{1000 \times 9.81 \, \text{N/m}^3} = 4.12 \, \text{mH}_2\text{O}$$

Specific speed refers to conditions of maximum overall efficiency, i.e. to maximum hydraulic efficiency if mechanical efficiency is constant. Then

$$\left(gH\right)^{5/4} = \frac{NP^{1/2}}{\varrho^{1/2} K_n} = \frac{\left(11 \, \text{rev/s}\right)\left(200 \times 10^3 \, \text{W}\right)^{1/2}}{\left(1000 \, \text{kg/m}^3\right)^{1/2} 0.09 \, \text{rev}}$$

whence $H = 39.67$ m.

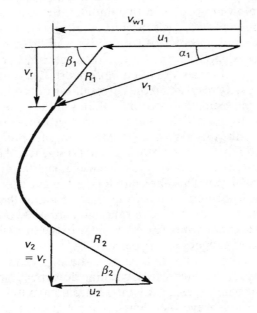

$$\therefore \text{ Hydraulic efficiency} = \frac{39.67 - 4.12}{39.67} = 0.896$$

$$\text{Overall efficiency} = \frac{P}{Q\varrho gH} = \frac{200 \times 10^3}{0.601 \times 1000 \times 9.81 \times 39.67}$$
$$= 0.855$$

$$\text{Euler head} = \left(39.67 - 4.12\right)\text{m} = 35.55\,\text{m}$$
$$= u_1 v_{w1}/g \text{ since } v_{w2} = 0$$
$$u_1 = \pi \times 11 \times 0.5\,\text{m/s} = 17.28\,\text{m/s}$$
$$\therefore v_{w1} = gH/u_1 = \left(9.81 \times 35.55/17.28\right)\text{m/s} = 20.18\,\text{m/s}$$
$$v_r = Q/A = \left\{0.601/\left(\pi \times 0.5 \times 0.06 \times 0.95\right)\right\}\text{m/s}$$
$$= 6.71\,\text{m/s}$$
$$\therefore \alpha_1 = \arctan\left(v_r/v_{w1}\right) = \arctan(6.71/20.18) = 18.4°$$
$$\text{and } \beta_1 = \arctan v_r/(v_{w1} - u_1) = \arctan(6.71/2.90) = 66.6°.$$
$$u_2 = \pi \times 11 \times 0.325\,\text{m/s} = 11.23\,\text{m/s}$$
$$\therefore \beta_2 = \arctan\left(v_r/u_2\right) = \arctan\left(6.71/11.23\right) = 30.9°$$

13.3.6 Cavitation

Non-uniformity of flow in machines may cause the pressure, even in a given cross-section, to vary widely. There may thus be, on the low-pressure side of the rotor, regions in which the pressure falls to values considerably below atmospheric. In a liquid, if the pressure at any point falls to the vapour pressure (at the temperature concerned), the liquid there boils and small bubbles of vapour form in large numbers. These bubbles are carried along by the flow, and on reaching a point where the pressure is higher they suddenly collapse as the vapour condenses to liquid again. A cavity results and the surrounding liquid rushes in to fill it. The liquid moving from all directions collides at the centre of the cavity, thus giving rise to very high local pressures (up to 1 GPa). Any solid surface in the vicinity is also subjected to these intense pressures, because, even if the cavities are not actually at the solid surface, the pressures are propagated from the cavities by pressure waves similar to those encountered in water hammer. This alternate formation and collapse of vapour bubbles may be repeated with a frequency of many thousand times a second. The intense pressures, even though acting for only a very brief time over a tiny area can cause severe damage to the surface. The material ultimately fails by fatigue, aided perhaps by corrosion, so the surface becomes badly scored and pitted. Parts of the surface may even be torn completely away. Associated with cavitating flow there may be considerable vibration and noise; when cavitation occurs in a turbine or pump it may sound as though gravel were passing through the machine.

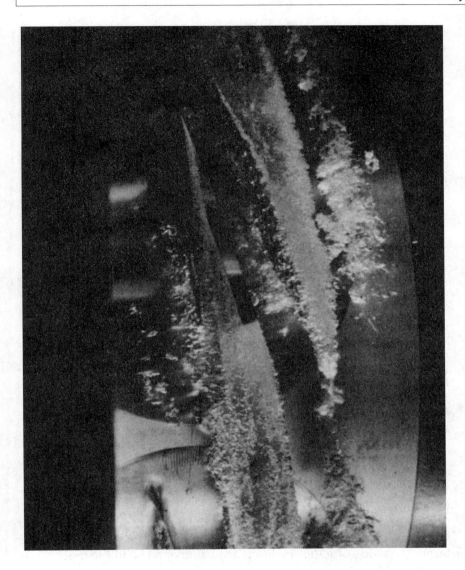

Fig. 13.21(a) (Crown Copyright: reproduced with permission).

Figure 13.21a shows cavitation occurring, and Fig. 13.21b illustrates the damage that a surface may soon suffer as a result. Not only is cavitation destructive: the larger pockets of vapour may so disturb the flow that the efficiency of a machine is impaired. Everything possible should therefore be done to eliminate cavitation in fluid machinery, that is, to ensure that at every point the pressure of the liquid is above the vapour pressure. When the liquid has air in solution this is released as the pressure falls and so *air cavitation* also occurs. Although air cavitation is less damaging than vapour cavitation to surfaces, it has a similar effect on the efficiency of the machine.

Since cavitation begins when the pressure reaches too low a value, it is likely to occur at points where the velocity or the elevation is high, and

Fig. 13.21(b) (Crown Copyright: reproduced with permission).

particularly at those where high velocity and high elevation are combined. In a reaction turbine the point of minimum pressure is usually at the outlet end of a runner blade, on the leading side. For the flow between such a point and the final discharge into the tail race (where the total head is atmospheric) the energy equation may be written

$$\frac{p_{\min}}{\varrho g} + \frac{v^2}{2g} + z - h_{\mathrm{f}} = \frac{p_{\mathrm{atm}}}{\varrho g} \tag{13.14}$$

Here h_{f} represents the head lost to friction in the draft tube, and the pressures are absolute.

Equation 13.14 incidentally shows a further reason why the outlet velocity v of the fluid from the runner should be as small as possible: the larger the value of v the smaller is the value of $p_{\min}$ and the more likely is cavitation.

Rearranging the equation gives

$$\frac{v^2}{2g} - h_{\mathrm{f}} = \frac{p_{\mathrm{atm}}}{\varrho g} - \frac{p_{\min}}{\varrho g} - z$$

For a particular design of machine operated under its design conditions, the left-hand side of this relation may be regarded as a particular proportion, say σ_{c}, of the net head H across the machine. Then

$$\sigma_{\mathrm{c}} = \frac{p_{\mathrm{atm}}/\varrho g - p_{\min}/\varrho g - z}{H}$$

For cavitation not to occur $p_{\min}$ must be greater than the vapour pressure of the liquid, p_v, i.e.

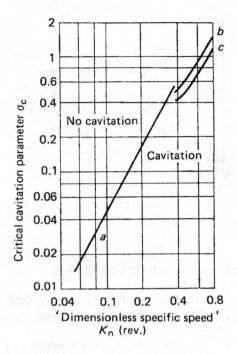

Fig. 13.22 Cavitation limits for reaction turbines (a) Francis, (b) fixed-blade propeller and (c) Kaplan.

$$\sigma > \sigma_c \quad \text{where} \quad \sigma = \frac{p_{atm}/\varrho g - p_v/\varrho g - z}{H} \qquad (13.15)$$

The expression 13.15 is known as Thoma's cavitation parameter, after the German engineer Dietrich Thoma* (1881–1943) who first advocated its use. If either z (the height of the turbine runner above the tail water surface) or H is increased, σ is reduced. To determine whether cavitation is likely in a particular installation, the value of σ may be calculated: if it is greater than the tabulated (empirical) value of σ_c for that design of turbine, cavitation should not be experienced.

In practice the expression is used to determine the maximum elevation z_{max} of the turbine above the tail water surface for cavitation to be avoided:

$$z_{max} = p_{atm}/\varrho g - p_v/\varrho g - \sigma_c H \qquad (13.16)$$

Equation 13.16 shows that the greater the net head H on which a given turbine operates, the lower it must be placed relative to the tail water level.

Figure 13.22 shows that turbines of high specific speed have higher values of σ_c and so they must be set much lower than those of smaller specific speed. For a high net head H it might be necessary to place the turbine *below* the tail water surface, thus adding considerably to the difficulties of construction and maintenance. This consideration restricts

* pronounced *Toh'-mah*.

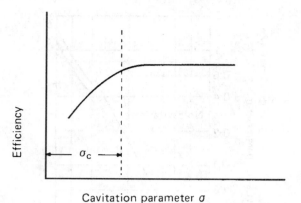

Fig. 13.23

the use of propeller turbines to low heads – to which, fortunately, they are best suited in other ways.

Figure 13.22, it should be realized, is no more than a useful general guide; in practice the incidence of cavitation depends very much on details of the design.

The general effect of cavitation on the efficiency of a turbine is indicated by Fig. 13.23.

Cavitation is a phenomenon by no means confined to turbines. Wherever it exists it is an additional factor to be considered if dynamic similarity is sought between one situation and another. Similarity of cavitation requires the cavitation number $(p - p_v)/\frac{1}{2}\varrho u^2$ (see Section 5.3.5) to be the same at corresponding points. Experiments suggest, however, that similarity of cavitation is difficult to achieve.

■ **Example 13.2** A vertical-shaft Francis water turbine runs at 6.25 rev/s under a net head of 27.5 m. The runner has, at inlet, a diameter of 0.75 m, a flow area of 0.2 m², and a blade angle of 75°. The guide-vane angle (at outlet) is 15°. At the entrance to the draft tube there is no whirl velocity and the absolute pressure is 35 kPa. Atmospheric pressure is 101.3 kPa. To design the draft tube, the critical cavitation parameter σ_c, based on the minimum pressure in the draft tube, is related to the 'dimensionless specific speed' K_n of the machine by $\sigma_c = 3.5(K_n/\text{rev})^{1.84}$. Calculate the overall efficiency of the turbine assuming that the mechanical efficiency is 0.97 and determine the limiting value for the height of the draft tube above the surface of the tail race.

Solution

$$u_1 = \pi ND = \pi 6.25 \times 0.75\,\text{m/s} = 14.73\,\text{m/s}$$
$$v_1 = u_1 \sin 105°/\sin 60° = 16.42\,\text{m/s}$$

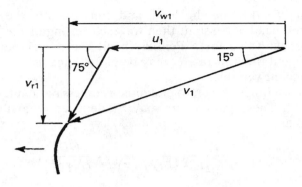

$$v_{r1} = v_1 \sin 15° = 4.25 \, \text{m/s}$$
$$v_{w1} = v_1 \cos 15° = 15.87 \, \text{m/s}$$
$$v_{w2} = 0$$

∴ Hydraulic efficiency $= u_1 v_{w1}/gH = 14.73 \times 15.87/(9.81 \times 27.5)$
$= 0.866$ and overall efficiency $= 0.97 \times 0.866 = 0.840$.

$$Q = A_1 v_{r1} = \left(0.2 \, \text{m}^2\right)\left(4.25 \, \text{m/s}\right) = 0.850 \, \text{m/s}$$

$$P = \eta_0 Q \varrho g H = 0.840\left(0.850 \, \text{m}^3/\text{s}\right)\left(1000 \, \text{kg/m}^3\right)\left(9.81 \, \text{N/kg}\right)\left(27.5 \, \text{m}\right)$$

$$= 192.7 \times 10^3 \, \text{W}$$

$$K_n = \frac{N}{(gH)^{5/4}}\left(\frac{P}{\varrho}\right)^{1/2} = \frac{6.25}{(9.81 \times 27.5)^{5/4}}\left(\frac{192.7 \times 10^3}{1000}\right)^{1/2} \text{rev}$$

$$= 0.0794 \, \text{rev}$$

$$\therefore \sigma > 3.5(0.0794)^{1.84} = 0.03306$$

i.e. $\left(\dfrac{p_{\text{atm}} - p_{\text{min}}}{\varrho g} - z_0\right)\Big/H > 0.03306$

$$\therefore \frac{(101.3 - 35)10^3}{1000 \times 9.81} \, \text{m} - z_0 > 0.03306 \times 27.5 \, \text{m}$$

whence $z_0 < 5.85 \, \text{m}$. ☐

13.3.7 The performance characteristics of turbines ✓

Although desirable, it is not always possible for a turbine to run at its maximum efficiency. Interest therefore attaches to its performance under conditions for which the efficiency is less than the maximum. In testing model machines it is usual for the head to be kept constant (or approximately so) while the load, and consequently the speed, are varied. If the head is constant then for each setting of the guide vane angle (or spear valve for a Pelton wheel) the power output P, the efficiency η and the flow rate Q may be plotted against the speed N as the independent variable.

It is more useful, however, to plot dimensionless parameters as shown in Fig. 13.24. These parameters may be deduced from the Π ratios (13.11) and are

$$\frac{P}{\varrho D^2 (gH)^{3/2}}, \quad \frac{Q}{D^2 (gH)^{1/2}}, \quad \frac{ND}{(gH)^{1/2}}$$

Thus one set of curves is applicable not just to the conditions of the test, but to any machine in the same homologous series, operating under any head.

More usually, however, the ϱ and g terms are dropped from these dimensionless forms. Often the D terms are omitted also, and the resulting ratios $P/H^{3/2}$, $Q/H^{1/2}$, $N/H^{1/2}$ are then referred to as 'unit power', 'unit flow' and 'unit speed'. Their *numerical* values correspond respectively to the power, volume flow rate and speed obtainable if the machine could be operated with unchanged efficiency under one unit of head (e.g. 1 m).

Most turbines are required to run at constant speed so that the electrical generators to which they are coupled provide a fixed frequency and voltage. For an impulse machine at constant speed under a given head, the vector diagrams are independent of the rate of flow. In theory, then, the hydraulic efficiency should be unaffected by the load, although in practice there is a small variation of the efficiency. For a reaction turbine, changes of load are dealt with by alteration of the guide vane angle. (True, the power output could be altered by throttling the flow through a partly closed valve in the supply line. This process, however, would wastefully dissipate a large part of the available energy in eddy formation at the valve.) Figure 13.25 shows the general effect of change of guide vane angle for a machine of the Francis type or fixed-blade propeller type. Only at the maximum efficiency point does the direction of the relative velocity at inlet conform with that of the inlet edges of the

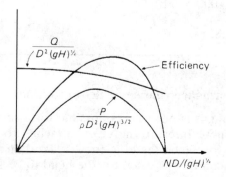

Fig. 13.24

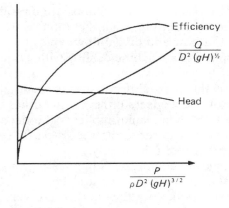

Fig. 13.25

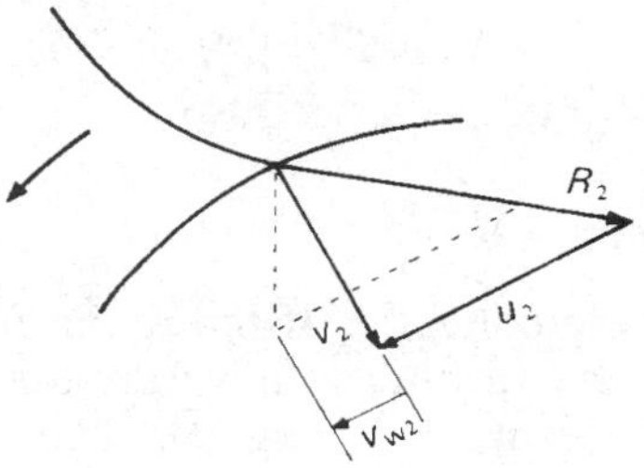

Fig. 13.26 Effect of change of flow rate on outlet vector triangle. (Full lines: max. efficiency conditions; dotted lines: altered conditions.)

runner blades. At other conditions these directions do not conform, and so the fluid does not flow smoothly into the passages in the runner. Instead, it strikes either the front or back surfaces of the blades; considerable eddy formation ensues and the consequent dissipation of energy reduces the efficiency of the machine.

In the Kaplan turbine the runner blade angle may be altered in addition to the guide vane angle. Thus it is possible to match the directions of the relative velocity at inlet and the inlet edges of the runner blades for a wide range of conditions. In consequence, the part-load efficiency of the Kaplan machine is superior to that of other types, as shown in Fig. 13.19.

A change of load also affects the conditions at outlet. A reduction in the rate of flow through the machine results in a decreased value of R_2. Consequently, if the blade velocity u_2 is unaltered, there is a departure from the ideal right-angled vector triangle at outlet (see Fig. 13.26); the resulting whirl component of velocity causes a spiral motion in the draft tube and hence a reduction of the draft-tube efficiency. The possibility of cavitation is also increased.

13.4 ROTODYNAMIC PUMPS

As we have already mentioned (Section 13.3) a rotodynamic pump is essentially a turbine 'in reverse'; mechanical energy is transferred from the rotor to the fluid. Like turbines, pumps are classified according to the direction of the fluid path through them: there are thus radial flow (or centrifugal), axial-flow and mixed-flow types. In general usage the word 'pump' is applied to a machine dealing with a liquid; a machine in which the working fluid is a gas is more usually termed a fan, blower or compressor. In fans the change of pressure is small and so changes of density may normally be neglected; this distinguishes them from compressors, in which – as their name implies – the density of the gas is considerably increased. The term 'blower' is partly synonymous with 'fan', but is used rather indiscriminately.

The density changes in compressors involve thermodynamic considerations and we shall not touch on these here. Fans, however, deal essentially with constant-density flow and so are very similar, both in construction and operation, to pumps. In what follows, then, statements about pumps may be taken to apply also to fans, except where specific reference is made to a liquid.

13.4.1   Centrifugal pumps ∨

This type of pump is the converse of the radial-flow (Francis) turbine. Whereas the flow in the turbine is inwards, the flow in the pump is outwards (hence the term 'centrifugal'). The rotor (usually called 'impeller') rotates inside a spiral casing as shown in Fig. 13.27. The inlet pipe is axial, and fluid enters the 'eye', that is, the centre, of the impeller with little, if any, whirl component of velocity. From there it flows outwards in the direction of the blades, and, having received energy from the impeller, is discharged with increased pressure and velocity into the casing. It then has a considerable tangential (whirl) component of velocity which is normally much greater than that required in the discharge pipe. The kinetic energy of the fluid leaving the impeller is largely dissipated in shock losses unless arrangements are made to reduce the velocity gradually. Figure 13.27 illustrates the simplest sort of pump, the 'volute' type. The volute is a passage of gradually increasing section which serves to reduce the velocity of the fluid and to convert some of the velocity head to static head. In this function the volute is often supplemented by a diverging discharge pipe.

A higher efficiency may be obtained by fitting a set of fixed guide vanes (a 'diffuser') round the outside of the impeller as shown in Fig. 13.28. These fixed vanes provide more opportunity for the gradual reduction of the velocity of the fluid so that less energy is wasted in eddies. However, unless the absolute velocity of the fluid leaving the impeller is in line with the entry edges of the diffuser blades, shock losses there will offset the gain in efficiency otherwise to be obtained. Thus the diffuser yields an improved efficiency over only a limited range of conditions for which the diffuser blade angles are approximately correct. It is possible to fit diffuser blades having adjustable angles, but the considerable extra complication is not normally warranted. The diffuser, which adds appreciably to the cost and bulk of the pump, is

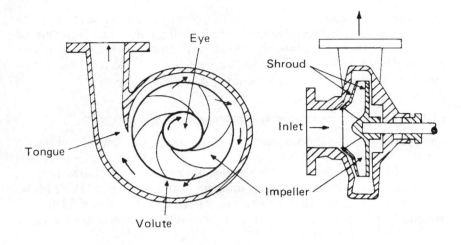

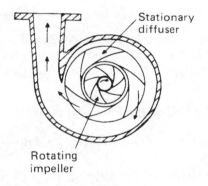

Fig. 13.27 Volute-type centrifugal pump.

Fig. 13.28 Diffuser-type centrifugal pump.

therefore not an unmixed blessing and, except for large machines where running costs are important, the gain in efficiency is not enough to justify its wide use.

The efficiency of a pump is in any case generally less than that of a turbine. Although the energy losses in the two types of machine are of the same kind, the flow passages of a pump are diverging, whereas those of a turbine are converging. The flow in a pump may therefore readily break away from the boundaries with consequent dissipation of energy in eddies. A modern diffuser pump (sometimes, unfortunately, called a 'turbine' pump because of its superficial resemblance to a reaction turbine) may have a maximum overall efficiency of well over 80%; the usual figure for the simpler volute pump is from 75% to 80%, although somewhat higher values are obtainable for large machines.

There are many variations on the basic arrangement shown in Fig. 13.27. This shows a 'single suction' pump in which the fluid enters from one side of the impeller. In a 'double suction' machine fluid enters from both sides. The impeller then often looks like two single suction impellers placed back to back. The symmetry of this arrangement has the advantage that the thrusts on each side, resulting from the change in direction of the flow, are balanced. To produce high pressures a multi-stage pump may be used. Two or more impellers may then be arranged on one shaft; these are connected in series, the fluid discharged from one impeller being led to the inlet of the next, so that the total head produced is the sum of the heads generated in each stage.

13.4.2 The basic equations applied to centrifugal pumps

The relations developed in Section 13.3.4 are applicable to pumps no less than to turbines. The assumptions on which they are founded are equally important and may be recalled here: we consider steady flow, with velocities at inlet and outlet uniform both in magnitude and in the angle made with the radius. The energy imparted to the fluid by the impeller is given by eqn 13.7 with the sign reversed since work is done *on* the fluid, not *by* it:

$$\text{Work done on fluid per unit mass} = u_2 v_{w2} - u_1 v_{w1} \quad (13.17)$$

Suffix 1 again refers to the inlet and suffix 2 to the outlet even though $r_2 > r_1$ for a pump. This expression may be transformed, with the aid of the trigonometric relations for the vector triangles, to the equivalent form (corresponding to eqn 13.10):

Work done on fluid per unit mass

$$= \frac{1}{2}\left\{\left(v_2^2 - v_1^2\right) + \left(u_2^2 - u_1^2\right) - \left(R_2^2 - R_1^2\right)\right\} \quad (13.18)$$

A centrifugal pump rarely has any sort of guide vanes at inlet. The fluid therefore approaches the impeller without appreciable whirl and so the inlet angle of the blades is designed to produce a right-angled vector triangle at inlet (as shown in Fig. 13.29). At conditions other than those for which the impeller was designed – for example, a smaller flow rate at the same shaft speed – the direction of the relative velocity R_1 does not coincide with that of a blade. Consequently the fluid changes direction abruptly on entering the impeller, eddies are formed and energy is dissipated. In addition, the eddies give rise to some back flow into the inlet pipe, thus causing the fluid to have some whirl before entering the impeller. Moreover, particularly if the pump is dealing with a fluid of high viscosity, some pre-whirl may be caused by viscous drag between the impeller and the incoming fluid.

Whatever the immediate cause of such pre-whirl, however, it will have come from the impeller. The initial angular momentum of the fluid may therefore be taken as zero. (This argument is equivalent to considering the inlet boundary of the 'control volume' further upstream of the impeller. Even in the enlarged control volume the impeller is the only thing providing torque.) We may therefore set $v_{w1} = 0$ in the Euler relation 13.17 to give

$$\text{Work done on fluid per unit mass} = u_2 v_{w2} \quad (13.19)$$

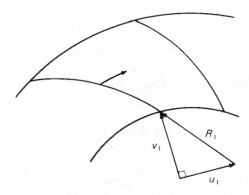

Fig. 13.29

It may be noted that since u_1 no longer enters the expression the work done is independent of the inlet radius.

The increase of energy received by the fluid in passing through the pump is most often expressed in terms of head H, i.e. energy/weight. It should not be forgotten that this quantity depends on g; indeed, a pump in an orbiting space-craft, for example, where the fluid would be weightless, would generate an infinite head! The use of energy/mass is therefore now sometimes advocated, but 'head' still seems to be preferred, if only for the brevity of the term.

The theoretical work done per unit weight is often termed 'the Euler head' and is given by dividing eqn 13.19 by g. The increase of total head across the pump is less than this, however, because of the energy dissipated in eddies and in friction. The gain in piezometric head, $p^*/\varrho g$, across the pump is known as the *manometric head* H_m: it is the difference of head that would be recorded by a manometer connected between the inlet and outlet flanges of the pump. (Accurate readings of inlet pressure, however, may be difficult to obtain if there is any swirling motion in the inlet pipe.) The ratio of the manometric head to the Euler head is gH_m/u_2v_{w2} (i.e. $\Delta p^*/\varrho u_2 v_{w2}$) and is known as the *manometric efficiency*: it represents the effectiveness of the pump in producing pressure from the energy given to the fluid by the impeller.

Except for fans, the velocity heads at inlet and outlet are usually similar, and small compared with the manometric head; thus the manometric head and the gain of total head are not greatly different. However, the overall efficiency – that is, the ratio of the power given to the fluid ($Q\varrho gH$, where H represents the difference of total head across the pump) to the shaft power – is appreciably lower than the manometric efficiency. This is because additional energy has to be supplied by the shaft to overcome friction in the bearings, and in the fluid in the small clearances surrounding the impeller. The energy required at the shaft (per unit mass of fluid) thus exceeds u_2v_{w2}.

The performance of a pump depends (among other things) on the outlet angle of the impeller blades. The blade outlet may face the direction of rotation (i.e. forwards), be radial, or face backwards. The

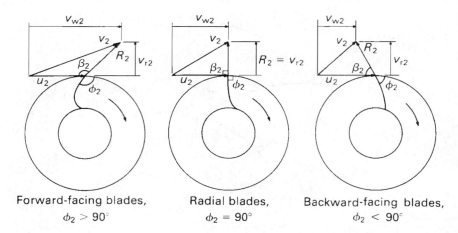

Forward-facing blades, $\phi_2 > 90°$

Radial blades, $\phi_2 = 90°$

Backward-facing blades, $\phi_2 < 90°$

Fig. 13.30

outlet blade angle ϕ_2 is usually defined as shown in Fig. 13.30. Thus for forward-facing blades $\phi_2 > 90°$ and for backward-facing blades $\phi_2 < 90°$. (American practice favours the definition of $180° - \phi_2$ as the blade angle.)

We assume for the moment that there is no discrepancy between the direction of the relative velocity R_2 and the outlet edge of a blade. Consequently the blade outlet angle ϕ_2 is identical with the angle β_2 in the diagram. Figure 13.30 shows the differences in the outlet vector diagrams of the three types of impeller for the same blade velocity u_2.

Ideally, that is, if the fluid were frictionless, the increase of total energy across the pump would be $u_2 v_{w2}$ per unit mass. Now, from Fig. 13.30, $v_{w2} = u_2 - v_{r2} \cot \beta_2$ where v_r represents the radial component of the fluid velocity (sometimes termed 'velocity of flow'). If Q represents the volume rate of flow through the pump and A_2 the outlet area perpendicular to v_{r2} (that is, the peripheral area of the impeller less the small amount occupied by the blades themselves) then for uniform conditions $v_{r2} = Q/A_2$. The ideal increase of energy per unit mass therefore equals

$$u_2 v_{w2} = u_2 \left(u_2 - v_{r2} \cot \beta_2 \right) = u_2 \left(u_2 - \frac{Q}{A_2} \cot \beta_2 \right) \quad (13.20)$$

The blade velocity u_2 is proportional to the rotational speed N, so the ideal energy increase gH equals

$$C_1 N^2 - C_2 N Q$$

where C_1 and C_2 are constants. Thus for a fixed speed N the variation of H with Q is ideally linear as shown in Fig. 13.31.

In practice, however, energy losses occur and some of the assumptions on which eqn 13.20 rests are not fulfilled. Consider the flow in the volute. Apart from frictional effects, no torque is applied to a fluid particle once it has left the impeller. The angular momentum of the particle is therefore constant, that is it follows a path along which

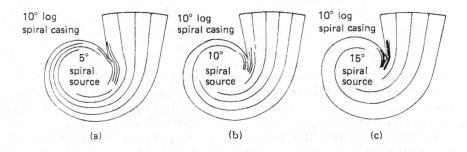

Fig. 13.31

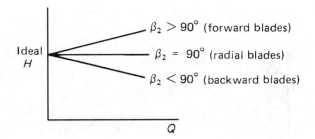

Fig. 13.32 (From Worster[1].)

$v_w r$ = constant. Ideally, the radial velocity from the impeller does not vary round the circumference. The combination of uniform radial velocity with the free vortex ($v_w r$ = constant) gives a pattern of spiral streamlines which should be matched by the shape of the volute. The latter is thus an important feature of the design of the pump.[1] At maximum efficiency about 10% of the energy increase produced by the impeller is commonly lost in the volute. Even a perfectly designed volute, however, can conform to the ideal streamline pattern at the design conditions only. At rates of flow greater or less than the optimum there are increased shock losses in the volute (Fig. 13.32); in addition there are variations of pressure and therefore of radial velocity v_{r2} round the impeller.

Shock losses are also possible at the inlet to the impeller. At conditions different from those of the design the fluid does not approach the blades in the designed direction. Consequently energy losses arise from the turbulence generated by the impact of the fluid against the blades. The energy (per unit mass) lost in this way may be shown to be approximately proportional to $(Q - Q_{ideal})^2$ for a given rotational speed N. Other, smaller, losses arise from friction between the fluid and the boundaries; there is also the recirculation of a small quantity of the fluid after leakage through the clearance spaces outside the impeller (Fig. 13.33).

All these effects modify the 'ideal' relation between head and discharge illustrated in Fig. 13.31, and the 'characteristic curves' for the pump take on the shapes shown in Fig. 13.34.

The pump, however, cannot be considered in isolation. The operating conditions at any moment are represented by a certain point on the

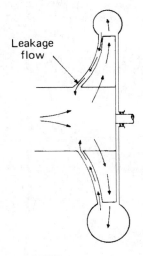

Leakage flow

Fig. 13.33

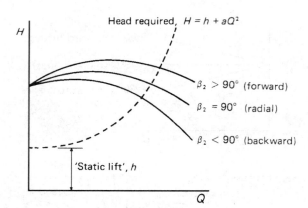

Head required, $H = h + aQ^2$

$\beta_2 > 90°$ (forward)

$\beta_2 = 90°$ (radial)

$\beta_2 < 90°$ (backward)

'Static lift', h

Fig. 13.34 Characteristic curves for fixed rotational speed N.

characteristic curve. This point is determined by the conditions in the external system (that is, the pipework, valves, and so on) to which the pump is connected. For example, a pump may be used to lift a liquid from a sump to a higher tank, the vertical distance involved being h. In addition to supplying this 'static lift' h, the pump must also provide sufficient head to overcome pipe friction and other losses (such as those at valves). As the flow is normally highly turbulent all these losses are proportional to the square of the velocity and thus of Q^2 (say aQ^2). The head required is therefore represented by the line $H = h + aQ^2$ in Fig. 13.34. The point of intersection of this line with the characteristic curve gives the values of H and Q at which the pump operates under these conditions. In other words, the pump offers one relation between H and Q, the external system offers another; these two relations must be satisfied simultaneously.

Impellers with backward-facing blades are often preferred although they give a smaller head for a given size and rate of flow. As shown in Fig. 13.30, this type of impeller gives the fluid a smaller absolute velocity v_2; thus the dissipation of energy in the volute is less and the efficiency correspondingly higher. Such impellers also have the advantage of a negative slope to the characteristic curve for nearly all values of Q: this enables two machines to be used in parallel without instability. For liquids the primary function of a pump is to increase the pressure, but fans are usually required to move a large quantity of air – or other gas – without much change in pressure. The volute is not then called upon to provide much conversion of velocity head to static head, and for such applications an impeller with forward-facing blades may be favoured.

13.4.3   The effects of non-uniform velocity distribution

The foregoing analysis has been based on the assumption that the velocities at inlet and outlet of an impeller are uniform. This condition might be approached if the volute were ideally matched to the impeller and if the impeller had a very large number of blades. In practice, however, the number of blades is limited, and the flow pattern in the passages between the blades becomes distorted. One important cause of the distortion is this. The function of the blades is to impart a whirl component of velocity to the fluid: this requires a force in the whirl direction to be exerted on the fluid by the blades; consequently at a given distance along a blade the pressure at the forward side (*xx* in Fig. 13.35) is greater than that at the other side *yy*. In other words, the individual blades act similarly to aerofoils and, as we saw in Section 9.9.2, a circulation is necessary to produce the transverse force. This circulation round a blade (indicated by the dotted loop in Fig. 13.35), when superimposed on the main flow, has the effect of reducing the velocity relative to the blade on the forward side, and increasing it on the other side. Thus the velocity pattern indicated in the left-hand blade passage is produced. In any diverging passage, rotating or not, there is a tendency for flow to break away from the surface: in a pump impeller this tendency is aggravated by the increased relative velocity at the back of the blades. Thus, instead of following the direction of the blades, the fluid flows rather in the direction of the dotted arrow in Fig. 13.35.

The inertia of the fluid particles also has an effect. The rotation of the impeller requires a change of the (absolute) direction taken by the fluid particles between two successive blades. But the particles, reluctant to do the bidding of the impeller blades, try to lag behind the blade movement. Those particles next to the forward side of a blade have no choice but to be hustled forward by it, but particles further from that side tend to deviate backwards from the prescribed path.

As a result of these two effects – the more ready separation along the back of a blade and the inertia of the fluid particles – the mean direction of the fluid leaving the impeller is modified as shown in Fig. 13.36. The velocity vector diagram that the outlet blade angle ϕ_2 leads us to expect is shown in full: the figure actually obtained is indicated by dotted lines. Conditions may vary somewhat round the impeller circumference, but the height of the 'average' diagram corresponds to the average value of v_{r2} (i.e. Q/A_2) and therefore, for a given rate of flow, is unchanged. The backward deviation of R_2 results in a backward shift of the vertex, a reduction in the whirl component v_{w2} and a consequent reduction in the head produced by the pump.

Methods have been suggested for estimating the deviation angle

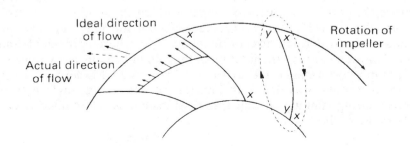

Fig. 13.35

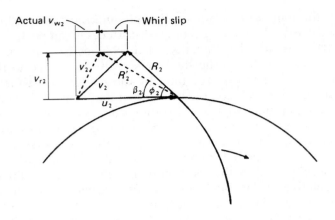

Fig. 13.36

$(\phi_2 - \beta_2)$ – which may be as high as 15° in some designs – but these are still semi-empirical[2,3].

The greater the number of blades in the impeller the more uniform is the flow direction, and for an infinite number of blades the deviation of R_2 and hence the 'whirl slip' (see Fig. 13.36) would be zero. Frictional losses, however, become greater as the number of blades is increased, and the number of blades used – from six to twelve if they are backward-facing – is a compromise between these two opposing considerations. Forward-facing blades, having a convex rear surface, greatly encourage the tendency of the flow to break away from that side. This type of impeller therefore usually has a much greater number of blades (sometimes as many as 60 in large fans).

Deviation of the flow also occurs in turbine runners, but to a smaller degree.

13.4.4 Axial- and mixed-flow pumps

The axial-flow or propeller pump is the converse of the propeller turbine and is very similar in appearance to it. The impeller consists of a central boss on which are mounted a number of blades. It rotates within a cylindrical casing, ideally of sufficient length to allow uniform flow on each side of the pump, and the clearance between blades and casing is as small as practicable. On the outlet side of the impeller a set of stationary guide vanes is usually fitted: these vanes are designed to remove the whirl component of velocity which the fluid receives from the impeller (see Fig. 13.37). Guide vanes on the inlet side also are sometimes provided but, except in applications where appreciable tangential motion exists in the inlet pipe, they do not significantly improve the pump's performance and so are usually omitted. Modern axial-flow pumps are often provided with impeller blades whose angle may be altered during running so that a high efficiency is maintained over a wide range of discharge.

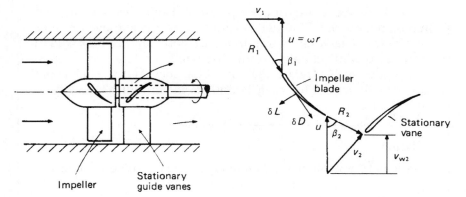

Fig. 13.37

The impeller of a mixed-flow pump is so shaped that the fluid enters axially but leaves with a substantial radial component.

The general formulae 13.7–13.10 apply, with reservations, no less to axial-flow machines than to centrifugal ones. In an axial machine a particle of fluid, in general, enters and leaves the impeller at the same radius. That is, $u_1 = u_2$ and, if we assume that v_w does not vary in the circumferential direction, the equations reduce to

$$\text{Work done on fluid per unit mass} = u\left(v_{w2} - v_{w1}\right)$$

$$= \frac{1}{2}\left\{\left(v_2^2 - v_1^2\right) - \left(R_2^2 - R_1^2\right)\right\} \tag{13.21}$$

It should be noted that the values of u, v, R and v_w vary (in general) with radius, and so the equations are applicable only to the fluid at a particular radius.

Impellers are commonly of 'free vortex design'; that is, the velocity of whirl at the outlet varies according to the relation $v_{w2}r = \text{constant}$. The product uv_{w2} is then constant since $u = \omega r$. With the further assumption that $v_{w1} = 0$, eqn 13.21 then applies regardless of radius (except perhaps close to the hub, where the free vortex design may be abandoned).

Since $v_{w2} = u - v_a \cot\beta_2$ and the axial component of velocity v_a is proportional to the volume flow rate Q, eqn 13.21 indicates that, for a fixed rotational speed and blade angle, the ideal head would decrease linearly with Q.

Example 13.3 A pair of contra-rotating axial-flow fans draw air into a duct from atmosphere, through a short entry section, and discharges the air to a ventilating system. A manometer connected to a pressure tapping in the duct wall just upstream of the first fan reads 11 mm H_2O below atmospheric pressure, and a manometer similarly connected just downstream of the second fan reads

75 mm H_2O above atmospheric pressure. Both fans have tip and hub diameters of 0.75 m and 0.4 m respectively, and each rotates at 25 rev/s. The total power input to both fans is 6.5 kW and there is no whirl at entry to the first fan. Both fans are of 'free-vortex' design. Calculate the total efficiency and the inlet and outlet blade angles for each fan at the mean radius. Density of air = 1.2 kg/m³.

Solution

Static head upstream = -11 mm $H_2O \equiv -11 \times 1000/1.2$ mm air

$$= -9.167 \text{ m air}$$

$\therefore$ Velocity upstream = $(2 \times 9.81 \times 9.167)^{1/2}$ m/s = 13.41 m/s

and volume flow rate = $\dfrac{\pi}{4}(0.75 \text{ m})^2 13.41 \text{ m/s} = 5.925 \text{ m}^3/\text{s}$

Total head rise across fans = $\{75 - (-11)\}$ mm H_2O = 86 mm H_2O which is equivalent to $1000 \times 9.81 \times 0.086$ Pa = 843.7 Pa.

$$\text{Fan total efficiency} = \frac{Q\Delta\varrho}{\text{Input power}} = \frac{\left(5.925 \text{ m}^3/\text{s}\right)\left(843.7 \text{ Pa}\right)}{6500 \text{ W}}$$

$$= 0.769$$

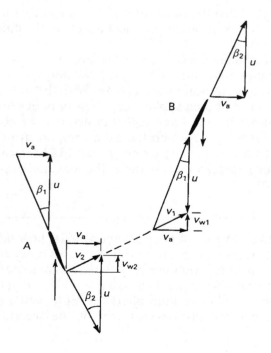

For no whirl at inlet to fan A or outlet from fan B (which is why contra-rotating fans are used) the theoretical head across both fans at a particular radius is

$$\left(\frac{uv_{w2} - 0}{g}\right)_A + \left(\frac{0 - uv_{w1}}{g}\right)_B$$

But as fans will be close together $|(v_{w2})_A| = |(v_{w1})_B|$ and as $(v_{w1})_B$ is in opposite direction to u_B it must be considered negative.

∴ Total theoretical head $= 2(uv_{w2})_A/g$ and theoretical pressure rise $= 2\varrho(uv_{w2})_A$. Since $uv_w = $ constant in 'free-vortex' flow these expressions are independent of radius.

But theoretical pressure rise $= 843.7\,\text{Pa}/0.769 = 1097\,\text{Pa}$

and at mean radius $u = \pi N \overline{D} = \pi 25\left(\frac{0.75 + 0.4}{2}\right)\text{m/s}$

$$= 45.16\,\text{m/s}$$

So, at mean radius, $\left(v_{w2}\right)_A = \dfrac{1097\,\text{Pa}}{2\left(1.2\,\text{kg/m}^3\right)\left(45.16\,\text{m/s}\right)}$

$$= 10.12\,\text{m/s}$$

Axial velocity through fans $\left(\text{assumed uniform}\right)$

$$= \frac{5.925\,\text{m}^3/\text{s}}{\frac{\pi}{4}\left(0.75^2 - 0.4^2\right)\text{m}^2}$$

$$= 18.74\,\text{m/s}$$

$$\therefore \left(\beta_1\right)_A = \arctan\left(v_a/u\right) = \arctan(18.74/45.16) = 22.5°$$

$$\left(\beta_2\right)_A = \arctan v_a/\left(u - v_{w2}\right) = \arctan\{18.74/(45.16 - 10.12)\}$$
$$= 28.1°$$

$$\left(\beta_1\right)_B = \arctan v_a/\left(u + v_{w1}\right) = \arctan\{18.74/(45.16 + 10.12)\}$$
$$= 18.7°$$

$$\left(\beta_2\right)_B = \arctan\left(v_a/u\right) = 22.5° \qquad\qquad \square$$

This simple analysis, however, is based on the assumption of uniform conditions at inlet and outlet, which, as we have already noted, would strictly hold only for blades of infinitesimal thickness with infinitesimal spacing. As an axial-flow machine normally has only a small number of fairly widely spaced blades, a better, though still simplified, analysis is often used in which each blade is considered to act as an isolated aerofoil.

Let us consider an axial-flow rotor with only a few blades (Fig. 13.37) and focus attention on a thin element of a blade at radius r from the axis and of thickness δr. We suppose for the moment that the flow past this element is unaffected by the presence of other blades or by the flow past other elements of the same blade (at different values of r). That is, we consider two-dimensional flow past the blade section. The force exerted by the fluid on this blade element depends on the magnitude and direction of the velocity R_1 of the fluid relative to the blade, and this force has components lift δL and drag δD, respectively perpendicular and parallel to the direction of R_1. The component of force acting *on the blade element* in the whirl direction is

$$\delta F_{\mathrm{w}} = -\delta L \sin \beta_1 - \delta D \cos \beta_1$$

Expressing δL and δD in terms of lift and drag coefficients (Section 9.9), we then have

$$\delta F_{\mathrm{w}} = \frac{1}{2}\varrho R_1^2 c \ \delta r \left(-C_{\mathrm{L}} \sin \beta_1 - C_{\mathrm{D}} \cos \beta_1 \right) \tag{13.22}$$

where c denotes the chord length of the aerofoil section.

The force in the whirl direction on the blade element contributes an amount $r \ \delta F_{\mathrm{w}}$ to the torque on the rotor. For the elements of n blades the work done per unit time interval is thus $n\omega r \ \delta F_{\mathrm{w}}$ where ω represents the angular velocity of the rotor. The work done (on the rotor) per unit mass of fluid passing the blade elements at radius r is therefore

$$\frac{n\omega r \ \delta F_{\mathrm{w}}}{\varrho v_{\mathrm{a}} 2\pi r \ \delta r}$$

where v_{a} denotes the axial component of velocity (assumed uniform) upstream of the rotor. Substituting for δF_{w} from eqn 13.22 and changing signs gives

Work done *on the fluid*/unit mass

$$= \frac{n\omega r}{v_{\mathrm{a}} 2\pi r} \frac{1}{2} R_1^2 c \left(C_{\mathrm{L}} \sin \beta_1 + C_{\mathrm{D}} \cos \beta_1 \right) = g(\Delta H) \tag{13.23}$$

where ΔH represents the increase of total head across the rotor at the radius considered. If there is no whirl at inlet, $v_{\mathrm{a}} = v_1 = \omega r \tan \beta_1$ and $R_1 = \omega r \sec \beta_1$. Equation 13.23 then yields

$$\Delta H = \frac{1}{2g}\left(\frac{nc}{2\pi r}\right)\omega^2 r^2 \left(C_{\mathrm{L}} \sec \beta_1 + C_{\mathrm{D}} \operatorname{cosec} \beta_1 \right) \tag{13.24}$$

The quantity $nc/2\pi r$ is termed the *solidity*. Since it equals $c \div$ (circumferential distance between corresponding points on adjacent blades) it is a measure of the proportion of the cross-section occupied by the blades.

In general, it varies along the length of the blades, usually being larger at the hub than at the blade tips. Of course, the greater the solidity the less trustworthy is the assumption that the flow past one blade does not affect that past another.

Since eqn 13.24 applies in general only to a single value of r, an integration along the blades must be performed to obtain results for the machine as a whole. It is commonly assumed for design purposes that ΔH and v_a are independent of r. Even so, analytical integration is not possible and a good deal of empirical information based on previous experience is required.

The values of lift and drag coefficients are normally taken from data obtained for single aerofoils in two-dimensional flow at constant angle of attack. The use of such values, however, is open to question. The relative velocity R_1 varies both in magnitude and direction along a blade; unless the blade is twisted appropriately, the angle of attack also varies with radius; in practice too there is some flow radially along the blade. Blade-element theory is thus no more than an approximation to the truth.

Nevertheless, we can deduce from it the general shape of the pump characteristic for a fixed-blade machine. For two-dimensional flow past isolated aerofoils of small camber (i.e. curvature of the centre line) and thickness at small angles of attack $C_L \simeq 2\pi \sin \alpha$. Here α denotes the angle between the upstream flow direction and that giving zero lift. For a particular blade element in the rotor $\alpha = \theta - \beta_1$, where θ denotes the (fixed) angle between the zero-lift direction and the whirl direction. So substituting $C_L = 2\pi \sin(\theta - \beta_1) = 2\pi(\sin \theta \cos \beta_1 - \cos \theta \sin \beta_1)$ in eqn 13.24 we obtain

$$\Delta H = \frac{1}{2g}\left(\frac{nc}{2\pi r}\right)\omega^2 r^2 \left\{2\pi\left(\sin \theta - \cos \theta \tan \beta_1\right) + C_D \operatorname{cosec} \beta_1\right\}$$

If the flow rate, and therefore the axial velocity component, is increased beyond the design value while the blade speed is kept constant, the angle β_1 is increased (see Fig. 13.37). Hence α decreases, C_D decreases (see Fig. 8.17) and ΔH decreases. Conversely, a reduction of flow rate causes an increase of ΔH. However, at a certain value of flow rate α will have been increased so much that the aerofoil stalls (as we recall from Section 8.8.6 the boundary layer then separates from much of the low-pressure side of the aerofoil), C_L falls markedly and the increase in ΔH is curtailed. Stall may not occur at all sections of the blades simultaneously, but when it is sufficiently widespread the overall performance of the machine declines. These effects are illustrated in Fig. 13.39c.

13.4.5   Similarity laws and specific speed for pumps

The relations developed in Section 13.3.5 are equally applicable to pumps. Just as the 'shape' of a turbine is characterized by its specific speed for maximum efficiency conditions, so is that of a pump. For a pump, however, the quantities of interest are N, H and Q rather than N, H and P and so the form of K_n used for a pump is $NQ^{1/2}/(gH)^{3/4}$. (This may be obtained directly from the relations 13.11 as $\Pi_1^{1/2}/\Pi_2^{3/4}$) Unfortunately g – being practically constant – has usually been disregarded. The resulting expression $N_s = NQ^{1/2}/H^{3/4}$ is thus not dimensionless and its numeric depends on the units adopted for N, Q and H. The unit for N has almost always been rev/min; for H the foot (or, in most of Europe, the metre) has been used; but the units for Q have been many. In Great Britain, Imperial gallons per minute has frequently been used but ft³/s has also found wide favour; in the United States, U.S. gallons per minute and ft³/s are used; in both countries ft³/min has been more usually employed for the discharge of fans. Litres/s and m³/s are both used in the metric system. In the face of such diversity it is essential that the units of specific speed be precisely stated, and it is much to be hoped that the 'dimensionless' form K_n will be universally adopted. (As for turbines, K_n is truly dimensionless only if the radian measure of angle is used in expressing N. Since, however, the revolution is usually preferred as a unit, this then remains as the unit of K_n.)

The effect of the shape of the rotor on the specific speed is broadly similar to that for turbines as indicated in Fig. 13.20. That is, radial-flow (i.e. centrifugal) impellers have the lower values of K_n and N_s and axial-flow designs have the higher values. The impeller, however, is not the entire pump; variations in the shape of the casing, especially of the volute in centrifugal pumps, may appreciably affect the specific speed. Nevertheless, in general, centrifugal pumps are best suited to providing high heads at moderate rates of flow, while axial-flow pumps are favoured for large rates of flow at low heads. As for turbines, the higher the specific speed the more compact the machine for given requirements.

For multi-stage pumps the specific speed refers to a single stage. For double-suction centrifugal impellers, half the discharge Q is generally used in calculating K_n or N_s.

Other useful results may be derived from the dimensionless parameters (13.11). For dynamically similar conditions Q/ND^3, for example, must be the same. Therefore if D is fixed, as in the same or an identical pump, Q is proportional to N so long as conditions are dynamically similar. From Π_2 and Π_4 it may be deduced that, with ϱ and g constant, $H \propto N^2$ and $P \propto N^3$ while conditions remain dynamically similar. If the flow conditions in the pump are similar the ratio between hydraulic head losses and the Euler head is unaltered and therefore the hydraulic efficiency remains unchanged. (The overall efficiency is not necessarily unchanged, because this takes account of mechanical losses such as bearing friction which may vary differently; these losses, however, are usually small.)

The relations $Q \propto N$, $H \propto N^2$, $P \propto N^3$ are often known as the *affinity laws* for pumps. They allow the performance characteristics of a pump at one speed to be predicted from the results of experiments carried out at a different speed (see Fig. 13.38). They may thus also be used for

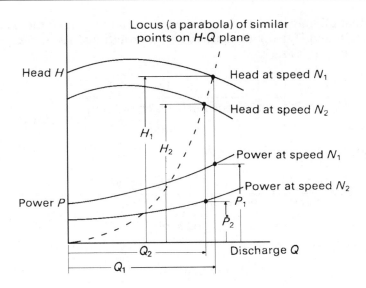

correcting individual experimental values of Q, H and P obtained at speeds slightly different from the intended one. Since, however, they apply only when dynamic similarity is maintained, they do not describe all possible changes in performance with a change of speed. For example, Q does not necessarily change in proportion to N if H is held constant.

When the range of speed is wide, the affinity laws are only approximately true because the modifying effect of Reynolds number ($\Pi_1 \times \Pi_3$) is neglected. It must also be remembered that cavitation can play havoc with dynamic similarity and so falsify the affinity laws.

Example 13.4 The results of tests on a water pump, 0.5 m in diameter and operating at 750 rev/min, are given in the table.

Q (m³ s⁻¹)	0	0.006	0.012	0.018	0.024	0.030	0.036	0.042	0.052
H (m)	15	16	16.5	16.5	15.5	13.5	10.5	7	0
η (%)	0	30	55	70	76	70	57	38	0

Use these data to deduce the pump characteristics when the speed is increased to 900 rev/min.

Solution
We use the similarity laws, with the dimensionless parameters

$$\Pi_1 = \frac{Q}{ND^3} \quad \text{and} \quad \Pi_2 = \frac{gH}{N^2 D^2}$$

So that all units are expressed in SI units, and to make the quantities Π_1 and Π_2 truly dimensionless, the rotational speed must be converted to rad s^{-1}.

$$\Pi_1 = \frac{Q_{750}}{ND^3} = \frac{Q_{750}\left(\text{m}^3/\text{s}\right)}{\dfrac{\left(750 \text{ rev/min}\right) \times \left(2\pi \text{ rad/rev}\right)}{60 \text{ s/min}} \times \left(0.5\,\text{m}\right)^3}$$

Also

$$\Pi_1 = \frac{Q_{900}}{ND^3} = \frac{Q_{900}\left(\text{m}^3/\text{s}\right)}{\dfrac{\left(900 \text{ rev/min}\right) \times \left(2\pi \text{ rad/rev}\right)}{60 \text{ s/min}} \times \left(0.5\,\text{m}\right)^3}$$

Hence

$$Q_{900} = \frac{900 \text{ rev/min}}{750 \text{ rev/min}} \times Q_{750} = 1.2 Q_{750}$$

Similarly

$$\Pi_2 = \frac{gH_{750}}{N^2 D^2} = \frac{9.81 \text{ m/s}^2 \times H_{750}\left(\text{m}\right)}{\left(\dfrac{\left(750 \text{ rev/min}\right) \times \left(2\pi \text{ rad/rev}\right)}{60 \text{ s/min}}\right)^2 \times \left(0.5\,\text{m}\right)^2}$$

and

$$\Pi_2 = \frac{gH_{900}}{N^2 D^2} = \frac{9.81 \text{ m/s}^2 \times H_{900}\left(\text{m}\right)}{\left(\dfrac{\left(900 \text{ rev/min}\right) \times \left(2\pi \text{ rad/rev}\right)}{60 \text{ s/min}}\right)^2 \times \left(0.5\,\text{m}\right)^2}$$

which yields after simplification

$$H_{900} = \left(\frac{900}{750}\right)^2 \times H_{750} = 1.44 H_{750}$$

Provided that the calculations are done carefully, it is seen that it is not necessary to convert the rotational speeds from rev/min to rad s^{-1}, as the conversion factors ultimately cancel out.

The table for the pump characteristics at 900 rev/min can now be drawn up.

Q (m³ s⁻¹)	0	0.0072	0.0144	0.0216	0.0288	0.036	0.0432	0.0504	0.0624
H (m)	21.6	23.0	23.8	23.8	22.3	19.4	15.1	10.1	0
η (%)	0	30	55	70	76	70	57	38	0

Example 13.5 Water is to be pumped at $0.04\,\text{m}^3/\text{s}$ upwards through a vertical distance of $28\,\text{m}$. The suction and delivery pipes will each be $150\,\text{mm}$ diameter with a friction factor, f, of 0.006, and the combined length will be $38\,\text{m}$. Losses at valves, etc. may be expected to total three times the velocity head in the pipes. The pump is to be driven by a constant-speed a.c. electric motor on a $50\,\text{Hz}$ supply and directly coupled to the shaft. Four single-stage, single-entry, centrifugal pumps are available:

1. $K_n = NQ^{1/2}(gH)^{-3/4} = 0.068\,\text{rev}$, impeller diameter, $D = 275\,\text{mm}$;
2. $K_n = 0.072\,\text{rev}$, $D = 200\,\text{mm}$;
3. $K_n = 0.099\,\text{rev}$, $D = 300\,\text{mm}$;
4. $K_n = 0.110\,\text{rev}$, $D = 250\,\text{mm}$.

In each pump the blades are backward-facing at $30°$ to the tangent, and the outlet width of the impeller passages is one-tenth of the diameter. Neglecting the blade thickness and whirl slip, and assuming a manometric efficiency of 75%, select the most suitable pump.

Solution

$$\text{Velocity in pipes} = 0.04\,\text{m}^3/\text{s} \div \frac{\pi}{4}(0.15\,\text{m})^2 = 2.264\,\text{m/s}$$

∴ Total head loss through pipes and valves

$$= \left(3 + \frac{4 \times 0.006 \times 38}{0.15}\right)\frac{2.264^2}{19.62}\,\text{m} = 2.371\,\text{m}$$

∴ Manometric head $= (28 + 2.371)\text{m} = 30.37\,\text{m}$

$N = (50/n)\,\text{rev/s}$ where n = number of pairs of poles.

$$K_n = \frac{NQ^{1/2}}{(gH)^{3/4}} = \frac{50(0.04)^{1/2}}{n(9.81 \times 30.37)^{3/4}}\,\text{rev} = \frac{0.1394}{n}\,\text{rev}$$

If $n = 2$, $K_n = 0.0697\,\text{rev}$, which suggests pump 1 or 2, and $N = 25\,\text{rev/s}$. Outlet flow area $= \pi D \times D/10$.

$$\therefore\ v_{r2} = \frac{0.04}{\pi D^2/10}\frac{\text{m}^3}{\text{s}} = \frac{0.1273}{D^2}\frac{\text{m}^3}{\text{s}}$$

$$u_2 = \pi ND = 78.54D\,\text{s}^{-1}$$

Manometric efficiency $= 0.75 = gH/u_2 v_{\text{w2}}$.

$$\therefore\ v_{\text{w2}} = \frac{9.81 \times 30.37}{0.75 \times 78.54D}\frac{\text{m}^2}{\text{s}} = \frac{5.06}{D}\frac{\text{m}^2}{\text{s}}$$

$$\tan 30° = \frac{v_{r2}}{u_2 - v_{w2}} = \frac{0.1273}{D^2\left(78.54D - \dfrac{5.06}{D}\right)}$$

$$\therefore\ 78.54D^3 - 5.06D = 0.1273 \cot 30° = 0.2205 \ \text{(in metre units)}$$

As a first approximation, neglect the 0.2205.

Then $D = \left(5.06/78.54\right)^{1/2}$ m $= 0.2538$ m

Next $D^3 = \left(0.2205 + 5.06 \times 0.2538\right)/78.54$, whence $D = 0.2678$ m

Then $D^3 = \left(0.2205 + 5.06 \times 0.2678\right)/78.54$, whence $D = 0.272$ m

That's near enough. So choose Pump 1.

13.4.6 Cavitation in centrifugal pumps

Cavitation is likely to occur on the inlet side of a pump particularly if the pump is situated at a level well above the surface of the liquid in the supply reservoir. For the sake of good efficiency and the prevention of damage to the impeller, cavitation should be avoided. A cavitation number for centrifugal pumps may be derived in a manner similar to that for turbines (see Section 13.3.6). Applying the energy equation between the surface of liquid in the supply reservoir and the entry to the impeller (where the pressure is a minimum) we have, for steady conditions,

$$\frac{p_0}{\varrho g} - h_f = \frac{p_{\min}}{\varrho g} + \frac{v_1^2}{2g} + z_1 \tag{13.25}$$

where v_1 represents the fluid velocity at the point where the static pressure has its least value $p_{\min}$, z_1 the elevation of this point above the surface of the liquid in the reservoir, and p_0 the pressure at that surface (usually, but not necessarily, atmospheric). Strainers and non-return valves are commonly fitted to intake pipes. The term h_f must therefore include the head loss occurring past these devices, in addition to losses caused by 'ordinary' pipe friction and by bends in the pipe.

For a given design of pump operating under specified conditions, $v_1^2/2g$ may be taken as a particular proportion of the head developed by the pump, say $\sigma_c H$. Then, rearranging eqn 13.25, we have

$$\sigma_c = \frac{p_0/\varrho g - p_{\min}/\varrho g - z_1 - h_f}{H}$$

For cavitation not to occur $p_{\min}$ must be greater than p_v, the vapour pressure of the liquid, i.e. $\sigma > \sigma_c$ where

$$\sigma = \frac{p_0/\varrho g - p_v/\varrho g - z_1 - h_f}{H} \tag{13.26}$$

and σ_c is the critical value of this parameter at which appreciable cavitation begins. Experiments show the σ_c is related to the specific speed of the pump.

The numerator of the expression 13.26 is known as the 'Net Positive Suction Head' (NPSH). ('Suction' is here simply a synonym for 'inlet'.)

In order that σ should be as large as possible, z_1 must be as small as possible. In some installations it may even be necessary to set the pump *below* the reservoir level (i.e. with z_1 negative) to avoid cavitation.

Conditions in axial-flow pumps are even more conducive to cavitation than those in centrifugal pumps. Since, however, the liquid does not enter an axial-flow machine at a single radius, the overall parameter of eqn 13.26 is not suitable for this type of pump, and more complicated analysis is necessary. Cavitation in an axial-flow pump usually begins on the backs of the blade-tips because that is where the pressure is least. However, breakaway of the flow from a blade may induce cavitation at other radii.

Recently, 'super-cavitating' machines have been developed in which the minimum pressure is unusually low, so the cavitation takes the form, not of small bubbles which collapse violently against the blade surfaces, but of large bubbles which are carried away from the surfaces. Without the restriction on speed that cavitation ordinarily imposes, a super-cavitating machine may be made smaller for a given flow rate. Also, since a lower minimum pressure is allowable, restrictions on the positioning of the machine are less severe. On the other hand, for conditions under which a conventional machine would be satisfactory, a super-cavitating one has a lower efficiency. Such machines are thus likely to find favour only in specialized applications where the advantages override the reduction of efficiency.

13.4.7 The performance characteristics of pumps

Pumps are normally run at constant speed. Interest therefore attaches to the variation of head H with discharge Q and also to the variations of efficiency and power required with Q. The results of a particular test may be made available for a different speed – or for a homologous pump of different diameter – by plotting the results in dimensionless form, that is, using the dimensionless parameters Q/ND^3, gH/N^2D^2 and

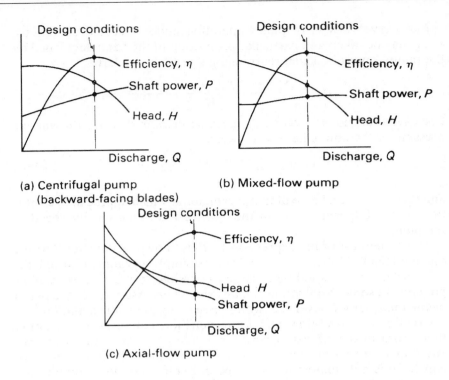

Fig. 13.39

(a) Centrifugal pump
(backward-facing blades)

(b) Mixed-flow pump

(c) Axial-flow pump

$P/\varrho N^3 D^5$ in place of Q, H and P respectively. Typical 'characteristic curves' for pumps are shown in Fig. 13.39.

A curve of H against Q which has a peak is termed an unstable characteristic. This is because the slope of the H–Q curve for an external system is normally always positive (see the dashed line in Fig. 13.34). If part of the pump characteristic also has a positive slope there is thus the possibility that, at the point of intersection, the pump characteristic could have a greater slope than the other curve. Any slight increase of Q, for example, would then result in the pump head rising more than the system head, and the excess head at the pump would cause Q to increase still further. Moreover, the two curves could intersect again at a second, higher, value of Q; if the two intersections were fairly close together the pump would tend to 'hunt' (or 'surge') from one to the other. Such instability is undesirable and so an unstable part of the characteristic should be outside the normal operating range of the pump. (The subject of the stability of pumping systems is a large one: a very thorough review, with 183 references, is given by Greitzer[4].)

For pumps handling liquids, the increase of total head is little different from the increase of piezometric head (i.e. the manometric head). For fans, however, kinetic energy often forms a substantial part of the total energy increase. Characteristic curves for fans are therefore normally based on the total head (static plus kinetic). The operating conditions of a fan are set by the intersection of its total head characteristic with the curve of the total head required by the external system (that is, the ducting and so on, to which the fan is connected). The latter head, it should be remembered, consists not only of that needed to overcome frictional resistance but also of the velocity head at the duct outlet.

(Where a fan is used to exhaust from a duct system to the atmosphere, the entire kinetic energy at outlet from the fan is wasted. If the total head at the inlet end of the duct is atmospheric and the fan exhausts directly to the atmosphere.

Fan Total Head = h_f for duct system + velocity head at fan outlet

i.e.

h_f for duct system = Fan Total Head − velocity head at fan outlet
= 'Fan Static Head'

Fan Static Head is conventionally defined in this way; it is *not* identical with the difference of static head between inlet and outlet. The intersection of curves of h_f for the duct system and Fan Static Head therefore gives the required matching of the fan to the external system. The operating conditions are commonly determined thus, rather than by the exactly equivalent process of matching Fan *Total* Head with h_f for the duct system *plus* velocity head at fan outlet.)

Example 13.6 A pump draws water from a large open sump through a short length of suction pipe. The water is delivered through a 21.1 m length of 9 cm diameter pipe and via a submerged exit into an open tank. A control valve is installed in the delivery pipe, for which $f = 0.0085$. Dissipation other than friction and the loss at the valve may be ignored. The bottom of the tank is 10 m above the free surface of the water in the sump.

(a) Express the head loss due to pipe friction in terms of flow rate Q.

The characteristics of the water pump, 0.5 m in diameter and operating at 750 rev/min, are given in the table.

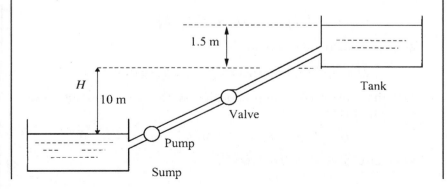

Q (m³ s⁻¹)	0	0.006	0.012	0.018	0.024	0.030	0.036	0.042	0.052
H (m)	15	16	16.5	16.5	15.5	13.5	10.5	7	0
η (%)	0	30	55	70	76	70	57	38	0

(b) Assuming the control valve is fully open, plot curves showing the variation of H with Q and η with Q, and use them to determine the discharge the pump will provide when the water in the delivery tank is 1.5 m deep. What power will be required?

Solution

(a) Combination of the head loss and continuity relations

$$h_l = h_f = \frac{4fl}{d}\frac{\bar{u}^2}{2g} \quad \text{and} \quad Q = \bar{u}A$$

yields

$$h_l = h_f = \frac{4fl}{d}\frac{16Q^2}{2\pi^2 d^4 g} = \frac{32 \times 0.0085 \times 21.1\,\text{m}}{\pi^2 \times 9.81\,\text{m/s}^2 \times (0.09\,\text{m})^5}Q^2\,(\text{m}^3/\text{s})^2$$

$$\approx (100Q)^2\,\text{m}$$

(b) Denote the total head at the free surfaces of the sump and tank by H_1 and H_2. Then

$$H_2 = H_1 - h_l + (H)_{\text{pump}}$$

Now

$$H_1 = \frac{p_1}{\varrho g} + \frac{u_1^2}{2g} + z_1 \quad \text{and} \quad H_2 = \frac{p_2}{\varrho g} + \frac{u_2}{2g} + z_2$$

Since

$$u_1 \approx u_2 \approx 0; \quad p_1 = p_2; \quad z_2 - z_1 = H_S$$
$$H_2 - H_1 = H_S$$

and

$$(H)_{\text{pump}} = H_S + h_l = (H)_{\text{pipe}}$$

The pipe system characteristic is

$$H = 10 + 1.5 + (100Q)^2\,\text{m} = 11.5 + (100Q)^2\,\text{m}$$

From the plot of the pump and pipe system characteristics, the intersection is at

$$H = 16\,\text{m}; \quad Q = 0.021\,\text{m}^3\text{s}^{-1}; \quad \eta = 74\%$$

Hence discharge $Q = 0.021\,\text{m}^3\text{s}^{-1}$.

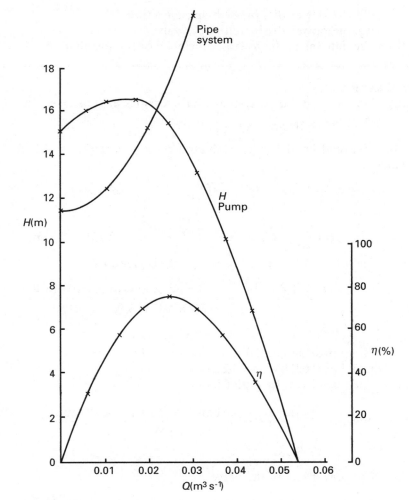

The power P is given by

$$P = \frac{\varrho g H Q}{\eta} = \frac{10^3\,\text{kg/m}^3 \times 9.81\,\text{m/s}^2 \times 16\,\text{m} \times 0.021\,\text{m}^3/\text{s}}{0.74 \times 10^3\,\text{W/kW}}$$

$$= 4.45\,\text{kW}$$

□

Example 13.7 The valve in Example 13.6 is now partially closed, reducing the flow rate to $0.015\,\text{m}^3\text{s}^{-1}$.

(a) If the pump rotational speed and depth of water in the tank remain unchanged, calculate:
  (i) the power consumption of the pump
  (ii) the power dissipated in the pump

■

(iii) the power dissipated by pipe friction

(iv) the power dissipated in the valve.

(b) Calculate, also, the overall efficiency of the installation.

Solution

(a) The pump and pipe system characteristics now intersect at

$$H = 16.5 \,\mathrm{m}; \quad Q = 0.015 \,\mathrm{m^3 s^{-1}}; \quad \eta = 63\%$$

The frictional head loss is evaluated as $h_f = (100Q)^2 = 2.25 \,\mathrm{m}$. Since

$$h_l = h_f + h_{\text{valve}} \quad \text{and} \quad \left(H\right)_{\text{pump}} = h_l + H_s = \left(H\right)_{\text{pipe}}$$

we deduce that

$$h_{\text{valve}} = \left(H\right)_{\text{pump}} - H_s - h_f = \left(16.5 - 11.5 - 2.25\right) = 2.75 \,\mathrm{m}$$

(i) The power P supplied to the pump is given by

$$P = \frac{\varrho g H Q}{\eta} = \frac{10^3 \,\mathrm{kg/m^3} \times 9.81 \,\mathrm{m/s^2} \times 16.5 \,\mathrm{m} \times 0.015 \,\mathrm{m^3/s}}{0.63 \times 10^3 \,\mathrm{W/kW}}$$

$$= 3.85 \,\mathrm{kW}$$

(ii) The power dissipated in the pump $= P(1 - \eta) = 3.85(1 - 0.63) \,\mathrm{kW} = 1.42 \,\mathrm{kW}$.

(iii) The power lost by pipe friction $=$

$$\varrho g h_f Q = \frac{10^3 \,\mathrm{kg/m^3} \times 9.81 \,\mathrm{m/s^2} \times 2.25 \,\mathrm{m} \times 0.015 \,\mathrm{m^3/s}}{10^3 \,\mathrm{W/kW}}$$

$$= 0.33 \,\mathrm{kW}$$

(iv) The power lost in the valve $=$

$$\varrho g h_{\text{valve}} Q = \frac{10^3 \,\mathrm{kg/m^3} \times 9.81 \,\mathrm{m/s^2} \times 2.75 \,\mathrm{m} \times 0.015 \,\mathrm{m^3/s}}{10^3 \,\mathrm{W/kW}}$$

$$= 0.40 \,\mathrm{kW}$$

(b) The useful rate of working to raise water $=$

$$\varrho g H_s Q = \frac{10^3 \,\mathrm{kg/m^3} \times 9.81 \,\mathrm{m/s^2} \times 11.5 \,\mathrm{m} \times 0.015 \,\mathrm{m^3/s}}{10^3 \,\mathrm{W/kW}}$$

$$= 1.69 \,\mathrm{kW}$$

At this stage we can check our working. The sum of the contributions in (ii)–(iv) plus (b) should equal that of (i). Thus $(1.42 + 0.33 + 0.40 + 1.69) \,\mathrm{kW} = 3.84 \,\mathrm{kW}$. The small difference between the two figures is due to rounding errors. The overall efficiency of the installation $= (1.69)/(3.85) = 44\%$.

13.5 HYDRODYNAMIC TRANSMISSIONS

Fundamentally a hydrodynamic transmission consists of a pump and a turbine connected in series. The fluid discharged from the pump drives the turbine; the discharge from the turbine is returned to the inlet of the pump. To minimize energy losses the pump and turbine rotors are enclosed in a single casing. The pump element is known as the 'primary' and receives energy from a prime mover such as an internal-combustion engine. Energy is delivered from the shaft of the turbine, or 'secondary' element. The output characteristics of the transmission are naturally those of a turbine (that is, the torque increases as the speed falls), and they are of particular value in such applications as vehicle and ship transmissions and machine tool drives.

There are two distinct types of hydrodynamic transmission – fluid couplings and torque converters.

13.5.1 The fluid coupling

The essential features of the coupling are illustrated in Fig. 13.40. The primary and secondary runners are the only elements involved. Fluid (usually an oil of low viscosity) flows directly from one runner to the other without passing through any intervening stationary passages. The casing is usually fixed to one of the runners and rotates with it, but the two runners are otherwise identical. Each has straight radial blades and so somewhat resembles half a grapefruit with the pulp removed from the segments. The flow is consequently radially outwards in the primary and radially inwards in the secondary. Flow occurs in this direction because the speed of the primary exceeds that of the secondary. The head produced in the primary is thus greater than the centrifugal head resisting flow through the secondary. If the two speeds were the same the heads would balance, and then no flow would occur and no torque would be transmitted.

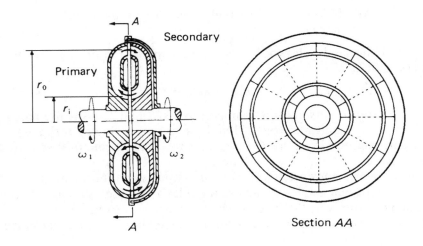

Section *AA* **Fig. 13.40** Fluid coupling.

For steady conditions, the input torque T_1 and output torque T_2 must be equal since there is no other member in the system to provide a torque reaction. Since power $= T\omega$, where ω represents angular velocity, the efficiency of the coupling $= T_2\omega_2/T_1\omega_1 = \omega_2/\omega_1 = 1 - s$ where s, the 'slip', $= (\omega_1 - \omega_2)/\omega_1$. The slip is usually expressed as a percentage, the value at normal operating conditions being about 2–3%. There are two reasons why the efficiency is less than 100%. One is that energy is dissipated by friction as the fluid moves relative to the solid surfaces. Secondly, turbulence is generated as the fluid from one runner strikes the blades of the other runner moving at slightly different velocity. Although such shock losses could be reduced by rounding the inlet edges of the blades, such a refinement is not worth the considerable extra expense.

The principal advantages of the coupling are found in unsteady operation. Torsional vibrations in either the primary or secondary shaft are not transmitted to the other; further, the full torque is developed only at full speed, thus easing the starting load on an internal-combustion engine or an electric motor. A so-called 'slip coupling' may also be used as a slipping clutch, that is, with ω_2 much less than ω_1: this is achieved by restricting the normal circulation of fluid either by reducing the quantity of fluid in the coupling or by throttling the flow. Although the efficiency suffers, the control of slip in this way is a useful temporary facility.

The relations derived for pump and turbine rotors (e.g. 13.7, 13.10) apply to the elements of a coupling – again with the provisos about uniformity of conditions at inlet and outlet of runners. For example, from eqn 13.7, the work done on the secondary runner per unit mass of fluid $= u_1 v_{w1} - u_2 v_{w2}$. If zero 'whirl slip' is assumed, then, since radial blades are used in both runners, the initial whirl component v_{w1} of the fluid entering the secondary is identical with the blade velocity of the primary at that radius, i.e. $\omega_1 r_o$, where r_o is the relevant radius (see Fig. 13.40). Moreover, v_{w2} for the secondary is identical with the blade velocity at outlet, $u_2 = \omega_2 r_i$. Therefore

Work done on the secondary per unit mass of fluid
$$= \omega_2 r_o \omega_1 r_o - \omega_2^2 r_i^2 = \omega_1 \omega_2 r_o^2 - \omega_2^2 r_i^2 \qquad (13.27)$$

Similarly it may be shown that the work done by the primary per unit mass of fluid is

$$\omega_1^2 r_o^2 - \omega_1 \omega_2 r_i^2 \qquad (13.28)$$

The difference between the expressions 13.27 and 13.28 is the energy dissipated per unit mass of fluid and, as the flow is highly turbulent, is proportional to Q^2, where Q represents the volume rate of flow round the circuit.

In practice, however, there is a significant variation of radius, and therefore of blade velocity, across the inlet and outlet sections of both runners. Moreover, the rate at which fluid passes from one runner to the other varies with the radius in a manner not readily determined.

Consequently, the expressions 13.27 and 13.28 can be regarded as no better than first approximations to the truth even if mean values are used for r_o and r_i.

Dimensional analysis indicates that the 'torque coefficient' of a coupling,

$$\frac{T}{\varrho \omega_1^2 D^5} = \phi\left(s, \ \frac{\varrho \omega_1 D^2}{\mu}, \ \frac{V}{D^3}\right)$$

where V represents the volume of fluid in the coupling and D the diameter. The term V/D^3 is thus proportional to the ratio of the volume of fluid to the total volume, and $\varrho \omega_1 D^2/\mu$ corresponds to the Reynolds number of the fluid flow. Since the flow is usually highly turbulent, however, the effect of Reynolds number on the torque is small, and that of slip s is far greater.

A fluid coupling used in a motor-car transmission is normally incorporated in the engine flywheel: it is then loosely known as a 'fluid flywheel'. The general characteristics of a coupling are illustrated in Fig. 13.41. The 'stall torque' is the torque transmitted when the secondary shaft is locked. The highest speed at which the primary can rotate under these conditions is given by the intersection of the two torque curves. For motor-car engines this speed is of the order of 15 rev/s.

13.5.2 The torque converter

The essential difference between a fluid coupling and a torque converter is that the latter has a set of stationary blades in addition to the primary and secondary runners. If the stationary blades are so designed that they change the angular momentum of the fluid passing through them, a torque is exerted on them: to prevent their rotation an opposing torque must be applied from the housing. With this additional torque on the assembly as a whole, the torque on the secondary runner no longer equals that on the primary. Appropriate design of the stationary

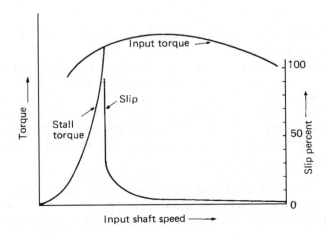

Fig. 13.41

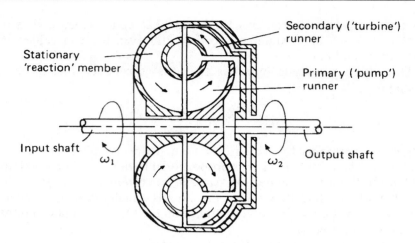

Fig. 13.42 Single-stage torque converter.

('reaction') blades allows the torque on the secondary to be as much as five times the input torque at the primary. Since, however, the reaction blades do not move, no work is done on them and the power output from the secondary runner is equal to the power input at the primary minus the power dissipated in turbulence. Many arrangements of the three elements of a torque converter are possible; one for a single-stage converter is shown in Fig. 13.42. For some purposes, particularly where a large torque multiplication is required, more complicated arrangements are used having two or even three 'turbine' stages.

A greater torque on the 'turbine' element than on the 'pump' element requires the fluid to undergo a greater change of angular momentum in the turbine element. The 'reaction' member is therefore so shaped as to increase the angular momentum of the fluid: the 'pump' impeller increases the angular momentum further: the 'turbine' part then removes – for steady conditions – all the angular momentum gained in the other two members. The output torque is thus the sum of the input torque and that on the reaction member:

$$T_2 = T_1 + T_{\text{reac}} \qquad (13.29)$$

Figure 13.43 shows the way in which torque ratio and efficiency vary for a typical converter.

The maximum efficiency of a converter is less than that of a fluid coupling because of the more complicated flow conditions. If the speed ratio is changed from that which gives maximum efficiency there is a corresponding change of the vector diagrams for each transition from one element to another, and much energy is dissipated in turbulence when the directions of the relative velocities of the fluid do not conform with the inlet edges of the blades.

Figure 13.43 shows that, for a converter designed to give a large increase of torque, the efficiency has a maximum value at a speed ratio much less than unity and falls off rapidly as the speed ratio approaches unity. This characteristic is a serious disadvantage in many applications.

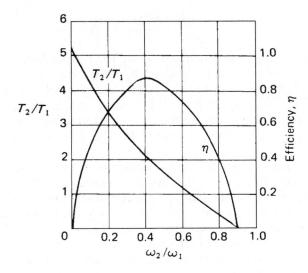

Fig. 13.43

One way of meeting the difficulty is this. At the higher speed ratios T_2/T_1 falls below unity, that is, the reaction torque becomes negative (eqn 13.29). If a ratchet device is fitted whereby the reaction member may rotate, but only in a forward direction, the reaction blades automatically begin to turn as the reaction torque changes sign. The entire device then behaves as a simple coupling; the efficiency is equal to the speed ratio, and the torque ratio remains at unity. Thus a combination of torque converter and coupling is obtained in which each is used in its best operating range. An alternative solution is to substitute a direct drive when the speed ratio reaches a certain value; this is achieved by a clutch that is required to slip until the speeds of the primary and secondary rotors are equalized.

The flow in torque converters may be studied by the use of the fundamental relations derived for pumps and turbines. For further details, however, more specialist works must be consulted.

13.6 THE EFFECT OF SIZE ON THE EFFICIENCY OF FLUID MACHINES

If the requirements laid down in Section 13.3.5 were exactly satisfied all fluid machines in the same homologous series would have the same efficiency when running at dynamically similar conditions. In practice, however, small machines, no matter how well designed and made, have lower efficiencies than larger members of the same homologous series.

One reason is that perfect dynamic similarity between a large machine and a small one cannot be achieved. For exact dynamic similarity each of the Π terms (13.11) must be the same for the large machine as for the small one. For example, if Π_2 is the same and the Reynolds number term Π_3 is the same, then

$$\Pi_2 \Pi_3^2 = \frac{gH}{N^2 D^2} \frac{N^2 D^4}{v^2} = \frac{g}{v^2} D^2 H$$

is also the same for both machines. Thus, for a given fluid, a reduction of D would have to be accompanied by an increase of H: a one-fifth scale model turbine would have to be tested at a head 25 times that for the prototype! The effect of Reynolds number on the efficiency is fortunately not large, so in practice models are tested under convenient low heads and empirical allowance made for the difference of Reynolds number.

More important, exact geometric similarity cannot be achieved. The *actual* roughness of surfaces in a small machine may differ little from that in a large machine; thus the *relative* roughness in the small machine is greater and the frictional losses are consequently more significant. The blades in the smaller machine may be *relatively* thicker. Clearances in the small machine cannot be reduced in the same proportion as other length measurements, and so leakage losses are relatively higher. Further departures from strict geometric similarity may be found in commercial pump impellers of nominally the same design.

All these effects reduce the hydraulic efficiency of the small machine. The overall efficiency is often reduced still further because the mechanical losses such as bearing friction and windage are relatively larger for the small machine.

Not all types of machine are affected similarly by changes of size. For example, a Pelton wheel does not suffer from leakage losses, so the effect of a reduction in size for such a machine is somewhat less than for a reaction machine.

When the results of tests on model machines are 'scaled up' for the prototype machine, empirical formulae are used to account for these differences of efficiency. A simple formula is that proposed by L. F. Moody (1880–1953) for reaction turbines:

$$\frac{1 - \eta}{1 - \eta_m} = \left(\frac{D_m}{D}\right)^n$$

where η represents the efficiency, D the diameter; suffix m refers to the model machine, and the exponent n is about 0.2.

REFERENCES

1. Worster, R. C. 'The flow in volutes and its effect on centrifugal pump performance', *Proc. Instn mech. Engrs*, **177**, 843–75 (1963) (see also corrigenda to Vol. 177).
2. Wislicenus, G. F. *Fluid Mechanics of Turbomachinery* (2nd edn), 267–81, Dover, New York (1965).
3. Moody, L. F. and Zowski, T. 'Hydraulic Machinery', Section 26 of *Handbook of Applied Hydraulics* (edited by C. V. Davis and K. E. Sorensen) (3rd edn), McGraw-Hill, New York (1969).

4. Greitzer, E. M. 'The stability of pumping systems', *J. Fluids Engng*, **103**, 193–242 (1981).

FURTHER READING

Anderson, H. *Centrifugal Pumps and Allied Machinery* (4th edn), Elsevier, Oxford (1994).

Danel, P. 'The hydraulic turbine in evolution', *Proc. Instn mech. Engrs*, **173**, 36–44 (1959).

Dean, R. C. 'On the necessity of unsteady flow in fluid machines', *J. basic Engng*, **81D**, 24–8 (1959).

Hammitt, F. G., *Cavitation and Multiphase Flow Phenomena*, McGraw-Hill, New York (1980).

Moss, S. A., Smith, C. W. and Foote, W. R. 'Energy transfer between a fluid and a rotor for pump and turbine machinery', *Trans. Am. Soc. mech. Engrs*, **64**, 567–97 (1942).

Norrie, D. H. *An Introduction to Incompressible Flow Machines*, Arnold, London (1963).

Pearsall, I. S. 'Cavitation', *Chartered mech. Engr*, **21**, No. 7, 79–85 (July 1974).

Stevens, J. C. and Davis, C. V. 'Hydroelectric Plants', Section 24 of *Handbook of Applied Hydraulics* (edited by C. V. Davis and K. E. Sorensen) (3rd edn), McGraw-Hill, New York (1969).

Wislicenus, G. F. *Fluid Mechanics of Turbomachinery* (2nd edn), Dover, New York (1965).

PROBLEMS

13.1 A single-acting reciprocating water pump, with a bore and stroke of 150 mm and 300 mm respectively, runs at 0.4 rev/s. Suction and delivery pipes are each 75 mm diameter. The former is 7.5 m long and the suction lift is 3 m. There is no air vessel on the suction side. The delivery pipe is 300 m long, the outlet (at atmospheric pressure) being 13.5 m above the level of the pump, and a large air vessel is connected to the delivery pipe at a point 15 m from the pump. Calculate the absolute pressure head in the cylinder at beginning, middle and end of each stroke. Assume that the motion of the piston is simple harmonic, that losses at inlet and outlet of each pipe are negligible, that the slip is 2%, and that f for both pipes is constant at 0.01. (Atmospheric pressure $\equiv$ 10.33 m water head.)

13.2 A reciprocating pump has two double-acting cylinders each 200 mm bore and 450 mm stroke, the cranks being at 90° to each other and rotating at 20 rev/min. The delivery pipe is

100 mm diameter, 60 m long. There are no air vessels. Assuming simple harmonic motion for the pistons determine the maximum and mean water velocities in the delivery pipe and the inertia pressure in the delivery pipe near the cylinders at the instant of minimum water velocity in the pipe.

13.3 In a hydro-electric scheme a number of Pelton wheels are to be used under the following conditions: total output required 30 MW; gross head 245 m; speed 6.25 rev/s; 2 jets per wheel; C_v of nozzles 0.97; maximum overall efficiency (based on conditions immediately before the nozzles) 81.5%; 'dimensionless specific speed' not to exceed 0.022 rev (for one jet); head lost to friction in pipe-line not to exceed 12 m. Calculate (a) the number of wheels required, (b) the diameters of the jets and wheels, (c) the hydraulic efficiency, if the blades deflect the water through 165° and reduce its relative velocity by 15%, (d) the percentage of the input power that remains as kinetic energy of the water at discharge.

13.4 The blading of a single-jet Pelton wheel runs at its optimum speed, 0.46 times the jet speed, with an overall efficiency of 0.85. Show that the 'dimensionless specific speed' is $0.1922 d/D$ rev, where d represents the jet diameter and D the wheel diameter. For the nozzle $C_v = 0.97$.

13.5 The following data refer to a Pelton wheel. Maximum overall efficiency 79%, occurring at a speed ratio of 0.46; C_v for nozzle = 0.97; jet turned through 165°. Assuming that the optimum speed ratio differs from 0.5 solely as a result of losses to windage and bearing friction which are proportional to the square of the rotational speed, obtain a formula for the optimum speed ratio and hence estimate the ratio of the relative velocity at outlet from the buckets to the relative velocity at inlet.

13.6 In a vertical-shaft inward-flow reaction turbine the sum of the pressure and kinetic heads at entrance to the spiral casing is 120 m and the vertical distance between this section and the tail-race level is 3 m. The peripheral velocity of the runner at entry is 30 m/s, the radial velocity of the water is constant at 9 m/s and discharge from the runner is without whirl. The estimated hydraulic losses are: (1) between turbine entrance and exit from the guide vanes, 4.8 m, (2) in the runner, 8.8 m, (3) in the draft tube, 790 mm, (4) kinetic head rejected to the tail race, 460 mm. Calculate the guide vane angle and the runner blade angle at inlet and the pressure heads at entry to and exit from the runner.

13.7 An inward-flow reaction turbine has an inlet guide vane angle of 30° and the inlet edges of the runner blades are at 120° to the direction of whirl. The breadth of the runner at inlet is a quarter of the diameter at inlet and there is no velocity of whirl at outlet. The overall head is 15 m and the speed 16.67 rev/s. The hydraulic and overall efficiencies may be assumed to be 88% and 85% respectively. Calculate the runner diameter at inlet and the power developed. (The thickness of the blades may be neglected.)

13.8   A vertical-shaft Francis turbine has an overall efficiency of 90% and runs at 7.14 rev/s with a water discharge of 15.5 m³/s. The velocity at the inlet of the spiral casing is 8.5 m/s and the pressure head at this point is 240 m, the centre-line of the casing inlet being 3 m above the tail-water level. The diameter of the runner at inlet is 2.23 m and the width at inlet is 300 mm. The hydraulic efficiency is 93%. Determine (a) the output power, (b) the 'dimensionless specific speed', (c) the guide vane angle, (d) the runner blade angle at inlet, (e) the percentage of the net head which is kinetic at entry to the runner. Assume that there is no whirl at outlet from the runner and neglect the thickness of the blades.

13.9   The runner of a vertical-shaft Francis turbine is 450 mm diameter and 50 mm wide at inlet and 300 mm diameter and 75 mm wide at outlet. The blades occupy 8% of the circumference. The guide vane angle is 24°, the inlet angle of the runner blades is 95° and the outlet angle is 30°. The pressure head at inlet to the machine is 55 m above that at exit from the runner, and of this head hydraulic friction losses account for 12% and mechanical friction 6%. Calculate the speed for which there is no shock at inlet, and the output power.

13.10   A quarter-scale turbine model is tested under a head of 10.8 m. The full-scale turbine is required to work under a head of 30 m and to run at 7.14 rev/s. At what speed must the model be run? If it develops 100 kW and uses 1.085 m³ of water per second at this speed, what power will be obtained from the full-scale turbine, its efficiency being 3% better than that of the model? What is the 'dimensionless specific speed' of the full-scale turbine?

13.11   A vertical-shaft Francis turbine is to be installed in a situation where a much longer draft tube than usual must be used. The turbine runner is 760 mm diameter and the circumferential area of flow at inlet is 0.2 m². The overall operating head is 30 m and the speed 6.25 rev/s. The guide vane angle is 15° and the inlet angle of the runner blades 75°. At outlet water leaves the runner without whirl. The axis of the draft tube is vertical, its diameter at the upper end is 450 mm and the (total) expansion angle of the tube is 16°. For a flow rate of Q m³/s the friction loss in the tube (of length l) is given by $h_f = 0.03\,Q^2 l$. If the absolute pressure head at the top of the tube is not to fall below 3.6 m of water, calculate the hydraulic efficiency of the turbine and show that the maximum permissible length of draft tube above the level of the tail water is about 5.36 m. (The length of the tube below tail-water level may be neglected. Atmospheric pressure ≡ 10.33 m water head.)

13.12 A Francis turbine develops 15 MW at 8.33 rev/s under a head of 180 m. For a water barometer height of 8.6 m estimate the maximum height of the bottom of the turbine runner above the tail water. [Use Fig. 13.22]

13.13 The impeller of a centrifugal pump has an outer diameter of 250 mm and runs at 25 rev/s. It has 10 blades each 5 mm thick; they are backward-facing at 30° to the tangent and the breadth of the flow passages at outlet is 12.5 mm. Pressure gauges are fitted close to the pump casing on the suction and discharge pipes and both are 2.5 m above the water level in the supply sump. The suction pipe is 120 mm diameter. When the discharge is 0.026 m³/s the gauge readings are respectively 4 m vacuum and 16.5 m. Assuming that there is no whirl at inlet and no whirl slip, estimate the manometric efficiency of the pump and the losses in the impeller if 50% of the velocity head at outlet from the impeller is recovered as static head in the volute.

13.14 The impeller of a centrifugal fan has an inner radius of 250 mm and width of 187.5 mm; the values at exit are 375 mm and 125 mm respectively. There is no whirl at inlet, and at outlet the blades are backward-facing at 70° to the tangent. In the impeller there is a loss by friction of 0.4 times the kinetic head corresponding to the *relative* outlet velocity, and in the volute there is a gain equivalent to 0.5 times the kinetic head corresponding to the absolute velocity at exit from the runner. The discharge of air is 5.7 m³/s when the speed is 13.5 rev/s. Neglecting the thickness of the blades and whirl slip, determine the head across the fan and the power required to drive it if the density of the air is sensibly constant at 1.25 kg/m³ throughout and mechanical losses account for 220 W.

13.15 A centrifugal fan, for which a number of interchangeable impellers are available, is to supply 4.5 m³ of air a second to a ventilating duct at a head of 100 mm water gauge. For all the impellers the outer diameter is 500 mm, the breadth 180 mm and the blade thickness negligible. The fan runs at 30 rev/s. Assuming that the conversion of velocity head to pressure head in the volute is counterbalanced by the friction losses there and in the impeller, that there is no whirl at inlet and that the air density is constant at 1.23 kg/m³, determine the most suitable outlet angle of the blades. (Neglect whirl slip.)

13.16 A centrifugal pump which runs at 16.6 rev/s is mounted so that its centre is 2.4 m above the water level in the suction sump. It delivers water to a point 19 m above its centre. For a flow rate of $Q\,\text{m}^3/\text{s}$ the friction loss in the suction pipe is $68Q^2\,\text{m}$ and that in the delivery pipe is $650Q^2\,\text{m}$. The impeller of the pump is 350 mm diameter and the width of the blade passages at outlet is 18 mm. The blades themselves occupy 5% of the circumference and are backward-facing at 35° to the tangent. At inlet the flow is radial and the radial component of velocity remains unchanged through the impeller. Assuming that 50% of the velocity head of the water leaving the impeller is converted to pressure head in the volute, and that friction and shock losses in the pump, the velocity heads in the suction and delivery pipes and whirl slip are all negligible, calculate the rate of flow and the manometric efficiency of the pump.

13.17 A single-stage centrifugal pump is to be used to pump water through a vertical distance of 30 m at the rate of 45 litres/s. Suction and delivery pipes will have a combined length of 36 m and a friction factor f of 0.006. Both will be 150 mm diameter. Losses at valves, etc. are estimated to total 2.4 times the velocity head in the pipes. The basic design of pump has a 'dimensionless specific speed' of 0.074 rev, forward-curved impeller blades with an outlet angle of 125° to the tangent and a width of impeller passages at outlet equal to one-tenth of the diameter. The blades themselves occupy 5% of the circumference. If a manometric efficiency (neglecting whirl slip) of 75% may be expected, determine a suitable impeller diameter.

13.18 The impeller of a centrifugal pump has an outer diameter of 250 mm and an effective outlet area of $17\,000\,\text{mm}^2$. The outlet blade angle is 32°. The diameters of suction and discharge openings are 150 mm and 125 mm respectively. At 24.2 rev/s and discharge $0.03\,\text{m}^3/\text{s}$ the pressure heads at suction and discharge openings were respectively 4.5 m below and 13.3 m above atmospheric pressure, the measurement points being at the same level. The shaft power was 7.76 kW. Water enters the impeller without shock or whirl. Assuming that the true outlet whirl component is 70% of the ideal, determine the overall efficiency and the manometric efficiency based on the true whirl component.

13.19 The following duties are to be performed by rotodynamic pumps driven by electric synchronous motors (speed $50/n$ rev/s, where n is an integer): (a) $14\,\text{m}^3/\text{s}$ of water against 1.5 m head; (b) 11.3 litres of oil (relative density 0.80) per second against 70 kPa pressure; (c) 5.25 litres of water per second against 5.5 MPa. Designs of pumps are available of which the 'dimensionless specific speeds' are 0.032, 0.096, 0.192, 0.45, 0.64 rev. Which design and speed should be used for each duty?

13.20 During a laboratory test on a water pump appreciable cavitation began when the pressure plus velocity head at inlet was reduced to 3.26 m while the total head change across the pump was 36.5 m and the discharge was 48 litres/s. Barometric pressure was 750 mm Hg and the vapour pressure of water 1.8 kPa. What is the value of σ_c? If the pump is to give the same total head and discharge in a location where the normal atmospheric pressure is 622 mm Hg and the vapour pressure of water 830 Pa, by how much must the height of the pump above the supply level be reduced?

13.21 A large centrifugal pump is to have a 'dimensionless specific speed' of 0.183 rev and is to discharge 2 m³ of liquid per second against a total head of 15 m. The kinematic viscosity of the liquid may vary between 3 and 6 times that of water. Determine the range of speeds and test heads for a one-quarter scale model investigation of the full-size pump, the model using water.

13.22 A 500 mm diameter fluid coupling containing oil of relative density 0.85 has a slip of 3% and a torque coefficient of 0.0014. The speed of the primary is 16.67 rev/s. What is the rate of heat dissipation when equilibrium is attained?

13.23 A fluid coupling is to be used to transmit 150 kW between an engine and a gear-box when the engine speed is 40 rev/s. The mean diameter at the outlet of the primary member is 380 mm and the cross-sectional area of the flow passage is constant at 0.026 m². The relative density of the oil is 0.85 and the efficiency of the coupling 96.5%. Assuming that the shock losses under steady conditions are negligible and that the friction loss round the fluid circuit is four times the mean velocity head, calculate the mean diameter at inlet to the primary member.

Units and conversion factors

Table A1.1 Prefixes for multiples and submultiples of SI units

Prefix	Symbol	Factor by which unit is multiplied
exa	E	10^{18}
peta	P	10^{15}
tera	T	10^{12}
giga (pronounced 'jyga')	G	10^{9}
mega (pronounced 'megga')	M	10^{6}
kilo	k	10^{3}
hecto	h	10^{2}
deca	da	10
deci	d	10^{-1}
centi	c	10^{-2}
milli	m	10^{-3}
micro	μ	10^{-6}
nano	n	10^{-9}
pico (pronounced 'peeko')	p	10^{-12}
femto	f	10^{-15}
atto	a	10^{-18}

Table A1.2 SI Units with internationally agreed names

Quantity	Unit	Symbol	Relationship
Force	newton	N	$1\,\mathrm{N} = 1\,\mathrm{kg\,m\,s^{-2}}$
Pressure (and stress)	pascal	Pa	$1\,\mathrm{Pa} = 1\,\mathrm{N\,m^{-2}} = 1\,\mathrm{kg\,m^{-1}\,s^{-2}}$
Energy and work	joule	J	$1\,\mathrm{J} = 1\,\mathrm{N\,m} = 1\,\mathrm{kg\,m^{2}\,s^{-2}}$
Power	watt	W	$1\,\mathrm{W} = 1\,\mathrm{J\,s^{-1}} = 1\,\mathrm{kg\,m^{2}\,s^{-3}}$
Frequency	hertz	Hz	$1\,\mathrm{Hz} = 1\,\mathrm{s^{-1}}$

Table A1.3 Conversion factors for mass, length and time

Mass

$$1 \equiv \frac{16\,\text{ozm}}{1\,\text{lbm}} \equiv \frac{112\,\text{lbm}}{1\,\text{cwt}} \equiv \frac{2240\,\text{lbm}}{1\,\text{tonm}} \equiv \frac{2000\,\text{lbm}}{1\,\text{US tonm}} \equiv \frac{0.4536\,\text{kg}}{1\,\text{lbm}} \equiv \frac{1016\,\text{kg}}{1\,\text{tonm}}$$

$$\equiv \frac{32.17\,\text{lbm}}{1\,\text{slug}} \equiv \frac{14.59\,\text{kg}}{1\,\text{slug}}$$

Length

$$1 \equiv \frac{12\,\text{in}}{1\,\text{ft}} \equiv \frac{25.4\,\text{mm}}{1\,\text{inch}} \equiv \frac{304.8\,\text{mm}}{1\,\text{ft}} \equiv \frac{914.4\,\text{mm}}{1\,\text{yd}} \equiv \frac{5280\,\text{ft}}{1\,\text{mile}} \equiv \frac{1.609\,\text{km}}{1\,\text{mile}}$$

$$\equiv \frac{1760\,\text{yd}}{1\,\text{mile}} \equiv \frac{6080\,\text{ft}}{1\,\text{British nautical mile}} \equiv \frac{1852\,\text{m}}{1\,\text{international nautical mile}}$$

Time

$$1 \equiv \frac{60\,\text{s}}{1\,\text{min}} \equiv \frac{60\,\text{min}}{1\,\text{h}} \equiv \frac{3600\,\text{s}}{1\,\text{h}} \equiv \frac{24\,\text{h}}{1\,\text{day}} \equiv \frac{1440\,\text{min}}{1\,\text{day}} \equiv \frac{86400\,\text{s}}{1\,\text{day}}$$

$$\equiv \frac{365.24\,\text{solar days}}{1\,\text{year}}$$

Table A1.4 Other conversion factors. Exact values are printed in bold type
Note: Only a limited number of the more important conversion factors are given here. It is possible to derive large numbers of other conversion factors from the information contained in Tables A1.1–A1.3

Area

$$1 \equiv \frac{144\,\text{in}^2}{1\,\text{ft}^2} \equiv \frac{0.0929\,\text{m}^2}{1\,\text{ft}^2} \equiv \frac{645\,\text{mm}^2}{1\,\text{inch}^2} \equiv \frac{0.836\,\text{m}^2}{1\,\text{yd}^2} \equiv \frac{4840\,\text{yd}^2}{1\,\text{acre}} \equiv \frac{640\,\text{acre}}{1\,\text{mile}^2}$$

$$\equiv \frac{0.4047\,\text{hectare}}{1\,\text{acre}} \equiv \frac{4.047 \times 10^3\,\text{m}^2}{1\,\text{acre}}$$

Angle (protractor measure only)

$$1 \equiv \frac{60\,\text{minutes of arc}}{1\,\text{degree}} \equiv \frac{60\,\text{seconds of arc}}{1\,\text{minute of arc}} \equiv \frac{0.9\,\text{degree}}{1\,\text{grade}}$$

[For conversion between protractor measure and radian measure the proportionality coefficient $\pi/(180\,\text{deg})$ is required.]

Density

$$1 \equiv \frac{16.02\,\text{kg/m}^3}{1\,\text{lbm/ft}^3} \equiv \frac{62.4\,\text{lbm/ft}^3}{1000\,\text{kg/m}^3} \equiv \frac{1000\,\text{kg/m}^3}{1\,\text{g/cm}^3}$$

Energy; Work

$$1 \equiv \frac{1.356\,\text{J}}{1\,\text{ft lbf}} \equiv \frac{10^{-7}\,\text{J}}{1\,\text{erg}} \equiv \frac{3.766 \times 10^{-7}\,\text{kW h}}{1\,\text{ft lbf}} \equiv \frac{3.6 \times 10^6\,\text{J}}{1\,\text{kW h}} \equiv \frac{1055\,\text{J}}{1\,\text{Btu}}$$

$$\equiv \frac{2.931 \times 10^{-4}\,\text{kW h}}{1\,\text{Btu}} \equiv \frac{252.0\,\text{cal}}{1\,\text{Btu}} \equiv \frac{1.8\,\text{Btu}}{1\,\text{CHU}} \equiv \frac{10^5\,\text{Btu}}{1\,\text{therm}} \equiv \frac{778\,\text{ft lbf}}{1\,\text{Btu}}$$

$$\equiv \frac{1401\,\text{ft lbf}}{1\,\text{CHU}} \equiv \frac{4.187\,\text{J}}{1\,\text{cal}} \equiv \frac{1.163 \times 10^{-6}\,\text{kW h}}{1\,\text{cal}} \equiv \frac{1.602 \times 10^{-19}\,\text{J}}{1\,\text{eV}}$$

Table A1.4 (contd.)

Force

$$1 \equiv \frac{32.17\,\text{pdl}}{1\,\text{lbf}} \equiv \frac{10^{-5}\,\text{N}}{1\,\text{dyne}} \equiv \frac{0.1383\,\text{N}}{1\,\text{pdl}} \equiv \frac{4.448\,\text{N}}{1\,\text{lbf}} \equiv \frac{0.4536\,\text{kgf}\dagger}{1\,\text{lbf}} \equiv \frac{9.81\,\text{N}}{1\,\text{kgf}\dagger}$$

$$\equiv \frac{9964\,\text{N}}{1\,\text{tonf}}$$

† Known in Germany and Eastern Europe as kilopond (kp).

Power

$$1 \equiv \frac{1.356\,\text{W}}{1\,\text{ft lbf/s}} \equiv \frac{550\,\text{ft lbf/s}}{1\,\text{hp}} \equiv \frac{75\,\text{m kgf/s}}{1\,\text{CV (metric horsepower)}}$$

$$\equiv \frac{0.986\,\text{hp}}{1\,\text{CV (metric horsepower)}} \equiv \frac{0.2931\,\text{W}}{1\,\text{Btu/h}} \equiv \frac{746\,\text{W}}{1\,\text{hp}}$$

Pressure; Stress

$$1 \equiv \frac{144\,\text{lbf/ft}^2}{1\,\text{lbf/in}^2} \equiv \frac{6895\,\text{Pa}}{1\,\text{lbf/in}^2} \equiv \frac{0.0703\,\text{kgf/cm}^2}{1\,\text{lbf/in}^2} \equiv \frac{1.544 \times 10^7\,\text{Pa}}{1\,\text{tonf/in}^2}$$

$$\equiv \frac{1.013 \times 10^5\,\text{Pa}}{1\,\text{atm}} \equiv \frac{14.70\,\text{lbf/in}^2}{1\,\text{atm}} \equiv \frac{2116\,\text{lbf/ft}^2}{1\,\text{atm}} \equiv \frac{10^5\,\text{Pa}}{1\,\text{bar}}$$

$$\equiv \frac{2.036\,\text{in Hg}\ddagger}{1\,\text{lbf/in}^2} \equiv \frac{29.92\,\text{in Hg}\ddagger}{1\,\text{atm}} \equiv \frac{7.50 \times 10^{-3}\,\text{mm Hg}\ddagger}{1\,\text{Pa}} \equiv \frac{0.491\,\text{lbf/in}^2}{1\,\text{in Hg}\ddagger}$$

$$\equiv \frac{3386\,\text{Pa}}{1\,\text{in Hg}\ddagger} \equiv \frac{133.3\,\text{Pa}}{1\,\text{mm Hg}\ddagger} \equiv \frac{2.307\,\text{ft water}\S}{1\,\text{lbf/in}^2} \equiv \frac{249.1\,\text{Pa}}{1\,\text{in water}\S}$$

$$\equiv \frac{0.03613\,\text{lbf/in}^2}{1\,\text{in water}\S} \equiv \frac{2989\,\text{Pa}}{1\,\text{ft water}\S}$$

‡ Strictly for mercury at 0°C and 1 atm pressure.
§ Strictly for pure water at 4°C and 1 atm pressure.

Velocity

$$1 \equiv \frac{0.3048\,\text{m/s}}{1\,\text{ft/s}} \equiv \frac{0.00508\,\text{m/s}}{1\,\text{ft/min}} \equiv \frac{1.467\,\text{ft/s}}{1\,\text{mile/h}} \equiv \frac{88\,\text{ft/min}}{1\,\text{mile/h}} \equiv \frac{1.609\,\text{km/h}}{1\,\text{mile/h}}$$

$$\equiv \frac{0.515\,\text{m/s}}{1\,\text{(British) knot}} \equiv \frac{0.447\,\text{m/s}}{1\,\text{mile/h}}$$

Viscosity (Absolute or dynamic)

$$1 \equiv \frac{47.9\,\text{Pa s}}{1\,\text{lbf s/ft}^2} \equiv \frac{0.1\,\text{Pa s}}{1\,\text{poise}} \equiv \frac{1.488\,\text{Pa s}}{1\,\text{pdl s/ft}^2}$$

Viscosity (kinematic) (μ/ρ)

$$1 \equiv \frac{0.0929\,\text{m}^2/\text{s}}{1\,\text{ft}^2/\text{s}} \equiv \frac{10^{-4}\,\text{m}^2/\text{s}}{1\,\text{stokes}}$$

continued overleaf

Table A1.4 (contd.)

Volume; Modulus of section

$$1 \equiv \frac{1728\,\text{in}^3}{1\,\text{ft}^3} \equiv \frac{2.832 \times 10^{-2}\,\text{m}^3}{1\,\text{ft}^3} \equiv \frac{28.32\,\text{litre}}{1\,\text{ft}^3} \equiv \frac{27\,\text{ft}^3}{1\,\text{yd}^3} \equiv \frac{1.639 \times 10^{-5}\,\text{m}^3}{1\,\text{in}^3}$$

$$\equiv \frac{0.7646\,\text{m}^3}{1\,\text{yd}^3} \equiv \frac{0.1605\,\text{ft}^3}{1\,\text{Imperial gallon}} \equiv \frac{4.546 \times 10^{-3}\,\text{m}^3}{1\,\text{Imperial gallon}} \equiv \frac{231\,\text{in}^3}{1\,\text{US gallon}}$$

$$\equiv \frac{3.785 \times 10^{-3}\,\text{m}^3}{1\,\text{US gallon}} \equiv \frac{0.833\,\text{Imperial gallon}}{1\,\text{US gallon}}$$

Volume flow rate

$$1 \equiv \frac{28.32 \times 10^{-3}\,\text{m}^3/\text{s}}{1\,\text{ft}^3/\text{s}} \equiv \frac{101.9\,\text{m}^3/\text{h}}{1\,\text{ft}^3/\text{s}} \equiv \frac{373.7\,\text{Imperial gallons/min}}{1\,\text{ft}^3/\text{s}}$$

$$\equiv \frac{2.242 \times 10^4\,\text{Imperial gallons/h}}{1\,\text{ft}^3/\text{s}} \equiv \frac{7.577 \times 10^{-5}\,\text{m}^3/\text{s}}{1\,\text{Imperial gallon/min}}$$

$$\equiv \frac{2.676 \times 10^{-3}\,\text{ft}^3/\text{s}}{1\,\text{Imperial gallon/min}}$$

Physical constants and properties of fluids

Approximate values of some physical constants
Standard acceleration due to gravity $\quad g = 9.81\,\mathrm{m\,s^{-2}}$
Universal gas constant $\quad R_0 = 8314\,\mathrm{J\,kg^{-1}K^{-1}}$
Standard temperature and pressure (s.t.p.) $\quad 273.15\,\mathrm{K}$ and $1.013 \times 10^5\,\mathrm{Pa}$

Approximate properties of liquid water
Density $\quad 10^3\,\mathrm{kg\,m^{-3}}$
Absolute viscosity at $20\,^\circ\mathrm{C}$ $\quad 10^{-3}\,\mathrm{Pa\,s} = 10^{-3}\,\mathrm{kg\,m^{-1}s^{-1}}$
Specific heat capacity (at $15\,^\circ\mathrm{C}$) $\quad 4.19\,\mathrm{kJ\,kg^{-1}K^{-1}}$
Bulk modulus (for moderate pressures) $\quad 2.05\,\mathrm{GPa}$

Approximate properties of air
Density at s.t.p. $\quad 1.29\,\mathrm{kg\,m^{-3}}$
Absolute viscosity at s.t.p. $\quad 1.7 \times 10^{-5}\,\mathrm{Pa\,s}$
Specific heat capacity at constant pressure $\quad 1\,\mathrm{kJ\,kg^{-1}K^{-1}}$
Specific heat capacity at constant volume $\quad 715\,\mathrm{J\,kg^{-1}K^{-1}}$
Ratio of specific heat capacities $\quad 1.4$
Gas constant $(R = p/\varrho T)$ $\quad 287\,\mathrm{J\,kg^{-1}K^{-1}}$

Density of mercury
Density at $15\,^\circ\mathrm{C}$ $\quad 13.56 \times 10^3\,\mathrm{kg\,m^{-3}}$

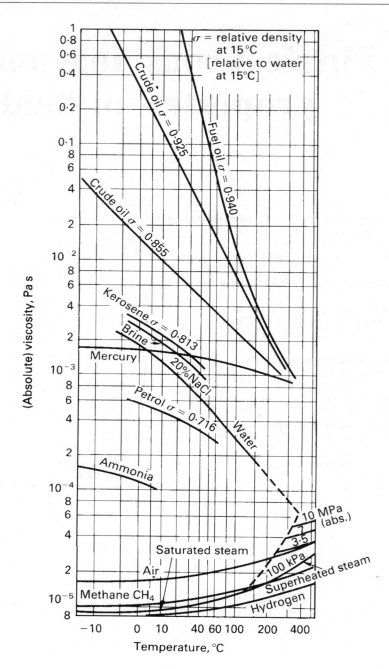

Fig. A2.1 Absolute viscosity of fluids.

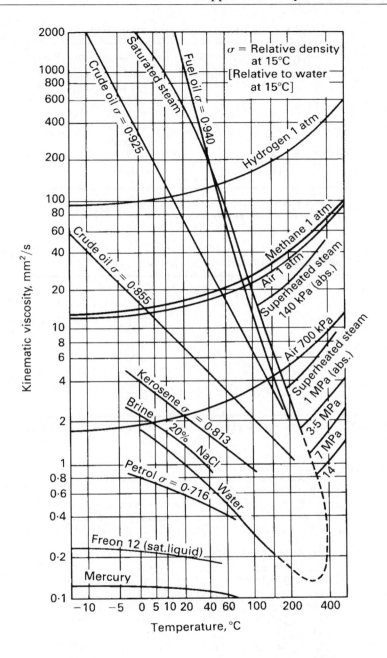

Fig. A2.2 Kinematic viscosity of fluids.

Table A2.1 Saturation vapour pressure of water

Temperature (°C)	Saturation vapour pressure (Pa)
0	615
10	1230
20	2340
40	7400
60	20000
80	47400
100	101500

The figures refer to plane interfaces between liquid and vapour. They are very slightly modified by curvature of the surface.

Table A2.2 Saturation vapour pressures at 20°C

	Saturation vapour pressure (Pa)
Mercury	0·16
Water	2340
Kerosene	3300
Ethanol	5900
Benzene	10000
Methanol	12500
Petrol	30400

(Here, too, the values refer to plane interfaces between liquid and vapour.)

Table A2.3 Properties of the international standard atmosphere

Altitude z (m)	Temperature T (K)	t (°C)	Pressure p (Pa)	Density ϱ (kg m^{-3})
0	288.15	15	101325	1.2252
1000	281.65	8.5	89875	1.1118
2000	275.15	2	79495	1.0067
3000	268.65	−4.5	70109	0.9093
4000	262.15	−11	61640	0.8193
5000	255.65	−17.5	54020	0.7362
6000	249.15	−24	47181	0.6598
7000	242.65	−30.5	41061	0.5896
8000	236.15	−37	35600	0.5252
9000	229.65	−43.5	30742	0.4664
10000	223.15	−50	26436	0.4127
11000	216.65	−56.5	22632	0.3639
12000	216.65	−56.5	19330	0.3108
13000	216.65	−56.5	16510	0.2655
14000	216.65	−56.5	14102	0.2268
15000	216.65	−56.5	12045	0.1937

Other values of atmospheric properties can be calculated as follows.

Range I: $0 \leq z \leq 11\,000\,\text{m}$

Defining sea level conditions:

$$T_0 = 288.15\,\text{K}; \quad p_0 = 1.01325 \times 10^5\,\text{Pa}; \quad \varrho_0 = 1.2252\,\text{kg m}^{-3}$$

At altitude z in metres:

Temperature $\quad\quad\quad T = (288.15 - 0.0065\,z)\,\text{K}$

Pressure $\quad\quad\quad\quad\, p = p_0(T/T_0)^{5.256}$

Density $\quad\quad\quad\quad\;\; \varrho = \varrho_0(T/T_0)^{4.256}$

Range II: $11\,000\,\text{m} \leq z \leq 20\,000\,\text{m}$

Defining conditions at 11 000 m:

$$T_1 = 216.65\,\text{K}; \quad p_1 = 22\,632\,\text{Pa}; \quad \varrho_1 = 0.3639\,\text{kg m}^{-3}$$

At altitude z in metres:

Temperature $\quad\quad\quad T = 216.65\,\text{K}$

Pressure $\quad\quad\quad\quad\, p = p_1\exp\left[-0.0001577(z - 11000)\right]$

Density $\quad\quad\quad\quad\;\; \varrho = \varrho_1\exp\left[-0.0001577(z - 11000)\right]$

Tables of gas flow functions

The following tables are provided as an aid in solving the problems in Chapter 11 and similar ones. Values of functions are therefore given only to three- or four-figure accuracy, and at intervals in the main variable for which simple linear interpolation is in general adequate. All are for a perfect gas with $\gamma = c_p/c_v = 1.4$. More detailed tables may be found in, for example:

Aeronautical Research Council *A Selection of Tables for Use in Calculations of Compressible Air Flow* (edited by L. Rosenhead), Clarendon Press, Oxford (1952).

Imrie, B. W. *Compressible Fluid Flow*, Butterworth, London (1973).

John, J. E. A. *Gas Dynamics*, Allyn and Bacon, Boston (1969).

Rotty, R. M. *Introduction to Gas Dynamics*, Wiley, New York (1962).

Shapiro, A. H. *The Dynamics and Thermodynamics of Compressible Fluid Flow*, Ronald Press, New York (1953, reprinted 1983).

Table A3.1 Plane normal shock

M_1	M_2	p_2/p_1	$\varrho_2/\varrho_1 = u_1/u_2$	T_2/T_1	$(p_0)_2/(p_0)_1$	$(p_0)_2/p_1$
1.00	1.000	1.000	1.000	1.000	1.000	1.893
1.02	0.981	1.047	1.033	1.013	1.000	1.938
1.04	0.962	1.095	1.067	1.026	1.000	1.984
1.06	0.944	1.144	1.101	1.039	1.000	2.032
1.08	0.928	1.194	1.135	1.052	0.999	2.082
1.10	0.912	1.245	1.169	1.065	0.999	2.133
1.12	0.897	1.297	1.203	1.078	0.998	2.185
1.14	0.882	1.350	1.238	1.090	0.997	2.239
1.16	0.868	1.403	1.272	1.103	0.996	2.294
1.18	0.855	1.458	1.307	1.115	0.995	2.350
1.20	0.842	1.513	1.342	1.128	0.993	2.407
1.22	0.830	1.570	1.376	1.141	0.991	2.466
1.24	0.818	1.627	1.411	1.153	0.988	2.526
1.26	0.807	1.686	1.446	1.166	0.986	2.588
1.28	0.796	1.745	1.481	1.178	0.983	2.650

Table A3.1 (contd.)

M_1	M_2	p_2/p_1	$\varrho_2/\varrho_1 = u_1/u_2$	T_2/T_1	$(p_0)_2/(p_0)_1$	$(p_0)_2/p_1$
1.30	0.786	1.805	1.516	1.191	0.979	2.714
1.32	0.776	1.866	1.551	1.204	0.976	2.778
1.34	0.766	1.928	1.585	1.216	0.972	2.844
1.36	0.757	1.991	1.620	1.229	0.968	2.912
1.38	0.748	2.055	1.655	1.242	0.963	2.980
1.40	0.740	2.120	1.690	1.255	0.958	3.049
1.42	0.731	2.186	1.724	1.268	0.953	3.120
1.44	0.723	2.253	1.759	1.281	0.948	3.191
1.46	0.716	2.320	1.793	1.294	0.942	3.264
1.48	0.708	2.389	1.828	1.307	0.936	3.338
1.50	0.701	2.458	1.862	1.320	0.930	3.413
1.52	0.694	2.529	1.896	1.334	0.923	3.489
1.54	0.687	2.600	1.930	1.347	0.917	3.567
1.56	0.681	2.673	1.964	1.361	0.910	3.645
1.58	0.675	2.746	1.998	1.374	0.903	3.724
1.60	0.668	2.820	2.032	1.388	0.895	3.805
1.62	0.663	2.895	2.065	1.402	0.888	3.887
1.64	0.657	2.971	2.099	1.416	0.880	3.969
1.66	0.651	3.048	2.132	1.430	0.872	4.05
1.68	0.646	3.126	2.165	1.444	0.864	4.14
1.70	0.641	3.205	2.198	1.458	0.856	4.22
1.72	0.635	3.285	2.230	1.473	0.847	4.31
1.74	0.631	3.366	2.263	1.487	0.839	4.40
1.76	0.626	3.447	2.295	1.502	0.830	4.49
1.78	0.621	3.530	2.327	1.517	0.822	4.58
1.80	0.617	3.613	2.359	1.532	0.813	4.67
1.82	0.612	3.698	2.391	1.547	0.804	4.76
1.84	0.608	3.783	2.422	1.562	0.795	4.86
1.86	0.604	3.870	2.454	1.577	0.786	4.95
1.88	0.600	3.957	2.485	1.592	0.777	5.05
1.90	0.596	4.045	2.516	1.608	0.767	5.14
1.92	0.592	4.13	2.546	1.624	0.758	5.24
1.94	0.588	4.22	2.577	1.639	0.749	5.34
1.96	0.584	4.32	2.607	1.655	0.740	5.44
1.98	0.581	4.41	2.637	1.671	0.730	5.54
2.00	0.577	4.50	2.667	1.688	0.721	5.64
2.02	0.574	4.59	2.696	1.704	0.712	5.74
2.04	0.571	4.69	2.725	1.720	0.702	5.85
2.06	0.567	4.78	2.755	1.737	0.693	5.95
2.08	0.564	4.88	2.783	1.754	0.684	6.06
2.10	0.561	4.98	2.812	1.770	0.674	6.17
2.12	0.558	5.08	2.840	1.787	0.665	6.27
2.14	0.555	5.18	2.868	1.805	0.656	6.38
2.16	0.553	5.28	2.896	1.822	0.646	6.49
2.18	0.550	5.38	2.924	1.839	0.637	6.60

continued overleaf

Table A3.1 (contd.)

M_1	M_2	p_2/p_1	$\varrho_2/\varrho_1 = u_1/u_2$	T_2/T_1	$(p_0)_2/(p_0)_1$	$(p_0)_2/p_1$
2.20	0.547	5.48	2.951	1.857	0.628	6.72
2.22	0.544	5.58	2.978	1.875	0.619	6.83
2.24	0.542	5.69	3.005	1.892	0.610	6.94
2.26	0.539	5.79	3.032	1.910	0.601	7.06
2.28	0.537	5.90	3.058	1.929	0.592	7.18
2.30	0.534	6.01	3.085	1.947	0.583	7.29
2.32	0.532	6.11	3.110	1.965	0.575	7.41
2.34	0.530	6.22	3.136	1.984	0.566	7.53
2.36	0.527	6.33	3.162	2.002	0.557	7.65
2.38	0.525	6.44	3.187	2.021	0.549	7.77
2.40	0.523	6.55	3.212	2.040	0.540	7.90
2.42	0.521	6.67	3.237	2.059	0.532	8.02
2.44	0.519	6.78	3.261	2.079	0.523	8.15
2.46	0.517	6.89	3.285	2.098	0.515	8.27
2.48	0.515	7.01	3.310	2.118	0.507	8.40
2.50	0.513	7.13	3.333	2.138	0.499	8.53
2.55	0.508	7.42	3.392	2.187	0.479	8.85
2.60	0.504	7.72	3.449	2.238	0.460	9.18
2.65	0.500	8.03	3.505	2.290	0.442	9.52
2.70	0.496	8.34	3.560	2.343	0.424	9.86
2.75	0.492	8.66	3.612	2.397	0.406	10.21
2.80	0.488	8.98	3.664	2.451	0.3895	10.57
2.85	0.485	9.31	3.714	2.507	0.3733	10.93
2.90	0.481	9.65	3.763	2.563	0.3577	11.30
2.95	0.478	9.99	3.811	2.621	0.3427	11.68
3.00	0.475	10.33	3.857	2.679	0.3283	12.06
3.10	0.470	11.05	3.947	2.799	0.3012	12.85
3.20	0.464	11.78	4.03	2.922	0.2762	13.65
3.30	0.460	12.54	4.11	3.049	0.2533	14.49
3.40	0.455	13.32	4.19	3.180	0.2322	15.35
3.50	0.451	14.13	4.26	3.315	0.2130	16.24
3.60	0.447	14.95	4.33	3.454	0.1953	17.16
3.70	0.444	15.81	4.39	3.596	0.1792	18.10
3.80	0.441	16.68	4.46	3.743	0.1645	19.06
3.90	0.438	17.58	4.52	3.893	0.1510	20.05
4.0	0.435	18.50	4.57	4.05	0.1388	21.07
4.5	0.424	23.46	4.81	4.88	0.0917	26.54
5.0	0.415	29.00	5.00	5.80	0.0617	32.65
∞	0.378	∞	6.00	∞	0	∞

Table A3.2 Isentropic flow

M	p/p_0	ϱ/ϱ_0	T/T_0	A/A_t	Prandtl–Meyer angle θ degrees
0	1.000	1.000	1.000	∞	
0.05	0.998	0.999	0.9995	11.59	
0.10	0.993	0.995	0.998	5.82	
0.15	0.984	0.989	0.996	3.91	
0.20	0.973	0.980	0.992	2.964	
0.25	0.957	0.969	0.988	2.403	
0.30	0.939	0.956	0.982	2.035	
0.35	0.919	0.941	0.976	1.778	
0.40	0.896	0.924	0.969	1.590	
0.42	0.886	0.917	0.966	1.529	
0.44	0.876	0.909	0.963	1.474	
0.46	0.865	0.902	0.959	1.425	
0.48	0.854	0.893	0.956	1.380	
0.50	0.843	0.885	0.952	1.340	
0.52	0.832	0.877	0.949	1.303	
0.54	0.820	0.868	0.945	1.270	Not applicable for $M < 1$
0.56	0.808	0.859	0.941	1.240	
0.58	0.796	0.850	0.937	1.213	
0.60	0.784	0.840	0.933	1.188	
0.62	0.772	0.831	0.929	1.166	
0.64	0.759	0.821	0.924	1.145	
0.66	0.746	0.812	0.920	1.127	
0.68	0.734	0.802	0.915	1.110	
0.70	0.721	0.792	0.911	1.094	
0.72	0.708	0.781	0.906	1.081	
0.74	0.695	0.771	0.901	1.068	
0.76	0.682	0.761	0.896	1.057	
0.78	0.669	0.750	0.892	1.047	
0.80	0.656	0.740	0.887	1.038	
0.82	0.643	0.729	0.881	1.030	
0.84	0.630	0.719	0.876	1.024	
0.86	0.617	0.708	0.871	1.018	
0.88	0.604	0.698	0.866	1.013	
0.90	0.591	0.687	0.861	1.009	
0.92	0.578	0.676	0.855	1.006	
0.94	0.566	0.666	0.850	1.003	
0.96	0.553	0.655	0.844	1.001	
0.98	0.541	0.645	0.839	1.000	

continued overleaf

Where A/A_t varies rapidly, say up to $M = 0.5$, interpolation of reciprocals is more satisfactory than linear interpolation. For example, for $M = 0.23$

$$\frac{A_t}{A} = \frac{1}{2.964} + \frac{0.23 - 0.20}{0.25 - 0.20}\left(\frac{1}{2.403} - \frac{1}{2.964}\right) = 0.3846$$

and hence $A/A_t = 1/0.3846 = 2.600$.

Table A3.2 (contd.)

M	p/p_0	ϱ/ϱ_0	T/T_0	A/A_t	Prandtl–Meyer angle θ degrees
1.00	0.528	0.634	0.833	1.000	0
1.05	0.498	0.608	0.819	1.002	0.49
1.10	0.468	0.582	0.805	1.008	1.34
1.15	0.440	0.556	0.791	1.017	2.38
1.20	0.412	0.531	0.776	1.030	3.56
1.25	0.3861	0.507	0.762	1.047	4.83
1.30	0.3609	0.483	0.747	1.066	6.17
1.35	0.3370	0.460	0.733	1.089	7.56
1.40	0.3142	0.437	0.718	1.115	8.99
1.45	0.2927	0.416	0.704	1.144	10.44
1.50	0.2724	0.3950	0.690	1.176	11.91
1.55	0.2533	0.3750	0.675	1.212	13.38
1.60	0.2353	0.3557	0.661	1.250	14.86
1.65	0.2184	0.3373	0.647	1.292	16.34
1.70	0.2026	0.3197	0.634	1.338	17.81
1.75	0.1878	0.3029	0.620	1.386	19.27
1.80	0.1740	0.2868	0.607	1.439	20.73
1.85	0.1612	0.2715	0.594	1.495	22.16
1.90	0.1492	0.2570	0.581	1.555	23.59
1.95	0.1381	0.2432	0.568	1.619	24.99
2.00	0.1278	0.2301	0.556	1.687	26.38
2.05	0.1182	0.2176	0.543	1.760	27.75
2.10	0.1094	0.2058	0.531	1.837	29.10
2.15	0.1011	0.1946	0.520	1.919	30.43
2.20	0.0935	0.1841	0.508	2.005	31.73
2.25	0.0865	0.1740	0.497	2.096	33.02
2.30	0.0800	0.1646	0.486	2.193	34.28
2.35	0.0740	0.1556	0.475	2.295	35.53
2.40	0.0684	0.1472	0.465	2.403	36.75
2.45	0.0633	0.1392	0.454	2.517	37.95
2.5	0.0585	0.1317	0.444	2.637	39.12
2.6	0.0501	0.1179	0.425	2.896	41.41
2.7	0.0429	0.1056	0.407	3.183	43.62
2.8	0.03685	0.0946	0.3894	3.500	45.75
2.9	0.03165	0.0849	0.3729	3.850	47.79
3.0	0.02722	0.0762	0.3571	4.23	49.76
3.1	0.02345	0.0685	0.3422	4.66	51.65
3.2	0.02023	0.0617	0.3281	5.12	53.47
3.3	0.01748	0.0555	0.3147	5.63	55.22
3.4	0.01512	0.0501	0.3019	6.18	56.91
3.5	0.01311	0.0452	0.2899	6.79	58.53
4.0	0.00659	0.02766	0.2381	10.72	65.78
4.5	0.00346	0.01745	0.1980	16.56	71.83
5.0	0.00189	0.01134	0.1667	25.00	76.92
6.0	0.000633	0.00519	0.1220	53.18	84.96
8.0	0.000102	0.00141	0.0725	190.1	95.62
10.0	0.000024	0.00050	0.0476	536	102.31
∞	0	0	0	∞	130.45

Table A3.3 Adiabatic flow with friction in duct of constant cross-section (Fanno flow)

M	p_c/p	$u/u_c = \varrho_c/\varrho$	T/T_c	$fl_{max}P/A$
0	0	0	1.200	∞
0.02	0.01826	0.02191	1.200	1778
0.04	0.03652	0.0438	1.200	440
0.06	0.05479	0.0657	1.199	193.0
0.08	0.07308	0.0876	1.198	106.7
0.10	0.09138	0.1094	1.198	66.9
0.12	0.1097	0.1313	1.197	45.4
0.14	0.1281	0.1531	1.195	32.51
0.16	0.1464	0.1748	1.194	24.20
0.18	0.1648	0.1965	1.192	18.54
0.20	0.1833	0.2182	1.190	14.53
0.22	0.2018	0.2398	1.188	11.60
0.24	0.2203	0.2614	1.186	9.39
0.26	0.2389	0.2829	1.184	7.69
0.28	0.2576	0.3043	1.181	6.36
0.30	0.2763	0.3257	1.179	5.30
0.32	0.2951	0.3470	1.176	4.45
0.34	0.3139	0.3682	1.173	3.752
0.36	0.3329	0.3893	1.170	3.180
0.38	0.3519	0.410	1.166	2.705
0.40	0.3709	0.431	1.163	2.308
0.42	0.3901	0.452	1.159	1.974
0.44	0.4094	0.473	1.155	1.692
0.46	0.4287	0.494	1.151	1.451
0.48	0.4482	0.514	1.147	1.245
0.50	0.4677	0.535	1.143	1.069
0.52	0.4874	0.555	1.138	0.917
0.54	0.507	0.575	1.134	0.787
0.56	0.527	0.595	1.129	0.674
0.58	0.547	0.615	1.124	0.576
0.60	0.567	0.635	1.119	0.491
0.62	0.587	0.654	1.114	0.417
0.64	0.608	0.674	1.109	0.3533
0.66	0.628	0.693	1.104	0.2979
0.68	0.649	0.713	1.098	0.2498

Note: The pressure ratio is here given as p_c/p, *not p/p_c.*

continued overleaf

Where $fl_{max}P/A$ varies rapidly, say up to $M = 0.6$ and above $M = 3.0$, linear interpolation is not satisfactory. To obtain the length parameter for a given value of M, direct calculation from eqn 11.76 is probably simplest. To obtain M from a given value of $fl_{max}P/A$, inverse-square interpolation is recommended; e.g. when $fl_{max}P/A = 13$,

$$\frac{1}{M^2} = \frac{1}{0.20^2} - \frac{14.53 - 13}{14.53 - 11.60}\left(\frac{1}{0.20^2} - \frac{1}{0.22^2}\right) = 27.27$$

whence $M = 0.2097$.

Table A3.3 (contd.)

M	p_c/p	$u/u_c = \varrho_c/\varrho$	T/T_c	$fl_{max}P/A$
0.70	0.670	0.732	1.093	0.2081
0.72	0.690	0.751	1.087	0.1722
0.74	0.712	0.770	1.082	0.1411
0.76	0.733	0.788	1.076	0.1145
0.78	0.754	0.807	1.070	0.0917
0.80	0.776	0.825	1.064	0.0723
0.82	0.797	0.843	1.058	0.0559
0.84	0.819	0.861	1.052	0.0423
0.86	0.841	0.879	1.045	0.03097
0.88	0.863	0.897	1.039	0.02180
0.90	0.886	0.915	1.033	0.01451
0.92	0.908	0.932	1.026	0.00892
0.94	0.931	0.949	1.020	0.00481
0.96	0.954	0.966	1.013	0.00206
0.98	0.977	0.983	1.007	0.00049
1.00	1.000	1.000	1.000	0
1.05	1.059	1.041	0.983	0.00271
1.10	1.119	1.081	0.966	0.00993
1.15	1.181	1.120	0.949	0.02053
1.20	1.243	1.158	0.932	0.03364
1.25	1.307	1.195	0.914	0.0486
1.30	1.373	1.231	0.897	0.0648
1.35	1.440	1.266	0.879	0.0820
1.40	1.508	1.300	0.862	0.0997
1.45	1.578	1.333	0.845	0.1178
1.50	1.649	1.365	0.828	0.1361
1.55	1.722	1.395	0.811	0.1543
1.60	1.796	1.425	0.794	0.1724
1.65	1.872	1.454	0.777	0.1902
1.70	1.949	1.482	0.760	0.2078
1.75	2.029	1.510	0.744	0.2250
1.80	2.109	1.536	0.728	0.2419
1.85	2.192	1.561	0.712	0.2583
1.90	2.276	1.586	0.697	0.2743
1.95	2.362	1.610	0.682	0.2899
2.0	2.449	1.633	0.667	0.3050
2.1	2.630	1.677	0.638	0.3339
2.2	2.817	1.718	0.610	0.3609
2.3	3.012	1.756	0.583	0.3862
2.4	3.214	1.792	0.558	0.410
2.5	3.423	1.826	0.533	0.432
2.6	3.640	1.857	0.510	0.453
2.7	3.864	1.887	0.488	0.472
2.8	4.096	1.914	0.467	0.490
2.9	4.335	1.940	0.447	0.507
3.0	4.583	1.964	0.429	0.522
3.5	5.935	2.064	0.3478	0.586
4.0	7.48	2.138	0.2857	0.633
5.0	11.18	2.236	0.2000	0.694
6.0	15.68	2.295	0.1463	0.730
∞	∞	2.449	0	0.822

Algebraic symbols

Table A4.1 lists a number of the algebraic symbols used in this book. Since the number of meanings exceeds the number of symbols available, even from both the Roman and Greek alphabets, many symbols have more than one meaning. Different meanings for the same symbol, however, occur in different parts of the book and so little confusion should arise. Some symbols used only once or twice are omitted from this general list.

Recommendations of the British Standards Institution (in B.S. 1991) have been mainly, although not slavishly, followed.

Dimensional, formulae are given in terms of the fundamental magnitudes of length [L], mass [M], time interval [T] and temperature [θ]. Numbers, that is, dimensionless magnitudes, are indicated by [1]. For this purpose, angles and angular velocities are assumed to be expressed in radian measure.

Table A4.1

Symbol	Definition	Dimensional formula
A	an area	$[L^2]$
Ak^2	second moment of area	$[L^4]$
a	an area	$[L^2]$
a	linear acceleration	$[LT^{-2}]$
a	radius of circular cylinder (Chapter 9)	$[L]$
a	maximum value of η for surface wave	$[L]$
a	sonic velocity	$[LT^{-1}]$
B	width of liquid surface in open channel	$[L]$
b	breadth of weir, boundary plane, etc.	$[L]$
b	span of aerofoil	$[L]$
b	base width of open channel	$[L]$
C	a constant	
C	Chézy's coefficient	$[L^{1/2}T^{-1}]$
C_c	coefficient of contraction	$[1]$
C_D	drag coefficient	$[1]$
C_{Di}	vortex or induced drag coefficient	$[1]$
C_d	coefficient of discharge	$[1]$
C_F	average skin friction coefficient	$[1]$
C_L	lift coefficient	$[1]$
C_v	coefficient of velocity	$[1]$

continued overleaf

Table A4.1 (contd.)

Symbol	Definition	Dimensional formula
c	chord length of aerofoil	[L]
c	distance between parallel surfaces	[L]
c	average clearance in journal bearing	[L]
c	velocity (celerity) of wave propagation	$[LT^{-1}]$
c_f	local skin friction coefficient	[1]
c_p	specific heat capacity at constant pressure	$[L^2T^{-2}\theta^{-1}]$
c_v	specific heat capacity at constant volume (i.e. at constant density)	$[L^2T^{-2}\theta^{-1}]$
D	drag force	$[MLT^{-2}]$
D	diameter, especially of piston or rotor	[L]
D_i	vortex or induced drag	$[MLT^{-2}]$
d	diameter, especially of pipe	[L]
E	elastic modulus	$[ML^{-1}T^{-2}]$
E	energy of system	$[ML^2T^{-2}]$
E	specific energy in open channel, $h + u^2/2g$	[L]
e	internal energy per unit mass	$[L^2T^{-2}]$
e	eccentricity of journal bearing	[L]
F	a force	$[MLT^{-2}]$
F	force per unit length in two-dimensional flow	$[MT^{-2}]$
Fr	Froude number, $u/(gl)^{1/2}$	[1]
f	friction factor	[1]
f	frequency of vortex shedding	$[T^{-1}]$
g	weight per unit mass	$[LT^{-2}]$
H	total head	[L]
H	upstream surface level above crest of notch or weir	[L]
H_m	manometric head of pump	[L]
h	depth below free surface of liquid	[L]
h	head of liquid	[L]
h	clearance in bearing	[L]
h	depth of flow in open channel	[L]
h	specific enthalpy $= e + p/\varrho$	$[L^2T^{-2}]$
h_f	head lost to friction	[L]
h_i	inertia head	[L]
h_1	head lost	[L]
h_0	value of clearance defined by Equation 6.48	[L]
i	energy gradient in open channel	[1]
K	a constant	
K	bulk modulus of elasticity	$[ML^{-1}T^{-2}]$
K'	effective bulk modulus of fluid in elastic pipe	$[ML^{-1}T^{-2}]$
K_n	dimensionless specific speed	[1]
K_s	isentropic bulk modulus	$[ML^{-1}T^{-2}]$
k	a constant	
k	head loss coefficient	[1]
k	Kozeny function	[1]
k	radius of gyration	[L]
k	roughness size	[L]

Table A4.1 (contd.)

Symbol	Definition	Dimensional formula
L	length of journal bearing	$[L]$
L	lift force	$[MLT^{-2}]$
l	a length	$[L]$
M	total mass	$[M]$
M	Mach number	$[1]$
M	relative molecular mass	$[1]$
M'	virtual mass	$[M]$
Mk^2	moment of inertia	$[ML^2]$
m	mass	$[M]$
m	mass flow rate (Chapter 11)	$[MT^{-1}]$
m	hydraulic mean depth, A/P	$[L]$
m	strength of source or sink (two-dimensional flow)	$[L^2T^{-1}]$
m	2π/wavelength	$[L^{-1}]$
N	rotational speed	$[T^{-1}]$
N_s	'specific speed' of turbine	$[M^{1/2}L^{-1/4}T^{-5/2}]$
N_s	'specific speed' of pump	$[L^{3/4}T^{-3/2}]$
n	a number	$[1]$
n	distance normal to streamline	$[L]$
n	Manning's roughness coefficient (a number)	$[1]$
P	perimeter in contact with fluid	$[L]$
P	power	$[ML^2T^{-3}]$
p	pressure (intensity)	$[ML^{-1}T^{-2}]$
p_a	atmospheric pressure	$[ML^{-1}T^{-2}]$
p_i	inertia pressure	$[ML^{-1}T^{-2}]$
p_v	vapour pressure	$[ML^{-1}T^{-2}]$
p^*	piezometric pressure, $p + \varrho gz$	$[ML^{-1}T^{-2}]$
Q	discharge (volume/time)	$[L^3T^{-1}]$
Q	heat added	$[ML^2T^{-2}]$
q	discharge per unit width, Q/b	$[L^2T^{-1}]$
q	velocity $= (u^2 + v^2)^{1/2}$ (Chapter 9)	$[LT^{-1}]$
q	heat added per unit mass of fluid	$[L^2T^{-2}]$
R	radius	$[L]$
R	gas constant, $p/\varrho T$	$[L^2T^{-2}\theta^{-1}]$
R	relative velocity in machines	$[LT^{-1}]$
Re	Reynolds number, ul/v (suffixes l, t, x to indicate $l = l$, $l =$ transition length, $l = x$ respectively)	$[1]$
r	radius	$[L]$
S	plan area of aerofoil	$[L^2]$
S	surface area of solid particle	$[L^2]$
s	a distance, especially in direction of flow	$[L]$
s	bed slope of open channel	$[1]$
s	specific entropy	$[L^2T^{-2}\theta^{-1}]$
T	torque	$[ML^2T^{-2}]$
T	thrust per unit width of bearing (Chapter 6)	$[MT^{-2}]$
T	wave period (Chapter 10)	$[T]$
T	absolute temperature	$[\theta]$
T_0	stagnation temperature (absolute)	$[\theta]$

continued overleaf

Table A4.1 (contd.)

Symbol	Definition	Dimensional formula
t	interval of time	$[T]$
t	thickness of pipe wall	$[L]$
U	velocity unaffected by obstacle	$[LT^{-1}]$
u	velocity	$[LT^{-1}]$
u	velocity parallel to x axis (Chapter 9)	$[LT^{-1}]$
u	blade velocity (Chapter 13)	$[LT^{-1}]$
u_m	main-stream velocity outside boundary layer	$[LT^{-1}]$
V	a volume	$[L^3]$
V	velocity of boundary plane	$[LT^{-1}]$
v	velocity parallel to y axis (Chapter 9)	$[LT^{-1}]$
v	'absolute' velocity of fluid in machine	$[LT^{-1}]$
v_i	downwash velocity	$[LT^{-1}]$
v_r	radial component of velocity in machine	$[LT^{-1}]$
v_w	velocity of whirl, i.e. circumferential component of 'absolute' velocity	$[LT^{-1}]$
W	weight	$[MLT^{-2}]$
W	'shaft work' done by system	$[ML^2T^{-2}]$
We	Weber number, $u\sqrt{(\varrho l/\gamma)}$	$[1]$
w	'shaft work' done per unit mass of fluid	$[L^2T^{-2}]$
x	coordinate	$[L]$
y	coordinate, especially distance from solid boundary	$[L]$
Z	height of crest of notch above upstream channel bed	$[L]$
z	coordinate, especially axial distance in journal bearing	$[L]$
z	height above an arbitrary datum level	$[L]$
$\mathcal{AR}$	aspect ratio of aerofoil	$[1]$

Greek Symbols

Symbol	Definition	Dimensional formula
α (alpha)	kinetic energy correction factor	$[1]$
α	an angle, especially angle of attack, angle between 'absolute' velocity of fluid and circumference of rotor	$[1]$
β (beta)	momentum correction factor	$[1]$
β	angle between fluid velocity and shock wave	$[1]$
β	angle between relative velocity and circumference of rotor	$[1]$
Γ (capital gamma)	circulation (in two-dimensional flow)	$[L^2T^{-1}]$
γ (gamma)	c_p/c_v	$[1]$
γ	surface tension	$[MT^{-2}]$
Δ (capital delta)	a particular value of $\eta = y/\delta$ in boundary layer or y/R in pipe	$[1]$
δ (delta)	nominal thickness of boundary layer	$[L]$
δ	tangent of angle between slipper and bearing plate	$[1]$
δ^*	displacement thickness of boundary layer	$[L]$

Table A4.1 (contd.)

Symbol	Definition	Dimensional formula
ε (epsilon)	kinematic eddy viscosity	$[L^2T^{-1}]$
ε	eccentricity ratio of journal bearing	$[1]$
ε	porosity	$[1]$
ζ (zeta)	vorticity (in two-dimensional flow)	$[T^{-1}]$
η (eta)	a coordinate perpendicular to constant-pressure plane	$[L]$
η	eddy viscosity	$[ML^{-1}T^{-1}]$
η	y/δ in boundary layer, y/R in pipe, or r/R in free jet	$[1]$
η	height of point on surface wave above equilibrium level	$[L]$
η	efficiency of machine	$[1]$
n_h	hydraulic efficiency	$[1]$
η_o	overall efficiency	$[1]$
θ (theta)	an angle	$[1]$
θ	temperature (on arbitrary scale)	$[\theta]$
θ	momentum thickness of boundary layer	$[L]$
λ (lambda)	temperature lapse rate, $-\partial T/\partial z$	$[\theta L^{-1}]$
λ	friction factor $= 4f$	$[1]$
λ	wavelength	$[L]$
λ	$(\theta^2/\nu)\mathrm{d}u_m/\mathrm{d}x$ for laminar boundary layer	$[1]$
μ (mu)	absolute viscosity; also known as dynamic viscosity	$[ML^{-1}T^{-1}]$
μ	Mach angle	$[1]$
ν (nu)	kinematic viscosity, μ/ϱ	$[L^2T^{-1}]$
ξ (xi)	a coordinate parallel to constant-pressure plane	$[L]$
ξ	dimensionless parameter $R(\tau_0/\varrho)^{1/2}/\nu$ (Chapter 8)	$[1]$
Π (capital pi)	a dimensionless parameter	$[1]$
π (pi)	$3.14159\ldots$	$[1]$
ϱ (rho)	density	$[ML^{-3}]$
σ (sigma)	relative density	$[1]$
σ	cavitation parameter	$[1]$
τ (tau)	shear stress	$[ML^{-1}T^{-2}]$
τ_0	shear stress at boundary	$[ML^{-1}T^{-2}]$
ϕ (phi)	velocity potential	$[L^2T^{-1}]$
ϕ	angle between blade and circumference of rotor	$[1]$
$\phi(x)$	a function of x	
ψ (psi)	stream function (in two-dimensional flow)	$[L^2T^{-1}]$
Ω (capital omega)	angular velocity of boundary surface	$[T^{-1}]$
ω (omega)	angular velocity	$[T^{-1}]$

continued overleaf

Table A4.1 (contd.)

Symbol	Definition
Other Symbols	
ln	natural logarithm, i.e., to base e
log	logarithm to base 10
Δ (capital delta)	an increment of
δ (delta)	a very small increase of
∂ ('curly d' or 'dif')	indicates partial derivative
Σ (capital sigma)	summation
$|x|$	the modulus of x, i.e., the magnitude of x without regard to its direction or sign
Superscript bar	the mean value of (e.g. $\bar{x}$ = mean value of x)

Suffixes
Suffixes are usually provided to meet particular needs but the following general usages have been adopted.

c	critical
s	suction (Chapter 13)
x, y, z	component of a vector quantity in x, y, z direction respectively
0	stagnation conditions (Chapter 11)
1, 2	at inlet, at outlet of control volume or machine rotor
∞	at a large distance upstream from body

Answers to problems

1.1 $56.2\,\text{m}^3$
1.2 $1.51\,\text{kg/m}^3$
1.3 $1061\,\text{kg/m}^3$
1.4   32.06
1.5   307.6 N
1.6   1.439 N
1.7   7.44 N m
1.8   36.5 Pa
1.9   5.95 mm
1.10 -1.563 mm
1.11 1508, 1.689 m/s
1.12 Laminar ($Re = 430$)

2.1   12.82 m
2.2   22.93 kPa, 40.2 kPa, 30.57 kPa, 211.2 kN
2.3   61.2 m
2.4   4.93 mm
2.5   44.2 kPa, 3.496 m, 35.83 m
2.6   171.1 Pa
2.7   38.3 kPa
2.8   2257 m
2.9   (a) 9.54 kN, 1.2 m from upper edge (b) 20.13 kN, 1.042 m (c) 645 kN, 0.904 m
2.10 On centre line at depth $3\pi d/32$
2.11 4.79 kN at each upper corner, 9.58 kN at bottom
2.12 999 N m
2.13 33.84 kN, 26.59 kN m, 1.111 m from top of aperture
2.14 1.252 MN, 8.42 m above base
2.15 5.29 MN/m, 42.57°, 30.31 m from face
2.16 8870 N, 651 mm
2.17 $b = \sigma c + a(1 - \sigma)/8$; $a^2(3 + \sigma)/48b$ below centre-line

2.18 1317 N, 1543 N, 2859 N
2.19 2.34 kg(f), 5.36 kg(f)
2.20 $0.1778\,l$
2.21 398 mm, 16.05 mm, 3747 Pa (gauge)
2.22 16.25 g
2.23 1.386
2.24 4.70°
2.25 0.641 kg and 0.663 kg
2.26 150 mm
2.27 1.308 m
2.28 3.822 s
2.29 3538 N
2.30 467 Pa

3.1   53.3 kPa gauge
3.2 $23.9\,\text{m}^3/\text{s}$
3.3   16.20 MW
3.5   63.43° to horizontal, 217.4 kPa
3.6 $0.02191\,\text{m}^3/\text{s}$
3.7   0.958
3.8   161.1 kPa gauge
3.9   34.53 kPa, 266.5 mm
3.10 162.2 mm
3.11 $0.0427\,\text{m}^3/\text{s}$
3.12 $0.0762\,\text{m}^3/\text{s}$

4.1 $0.0475\,\text{m}^3/\text{s}$
4.2   13.12 kN at 12.76° upwards from inlet axis
4.3   440 N
4.4 $2.22\,\text{m}^2/\text{s}^2$, $-1.278\,\text{m}^2/\text{s}^2$; 311.1 W
4.5 $0.01045\,\text{m}^2$, 30.92 kW
4.6   8.82 g/s
4.7   447 kg
4.8   25.77 kN, 79.8%, 387 kW

4.9   2.63 m, 916 kW

4.10   8460 N, 11.33 m/s, 41.8 Pa, -32.99 Pa, 95.8 kW

5.1   2.059 litres/s, 1 : 4.74

5.2   225.2 mm

5.3   39.22 rev/s, 0.933 N m

5.4   2.797 m^3/s

5.5   7.82 m/s, 322 N

5.6   13.41 m/s, 92.0 N m

5.8   3.91 m/s, 42.96 N

5.9   7.61, 1236 N

5.10   821 kPa, 450 m/s

5.11   1.24 h

5.12   10.68 kPa, 14.03 litres/s

5.13   7.5 m/s, 410 kN

6.1 $Re = 357$, 116 kPa, 183.3 Pa

6.3   192.5 kPa

6.5   0.689 mm

6.7   62.8 s

6.8   27.13 μm, 0.0549 m/s

6.9   0.934 Pa s, $Re = 0.01362$

6.10   0.1505 Pa s

6.11   0.0626 Pa s, 589 W, 5.63 MPa, 91.1 mm from toe

6.12 $x = 0.689 l$

6.13   0.01771 Pa s, 0.02895 Pa s, 30.02 W, 130.3° forward of load line, yes – it is almost in the centre of the reduced pressure region

7.1   6.12 kW

7.2   10.08 kW, 157 W

7.3   25.06 mm, 25.85 kW

7.4   692 kPa, 0.0547 m^3/s

7.5   141.5 mm, say 150 mm

7.6   100 kPa

7.7   8.35 litres/s

7.8   116.5 mm, say 120 mm

7.9   approx. 12.7 litres/s

7.10   11.34 kg/s

7.11   26.5%

7.12   6.47 m above pipe inlet

7.13   2.571 m, 1.156 kW

7.14   569 mm, increased in ratio $n^{10/3}$

7.15   9.64 m

7.16   24.6 litres/s, 16.0 litres/s

7.17 $d_B = 512$ mm, $d_C = 488$ mm

7.18   0.939 m^3/s, 44.1 Pa, 0.826 m^3/s

7.19   1719 s

7.20   2100 s

7.21   201.7 mm

7.22   40.4 mm, say 40 mm

7.23   about 1625 s

7.24   980 s, 3.072 kWh

7.25   >62.1 mm

7.26   1.110 mPa s

7.27   12.46 Pa, 4.98 Pa

8.1   0.3634, 0.1366

8.2   11.99 mm, 0.0208 Pa, 0.763 W, 2.654 W

8.3   55.69 N, 167.1 W

8.4   1.192

8.5   87.9 kW, about 165 mm

8.6   0.00293

8.7   0.546 mm, 1.968 m/s, 317.2 Pa

8.8   1445 Hz

8.9   129.2°, skin friction neglected

8.11   approx. 3.08 N, 42.0 N (using Fig. 8.14)

8.12   6.23 m

8.13   about 1.71 mm, about 0.26 m/s

8.14   0.002955, 2.866 m/s, 18.54 Pa

8.15   0.204 m^3/s, 0.03066 Pa

9.1   yes, no

9.2   (a), (e)

9.3 $\phi = -y - 4xy + \text{constant}$, 3.761 kPa

9.4 $\psi = A\theta + \text{constant}$, flow to a sink

9.5   0.3 m/s, 28.8 Pa/m, -0.036 m/s^2

9.7   28.01 rad/s

9.8   62.5 kPa, 1533 N

9.9 $r_1\sqrt{2}$

9.10 $\omega^2 R^2/g$

9.11   52.8 kPa

9.12   9250 N

9.13   0.429 m/s at 291.8°, (6/7 m, 0)

9.14 $\dfrac{1}{2}\varrho U^2 \left(\dfrac{X^2 + Y^2}{Y^2 + b^2} \right)^2$

9.15 $\pm 0.471\,\text{m}^3/\text{s per m}$, 156.2 mm apart on long axis, 4.36 m/s

9.16 $1.5\,\text{m}^3/\text{s per m}$, 15.92 mm, $x = -y \cot(20\pi y)$ (m units), 194.5 Pa

9.17 $\psi = -Uy$

$$-\frac{m}{2\pi}\arctan\left(\frac{2xy}{x^2 - y^2 + a^2}\right),$$

$$\left(-\frac{m}{2\pi U} - \left\{\frac{m^2}{4\pi^2 U^2} - a^2\right\}^{1/2}, 0\right),$$

contour is a half-body,

$$U\left\{1 + \left(4m^4/\pi^2\right) \times \left(m^2 - 4U^2 a^2\right)^{-2}\right\}^{1/2}$$

9.18 1296 N

9.19 2.580

9.20 $2\pi\sqrt{3}$

9.21 5.84 rev/s, clockwise, 3.69° E of N and 47.3° W of N, 2058 Pa

9.23 155.6 kPa

9.24 0.558, 0.0685

9.25 0.790, 0.0405, 0.02650, 7.96°, 0.790, 0.0537

10.1 $0.3827\,\text{m}^3/\text{s}$

10.2 $2.216\,\text{m}^3/\text{s}$

10.3 4.456 m

10.4 depth 520 mm, base width 601 mm

10.5 $48.2\,\text{m}^{1/3}/\text{s}$, $0.532\,\text{m}^3/\text{s}$

10.6 460 mm, 0.001588

10.7 2.41 m

10.8 $0.554\,\text{m}^3/\text{s}$, 1.185 m/s, 0.659, 0.277

10.9 1.027 m, 1.357 m, 1.226, 0.808, 0.00394

10.10 $(gE/2)^{1/2}$, $3E/4$

10.11 4.15 m

10.12 $75.4\,\text{m}^{1/3}/\text{s}$, 4.75 Pa

10.13 956 mm, 6.97 kW

10.14 0.0168

10.15 $71.3\,\text{m}^3/\text{s}$, 192.5 kN

10.16 $5.58\,\text{m}^3/\text{s}$

10.17 222 mm

10.18 $0.337\,\text{m}^3/\text{s}$

10.19 294 mm, 74.6 mm, 305.5 N

10.20 202 m

10.21 about 88 m

10.22 95.6 m, 1.281 m

10.23 0.038 N/m, 0.1096 m/s, 0.260 m/s

10.24 5.589 m/s, 27.95 m, 3.407 m/s, 7.71 kW, 0.1424 m/s, 19.66 kPa, 0.2688 m, 0.1136 m

10.25 452 kW, 0.614 m

11.1 310.6 m/s, $0.777\,\text{kg/m}^3$, 761 m/s

11.2 23.88 kPa, 443 m/s, 69.6 kg/s, $0.288\,\text{m}^2$

11.3 147°C

11.4 1.386, 1.665, 280.7 m/s, 13.2°C

11.5 1.967, 24.43 kPa, 456 m/s

11.6 40.7° to wall, 1.250

11.7 13.34°

11.8 404 m/s

11.9 1059 km/h, 6.1°C

11.10 (a) 400 kPa, −6°C, 1.035 kg/s
(b) 369.6 kPa, −12.2°C, 1.039 kg/s

11.11 $1338\,\text{mm}^2$, $4680\,\text{mm}^2$, 380 kPa, 200.6 kPa, −159°C, 600 m/s

11.12 1.692

11.13 0.60 kg/s

11.14 0.413, 1.535 times throat area, 1.840, 106.3 kPa, 940 kPa

11.15 4.75 m, 9.2°C, 82.7 kPa, 44.2°C, 150.9 kPa

11.16 241.4 kPa, 240.8 kPa

11.17 44.1 mm

11.18 15.97 kW

11.19 0.292 kg/s, 435 kPa

12.2 415 m, 1.517 m/s

12.3 3.37 m/s

12.4 29.29 litres/s

12.5 85.7 mm, 1.432 m/s

12.6 0.924 s

12.7 15.18, 0.759 s

12.8 45.2%, less than 4.97 s

12.9 162.9 m, 167.1 m, 167.7 m

12.10 $0.0836\,\text{m}^{-1}$, $3.024 \times 10^{-4}\,\text{s}^{-2}$

12.11 17.89 m

13.1 Suction: 4.43, 6.87, 10.23 m. Delivery: 31.34, 26.47, 19.75 m

13.2 2.666 m/s, 2.40 m/s, ±236.9 kPa

13.3 5, 177 mm, 1.536 m, 85.1%, 1.35% – based on optimum speed ratio 0.46

13.5 0.854

13.6 14.28°, 59.2°, 47.3 m, −5.88 m

13.7 251 mm, 35.2 kW

13.8 33.76 MW, 0.0773 rev, 9.31°, 124.29°, 42.9%

13.9 14.65 rev/s, 235.5 kW

13.10 17.14 rev/s, 7.66 MW, 0.513 rev

13.11 81.5%

13.12 1.94 m

13.13 71.2%, 3.05 m

13.14 58.1 mm water, 5.83 kW

13.15 backward-facing at 39.5°

13.16 38.23 litres/s, 78.6%

13.17 238 mm, say 250 mm

13.18 68.1%, 81.1%

13.19 (a) K_n 0.64 rev at 1.28 rev/s
(b) K_n 0.192 rev at 50 rev/s
(or K_n 0.096 rev at 25 rev/s)
(c) K_n 0.032 rev, 10 stages, at 50 rev/s

13.20 0.0843, 1.637 m

13.21 14.58 to 29.16 rev/s, 6.67 to 26.67 m

13.22 1.282 kW

13.23 238.4 mm

Index